CHEMISTRY
STRUCTURE
AND DYNAMICS

JAMES N. SPENCER
Franklin and Marshall College

GEORGE M. BODNER
Purdue University

LYMAN H. RICKARD
Millersville University

JOHN WILEY & SONS, INC.
New York / Chichester / Weinheim / Brisbane / Singapore / Toronto

ACQUISITIONS EDITOR Nedah Rose
SENIOR MARKETING MANAGER Charity Robey
SENIOR PRODUCTION EDITOR Elizabeth Swain
DESIGNER Madelyn Lesure
PHOTO EDITOR Hilary Newman
ILLUSTRATION COORDINATOR Sigmund Malinowski

Cover photo: John Lund/Tony Stone Images, New York

This book was set in 11/13 Times Ten Roman by York Graphics and printed and bound by Courier / Westford. The cover was printed by Phoenix.

This book is printed on acid-free paper.

The paper in this book was manufactured by a mill whose forest management programs include sustained yield harvesting of its timberlands. Sustained yield harvesting principles ensure that the number of trees cut each year does not exceed the amount of new growth.

Library of Congress Cataloging in Publication Data
Spencer, James N. (James Nelson), 1941–
    Chemistry : structure and dynamics : core text / James N. Spencer,
George M. Bodner, Lyman H. Rickard.
        p.   cm.
    Includes index.
    ISBN 0-471-05387-2 (pbk. : alk. paper)
    1. Chemistry.   I. Bodner, George M.   II. Rickard, Lyman H.
III. Title.
QD33.S73   1998
540–dc21   98-10375
                                                        CIP

Printed in the United States of America
10 9 8 7 6 5 4 3 2 1

P R E F A C E

In 1989 the Division of Chemical Education of the American Chemical Society recognized the need to foster the development of alternative introductory chemistry curricula. The Task Force on the General Chemistry Curriculum was created to meet the need. We were members of the Task Force and our book, *Chemistry: Structure and Dynamics,* is one of the products of the work of the Task Force.

This text encourages innovation in the teaching of general chemistry by providing flexible materials that allow instructors to build a custom curriculum appropriate for their students. We achieved this flexibility through the use of a core/modular structure. The text contains the core and chapter appendices. The modules are found on the student CD that accompanies the text. The core provides the fundamentals that all students need to prepare them for the degree programs in which an introductory chemistry course is a prerequisite. By selecting topics from the chapter appendices and modules instructors can customize their course to fit the needs of their particular students.

The text is written for students taking introductory chemistry for science and math majors. Its flexibility, however, makes it appropriate for most introductory chemistry courses. We decided which concepts to include in the core based on two criteria. First, the concepts should be the fundamental building blocks for understanding chemistry—concepts that provide the basis for both the core and the modules. Second, these concepts should be perceived by *the students* as being directly applicable to their majors or careers.

Chemistry: Structure and Dynamics is characterized by the following key features:

- **Brevity.** The core contains approximately 600 pages with 75 pages of chapter appendices.
- **Flexibility.** Modules and chapter appendices can be added to the core to provide a curriculum that meets the needs of the students at a particular institution.
- **Currency.** New models and current methods of understanding chemical concepts not yet found in traditional texts are used to introduce material.
- **Unifying Themes.** Major themes are used to link the core material into a unified whole.
- **Balanced Coverage.** Organic and biochemical examples are used throughout the text to provide a more balanced coverage of all areas of chemistry.
- **Conceptual Nature.** There is an increased emphasis on conceptual questions and problems at the end of the chapters. A large selection of traditional problems are provided, but these are supplemented with conceptual problems and discussion questions that ask students to explain, describe, or suggest experiments.
- **Case Studies.** A chapter on chemical analysis that is unique to introductory texts has been included in the core. This chapter uses case studies to introduce students to the methods and tools that chemists use to solve real-world problems.

THE CORE

Traditional texts present concepts and principles in isolation, with little, if any, connection between each concept or principle and the rest of the material. In *Chemistry: Structure and Dynamics* unifying themes are used to integrate the core topics.

- The *process of science* theme incorporates experimental data in discussions that traditional texts present as the *product* of scientific process. Our goal is to provide students with the evidence that will allow them to understand why chemists believe what they do.
- A second theme is the interrelationships between chemistry on the macroscopic and microscopic scale. This theme is developed to help students understand both how and why chemists make observations on the macroscopic scale so as to comprehend the microscopic world of atoms and molecules, and then use the resulting understanding of microscopic structure and properties to explain, predict, and—most importantly—control the macroscopic properties of matter.
- Atomic and molecular structure are developed early in the text and then repeatedly used to help students understand the physical and chemical properties of matter.

The final chapter of the core introduces students to the types of real problems that chemists can help solve and to the instruments they use to do so. Although most of the students in the typical introductory chemistry course are not chemistry majors, many of them will have careers that will require them to interact with chemists. Most of these interactions will involve analyses of samples. This chapter introduces information on how samples are analyzed and some of the major tools that chemists use. Case studies representing problems from a variety of fields have been used to present real-world problems. This chapter is not meant to emphasize instrumental analysis or spectral interpretation. Instead it is meant to make instruments seem less like mysterious black boxes and the interpretation of spectra seem less like magic.

THE CHAPTER APPENDICES AND MODULES

The chapter appendices extend the core material. The core chapter on bonding, for example, covers Lewis structures, molecular geometry, and the concept of polarity. A chapter appendix is then available that includes hybridization, valence bond theory, and molecular orbital theory. The modules are designed to introduce new topics, such as biochemistry, polymer chemistry, nuclear chemistry, and coordination chemistry.

The core/modular approach has advantages over both the traditional 1000-page texts and new shorter texts. When traditional texts are used, sections or even whole chapters are skipped. This can be frustrating to those students who depend heavily on the text for understanding new material because later topics in the text are often explained using concepts that have been omitted. *Chemistry: Structure and Dynamics* avoids this problem by building the modules on the concepts presented in the core. The core covers all concepts that are prerequisite to the modules.

NEW MODELS

The new models used in this text can be divided into three types: (1) data-driven models, (2) models that reflect current understanding of chemical theories, and (3) models that make it easier for students to understand traditional concepts. The development of electron configurations from experimental photoelectron (PES) data is an example of a data-driven model that supports the unifying theme of the process of science by demonstrating to students how experimental data can be used to construct models. This approach gives students a more concrete, and still scientifically correct, foundation on which to base their understanding of electron configuration than does use of the more abstract quantum numbers. The use of experimental data and the graphical representation of that data to develop the gas laws is another example of this type of model. Once the kinetic molecular theory is developed, it is used throughout the text to provide a consistent background for understanding temperature, heat, and equilibrium processes.

Models that reflect current understanding of chemical theories include the replacement of the valence shell electron pair repulsion (VSEPR) theory for predicting molecular geometry with Gillespie's more recent electron domain (ED) model and the use of bond-type triangles to explain the interrelationship of covalent, ionic, and metallic bonding. Bond-type triangles are then used in the text to help explain physical and chemical properties of compounds and to predict properties of new materials.

New models in the text are also used to present familiar concepts in innovative ways that make it easier for students to understand. For example, enthalpies of atom combination replace enthalpies of formation. Because chemistry is concerned with the making and breaking of bonds, thermochemical calculations are done using the energetics of breaking reactant bonds to produce atoms in the gas phase and then allowing these atoms to combine to form product molecules. In the enthalpy of formation approach found in most traditional texts the standard states are the elements in their stable states of aggregation. This construct places another concept between the student and the idea that chemical reactions liberate or absorb heat through the making and breaking of bonds. The major strength of the atom combination method is the use of the powerful visual model of reactants being broken down into gaseous atoms and then recombining to form products. The advantage to this method is that bond breaking always requires an input of energy, whereas bond making produces energy. Instructors accustomed to thinking of enthalpy changes in terms of enthalpies of formation may at first find this method more awkward. However, our experience indicates that students find this approach much easier to visualize than enthalpies of formation.

A major advantage of the atom combination approach is that the same standard states and reaction diagrams are used for all thermodynamic parameters. Traditionally, third law entropies are used in conjunction with enthalpies based on elemental standard states to introduce free energy. The use of third law entropies that are based on yet a different concept and standard state further confuses students. The atom combination approach clearly shows students the origin of the entropies of substances and, because of the direct relationship with enthalpies and free energies, makes all these concepts more accessible to students.

Another example of the use of models that reflect current understanding of chemical theories is the introduction of average valence electron energies (AVEE), calculated from PES data, which describe how tightly an atom holds on to its electrons. AVEE values are used to predict which elements will form cations and which will form anions and to explain the diagonal line in the periodic table that separates metals from nonmetals. AVEE is then used to develop the concept of electronegativity, thus giving electronegativity a more concrete meaning for students. Not only do the students develop the electronegativity concept from the same data used to determine the electronic structure of atoms, but they also see how electronegativities relate to oxidation numbers, partial charges, and formal charges. This theme is carried through the text and provides a unifying thread for oxidation–reduction, resonance structures, and polarity.

USING THE CORE TEXT AND MODULES

The core text consists of 15 chapters. With the exception of Chapter 2, the core chapters are designed to be covered in order. Chapter 2 covers the mole, chemical equations, molarity, and stoichiometry. This information is provided early in the text for those instructors who traditionally cover these topics at the beginning of the course. However, the mole concept is used only qualitatively in Chapters 3–5. Mole calculations and stoichiometry are not required until Chapter 6. Molarity and solution stoichiometry are not required until Chapter 8. Instructors can choose to cover portions of Chapter 2 as fits their curriculum.

The core may be supplemented using the chapter appendices, called Special Topics (found at the end of Chapters 1, 3, 4, 6, 7, 8, 10, 11, 12, and 14), or modules (found on the student CD). The chapter appendices extend the core topics found in the chapters they accompany. The material in the chapter appendices may be skipped at the discretion of the instructor because subsequent core chapters depend only on the concepts covered in the core chapters. Topics covered in the core chapters and their accompanying appendices are listed in the table of contents.

Although the modules are located in the student CD, they may be inserted at various points in the text depending on the core material prerequisite to the modules. The table below lists the core chapters that are prerequisite for each module.

Prerequisites for Modules

Module	*Prerequisite Core Chapters*
Biochemistry	Chapter 11 and Organic Chemistry: Functional Groups
Complex Ion Equilibria	Chapter 11
Materials Science	Chapter 9
Chemistry of the Nonmetals	Chapter 12
Nuclear Chemistry	Chapter 14
Organic Chemistry: Structure and Nomenclature of Hydrocarbons	Chapter 7, Chapters 3 and 4 Appendices
Organic Chemistry: Functional Groups	Structure and Nomenclature Module
Organic Chemistry: Reaction Mechanisms	Functional Groups Module
Polymer Chemistry	Chapter 7, and Organic Chemistry Module: Functional Groups, Chapters 3 and 4 Appendices
Solubility Equilibria	Chapter 11
Transition-Metal Chemistry	Chapter 5, Chapters 3 and 4 Appendices

TO THE STUDENT

This text, *Chemistry: Structure and Dynamics,* differs from most texts currently available for introductory chemistry. The text contains a core of material that provides a basis for understanding the fundamental concepts of chemistry, and presents new models to teach familiar concepts in ways that should make the concepts easier to grasp. Consistent themes throughout the text tie together many seemingly isolated topics.

The chapters in this text are designed to be read linearly, from the beginning to the end, rather than reading individual topics within the chapter. It is important to answer the Checkpoints and to work through the Exercises as you encounter them in your reading. Important new concepts are sometimes introduced in the Checkpoints and Exercises. In addition, answering the Checkpoints and working the Exercises will change your reading of the chapter from passive to active reading. The checkpoints will assist you in determining if you have understood what you have just read. Answers to the checkpoints can be found in Appendix D. The exercises are problems similar to the problems found at the end of the chapter. Solutions to selected end of chapter problems are found in Appendix C. Work the assigned end of chapter problems and then check your answer using Appendix C.

SUPPLEMENTS

Chemistry; A Guided Inquiry, written by Richard S. Moog and John J. Farell, of Franklin and Marhsall College. This supplement facilitates implementation of cooperative learning in general chemistry and can be used as part of a recitation section, as a workbook, or as a means of interacting with students in the classroom. This book uses all of the new approaches featured in *Chemistry: Structure and Dynamics,* and features problem assignments from the Core text. The book is a guided inquiry in which data, written descriptions, models and figures are used to develop chemical concepts.

Discovering Chemistry, written by Ram Lamba, Jesus Monzon, George Bodner, George Lisensky, Brock Spencer, Laura Parmentier, and Nancy Kerner. This laboratory manual contains a series of discovery-based labs for use in general chemistry that differ significantly from the traditional labs. The goal of these experiments is to achieve a balance between the process and content of chemistry by encouraging students to design and carry out experiments while thinking about and discussing the implications of the experiments performed or the data collected. This lab manual allows students to experience chemistry as it is done by practicing chemists, within an environment in which they also build conceptual knowledge.

Instructor's Manual, prepared by Jim Spencer, Lyman Rickard, and Jim Hovick. The Instructor's Manual highlights material in the Core text that differs from more traditional texts, focusing on how the text's method of presentation can best be utilized in the course. The manual contains complete solutions to all text problems as well as additional case studies. Further information is given on PES, atom combination parameters, bond type triangles, AVEE, and partial charge calculations.

Test Bank, prepared by George Bodner. This Test Bank contains over 600 multiple-choice and short-answer questions designed for an introductory chemistry course.

Computerized Test Bank (Mac and IBM versions available). Computerized version of the printed test bank.

Website (http://www.wiley.com/spencer) The website will include ancillaries for the instructor, including additional information about the text, the Instructor's Manual, and the Test Bank. Exciting features for the students include on-line quizzing with feedback and interactive tutorials which focus on more difficult concepts in the course.

Instructor's CD, will contain four-color photos, illustrations from the text, QuickTime movies of lab techniques, the Instructor's Manual, and Test Bank.

Student CD, will contain the text of the Modular Chapters, full color photos, illustrations, and a glossary of important terms. This CD is located on the inside back cover of the Core Text.

ACKNOWLEDGMENTS

There are many people who contributed to this textbook. First are all the members of the Task Force on the General Chemistry Curriculum, without whose discussions and ideas this project would never have been initiated. We are also indebted to the many reviewers:

Linda Allen	Louisiana State University
Dennis M. Anjo	Cal State, Long Beach
Chris Bailey	Wells College
Jay Bardole	Vincennes University
Elisabeth Bell-Loncella	University of Pittsburgh, Johnstown
Sheila Cancella	Raritan Valley Community College
David A. Cleary	Gonzaga University
Martin Cowie	University of Alberta
Michael Doyle	Trinity University
Robert Eierman	University of Wisconsin—Eau Claire
William Evans	University of California, Irvine
Keith Hansen	Lamar University
Lee Hansen	Brigham Young University
Tom Gilbert	Northeastern University
L. Peter Gold	Pennsylvania State University
Stan Grenda	University of Nevada, Las Vegas
Eugene Grimley III	Elon College
Mark Iannone	Millersville University
Keith Kester	Colorado College
Leslie Kinsland	University of Southwestern Louisiana
Nancy Konigsberg-Kerner	University of Michigan, Ann Arbor
David Lewis	Colgate University
Robert Loeschen	California State University, Long Beach
Baird Lloyd	Miami University of Ohio
David MacInnes Jr.	Guilford College
Doug Martin	Sonoma State University
Claude Mertzenich	Luther College
Patricia Metz	Texas Tech University
Susan Morante	Mont Royal College
Edward Paul	Stockton College
E. B. Robertson	University of Calgary
Doug Rustad	Sonoma State University
William Stanclift	Northern Virginia Community College, Annandale
Wayne E. Steinmetz	Pomona College
Robert L. Swofford	Wake Forest University
Sandra Turchi	Millersville University
John Woolcock	Indiana University of Pennsylvania
Andrew Zanella	Claremont College

Particular contributions were made by John Farrell and Rick Moog of Franklin and Marshall College; Ron Gillespie of McMaster University; Dudley Herschbach of Harvard University; Lee Allen of Princeton University; Gordon Sproul of the University of South Carolina at Beaufort; Bert Schiffer, a Franklin and Marshall student; and Jim Hovick at the University of North Carolina at Charlotte.

It is a particular pleasure to acknowledge the support and assistance of the staff at John Wiley and Sons, Inc. Nedah Rose, our editor, guided and encouraged us during the writing of this text. We appreciate her support for a new and different text for introductory chemistry. Jennifer Yee and Alicia Solis also attended to countless details.

Finally, to Kathy and Lynette and Killer, we owe a debt for their patience and encouragement.

James N. Spencer   George M. Bodner   Lyman H. Rickard

CONTENTS

C H A P T E R
1
ELEMENTS AND COMPOUNDS

1.1 CHEMISTRY: A DEFINITION

It seems logical to start a book of this nature with the question: What is chemistry? Most dictionaries define chemistry as *the science that deals with the composition, structure, and properties of substances and the reactions by which one substance is converted into another.* Knowing the definition of chemistry, however, is not the same as understanding what it means.

Perhaps the best way to understand the nature of chemistry is to examine examples of what it isn't. In 1921, a group from the American Museum of Natural History began excavations at an archaeological site on Dragon-Bone Hill, near the town of Chou-k'outien, 34 miles southwest of Beijing, China. Fossils found at this site were assigned to a new species, *Homo erectus pekinensis,* commonly known as Peking man. The excavations suggested that for at least 500,000 years, people have known enough about the properties of stone to make tools, and they have been able to take advantage of the chemical reactions involved in combustion in order to cook food. But even the most liberal interpretation would not allow us to call this chemistry because of the absence of any evidence of control over these reactions or processes.

The ability to control the transformation of one substance into another can be traced back to the origin of two different technologies: brewing and metallurgy. People have been brewing beer for at least 12,000 years, since the time when the first cereal grains were cultivated, and the process of extracting metals from ores has been practiced for at least 6000 years, since copper was first produced by heating the ore malachite.

But brewing beer by burying barley until it germinates and then allowing the barley sprouts to ferment in the open air was not chemistry. Neither was extracting copper metal from one of its ores, because this process was carried out without any understanding of what was happening, or why. Even the discovery at around 3500 B.C. that copper mixed with 10% to 12% tin gave a new metal that was harder than copper, and yet easier to melt and cast, was not chemistry. The preparation of bronze was a major breakthrough in metallurgy, but it didn't provide us with an understanding of how to make other metals.

Between the sixth and the third centuries B.C., the Greek philosophers tried to build a theoretical model for the behavior of the natural world. They argued that the world was made up of four primary, or *elementary*, substances: fire, air, earth, and water. These substances differed in two properties: hot versus cold, and dry versus wet. Fire was hot and dry; air was hot and wet; earth was cold and dry; water was cold and wet.

This model was the first step toward the goal of understanding the properties and compositions of different substances and the reactions that convert one substance to another. But some elements of modern chemistry were still missing. This model could explain certain observations of how the natural world behaved, but it couldn't predict new observations or behaviors. It was also based on pure speculation. In fact, its proponents were not interested in using the results of experiments to test the model.

Modern chemistry is based on certain general principles.

- **One of the goals of chemistry is to recognize patterns in the behaviors of different substances.** An example of this might be observations in 1794 by the French chemist Antoine Lavoisier that many substances which burn in air gain weight.

- **Once a pattern is recognized, it should be possible to develop a model that explains these observations.** Lavoisier concluded that substances that burn in air combine with the oxygen in the air to form products that weigh more than the starting material.

- **These models should allow us to predict the behavior of other substances.** In 1869, Dmitri Mendeléeff used his model for the behavior of the known elements to predict the properties of elements that had not yet been discovered.

- **When possible, the models should be quantitative.** They should not only predict what happens, but by how much.

- **The models should be able to make predictions that can be tested experimentally.** Mendeléeff's periodic table was accepted by other chemists because of the agreement between his predictions and the results of experiments based on the predictions.

In essence, *chemistry is an experimental science.* Experiment serves two important roles. It forms the basis of observations that define the problems which theories must explain, and it provides a way of checking the validity of new theories. This text emphasizes an experimental approach to chemistry. As often as possible, it presents the experimental basis of chemistry before the theoretical explanations of these observations.

1.2   PHYSICAL AND CHEMICAL PROPERTIES

The **composition** of a substance can be described in terms of the number and types of atoms found in that substance. The arrangement of atoms in three-dimensional space is referred to as the **structure** of the substance. The substance known as vitamin B_{12} is described in terms of the formula $C_{63}H_{88}N_{14}O_{14}PCo$, which suggests that each molecule contains 63 carbon atoms, 88 hydrogen atoms, 14 nitrogen atoms, 14 oxygen atoms, and one atom each of phosphorus and cobalt. The arrangement of the atoms is described in terms of structures such as the one shown in Figure 1.1.

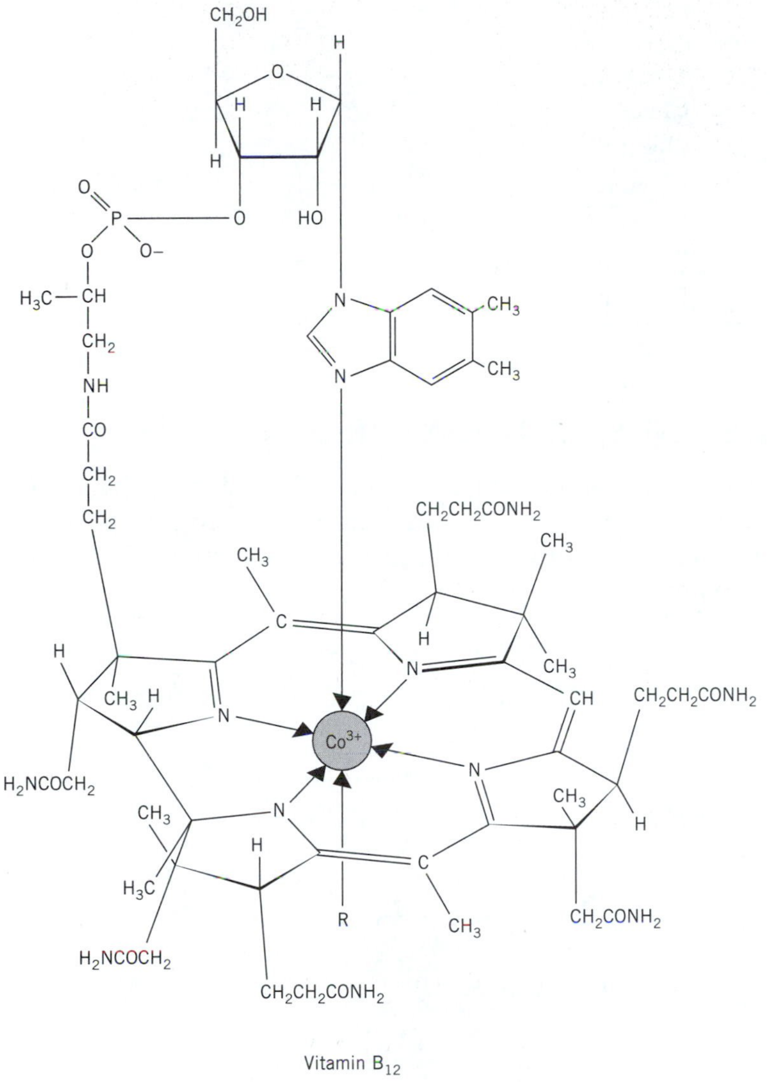

Vitamin B_{12}

FIGURE 1.1 The structure of vitamin B_{12}.

The **properties** of a substance are its characteristic traits. Butane, for example, is a liquid when found in a pressurized cigarette lighter but is a gas at room temperature and atmospheric pressure. In the presence of a spark butane will burst into a flame. The characteristic properties of a substance are divided into two categories: physical properties and chemical properties. The **physical properties** of a substance are the characteristic properties that can be observed without a change in the composition of the substance, such as the fact that butane is a gas at room temperature and atmospheric pressure. Other physical properties include the color, hardness, density, vapor pressure, melting point, and boiling point of the substance. Although the appearance of the substance changes when it melts or boils, the composition of the substance doesn't change. The same number of atoms is still present in the same ratio when a solid turns into a liquid, or a liquid into a gas.

The **chemical properties** of a substance are observed when the substance undergoes a chemical reaction. The fact that butane burns in the presence of oxygen and a spark, for example, is one of the most important chemical properties of this substance. In the course of the reaction, the butane is transformed into two new substances, carbon dioxide and water. Table 1.1 lists several common chemical and physical properties.

TABLE 1.1   Examples of Physical and Chemical Properties

Lead has a density of 11.34 g/cm^3	Physical
Milk sours	Chemical
Sulfur is yellow	Physical
Gasoline is flammable	Chemical
Iron rusts	Chemical
Sugar melts at 185°C	Physical
Diamonds are hard	Physical

1.3   ELEMENTS, COMPOUNDS, AND MIXTURES

Matter is defined as anything that has mass and occupies space. All substances that we encounter—whether natural or synthetic—are matter. Matter can be divided into pure substances and mixtures, as shown in Figure 1.2. Pure substances can be further divided into the categories of elements and compounds.

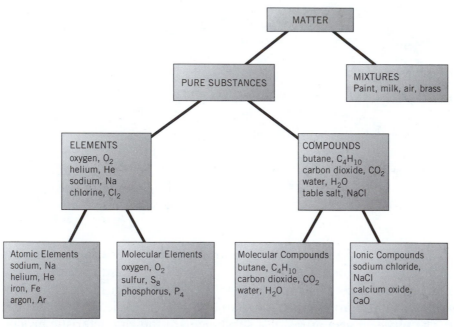

FIGURE 1.2 The classification of matter.

A **pure substance** has a constant composition. Water, for example, is always 88.81% oxygen by weight. Pure substances also have constant chemical and physical properties. Water freezes at exactly 0°C and boils at exactly 100°C at atmospheric pressure. The composition and properties of a pure substance are the same regardless of where it is found. When pure, the salt found in a salt shaker on the dining room table has the same composition as the salt dug from mines deep beneath the earth or the salt obtained by evaporating seawater. No matter where salt comes from, it always contains 1.54 times as much chlorine by weight as sodium. Since it always has the same composition, salt has the same chemical and physical properties.

Mixtures, such as a cup of coffee, have different compositions from sample to sample, and therefore varying properties. If you are a coffee drinker you will have noted that cups of coffee from your home, the college cafeteria, and a gourmet coffeehouse are not the same. They will vary in appearance, aroma, and flavor because of differences in the composition of the coffee.

Pure substances can be divided into elements and compounds. **Elements** are substances that contain only one kind of atom. As of this textbook edition, 109 elements have been discovered. They include a number of substances with which you are familiar, such as the oxygen in the atmosphere, the aluminum metal in aluminum foil, the iron in nails, the copper in electrical wires, and the mercury in thermometers. Elements are the fundamental building blocks from which all other substances are made. Imagine cutting a piece of gold metal in half and then repeating this process again and again and again. In theory, we should eventually end up with a single gold atom. If we tried to split this atom in half we would end up with something that no longer retains any of the properties of gold. An **atom**, in other words, can be defined as the smallest particle of an element that has any of the properties of that element.

Compounds (such as water or table salt) contain more than one element chemically combined in fixed proportions. Water is composed of the elements hydrogen and oxygen. Table salt is composed of the elements sodium and chlorine. The 109 elements are used in various combinations to make over 6 million known compounds and may be used to make millions more that are yet unknown. If we tried to divide a compound, such as water, into infinitesimally small portions eventually we would end up with a single molecule of water containing two hydrogen atoms and one oxygen atom. This molecule if further divided would be broken into its individual atoms. At this point we would no longer have water. A **molecule** is a group of atoms bound together that exists as a discrete particle. A molecular compound is composed of electrically neutral particles. The composition of a molecular compound can be represented by a **chemical formula** as shown in Figure 1.3. The subscripts in a chemical formula represent the relative numbers of atoms present in the compound. When there is no subscript, as in the case of carbon in CO_2, a value of one is assumed. Thus, the formula CO_2 represents a molecule that contains one carbon atom and two oxygen atoms.

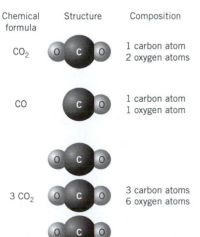

Chemical formula | Structure | Composition

CO_2 — 1 carbon atom / 2 oxygen atoms

CO — 1 carbon atom / 1 oxygen atom

$3 CO_2$ — 3 carbon atoms / 6 oxygen atoms

FIGURE 1.3 The composition of a compound is represented by a formula. The formula CO_2 represents a molecule that contains one carbon atom and two oxygen atoms. The formula CO tells us that this molecule consists of one carbon and one oxygen atom. The coefficient, 3, that appears in front of CO_2 represents three CO_2 molecules for a total of three carbon atoms and six oxygen atoms.

The same elements can combine in different ratios to give completely different compounds. Carbon dioxide, CO_2, is the gas that results from respiration (we exhale CO_2). Carbon and oxygen can combine to also form carbon monoxide, CO, a very toxic gas. In this compound carbon and oxygen are in a 1:1 ratio.

We use coefficients in front of chemical formulas to represent the number of units of a compound present. If we wish to describe the three molecules of carbon dioxide shown in Figure 1.3, we would write 3 CO_2.

A second type of compound, **an ionic compound,** is composed of both positive and negative electrically charged particles, which form a three-dimensional network rather than discrete particles as in molecules. We can also represent an ionic compound with a chemical formula. The chemical formula of an ionic compound consists of the smallest whole number ratio of positive and negative particles in the three-dimensional network. The NaCl and CaO listed in Figure 1.2 are examples of chemical formulas of ionic compounds.

Elements can also exist in the form of molecules, but these molecules are composed of identical atoms. The oxygen we breathe, for example, consists of molecules that contain two oxygen atoms, O_2 (Figure 1.4). Elemental phosphorus molecules are composed of four phosphorus atoms (P_4), and elemental sulfur contains molecules composed of eight sulfur atoms (S_8). Figure 1.2 illustrates the relationships between the classifications of matter.

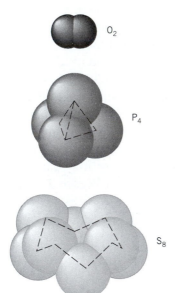

O_2

P_4

S_8

FIGURE 1.4 Some nonmetallic elements also form molecules. At room temperature, oxygen exists as O_2 molecules, phosphorus forms P_4 molecules, and sulfur forms cyclic S_8 molecules.

Checkpoint
Describe the difference between the symbols 8 S and S_8.

Elements and compounds are classified as pure substances because they have constant properties and composition. As a result, a substance can't be classified as an element or compound on the basis of its physical properties (Figure 1.5). The only way to determine whether a substance is an element or a compound is to try to break it down into simpler substances. Elements can be broken down into only one kind of atom.

If a substance can be decomposed into more than one kind of atom, it is a compound. Water, for example, can be decomposed into the elements hydrogen and oxygen by passing an electric current through the liquid, as shown in Figure 1.6. In a similar fashion, salt can be decomposed into its elements—sodium and chlorine—by passing an electric current through a molten sample. See Table 1.2 for examples of common elements, compounds, and mixtures.

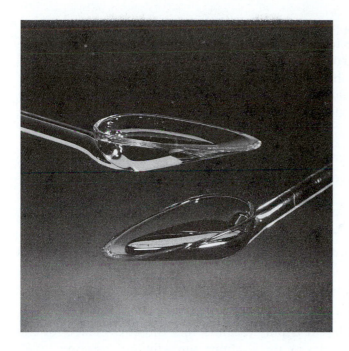

FIGURE 1.5 It isn't obvious from their appearances that the liquid on the left is a compound (water) and the liquid on the right (mercury metal) is an element.

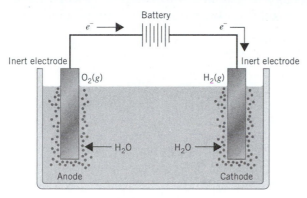

FIGURE 1.6 Electrolysis of water results in the production of oxygen gas and hydrogen gas.

TABLE 1.2   Examples of Elements, Compounds, and Mixtures

Oxygen gas	Element
Carbon dioxide gas	Compound
Gasoline	Mixture
Distilled water	Compound
Tap water	Mixture
Sugar	Compound
Raisin bran cereal	Mixture

1.4   EVIDENCE FOR THE EXISTENCE OF ATOMS

Most students in a beginning chemistry course already believe in atoms. If asked to describe the evidence on which they base this belief, however, they hesitate. Our senses argue against the existence of atoms. The atmosphere in which we live feels like a continuous fluid. We don't feel bombarded by collisions with individual particles in the air. The water we drink looks like a continuous fluid. We can take a glass of water, pour out half, divide the remaining water in half, and repeat this process again and again, without ever appearing to reach the point at which it is impossible to divide it one more time. Because our senses suggest that matter is continuous, it isn't surprising that the debate about the existence of atoms goes back to the ancient Greeks and continued well into the twentieth century.

Experiments with gases that first became possible at the turn of the nineteenth century led John Dalton in 1803 to propose a model for the atom based on the following assumptions:

- Matter is made up of atoms that are indivisible and indestructible.
- All atoms of an element are identical.
- Atoms of different elements have different weights and different chemical properties.
- Atoms of different elements combine in simple whole number ratios to form compounds.
- Atoms cannot be created or destroyed. When a compound is decomposed, the atoms are recovered unchanged.

Dalton's assumptions form the basis of modern atomic theory. However, modern experiments have shown that not all atoms of an element are exactly alike and that atoms can be broken down into subatomic particles. It is only recently that direct evidence for the existence of atoms has become available. Using the scanning tunneling microscope developed in the 1980s, scientists have finally been able to observe and even manipulate individual atoms.

Research in the 1990s

Scanning Tunneling Microscopy

In 1982 a paper describing a new technique known as *scanning tunneling microscopy* (STM) was published by a group of scientists from the IBM Research Laboratory in Zurich [G. Binnig, H. Rohrer, Ch. Geber, and E. Weibel, *Physical Review Letters*, **49,** 57 (1982)].[1] This paper was built on prior work in the same laboratory that showed that the magnitude of the current that flowed between the tip of a sharp piece of tungsten metal and the surface of platinum metal over which it was moved was very sensitive to the distance between the tip and the metal surface. In their 1982 paper, the IBM researchers showed how this information could be used to study the surface of the metal. As the tungsten tip moves over the surface being studied, it is raised or lowered as needed to give a constant current. Measurements of the motion of the tip are then recorded and analyzed to yield an image of individual atoms or molecules.

STM allows scientists to study the physical nature and chemical composition of the surface of solids on the atomic level. Of particular commercial importance to the microelectronics industry is the ability of STM to observe crystal defects on the surface of silicon. With the ever-decreasing size of integrated circuit components these defects become very important.

Experiments have demonstrated that the STM probe can be used to move individual atoms or molecules. The "molecular man" shown in Figure 1.7 was formed by moving 28 carbon monoxide molecules into position on a platinum surface. The ability to manipulate individual atoms has the potential to allow scientists to control reactions of single atoms and molecules. This could lead to the production of new chemical substances that are not possible using normal chemical methods.

---

[1]Binnig and Rohrer received the Nobel Prize in physics in 1986 for the development of STM.

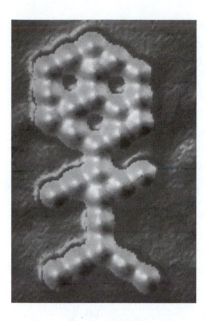

FIGURE 1.7 Scientists now have the ability to manipulate individual atoms and molecules. This "molecular man" was formed from carbon monoxide molecules moved into position on a platinum surface.

1.5   THE STRUCTURE OF ATOMS

Contrary to Dalton's model of the atom, atoms are not indivisible. Today, we recognize that atoms are composed of the three fundamental subatomic particles listed in Table 1.3: **electrons, protons,** and **neutrons.** Chemists normally refer to these particles as fundamental because they are the building blocks of all atoms.[2] Although gold atoms and oxygen atoms are quite different from one another, the electrons, protons, and neutrons found within gold atoms are indistinguishable from the electrons, protons, and neutrons found within oxygen atoms.

TABLE 1.3   Fundamental Subatomic Particles

Particle	Symbol	Absolute Charge (C)[a]	Relative Charge	Absolute Mass (g)	Relative Mass
Electron	e^-	-1.60×10^{-19}	-1	9.11×10^{-28}	0
Proton	p^+	1.60×10^{-19}	$+1$	1.673×10^{-24}	1
Neutron	n^0	0	0	1.675×10^{-24}	1

[a]See the discussion of scientific notation in Appendix A.4 if you wish to review numbers such as 1.60×10^{-19}.

The electric charge on a single electron and a single proton are given in columns three and four of Table 1.3. In column three the charge on these particles is reported as an **absolute** value in units of coulombs (C), the fundamental unit of electric charge. Quite often in chemistry, instead of being concerned with an absolute measurement we are more interested in the relationship between the sizes of two measurements. This relationship can be expressed as a ratio between the two measurements. Because the magnitude of the charge on an electron and a proton is the same—they differ only in the sign of the charge— the **relative** charge on these particles is -1 or $+1$.

[2]Protons and neutrons are not fundamental particles in the strictest sense because they are composed of still smaller particles, the so-called up and down quarks.

Because the charge on a proton has the same magnitude as the charge on an electron, the charge on one proton exactly balances the charge on an electron, and vice versa. Thus, atoms are electrically neutral when they contain the same number of electrons and protons.

The masses of the three subatomic particles are also described in Table 1.3 in terms of both absolute and relative values. The first column for mass gives the experimental values in units of grams (g), the fundamental unit for measurements of mass. The second column for mass gives the relative values. Because the mass of a proton is almost the same as that of a neutron, both particles are assumed to have a relative mass of 1. Because the ratio of the mass of an electron to that of a proton is so very small it is considered negligible, and the electron is assigned a relative mass of zero.

Most of the mass of an atom is concentrated in the **nucleus,** which contains all of the protons and neutrons in the atoms; 99.97% of the mass of a carbon atom, for example, can be found in the nucleus of that atom. The term *nucleus* comes from a Latin word meaning "little nut." This term was originally chosen to convey the image that the nucleus of an atom occupies an infinitesimally small fraction of the volume of an atom. The radius of an atom is approximately 10,000 times larger than its nucleus. The relative size of an atom, its nucleus, and its electrons can be appreciated by using the New Orleans Superdome as an analogy for the size of an atom. In this model the nucleus would be the size of a small pea suspended above the 50-yard line, and the electrons would be microscopic mosquitos flying inside the arena. Most of the volume of an atom consists of empty space through which the electrons move.

It is impossible to determine the exact position or path of an electron. Because of this chemists often visualize electrons in two ways. Electrons can be described as very small particles or as a cloud of negative charge spread out through the volume of space surrounding the nucleus that corresponds to the size of the atom (Figure 1.8).

FIGURE 1.8 The hydrogen atom. Hydrogen has a nucleus with one proton, whose positive charge is balanced by one electron. The exact position of an electron in an atom cannot be determined. Often electrons are described as a cloud of negative charge spread out in the space surrounding the nucleus. The boundary of the atom, defined by the dotted line, is not a physical boundary but instead is a volume that contains most of the electron density of the atom.

1.6 ATOMIC SYMBOLS

Chemists use a shorthand notation to save both time and space when describing atoms. Each element is represented by a unique symbol. Most of these symbols make sense because they are derived from the name of the element.

H = hydrogen	B = boron
C = carbon	N = nitrogen
O = oxygen	P = phosphorus
Se = selenium	Si = silicon
Mg = magnesium	Br = bromine
Al = aluminum	Ca = calcium
Cr = chromium	Zn = zinc

Symbols that don't seem to make sense come from the Latin or German names of the elements. Fortunately, there are only a handful of elements in this category.

Ag = silver   Na = sodium
Au = gold   Pb = lead
Cu = copper   Sb = antimony
Fe = iron   Sn = tin
Hg = mercury   W = tungsten
K = potassium

Additional symbols and names of the elements can be found on the inside back cover of the text.

1.7   ATOMIC NUMBER AND MASS NUMBER

Each element has been assigned an **atomic number (Z)** between 1 and 109 that describes the number of protons in the nucleus of an atom of that element. **The number of protons in the nucleus of an atom determines the identity of the atom.** Every carbon atom ($Z = 6$) has six protons in the nucleus of the atom, whereas sodium atoms ($Z = 11$) have eleven (Figure 1.9). Since atoms are electrically neutral, the nucleus of the carbon atom must be surrounded by six electrons, and the sodium atom must contain 11 electrons.

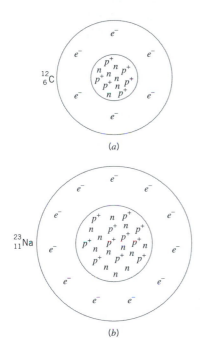

FIGURE 1.9 (a) A $^{12}_{6}$C atom consists of a nucleus of six protons and six neutrons. Six electrons are required to balance the positive nuclear charge. (b) A $^{23}_{11}$Na atom contains 11 protons, 12 neutrons, and 11 electrons.

The nucleus of an atom is also described by a **mass number (A),** which is the sum of the number of protons and neutrons in the nucleus. The difference between the mass number and the atomic number of an atom is therefore equal to the number of neutrons in the nucleus of that atom. The carbon atom shown in Figure 1.9 has a mass number of 12 because it contains six protons and six neutrons. The sodium atom in this figure has a mass number of 23 because it contains 11 protons and 12 neutrons.

A shorthand notation has been developed to describe the number of neutrons and protons in the nucleus of an atom. The atomic number is written in the bottom left corner of the symbol for the element and the mass number is written in the top left corner: $^{A}_{Z}X$. The atoms in Figure 1.9 would therefore be given the symbol $^{12}_{6}$C and $^{23}_{11}$Na. Each element, however, has a unique atomic number, and each element has its own symbol. It is therefore redundant to give both the symbol for the element and its atomic number. The symbols for the atoms in Figure 1.9 are therefore usually written as ^{12}C and ^{23}Na.

1.8   ISOTOPES

The number of protons in the nucleus of an atom determines the identity of the atom. As a result, all atoms of an element must have the same number of protons. But they don't have to contain the same number of neutrons.

Atoms with the same atomic number but different mass numbers are called **isotopes.** Carbon, for example, has the three naturally occurring isotopes shown in Figure 1.10: ^{12}C, ^{13}C, and ^{14}C. Carbon-12 (^{12}C) has six protons and six neutrons, ^{13}C has six protons and seven neutrons, and ^{14}C has six protons and eight neutrons.

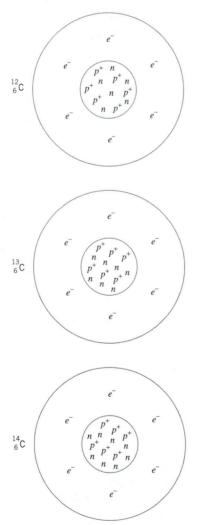

FIGURE 1.10 The three naturally occurring isotopes of carbon: ^{12}C, ^{13}C, ^{14}C.

Each element occurs in nature as a mixture of its isotopes. Consider a "lead" pencil, for example. These pencils don't contain the element lead, which is fortunate because many people chew their pencils and lead can be very toxic. They contain a substance that was once known as "black lead" and is now known as graphite.[3] Because it is almost pure carbon, the graphite

---

[3]The graphite in most pencils is mixed with clay; the more clay, the harder the pencil.

in a pencil contains a mixture of ^{12}C, ^{13}C, and ^{14}C atoms. The three isotopes, however, do not occur to the same extent in nature. Most of the atoms (98.892%) are ^{12}C, a small percentage (1.108%) are ^{13}C, and only 1 in about 10^{12} is the radioactive isotope of carbon, ^{14}C. The percentage of atoms occurring as a given isotope is referred to as the natural abundance of that isotope. Some elements, such as fluorine, have only one naturally occurring isotope, ^{19}F, whereas other elements have several, as shown in Table 1.4. In this table the mass of a particular isotope is given in grams (g) as well as in a unit known as the atomic mass unit (amu). The amu also gives the mass of an atom and is often more convenient for use than the mass in grams.

TABLE 1.4 Common Isotopes of Some of the Lighter Elements

Isotope	Natural Abundance (%)	Mass (g)	Mass (amu)
^1H	99.985	1.6735×10^{-24}	1.007825
^2H	0.015	3.3443×10^{-24}	2.01410
^6Li	7.42	9.9883×10^{-24}	6.01512
^7Li	92.58	1.1650×10^{-23}	7.01600
^{10}B	19.7	1.6627×10^{-23}	10.0129
^{11}B	80.3	1.8281×10^{-23}	11.00931
^{12}C	98.892	1.9926×10^{-23}	12.00000
^{13}C	1.108	2.1592×10^{-23}	13.003
^{16}O	99.76	2.6560×10^{-23}	15.99491
^{17}O	0.04	2.8228×10^{-23}	16.99913
^{18}O	0.20	2.9888×10^{-23}	17.99916
^{20}Ne	90.51	3.3198×10^{-23}	19.99244
^{21}Ne	0.27	3.4861×10^{-23}	20.99384
^{22}Ne	9.22	3.6518×10^{-23}	21.99138

Checkpoint

There are two naturally occurring isotopes of lithium, ^6Li and ^7Li. If you selected 10,000 lithium atoms at random, how many would be ^6Li? How many would be ^7Li?

Exercise 1.1

What is the ratio of the mass in grams of a ^{12}C atom to that of a ^1H atom? What is the ratio of the mass in amu of a ^{12}C atom to that of a ^1H atom? How much heavier is an atom of ^{12}C than an atom of ^1H? Use both the mass in grams and the mass in amu to obtain your answer.

Solution

$$\text{ratio in grams} \quad \frac{^{12}C}{^1H} = \frac{1.9926 \times 10^{-23}}{1.6735 \times 10^{-24}} = 11.907$$

$$\text{ratio in amu} \quad \frac{^{12}C}{^1H} = \frac{12.000}{1.0078} = 11.907$$

Using either answer, the mass of a ^{12}C atom is 11.907 times heavier than a ^1H atom.

When performing calculations in science it is important to maintain the proper number of significant figures. For a review of significant figures see the Special Topics at the end of this chapter and Appendix A.3 at the end of the text.

1.9   IONS

Imagine that you have a piece of copper metal. Would it be easier to change the number of electrons on some of the copper atoms in this piece of metal or to change the number of protons?

It is much easier to change the number of electrons on an atom than the number of protons in the nucleus. The best evidence for this is the fact that copper can conduct an electric current. Copper can conduct an electric current because electrons are transferred from atom to atom. Since the number of protons determines the identity of an atom and the copper wire does not change to some other element when conducting electricity, the moving charged particles must be electrons not protons.

The electrically charged particles formed when electrons are added to or removed from a neutral atom are called **ions**.[4] Neutral atoms are turned into positively charged ions called **cations** by removing one or more electrons (Figure 1.11). By removing an electron from a sodium atom—which contains 11 electrons and 11 protons—a Na^+ ion is produced that has 10 electrons and 11 protons. Ions with larger positive charges can be produced by removing more electrons. A neutral aluminum atom, for example, has 13 electrons and 13 protons. If we remove three electrons from this atom, we get a positively charged Al^{3+} ion that has 10 electrons and 13 protons, for a net charge of +3.

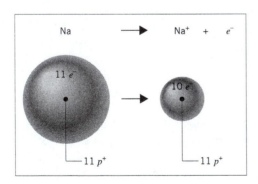

FIGURE 1.11 Removing an electron from a neutral sodium atom produces a Na^+ ion that has a net charge of +1.

Atoms that gain extra electrons become negatively charged ions called **anions,** as shown in Figure 1.12. A neutral chlorine atom, for example, has 17 protons and 17 electrons. By adding one more electron to this atom, a Cl^- ion is produced that has 18 electrons and 17 protons, for a net charge of −1.

The gain or loss of electrons by an atom to form negative or positive ions has an enormous impact on the chemical and physical properties of the atom. Sodium metal, which consists of neutral sodium atoms, bursts into flame when it comes in contact with water. But positively charged Na^+ ions are so unreactive with water they are essentially inert.[5] Neutral chlorine atoms instantly combine to form Cl_2 molecules, which are so reactive that

---

[4]The force of attraction that holds protons within the nucleus of an atom is at least 10^6 times as large as the force of attraction between the nucleus and the electrons that surround it. As a result, ions can be formed only by the gain or loss of electrons.
[5]A substance that seldom undergoes a chemical reaction is said to be *inert.*

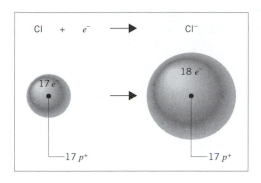

FIGURE 1.12 Adding an extra electron to a neutral chlorine atom produces a Cl^- ion that has a net charge of -1.

entire communities are evacuated when trains carrying chlorine gas derail. However, chloride ions do not react with one another.

The enormous difference between the chemistry of neutral atoms and that of their ions means that you will have to pay close attention to the symbols you read, to make sure that you never confuse one with the other.

Exercise 1.2

Find the number of protons, electrons, and neutrons in each of the following atoms and ions.

(a) ^7_3Li (b) $^{24}_{12}Mg^{2+}$ (c) $^{79}_{35}Br^-$

Solution

(a) Lithium has an atomic number of 3, which means that the nucleus of any atom of this element must contain 3 protons. Since the symbol for lithium is written to describe a neutral atom, the number of protons and electrons in the atom must be the same. This atom therefore contains 3 electrons. The mass number is equal to the sum of the neutrons and protons. This atom has a mass number of 7 and the nucleus contains 3 protons. It therefore contains 4 neutrons.

(b) The atomic number of magnesium is 12, which means that this ion contains 12 protons. Because the ion carries a charge of $+2$, there must be two more protons (positive charges) than electrons (negative charges). This ion therefore contains 10 electrons. The mass number is 24 and there are 12 protons, which means there must also be 12 neutrons.

(c) Bromine has an atomic number of 35 and therefore has 35 protons. Since the ion has a -1 charge the ion must have one more electron than a neutral atom. This ion therefore contains 36 electrons. Because the mass number of the ion is 79 and it contains 35 protons, there must be 44 neutrons as well.

1.10 POLYATOMIC IONS

Simple ions, such as Mg^{2+} and N^{3-} ions, are formed by adding or subtracting electrons from neutral atoms. **Polyatomic ions** are electrically charged substances composed of more than one atom. There are only two polyatomic cations that you will commonly encounter. These are NH_4^+ and H_3O^+. There are many more polyatomic anions. A few of the more common anions are listed in Table 1.5.

TABLE 1.5   Common Polyatomic Negative Ions

-1 ions

HCO_3^-	hydrogen carbonate	OH^-	hydroxide
$CH_3CO_2^-$	acetate	ClO_4^-	perchlorate
NO_3^-	nitrate	ClO_3^-	chlorate
NO_2^-	nitrite	ClO_2^-	chlorite
MnO_4^-	permanganate	ClO^-	hypochlorite
CN^-	cyanide		

-2 ions

CO_3^{2-}	carbonate	O_2^{2-}	peroxide
SO_4^{2-}	sulfate	CrO_4^{2-}	chromate
SO_3^{2-}	sulfite	$Cr_2O_7^{2-}$	dichromate
$S_2O_3^{2-}$	thiosulfate		

-3 ions

PO_4^{3-}	phosphate	AsO_4^{3-}	arsenate
BO_3^{3-}	borate		

1.11   PREDICTING THE FORMULAS OF IONIC COMPOUNDS

A famous Sherlock Holmes story that revolves around a dog that *didn't bark* reminds us that we often forget to notice things that don't happen. As a chemical equivalent of the dog that doesn't bark, consider what happens when you pour a few crystals of the ionic compound, table salt, into the palm of your hand. Even though table salt is composed of electrically charged ions, you don't feel the same thing that you would feel if you made the mistake of touching a bare wire that had been plugged into an electric socket. In other words, your senses tell you something, that ionic compounds carry no net electric charge. This means that they must contain just as many positive charges as negative charges.

The formulas of compounds that contain ions can be predicted by assuming that the total charge on the positive ions must balance the total charge on the negative ions. The charges on the ions, therefore, dictate the ionic compound that is formed. Chemical formulas for ionic compounds are represented by the simplest whole number ratio of positive to negative ions that will yield an electric charge of zero on the compound.

Checkpoint

What are the ions that compose each of the following ionic compounds: NaOH, K_2SO_4, $BaSO_4$, and $Be_3(PO_4)_2$?

Exercise 1.3

Predict the formulas of the compounds formed by the following sets of ions.

(a) Calcium fluoride is found in the mineral fluorite. It is composed of calcium $+2$ ions, Ca^{2+}, and fluoride ions, F^-. Predict the formula of calcium fluoride.

(b) Rubies are composed primarily of aluminum oxide with traces of chromium (III) oxide that give it a red color. Predict the formula of the compound chromium (III) oxide composed of Cr^{3+} and O^{2-}.

Solution

(a) In order for a compound to have an electrically neutral charge the total positive charges must be balanced by the total negative charges. The charge on each calcium ion is $+2$, whereas the charge on each fluoride ion is -1. In order for the charges to cancel one another there must be two fluoride ions for every calcium ion. This provides a total positive charge of $+2$ from the calcium ion and a total charge of -2 from the two fluoride ions. The formula for this compound is therefore CaF_2.

(b) Balancing the charges in chromium (III) oxide can be done as if finding the lowest common denominator for two fractions. If the two charges, $+3$ and -2, were denominators in fractions, the common fraction would have a common denominator of 6. If we make sure that we have a total charge from the Cr^{3+} ions of $+6$, we can balance it with a total charge from the O^{2-} ions of -6. Two Cr^{3+} ions would provide the charge of $+6$. Three O^{2-} ions are required to provide a charge of -6. This compound therefore has the formula Cr_2O_3.

Exercise 1.4

Predict the formula of the following ionic compounds, which contain polyatomic ions.

(a) Calcium carbonate is the major component in chalk, limestone, and marble. It is composed of calcium ions, Ca^{2+}, and the polyatomic anion carbonate, CO_3^{2-}. Predict the chemical formula of calcium carbonate.

(b) Strontium nitrate is a component of fireworks that produces a red color. This ionic compound is composed of Sr^{2+} and NO_3^-. Predict the chemical formula.

Solution

(a) The total charge from the positive ions must balance the charge from the negative ions. Since the calcium ion has a charge of $+2$ and the carbonate ion has a charge of -2, the charge of one calcium ion cancels the charge of one carbonate ion. The ionic formula of calcium carbonate is $CaCO_3$.

(b) The $+2$ charge on the strontium requires two negative nitrate ions in order to balance the charge. The chemical formula is written as $Sr(NO_3)_2$. When more than one polyatomic ion is required to balance the charge, parentheses are used to indicate that the subscript 2 represents two entire nitrate ions, not just two nitrogens or two oxygens. The chemical formula, $Sr(NO_3)_2$, represents one strontium, two nitrogen, and six oxygen atoms.

This procedure is used for predicting the formulas of ionic compounds when the charges of the individual ions are known. A different method is used when predicting the chemical formula of a molecule that consists of electrically neutral atoms rather than charged ions. This will be described in Chapter 4.

1.12   THE PERIODIC TABLE

While trying to organize a discussion of the properties of the elements for a chemistry course at the Technological Institute in Petrograd, Dmitri Ivanovitch Mendeléeff[6] listed

---

[6]There are at least a half dozen ways of spelling Mendeléeff's name because of disagreements about translations from the Cyrillic alphabet. The version used here is the spelling that Mendeléeff himself used when he visited England in 1887.

the properties of each element on a different card. As he arranged the cards in different orders, he noticed that the properties of the elements repeat in a periodic fashion when the elements are listed more or less in order of increasing atomic weight. In 1869 Mendeléeff published the first of a series of papers outlining a table of the elements in which the properties of the elements repeated in a periodic fashion. A copy of the table Mendeléeff published in 1871 is shown in Figure 1.13. This table lists elements with similar chemical properties in the same column.

Reihen	Gruppe I. — R^2O	Gruppe II. — RO	Gruppe III. — R^2O^3	Gruppe IV. RH^4 RO^2	Gruppe V. RH^3 R^2O^5	Gruppe VI. RH^2 RO^3	Gruppe VII. RH R^2O^7	Gruppe VIII. — RO^4
1	H = 1							
2	Li = 7	Be = 9,4	B = 11	C = 12	N = 14	O = 16	F = 19	
3	Na = 23	Mg = 24	Al = 27,3	Si = 28	P = 31	S = 32	Cl = 35,5	
4	K = 39	Ca = 40	— = 44	Ti = 48	V = 51	Cr = 52	Mn = 55	Fe = 56, Co = 59, Ni = 59, Cu = 63.
5	(Cu = 63)	Zn = 65	— = 68	— = 72	As = 75	Se = 78	Br = 80	
6	Rb = 85	Sr = 87	?Yt = 88	Zr = 90	Nb = 94	Mo = 96	— = 100	Ru = 104, Rh = 104, Pd = 106, Ag = 108.
7	(Ag = 108)	Cd = 112	In = 113	Sn = 118	Sb = 122	Te = 125	J = 127	
8	Cs = 133	Ba = 137	?Di = 138	?Ce = 140	—	—	—	— — — —
9	(—)	—	—	—	—	—	—	
10	—	—	?Er = 178	?La = 180	Ta = 182	W = 184	—	Os = 195, Ir = 197, Pt = 198, Au = 199.
11	(Au = 199)	Hg = 200	Tl = 204	Pb = 207	Bi = 208	—	—	
12	—	—	—	Th = 231	—	U = 240	—	— — — —

FIGURE 1.13 This version of Mendeléeff's periodic table was published in the journal *Annalen der Chemie* in 1871.

More than 700 versions of the periodic table were proposed in the first 100 years after the publication of Mendeléeff's table. A modern version of the table is shown in Figure 1.14. In this version the elements are arranged in order of increasing atomic number. The atomic number is given above the atomic symbol. The vertical columns are known as **groups,** or families. Traditionally these groups have been distinguished by a **group number** consisting of a Roman numeral followed by either an A or a B. In the United States, the elements in the first column on the left-hand side of the table are Group IA. The next column is IIA, then IIIB, and so on across the periodic table to VIIIA.

Checkpoint

What are the atomic numbers of F and Pb? What are the atomic symbols of the elements with atomic numbers 24 and 74?

Unfortunately the same notation is not used in all countries. The elements known as Group VIA in the United States are Group VIB in Europe. A new convention for the periodic table has been proposed that numbers the columns from 1 to 18, reading from left to right. This convention has obvious advantages. It is perfectly regular and therefore unambiguous. The advantages of the old format are less obvious, but they are equally real. This book therefore introduces the new convention but retains the old.

The elements in a column of the periodic table have similar chemical properties. Elements in the first column, for example, combine in similar ways with chlorine to form compounds.

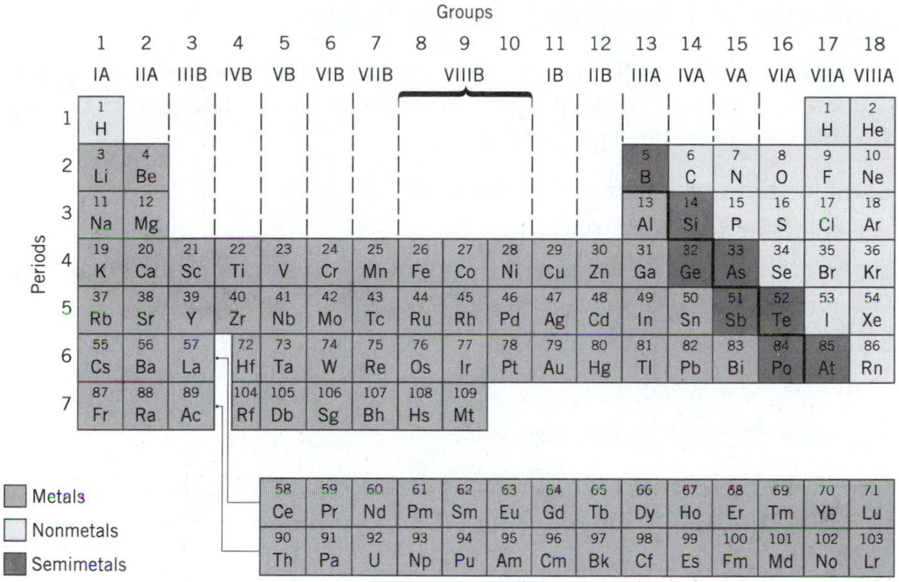

Groups

		1	2	3	4	5	6	7	8	9	10	11	12	13	14	15	16	17	18
		IA	IIA	IIIB	IVB	VB	VIB	VIIB		VIIIB		IB	IIB	IIIA	IVA	VA	VIA	VIIA	VIIIA

FIGURE 1.14 The elements can be organized into a periodic table in which elements with similar chemical properties are placed in vertical columns or groups. More than 75% of the known elements are metals. Another 15%, clustered primarily in the upper right corner of the table, are nonmetals. Along the dividing line between these categories is a handful of semimetals, which have properties that lie between the extremes of metals and nonmetals.

The horizontal rows in the periodic table are called **periods.** The first period contains only two elements: hydrogen (H) and helium (He). The second period contains eight elements (Li, Be, B, C, N, O, F, and Ne). Although there are nine horizontal rows in the periodic table in Figure 1.14, there are only seven periods. The two rows at the bottom of the table belong in the sixth and seventh periods. These rows are listed at the bottom to prevent the table from becoming so wide that it is unwieldy.

The 109 elements in the periodic table can be divided into three categories: **metals, nonmetals,** and **semimetals.** The dividing line between the metals and the nonmetals in Figure 1.14 is marked with a heavy stair-step line. As you can see from Figure 1.14, more than 75% of the elements are metals. These elements are found toward the bottom and left-hand side of the table.

Only 17 elements are nonmetals. With only one exception—hydrogen, which appears on both sides of the table—these elements are clustered in the upper right corner of the periodic table. A cluster of elements that are neither metals nor nonmetals can be found on either side of the heavy diagonal line. These elements are called the semimetals, or metalloids.

Exercise 1.5

Classify the elements in Group IVA as either a metal, a nonmetal, or a semimetal.

Solution

Group IVA contains five elements: carbon, silicon, germanium, tin, and lead. According to Figure 1.14, these elements fall into the following categories.

Nonmetal: C
Semimetal: Si and Ge
Metal: Sn and Pb

1.13   METALS, NONMETALS, AND SEMIMETALS

The tendency to divide elements into metals, nonmetals, and semimetals is based on differences between the chemical and physical properties of these three categories of elements. **Metals** have the following physical properties:

- They have a metallic shine or luster. (They look like metals!)
- They are usually solids at room temperature.
- They are *malleable*. They can be hammered, pounded, or pressed into different shapes without breaking.
- They are *ductile*. They can be drawn into thin sheets or wires without breaking.
- They conduct heat and electricity.

Nonmetals have the opposite physical properties:

- They seldom have a metallic luster. They tend to be colorless, like the oxygen and nitrogen in the atmosphere, or brilliantly colored, like bromine.
- They are often gases at room temperature.
- The nonmetallic elements that are solids at room temperature (such as carbon, phosphorus, sulfur, and iodine) are neither malleable nor ductile. They cannot be shaped with a hammer or drawn into sheets or wires.
- They are poor conductors of either heat or electricity. Nonmetals tend to be insulators, not conductors.

The **semimetals** have properties that lie between these extremes. They often look like metals but are brittle like nonmetals. They are neither conductors nor insulators but make excellent semiconductors.

KEY TERMS

Absolute measurement	Group number	Period
Atom	Ion	Periodic table
Atomic mass unit	Isotope	Physical properties
Atomic number	Mass number	Properties
Chemical formula	Metal	Pure substance
Chemical properties	Mixture	Relative measurement
Compound	Molecule	Semimetal
Element	Nonmetal	Structure
Group	Nucleus	

PROBLEMS

Elements and Compounds and Mixtures

1. Define the following terms: *element, compound,* and *mixture.* Give an example of each.
2. Describe the difference between elements and compounds on the macroscopic (objects are visible to the naked eye) scale and on the atomic scale.

3. Classify the following substances into the categories of elements, compounds, mixtures, metals, nonmetals, and semimetals. Use as many labels as necessary to classify each substance. Use whatever reference books you need to identify each substance.
   (a) diamond   (b) brass   (c) soil   (d) glass   (e) cotton   (f) milk of magnesia
   (g) salt   (h) iron   (i) steel

4. Granite consists of four minerals in varying composition: feldspar, mica, and quartz. Is granite an element, a compound, or a mixture?

5. List the symbols for the following elements.
   (a) antimony   (b) gold   (c) iron   (d) mercury   (e) potassium   (f) silver
   (g) tin   (h) tungsten

6. Name the elements with the following symbols.
   (a) Na   (b) Mg   (c) Al   (d) Si   (e) P   (f) Cl   (g) Ar

7. Name the elements with the following symbols.
   (a) Ti   (b) V   (c) Cr   (d) Mn   (e) Fe   (f) Co   (g) Ni   (h) Cu   (i) Zn

8. Name the elements with the following symbols.
   (a) Mo   (b) W   (c) Rh   (d) Ir   (e) Pd   (f) Pt   (g) Ag   (h) Au   (i) Hg

Evidence for the Existence of Atoms

9. Describe some of the evidence for the existence of atoms and some of the evidence from our senses that seems to deny the existence of atoms.

10. Describe what the formula P_4S_3 tells us about this compound.

11. Describe the difference between the following pairs of symbols.
    (a) Co and CO   (b) Cs and CS_2 (c) Ho and H_2O (d) 4 P and P_4

12. Describe the difference between the following pairs of symbols.
    (a) H and H^+ (b) H and H^- (c) 2 H and H_2 (d) H^+ and H^-

13. Explain the difference between H^+ ions, H atoms, and H_2 molecules on the atomic scale.

The Structure of the Atom

14. Describe the differences between a hydrogen atom and a proton, between a hydrogen atom and a neutron, and between a proton and a neutron.

15. Describe the relationship between the atomic number, mass number, number of protons, number of neutrons, and number of electrons in a calcium atom, ^{40}Ca, and its positively charged ion, $^{40}Ca^{2+}$.

16. Write the symbol for the atom or ion that contains 24 protons, 21 electrons, and 28 neutrons.

17. Calculate the number of protons and neutrons in the nucleus and the number of electrons surrounding the nucleus of a $^{39}K^+$ ion. What are the atomic number and the mass number of this ion?

18. Calculate the number of protons and neutrons in the nucleus and the number of electrons surrounding the nucleus of a $^{127}I^-$ ion. What are the atomic number and the mass number of this ion?

19. Identify the element that has atoms with mass numbers of 20 that contain 11 neutrons.

20. Give the symbol for the atom or ion that has 34 protons, 45 neutrons, and 36 electrons.

21. Calculate the number of electrons in a $^{134}Ba^{2+}$ ion.

22. Complete the following table.

Isotope	Atomic Number (Z)	Mass Number (A)	Number of Electrons
^{31}P	15	—	—
^{18}O	—	—	8
—	19	39	18
$^{58}Ni^{2+}$	—	58	—

23. How many isotopes of oxygen occur naturally on earth?

24. What do all isotopes of oxygen have in common? What is different?

25. If you select one carbon atom at random, what is the mass of that atom likely to be (in grams and in amu)?

26. What is the mass (in amu) of 100 ^{12}C atoms? Of 100 ^{13}C atoms?

27. If you select 100 carbon atoms at random, will the total mass be
    (a) 1200.00 amu?
    (b) slightly more than 1200.00 amu?
    (c) slightly less than 1200.00 amu?
    (d) 1300.3 amu?
    (e) slightly less than 1300.3 amu?
    Explain your reasoning.

Predicting the Formulas of Common Ionic Compounds

28. Fluoride toothpastes convert the mineral apatite in tooth enamel into fluorapatite, $Ca_5(PO_4)_3F$. If fluorapatite contains Ca^{2+} and PO_4^{3-} ions, what is the charge on the fluoride ion in this compound?

29. Verdigris is a green pigment used in paint. The simplest formula for the compound is $Cu_3(OH)_2(CH_3CO_2)_4$. What is the charge on the copper ions in this compound, if the other ions both carry a charge of -1?

30. Predict the formulas for neutral compounds containing the following pairs of ions.
    (a) Mg^{2+} and NO_3^- (b) Fe^{3+} and SO_4^{2-} (c) Na^+ and CO_3^{2-}

31. Predict the formulas for neutral compounds containing the following pairs of ions.
    (a) Na^+ and O_2^{2-} (b) Zn^{2+} and PO_4^{3-} (c) K^+ and $PtCl_6^{2-}$

32. Predict the formulas for sodium nitride and aluminum nitride if the formula for magnesium nitride is Mg_3N_2.

33. Compounds that contain the O^{2-} ion are called oxides. Those that contain the O_2^{2-} ion are called peroxides. If the formula for potassium oxide is K_2O, what is the formula for potassium peroxide?

34. What is the value of x in the $[Co(NO_2)_x]^{3-}$ ion if this complex ion contains Co^{3+} and NO_2^- ions?

35. Magnetic iron oxide has the formula Fe_3O_4. Explain this formula by assuming that Fe_3O_4 contains both Fe^{2+} and Fe^{3+} ions combined with O^{2-} ions.

The Periodic Table

36. Describe the differences among families, periods, and groups of elements in the periodic table.

37. Mendeléeff placed both silver and copper in the same group as lithium and sodium. Look up the chemistry of these four elements in the *CRC Handbook of Chemistry and*

Physics.[7] Describe some of the similarities that allow these elements to be classified in a single group on the basis of their chemical properties.

Metals, Nonmetals, and Semimetals

38. Classify each of the following elements as a metal, nonmetal, or semimetal.
    (a) Na   (b) Mg   (c) Al   (d) Si   (e) P   (f) S   (g) Cl   (h) Ar
39. Classify each of the following elements as a metal, nonmetal, or semimetal.
    (a) N   (b) Sb   (c) Sc   (d) Se   (e) Ge   (f) Sm   (g) Sn   (h) Sr
40. List three molecular elements and three atomic elements.
41. Classify the elements in Group VA as either metals, nonmetals, or semimetals. Describe what happens to the properties of the elements as we go down this column of elements.
42. Classify the elements in the third period as either metals, nonmetals, or semimetals. Describe what happens to the properties of the elements as we go from left to right across this period.

Integrated Problems

43. The mass number of the atom in Group IIA from which the ion is formed is 40. The formula of the ionic compound formed with the carbonate ion is XCO_3. How many electrons, protons, and neutrons does the ion X have? What is the chemical symbol for X?
44. There are two naturally occurring isotopes of element X. One of these isotopes has a natural abundance of 80.3% and a mass of 11.00931 amu. The second isotope is lighter. Identify element X. Give the number of neutrons, protons, and electrons in each isotope.

[7]CRC Press, Boca Raton, Florida.

C H A P T E R
1
SPECIAL TOPICS

1A.1   Significant Figures

1A.2   Unit Conversions

1A.1   SIGNIFICANT FIGURES

One way in which a scientist can indicate the quality of a measurement or the degree to which a piece of equipment allows a measurement to be made is by the use of significant figures. Consider the measurement 32.45 grams. The digits in 32.4 are known with certainty while there is uncertainty associated with the 0.05. The actual value might be 32.47 g, 32.44 g, or some other value that deviates in the second digit after the decimal. The actual value of the last reported significant figure is uncertain. All other significant figures in a measurement are assumed to be known with certainty. Thus in 32.45 there are four significant figures and the last digit is uncertain. You should use correct significant figures when recording data and reporting results in the laboratory and when working homework and test questions.

The uncertainty associated with many measurements results from estimating between the smallest divisions on a measuring device. Measurements should always be recorded to show the digits that are known with certainty (those digits corresponding to a division on the measuring device) plus one digit which is uncertain (corresponding to the estimation between the smallest divisions).

1. Figure 1A.1 shows a diagram of three volumes of water measured in 100 mL graduated cylinders. Record each measurement to the correct number of significant figures. A graduated cylinder is always read from the bottom of the meniscus.

Always read from the bottom of the meniscus.   **FIGURE 1A.1** Read the volume of liquid in the graduated cylinder.

Counting Significant Figures in a Measurement

Expressing numbers in scientific notation can help you count the number of significant figures. The exponential portion of a number expressed in scientific notation is not used in counting significant figures.

If all of the numbers are nonzero the number of digits equals the number of significant figures.

Measurement	Number of Significant Figures
1426.5	5
32.4561	6

Leading zeros are never significant. They are used only to set the position of the decimal point.

Measurement	Scientific Notation	Significant Figures
0.369	3.69×10^{-1}	3
0.00029	2.9×10^{-4}	2
0.008957	8.957×10^{-3}	4

Zeros between significant figures are always significant.

Measurement	Scientific Notation	Significant Figures
104.56	1.0456×10^{2}	5
0.002305	2.305×10^{-3}	4

Trailing zeros that are **not** needed to set the position of the decimal point are significant. Their function is to show how accurately the measurement is known.

Measurement	Scientific Notation	Significant Figures
12.30	1.230×10^{1}	4
5.00	5.00×10^{0}	3
230.	2.30×10^{2}	3

Trailing zeros that have no decimal point do not clearly show the number of significant figures. These measurements are best written in scientific notation. For example the measurement 700 milliters (mL; see Appendix A.1) could represent 7×10^{2} mL, 7.0×10^{2} mL, or 7.00×10^{2} mL (i.e., one, two, or three significant figures depending on how the measurement was made).

2. Express each of the following measurements in scientific notation and count the number of significant figures.

Measurement	Scientific Notation	Significant Figures
98.30		
104.02		
0.00285		
100.03		
0.04340		

Significant Figures in Calculations

It is important to maintain the proper number of significant figures when doing calculations. The number of digits shown in the result must correctly reflect the significant figures associated with each measurement used in the calculation.

Addition and Subtraction: In addition and subtraction the answer is restricted to the number of digits **after the decimal.** The measurement with the least number of digits **after the decimal** determines the number of digits after the decimal in the answer.

$$
\begin{array}{ll}
140.15 & \text{(two digits after the decimal)} \\
34.4129 & \text{(four digits after the decimal)} \\
\underline{2032.1} & \text{(one digit after the decimal)} \\
2206.6629 & \text{(calculator answer)}
\end{array}
$$

The answer in correct significant figures is 2206.7 (limited to one digit after the decimal). Note that the last digit needed to be rounded up to 7 since the portion that was dropped (0.0629) was greater than 0.0500.

Multiplication and Division: In multiplication and division the answer can have no more total significant figures than the measurement with the least number of significant figures.

$$12.34 \times 0.0203 \times 25.673 = 6.431137846$$
$$\text{Significant figures} \quad (4) \quad\;\; (3) \quad\;\;\; (5)$$

The answer in correct significant figures is 6.43 (limited to three significant figures).

Calculations with Multiple Steps: In calculations with multiple steps it is necessary to look at the number of significant figures in the answer to each step.

3. Explain why there are only three significant figures in the answer for the following calculation.

$$\frac{(123.4 + 0.42)}{(17.48 - 2.8)} = \frac{123.8}{14.7} = 8.42$$

4. Explain why there are only two significant figures in the answer.

$$\frac{(123.4 + 0.42)}{(17.48 - 17.00)} = \frac{123.8}{0.48} = 2.6 \times 10^2$$

Measurements versus Definitions

5. Use a metric rule to find the diameter of the following circle to two significant figures (for this measurement do not estimate between the mm marks).

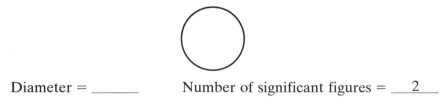

Diameter = _____   Number of significant figures = ___2___

6. Now use the definition of radius (radius = diameter/2) to find the radius of the circle using the measurement you made of diameter. Maintain significant figures as described in the previous sections.

Radius =   _____
Number of significant figures in numerator =   _____

Number of significant figures in denominator = _____
Number of significant figures in radius =   _____

7. Repeat the measurement of the diameter of the circle. This time, record the measurement to three significant figures (estimate between the millimeter marks). Now determine the radius of the circle while maintaining significant figures as described above.

Diameter =   _____
Number of significant figures in numerator =   _____
Number of significant figures in denominator = _____
Number of significant figures in radius =   _____
Radius =   _____

8. Compare the number of significant figures in the two values of the radius. Does the number of significant figures in the radii that you calculated in steps A1.6 and A1.7 differentiate between the number of significant figures in the two measurements that you made?

The purpose of significant figures is to reflect the quality of a **measurement.** The problem with the radii that you calculated is that the 2 in the denominator of the radius formula is not a measurement. It instead is a definition of radius. The 2 can be assumed to have an infinite number of significant figures. This also applies to counting numbers such as 24 people in a class. There could not be 24.5 or 23.9 people in a class. Numbers in a calculation that are not measurements (definitions and counting numbers) can be considered to have an infinite number of significant figures.

9. Now determine the radius of the circle for both measurements (steps A1.6 and A1.7) assuming that the 2 in the denominator has an infinite number of significant figures.

Measurement 1:
Radius = _____   Number of significant figures = _____
Measurement 2:
Radius = _____   Number of significant figures = _____

1A.2   UNIT CONVERSIONS

When conversion factors are given (as in Table B.2 in Appendix B of your text), you must determine the appropriate number of significant figures to use in converting a measurement to another unit.

$$1 \text{ kg} = 2.2046 \text{ lb}$$

This relationship is the result of a measurement. It therefore has a limited number of significant figures. You may assume that the number of significant figures associated with the 1 kg is equal to the number of significant figures in the pound measurement. This relationship can be determined by determining the weight in pounds of a 1.00000-kg mass using a balance or scale.

$$1.00 \text{ kg} = 2.20 \text{ lb}$$

Using a better balance or scale can give a better relationship.

$$1.0000 \text{ kg} = 2.2046 \text{ lb}$$

Some conversion factors are definitions and therefore have an infinite number of significant figures. Below are examples of two relationships between units that are based on definitions not measurements.

$$1 \text{ ft} = 12 \text{ inches} \quad \text{or} \quad 1.0000 \ldots \text{ ft} = 12.0000 \ldots \text{ inches}$$
$$1 \text{ inch} = 2.54 \text{ cm (by definition)} \quad \text{or} \quad 1.000 \ldots \text{ in.} = 2.54000 \ldots \text{ cm}$$

To convert 14 lb into kilograms:

$$14 \text{ } lb \times \frac{1 \text{ } kg}{2.2046 \text{ } lb} = 6.4 \text{ } kg$$

To convert 14.0000 lb into kilograms:

$$14.0000 \text{ } lb \times \frac{1 \text{ } kg}{2.2046 \text{ } lb} = 6.3504 \text{ } kg$$

To convert 58 cm into inches:

$$58 \text{ } cm \times \frac{1 \text{ } inch}{2.54 \text{ } cm} = 23 \text{ } inches$$

To convert 58.0000 cm into inches:

$$58.0000 \text{ } cm \times \frac{1 \text{ } inch}{2.54 \text{ } cm} = 22.8346 \text{ } inches$$

10. Explain the difference between the number of significant figures found in each of the previous four conversions.
11. Perform the following calculations and express the result in correct significant figures.
    (a) Convert 18.9 inches into centimeters
    (b) Convert 47.50 kg into pounds.
    (c) Convert 35.56 inches into centimeters.
    (d) Convert 92.3 cm into feet.

2

THE MOLE: THE LINK BETWEEN THE MACROSCOPIC AND THE ATOMIC WORLD OF CHEMISTRY

2.1   THE MACROSCOPIC, ATOMIC, AND SYMBOLIC WORLDS OF CHEMISTRY

Chemists work in three very different worlds (Figure 2.1). Most measurements are done on the **macroscopic** scale—with objects visible to the naked eye. When you walk into a chemical laboratory, you find a variety of bottles, tubes, flasks, and beakers designed to study samples of liquids and solids large enough to be seen. You may also find sophisticated instruments that can be used to analyze very small quantities of materials, but even these samples are visible to the naked eye.

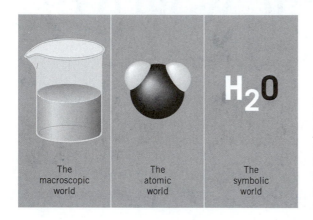

FIGURE 2.1 Chemists simultaneously work in three different worlds: (1) the macroscopic world of objects visible to the naked eye, which may be represented by a beaker of water; (2) the atomic world, in which water is thought of as molecules that contain two hydrogen atoms bound to an oxygen atom; and (3) the symbolic world, in which water is represented as H_2O.

Although they perform experiments on the macroscopic scale, chemists think about the behavior of matter in terms of a world of atoms and molecules. In this **atomic world,** water is no longer a liquid that freezes at 0°C and boils at 100°C but individual molecules that contain two hydrogen atoms and an oxygen atom.

One of the challenges students face when they encounter chemistry for the first time is understanding the process by which chemists perform experiments on the macroscopic scale that can be interpreted in terms of the structure of matter on the atomic scale. The task of bridging the gap between the atomic and macroscopic worlds is made more difficult by the fact that chemists also work in a **symbolic world,** in which they represent water as H_2O and write equations such as the following to represent what happens when hydrogen and oxygen react to form water.

$$2\,H_2 + O_2 \longrightarrow 2\,H_2O$$

The problem with the symbolic world is that chemists use the same symbols to describe what happens on both the macroscopic and the atomic scales. The symbol "H_2O," for example, is used to represent both a single water molecule and a beaker full of water.

It is easy to forget the link between the symbols chemists use to represent reactions and the particles involved in these reactions. Figure 2.2 provides an example of how you might envision the reaction described in the chemical equation written above. The reaction starts with a mixture of H_2 and O_2 molecules, each containing a pair of atoms. It produces water molecules, which contain two hydrogen atoms and an oxygen atom.

2.2   THE MASS OF AN ATOM

Atoms are so small that a sliver of copper metal just big enough to be detected on a good analytical balance contains about 1×10^{17} atoms. As a result, it is impossible to measure

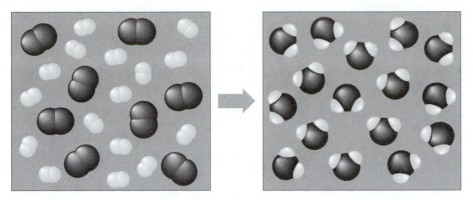

FIGURE 2.2 Mechanical models such as this make it easier to remember that chemical reactions involve particles in a state of constant motion that collide and then react.

the *absolute mass* of a single atom. We can, however, measure the *relative masses* of different atoms.

Figure 2.3 shows a diagram of a **mass spectrometer** that can be used to determine the relative mass of an atom or molecule. The sample is injected into an evacuated chamber. The particles in the sample flow past a filament, where they collide with high-energy electrons. As a result of these collisions, the neutral atoms or molecules in the sample lose electrons to form positively charged ions. As they pass between the poles of a magnet, the ions interact with the magnetic field. The interaction between the magnetic field and the charges on the ions bends the path along which the ions travel. The larger the mass of the ion, the smaller the angle through which its path is bent before it enters the detector.

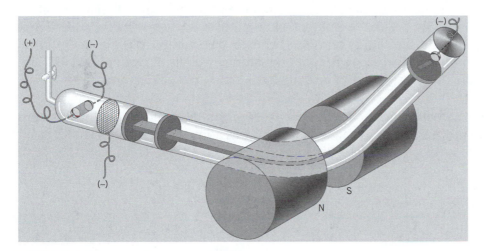

FIGURE 2.3 Diagram of a mass spectrometer.

Because the mass spectrometer can only tell us the relative mass of an atom, we need a standard with which our measurement can be compared. The standard used to calibrate these measurements is the ^{12}C isotope of carbon. Measurements of the masses of atoms can be made in grams or **atomic mass units** (amu; see Table 1.4). The mass of a single atom in grams is such a small number that the atomic mass unit is typically the preferred unit. By definition, the mass of a single atom of the ^{12}C isotope is exactly 12 atomic mass units or 12 amu. Thus, the mass number for this particular isotope, ^{12}C, is exactly equal to its mass in amu.

The masses of the atoms of the common isotopes of the lighter elements reported in Table 1.4 are all measured relative to ^{12}C. The following example shows how the mass of an atom can be determined from experimental data. Measurements taken with a mass spectrometer show that the mass of a ^{16}O atom is 1.3329 times as heavy as a ^{12}C atom.

$$\frac{\text{Mass of } ^{16}O}{\text{Mass of } ^{12}C} = 1.3329$$

We can therefore calculate the mass of a ^{16}O atom in units of atomic mass units by multiplying the known mass of a single ^{12}C atom by 1.3329.

$$\text{Mass of } ^{16}O = 1.3329 \times 12.000 \text{ amu} = 15.995 \text{ amu}$$

Most elements exist in nature as mixtures of isotopes. As we have seen, the graphite in a lead pencil is composed of a mixture of ^{12}C (98.892%, 12.000 amu), ^{13}C (1.108%, 13.003 amu), and an infinitesimally small amount of ^{14}C. It is therefore useful to calculate the **average mass** of a sample of carbon atoms. Because there is a large difference in the natural abundance of these isotopes, the average mass of a carbon atom must be a weighted average of the masses of the different isotopes. Because the amount of ^{14}C is so small, the average mass of a carbon atom is calculated using only the two most abundant isotopes of the element.

$$\left(12.000 \text{ amu} \times \frac{98.892}{100}\right) + \left(13.003 \text{ amu} \times \frac{1.108}{100}\right) = 12.011 \text{ amu}$$

The average mass of a carbon atom is much closer to the mass of a ^{12}C atom than a ^{13}C atom because the vast majority of the atoms in a sample of carbon are ^{12}C. This weighted average of all the naturally occurring isotopes of an atom is known as the **atomic weight** of the element. It is this value that is reported beneath the symbol of the element in the periodic table. It is important to recognize that the atomic weight of carbon is 12.011 amu even though no individual carbon atom actually has a mass of 12.011 amu.

EXAMPLE   The characteristics of a weighted average can be demonstrated with a common example from the classroom. Assume that a test was administered to ten students and the following grades were obtained: 84, 87, 84, 87, 92, 96, 92, 87, 84, 87. The average score on this exam is a weighted average of the results obtained by the individual students. The first step in determining the average score could involve calculating the percentage of the students who received each score.

Grade	Number of Grades	Percent of Grades
96	1	10%
92	2	20%
87	4	40%
84	3	30%

The average score on this exam can then be calculated from the grades and the percentage of the students receiving each grade.

$$\left[\left(96 \times \frac{10}{100}\right) + \left(92 \times \frac{20}{100}\right) + \left(87 \times \frac{40}{100}\right) + \left(84 \times \frac{30}{100}\right)\right] = 88$$

The average score on this exam is 88 even though no individual student received a grade of 88. In a similar fashion, the atomic weight of carbon is 12.011 amu even though no individual atom has this mass.

Exercise 2.1

Calculate the atomic weight of chlorine if 75.77% of the atoms have a mass of 34.97 amu and 24.23% have a mass of 36.97 amu.

Solution

Percent literally means "per hundred." Chlorine is therefore a mixture of atoms for which 75.77 parts per hundred have a mass of 34.97 amu and 24.23 parts per hundred have a mass of 36.97 amu. The atomic weight of chlorine is therefore 35.45 amu.

$$\left[\left(34.97 \text{ amu} \times \frac{75.77}{100}\right) + \left(36.97 \text{ amu} \times \frac{24.23}{100}\right)\right] = 35.45 \text{ amu}$$

2.3   THE MOLE AS THE BRIDGE BETWEEN THE MACROSCOPIC AND ATOMIC SCALES

Imagine that you pick the following items off the shelves of a grocery store: a dozen eggs, a 2-lb bag of sugar, a 5-lb bag of flour, and a quart of milk. When you open the egg carton, you know exactly how many eggs it should contain—a dozen. But the same cannot be said about the sugar, flour, and milk. A recipe may call for 1 egg, but it never calls for 1 grain of sugar because a grain of sugar is too small to be useful. By the time you get to adding the flour or milk, the problem becomes even more serious; it is physically impossible to add a single grain of flour. Recipes therefore call for a half cup of sugar or two cups of flour or a cup of milk.

Chemists face a similar problem because it takes an enormous number of atoms to give a sample large enough to be seen with the naked eye. (A dot of graphite from a pencil that is just large enough to be weighed on an analytical balance contains approximately 5×10^{19} atoms.) Chemists therefore created a unit known as the **mole** (from Latin, meaning "a huge pile") that can serve as the bridge between chemistry on the macroscopic and atomic scales. The mole is defined as follows.

> **A mole of any substance contains the same number of particles as the number of atoms in exactly 12.000 . . . grams of the ^{12}C isotope of carbon.**

Note that a single ^{12}C atom has a mass of exactly 12 amu and a mole of these atoms has a mass of exactly 12 grams.

$$1 \; ^{12}C \text{ atom} = 12.000 \ldots \text{amu}$$
$$1 \text{ mol of } ^{12}C \text{ atoms} = 12.000 \ldots \text{g}$$

The mole is the most fundamental unit of chemistry because it allows us to determine the number of elementary particles in a sample of a pure substance by simply determining the mass of the sample. Assume, for example, that we wanted to obtain a sample of aluminum metal that contained the same number of atoms as a mole of ^{12}C atoms. We can start by looking up the atomic weight of aluminum in the periodic table. When using atomic weights it is only necessary to retain as many digits in the atomic weight as are needed to maintain the appropriate number of significant digits (see Appendix A.3).

$$1 \text{ Al atom} = 26.982 \text{ amu}$$

Aluminum has only one naturally occurring isotope, therefore the weight that appears on the periodic chart is the mass of an aluminum atom. We then note that an aluminum atom is a little more than twice as heavy as a ^{12}C atom

$$\frac{1 \text{ Al atom}}{1 \ ^{12}\text{C atom}} = \frac{26.982 \text{ amu}}{12.000 \text{ amu}} = 2.2485$$

If a mole of aluminum contains exactly the same number of atoms as a mole of ^{12}C, then a mole of aluminum must have a mass that is 2.2485 times the mass of a mole of ^{12}C atoms.

$$\frac{1 \text{ mol Al}}{1 \text{ mol } ^{12}\text{C}} = \frac{26.982 \text{ g}}{12.000 \text{ g}} = 2.2485$$

Thus, a mole of aluminum would have a mass of 26.982 g.

Figure 2.4 shows two beakers. The beaker on the left contains 12.011 g of carbon (natural mixture of the isotopes of carbon), and the beaker on the right contains 26.982 g of aluminum. Therefore, both beakers contain 1 mole of the respective elements. There are the same number of atoms in both beakers.

12.011 g   26.982 g

Carbon   Aluminum

FIGURE 2.4 Each beaker contains 1 mole of the element, either carbon or aluminum. Both beakers contain the same number of atoms.

The results of the above relationship lead to the following generalization:

A mole of atoms of any element will have a mass in grams equal to the atomic weight of the element.

Therefore, the mass of a mole of any element can be read directly from the periodic table. The mass of a mole of a substance is often called its **molar mass.** The molar mass of ^{12}C, for example, is 12 g per mole (abbreviated mol). The molar mass of a sample of carbon that contains both ^{12}C and ^{13}C atoms in their natural abundances would be 12.011 g/mol. The relationship between atomic weight and molar mass exists for all elements.

Element	Atomic Weight	Molar Mass
Carbon	12.011 amu	12.011 g
Aluminum	26.982 amu	26.982 g
Iron	55.847 amu	55.847 g

The key to understanding the concept of the mole is recognizing that 12.011 grams of carbon contains the same number of atoms as 26.982 grams of aluminum or 55.847 grams of iron.

Checkpoint

What are the atomic weight and molar mass of potassium and uranium?

2.4 THE MOLE AS A COLLECTION OF ATOMS

If we knew how many ^{12}C atoms could be found in a mole of carbon, we would also know the number of elementary particles in a mole of any pure substance. Knowing the number of elementary particles in a mole would allow us to calculate the number of elementary particles in any mass of a pure substance simply by measuring the mass of the sample.

Chemists often need to know the relative number of particles of each reactant in a chemical reaction. For many years, chemists struggled to determine the number of elementary particles in a mole. The only way to achieve this goal is to measure the same quantity on both the atomic and the macroscopic scales. In 1910 Robert Millikan measured the charge in coulombs, C, on a single electron for the first time. Because the charge on a mole of electrons was already known from experimental measurements, it was possible to estimate the number of particles in a mole for the first time. Using more recent—and more accurate—data, we get the following results.

$$\frac{96,485 \text{ C}}{1 \text{ mol}} \times \frac{1 \text{ electron}}{1.60217733 \times 10^{-19} \text{ C}} = 6.0221 \times 10^{23} \frac{\text{electrons}}{\text{mol}}$$

This quantity is known as **Avogadro's number,** or more accurately **Avogadro's constant.**

Avogadro's number is so large it is difficult to comprehend. It would take 6 million million galaxies the size of the Milky Way to yield 6×10^{23} stars. At the speed of light, it would take 102 billion years to travel 6.02×10^{23} miles. There are only about 40 times this number of drops of water in all of the oceans on earth.

In everyday life units such as dozen (12) and gross (144) are used to describe a collection of items. The mole is sometimes referred to as the "chemist's dozen." The concept of the mole can be applied to any particle. In addition to talking about a mole of ^{12}C atoms, we can talk about a mole of Mg atoms, a mole of Na^+ ions, or a mole of electrons. Each time we use the term, we refer to Avogadro's number of items.

1 mole of ^{12}C atoms is composed of 6.022×10^{23} atoms of ^{12}C

1 mole of Al atoms is composed of 6.022×10^{23} atoms of Al

1 mole of Na^+ ions is composed of 6.022×10^{23} ions of Na^+

1 mole of electrons is composed of 6.022×10^{23} electrons

1 mole of molecules is composed of 6.022×10^{23} molecules

Once we know the number of elementary particles in a mole, we can determine the number of particles in a sample of a pure substance by weighing the sample. To see how this is done, let's consider a concrete example of the process by which objects of known mass are counted by weighing a sample.

The relationship between counting and weighing can be demonstrated with the following example. A dozen balls are placed on a balance, as shown in Figure 2.5. The mass of the dozen balls is found to be 107 grams. Now assume that an unknown number of balls has a mass of 178 grams. How many balls are in the unknown sample?

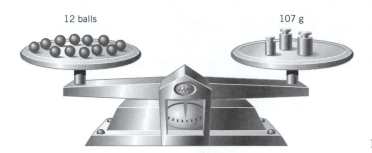

FIGURE 2.5 One dozen balls weighs 107 g.

We can build two unit factors from our knowledge of the mass of a dozen balls.

$$\frac{107 \text{ g}}{1 \text{ dozen balls}} \quad \text{or} \quad \frac{1 \text{ dozen balls}}{107 \text{ g}}$$

The question is: Which unit factor should we use? A technique known as **dimensional analysis** can guide us to the correct conversion factor. All we have to do is keep track of what happens to the units during the calculation. If the units cancel as expected, the calculation has been set up properly.

In this case, we know the mass of the unknown sample and the mass of a dozen balls. We therefore set up the calculation as follows:

$$178 \text{ g} \times \frac{1 \text{ dozen balls}}{107 \text{ g}} = 1.66 \text{ dozen balls}$$

We can now calculate the number of balls in the sample from the fact that there are 12 balls in a dozen.

$$166 \text{ dozen} \times \frac{12 \text{ balls}}{1 \text{ dozen}} = 20 \text{ balls}$$

In this example a dozen is analogous to a mole and 107 g/dozen is analogous to the molar mass of an element. You may well be asking yourself, wouldn't it be easier to just count the 20 balls? In the case of the balls it would be much easier just to count them rather than weighing them and then doing the above calculation. However, the balls represent atoms, which are too small and too numerous to count. Counting atoms would be an impossible task. The only method to determine the number of atoms in a pure sample is to weigh the sample and calculate the number of atoms.

We can use the same logic as in the above example to calculate the number of carbon atoms in a one-carat diamond. All we need to know is that a diamond can be thought of as a single crystal that contains only carbon atoms and that the mass of a carat is defined as 200.0 milligrams (mg; see Appendix A.1 for a discussion of unit prefixes and unit conversions). A one-carat diamond therefore has a mass of 0.2000 grams.

$$1 \text{ carat} \times \frac{200.0 \text{ mg}}{1 \text{ carat}} \times \frac{1 \text{ g}}{1000 \text{ mg}} = 0.2000 \text{ g}$$

The atomic weight of carbon is 12.011 amu, which means that the molar mass of carbon is 12.011 g/mol.

$$6.0221 \times 10^{23} \text{ atoms of C} = 1 \text{ mol C} = 12.011 \text{ g}$$

We therefore start the calculation by determining the number of moles of carbon in the diamond.

$$0.2000 \text{ g C} \times \frac{1 \text{ mol C}}{12.011 \text{ g C}} = 0.01665 \text{ mol C}$$

We then use Avogadro's number to calculate the number of carbon atoms in the diamond.

$$0.01665 \text{ mol C} \times \frac{6.022 \times 10^{23} \text{ atoms}}{1 \text{ mol C}} = 1.003 \times 10^{22} \text{ C atoms}$$

Checkpoint

How many aluminum atoms are in 1.0 g of aluminum?

2.5 CONVERTING GRAMS INTO MOLES AND NUMBER OF ATOMS

The mole is the bridge between chemistry on the macroscopic scale, where we do experiments, and the atomic scale, where we think about the implications of these experiments. As a result, one of the most common calculations in chemistry involves converting measurements of the mass of a sample into the number of moles of the substance it contains. To show how this is done, consider the following question: How many moles of sulfur atoms can be found in a sample of sulfur that has a mass of 45.5 grams?

What information do we need to convert grams of a substance into moles?

$$\text{grams} \longrightarrow \text{moles}$$

It seems reasonable to try to find out the number of *grams per mole* of sulfur. According to the periodic table, the atomic weight of sulfur is 32.07 amu. This means that a mole of sulfur atoms would have a mass of 32.07 grams.

$$1 \text{ mol S} = 32.07 \text{ g}$$

This equality can be transformed into two unit factors.

$$\frac{1 \text{ mol S}}{32.07 \text{ g S}} \quad \text{and} \quad \frac{32.07 \text{ g S}}{1 \text{ mol S}}$$

Multiplying the size of the sample in grams by the unit factor on the left gives us the number of moles of sulfur atoms in the sample.

$$45.5 \text{ g S} \times \frac{1 \text{ mol S}}{32.07 \text{ g S}} = 1.42 \text{ mol S}$$

Once we know the number of moles of sulfur atoms, we can use Avogadro's number to calculate the number of atoms in the sample.

$$1.42 \text{ mol S} \times \frac{6.02 \times 10^{23} \text{ S atoms}}{1 \text{ mol S}} = 8.55 \times 10^{23} \text{ S atoms}$$

In general, you need two pieces of information to do calculations of this nature. You need to know the mass of a mole of the substance, and you need to know the number of particles in a mole.

$$\text{Mass} \xrightarrow[\text{Macroscopic Scale}]{\overset{Molar\ mass}{\longleftarrow\!\!\longrightarrow}} \text{Moles} \xrightarrow[\text{Atomic Scale}]{\overset{Avogadro's\ number}{\longleftarrow\!\!\longrightarrow}} \text{Atoms}$$

Exercise 2.2

Calculate the mass of a sample of iron metal that would contain 0.250 moles of iron atoms.

Solution

According to the periodic table, the molar mass of iron is 55.85 g/mol. This can be represented in terms of either of the following units factors.

$$\frac{1 \text{ mol Fe}}{55.85 \text{ g Fe}} \quad \text{or} \quad \frac{55.85 \text{ Fe}}{1 \text{ mol Fe}}$$

In order to convert from moles to grams we need the unit factor that tells us how many grams of iron can be found in one mole of this metal.

$$0.250 \text{ mol Fe} \times \frac{55.85 \text{ g Fe}}{1 \text{ mole Fe}} = 14.0 \text{ g}$$

Exercise 2.3

Calculate the number of atoms in a 0.123-gram sample of aluminum foil.

Solution

Before we can do anything else, we need to know the number of moles of aluminum metal in the sample. This can be calculated from the mass of the sample and the molar mass of aluminum, which is 26.982 g/mol.

$$0.123 \text{ g Al} \times \frac{1 \text{ mol Al}}{26.982 \text{ g Al}} = 4.56 \times 10^{-3} \text{ mol Al}$$

Now we can use Avogadro's number to calculate the number of atoms in the sample.

$$4.56 \times 10^{-3} \text{ mol Al} \times \frac{6.02 \times 10^{23} \text{ Al atoms}}{1 \text{ mol Al}} = 2.75 \times 10^{21} \text{ Al atoms}$$

2.6   THE MOLE AS A COLLECTION OF MOLECULES

Before we can apply the concept of the mole to compounds, such as carbon dioxide (CO_2) or the sugar known as glucose ($C_6H_{12}O_6$), we have to be able to calculate the molecular weight of these compounds. As might be expected, the molecular weight of a compound is the sum of the atomic weights of the atoms in the formula of the compound.

Exercise 2.4

Calculate both the average mass of a single molecule of carbon dioxide and glucose and the molecular weight of these compounds.

Solution

The average mass of a *molecule of carbon dioxide* would be equal to the sum of the atomic weights of the three atoms in a CO_2 molecule.
  Mass of a single CO_2 molecule:

$$
\begin{aligned}
1 \text{ C atom } &= 1(12.011 \text{ amu}) = 12.011 \text{ amu} \\
2 \text{ O atoms } &= 2(15.999 \text{ amu}) = \underline{31.998 \text{ amu}} \\
&\phantom{= 2(15.999 \text{ amu}) = } 44.009 \text{ amu}
\end{aligned}
$$

The mass of a *mole of carbon dioxide* would be 44.009 grams.
  The average mass of a molecule of glucose is equal to the sum of the atomic weights of the 24 atoms in a $C_6H_{12}O_6$ molecule.
  Mass of a single $C_6H_{12}O_6$ molecule:

$$
\begin{aligned}
6 \text{ C atoms } &= 6(12.011 \text{ amu}) = 72.066 \text{ amu} \\
12 \text{ H atoms } &= 12(1.0079 \text{ amu}) = 12.095 \text{ amu} \\
6 \text{ O atoms } &= 6(15.999 \text{ amu}) = \underline{95.994 \text{ amu}} \\
&\phantom{= 6(15.999 \text{ amu}) = 0} 180.155 \text{ amu}
\end{aligned}
$$

The molecular weight of the compound is therefore 180.155 grams/mol.

For many years, chemists referred to the results of the calculations in the previous exercise as the **molecular weight** of the compound. This term is misleading for several reasons. First, no $C_6H_{12}O_6$ molecule ever has a mass equal to 180.155 amu. This is the *average mass* of the sugar molecules, most of which contain only ^{12}C and others of which contain one or perhaps two ^{13}C atoms. Second, some compounds, as we shall see, don't exist as molecules, so it is misleading to talk about their "molecular" weight. Some chemists therefore recommend that we describe the results of these calculations as the **mass of a mole** or the **molar mass** of a compound. Because the term molecular weight is so extensively used by chemists and is widely found in the chemical literature, however, we shall continue to use it in this text.

Exercise 2.5

Describe the difference between the mass of a mole of oxygen atoms (O) and the mass of a mole of oxygen molecules (O_2).

Solution

Because the atomic weight of oxygen is 15.999 amu, a mole of oxygen atoms has a mass of 15.999 grams. Each O_2 molecule has two atoms, however, so the molecular weight of O_2 molecules is twice as large as the atomic weight of the atom.

$$1 \text{ mol O} = 15.999 \text{ g} \qquad 1 \text{ mol } O_2 = 31.998 \text{ g}$$

The diagram we used to summarize mole–mass conversions for elements can be used for chemical compounds. In this case, however, we can take the calculation one step further by using the formula of the compound to calculate the number of atoms of a given element in the sample.

$$\text{Mass} \xrightarrow[\text{Macroscopic Scale}]{\textit{Molecular weight}} \text{Moles} \xleftrightarrow{\textit{Avogadro's number}} \text{Molecules} \xleftrightarrow[\text{Atomic Scale}]{\textit{Chemical formula}} \text{Atoms}$$

To illustrate the power of the mole concept, consider the following question:

What is the formula of carbon dioxide if 2.73 grams of carbon combine with 7.27 grams of oxygen molecules (O_2) when the carbon burns?

The first step toward the answer of an unfamiliar problem often involves drawing a diagram that helps us organize the information in the problem and visualize the process taking place. We could start, for example, with the simple diagram in Figure 2.6, which summarizes the relationship between the mass of carbon and the oxygen gas consumed in this reaction.

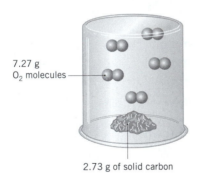

7.27 g
O_2 molecules

2.73 g of solid carbon

FIGURE 2.6 2.73 g of carbon is required to react with 7.27 g of O_2 molecules.

The next step in any problem of this kind is to convert the grams into moles. To do this, we need to know the relationship between the number of grams and the number of moles of the substance. It doesn't matter which element we start with because we eventually have to work with both, so let's arbitrarily start with carbon.

The atomic weight of carbon is 12.011 amu, which means that a mole of carbon has a mass of 12.011 grams. We can use this information to construct two unit factors.

$$\frac{1 \text{ mol C}}{12.011 \text{ g C}} \qquad \text{and} \qquad \frac{12.011 \text{ g C}}{1 \text{ mol C}}$$

Converting grams of carbon into moles requires a unit factor that has units of moles in the numerator and grams in the denominator. Dimensional analysis therefore suggests that the problem should be set up as follows.

$$2.73 \text{ g C} \times \frac{1 \text{ mol C}}{12.011 \text{ g C}} = 0.227 \text{ mol C}$$

The same format can be used to convert grams of oxygen into moles of oxygen atoms.

$$7.27 \text{ g O}_2 \times \frac{1 \text{ mol O}_2}{31.998 \text{ g O}_2} = 0.227 \text{ mol O}_2 \text{ molecules}$$

$$0.227 \text{ mol O}_2 \text{ molecules} \times \frac{2 \text{ O atoms}}{1 \text{ O}_2 \text{ molecule}} = 0.454 \text{ mol O atoms}$$

So far we have found that the reaction of 2.73 g of carbon and 7.27 g of oxygen corresponds to 0.227 mol of carbon reacting with 0.454 mol of oxygen atoms. Because atoms are neither created nor destroyed in a chemical reaction, the same number of atoms of each element must be found on both sides of the equation used to describe the reaction. The product of this reaction must have a mass of 10.00 g (2.73 + 7.27), and must contain 0.227 mol of carbon atoms and 0.454 mol of oxygen atoms.

We can then reread the question and ask: "Have we made any progress toward the answer?" In this case, we are trying to find the chemical formula for carbon dioxide, which gives the ratio of carbon atoms to oxygen atoms. (If the formula were CO, there would be just as many carbon atoms as oxygen atoms in the compound. If the formula were CO_2, there would be twice as many oxygen atoms in the compound.) The next step in the problem may therefore involve determining the relationship between the number of moles of carbon atoms and moles of oxygen atoms in our sample.

$$\frac{0.454 \text{ mol O}}{0.227 \text{ mol C}} = 2.00$$

There are twice as many moles of oxygen atoms as there are moles of carbon atoms in this sample. Because a mole of atoms always contains the same number of atoms, the only possible conclusion is that there are twice as many oxygen atoms as carbon atoms in the compound. In other words, the formula for carbon dioxide must be CO_2. The mole concept is the bridge between macroscopic measurements (the mass of carbon and oxygen) and the microscopic atomic world (the number of carbon and oxygen atoms in a carbon dioxide molecule).

Exercise 2.6

Determine the number of carbon atoms in 0.500 grams of carbon dioxide, CO_2.

Solution

The first step in this calculation involves converting the mass of the sample into the number of moles of CO_2. We will need the molecular weight of CO_2, which was determined in Exercise 2.4.

$$0.500 \text{ g CO}_2 \times \frac{1 \text{ mol CO}_2}{44.009 \text{ g CO}_2} = 1.14 \times 10^{-2} \text{ moles of CO}_2$$

Once we know the number of moles, we can use Avogadro's number to calculate the number of CO_2 molecules in the sample.

$$1.14 \times 10^{-2} \text{ mol CO}_2 \times \frac{6.02 \times 10^{23} \text{ CO}_2 \text{ molecules}}{1 \text{ mol CO}_2} = 6.86 \times 10^{21} \text{ CO}_2 \text{ molecules}$$

We can now use the chemical formula for carbon dioxide to determine the number of carbon atoms in the sample. The formula suggests that there is a single carbon atom for each CO_2 molecule.

$$6.86 \times 10^{21} \text{ CO}_2 \text{ molecules} \times \frac{1 \text{ C atom}}{1 \text{ CO}_2 \text{ molecule}} = 6.86 \times 10^{21} \text{ C atoms}$$

Checkpoint
How many atoms of carbon are there in one molecule of C_2H_2? How many moles of carbon atoms are in one mole of C_2H_2? How many atoms of carbon are in one mole of C_2H_2?

2.7   PERCENT MASS

Chemists are often interested in what portion of the mass of a compound is due to a specific element. This can be expressed in terms of **percent mass.** Percent represents parts per hundred. If you take a quiz with 25 problems and get 20 of them correct, the fraction correct is 20/25. To express this in terms of percent (the number correct per hundred problems) the fraction is multiplied by 100.

$$\% \text{ correct} = \frac{20 \text{ correct problems}}{25 \text{ total problems}} \times 100 = 80\%$$

The percent mass of an element in a compound may be determined either from experimental data (discussed in Section 2.9) or from the chemical formula. When the chemical formula is known, the percent mass can be calculated based on one mole of the compound. The percent mass of carbon in ethanol (CH_3CH_2OH) is determined by first calculating the mass of one mole of ethanol (the molecular weight).

$$
\begin{array}{l}
(2 \text{ carbons})(12.011 \text{ g/mol}) = 24.022 \text{ g of C/mole of ethanol} \\
(6 \text{ hydrogens})(1.0079 \text{ g/mol}) = 6.0474 \text{ g of H/mole of ethanol} \\
(1 \text{ oxygen})(15.999 \text{ g/mol}) = 15.999 \text{ g of O/mole of ethanol} \\
\hline
 46.068 \text{ g of ethanol/mole of ethanol}
\end{array}
$$

$$\% \text{ C in ethanol} = \frac{\text{mass of C in one mole of ethanol}}{\text{mass of one mole of ethanol}} \times 100$$

$$= \frac{24.022 \text{ g/mol}}{46.068 \text{ g/mol}} \times 100 = 52.145\% \text{ C}$$

The percent masses of hydrogen and oxygen in ethanol can be calculated in a similar fashion.

$$\% \text{ H in ethanol} = \frac{6.0474 \text{ g/mol}}{46.068 \text{ g/mol}} \times 100 = 13.127\% \text{ H}$$

$$\% \text{ O in ethanol} = \frac{15.999 \text{ g/mol}}{46.068 \text{ g/mol}} \times 100 = 34.729\% \text{ O}$$

2.8   DETERMINING THE FORMULA OF A COMPOUND

Section 2.6 showed one way to determine the formula of a compound. By carefully measuring the amount of carbon and oxygen that combine to form carbon dioxide, it was possible to show that the formula for this compound is CO_2. Let's look at another way to approach this problem using percent mass data. This time we will examine the compound methane, once known as "marsh gas" because it was first collected above certain swamps, or marshes, in Britain.

Marsh gas is 74.9% carbon and 25.1% hydrogen by mass. A 100-g sample of the gas therefore contains 74.9 g of carbon and 25.1 g of hydrogen.

$$100 \text{ g methane} \times \frac{74.9 \text{ g C}}{100 \text{ g methane}} = 74.9 \text{ g C}$$

$$100 \text{ g methane} \times \frac{25.1 \text{ g H}}{100 \text{ g methane}} = 25.1 \text{ g H}$$

This is useful information because we can use the atomic weights of these elements to convert the grams of carbon and hydrogen in this sample into moles.

$$74.9 \text{ g C} \times \frac{1 \text{ mol C}}{12.011 \text{ g C}} = 6.24 \text{ mol C}$$

$$25.1 \text{ g H} \times \frac{1 \text{ mol H}}{1.0079 \text{ g H}} = 24.9 \text{ mol H}$$

We now know the number of moles of carbon atoms and the number of moles of hydrogen atoms in a sample. We also know that there are always 6.022×10^{23} atoms in a mole of atoms of any element. It therefore might be useful to look at the ratio of the moles of these elements in the sample.

$$\frac{24.9 \text{ mol H}}{6.24 \text{ mol C}} = 3.99$$

The 100-g sample of marsh gas contains four times as many moles of hydrogen atoms as moles of carbon atoms. This means that there are four times as many hydrogen atoms as carbon atoms in this sample.

This experiment tells us the simplest or empirical formula but not the molecular formula of a marsh gas molecule. These results are consistent with molecules that contain one carbon atom and four hydrogen atoms: CH_4. But they are also consistent with formulas of C_2H_8, C_3H_{12}, C_4H_{16}, and so on. All we know at this point is that the molecular formula for the molecule is some multiple of the empirical formula, CH_4. Marsh gas is now known to have the formula CH_4.

Exercise 2.7

Calculate the empirical chemical formula for vitamin C, which is 40.9% C, 54.5% O, and 4.58% H by mass.

Solution

We start by calculating the number of grams of each element in a 100-g sample of vitamin C.

A diet that contains natural sources of vitamin C such as citrus fruits can be supplemented with vitamin C tablets.

$$100 \text{ g} \times 40.9\% \text{ C} = 40.9 \text{ g C}$$
$$100 \text{ g} \times 54.5\% \text{ O} = 54.5 \text{ g O}$$
$$100 \text{ g} \times 4.58\% \text{ H} = 4.58 \text{ g H}$$

We then convert the number of grams of each element into the number of moles of atoms of that element.

$$40.9 \text{ g C} \times \frac{1 \text{ mol C}}{12.011 \text{ g C}} = 3.41 \text{ mol C}$$

$$54.5 \text{ g O} \times \frac{1 \text{ mol O}}{15.999 \text{ g O}} = 3.41 \text{ mol O}$$

$$4.58 \text{ g H} \times \frac{1 \text{ mol H}}{1.0079 \text{ g H}} = 4.54 \text{ mol H}$$

Because we are interested in the simplest whole number ratio of these elements, we now divide through by the element with the smallest number of moles of atoms.

$$\frac{3.41 \text{ mol O}}{3.41 \text{ mol C}} = 1.00 \qquad \frac{4.54 \text{ mol H}}{3.41 \text{ mol C}} = 1.33$$

The ratio of C to H to O atoms in vitamin C is therefore $1:1\frac{1}{3}:1$. It doesn't make sense to write the ratio of atoms as $CH_{1\frac{1}{3}}O$, because there is no such thing as one-third of a hydrogen atom. We therefore multiply this ratio by small whole numbers until we get a formula in which all of the coefficients are integers

$$2(CH_{1\frac{1}{3}}O) = C_2H_{2\frac{2}{3}}O_2$$
$$3(CH_{1\frac{1}{3}}O) = C_3H_4O_3$$

Multiplying the ratio by three gives an empirical formula of $C_3H_4O_3$ for vitamin C.

The mass of a mole of $C_3H_4O_3$ molecules would be 88.062 g.

$$
\begin{array}{lll}
3 \text{ moles of C atoms} & = 3(12.011 \text{ g/mol}) & = 36.033 \text{ g} \\
4 \text{ moles of H atoms} & = 4(\ 1.0079 \text{ g/mol}) & = \ \ 4.0316 \text{ g} \\
3 \text{ moles of O atoms} & = 3(15.999 \text{ g/mol}) & = \underline{47.997 \text{ g}} \\
& & \ \ \ 88.062 \text{ g}
\end{array}
$$

This is as far as we can go with percent mass data. A separate experiment may be used to determine that the molecular weight of vitamin C is 176 g/mol. The molecular weight is therefore twice as large as the mass of a mole of $C_3H_4O_3$ molecules.

$$
\frac{176 \text{ g/mol}}{88.062 \text{ g/mol}} = 2.00
$$

The only possible conclusion is that a molecule of vitamin C is twice as large as the empirical formula for this compound. In other words, the **molecular formula** of vitamin C is $C_6H_8O_6$.

Checkpoint

A compound contains 24 g of carbon and 1.0 mole of calcium. What is the empirical formula of the compound? What additional information is necessary to determine the molecular formula?

2.9 ELEMENTAL ANALYSIS

Percent mass data for a compound are obtained by a process known as **elemental analysis.** The analysis is carried out by burning a small sample in the microanalysis apparatus shown in Figure 2.7. A few milligrams of the compound are added to a tiny platinum boat, which is placed in a furnace heated to about 850°C, and a stream of oxygen (O_2) gas is passed over the sample.

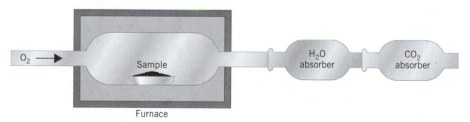

FIGURE 2.7 Diagram of the microanalysis apparatus used to determine the percentage by mass of carbon and hydrogen in a compound.

Compounds that contain only carbon and hydrogen burn to form a mixture of CO_2 and H_2O. If elements besides carbon and hydrogen are present, other gases may be formed as well. The CO_2 and H_2O produced in this combustion reaction are swept out of the furnace by the stream of oxygen gas and trapped in a pair of absorbers. The water vapor is absorbed onto a sample of magnesium perchlorate [$Mg(ClO_4)_2$] of known mass. The carbon dioxide is absorbed onto a known mass of the mineral ascharite ($Mg_2Br_2O_4 \cdot 2H_2O$). Ascharite contains bound water molecules and is called a hydrate. The water is designated by a dot followed by the number of bound water molecules.

A 3.00-mg sample of aspirin was analyzed in the apparatus shown in Figure 2.7. Aspirin is known to contain three elements: carbon, hydrogen, and oxygen. The results of the analysis found that 6.60 mg of CO_2 and 1.20 mg of H_2O were formed by burning the aspirin.

Our goal is to convert this information into data that describe the percent mass of carbon and hydrogen in aspirin. The logical place to start is by converting the number of grams of CO_2 and H_2O given off in this reaction into moles of these compounds.

$$0.00660 \text{ g } CO_2 \times \frac{1 \text{ mol } CO_2}{44.009 \text{ g } CO_2} = 1.50 \times 10^{-4} \text{ mol } CO_2$$

$$0.00120 \text{ g } H_2O \times \frac{1 \text{ mol } H_2O}{18.015 \text{ g } H_2O} = 6.66 \times 10^{-5} \text{ mol } H_2O$$

It would now be useful to know how many moles of carbon atoms and hydrogen atoms are present in the gases given off in this reaction. We therefore start by noting that there is one carbon atom in each CO_2 molecule, which means there is a mole of carbon atoms in a mole of CO_2 molecules.

$$1.50 \times 10^{-4} \text{ mol } CO_2 \times \frac{1 \text{ mol C}}{1 \text{ mol } CO_2} = 1.50 \times 10^{-4} \text{ mol C}$$

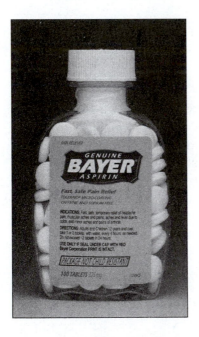

Elemental analysis can be used to determine the empirical formula for compounds such as aspirin.

Because all of the carbon came from the aspirin, the aspirin must have contained 1.50×10^{-4} mol of carbon atoms.

There are two hydrogen atoms in each H_2O molecule and therefore 2 mol of hydrogen atoms in a mole of H_2O molecules.

$$6.66 \times 10^{-5} \text{ mol } H_2O \times \frac{2 \text{ mol H}}{1 \text{ mol } H_2O} = 1.33 \times 10^{-4} \text{ mol H}$$

The aspirin therefore contained 1.33×10^{-4} moles of hydrogen atoms.

We now know the number of moles of carbon atoms and hydrogen atoms in the original sample, so we can calculate the number of grams of each element in the sample.

$$1.50 \times 10^{-4} \text{ mol C} \times \frac{12.011 \text{ g C}}{1 \text{ mol C}} = 1.80 \times 10^{-3} \text{ g C}$$

$$1.33 \times 10^{-4} \text{ mol H} \times \frac{1.0079 \text{ g H}}{1 \text{ mol H}} = 1.34 \times 10^{-4} \text{ g H}$$

According to this calculation, a 3.00-mg sample of aspirin contains 1.80 mg of carbon and 0.134 mg of hydrogen. Aspirin is therefore 60.0% C and 4.47% H by mass.

$$\frac{1.80 \text{ mg C}}{3.00 \text{ mg aspirin}} \times 100 = 60.0\% \text{ C}$$

$$\frac{0.134 \text{ mg H}}{3.00 \text{ mg aspirin}} \times 100 = 4.47\% \text{ H}$$

Carbon and hydrogen add up to only 64.5% of the total mass of the aspirin. The remaining mass (35.5%) must be due to the third element: oxygen. Microanalysis therefore suggests that aspirin is 60.0% C, 4.47% H, and 35.5% O by mass. As we saw in Exercise 2.7, this method can be used to determine that the empirical formula for aspirin is $C_9H_8O_4$.

2.10   SOLUTE, SOLVENT, AND SOLUTION

In the chemistry laboratory we often work with solutions rather than with individual pure chemicals. A **solution** is a uniform mixture. This means that the composition is the same throughout the mixture.

Suppose we dissolve a solid such as copper(II) sulfate pentahydrate ($CuSO_4 \cdot 5H_2O$) in water, as shown in Figure 2.8. The substance that dissolves ($CuSO_4 \cdot 5H_2O$) is called the **solute.** The substance in which the solute dissolves (water) is called the **solvent.** The mixture of solute and solvent is the solution. The reagent present in the smallest quantity is the solute.

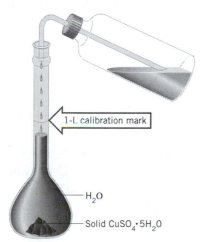

1-L calibration mark

H_2O

Solid $CuSO_4 \cdot 5H_2O$

FIGURE 2.8 A *solution* of $CuSO_4 \cdot 5 H_2O$ in water is made by dissolving the *solute* ($CuSO_4 \cdot 5 H_2O$) in a *solvent* (H_2O).

Not all solutions consist of a solid solute and liquid solvent. Table 2.1 gives examples of different kinds of solutions. $CuSO_4 \cdot 5H_2O$ is a dark blue solid which forms a blue solution when dissolved in water. When H_2 gas dissolves in platinum metal to form a solid solution, H_2 is the solute. When liquid mercury dissolves in sodium metal, mercury is the solute. Wine that is 12% ethanol (CH_3CH_2OH) by volume is a solution of a small quantity of liquid ethanol (the solute) in a larger volume of liquid water (the solvent). In a 50:50 mixture either component of the mixture can be thought of as the solute.

TABLE 2.1   Examples of Solutions

Solute	Solvent	Solution
$CuSO_4 \cdot 5H_2O(s)$	$H_2O(l)$	$CuSO_4 \cdot 5H_2O(aq)$
$H_2(g)$	$Pt(s)$	$H_2/Pt(s)$
$Hg(l)$	$Na(s)$	$Na/Hg(s)$
$CH_3CH_2OH(l)$	$H_2O(l)$	Wine

(s) = solid; (l) = liquid; (g) = gas; (aq) = aqueous.

2.11   SOLUTION CONCENTRATION

The ratio of the amount of solute to the amount of solution is known as the **concentration** of the solution.

$$\text{Concentration} = \frac{\text{amount of solute}}{\text{amount of solution}}$$

The concept of concentration is a common one. We talk about *concentrated* orange juice, which must be *diluted* with water. We even describe certain laundry products as *concentrated*, which means that we don't have to use as much of them.

2.12   MOLARITY AS A WAY OF COUNTING PARTICLES IN SOLUTION

Chemists use one concentration unit more than any other: **molarity (M)**. The molarity of a solution is defined as the number of moles of solute per liter of solution. Molarity is calculated by dividing the number of moles of solute in the solution by the volume of the solution in liters.

$$\text{Molarity} = \frac{\text{moles of solute}}{\text{liters of solution}}$$

Exercise 2.8

Copper sulfate is available as blue crystals that contain water molecules coordinated to the Cu^{2+} ions in the crystal. Because the crystals contain five water molecules per Cu^{2+} ion, the compound is called a pentahydrate, and the formula is written as $CuSO_4 \cdot 5H_2O$. Calculate the molarity of a solution prepared by dissolving 1.25 g of this compound in enough water to give 50.0 mL of solution.

Solution

A useful strategy for solving problems involves looking at the goal of the problem and asking: What information do we need to answer the question? The molarity of a solution is calculated by dividing the number of moles of solute by the volume of the solution. We therefore need two pieces of information: the number of moles of solute and the volume of the solution in liters.

The volume of the solution needs to be expressed in liters.

$$50.0 \text{ mL} \times \frac{1 \text{ L}}{1000 \text{ mL}} = 0.0500 \text{ L}$$

The number of moles of solute can be calculated from the mass of solute used to prepare the solution and the mass of a mole of this compound.

$$1.25 \text{ g CuSO}_4 \cdot 5\text{H}_2\text{O} \times \frac{1 \text{ mol}}{249.68 \text{ g CuSO}_4 \cdot 5\text{H}_2\text{O}} = 0.00501 \text{ mol CuSO}_4 \cdot 5\text{H}_2\text{O}$$

The molarity of the solution is then calculated by dividing the number of moles of solute in the solution by the volume of the solution.

$$\frac{0.00501 \text{ mol CuSO}_4 \cdot 5\text{H}_2\text{O}}{0.0500 \text{ L}} = 0.100 \text{ M CuSO}_4 \cdot 5\text{H}_2\text{O}$$

Sections 2.5 and 2.6 described how to use the mass of a pure chemical to count the number of moles that were present in a sample. Molarity can be used to count the number of moles of solute in a given volume of a solution. Note that molarity has units of moles per liter. The product of the molarity of a solution times its volume in liters is therefore equal to the number of moles of solute dissolved in the solution.

$$\frac{\text{mol}}{\text{L}} \times \text{L} = \text{mol}$$

We can write this relationship in terms of the following generic equation:

$$M \times V = n$$

where M is molarity, V is volume in liters, and n is number of moles.

Exercise 2.9

How many moles of sodium sulfate, Na_2SO_4, are present in 250 mL of a 0.150 M solution of sodium sulfate?

Solution

We use the molarity relationship to calculate the answer as follows:

$$M \times V = n$$
$$(0.150 \ M) \times (0.250 \text{ L}) = n$$
$$n = 0.0375 \text{ moles of Na}_2\text{SO}_4$$

2.13   DILUTIONS

Adding more solvent to a solution to decrease the solute concentration is known as **dilution.** Starting with a known volume of a solution of known molarity, we can prepare a more dilute solution of any desired concentration.

Exercise 2.10

Describe how you would prepare 2.50 L of an 0.360 M solution of sulfuric acid (H_2SO_4) starting with concentrated sulfuric acid that is 18.0 M.

Solution

We have one piece of information about the concentrated H_2SO_4 solution (the concentration is 18.0 moles per liter) and two pieces of information about the dilute solution (the volume is 2.50 liters and the concentration is 0.360 mole per liter), as shown in Figure 2.9. It seems reasonable to start with the solution about which we know the most. If we know the concentration (0.360 M) and the volume (2.50 liters) of the dilute sulfuric acid solution we are trying to prepare, we can calculate the number of moles of H_2SO_4 it must contain.

$$\frac{0.360 \text{ mol sulfuric acid}}{1 \cancel{L}} \times 2.50 \cancel{L} = 0.900 \text{ mol sulfuric acid}$$

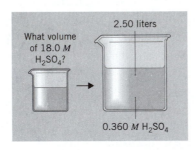

FIGURE 2.9 The volume and concentration of the dilute H_2SO_4 solution plus the concentration of the concentrated solution are all known in Exercise 2.10.

What volume of concentrated H_2SO_4 contains the same number of moles of H_2SO_4? We can start with the equation that describes the relationship between the molarity (M) of a solution, the volume of the solution (V) and the number of moles of solute in the solution (n).

$$M \times V = n$$

We then substitute into this equation the molarity of the concentrated sulfuric acid solution and the number of moles of sulfuric acid molecules needed to prepare the dilute solution.

$$\frac{18.0 \text{ mol sulfuric acid}}{1 \text{ L}} \times V = 0.900 \text{ mol sulfuric acid}$$

We then solve this equation for the volume of the solution.

$$V = 0.0500 \text{ L}$$

According to this calculation, we can prepare 2.50 L of 0.360 M H_2SO_4 solution by adding 50.0 mL of concentrated sulfuric acid to enough water to give a total volume of 2.50 L.

2.14   CHEMICAL REACTIONS AND THE LAW OF CONSERVATION OF ATOMS

We have focused so far on individual compounds such as carbon dioxide (CO_2) and glucose ($C_6H_{12}O_6$). Much of the fascination of chemistry, however, revolves around chemical reactions. The first breakthrough in the study of chemical reactions resulted from the work of the French chemist Antoine Lavoisier between 1772 and 1794. Lavoisier confirmed experimentally that the total mass of the products of a chemical reaction is always the same as the total mass of the starting materials consumed in the reaction. His results led to one of the fundamental laws of chemical behavior: the **law of conservation of matter,** which states that matter is conserved in a chemical reaction.

We now understand why matter is conserved—atoms are neither created nor destroyed in a chemical reaction. Hydrogen atoms in an H_2 molecule can combine with oxygen atoms in an O_2 molecule to form H_2O. But the number of hydrogen and oxygen atoms before and after the reaction is the same. The total mass of the products of a reaction therefore must be the same as the total mass of the reactants that undergo reaction.

2.15   CHEMICAL EQUATIONS AS A REPRESENTATION OF CHEMICAL REACTIONS

It is possible to describe a chemical reaction in words, but it is much easier to describe it with a **chemical equation.** The formulas of the starting materials, or **reactants,** are written on the left side of the equation, and the formulas of the **products** are written on the right. Instead of an equal sign, the reactants and products are separated by an arrow. The reaction between hydrogen and oxygen to form water, which is shown in Figure 2.10, is represented by the following equation.

$$2\,H_2 + O_2 \longrightarrow 2\,H_2O$$

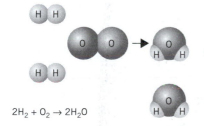

$2H_2 + O_2 \rightarrow 2H_2O$

FIGURE 2.10 The reaction between hydrogen, H_2, and oxygen, O_2, to form water, H_2O. The number and kinds of atoms in the reactants are the same as in the products; that is, the atoms are conserved.

It is often useful to indicate whether the reactants or products are solids, liquids, or gases by writing an s, l, or g in parentheses after the symbol for the reactants or products.

$$2\,H_2(g) + O_2(g) \longrightarrow 2\,H_2O(g)$$

Most of the reactions you will encounter in this course will occur when solutions of two substances dissolved in water are mixed. These **aqueous** solutions (from the Latin word *aqua,* "water") are so important we use the special symbol *aq* to describe them. This way we can distinguish between glucose as a solid, $C_6H_{12}O_6(s)$, and solutions of this sugar dissolved in water, $C_6H_{12}O_6(aq)$. Or between salt as an ionic solid, $NaCl(s)$, and solutions of salt dissolved in water, $NaCl(aq)$. The process by which a sample dissolves in water will be indicated by equations such as the following.

$$C_6H_{12}O_6(s) \xrightarrow{H_2O} C_6H_{12}O_6(aq)$$

$$NaCl(s) \xrightarrow{H_2O} NaCl(aq)$$

The above equations are referred to as **molecular equations** because all of the reactants and products are written as electrically neutral (no electric charge). Ionic compounds and some molecular compounds break up into their component ions when they dissolve in water. Therefore, the aqueous forms of these compounds may be written as aqueous ions. Since salt is an ionic compound the chemical equation describing the dissolution of salt can be written as follows.

$$NaCl(s) \xrightarrow{H_2O} Na^+(aq) + Cl^-(aq)$$

This form of a chemical equation is referred to as an **ionic equation.**

Checkpoint

The molecular equation for the reaction of $HCl(aq)$ with $NaOH(aq)$ is written below. $HCl(aq)$, $NaOH(aq)$, and $NaCl(aq)$ all break up into their ions when in aqueous solution. Write the ionic equation for the reaction.

$$HCl(aq) + NaOH(aq) \longrightarrow NaCl(aq) + H_2O(l)$$

Chemical equations are such a powerful shorthand for describing chemical reactions that we tend to think about reactions in terms of these equations. It is important to remember that a chemical equation is a statement of what *can happen,* not necessarily what *will happen.* The following equation, for example, does not guarantee that hydrogen will react with oxygen to form water.

$$2 H_2(g) + O_2(g) \longrightarrow 2 H_2O(g)$$

It is possible to fill a balloon with a mixture of hydrogen and oxygen and find that no reaction occurs until the balloon is touched with a flame. All the equation tells us is what would happen if, or when, the reaction occurs.

2.16   TWO VIEWS OF CHEMICAL EQUATIONS: MOLECULES VERSUS MOLES

Chemical equations such as the following can be used to represent what happens on either the atomic or macroscopic scale.

$$2 H_2(g) + O_2(g) \longrightarrow 2 H_2O(g)$$

This equation can be read in either of the following ways.

- When hydrogen reacts with oxygen, two molecules of hydrogen and one molecule of oxygen are consumed for every two molecules of water produced.
- When hydrogen reacts with oxygen, 2 mol of hydrogen and 1 mol of oxygen are consumed for every 2 mol of water produced.

Regardless of whether we think of the reaction in terms of molecules or moles, chemical equations must be balanced—they must have the same number of atoms of each element on both sides of the equation. As a result, the mass of the reactants will be equal to the mass of the products of the reaction. On the atomic scale, the following equation is balanced because the total mass of the reactants in atomic mass units is equal to the mass of the products.

$$2\ H_2(g) + O_2(g) \longrightarrow 2\ H_2O(g)$$
$$2 \times 2\ \text{amu} + 32\ \text{amu} \qquad 2 \times 18\ \text{amu}$$
$$36\ \text{amu} \qquad\qquad 36\ \text{amu}$$

On the macroscopic scale, it is balanced because the mass of 2 mol of hydrogen and 1 mol of oxygen is equal to the mass of 2 mol of water.

$$2\ H_2(g) + O_2(g) \longrightarrow 2\ H_2O(g)$$
$$2 \times 2\ \text{g} + 32\ \text{g} \qquad 2 \times 18\ \text{g}$$
$$36\ \text{g} \qquad\qquad 36\ \text{g}$$

The following diagram is a useful way of visualizing the relationship between the mass of the starting materials and products of the reaction. The box on the left shows the reactants, and the box on the right shows the products. The box centered above the arrow of the reaction represents all of the atoms found in the products and reactants.

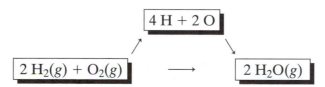

If we think about this reaction in terms of H_2 and O_2 molecules combining to form H_2O molecules, the equation is balanced because we have four hydrogen atoms and two oxygen atoms on both sides of the equation. If we think about the reaction in terms of moles of starting materials and products, the equation must be balanced because we have 4 mol of hydrogen atoms and 2 mol of oxygen atoms on both sides of the arrow.

It is important to recognize that reactions usually don't occur by passing through an intermediate stage in which they form isolated atoms. But this approach can be a useful way to emphasize the fact that atoms are conserved in a chemical reaction. Each and every atom among the starting materials must be found in one of the products of the reaction.

2.17   BALANCING CHEMICAL EQUATIONS

There is no sequence of rules that can be followed blindly to get a balanced chemical equation. All we can do is manipulate the coefficients written in front of the formulas of the reactants and products until the number of atoms of each element on both sides of the equation are the same. The subscripts in the chemical formulas cannot be changed in balancing the chemical equation because that would change the identity of the products and reactants. Persistence is required to balance chemical equations. The equation must be explored until the same number of atoms of each element is on both sides of the equation.

While doing this it is usually a good idea to tackle the easiest part of a problem first. Consider, for example, the equation for the combustion of glucose ($C_6H_{12}O_6$). Everything that we digest, at one point or another, gets turned into a sugar that is oxidized to give the energy that fuels our bodies. Although there are a variety of sugars that can be used as fuels, the primary source of energy that drives our bodies is glucose, or *blood sugar* as it is also known. The bloodstream delivers both glucose and oxygen to tissues, where they react to give a mixture of carbon dioxide and water.

$$C_6H_{12}O_6(aq) + O_2(g) \longrightarrow CO_2(g) + H_2O(l)$$

If you look at this equation carefully, you might conclude that it is going to be easier to balance the carbon and hydrogen atoms than the oxygen atoms in this reaction. All of the carbon atoms in glucose end up in CO_2 and all of the hydrogen atoms end up in H_2O, but there are two sources of oxygen among the starting materials and two compounds that contain oxygen among the products. This means that there is no way to predict the number of O_2 molecules that are consumed in this reaction until we know how many CO_2 and H_2O molecules are produced.

We can start the process of balancing this equation by noting that there are six carbon atoms in each $C_6H_{12}O_6$ molecule. Thus, six CO_2 molecules are formed for every $C_6H_{12}O_6$ molecule consumed.

$$1\ C_6H_{12}O_6 + \underline{\quad} O_2 \longrightarrow 6\ CO_2 + \underline{\quad} H_2O$$

There are 12 hydrogen atoms in each $C_6H_{12}O_6$ molecule, which means there must be 12 hydrogen atoms, or 6 H_2O molecules, on the right-hand side of the equation.

$$1\ C_6H_{12}O_6 + \underline{\quad} O_2 \longrightarrow 6\ CO_2 + 6\ H_2O$$

Now that the carbon and hydrogen atoms are balanced, we can try to balance the oxygen atoms. There are 12 oxygen atoms in 6 CO_2 molecules and 6 oxygen atoms in 6 H_2O molecules. To balance the 18 oxygen atoms in the products of this reaction we need a total of 18 oxygen atoms among the starting materials. But each $C_6H_{12}O_6$ molecule already contains 6 oxygen atoms. We therefore need 6 O_2 molecules among the reactants.

$$6\ C + 12\ H + 18\ O$$

$$1\ C_6H_{12}O_6 + 6\ O_2 \qquad \longrightarrow \qquad 6\ CO_2 + 6\ H_2O$$

There are now 6 carbon atoms, 12 hydrogen atoms, and 18 oxygen atoms on each side of the equation. The balanced equation for this reaction, which is shown in Figure 2.11, is therefore written as follows.

$$C_6H_{12}O_6(aq) + 6\ O_2(g) \longrightarrow 6\ CO_2(g) + 6\ H_2O(l)$$

$$C_6H_{12}O_6 + 6O_2 \rightarrow 6CO_2 + 6H_2O$$

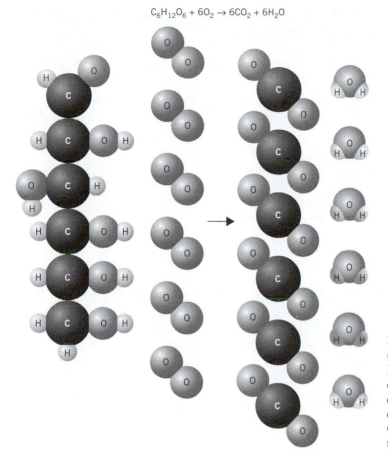

FIGURE 2.11 When glucose ($C_6H_{12}O_6$) burns to form CO_2 and H_2O, the 6 carbon atoms and 12 hydrogen atoms in the $C_6H_{12}O_6$ molecule must end up in the product molecules. Oxygen atoms from both $C_6H_{12}O_6$ and the 6 O_2 must appear in the product molecules.

Exercise 2.11

Write a balanced equation for the reaction that occurs when ammonia burns in air to form nitrogen oxide and water.

$$\underline{\quad} NH_3 + \underline{\quad} O_2 \longrightarrow \underline{\quad} NO + \underline{\quad} H_2O$$

Solution

We might start by balancing the nitrogen atoms. If we start with one molecule of ammonia and form one molecule of NO, the nitrogen atoms are balanced.

$$1\, NH_3 + \underline{\quad} O_2 \longrightarrow 1\, NO + \underline{\quad} H_2O$$

We can then turn to the hydrogen atoms. We have three hydrogen atoms on the left and two hydrogen atoms on the right in this equation. One way of balancing the hydrogen atoms is to look for the lowest common denominator: $2 \times 3 = 6$. We therefore set up the equation so that there are six hydrogen atoms on both sides. Doing this doubles the amount of NH_3 consumed in the reaction, so we have to double the amount of NO produced.

$$2\, NH_3 + \underline{\quad} O_2 \longrightarrow 2\, NO + 3\, H_2O$$

Because the nitrogen and hydrogen atoms are both balanced, the only task left is to balance the oxygen atoms. There are five oxygen atoms on the right side of this equation, so we need five oxygen atoms on the left. This could be accomplished by using a coefficient of $2\frac{1}{2}$ in front of oxygen as in the following equation:

$$2\text{ NH}_3 + 2\tfrac{1}{2}\text{ O}_2 \longrightarrow 2\text{ NO} + 3\text{ H}_2\text{O}$$

You will encounter some chemical equations written with fractional coefficients. However, such a reaction does not have meaning on the atomic scale, only on the molar scale. One-half molecule of O_2 would no longer be an oxygen molecule, but we could have one-half mole of oxygen molecules. If we insist that chemical equations work on both the atomic and macroscopic scale we must multiply the equation by 2.

$$\boxed{4\text{ N} + 12\text{ H} + 10\text{ O}}$$

$$4\text{ NH}_3 + 5\text{ O}_2 \qquad \longrightarrow \qquad 4\text{ NO} + 6\text{ H}_2\text{O}$$

The balanced equation for this reaction, which is shown in Figure 2.12, is therefore written as follows.

$$4\text{ NH}_3(g) + 5\text{ O}_2(g) \longrightarrow 4\text{ NO}(g) + 6\text{ H}_2\text{O}(g)$$

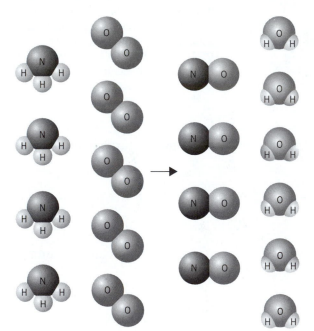

FIGURE 2.12 All atoms in the reactants must be accounted for in the products. The 12 hydrogen atoms in 4 NH_3, for example, are found in the 6 water molecules.

Checkpoint

What must be the value of x in the equation below in order to balance the equation?

$$x\text{ SO}_3(g) \longrightarrow 2\text{ SO}_2(g) + \text{O}_2(g)$$

2.18   MOLE RATIOS AND CHEMICAL EQUATIONS

There are two fundamental goals of science: (1) explaining observations about the world around us and (2) predicting what will happen under a particular set of conditions. Any chemical equation explains something about the world, but a balanced chemical equation has the added advantage of allowing us to predict what happens when the reaction takes place.

We describe a chemical reaction with a balanced chemical equation just as we might describe the preparation of oatmeal cookies with a recipe. A recipe for 5 dozen oatmeal cookies calls for 1 egg, 1 cup of flour, $\frac{1}{4}$ cup of water, 3 cups of oatmeal, and 1 cup of brown sugar. However, you are not restricted to preparing 5 dozen cookies. By reducing all the ingredients by two-fifths you could prepare only two dozen, and by doubling the amounts of the ingredients you could prepare ten dozen. In the same way in the laboratory we are not restricted to only working with the number of moles listed in the balanced chemical equation. Using the ratio of moles in the balanced chemical equation we can relate the moles of one compound to another in the chemical equation.

Consider the following question, for example:

Hydrazine, N_2H_4, can act as a rocket fuel. How many moles of oxygen are required to completely react with 5.20 mol of hydrazine in the following reaction, which is shown in Figure 2.13?

$$N_2H_4(l) + O_2(g) \longrightarrow N_2(g) + 2\ H_2O(l)$$

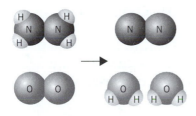

$$N_2H_4 + O_2 \rightarrow N_2 + 2H_2O$$

FIGURE 2.13 The reaction of $N_2H_4(l)$ with oxygen gas. The coefficients show that 1 mol of hydrazine reacts with 1 mol of O_2.

The coefficients in front of hydrazine and oxygen in this equation imply that 1 mol of hydrazine is consumed in this reaction for each mole of oxygen. We can express this information in terms of the following **mole ratios.** Stoichiometric coefficients are counting numbers (Chapter 1 Special Topics) and therefore have an infinite number of significant figures.

$$\frac{1\ \text{mol}\ O_2}{1\ \text{mol}\ N_2H_4} \quad \text{or} \quad \frac{1\ \text{mol}\ N_2H_4}{1\ \text{mol}\ O_2}$$

To determine the moles of oxygen required to react with 5.20 mol of N_2H_4, we must decide which of these mole ratios to use. Dimensional analysis suggests that we should use the mole ratio on the left because it allows us to convert moles of hydrazine to moles of oxygen.

$$5.20\ \text{mol}\ N_2H_4 \times \frac{1\ \text{mol}\ O_2}{1\ \text{mol}\ N_2H_4} = 5.20\ \text{mol}\ O_2$$

Exercise 2.12

How many moles of water are formed when 7.34 mol of hydrazine reacts with excess oxygen?

Solution

Since oxygen is in excess this reaction will occur until all of the hydrazine reacts and then stop. We now need the mole ratio of water to hydrazine. According to the balanced equation 2 mol of water are formed for every mole of hydrazine consumed in this reaction.

$$\frac{2 \text{ mol } H_2O}{1 \text{ mol } N_2H_4} \quad \text{or} \quad \frac{1 \text{ mol } N_2H_4}{2 \text{ mol } H_2O}$$

We can therefore calculate the number of moles of water produced in this reaction from the following equation:

$$7.34 \text{ mol } N_2H_4 \times \frac{2 \text{ mol } H_2O}{1 \text{ mol } N_2H_4} = 14.7 \text{ mol } H_2O$$

Let's now use what we have learned to see how a balanced chemical equation can be used to predict how much O_2 we have to breathe to digest 10.00 g of sugar. We start with the balanced equation for the reaction.

$$C_6H_{12}O_6(aq) + 6 \text{ } O_2(g) \longrightarrow 6 \text{ } CO_2(g) + 6 \text{ } H_2O(l)$$

We then ask the fundamental question: How many moles of $C_6H_{12}O_6$ molecules can be found in 10.00 g of this compound?

The only way we can convert grams of a substance into moles is to know something about the ratio of grams per mole in a sample of this substance. In other words, we need to know the molecular weight of the substance. The molecular weight of glucose calculated in Exercise 2.4 can be used to construct a pair of unit factors.

$$\frac{1 \text{ mol } C_6H_{12}O_6}{180.16 \text{ g } C_6H_{12}O_6} \quad \text{or} \quad \frac{180.16 \text{ g } C_6H_{12}O_6}{1 \text{ mol } C_6H_{12}O_6}$$

By paying attention to the units during the calculation, it becomes easy to choose the correct unit factor to convert grams of sugar into moles.

$$10.00 \text{ g } C_6H_{12}O_6 \times \frac{1 \text{ mol } C_6H_{12}O_6}{180.16 \text{ g } C_6H_{12}O_6} = 0.05551 \text{ mol } C_6H_{12}O_6$$

We now turn to the balanced equation for the reaction.

$$C_6H_{12}O_6(aq) + 6 \text{ } O_2(g) \longrightarrow 6 \text{ } CO_2(g) + 6 \text{ } H_2O(g)$$

This equation can be used to construct two mole ratios that describe the relationship between the moles of sugar and moles of oxygen consumed in the reaction.

$$\frac{6 \text{ mol } O_2}{1 \text{ mol } C_6H_{12}O_6} \quad \text{or} \quad \frac{1 \text{ mol } C_6H_{12}O_6}{6 \text{ mol } O_2}$$

By focusing on the units of this problem, we can select the correct mole ratio to convert moles of sugar into an equivalent number of moles of oxygen.

$$0.05551 \text{ mol } C_6H_{12}O_6 \times \frac{6 \text{ mol } O_2}{1 \text{ mol } C_6H_{12}O_6} = 0.3331 \text{ mol } O_2$$

We now need only one more step to complete our calculation—we need to convert the number of moles of O_2 consumed in the reaction into grams of oxygen. In Exercise 2.5 we concluded that the molecular weight of O_2 is exactly twice the atomic weight. The next step in the calculation therefore involves multiplying the number of moles of O_2 needed by the molecular weight.

$$0.3331 \text{ mol } O_2 \times \frac{31.998 \text{ g } O_2}{1 \text{ mol } O_2} = 10.66 \text{ g } O_2$$

We now have the answer to our original question. We need to breathe 10.66 g of oxygen to digest 10.00 g of the glucose that we carry through our bloodstream as the source of the energy needed to fuel our bodies.

2.19 STOICHIOMETRY

By now, you have encountered all the steps necessary to do calculations of the sort that are grouped under the heading **stoichiometry.** The goal of these calculations is to use a balanced chemical equation to predict the relationships between the amounts of the reactants and products of a chemical reaction. There are three steps in these calculations.

- Find the starting material or product of the reaction for which you know both the mass of the sample and the formula of the substance. Use the molecular weight of this substance to convert the number of grams in the sample into an equivalent number of moles. This step is a straightforward mass to moles calculation. You do not need to consider the number of moles of the substance used in the chemical reaction. That will be done in the next step.

- Use the balanced equation for the reaction to create a mole ratio that can convert the number of moles of the substance into moles of one of the other components of the reaction.

- Use the molecular weight of the other component of the reaction to convert the number of moles involved in the reaction into grams of that substance. This is a calculation of the mass of a substance from the number of moles, which is the converse of the first step. The stoichiometry of the chemical equation is not used here.

Exercise 2.13

Calculate the number of moles and mass of ammonia needed to prepare 3.00 grams of nitrogen oxide (NO) by the following reaction.

$$4 \text{ NH}_3(g) + 5 \text{ O}_2(g) \longrightarrow 4 \text{ NO}(g) + 6 \text{ H}_2O(g)$$

Solution

The only component of this reaction about which we know both the formula of the compound and the mass of the sample is nitrogen oxide. We therefore start by converting 3.00 g of NO into an equivalent number of moles of the compound. To do this, we need to calculate the molecular weight of NO, which is 30.006 g/mol. The number of moles of NO formed in this reaction can therefore be calculated as follows.

$$3.00 \text{ g NO} \times \frac{1 \text{ mol NO}}{30.006 \text{ g NO}} = 0.100 \text{ mol NO}$$

We now use the balanced equation for the reaction to determine the mole ratio that allows us to calculate the number of moles of NH_3 needed to produce 0.100 mol of NO.

$$0.100 \text{ mol NO} \times \frac{4 \text{ mol NH}_3}{4 \text{ mol NO}} = 0.100 \text{ mol NH}_3$$

We then use the molecular weight of NH_3 to calculate the mass of ammonia consumed in the reaction.

$$0.100 \text{ mol NH}_3 \times \frac{17.031 \text{ g NH}_3}{1 \text{ mol NH}_3} = 1.70 \text{ g NH}_3$$

According to this calculation, we need to start with 1.70 grams of ammonia to obtain 3.00 grams of nitrogen oxide.

Checkpoint

How many moles of H_2 are required to react completely with 2 moles of CO in the following reaction?

$$CO(g) + 2 H_2(g) \longrightarrow CH_3OH(g)$$

How many molecules of hydrogen, H_2, would this be? How many atoms of hydrogen, H, would this be?

Research in the 1990s

The Stoichiometry of the Breathalyzer

A patent was issued to R. F. Borkenstein in 1958 for the Breathalyzer, which remains the method of choice for determining whether an individual is DUI—driving under the influence—or DWI—driving while intoxicated. The chemistry behind the Breathalyzer is described by the following equation.

$$3 CH_3CH_2OH(g) + 2 Cr_2O_7{}^{2-}(aq) + 16 H^+(aq) \longrightarrow$$
$$3 CH_3CO_2H(aq) + 4 Cr^{3+}(aq) + 11 H_2O(l)$$

The instrument contains two ampules that hold 0.75 mg potassium dichromate ($K_2Cr_2O_7$) each, dissolved in sulfuric acid (H_2SO_4). One of the ampules is used as a reference. The other is opened and the breath sample to be analyzed is added. If alcohol is present in the breath, it reacts with the yellow-orange $Cr_2O_7^{2-}$ ion to form a green Cr^{3+} ion. The potassium dichromate in the ampule is in excess of the maximum amount of ethanol that might be expected to be present in someone's breath. The amount of green Cr^{3+} that is produced by the reaction is therefore directly dependent on the amount of ethanol that is present. For every 3 mol of ethanol that reacts 2 mol of dichromate will be used up, causing the yellow-orange color to decrease, and 4 mol of Cr^{3+} will be produced, thus increasing the green color of the solution in the test ampule. The extent to which the color balance between the two ampules is disturbed is a direct measure of the amount of alcohol in the breath sample.

Measurements of the alcohol on the breath are then converted into estimates of the concentration of alcohol in the blood. The link between these quantities is the assumption that 2100 mL of air exhaled from the lungs contains the same amount of alcohol as 1 mL of blood.

Measurements taken with the Breathalyzer are reported in units of percent blood-alcohol concentration (BAC) from 0 to 0.40%. In most U.S. states, a BAC of 0.10% is sufficient for a DWI conviction. (This corresponds to a blood-alcohol concentration of 0.10 g of alcohol per 100 mL of blood.)

Between January 1989 and December 1990, almost 50 papers were published that described research related to the measurement of blood-alcohol concentration. Several studies probed the implications of the fact that the ratio of alcohol in the breath to the blood-alcohol concentration varies from one individual to another. The research has shown that Breathalyzers are far more likely to *underestimate* blood-alcohol concentration than overestimate BAC.[1] Other research has shown that there is no significant difference in the rate at which an individual metabolizes alcohol with age.[2] This research has also led to the development of more accurate methods for determining blood-alcohol concentration, particularly during autopsies that are done on those who drink and drive.[3]

Work with general chemistry students has revealed an interesting misconcept. Some students believed they could "cheat" on a Breathalyzer test by placing a copper penny in their mouth. (Modern folklore apparently suggests that this will decrease the amount of alcohol on the breath.) Copper metal will, in fact, catalyze the following reaction, in which ethyl alcohol is oxidized to acetaldehyde.

$$CH_3CH_2OH \xrightarrow{\;Cu\;} Cu_3CHO + H_2$$

There is only one minor problem—*the copper penny has to be heated until it glows red-hot before it will do this!*

2.20   THE NUTS AND BOLTS OF LIMITING REAGENTS

According to Exercise 2.13, we need 1.70 g of ammonia to make 3.00 g of nitrogen oxide by the following reaction.

$$4\,NH_3(g) + 5\,O_2(g) \longrightarrow 4\,NO(g) + 6\,H_2O(g)$$

But we also need something else—we need enough oxygen for the reaction to take place.

[1]G. Simpson, *Journal of Analytical Toxicology*, **13**(2), 120 (1989).
[2]P. M. Hein and R. Volk, *Blutalkohol*, **26**(4), 276 (1990).
[3]J. V. Maracini, T. Carroll, S. Grant, S. Halleran, and J. A. Benz, *Journal of Forensic Science*, **41**, 181 (1989).

Figure 2.14 shows the amount of NO produced when 1.70 g of ammonia is allowed to react with different amounts of oxygen. At first, the amount of NO produced is directly proportional to the amount of O_2 present when the reaction begins. At some point, however, the yield of the reaction reaches a maximum. No matter how much more O_2 we add to the system, no more NO is produced.

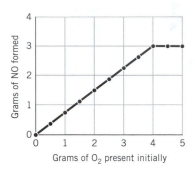

FIGURE 2.14 Grams of NO formed by the reaction of 1.70 g ammonia with oxygen according to $4 NH_3(g) + 5 O_2(g) \rightarrow 4 NO(g) + 6 H_2O(g)$. Addition of oxygen up to 4.0 g produces more NO, but from this point on no matter how much O_2 is added no more NO is produced from the same 1.70 g of NH_3.

We eventually reach a point at which the reaction runs out of NH_3 before all of the O_2 is consumed. When this happens, the reaction must stop. No matter how much O_2 is added to the system, we can't get more than 3.00 g of NO from 1.70 g of NH_3.

When there isn't enough NH_3 to consume all the O_2 in the reaction, the amount of NH_3 limits the amount of NO that can be produced. Ammonia is therefore the **limiting reagent** in this reaction. Because there is more O_2 than we need, it is the **excess reagent.**

The concept of limiting reagent is important because chemists frequently run reactions in which only a limited amount of one of the reactants is present. An analogy might help clarify what goes on in limiting reagent problems.

Let's start with exactly 10 nuts and 10 bolts, as shown in Figure 2.15. How many NB "molecules" can be made by screwing one nut (N) onto each bolt (B)? The answer is obvious—we can make exactly 10. After that, we run out of both nuts and bolts. Because we run out of both nuts and bolts at the same time, neither is a limiting reagent.

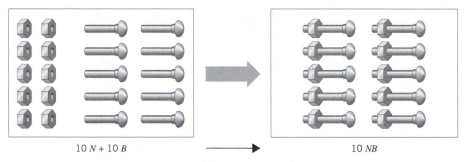

$10 N + 10 B$ $10 NB$

FIGURE 2.15 Starting with 10 nuts (N) and 10 bolts (B) we can make 10 NB molecules, with no nuts or bolts left over.

Now let's assemble N_2B molecules by screwing two nuts onto each bolt. Starting with 10 nuts and 10 bolts, we can make only five N_2B molecules, as shown in Figure 2.16. Because we run out of nuts, they must be the limiting reagent. Because five bolts are left over, they are the excess reagent.

It is also possible to put two bolts into a single nut. Starting with 10 nuts and 10 bolts, we can make only five NB_2 molecules, as shown in Figure 2.17. This time, the bolts are the limiting reagent and the nuts are present in excess.

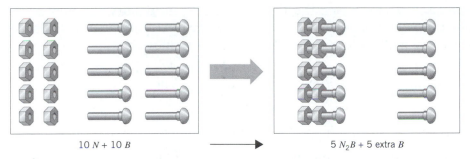

10 N + 10 B $\longrightarrow$ 5 N_2B + 5 extra B

FIGURE 2.16 Starting with 10 nuts and 10 bolts we can make only five N_2B molecules, and we will have five bolts left over. Because the number of N_2B molecules is limited by the number of nuts, the nuts are the limiting reagent in this analogy.

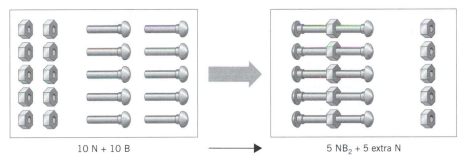

10 N + 10 B $\longrightarrow$ 5 NB_2 + 5 extra N

FIGURE 2.17 Starting with 10 nuts and 10 bolts we can make only five NB_2 molecules, and we will have five nuts left over. In this case, the bolts are the limiting reagent.

Now let's extend the analogy to a slightly more difficult problem in which we assemble as many N_2B molecules as possible from a collection of 30 nuts and 20 bolts. There are three possibilities: (1) we have too many nuts and not enough bolts, (2) we have too many bolts and not enough nuts, or (3) we have just the right number of both nuts and bolts.

One way to approach the problem is to pick one of these alternatives and test it. Because there are more nuts (30) than bolts (20), let's assume that we have too many nuts and not enough bolts. In other words, let's assume that bolts are the limiting reagent in this problem. Now let's test that assumption. According to the formula, N_2B, we need two nuts for every bolt. Thus, we need 40 nuts to consume 20 bolts.

$$20 \text{ bolts} \times \frac{2 \text{ nuts}}{1 \text{ bolt}} = 40 \text{ nuts}$$

According to this calculation, we need more nuts (40) than we have (30). Thus, our original assumption is wrong. We don't run out of bolts, we run out of nuts.

Because our original assumption is wrong, let's turn it around and try again. Now let's assume that nuts are the limiting reagent and calculate the number of bolts we need.

$$30 \text{ nuts} \times \frac{1 \text{ bolt}}{2 \text{ nuts}} = 15 \text{ bolts}$$

Do we have enough bolts to use up all the nuts? Yes, we need only 15 bolts yet we have 20 bolts to choose from.

Our second assumption was correct. The limiting reagent in this case is nuts, and the excess reagent is bolts. We can now calculate the number of N_2B molecules that can be assembled from 30 nuts and 20 bolts. Because the limiting reagent is nuts, the number of nuts limits the number of N_2B molecules we can make. Because we get one N_2B molecule for every two nuts, we can make a total of 15 of the N_2B molecules.

$$30 \text{ nuts} \times \frac{1 \ N_2B \text{ molecule}}{2 \text{ nuts}} = 15 \ N_2B \text{ molecules}$$

The following sequence of steps is helpful in working limiting reagent problems.

- Recognize that you have a limiting reagent problem, or at least consider the possibility that there might be a limiting amount of one of the reactants.
- Assume that one of the reactants is the limiting reagent.
- See if you have enough of the other reactant to consume the material you have assumed to be the limiting reagent.
- If you do, your original assumption was correct.
- If you don't, assume that another reagent is the limiting reagent and test this assumption.
- Once you have identified the limiting reagent, calculate the amount of product formed.

Exercise 2.14

Magnesium metal burns rapidly in air to form magnesium oxide. This reaction gives off an enormous amount of energy in the form of light and is used in both flares and fireworks. How many moles and what mass of magnesium oxide (MgO) is formed when 10.0 g of magnesium reacts with 10.0 g of O_2?

The white light emitted during firework displays is produced by burning magnesium metal.

Solution

It might be useful to start with a simple diagram, such as Figure 2.18, that summarizes the relevant information in the problem. The next step toward solving the problem involves writing a balanced equation for the reaction.

$$2 \text{ Mg}(s) + O_2(g) \longrightarrow 2 \text{ MgO}(s)$$

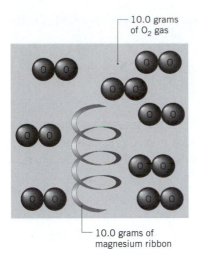

10.0 grams
of O_2 gas

10.0 grams of
magnesium ribbon

FIGURE 2.18 Reactants for Exercise 2.14.

We then pick one of the reactants and assume it is the limiting reagent. For the sake of argument, let's assume that magnesium is the limiting reagent and O_2 is present in excess. Our immediate goal is to test the validity of this assumption. If it is correct, we will have more O_2 than we need to burn 10.0 g of magnesium. If it is wrong, O_2 is the limiting reagent.

We start by converting grams of magnesium into moles of magnesium.

$$10.0 \text{ g Mg} \times \frac{1 \text{ mol Mg}}{24.31 \text{ g Mg}} = 0.411 \text{ mol Mg}$$

We then use the balanced equation to predict the number of moles of O_2 needed to burn this much magnesium. According to the equation for the reaction, it takes 1 mol of O_2 to burn 2 mol of magnesium. We therefore need 0.206 mol of O_2 to consume all of the magnesium.

$$0.411 \text{ mol Mg} \times \frac{1 \text{ mol } O_2}{2 \text{ mol Mg}} = 0.206 \text{ mol } O_2$$

We now calculate the mass of the O_2.

$$0.206 \text{ mol } O_2 \times \frac{32.00 \text{ g } O_2}{1 \text{ mol } O_2} = 6.59 \text{ g } O_2$$

According to this calculation, we need 6.59 g of O_2 to burn all the magnesium. Because we have 10.0 g of O_2, our original assumption was correct. We have more than enough O_2 and only a limited amount of magnesium.

We can now calculate the amount of magnesium oxide formed when all of the limiting reagent is consumed. The balanced equation suggests that 2 mol of MgO is produced for every 2 mol of magnesium consumed. Thus, 0.411 mol of MgO can be formed in this reaction.

$$0.411 \text{ mol Mg} \times \frac{2 \text{ mol MgO}}{2 \text{ mol Mg}} = 0.411 \text{ mol MgO}$$

We can use the molecular weight of MgO to calculate the number of grams of MgO that can be formed.

$$0.411 \text{ mol MgO} \times \frac{40.30 \text{ g MgO}}{1 \text{ mol MgO}} = 16.6 \text{ g MgO}$$

We can check the result of our calculations by noting that 10.0 grams of magnesium combine with 6.59 grams of O_2 to form 16.6 grams of MgO. (3.4 g of O_2 remains unused.) Mass is therefore conserved and we can feel confident that our calculations are correct.

Checkpoint

If 4 molecules of $I_2(s)$ are reacted with 4 atoms of Mg(s) what will be the limiting reagent for the following chemical reaction?

$$Mg(s) + I_2(s) \longrightarrow MgI_2(s)$$

2.21   SOLUTION STOICHIOMETRY

Many chemical reactions are performed in the laboratory using solutions that contain the reactants. It is necessary, therefore, to determine the moles of solute to be able to do stoichiometric calculations. The relationship between volume and molar concentration can be used to determine the moles of a reacting species in solution just as mass and molar mass were used in the previous sections. The following examples illustrate the use of volumes and concentrations to relate reacting species.

Exercise 2.15

The reaction between oxalic acid ($H_2C_2O_4$) and sodium hydroxide (NaOH) can be described by the following equation.

$$H_2C_2O_4(aq) + 2 \text{ NaOH}(aq) \longrightarrow 2 \text{ Na}^+(aq) + C_2O_4{}^{2-}(aq) + 2 \text{ H}_2O(l)$$

Calculate the concentration of an oxalic acid solution if it takes 34.0 mL of a 0.200 M NaOH solution to consume the acid in a 25.0-mL sample of the solution.

Solution

The above chemical equation can be rewritten to simplify it. The complete ionic equation can be written as follows.

$$H_2C_2O_4(aq) + 2 \text{ Na}^+(aq) + 2 \text{ OH}^-(aq) \longrightarrow 2 \text{ Na}^+(aq) + C_2O_4{}^{2-}(aq) + 2 \text{ H}_2O(l)$$

Because 2 $Na^+(aq)$ appears on both sides of the equation we can write

$$H_2C_2O_4(aq) + 2 \text{ OH}^-(aq) \longrightarrow C_2O_4{}^{2-}(aq) + 2 \text{ H}_2O(l)$$

This is called the **net ionic equation.**

We only know the volume of the oxalic acid solution, but we know both the volume and the concentration of the NaOH solution, as shown in Figure 2.19. It therefore seems

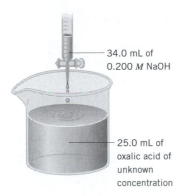

34.0 mL of
0.200 M NaOH

25.0 mL of
oxalic acid of
unknown
concentration   **FIGURE 2.19** Addition of sodium hydroxide to oxalic acid in Exercise 2.15.

reasonable to start by calculating the number of moles of NaOH in the given amount of solution.

$$\frac{0.200 \text{ mol NaOH}}{1 \text{ L}} \times 0.0340 \text{ L} = 6.80 \times 10^{-3} \text{ mol NaOH}$$

We now know the number of moles of NaOH consumed in the reaction, and we have a balanced chemical equation for the reaction that occurs when the two solutions are mixed. We can therefore calculate the number of moles of $H_2C_2O_4$ needed to consume this much NaOH.

$$6.80 \times 10^{-3} \text{ mol NaOH} \times \frac{1 \text{ mol } H_2C_2O_4}{2 \text{ mol NaOH}} = 3.40 \times 10^{-3} \text{ mol } H_2C_2O_4$$

We now know the number of moles of $H_2C_2O_4$ in the original oxalic acid solution. Using the volume of solution given in the statement of the problem, we can calculate the number of moles of oxalic acid per liter, or the molarity of the solution.

$$\frac{3.40 \times 10^{-3} \text{ mol } H_2C_2O_4}{0.0250 \text{ L}} = 0.136 \text{ M } H_2C_2O_4$$

The oxalic acid solution therefore has a concentration of 0.136 moles per liter.

Exercise 2.16

Calculate the volume of 1.50 *M* HCl that would react completely with 25.0 g of $CaCO_3$ according to the following balanced equation.

$$CaCO_3(s) + 2 \text{ HCl}(aq) \longrightarrow CaCl_2(aq) + CO_2(g) + H_2O(l)$$

Solution

We have only one piece of information about the HCl solution (its concentration is 1.50 *M*) and only one piece of information about $CaCO_3$ (it has a mass of 25.0 g), as shown in Figure 2.20. Which of these numbers do we start with? When relating two compounds in a chemical equation, we need to compare them on a mole basis. There is nothing we can do with the concentration of the HCl solution unless we know either the number of moles of HCl consumed in the reaction or the volume of the solution. We can work with the mass

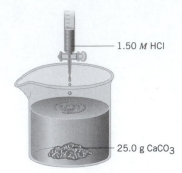

— 1.50 *M* HCl

— 25.0 g CaCO3

FIGURE 2.20 The concentration of the HCl solution and the number of grams of $CaCO_3$ are the only two pieces of information given in Exercise 2.16.

of $CaCO_3$ consumed in the reaction, however. We can start the calculation by converting grams of $CaCO_3$ into moles of $CaCO_3$.

$$25.0 \text{ g } CaCO_3 \times \frac{1 \text{ mol } CaCO_3}{100.09 \text{ g } CaCO_3} = 0.250 \text{ mol } CaCO_3$$

We then ask: Does this information get us any closer to our goal of calculating the volume of HCl consumed in this reaction? We know the number of moles of $CaCO_3$ present initially, and we have a balanced equation that states that 2 mol of HCl is consumed for every mole of $CaCO_3$. We can therefore calculate the number of moles of HCl consumed in the reaction.

$$0.250 \text{ mol } CaCO_3 \times \frac{2 \text{ mol HCl}}{1 \text{ mol } CaCO_3} = 0.500 \text{ mol HCl}$$

We now have the number of moles of HCl (0.500 mol) and the molarity of the solution (1.50 *M*). The number of moles and the molarity are related to volume by the following equation.

$$M \times V = n$$

Substituting the known values of the concentration of the HCl solution and the number of moles of HCl into the equation gives the following result.

$$\frac{1.50 \text{ mol}}{1 \text{ L}} \times V = 0.500 \text{ mol HCl}$$

We then solve the equation for the volume of the solution that would contain this amount of HCl.

$$V = 0.333 \text{ L}$$

According to this calculation, we need 333 mL of 1.50 M HCl to consume 25.0 g of $CaCO_3$.

Checkpoint
Write the ionic and the net ionic equation for the molecular reaction of Exercise 2.16.

KEY TERMS

Aqueous

Chemical equation

Concentration

Dilution

Elemental analysis

Excess reagent

Ionic equation

Law of conservation of
  matter

Limiting reagent

Molarity

Molecular equation

Mole ratio

Net ionic equation

Percent mass

Product

Reactant

Solute

Solution

Solvent

Stoichiometry

PROBLEMS

The Mass of an Atom

1. What would be the atomic weight of neon on a scale on which the mass of a ^{12}C atom is defined as exactly 1 amu?

2. Identify the element that has an atomic weight 4.33 times as large as carbon.

3. Calculate the atomic weight of bromine if naturally occurring bromine is 50.69% ^{79}Br atoms that have a mass of 78.9183 amu and 49.31% ^{81}Br atoms that have a mass of 80.9163 amu.

4. Naturally occurring zinc is 48.6% ^{64}Zn atoms (63.9291 amu), 27.9% ^{66}Zn atoms (65.9260 amu), 4.1% ^{67}Zn atoms (66.9721 amu), 18.8% ^{68}Zn atoms (67.9249 amu), and 0.6% ^{70}Zn atoms (69.9253 amu). Calculate the atomic weight of zinc.

The Mole as a Collection of Atoms

For problems 5 through 9 refer back to Table 1.4 in Chapter 1 for mass values.

5. What is the mass (in grams) of one ^{1}H atom? Of one ^{12}C atom?

6. What is the mass (in grams) of 4.35×10^6 atoms of ^{12}C?

7. What is the mass (in grams) of 6.022×10^{23} atoms of ^{12}C?

8. What is the mass (in grams) of one molecule of carbon dioxide which has one ^{12}C atom and two ^{16}O atoms, $^{12}C^{16}O_2$.

9. For 100,000 carbon atoms selected at random, what is the total mass (in amu)? What is the mass, on average, of a carbon atom? Does any carbon atom have this mass?

10. For any large collection of carbon atoms, what is the average mass of a carbon atom (in amu)? In grams?

11. For a large collection of magnesium (Mg) atoms, what is the average mass of a magnesium atom (in amu)?

12. What is the mass (in grams) of 6.022×10^{23} random magnesium atoms?

13. Examine the periodic table and find magnesium. What is the significance of the number given just below the symbol for magnesium?

14. Identify the element that contains atoms that have an average mass of 4.48×10^{-23} grams.

15. If the average mass of a chromium atom is 51.996 amu, what is the mass of a mole of chromium atoms?

16. If the average sulfur atom is approximately twice as heavy as the average oxygen atom, what is the ratio of the mass of a mole of sulfur atoms to the mass of a mole of oxygen atoms?

17. Calculate the mass in grams of a mole of atoms of the following elements.
    (a) C   (b) Ni   (c) Hg
18. If eggs sell for $0.90 a dozen, what does it cost to buy 2.5 dozen eggs? If the molar mass of carbon is 12.011 g, what is the mass of 2.5 mol of carbon atoms?

Avogadro's Number

19. What would be the value of Avogadro's number if a mole were defined as the number of ^{12}C atoms in 12 pounds of ^{12}C?
20. Calculate the number of atoms in 16 g of O_2, 31 g of P_4, and 32 g of S_8.
21. Calculate the number of chlorine atoms in 0.756 g of K_2PtCl_6.
22. Calculate the number of oxygen atoms in each of the following.
    (a) 0.100 mol of potassium permanganate, $KMnO_4$
    (b) 0.25 mol of dinitrogen pentoxide, N_2O_5
    (c) 0.45 mol of penicillin, $C_{16}H_{17}N_2O_5SK$
23. Benzaldehyde has the pleasant, distinctive odor of almonds. What is the molecular weight of benzaldehyde if a single molecule has a mass of 1.762×10^{-22} gram?

The Mole as a Collection of Molecules

24. How many carbon atoms are found in one dozen methane, CH_4, molecules? Give your answer in terms of a number (such as 17 or 3.25×10^{15} atoms).
25. How many hydrogen atoms are found in one dozen methane molecules?
26. How many carbon atoms are found in 1 mol of methane molecules?
27. How many hydrogen atoms are found in 1 mol of methane molecules?
28. What is the average mass (in amu) of one methane molecule?
29. What is the mass (in grams) of one mole of methane?
30. Consider 1.00 mol of hydrogen gas, H_2. How many hydrogen molecules are present? How many hydrogen atoms are present? What is the mass of the sample?
31. Indicate whether each of the following statements is true or false, and explain your reasoning.
    (a) One mole of NH_3 weighs more than 1 mol of H_2O.
    (b) There are more carbon atoms in 48 g of CO_2 than in 12 g of diamond (a pure form of carbon).
    (c) There are equal numbers of nitrogen atoms in 1 mol of NH_3 and 1 mol of N_2.
    (d) The number of Cu atoms in 100 g of Cu(s) is the same as the number of Cu atoms in 100 g of copper(II) oxide, CuO.
    (e) The number of Ni atoms in 100 moles of Ni(s) is the same as the number of Ni atoms in 100 mol of nickel(II) chloride, $NiCl_2$.
32. Which pair of samples contains the same number of hydrogen atoms?
    (a) 1 mol of NH_3 and 1 mol of N_2H_4 (b) 2 mol of NH_3 and 1 mol of N_2H_4
    (c) 2 mol of NH_3 and 3 mol of N_2H_4 (d) 4 mol of NH_3 and 3 mol of N_2H_4
33. Which of the following contains the largest number of carbon atoms?
    (a) 0.10 mol of acetic acid, CH_3CO_2H (b) 0.25 mol of carbon dioxide, CO_2
    (c) 0.050 mol of glucose, $C_6H_{12}O_6$ (d) 0.0010 mol of sucrose, $C_{12}H_{22}O_{11}$
34. Calculate the molecular weights of formic acid, HCO_2H, and formaldehyde, H_2CO.
35. Calculate the molecular weights of the following compounds.
    (a) methane, CH_4 (b) glucose, $C_6H_{12}O_6$
    (c) diethyl ether, $(CH_3CH_2)_2O$ (d) thioacetamide, CH_3CSNH_2

36. Calculate the molecular weight of the following compounds.
    (a) tetraphosphorus decasulfide, P_4S_{10} (b) nitrogen dioxide, NO_2
    (c) zinc sulfide, ZnS   (d) potassium permanganate, $KMnO_4$

37. Calculate the molecular weight of the following compounds.
    (a) chromium hexacarbonyl, $Cr(CO)_6$ (b) iron(III) nitrate, $Fe(NO_3)_3$
    (c) potassium dichromate, $K_2Cr_2O_7$ (d) calcium phosphate, $Ca_3(PO_4)_2$

38. Root beer hasn't tasted the same since the FDA outlawed the use of sassafras oil as a food additive because sassafras oil is 80% safrole, which has been shown to cause cancer in rats and mice. Calculate the molecular weight of safrole, $C_{10}H_{10}O_2$.

39. MSG ($C_5H_8NNaO_4$) is a spice commonly used in Chinese cooking that causes some people to feel light-headed (a disorder commonly known as Chinese restaurant syndrome). Calculate the molecular weight of MSG.

40. Calculate the molecular weights of the active ingredients in the following prescription drugs.
    (a) Darvon, $C_{22}H_{30}ClNO_2$ (b) Valium, $C_{16}H_{13}ClN_2O$
    (c) tetracycline, $C_{22}H_{24}N_2O_8$

Using Measurements of Mass to Count Atoms and Molecules

41. Calculate the number of moles of carbon atoms in 0.244 g of calcium carbide, CaC_2.

42. Calculate the number of moles of phosphorus in 15.95 g of tetraphosphorus decaoxide, P_4O_{10}.

43. Calculate the atomic weight of platinum, if 0.8170 mol of the metal has a mass of 159.4 g.

44. Calculate the mass of 0.0582 mol of carbon tetrachloride, CCl_4.

45. Calculate the density of carbon tetrachloride, if 1.000 mol of CCl_4 occupies a volume of 96.94 cm^3. Density is a physical property of substances and is defined as mass divided by volume.

46. Calculate the volume of a mole of aluminum, if the density of the metal is 2.70 g/cm^3. Density is defined in problem 45.

Percent Mass Calculations

47. Calculate the percent mass of chromium in each of the following oxides.
    (a) CrO (b) Cr_2O_3 (c) CrO_3

48. Calculate the percent mass of nitrogen in the following fertilizers.
    (a) $(NH_4)_2SO_4$ (b) KNO_3 (c) $NaNO_3$ (d) $(H_2N)_2CO$

49. Calculate the percent mass of carbon, hydrogen, and chlorine in DDT, $C_{14}H_9Cl_5$.

50. Emeralds are gem-quality forms of the mineral beryl, $Be_3Al_2(SiO_3)_6$. Calculate the percent mass of silicon in beryl.

51. Osteoporosis is a disease common in older women who have not had enough calcium in their diets. Calcium can be added to the diet by tablets that contain either calcium carbonate ($CaCO_3$), calcium sulfate ($CaSO_4$), or calcium phosphate [$Ca_3(PO_4)_2$]. On a per gram basis, which is the most efficient way of getting Ca^{2+} ions into the body?

Empirical Formulas

52. A molecule containing only nitrogen and oxygen contains (by mass) 36.8% N.
    (a) How many grams of N would be found in a 100-g sample of the compound? How many grams of O would be found in the same sample?

(b) How many moles of N would be found in a 100-g sample of the compound? How many moles of O would be found in the same sample?

(c) What is the ratio of the number of moles of O to the number of moles of N?

(d) What is the empirical formula of the compound?

53. Stannous fluoride, or "Fluoristan," is added to toothpaste to help prevent tooth decay. What is the empirical formula for stannous fluoride if the compound is 24.25% F and 75.75% Sn by mass?

54. Iron reacts with oxygen to form three compounds: FeO, Fe_2O_3, and Fe_3O_4. One of these compounds, known as magnetite, is 72.36% Fe and 27.64% O by mass. What is the formula of magnetite?

55. The chief ore of manganese is an oxide known as pyrolusite, which is 36.8% O and 63.2% Mn by mass. Which of the following oxides of manganese is pyrolusite?
(a) MnO (b) MnO_2 (c) Mn_2O_3 (d) MnO_3 (e) Mn_2O_7

56. Nitrogen combines with oxygen to form a variety of compounds, including N_2O, NO, NO_2, N_2O_3, N_2O_4, and N_2O_5. One of these compounds is called nitrous oxide, or "laughing gas." What is the formula of nitrous oxide if this compound is 63.65% N and 36.35% O by mass?

57. Chalcopyrite is a bronze-colored mineral that is 34.59% Cu, 30.45% Fe, and 34.96% S by mass. Calculate the empirical formula for the mineral.

58. A compound of xenon and fluorine is found to be 53.5% xenon by mass. What is the empirical formula of the compound?

59. In 1914, E. Merck and Company synthesized and patented a compound known as MDMA as an appetite suppressant. Although it was never marketed, it has reappeared in recent years as a street drug known as Ecstasy. What is the empirical formula of the compound if it contains 68.4% C, 7.8% H, 7.2% N, and 16.6% O by mass?

60. What is the empirical formula of the compound that contains 0.483 g of nitrogen and 1.104 g of oxygen?
(a) N_2O (b) NO (c) NO_2 (d) N_2O_3 (e) N_2O_4

61. What is the empirical formula of the compound formed when 9.33 g of copper metal reacts with excess chlorine to give 14.54 g of the compound?

Empirical and Molecular Formulas

62. Is it possible to determine the molecular formula of a compound solely from its percent composition? Why or why not?

63. β-Carotene is the protovitamin from which nature builds vitamin A. It is widely distributed in the plant and animal kingdoms, always occurring in plants together with chlorophyll. Calculate the molecular formula for β-carotene if the compound is 89.49% C and 10.51% H by mass and its molecular weight is 536.89 g/mol.

64. The phenolphthalein used as an indicator in acid–base titrations is also the active ingredient in laxatives such as ExLax. Calculate the molecular formula for phenolphthalein if the compound is 75.46% C, 4.43% H, and 20.10% O by mass and has a molecular weight of 318.31 grams per mole.

65. Caffeine is a central nervous stimulant found in coffee, tea, and cola nuts. Calculate the molecular formula of caffeine if the compound is 49.48% C, 5.19% H, 28.85% N, and 16.48% O by mass and has a molecular weight of 194.2 grams per mole.

66. Aspartame, also known as NutraSweet, is 160 times sweeter than sugar when dissolved in water. The true name for this artificial sweetener is N-L- α-aspartyl-L-phenyl-

alanine methyl ester. Calculate the molecular formula of aspartame if the compound is 57.14% C, 6.16% H, 9.52% N, and 27.18% O by mass and has a molecular weight of 294.30 grams per mole.

Elemental Analysis: Experimental Results

67. How accurately would you have to measure the percent by mass of carbon and hydrogen to tell the difference between diazepam (Valium), with the formula $C_{16}H_{13}ClN_2O$, and chlordiazepoxide (Librium), with the formula $C_{16}H_{14}ClN_3O$?

68. The methane in natural gas, the propane used in camping stoves, and the butane used in butane lighters are all members of a family of compounds known as the alkanes, which have the generic formula C_nH_{2n+2}. Calculate the value of n for butane if 3.15 mg of butane burns in air to form 9.54 mg of CO_2 and 4.88 mg of H_2O.

69. In small quantities, the nicotine in tobacco is addictive. In large quantities, it is a deadly poison. Calculate the molecular formula of nicotine, $C_xH_yN_z$, if the molecular weight of nicotine is 162.2 g/mol and 4.38 mg of the compound burns to form 11.9 mg of CO_2 and 3.41 mg of water.

Solutions and Concentration

70. Define the terms *solution, solvent,* and *solute* and give an example of each.

71. Describe in detail the steps you would take to prepare 125 mL of 0.745 M oxalic acid, starting with solid oxalic acid dihydrate ($H_2C_2O_4 \cdot 2\ H_2O$). Describe the glassware you would need, the chemicals, the amounts of each chemical, and the sequence of steps you would take.

72. Muriatic acid is sold in many hardware stores for cleaning bricks and tile. What is the molarity of the commercial solution if 125 mL contains 27.3 g of HCl?

73. Silver chloride is only marginally soluble in water; only 0.00019 g of AgCl dissolves in 100 mL of water. Calculate the molarity of this solution.

74. Ammonia (NH_3) is relatively soluble in water. Calculate the molarity of a solution that contains 252 g of NH_3 per liter.

75. During a physical exam, one of the authors was found to have a cholesterol level of 160 milligrams per deciliter (0.100 L). If the molecular weight of cholesterol is 386.67 grams per mole, what is the cholesterol level in his blood in units of moles per liter?

76. At 25°C, 5.77 g of chlorine gas dissolves in 1.00 liter of water. Calculate the molarity of Cl_2 in this solution.

77. Calculate the mass of Na_2SO_4 needed to prepare 0.500 L of a 0.150 M solution.

78. When asked to prepare a liter of 1.00 M K_2CrO_4, a student weighed out exactly 1.00 mol of K_2CrO_4 and added this solid to 1.00 L of water in a volumetric flask. What did the student do wrong? Did the student get a solution that was more concentrated than 1.00 M or less concentrated than 1.00 M? How would you prepare the solution?

79. Benzene, C_6H_6, was once used as a common solvent in chemistry. Its use is now restricted because of fear that it might cause cancer. The recommended limit of exposure to benzene is 3.2 milligrams per cubic meter of air (3.2 mg/m^3). Calculate the molarity of the solution. Calculate the number of parts per million (mg/kg) by mass of benzene in the solution. (See Table A.6.) The density of air is 1.29 g/L.

80. You can make the chromic acid bath commonly used to clean glassware in the lab by dissolving 92 g of sodium dichromate ($Na_2Cr_2O_7 \cdot 2\ H_2O$) in enough water to give 458

mL of solution and then adding 800 mL of concentrated sulfuric acid. Calculate the molarity of the $Cr_2O_7^{2-}$ ion in the solution.

81. People who smoke marijuana can be detected by looking for the tetrahydrocannabinols (THC) that are the active ingredient in marijuana. The present limit on detection of THC in urine is 20 nanograms of THC per milliliter of urine (20 ng/mL). Calculate the molarity of the solution at the limit of detection if the molecular weight of THC is 315 g/mol.

Dilution Calculations

82. Calculate the volume of 17.4 M acetic acid needed to prepare 1.00 L of 3.00 M acetic acid.

83. Calculate the concentration of the solution formed when 15.0 mL of 6.00 M HCl is diluted with 25.0 mL of water.

84. Describe how you would prepare 0.200 liter of 1.25 M nitric acid from a solution that is 5.94 M HNO_3.

Balancing Chemical Equations

85. Balance the following chemical equations.
    (a) $Cr(s) + O_2(g) \rightarrow Cr_2O_3(s)$
    (b) $SiH_4(g) \rightarrow Si(s) + H_2(g)$
    (c) $SO_3(g) \rightarrow SO_2(g) + O_2(g)$

86. Balance the following chemical equations.
    (a) $Pb(NO_3)_2(s) \rightarrow PbO(s) + NO_2(g) + O_2(g)$
    (b) $NH_4NO_2(s) \rightarrow N_2(g) + H_2O(g)$
    (c) $(NH_4)_2Cr_2O_7(s) \rightarrow N_2(g) + Cr_2O_3(s) + H_2O(g)$

87. Balance the following chemical equations.
    (a) $CH_4(g) + O_2(g) \rightarrow CO_2(g) + H_2O(g)$
    (b) $H_2S(g) + O_2(g) \rightarrow H_2O(g) + SO_2(g)$
    (c) $B_5H_9(g) + O_2(g) \rightarrow B_2O_3(s) + H_2O(g)$

88. Balance the following chemical equations.
    (a) $PF_3(g) + H_2O(l) \rightarrow H_3PO_3(aq) + HF(aq)$
    (b) $P_4O_{10}(s) + H_2O(l) \rightarrow H_3PO_4(aq)$

89. Balance the following chemical equations.
    (a) $C_3H_8(g) + O_2(g) \rightarrow CO_2(g) + H_2O(g)$
    (b) $C_2H_5OH(l) + O_2(g) \rightarrow CO_2(g) + H_2O(g)$
    (c) $C_6H_{12}O_6(s) + O_2(g) \rightarrow CO_2(g) + H_2O(l)$

Mole Ratios and Chemical Equations

90. For the reaction $2\ CO(g) + O_2(g) \rightarrow 2\ CO_2(g)$, how many moles of CO_2 are produced when 5 mol of O_2 is consumed?

91. For the reaction $2\ CO(g) + O_2(g) \rightarrow 2\ CO_2(g)$, does the number of moles of gas present increase, decrease, or remain the same as the reaction occurs?

92. For the reaction $CuO(s) + H_2(g) \rightarrow Cu(s) + H_2O(g)$, how many moles of CuO react to produce 12 mol of Cu?

93. Carbon disulfide burns in oxygen to form carbon dioxide and sulfur dioxide.

$$CS_2(l) + 3\ O_2(g) \longrightarrow CO_2(g) + 2\ SO_2(g)$$

Calculate the number of O_2 molecules it would take to consume 500 molecules of CS_2. Calculate the number of moles of O_2 molecules it would take to consume 5.00 mol of CS_2.

94. Calculate the number of moles of oxygen produced when 6.75 mol of manganese dioxide decomposes into Mn_3O_4 and O_2.

$$3\ MnO_2(s) \longrightarrow Mn_3O_4(s) + O_2(g)$$

95. Calculate the number of moles of carbon monoxide needed to reduce 3.00 mol of iron(III) oxide to iron metal.

$$Fe_2O_3(s) + 3\ CO(g) \longrightarrow 2\ Fe(s) + 3\ CO_2(g)$$

Stoichiometry

96. Describe in grammatically correct English sentences the steps taken to calculate the number of grams of CO_2 produced in the reaction below given that x grams of CO is consumed.

$$2\ CO(g) + O_2(g) \longrightarrow 2\ CO_2(g)$$

97. Calculate the mass of oxygen released when enough mercury(II) oxide decomposes to give 25 g of liquid mercury.

$$2\ HgO(s) \longrightarrow 2\ Hg(l) + O_2(g)$$

98. Calculate the mass of CO_2 produced and the mass of oxygen consumed when 10.0 g of methane (CH_4) is burned in oxygen to produce CO_2 and H_2O.

99. How many pounds of sulfur will react with 10.0 pounds of zinc to form zinc sulfide, ZnS?

100. Calculate the mass of oxygen that can be prepared by decomposing 25.0 g of potassium chlorate.

$$2\ KClO_3(s) \longrightarrow 2\ KCl(s) + 3\ O_2(g)$$

101. Predict the formula of the compound produced when 1.00 g of chromium metal reacts with 0.923 g of oxygen, O_2.

102. Ethanol, or ethyl alcohol, is produced by the fermentation of sugars such as glucose.

$$C_6H_{12}O_6(aq) \longrightarrow 2\ C_2H_5OH(aq) + 2\ CO_2(g)$$

Calculate the number of kilograms of alcohol that can be produced from 1.00 kilogram of glucose.

103. Calculate the number of pounds of aluminum metal that can be obtained from 1.000 ton of bauxite, $Al_2O_3\cdot 2\ H_2O$.

104. Calculate the mass of phosphine, PH_3, that can be prepared when 10.0 g of calcium phosphide, Ca_3P_2, reacts with excess water.

$$Ca_3P_2(s) + 6\ H_2O(l) \longrightarrow 3\ Ca(OH)_2(aq) + 2\ PH_3(g)$$

105. Hydrogen chloride can be obtained by reacting phosphorus trichloride with excess water and then boiling the HCl gas out of the solution into a collection flask.

$$PCl_3(g) + 3\ H_2O(l) \longrightarrow 3\ HCl(aq) + H_3PO_3(aq)$$

Calculate the mass of HCl gas that can be prepared from 15.0 g of PCl_3.

106. Nitrogen reacts with hydrogen to form ammonia:

$$N_2(g) + 3\ H_2(g) \longrightarrow 2\ NH_3(g)$$

which burns in the presence of oxygen to form nitrogen oxide

$$4\ NH_3(g) + 5\ O_2(g) \longrightarrow 4\ NO(g) + 6\ H_2O(l)$$

which reacts with excess oxygen to form nitrogen dioxide

$$2\ NO(g) + O_2(g) \longrightarrow 2\ NO_2(g)$$

which dissolves in water to give nitric acid.

$$3\ NO_2(g) + H_2O(l) \longrightarrow 2\ HNO_3(aq) + NO(g)$$

Calculate the mass of nitrogen needed to make 150 g of nitric acid assuming an excess of all other reactants.

Limiting Reagents

107. Calculate the number of water molecules that can be prepared from 500 H_2 molecules and 500 O_2 molecules.

$$2\ H_2(g) + O_2(g) \longrightarrow 2\ H_2O(l)$$

What would happen to the potential yield of water molecules if the amount of O_2 were doubled? What if the amount of H_2 were doubled?

108. Calculate the number of moles of P_4S_{10} that can be produced from 0.500 mol of P_4 and 0.500 mol of S_8.

$$4\ P_4(s) + 5\ S_8(s) \longrightarrow 4\ P_4S_{10}(s)$$

What would happen to the potential yield of P_4S_{10} if the amount of P_4 were doubled? What if the amount of S_8 were doubled?

109. Calculate the number of moles of nitrogen dioxide, NO_2, that could be prepared from 0.35 mol of nitrogen oxide and 0.25 mol of oxygen.

$$2\ NO(g) + O_2(g) \longrightarrow 2\ NO_2(g)$$

Identify the limiting reagent and the excess reagent in the reaction. What would happen to the potential yield of NO_2 if the amount of NO were increased? What if the amount of O_2 were increased?

110. Calculate the mass of hydrogen chloride that can be produced from 10.0 g of hydrogen and 10.0 g of chlorine.

$$H_2(g) + Cl_2(g) \longrightarrow 2\ HCl(g)$$

What would have to be done to increase the amount of hydrogen chloride produced in the reaction?

111. Calculate the mass of calcium nitride, Ca_3N_2, that can be prepared from 54.9 g of calcium and 43.2 g of nitrogen.

$$3\ Ca(s) + N_2(g) \longrightarrow Ca_3N_2(s)$$

112. PF_3 reacts with XeF_4 to give PF_5.

$$2\ PF_3(g) + XeF_4(s) \longrightarrow 2\ PF_5(g) + Xe(g)$$

How many moles of PF_5 can be produced from 100.0 g of PF_3 and 50.0 g of XeF_4?

113. Trimethyl aluminum, $Al(CH_3)_3$, must be handled in an apparatus from which oxygen has been rigorously excluded, because the compound bursts into flame in the presence of oxygen. Calculate the mass of trimethyl aluminum that can be prepared from 5.00 g of aluminum metal and 25.0 g of dimethyl mercury.

$$2\ Al(s) + 3\ Hg(CH_3)_2(l) \longrightarrow 2\ Al(CH_3)_3(l) + 3\ Hg(l)$$

114. The thermite reaction, used to weld rails together in the building of railroads, is described by the following equation.

$$Fe_2O_3(s) + 2\ Al(s) \longrightarrow Al_2O_3(s) + 2\ Fe(l)$$

Calculate the mass of iron metal that can be prepared from 150 grams of aluminum and 250 grams of iron(III) oxide.

Solution Stoichiometry

115. Calculate the concentration of an aqueous KCl solution if 25.00 mL of the solution gives 0.430 g of AgCl when treated with excess $AgNO_3$. Write the ionic and net ionic equations for the reaction.

$$KCl(aq) + AgNO_3(aq) \longrightarrow AgCl(s) + KNO_3(aq)$$

116. Calculate the volume of 0.25 M NaI that would be needed to precipitate all of the Hg^{2+} ion from 45 mL of a 0.10 M $Hg(NO_3)_2$ solution. Write the ionic and net ionic equations for this reaction.

$$2\ NaI(aq) + Hg(NO_3)_2(aq) \longrightarrow HgI_2(s) + 2\ NaNO_3(aq)$$

117. Calculate the molarity of an acetic acid solution if 34.57 mL of the solution is needed to neutralize 25.19 mL of 0.1025 M sodium hydroxide. CH_3CO_2H does not break up into its ions in solution, but all other aqueous reactants and products in the reaction do. Write the ionic and net ionic equations for the reaction.

$$CH_3CO_2H(aq) + NaOH(aq) \longrightarrow Na^+(aq) + CH_3CO_2^-(aq) + H_2O(l)$$

118. Calculate the molarity of a sodium hydroxide solution if 10.42 mL of the solution is needed to neutralize 25.00 mL of 0.2043 M oxalic acid. $H_2C_2O_4$ does not break up into its ions in solution, but all other aqueous reactants and products in the reaction do. Write the ionic and net ionic equations for the reaction.

$$H_2C_2O_4(aq) + 2\,NaOH(aq) \longrightarrow Na_2C_2O_4(aq) + 2\,H_2O(l)$$

119. Calculate the volume of 0.0985 M sulfuric acid that would be needed to neutralize 10.89 mL of a 0.01043 M aqueous ammonia solution. NH_3 does not break up into its ions in solution, but all other aqueous reactants and products in the reaction do. Write the ionic and net ionic equations for the reaction.

$$H_2SO_4(aq) + 2\,NH_3(aq) \longrightarrow (NH_4)_2SO_4(aq)$$

120. α-D-Glucopyranose reacts with the periodate ion as follows.

$$C_6H_{12}O_6(aq) + 5\,IO_4^-(aq) \longrightarrow 5\,IO_3^-(aq) + 5\,HCO_2H(aq) + H_2CO(aq)$$

Calculate the molarity of the glucopyranose solution if 25.0 mL of 0.750 M IO_4^- is required to consume 10.0 mL of the solution.

121. Oxalic acid reacts with the chromate ion in acidic solution as follows.

$$3\,H_2C_2O_4(aq) + 2\,CrO_4^{2-}(aq) + 10\,H^+(aq) \longrightarrow$$
$$6\,CO_2(g) + 2\,Cr^{3+}(aq) + 8\,H_2O(l)$$

Calculate the molarity of the oxalic acid solution if 10.0 mL of the solution consumes 40.0 mL of 0.0250 M CrO_4^{2-}.

Integrated Problems

122. Calculate the atomic weight of the metal, M, that forms a compound with the formula MCl_2 that is 74.5% Cl by mass.

123. Halothane is an anesthetic that is 12.17% C, 0.51% H, 40.48% Br, 17.96% Cl, and 28.87% F by mass. What is the molecular formula of the compound if each molecule contains one hydrogen atom?

124. A compound that is 31.9% K and 28.9% Cl by mass decomposes when heated to give O_2 and a compound that is 52.4% K and 47.6% Cl by mass. Write a balanced chemical equation for the reaction.

125. A compound that combines in fixed amounts with one or more molecules of water is known as a hydrate. For example, the hydrate of copper sulfate is composed of five water molecules and has the formula $CuSO_4 \cdot 5H_2O$. In the lab a 5.00-gram sample of the hydrate of barium chloride, $BaCl_2 \cdot xH_2O$, is heated to drive off the water. After heating, 4.26 g of anhydrous barium chloride, $BaCl_2$, remains. What mass of water was driven off during heating? How many moles of water were driven off during heating? How many moles of anhydrous $BaCl_2$ remain after heating? What is the value of x in the formula of the hydrate, $BaCl_2 \cdot xH_2O$?

126. Predict the formula of the compound produced when 1.00 g of chromium metal reacts with 0.923 g of oxygen atoms, O.

127. A 3.500-g sample of an oxide of manganese contains 1.288 grams of oxygen. What is the empirical formula of the compound?

128. Cocaine is a naturally occurring substance that can be extracted from the leaves of the *coca* plant, which grows in South America (and is not to be confused with chocolate, or *cocoa*, which is extracted from the seeds of another South American plant). If the chemical formula for cocaine is $C_{17}H_{21}O_4N$, what is the percentage by mass of carbon, hydrogen, oxygen, and nitrogen in the compound? Comment on the ease with which elemental analysis of the carbon and hydrogen in a compound can be used to distinguish between the white, crystalline powder known as aspirin ($C_9H_8O_4$), which is used to cure headaches, and the white, crystalline powder known as cocaine, which is more likely to cause headaches.

129. The oxygen-carrying protein known as hemoglobin is 0.335% Fe by mass. Calculate the molecular weight of the protein. There are four Fe atoms in each hemoglobin molecule.

130. Metal carbonates decompose when they are heated to form metal oxides and carbon dioxide.

$$MCO_3(s) \longrightarrow MO(s) + CO_2(g)$$

Which of the following metal carbonates would lose 35.1% of its mass when it decomposes?
(a) Li_2CO_3 (b) $MgCO_3$ (c) $CaCO_3$ (d) $ZnCO_3$ (e) $BaCO_3$

131. A crucible and sample of $CaCO_3$ weighing 42.670 g were heated until the compound decomposed to form CaO and CO_2.

$$CaCO_3(s) \longrightarrow CaO(s) + CO_2(g)$$

The crucible had a mass of 35.351 g. What is the theoretical mass of the crucible and residue after the decomposition is complete?

132. Nitrogen reacts with red-hot magnesium to form magnesium nitride

$$3\ Mg(s) + N_2(g) \longrightarrow Mg_3N_2(s)$$

which reacts with water to form magnesium hydroxide and ammonia.

$$Mg_3N_2(s) + 6\ H_2O(l) \longrightarrow 3\ Mg(OH)_2(aq) + 2\ NH_3(aq)$$

Calculate the number of grams of magnesium that would be needed to prepare 15.0 g of ammonia.

133. A sealed bottle contains oxygen gas (O_2) and liquid butyl alcohol ($C_4H_{10}O$). There is sufficient oxygen to react completely with the butyl alcohol to produce carbon dioxide (CO_2) and water (H_2O) gas. Write a chemical equation to describe this reaction. The bottle remains sealed during the reaction. Compare the number of molecules in the bottle originally ($C_4H_{10}O$ and O_2) with the number of molecules present in the bottle after the reaction (CO_2 and H_2O). Will the number of molecules in the bottle increase, decrease, or remain the same as the reaction takes place?

134. Criticize the following erroneous statements or conclusions:
    (a) A carbon-12 atom weighs 12.0 g.
    (b) There are 6.02×10^{23} atoms of water in one mole of water.
    (c) By making careful measurements in the laboratory, a student has determined the molecular weight of an unknown gas to 0.045 g/mol.
    (d) The limiting reagent is the reactant present in the smallest amount.

135. A 2.50-g sample of bronze was dissolved in sulfuric acid. The copper in the alloy reacted with sulfuric acid as follows.

$$Cu(s) + 2\ H_2SO_4(aq) \longrightarrow CuSO_4(aq) + SO_2(g) + 2\ H_2O(l)$$

The $CuSO_4$ formed in the reaction was mixed with KI to form CuI.

$$2\ CuSO_4(aq) + 5\ I^-(aq) \longrightarrow 2\ CuI(s) + I_3^-(aq) + 2\ SO_4^{2-}(aq)$$

The I_3^- formed in that reaction was then titrated with $S_2O_3^{2-}$.

$$I_3^-(aq) + 2\ S_2O_3^{2-}(aq) \longrightarrow 3\ I^-(aq) + S_4O_6^{2-}(aq)$$

Calculate the percentage by mass of copper in the original sample if 31.5 mL of 1.00 $M\ S_2O_3^{2-}$ was consumed in the titration.

136. Assume that you start with a glass of water, a glass of methanol, and a teaspoon. Exactly one teaspoon of water is removed from the glass of water and added to the glass of methanol. The resulting mixture of methanol is stirred until the two liquids are thoroughly mixed. Exactly one teaspoon of this mixture is then transferred back to the water. Which of the following statements is true?
    (a) The volume of water that ends up in the methanol is larger than the volume of methanol that ends up in the water.
    (b) The net volume of water transferred to the methanol is smaller than the net volume of methanol transferred to the water.
    (c) The net volume of water added to the methanol is exactly the same as the net volume of methanol added to the water.

137. Iron can react with O_2 to produce two different oxides, $Fe_2O_3(s)$ or $Fe_3O_4(s)$. Write the chemical equations that describe both reactions. If 167.6 g of Fe reacts completely with excess $O_2(g)$ to produce 231.6 g of product, which oxide was formed?

138. Draw and label a diagram like Figures 2.11, 2.12, and 2.13 to represent this chemical reaction:

$$CS_2(g) + 3\ Cl_2(g) \longrightarrow S_2Cl_2(g) + CCl_4(g)$$

139. The following two experiments are performed for the chemical reaction given below. The reaction goes to completion.

$$2\ X(aq) + Y(aq) \longrightarrow Z(aq)$$
$$\text{colorless} \quad \text{colorless} \quad \quad \text{red}$$

Experiment 1: 100 mL of a 0.100 M solution of x is added to 100 mL of a 0.200 M solution of Y.

Experiment 2: 100 mL of a 0.100 M solution of X is added to 100 mL of a 0.500 M solution of Y.

Which of the following would you expect to observe after mixing the two solutions? Explain your reasoning.
    (a) The solution formed in experiment 1 will be a darker red.
    (b) The solution formed in experiment 2 will be a darker red.
    (c) The solutions formed in both experiments will have the same shade of red.

3
THE STRUCTURE OF THE ATOM

3.1 RUTHERFORD'S MODEL OF THE ATOM

Shortly after radioactivity was discovered at the turn of the twentieth century, Ernest Rutherford became interested in α particles (positively charged helium nuclei) emitted by uranium metal and its compounds. Rutherford found that α particles were absorbed by a thin sheet of metal, a few tenths of a centimeter thick, but they could pass through metal foil that was thin enough.

Rutherford observed that a narrow beam of α particles was broadened as it passed through the metal foil. Working with his assistant Hans Geiger, Rutherford measured the angle through which the α particles were scattered by a thin piece of gold foil. Because it is unusually ductile, gold can be made into a foil that is only 0.00004 cm—or about 1400 atoms—thick. When the foil was bombarded with α particles, Rutherford and Geiger found that the angle of scattering was small, on the order of 1°.

These results were consistent with Rutherford's expectations. He knew that the α particle had a considerable mass (for a subatomic particle) and that it moved quite rapidly. Although the α particles would be scattered slightly by collisions with the atoms through which they passed, Rutherford expected the α particles to pass through the metal foil much the same way a rifle bullet would pass through a bag of sand.

J. J. Thompson (left), who showed that electrons were subatomic particles, and his student, Ernest Rutherford (right), who proposed that atoms were composed of negatively charged electrons orbiting an infinitesimally small positively charged nucleus.

One day, Geiger suggested that a research project should be given to Ernest Marsden, who was working in Rutherford's laboratory. Rutherford responded, "Why not let him see whether any α particles can be scattered through a large angle?" When this experiment was done, Marsden found that a small fraction (perhaps 1 in 20,000) of the α particles were scattered through angles larger than 90° (see Figure 3.1a). Many years later, reflecting on his reaction to these results, Rutherford said, "It was quite the most incredible event that has ever happened to me in my life. It was almost as incredible as if you fired a 15-inch shell at a piece of tissue paper and it came back and hit you."

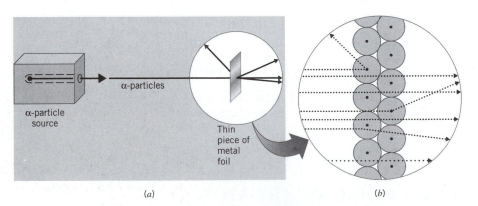

(a)   (b)

FIGURE 3.1 (a) Block diagram of the Rutherford–Marsden–Geiger experiment. (b) Most α particles pass through empty space between nuclei. A few come close enough to be repelled by the nucleus and are deflected through small angles. Occasionally, an α particle travels along a path that would lead to a direct hit with the nucleus. These particles are deflected through large angles by the force of repulsion between the α particle and the positively charged nucleus of the atom.

Rutherford found that he could explain Marsden's results by assuming that the positive charge and most of the mass of an atom are concentrated in a small fraction of the total volume, which he called the nucleus. When he derived mathematical equations for the scattering that would occur, his equations predicted that the number of α particles scattered through a given angle should be proportional to the thickness of the foil and the square of the charge on the nucleus, and inversely proportional to the velocity with which the α particles moved raised to the fourth power. In a series of experiments, Geiger and Marsden verified each of these predictions.

Most of the α particles were able to pass through the gold foil without encountering anything large enough to significantly deflect their path. A small fraction of the α particles came close to the nucleus of a gold atom as they passed through the foil. When this happened, the force of repulsion between the positively charged α particle and the nucleus deflected the α particle by a small angle, as shown in Figure 3.1b. Occasionally, an α particle traveled along a path that would eventually lead to a direct collision with the nucleus of one of the 2000 or so atoms it had to pass through. When this happened, repulsion between the nucleus and the α particle deflected the α particle through an angle of 90° or more.

By carefully measuring the fraction of the α particles deflected through large angles, Rutherford was able to estimate the size of the nucleus. According to his calculations, the radius of the nucleus is at least 10,000 times smaller than the radius of the atom. The vast majority of the volume of an atom is therefore empty space.

Checkpoint

If a circle 1 cm in diameter represents the nucleus of an atom, how far away would the circle that represents the radius of the atom have to be drawn?

3.2 PARTICLES AND WAVES

Rutherford's model assumes that most of the mass and all of the positive charge of an atom are concentrated in an infinitesimally small nucleus that is surrounded by a sea of lightweight, negatively charged electrons. Our next goal is to develop a picture of how these electrons are distributed around the nucleus. Much of what we know about the arrangement of electrons in an atom has been obtained by studying the interaction between matter and different forms of **electromagnetic radiation.** First, therefore, we need to understand what we mean when we say that electromagnetic radiation has some of the properties of both a particle and a wave.

Scientists have divided matter into two categories: particles and waves. **Particles** are easy to understand because they have a definite mass and they occupy space. **Waves** are more challenging. They have no mass and yet they carry energy as they travel through space. The best way to demonstrate that waves carry energy is to watch what happens when a pebble is tossed into a lake. As the waves produced by this action travel across the lake, they set in motion any leaves that lie on the surface.

In addition to their ability to carry energy, waves have four other characteristic properties: speed, frequency, wavelength, and amplitude. As we watch waves travel across the surface of a lake, we can see that they move at a certain **speed.** By watching waves strike a pier at the edge of the lake, we can see that they are also characterized by a **frequency** (ν)**,** which is the number of wave cycles that hit the pier per unit of time. The frequency of a wave is therefore reported in units of cycles per second (s^{-1}) or hertz (Hz).

The idealized waves in Figure 3.2 illustrate the definitions of amplitude and wavelength. The **wavelength (λ)** is the distance between two highest adjacent points or two lowest adjacent points.

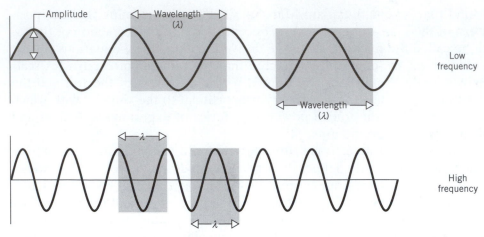

FIGURE 3.2 The wavelength (λ) of a wave is the distance between two highest adjacent points or two lowest adjacent points. The *amplitude* is the difference between the height of the highest point on the wave and its midline. The *frequency* (ν) is the number of waves (or cycles) that pass a fixed point per unit time. The product of frequency times wavelength is the speed at which the wave moves through space.

If we measure the frequency (ν) of any type of wave in cycles per second and the wavelength (λ) in meters, the product of the two numbers has the units of meters per second. The product of the frequency (ν) times the wavelength (λ) of a wave is therefore the speed (v) at which the wave travels through space.

$$\nu\lambda = v$$

To understand the relationship between the speed, frequency, and wavelength of a wave, it may be useful to consider a concrete example. Imagine that you are at a railroad crossing, watching a train that consists of 45-foot-long boxcars go by. Assume that it takes 3.0 seconds for each boxcar to pass in front of your car. The "frequency" of the train is 1 car every 3.0 seconds, or $1/3.0 \text{ s}^{-1}$. How fast is the train moving? The speed at which the train moves through space is the product of the "frequency" of this phenomenon times its "wavelength."

$$\frac{1}{3.0 \text{ s}} \times 45 \text{ ft} = 15 \frac{\text{ft}}{\text{s}} \approx 10 \frac{\text{mi}}{\text{h}}$$

Exercise 3.1

Orchestras in the United States tune their instruments to an A that has a frequency of 440 cycles per second, or 440 Hz. If the speed of sound is 1116 feet per second, what is the wavelength of this note?

Solution

The product of the frequency times the wavelength of any wave is equal to the speed with which the wave travels through space: $\nu\lambda = v$. Substituting the speed with which the note travels through space and the frequency of the note gives the following result:

$$(440 \text{ s}^{-1})(\lambda) = 1116 \text{ ft/s}$$

Solving the equation for λ gives a wavelength of 2.54 feet for the note.

3.3 LIGHT AND OTHER FORMS OF ELECTROMAGNETIC RADIATION

In 1865 James Clerk Maxwell proposed that light is a wave with both *electric* and *magnetic* components. Light is therefore a form of **electromagnetic radiation.** Because it is a wave, light is bent when it enters a glass prism. When white light, which is electromagnetic radiation made up of many different wavelengths, is focused on a prism, the light rays of different wavelengths are bent by differing amounts and the light is transformed into a band of colors. Radiation separated into its different wavelength components is called a **spectrum.** Starting from the side of the spectrum where the light is bent by the smallest angle, the colors are red, orange, yellow, green, blue, and violet.

Visible light contains the narrow band of frequencies and wavelengths in the small portion of the electromagnetic spectrum that our eyes can detect. It includes radiation with wavelengths between about 400 nm (violet) and 700 nm (red). Because the wavelength of electromagnetic radiation can be as long as 40 m or as short as 10^{-5} nm, the visible spectrum is only a tiny portion of the total range of electromagnetic radiation.

The electromagnetic spectrum includes radio and TV waves, microwaves, infrared radiation, visible light, ultraviolet radiation, X rays, γ rays, and cosmic rays, as shown in Figure 3.3. These different forms of radiation all travel at the speed of light in a vacuum (c). They differ, however, in their frequencies and wavelengths. The product of the frequency times the wavelength of electromagnetic radiation is always equal to the speed of light.

$$\nu \lambda = c$$

As a result, electromagnetic radiation that has a long wavelength has a low frequency, and radiation with a high frequency has a short wavelength.

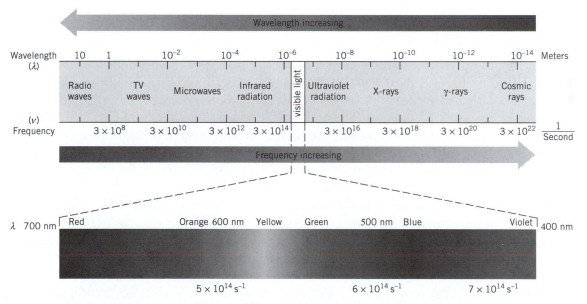

FIGURE 3.3 The visible spectrum is the small portion of the total electromagnetic spectrum that our eyes can detect. Other forms of electromagnetic radiation include radio and TV waves, microwaves, infrared and ultraviolet radiation, as well as X rays, γ rays, and cosmic rays.

Checkpoint
What is the frequency range of the waves used in a microwave oven? Why can't we see these waves?

Exercise 3.2

Calculate the frequency of red light that has a wavelength of 700.0 nm if the speed of light in a vacuum is 2.998×10^8 m/s.

Solution

The product of the frequency times the wavelength of any wave is equal to the speed with which the wave travels through space. In this case, the wave travels at the speed of light: 2.998×10^8 m/s. Before we can calculate the frequency of the radiation we have to convert the wavelength into units of meters.

$$700.0 \text{ nm} \times \frac{1 \text{ m}}{10^9 \text{ nm}} = 7.000 \times 10^{-7} \text{ m}$$

We can then substitute this wavelength and the speed of light into the following equation: $\nu\lambda = c$.

$$\nu(7.000 \times 10^{-7} \text{ m}) = 2.998 \times 10^8 \text{ m/s}$$

We then solve the equation to obtain the frequency of the light in units of cycles per second.

$$\nu = 4.283 \times 10^{14} \text{ s}^{-1}$$

3.4   ATOMIC SPECTRA

For more than 200 years chemists have known that sodium salts produce a yellow color when added to a flame. Robert Bunsen, however, was the first to systematically study this phenomenon. (Bunsen went so far as to design a new burner that would produce a colorless flame for the work.) Between 1855 and 1860, Bunsen and colleague Gustav Kirchhoff developed a spectroscope that focused the light from the burner flame onto a prism that separated the light into its spectrum. Using this device, Bunsen and Kirchhoff were able to show that the **emission spectrum** of sodium salts contains two narrow bands of radiation in the yellow portion of the spectrum.

Chemists and physicists soon began using the spectroscope to catalog the wavelengths of light emitted or absorbed by a variety of compounds. The data were then used to detect the presence of certain elements in everything from mineral water to sunlight. No obvious patterns were discovered in the data until 1885, however, when Johann Jacob Balmer analyzed the spectrum of hydrogen.

When an electric current is passed through a glass tube that contains hydrogen gas at low pressure, hydrogen atoms are produced and the tube gives off blue light. When the light is passed through a prism (as shown in Figure 3.4), four narrow lines of bright light are observed against a black background. Table 3.1 gives the visible spectrum, the characteristic wavelengths, and the corresponding colors of the emitted light from the hydrogen atoms formed in the glass tube. Conversely if visible light is passed through a gas of hydrogen atoms, four wavelengths of light would be absorbed. These wavelengths are exactly the same as those emitted in the gas discharge tube of Figure 3.4.

TABLE 3.1   Characteristic Lines in the Visible Spectrum of Hydrogen

Wavelength (nm)	Color
656.3	Red
486.1	Blue-green
434.0	Blue-violet
410.2	Violet

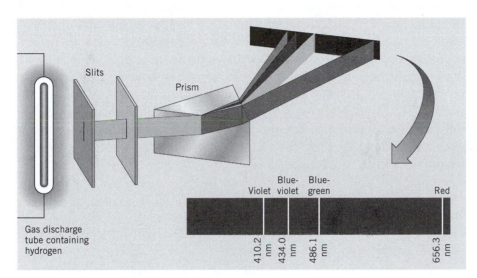

FIGURE 3.4 The blue light given off when a tube filled with H_2 gas is excited with an electric discharge can be separated into four narrow lines of visible light when it is passed through a prism.

3.5   QUANTIZATION OF ENERGY

The spectrum of the hydrogen atom raises several important questions.

- Why do hydrogen atoms give off only a handful of narrow lines of radiation when they emit light?
- Why do hydrogen atoms emit light when excited that has the same wavelengths as the light they absorb at room temperature?

Answers to these questions came from work on a related topic.

It is common knowledge that objects give off light when heated. Examples range from the gentle red glow of an electric burner on a stove to the bright light emitted when the tungsten wire in a light bulb is heated by passing an electric current through the wire. What is less well known is a phenomenon discovered by Thomas Wedgwood in 1792. Wedgwood, whose father started the famous porcelain factory, noticed that many objects give off a red glow when heated to the same temperature. The bottom of a crucible and the iron triangle on which the crucible rests, for example, both glow red when heated with a Bunsen burner.

Wedgwood also noticed that the color of the light emitted by an object changes as it is heated to higher temperatures until the object glows white-hot. But the spectrum of light given off at a particular temperature is the same for any object. The fact that sunlight is equivalent to the light emitted by an object at 6000°C, for example, has led to the assumption that this is the temperature of the surface of the sun.

The spectrum of radiation that an object emits when it is heated was eventually understood through the work of Max Planck and Albert Einstein. Planck was able to explain the spectrum of a radiating body by hypothesizing that the energy of the electromagnetic spectrum was made up of small energy packets. He argued that there were only a limited number of values for the energy of the system. In theory, the number of possible values for the energy of the radiating body was countable, which means that its energy was **quantized.**

Planck introduced the notion of quantization to explain how light was emitted. In 1905 Albert Einstein further developed Planck's theory when he concluded that not only was radiation emitted in discrete packets of energy but indeed light, no matter what its origin, always consists of packets of energy. Thus sunlight or the light from a light bulb may be thought of as a stream of individual energy packets. These suggestions were made at a time when everyone agreed that light was a wave (and therefore continuous). These small bundles, or packets, of energy were called **photons.** This means that light is also quantized. Because light isn't continuous, it is possible to count the number of photons that an object emits or absorbs. Einstein concluded that the energy (E) carried by one photon is directly proportional to the frequency of the radiation.

$$E = h\nu$$

In this equation, h is Planck's constant, which is equal to 6.626×10^{-34} joule · second (J · s).

Exercise 3.3

Calculate both the energy (E) of a single photon of red light with a wavelength of 700.0 nm and the energy of a mole of such photons.

Solution

In Exercise 3.2 we found that red light with a wavelength of 700.0 nm has a frequency of 4.283×10^{14} s^{-1}. Substituting the frequency into the Planck–Einstein equation gives the following result, in the energy unit joules (J).

$$E = (6.626 \times 10^{-34} \text{ J · s})(4.283 \times 10^{14} \text{ s}^{-1}) = 2.838 \times 10^{-19} \text{ J}$$

A single photon of red light carries an insignificant amount of energy. But a mole of the photons carries 1.709×10^5 J of energy.

$$\frac{2.838 \times 10^{-19} \text{ J}}{1 \text{ photon}} \times \frac{6.022 \times 10^{23} \text{ photons}}{1 \text{ mol}} = 1.709 \times 10^5 \text{ J/mol}$$

This amount of energy would raise the temperature of 1 liter of water by more than 40°C.

As we have seen, atoms emit or absorb radiation at only a limited number of frequencies or wavelengths. If the energy of the radiation is directly proportional to its frequency, this means that the atoms emit or absorb radiation with only a limited number of energies. This suggests that there are only a limited number of stable energy states within an atom. If this is true, these energy levels are countable. In other words, the energy levels of an atom are *quantized*.

This hypothesis explains why an object gives off red light when heated until it just starts to glow. Red light has the longest wavelength, and therefore the smallest frequency, of any form of visible radiation. Because the energy of electromagnetic radiation is proportional to its frequency, red light carries the smallest amount of energy of any form of visible radiation. The light given off when an object just starts to glow, therefore, has the color characteristic of the lowest energy form of electromagnetic radiation our eyes can detect. As the object becomes hotter it also emits light at higher frequencies, until it eventually appears to give off white light.

Checkpoint
Could the color of a heated metal bar be used to determine its temperature? Explain.

3.6   THE BOHR MODEL OF THE ATOM

Although Rutherford was never able to incorporate electrons into his model of the atom, one of his students, Niels Bohr, proposed a model for the hydrogen atom that accounted for its spectrum. The Bohr model assumed that the negatively charged electron and the positively charged nucleus of a hydrogen atom were held together by the force of attraction between oppositely charged particles. This force is directly proportional to the charge on the electron (q_e) and the charge on the nucleus (q_p) and inversely proportional to the square of the distance between the particles (r^2).

$$F = \frac{q_e \times q_p}{r^2}$$

Bohr then derived an equation that reproduced the absorption and emission spectra of the hydrogen atom by including the following assumptions.

*   The energy of the electron is proportional to its distance from the nucleus. (It takes energy to move an electron from a region close to the nucleus to one that is farther away.)
*   Only a limited number of regions with certain energies are allowed. As a result, the energy of the electron in a hydrogen atom is quantized.
*   When light is absorbed an electron moves from a lower energy state to a higher energy state. The energy of the radiation is equal to the difference between the energies of the two states.
*   Light is emitted when an electron falls from a higher energy state into a lower energy state. Once again, the energy of the radiation is equal to the difference between the energies of the two states.

Although it was remarkably successful for the hydrogen atom, the Bohr model was unable to explain the properties of atoms more complex than hydrogen and was eventually abandoned for a **quantum mechanical model** of the atom. The quantum mechanical model has the advantage that it is more powerful, but it achieves this power at a significant cost in terms of the ease with which it can be visualized.

We will retain four ideas from the Bohr model as we try to visualize the quantum mechanical model of the atom, although these ideas will be modified slightly.

- Electrons are attracted to the nucleus of an atom by the electrostatic force of attraction between oppositely charged objects.
- Electrons reside in regions in space that are at different distances from the nucleus.
- There are only a limited number of regions in space in which an electron can reside. As a result, the energy of an electron in an atom is quantized.
- Atoms emit or absorb radiation when an electron moves from one energy state to another.

3.7   THE ENERGY STATES OF THE HYDROGEN ATOM

For the four bands in the visible spectrum of the hydrogen atom reported in Table 3.1, the only information that was given was the wavelength and the color of the radiation emitted or absorbed. Table 3.2 includes both the frequency and the energy of this radiation as well.

TABLE 3.2   Characteristic Lines in the Visible Spectrum of the Hydrogen Atom

Wavelength (nm)	Color	Frequency (s^{-1})	Energy (kJ/mol)
656.3	Red	4.568×10^{14}	182.2
486.1	Blue-green	6.167×10^{14}	246.0
434.0	Blue-violet	6.908×10^{14}	275.5
410.2	Violet	7.309×10^{14}	291.6

The data in Table 3.2 only describe the spectrum of the hydrogen atom in the visible portion of the electromagnetic spectrum. Between 1908 and 1924, additional bands were observed in the infrared and ultraviolet portions of the spectrum, which indicate that there are additional energy states for the electron in the hydrogen atom.

Figure 3.5 displays a more complete energy diagram of the hydrogen atom, in which each energy state is represented by a horizontal line. The energy states are labeled with a variable n. The lowest energy state in which the electron is closest to the nucleus is represented by $n = 1$. Subsequent higher energy states are labeled $n = 2, 3, 4, \ldots$, respectively. The energy and wavelength of the photon absorbed or emitted when an electron moves between different energy states are indicated next to the vertical lines in the diagram. When the electron moves from one of the energy states toward the top of the diagram into one that lies toward the bottom, radiation is emitted with the indicated energy and wavelength. When radiation with this energy and wavelength is absorbed, the electron moves in the opposite direction.

You will notice in Figure 3.5 that the energy value listed for each energy state is a negative number. We have described electrons closest to the nucleus as being in a low (more stable) energy state, and we stated that an electron must absorb (gain) energy to move to a higher energy state. If an electron absorbs (gains) enough energy it can be completely removed from the atom (the atom is ionized). When an electron is completely removed from the atom, the electron is said to be in an energy state corresponding to an infinite distance from the nucleus, $n = \infty$. An electron that has been removed from an atom no longer has any force of attraction toward the nucleus, and therefore the energy associated with the electron in relation to the nucleus is zero. The energy state $n = \infty$ (zero energy) is used as a reference point for the

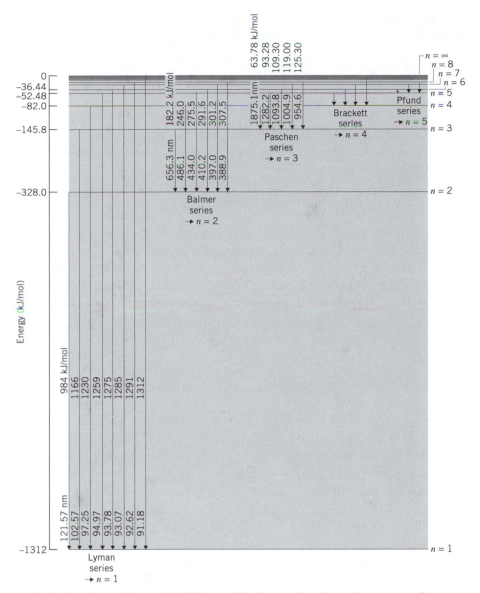

FIGURE 3.5 According to the Bohr model, the wavelengths and corresponding energies in the ultraviolet spectrum of the hydrogen atom discovered by Lyman were the result of electrons dropping from higher energy states into the $n = 1$ state. The Balmer, Paschen, Brackett, and Pfund series result from electrons falling from high energy states into the $n = 2$, $n = 3$, $n = 4$, and $n = 5$ states, respectively.

measurement of the energy states within the atom. Since all of the energy states are lower (more stable) than $n = \infty$, these states must have less energy than the reference value of zero. Therefore, all energy states within an atom have negative energy values.

It may be useful to view the energy states for the hydrogen atom on a number line:

$\longleftarrow$ —————————————————————————— $\longrightarrow$ (not drawn to scale)

−1312 kJ/mol   −328   −145   −82   0

Energy increases as you move to the right along the number line from the minimum energy of −1312 kJ/mol to a maximum energy of 0 kJ/mol. An electron in the lowest energy state (−1312 kJ/mol) must gain energy in order to be promoted to the second energy state (−328 kJ/mol).

The absorption of a photon of precisely the right energy to completely remove the electron from the lowest energy state of the hydrogen atom is described by the following equation.

$$H(g) + h\nu \longrightarrow H^+(g) + e^-$$

According to the energy scale on the left side of Figure 3.5, it takes 1312 kJ/mol of energy to remove an electron from a neutral hydrogen atom in the gas phase. If an electron is brought from a great distance to the nucleus of a hydrogen ion, H^+, we have to switch the reactants and products of the reaction.

$$H^+(g) + e^- \longrightarrow H(g) + h\nu$$

The magnitude of the energy associated with this process is exactly the same, 1312 kJ/mol. But now the energy is released in the form of electromagnetic radiation.

The data in Figure 3.5 can be used to calculate the energy associated with the radiation that accompanies the movement of an electron between any of the energy states of the hydrogen atom. Assume, for the moment, that we wanted to excite an electron from the lowest energy state into the second energy state. To move an electron from the first energy state to the second requires an amount of energy equal to the difference in energy between the two states.

$$E = 1312 \text{ kJ/mol} - 328.0 \text{ kJ/mol} = 984 \text{ kJ/mol}$$

According to the data in Figure 3.5, the wavelength of the radiation that must be absorbed to achieve this is 122 nm. If an electron in the second energy level dropped into the first energy level, the photon that would be emitted would have a wavelength of 122 nm and it would carry 984 kJ/mol of energy. Absorption or emission of a photon is possible, but the energy absorbed or emitted for movement between two given energy states is the same.

Checkpoint

If electromagnetic radiation with an energy of 1166 kJ/mol were absorbed by a mole of hydrogen atoms in their lowest energy state, what would be the resulting energy of the hydrogen atoms? What would be the energy level (value of n)?

Exercise 3.4

Identify the transitions between different energy states that give rise to the four lines in the spectrum of the hydrogen atom given in Table 3.2.

Solution

Table 3.2 contains the following values for the energies of the four lines in the visible spectrum of the hydrogen atom: 182.2, 246.0, 275.5, and 291.6 kJ/mol. Figure 3.5 suggests that each of these lines corresponds to a transition between the second energy state and a higher energy state.

182.2 kJ/mol	$3 \longleftrightarrow 2$
246.0 kJ/mol	$4 \longleftrightarrow 2$
275.5 kJ/mol	$5 \longleftrightarrow 2$
291.6 kJ/mol	$6 \longleftrightarrow 2$

Because of interactions between the electrons of an atom, the spectra for atoms that contain more than one electron can be far more complex than the spectrum for the hydrogen atom. As we'll see, however, the same basic postulates about the process by which radiation is emitted or absorbed by the hydrogen atom can be applied to investigate the energy levels of other atoms.

3.8   THE FIRST IONIZATION ENERGY

An electron can be completely removed from an atom if sufficient energy is supplied to overcome the attraction between the nucleus and the electron. The amount of energy that must be supplied depends on the energy state of the electron. The **first ionization energy** of an atom is defined as the minimum energy required to completely remove an electron from a ground state atom in the gas phase. In a ground state atom all electrons are in their lowest possible energy states. The first ionization energy of hydrogen, for example, is represented by the equation:

$$H(g) \longrightarrow H^+(g) + e^-$$

The ionization energy of a single atom is a very small quantity. It takes only 2.179×10^{-18} J to remove an electron from a single hydrogen atom in the gas phase. We can calculate the frequency of the radiation that carries this energy from the Planck–Einstein equation.

$$E = h\nu$$
$$2.179 \times 10^{-18} \text{ J} = (6.626 \times 10^{-34} \text{ J} \cdot \text{s})(\nu)$$
$$\nu = 3.288 \times 10^{15} \text{ s}^{-1}$$

We can then use the relationship between the frequency and wavelength of electromagnetic radiation to calculate the wavelength of the radiation.

$$\nu\lambda = c$$
$$(3.288 \times 10^{15} \text{ s}^{-1})(\lambda) = 2.998 \times 10^8 \text{ m/s}$$
$$\lambda = 9.118 \times 10^{-8} \text{ m}$$
$$\lambda = 91.18 \text{ nm}$$

This frequency and wavelength correspond to electromagnetic radiation in the ultraviolet region of the spectrum.

We can therefore measure the first ionization energy of an atom by shining ultraviolet light or X rays of known frequencies on a sample of atoms. Some of the photons in this portion of the electromagnetic spectrum have enough energy to knock an electron out of the atom. The first ionization energy of the element is measured by finding the smallest amount of electromagnetic energy required to remove an electron from the atom.

Although it takes relatively little energy to remove an electron from a single atom, the ionization energy becomes significant when we consider a mole of atoms. The first ionization energy (IE) for a mole of hydrogen atoms, for example, is 1312 kJ/mol.

$$IE = 2.179 \times 10^{-18} \frac{J}{\text{atom}} \times 6.022 \times 10^{23} \frac{\text{atoms}}{\text{mol}} = 1312 \frac{kJ}{\text{mol}}$$

We can bring the magnitude of the first ionization energy for hydrogen into perspective by noting that it is more than one and a half times as large as the energy released when we ignite one mole of methane that fuels the Bunsen burners in a chemistry laboratory.

$$CH_4(g) + 2\ O_2(g) \longrightarrow CO_2(g) + 2\ H_2O(g)$$

So much energy is consumed when the electrons are removed from a mole of hydrogen atoms that an equivalent amount of energy in the form of heat would be able to raise the temperature of 4 L of water by more than 75°C!

Experimental values of the first ionization energies for the gas phase atoms of the first 20 elements are given in Table 3.3 and Figure 3.6. There are several clear patterns in the data.

TABLE 3.3　First Ionization Energies for Gas Phase Atoms of the First 20 Elements

Symbol	Z	IE (kJ/mol)	Symbol	Z	IE (kJ/mol)
H	1	1312.0	Na	11	495.8
He	2	2372.3	Mg	12	737.7
Li	3	520.2	Al	13	577.6
Be	4	899.4	Si	14	786.4
B	5	800.6	P	15	1011.7
C	6	1086.4	S	16	999.6
N	7	1402.3	Cl	17	1251.1
O	8	1313.9	Ar	18	1520.5
F	9	1681.0	K	19	418.8
Ne	10	2080.6	Ca	20	589.8

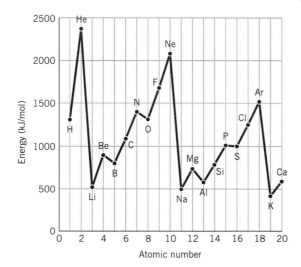

FIGURE 3.6 Plot of the first ionization energies of the atoms of the first 20 elements in the periodic table. There is a gradual increase in the first ionization energy across a row of the periodic table from H to He, from Li to Ne, and from Na to Ar. K and Ca appear to start a similar trend. There is a gradual decrease in the first ionization energy down a column of the periodic table, as can be seen by comparing the values for He, Ne, and Ar.

- The first ionization energy increases drastically as we go across a row of the periodic table. It doesn't matter whether we compare hydrogen and helium, lithium and neon, or sodium and argon; the first ionization energy increases by a factor of between 2 and 4 from the left side of the periodic table to the right.

- There is a dramatic drop in the first ionization energy as we go from the end of one row to the beginning of the next. It doesn't matter whether we compare helium and lithium, or neon and sodium, the first ionization energy decreases by more than a factor of 4.

- There is a gradual decrease in the first ionization energy as we go down a column of the periodic table. Consider going from He (IE = 2372 kJ/mol) to Ne (IE = 2081 kJ/mol) to Ar (IE = 1521 kJ/mol), for example.
- There are minor exceptions to the gradual increase in the first ionization energy across a row of the periodic table. The increase from Li (IE = 520 kJ/mol) to Be (IE = 899 kJ/mol), for example, is followed by a small decrease as we continue to B (IE = 801 kJ/mol).

Checkpoint

Can the first ionization energies of the 20 elements shown in Table 3.3 be used to arrange these elements in their appropriate positions in the periodic table? Explain.

3.9   THE SHELL MODEL

The data in Table 3.3 are consistent with the idea that electrons are attracted more strongly as nuclear charge increases and less strongly as the distance from the nucleus increases. Consider the first and second elements in the periodic table, for example. The first ionization energy increases by a factor of about 2 as we go from H (IE = 1312 kJ/mol) to He (IE = 2372 kJ/mol). This is consistent with the hypothesis that the helium electrons feel roughly twice the force of attraction for the nucleus because the positive charge on the helium nucleus is twice as large as the charge on the hydrogen nucleus.

FIGURE 3.7 The helium atom has a nucleus that has twice the positive charge of a hydrogen atom. The first ionization energy of He is about twice that of H. The shell model assumes that the energy of an electron depends on nuclear charge and distance from the nucleus.

The most easily removed electron on a lithium atom would be expected to have a first ionization energy about 1.5 times as large as the first ionization energy of helium, and three times as large as the first ionization energy of hydrogen if the distance of the electron from the nucleus was about the same as for H and He. Nothing could be further from the truth. The first ionization energy of lithium (IE = 520 kJ/mol) is only about 20% as large as that of helium (IE = 2372 kJ/mol), and it is less than half as large as hydrogen (IE = 1312 kJ/mol).

The only way to explain the data is to assume that the electron removed when the first ionization energy of lithium is measured is at a different distance from the nucleus than the two electrons on a helium atom. Because this electron is further away from the nucleus, it is much easier to remove from the atom.

The data in Table 3.3 for the first three elements therefore suggest that the electrons in an atom are arranged in shells. Within the **shell model** of the atom, we assume that the hydrogen and helium atoms consist of a nucleus surrounded by either one or two electrons in a single shell. Lithium then contains two electrons that lie close to the nucleus of the atom, in the same shell as the electrons of the first two elements, and a single electron in a shell that is farther from the nucleus, as shown in Figure 3.8.

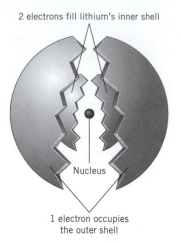

2 electrons fill lithium's inner shell

Nucleus

1 electron occupies
the outer shell

FIGURE 3.8 The lithium atom consists of a nucleus with a positive charge of +3, two electrons in a shell close to the nucleus, and one electron in a shell farther from the nucleus. Reprinted from Carl Snyder, *The Extraordinary Chemistry of Ordinary Things,* John Wiley & Sons, Inc., New York, 1992, p. 24.

The nucleus and the two inner electrons constitute the core of the lithium atom. The outermost electron in lithium does not experience the full +3 nuclear charge but rather a charge reduced by the underlying electrons. It is convenient to define a **core charge,** which, although it does not give the actual charge felt by an outer shell electron, is useful for organizing atomic properties. The core charge of lithium is the sum of the positive charge on the nucleus and the negative charge on the two inner shell core electrons:

Li: (+3 nuclear charge) + (−2 inner electron charge) = +1 core charge

Because the outer electron in lithium is at a greater distance from the nucleus and experiences a smaller nuclear attraction than the electrons in an He atom, the outer electron in Li has a lower first ionization energy than the He electron.

Lithium is the start of a series of eight elements that are characterized by a general increase in the first ionization energy with atomic number. Although there are minor variations in this trend, it is consistent with the idea that, in all, eight electrons are added to the atom in the second shell of electrons.

The shell that lies closest to the nucleus is described as the $n = 1$ shell of electrons. The data in Table 3.3 suggest that hydrogen has a single electron in the $n = 1$ shell and that helium has two electrons in this shell. Lithium has two electrons in the $n = 1$ shell and a third electron in the $n = 2$ shell. The next element Be has an ionization energy (899 kJ/mol) that is larger than that for Li and reflects the increased core charge:

Be: (+4 nuclear charge) + (−2 inner electron charge) = +2 core charge

Note that for the period 1 elements (first horizontal row in the periodic table) the core charge is the same as the atomic number Z and that for period 2 elements the core charge is $Z - 2$. Beryllium has two electrons in both the $n = 1$ and $n = 2$ shells, and neon has two electrons in the $n = 1$ shell and eight electrons in the $n = 2$ shell, as shown in Figure 3.9.

Checkpoint
What are the core charges for carbon and fluorine?

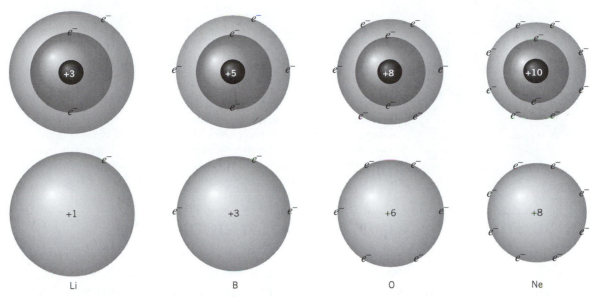

FIGURE 3.9 The top row shows the shell model for four atoms from the second period of the periodic table. The bottom row shows the model for the core charge of the atoms.

Once the eighth electron has been added to the $n = 2$ shell, there is a precipitous drop in the first ionization energy as we go from neon (IE = 2081 kJ/mol) to sodium (IE = 496 kJ/mol). This drop is similar to the one we observed from helium to lithium. It even has a similar magnitude: the first ionization energy for sodium ($Z = 11$) is slightly more than 20% of the ionization energy for neon ($Z = 10$). This suggests that the eleventh electron in the sodium atom is placed in a third shell ($n = 3$), at an even greater distance from the nucleus than the second shell of electrons used for the elements between Li and Ne, as shown in Figure 3.10.

FIGURE 3.10 Shell model of the sodium atom. There are three shells and a core charge of +1. Adapted from Carl Snyder, *The Extraordinary Chemistry of Ordinary Things,* John Wiley & Sons, Inc., New York, 1992.

Checkpoint
Why are the first ionization data (Table 3.3) not consistent with having nine electrons in the second shell of sodium?

The data in Table 3.3 for the elements between Na ($Z = 11$) and Ar ($Z = 18$) mirror the pattern observed between Li ($Z = 3$) and Ne ($Z = 10$). With slight variations, there is a gradual increase in the first ionization energy of these elements. This increase is then followed by another precipitous drop in the first ionization energy as we go from Ar (IE = 1521 kJ/mol) to potassium (IE = 419 kJ/mol). The data suggest that we leave the third shell at this point and enter a fourth shell ($n = 4$), as shown in Figure 3.11.

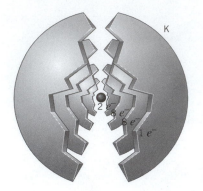

FIGURE 3.11 Shell model of the potassium atom. Now there are four shells. The first shell is complete with two electrons, the $n = 2$ and $n = 3$ shells hold eight electrons, and the fourth shell has only one electron. Adapted from Carl Snyder, *The Extraordinary Chemistry of Ordinary Things,* John Wiley & Sons, Inc., New York, 1992.

3.10   THE SHELL MODEL AND THE PERIODIC TABLE

When the data in Table 3.3 are arranged in terms of the periodic table, many of the trends we observe are reinforced. There are two elements in the first row of the periodic table, and two electrons occupy the first shell in our shell model. There are eight electrons in the second and third shells, and eight elements in the second and third rows of the periodic table.

Furthermore, the number of electrons in each of the outermost shells on an atom is consistent with the group in which the element is found. We have concluded that H, Li, Na, and K each have one electron in their outermost shell, and each of these elements is in Group IA of the periodic table. In addition, as this group is descended from H to K the ionization energy decreases. This is consistent with all of these atoms having a core charge of $+1$, with the outermost electron in a shell that is progressively farther and farther from the nucleus. In fact, the number of electrons in the outermost shell of each atom among the first 20 elements in the periodic table corresponds exactly with the group number for that element.

As we shall see in the next chapter, the electrons in the outermost shell of an atom are involved in the formation of bonds between atoms. Because these bonds are relatively strong, the electrons that form them are often called the **valence electrons** (from the Latin stem *valens,* "to be strong"). A more complete discussion of valence electrons is found in the next chapter.

The periodic table was originally created to group elements that had similar chemical and physical properties. The arrangement of the electrons in the atom deduced from the shell model is reflected in the arrangement of elements in the periodic table. This suggests that many of the chemical and physical properties of the elements are related to the number of electrons in the outermost shell, which are the valence electrons in these atoms. Also note that the core charge and the number of electrons in the outermost shell correspond to the group numbers IA through VIIIA.

Checkpoint

Is the radius of the valence shell of Na larger, smaller, or the same as the radius of the valence shell of Li?

3.11   PHOTOELECTRON SPECTROSCOPY AND THE STRUCTURE OF ATOMS

The shell model of the atom that we have deduced from first ionization energy data assumes that the electrons in an atom are arranged in shells about the nucleus, with successive shells being farther and farther from the nucleus of the atom. The data we used to derive this model represented the *minimum* energy needed to remove an electron from the atom. Thus, in each case, the first ionization energy reflects the ease with which the outermost electron can be removed from the atom.

For atoms that have many electrons, we would expect that it would take more energy to remove an electron from an inner shell than it does to remove the electron from the valence shell. It takes more energy to remove an electron from the $n = 2$ shell than from the $n = 3$ shell, for example, and even more energy to remove an electron from the $n = 1$ shell.

This raises an interesting question: Do all of the electrons in a given shell have the same energy? This question can be answered by using a slightly different technique to measure the energy required to remove an electron from a neutral atom in the gas phase to form a positively charged ion.

This time, we will shine radiation on the sample that carries enough energy to excite an atom to the point that one of its electrons from any shell is ejected from the atom to form a positively charged ion. The experiment is known as **photoelectron spectroscopy (PES),** as shown in Figure 3.12. This experiment measures the energy required to remove one electron from a neutral atom in the gas phase to form a positively charged ion. PES differs from the experiment used to obtain the first ionization energies given in Table 3.3 by its ability to remove electrons *from different shells in the atom.* Not only can an electron from the outermost shell be removed, but an electron from one of the shells that are deep within the core of electrons that surround the nucleus may be ejected. Many atoms will be in the path of the electromagnetic radiation, but only a single electron is ejected from each atom. However, each atom may have many electrons in different shells. Any one of these electrons may be ejected. Thus PES allows measurement of the energy required to remove any one of the electrons in an atom.

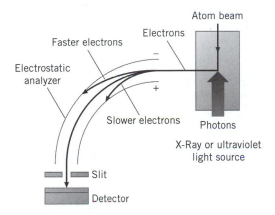

FIGURE 3.12 Photoelectron spectroscopy. An electron absorbs a high energy photon and is ejected from an atom. The kinetic energy, KE, of the ejected electron is measured, and the energy required to remove the electron from the atom, IE, is calculated from IE = KE − $h\nu$ where $h\nu$ is the energy of the photon. Electrons may be removed from any shell in the atom. This is different from measurements of first ionization energies, which remove electrons from only the outermost shell. Reprinted from R. J. Gillespie et al., *Atoms, Molecules and Reactions,* Prentice-Hall, Englewood Cliffs, New Jersey, 1994, p. 198.

An atom in the gas phase absorbs a high energy photon from the ultraviolet (UV) or X-ray region of the electromagnetic spectrum. The energy of this photon ($h\nu$) is larger than the energy required to remove an electron from the atom (IE). The excess energy is carried off in the form of kinetic energy (KE) by the electron ejected from the atom.

$$h\nu = \text{IE} + \text{KE}$$

When this experiment is done, we know the energy of the radiation ($h\nu$) used to excite the atom. If we measure the energy of the so-called photoelectron that is ejected (KE) when this radiation is absorbed, we can calculate the energy required to remove this electron from the atom (IE), as shown in Figure 3.13.

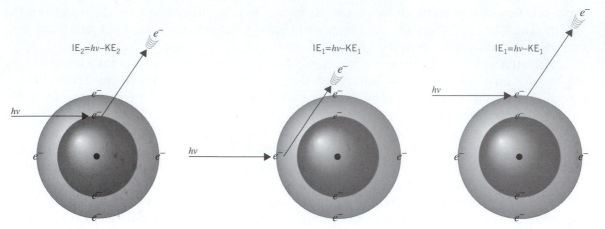

FIGURE 3.13 A high energy photon can eject an electron from an atom. If the kinetic energy of the ejected electron is calculated, and the energy of the photon is known, the energy required to remove the electron can be found. The ionization energy of the inner shell electrons, IE_2, will be greater than that of the outer shell electrons, IE_1.

Data from PES experiments are obtained as peaks in a spectrum that plots the energy to eject an electron (IE) versus the intensity of the signal observed as shown in Figure 3.14. The intensity of the signal measures the relative number of electrons of equivalent energy ejected during the experiment. If we see two peaks, for example, that have a relative intensity of 2:1 we can conclude that one of the energy levels from which electrons are removed in this experiment contains twice as many electrons as the other.

As we examine PES data in the next section, it is important to remember that the electromagnetic radiation energy supplied may remove an electron from the outermost shell, or remove an electron from one of the shells that are deep within the core of the atom.

Checkpoint
What determines the number of peaks in a photoelectron spectrum of an atom?

3.12 ELECTRON CONFIGURATIONS FROM PHOTOELECTRON SPECTROSCOPY

Hydrogen can only have one peak in the photoelectron spectrum because it contains only a single electron. As expected, this peak comes at an energy of 1312 kJ/mol, or, as we'll express it from now on in megajoules, 1.31 MJ/mol. This is the energy required to eject the electron from the hydrogen atom. The resulting PES spectrum is shown in Figure 3.14. The energy to eject the electron is plotted along the x axis, and the relative number of electrons ejected is plotted along the y axis.

Helium also has only one peak in the PES experiment, which occurs at an energy of 2372 kJ/mol or 2.37 MJ/mol. The PES spectrum for He is also shown in Figure 3.14. Note that the helium peak is shifted to the left on the x axis relative to the peak in the hydrogen spectrum. In the PES spectra energy *increases* as you move to the *left* on the x axis. This means that the energy to remove an electron from a helium atom is greater than the energy required to remove an electron from hydrogen. The intensity of the peak (relative number of electrons) is twice as great as that of the hydrogen peak. This is consistent with our hypothesis that the two electrons in this atom both occupy the $n = 1$ shell.

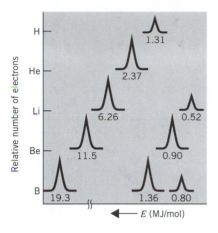

FIGURE 3.14 Photoelectron spectra of several atoms. The energy required to remove an electron from an atom is plotted as increasing from right to left. The energy required to remove an electron from the hydrogen atom is 1.31 MJ/mol and that for the helium atom is 2.37 MJ/mol. The height of the He spectrum is twice that of the H spectrum because He has twice as many electrons as H. As the electrons are more tightly held by the nucleus, the energy required to remove an electron increases. The computer-generated spectra have been adjusted so that the peak heights of the spectra of different atoms are directly comparable.

Checkpoint

Why is there only one peak in the photoelectron spectrum of He even though there are two electrons in each He atom?

Our shell model of the atom leads us to expect two peaks in the PES spectrum for lithium, which is exactly what is observed. These peaks occur at ejection energies (IE) of 6.26 and 0.52 MJ/mol, and they have a relative intensity of 2:1. The outermost electron in the Li atom is relatively easy to remove because it is in the $n = 2$ shell. But it takes a great deal of energy to reach into the $n = 1$ shell, as shown in Figure 3.14, because the electrons in that shell lie close to a nucleus that carries a charge of +3.

Two peaks are also observed in the PES spectrum of beryllium, with a relative intensity of 1:1. In this case, it takes an enormous amount of energy to reach into the $n = 1$ shell (IE = 11.5 MJ/mol) to remove one of the electrons that lie close to the nucleus with its charge of +4. It takes quite a bit less energy (IE = 0.90 MJ/mol) to remove one of the electrons in the $n = 2$ shell, as shown in Figure 3.14.

An interesting phenomenon occurs when we compare the photoelectron spectrum of boron with the spectra for the first four elements, as shown in Figure 3.14. There are now three distinct peaks in the spectrum, at energies of 19.3, 1.36, and 0.80 MJ/mol. There is a peak in the PES spectrum for removing one of the electrons from the $n = 1$ shell (IE = 19.3 MJ/mol). But there are also two additional peaks, with a relative intensity of 2:1, that correspond to removing an electron from the $n = 2$ shell (IE = 1.36 and 0.80 MJ/mol).

The same phenomenon occurs in the PES spectra for carbon, nitrogen, oxygen, fluorine, and neon, as shown in Table 3.4. In each case, we see three peaks. As the charge on the nucleus increases, it takes more and more energy to remove an electron from the $n = 1$ shell, until, by the time we get to neon, it takes 84.0 MJ/mol to remove an electron from this shell. (This is more than 100 times the energy given off in a typical chemical reaction.)

The PES spectra for B, C, N, O, F, and Ne contain a second peak, of gradually increasing energy because of the increasing nuclear charge, that has the same intensity as the peak for the $n = 1$ shell. And, in each case, we get a third peak, of gradually increasing energy, that corresponds to the electrons which are the easiest to remove from the atoms. The intensity of the third peak increases from element to element, representing a single electron for boron up to a total of six electrons for neon, as shown in Figures 3.14 and 3.15.

The PES data for the first 10 elements reinforce our belief in the shell model of the atom. But they suggest that we have to refine the model to explain the fact that the electrons in the $n = 2$ shell seem to occupy different energy levels. In other words, we have to introduce the concept of **subshells** within the shells of electrons.

TABLE 3.4   Ionization Energies for Gas Phase Atoms of the First 10 Elements Obtained from Photoelectron Spectra

Element	IE (MJ/mol)		
	First Peak	Second Peak	Third Peak
H	1.31		
He	2.37		
Li	6.26	0.52	
Be	11.5	0.90	
B	19.3	1.36	0.80
C	28.6	1.72	1.09
N	39.6	2.45	1.40
O	52.6	3.12	1.31
F	67.2	3.88	1.68
Ne	84.0	4.68	2.08

Source: D. A. Shirley et al., *Physical Review B* (15), 544–552 (1977).

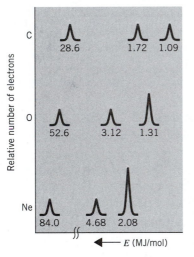

FIGURE 3.15 Photoelectron spectra of C, O, and Ne. The nuclear charge increases from carbon to oxygen to neon, and the spectra shift to the left to higher energies. The peak farthest to the right shows the electrons most easily removed and these peak heights (relative number of electrons) increase in a ratio of 2:4:6. The computer-generated spectra have been adjusted so that the peak heights for different atoms are directly comparable.

When the first evidence for subshells was discovered shortly after the turn of the twentieth century, a shorthand notation was introduced in which these subshells were described as either *s, p, d,* or *f*. Within any shell of electrons, the *s* subshell always corresponds to the largest ejection energy.

The PES data in Table 3.4 suggest that there is only one subshell in the $n = 1$ shell. Chemists usually represent this by writing 1*s*, where the number represents the shell and the letter represents the subshell. These data also suggest that the 1*s* subshell can hold a maximum of two electrons. This hypothesis is confirmed by the data for the next three elements. The third and fourth electrons on an atom seem to be added to the 2*s* subshell. Once we have five electrons, however, the 2*s* subshell seems to be filled, and we have to add the fifth electron to the next subshell: 2*p*.

By convention, the information we have deduced from the PES data is referred to as the atom's **electron configuration** and is written as follows.

$$
\begin{array}{ll}
\text{H } (Z = 1) & 1s^1 \\
\text{He } (Z = 2) & 1s^2 \\
\text{Li } (Z = 3) & 1s^2\,2s^1 \\
\text{Be } (Z = 4) & 1s^2\,2s^2 \\
\text{B } (Z = 5) & 1s^2\,2s^2\,2p^1
\end{array}
$$

The superscripts in the electron configurations designate the number of electrons in each subshell. The existence of two subshells within the $n = 2$ shell explains the minor inversion in the first ionization energies of boron and beryllium shown in Figure 3.6. It is slightly easier to remove the outermost electron from B (IE = 0.80 MJ/mol) than from Be (IE = 0.90 MJ/mol), in spite of the greater charge on the nucleus of the boron atom, because the outermost electron on B is in the $2p$ subshell, whereas the outermost electron on Be has to be removed from the $2s$ subshell (Figure 3.16).

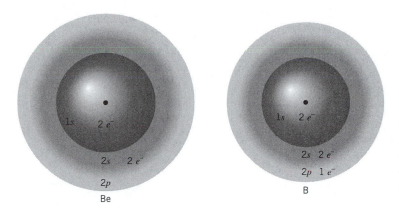

FIGURE 3.16 The outermost electron in B is easier to remove than the outermost electron in Be because the electron is in the $2p$ subshell. The easiest electron to remove from Be is in the $2s$ subshell.

As we continue across the second row of the periodic table, from B to Ne, the number of electrons in the $2p$ subshell gradually increases to 6. The best evidence for this is the growth in the intensity of the peak corresponding to the $2p$ subshell. In boron, this peak has half the area of the $1s$ or $2s$ peaks. When we get to neon, we find it has three times the area of the $1s$ or $2s$ peaks. This suggests that the $2p$ subshell can hold a maximum of six electrons. We can therefore continue the process of translating PES data into the electron configurations for the atoms as follows (see also Figure 3.17):

FIGURE 3.17 Subshells of the elements carbon through neon. PES data show that the $n = 2$ shell has two subshells, the $2s$ and $2p$ subshells.

$$
\begin{array}{ll}
\text{C } (Z = 6) & 1s^2\,2s^2\,2p^2 \\
\text{N } (Z = 7) & 1s^2\,2s^2\,2p^3 \\
\text{O } (Z = 8) & 1s^2\,2s^2\,2p^4 \\
\text{F } (Z = 9) & 1s^2\,2s^2\,2p^5 \\
\text{Ne } (Z = 10) & 1s^2\,2s^2\,2p^6
\end{array}
$$

PES data for the next 11 elements in the periodic table are given in Table 3.5. In Table 3.5, the columns are labeled in terms of the representations for the subshells: $1s$, $2s$, $2p$, $3s$, $3p$, $3d$, and $4s$.

TABLE 3.5   Ionization Energies for Gas Phase Atoms of Elements 11 through 21 Obtained from Photoelectron Spectra

Element	IE (MJ/mol)						
	1s	2s	2p	3s	3p	3d	4s
Na	104	6.84	3.67	0.50			
Mg	126	9.07	5.31	0.74			
Al	151	12.1	7.79	1.09	0.58		
Si	178	15.1	10.3	1.46	0.79		
P	208	18.7	13.5	1.95	1.01		
S	239	22.7	16.5	2.05	1.00		
Cl	273	26.8	20.2	2.44	1.25		
Ar	309	31.5	24.1	2.82	1.52		
K	347	37.1	29.1	3.93	2.38		0.42
Ca	390	42.7	34.0	4.65	2.90		0.59
Sc	433	48.5	39.2	5.44	3.24	0.77	0.63

Source: D. A. Shirley et al., *Physical Review B* (15), 544–552 (1977).

Checkpoint

For neon, why is it expected that 2/10 of the ejected electrons will come from the $n = 1$ shell and 8/10 from the $n = 2$ shell?

As we might expect, sodium has four peaks in the photoelectron spectrum, corresponding to the loss of electrons from the $1s$, $2s$, $2p$, and $3s$ subshells, as shown in Figure 3.18.

$$\text{Na } (Z = 11) \qquad 1s^2\ 2s^2\ 2p^6\ 3s^1$$

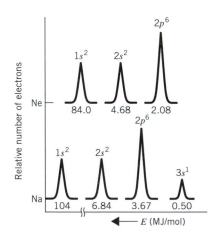

FIGURE 3.18 Photoelectron spectra of Ne and Na. Sodium has four peaks corresponding to loss of electrons from the $1s$, $2s$, $2p$, and $3s$ subshells. The heights of the peaks correspond to the number of electrons in a subshell. The computer-generated spectra have been adjusted so that the peak heights for different atoms are directly comparable.

It is easier to remove the electron in the $n = 3$ shell (IE = 0.50 MJ/mol) than the electrons in the $n = 2$ shell (IE = 6.84 and 3.67 MJ/mol), which in turn are easier to remove than the electrons in the $n = 1$ shell (IE = 104 MJ/mol).

Magnesium also gives four peaks in the PES experiment, which is consistent with the following electron configuration:

$$\text{Mg } (Z = 12) \qquad 1s^2 \, 2s^2 \, 2p^6 \, 3s^2$$

Aluminum and each of the subsequent elements up to argon have five peaks in the PES spectrum. The new peak corresponds to the $3p$ subshell, and these elements have the following electron configurations.

$$
\begin{array}{ll}
\text{Al } (Z = 13) & 1s^2 \, 2s^2 \, 2p^6 \, 3s^2 \, 3p^1 \\
\text{Si } (Z = 14) & 1s^2 \, 2s^2 \, 2p^6 \, 3s^2 \, 3p^2 \\
\text{P } (Z = 15) & 1s^2 \, 2s^2 \, 2p^6 \, 3s^2 \, 3p^3 \\
\text{S } (Z = 16) & 1s^2 \, 2s^2 \, 2p^6 \, 3s^2 \, 3p^4 \\
\text{Cl } (Z = 17) & 1s^2 \, 2s^2 \, 2p^6 \, 3s^2 \, 3p^5 \\
\text{Ar } (Z = 18) & 1s^2 \, 2s^2 \, 2p^6 \, 3s^2 \, 3p^6
\end{array}
$$

The electron configurations of the elements in the third row of the periodic table therefore follow the same pattern as the corresponding elements in the second row.

By the time we get to potassium and calcium, we find six peaks in the PES spectrum, with the smallest ionization energy comparable to, but slightly less than, the ionization energies of the $3s$ electrons on sodium and magnesium. We therefore write the electron configurations of potassium and calcium as follows.

$$
\begin{array}{ll}
\text{K } (Z = 19) & 1s^2 \, 2s^2 \, 2p^6 \, 3s^2 \, 3p^6 \, 4s^1 \\
\text{Ca } (Z = 20) & 1s^2 \, 2s^2 \, 2p^6 \, 3s^2 \, 3p^6 \, 4s^2
\end{array}
$$

First ionization data and photoelectron spectra are used to construct a model of the atom in which electrons have energy shells and subshells. Figure 3.19 shows a representation of the relative energies of the shells and subshells in H, Be, Ne, and Ca atoms. The same *labels* ($1s$, $2s$, $2p$) are used to describe the energy states in all atoms. However, the energy associated with a given label will change from one atom to another. As we saw with the energy states in the Bohr atom in Figure 3.5, the energy values listed beneath each energy state are negative values. In a mole of hydrogen atoms 1.31 MJ of energy is required to remove a mole of electrons from the $1s$ energy state. When an electron is completely removed from an atom, the energy of the electron relative to the atom is zero. Therefore, the energy that a mole of hydrogen electrons in the $1s$ energy state possesses is -1.31 MJ. In Be the $1s$ electrons are held more tightly, which makes the $1s$ energy state more stable and therefore lower on the diagram. An energy of 11.5 MJ is required to ionize $1s$ electrons from a mole of Be atoms, giving the $1s$ energy level an energy state of -11.5 MJ/mol. In Ne and Ca the electrons in the $1s$ state are held even more tightly. The energies of a mole of $1s$ electrons are -84.0 and -390 MJ, respectively. Even though the energy associated with a given state, for example, $2s$, varies from one atom to another, the energy states for any given atom increase from the most stable lowest energy state, $1s$ (most tightly held electrons), to less stable higher energy states (less tightly held electrons), namely, $2s$, $2p$, $3s$, $3p$, and $4s$. However, as we go to higher energy levels the order of the energy levels can vary from one type of atom to another.

An interesting phenomenon occurs in the next element in the periodic table, scandium (Sc). Our shell model predicts that the subshells used to hold electrons in calcium are all filled. The twenty-first electron therefore has to go into a new subshell. But the ionization energy for the new peak that appears in the PES spectrum doesn't occur at a lower energy than the subshells used previously, which has been observed for every other subshell as it has appeared. The new peak appears at a higher energy than the $4s$ subshell on scandium.

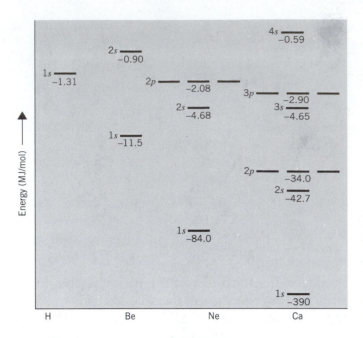

FIGURE 3.19 Relative energies of the electrons within the shells and subshells in H, Be, Ne, and Ca atoms in units of megajoules per mole (MJ/mol). The $1s$ electrons are more tightly held as the nuclear charge increases from H through Ca. The labels $1s$, $2s$, $2p$, $3s$, $3p$, and $4s$ are used to describe the energy states.

Evidence from other forms of spectroscopy suggest that the twenty-first electron on the scandium atom goes into the $n = 3$ shell, not the $n = 4$ shell. The subshell used to hold this electron is the $3d$ subshell. As we can see from the data in Table 3.5, the $3d$ subshell is very close in energy to the $4s$ subshell. For scandium, it is easier to remove a $4s$ electron than a $3d$ electron. The electron configuration of scandium may be written as follows:

$$\text{Sc } (Z = 21) \qquad 1s^2 \, 2s^2 \, 2p^6 \, 3s^2 \, 3p^6 \, 4s^2 \, 3d^1$$

This way of writing electron configurations is convenient because it follows the arrangement of the elements on the periodic table. Another way of writing electron configurations is to list the subshells in order of decreasing ionization energy.

$$\text{Sc } (Z = 21) \qquad 1s^2 \, 2s^2 \, 2p^6 \, 3s^2 \, 3p^6 \, 3d^1 \, 4s^2$$

In this text the first method is employed.

Scandium is followed by nine elements—known as the **transition metals**—that have no analogs in the second and third rows of the periodic table. In these elements, electrons continue to fill the $3d$ subshell until we reach zinc.

$$\begin{array}{ll} \text{Ti} & 1s^2 \, 2s^2 \, 2p^6 \, 3s^2 \, 3p^6 \, 4s^2 \, 3d^2 \\ \text{V} & 1s^2 \, 2s^2 \, 2p^6 \, 3s^2 \, 3p^6 \, 4s^2 \, 3d^3 \\ \vdots & \\ \text{Zn} & 1s^2 \, 2s^2 \, 2p^6 \, 3s^2 \, 3p^6 \, 4s^2 \, 3d^{10} \end{array}$$

Thus, the $3d$ subshell can contain a maximum of 10 electrons, giving a total of up to 18 electrons in the $n = 3$ shell.

The next element is gallium, at which point the $4p$ level begins to fill, giving six elements (gallium to krypton) that have electron configurations analogous to those of the elements in the second and third rows of the periodic table.

Checkpoint

Using data from Table 3.5, sketch the photoelectron spectrum of scandium, Sc. Give the relative intensities of the peaks and assign energies to each peak. Use the photo-electron spectrum to list the order in which electrons are removed from subshells of scandium. Use the electron configuration of Sc to list the order in which its sub-shells are filled with electrons. Is there a difference in the orders in which electrons are added to and taken away from Sc?

When forming a positive ion, we find that $4s$ electrons are easier to remove than the $3d$ electrons for all of the first row transition metals. The electron configuration of the vanadium ion, V^{2+}, is

$$V^{2+} \qquad 1s^2\, 2s^2\, 2p^6\, 3s^2\, 3p^6\, 3d^3$$

The two electrons are lost from the outermost shell, $4s$, to give the V^{2+} ion.

3.13   ORBITALS AND THE PAULI EXCLUSION PRINCIPLE

So far, we've assumed that the characteristic properties of an electron are its charge and its mass. A third property of the electron was discovered in 1920 by the German physicists Otto Stern and Walter Gerlach.

The Stern–Gerlach experiment began by heating a piece of silver metal until silver atoms evaporated from the surface. These atoms were passed through slits into an evacuated chamber where they were allowed to pass through a magnetic field. Because the magnetic field was inhomogeneous, the atoms felt a net force pushing them in a direction perpendicular to their path as they passed through the magnetic field. As a result, the path of the silver atoms was bent or deflected by the magnetic field.

Once they passed through the magnetic field, the atoms struck a glass plate, where they were deposited as silver metal. When the experiment was done at very low residual pressures and long exposure times, Stern and Gerlach found that the silver metal was deposited in two closely spaced areas, suggesting that silver atoms interact with a magnetic field in two distinctly different ways.

The results of this experiment were interpreted by assuming that an electron has a third characteristic property, besides its charge and mass. The electron behaves as if it were spinning in either a clockwise or counterclockwise direction on its axis. Because of its charge, the "spinning" electron produces a magnetic field. The Stern–Gerlach experiment shows that electrons do indeed produce a magnetic field. The "spin" of an electron is a model proposed to explain the origin of this field. Two electrons which have opposite spin are said to be **paired.** The presence of an unpaired electron in an atom causes the atom to be-have as a magnet; in other words, the atom is said to be magnetic. If there are no unpaired electrons in an atom, the atom has no magnetic properties. Thus, pairing of one electron with another cancels their magnetic fields.

Silver has an odd number of electrons ($Z = 47$). Therefore, even if most of the elec-trons pair so that the spins on these electrons cancel, there must be at least one odd elec-tron on each silver atom that is unpaired. This unpaired electron interacts with an exter-nal magnetic field to cause the paths of moving silver atoms to be deflected in one direction if the electron spins clockwise on its axis, and in the other direction if it spins counter-clockwise.

Checkpoint

Predict the results of a Stern–Gerlach experiment on a beam of

(a) Li atoms   (b) Be atoms   (c) B atoms   (d) N atoms

A further consequence of an electron having a spin is that electrons having the same spin (both electrons spinning in a clockwise direction, for example) have a low probability of being close together and a high probability of being far apart. There is no restriction on electrons of opposite spin being close together.

Because electrons of the same spin keep apart, an electron tends to exclude all other electrons of the same spin from the space that it occupies. However, an electron of opposite spin may enter this space. This space from which other electrons of the same spin tend to be excluded is called the **orbital** of the electron.

Because electrons have a negative charge, an electrostatic repulsion between electrons will also be present. The charge and spin reinforce one another to keep electrons of the same spin in separate orbitals. Electrons of opposite spin also undergo electrostatic repulsions, but exclusion of an electron having one spin from the space occupied by an electron of another spin does not come into play. Repulsion of like charges also tends to prevent the orbitals of electrons of opposite spins from coinciding. Thus, unless some other factor is present all electrons tend to avoid one another.

In 1924 the Austrian physicist Wolfgang Pauli proposed the following hypothesis, which has become known as the Pauli exclusion principle.

No more than two electrons can occupy an orbital, and, if there are two electrons in an orbital, the spins of the electrons must be paired.

The concept of the spin of an electron and the Pauli exclusion principle provide the last step needed to complete our model of the structure of the atom.

The electrons in a shell or a subshell occupy three-dimensional orbitals in space. An orbital can contain either one or two electrons. If there are two electrons in the orbital, they must have different spins.

Thus, we assume that there is a single orbital in the 1s, 2s, 3s, and 4s subshells. Because the 2p and 3p subshells can contain up to six electrons, there must be three orbitals in these subshells. Since the 3d orbitals can hold up to 10 electrons, there must be five orbitals in these subshells. For free atoms there is no difference between the orbitals in a subshell. The orbitals within a single subshell of free atoms have the same energy and are therefore indistinguishable.

3.14   PREDICTING ELECTRON CONFIGURATIONS

It is clear from the electron configurations of the first 36 elements in the periodic table that they follow a regular pattern. This pattern can be summarized by a set of simple rules that can be used to predict the electron configurations for most of the elements.

- The number of subshells in a shell is equal to the shell number, n.

$$
\begin{array}{ll}
n = 1 & 1s \\
n = 2 & 2s \text{ and } 2p \\
n = 3 & 3s, 3p, \text{ and } 3d \\
n = 4 & 4s, 4p, 4d, \text{ and } 4f
\end{array}
$$

- In order of increasing energy, the subshells are designated *s, p, d,* and *f,* respectively.

$$4s < 4p < 4d < 4f$$

- There are always an odd number of orbitals in a subshell, and the number of orbitals in a subshell increases in the same order as the energy of these subshells.

$$4s = 1 \text{ orbital}$$
$$4p = 3 \text{ orbitals}$$
$$4d = 5 \text{ orbitals}$$
$$4f = 7 \text{ orbitals}$$

- Because each orbital can hold up to two electrons, the maximum number of electrons in a subshell is equal to twice the number of orbitals.

s	2 electrons
p	6 electrons
d	10 electrons
f	14 electrons

- Electrons are added to an atom, one at a time, starting with the lowest available orbital.

A device for remembering the order in which orbitals are filled is shown in Figure 3.20. The order of filling of the orbitals is read by following the arrows, starting at the top of the first line and proceeding on to the second, third, and fourth lines, and so on.

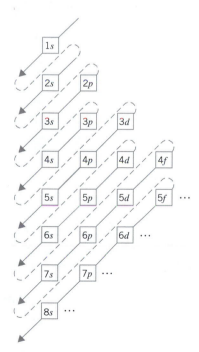

FIGURE 3.20 The order of filling for atomic orbitals can be predicted by following the arrows in this diagram.

Exercise 3.5

Predict the electron configuration for a neutral tin atom (Sn, $Z = 50$).

Solution

We start by predicting the order of filling of atomic orbitals from the diagram in Figure 3.20. We then add electrons to these orbitals, starting with the lowest orbital, until all 50 electrons have been included.

The electron configuration for an atom can be written by remembering that each orbital can hold two electrons. Because an s subshell contains only one orbital, only two electrons can be added to this subshell. There are three orbitals in a p subshell, however, so a set of p orbitals can hold up to six electrons. There are five orbitals in a d subshell, so a set of d orbitals can hold up to 10 electrons. The following is the complete electron configuration for tin.

$$\text{Sn } (Z = 50) \qquad 1s^2\ 2s^2\ 2p^6\ 3s^2\ 3p^6\ 4s^2\ 3d^{10}\ 4p^6\ 5s^2\ 4d^{10}\ 5p^2$$

3.15   ELECTRON CONFIGURATIONS AND THE PERIODIC TABLE

There is something unusually stable about atoms, such as He, Ne, and Ar, that have electron configurations with filled shells of orbitals. By convention, we therefore write abbreviated electron configurations in terms of the number of electrons beyond the previous element with a filled-shell electron configuration. The electron configuration of lithium, for example, could be written as follows.

$$\text{Li } (Z = 3) \qquad \text{[He] } 2s^1$$

When the electron configurations of the elements are arranged so that we can compare elements in one of the horizontal rows of the periodic table, we find that these rows correspond to the filling of subshells. The second row, for example, contains elements in which the subshells in the $n = 2$ shell are filled.

Li $(Z = 3)$	[He] $2s^1$
Be $(Z = 4)$	[He] $2s^2$
B $(Z = 5)$	[He] $2s^2\ 2p^1$
C $(Z = 6)$	[He] $2s^2\ 2p^2$
N $(Z = 7)$	[He] $2s^2\ 2p^3$
O $(Z = 8)$	[He] $2s^2\ 2p^4$
F $(Z = 9)$	[He] $2s^2\ 2p^5$
Ne $(Z = 10)$	[He] $2s^2\ 2p^6$

There is an obvious pattern within the vertical columns, or groups, of the periodic table as well. The elements in a group have similar configurations for their outermost electrons. This relationship can be seen by looking at the electron configurations of elements in columns on either side of the periodic table.

Group IA		*Group VIIA*	
H	$1s^1$		
Li	[He] $2s^1$	F	[He] $2s^2\ 2p^5$
Na	[Ne] $3s^1$	Cl	[Ne] $3s^2\ 3p^5$
K	[Ar] $4s^1$	Br	[Ar] $4s^2\ 3d^{10}\ 4p^5$
Rb	[Kr] $5s^1$	I	[Kr] $5s^2\ 4d^{10}\ 5p^5$
Cs	[Xe] $6s^1$	At	[Xe] $6s^2\ 4f^{14}\ 5d^{10}\ 6p^5$

Figure 3.21 shows the relationship between the periodic table and the orbitals being filled during the process by which the electron configuration is generated. The two columns on the left side of the periodic table correspond to the filling of an s orbital. The next 10 columns include elements in which the five orbitals in a d subshell are filled. The six columns on the right represent the filling of the three orbitals in a p subshell (except for He which has only one occupied s orbital). Finally, the 14 columns at the bottom of the table correspond to the filling of the seven orbitals in an f subshell.

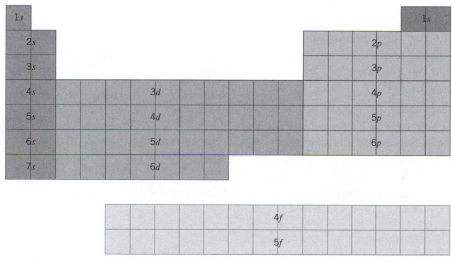

FIGURE 3.21 The periodic table reflects the order in which atomic orbitals are filled. The s orbitals are filled in the two columns on the far left, and the p orbitals are filled in the six columns on the right. The d orbitals are filled along the transition between the s and p orbitals. The f orbitals are filled in the two long rows of elements at the bottom of the table.

Exercise 3.6

Predict the electron configurations for calcium ($Z = 20$) and zinc ($Z = 30$) from their positions in the periodic table.

Solution

Calcium is in the second column and the fourth row of the table. The second column corresponds to the filling of an s orbital. The $1s$ orbital is filled in the first row, the $2s$ orbital in the second row, and so on. By the time we get to the fourth row, we are filling the $4s$ orbital. Calcium therefore has all of the electrons of argon, plus a filled $4s$ orbital.

$$\text{Ca } (Z = 20) \qquad [\text{Ar}] \, 4s^2$$

Zinc is the tenth element in the region of the periodic table where d orbitals are filled. Zinc therefore has a filled subshell of d orbitals. The only question is: Which set of d orbitals is filled? Although zinc is in the fourth row of the periodic table, the first time d orbitals occur is in the $n = 3$ shell. The following is therefore the abbreviated electron configuration for zinc.

$$\text{Zn } (Z = 30) \qquad [\text{Ar}] \, 4s^2 \, 3d^{10}$$

3.16   ELECTRON CONFIGURATIONS AND HUND'S RULES

It is sometimes useful to illustrate the electron configuration of an atom in terms of an **orbital diagram** that shows the spins of the electrons. By convention, electron spins are represented by arrows pointing up or down. Each orbital is represented by a single line or box.

Consider the electrons in the $n = 2$ shell of a boron atom, for example.

$$B \qquad [He]\ 2s^2\ 2p^1 \qquad \underset{2s}{\uparrow\downarrow} \qquad \underset{2p}{\uparrow \ \_\_ \ \_\_}$$

The $2s$ orbital is designated by a single line, whereas three lines are needed to represent the three $2p$ orbitals. There is only one electron in the $2p$ subshell, so we add a single electron to one of the three orbitals in this subshell, with the spin of that electron shown in an arbitrarily chosen direction.

A problem arises when we try to adapt the diagram to the next element, carbon. Where do we put the second electron? Do we put it in the same orbital, to form a pair of electrons of opposite spin? Do we put it in a different orbital in the subshell with the same spin as the first electron? Or do we put it in a different orbital, but with the spins of the two electrons paired? The German physicist Friedrich Hund found that the most stable arrangement of electrons can be predicted from the following rules.

- One electron is added to each orbital in a subshell before two electrons are added to any orbital in the subshell.
- Electrons are added to a subshell with the same spin until each orbital in the subshell has at least one electron.

According to **Hund's rules,** the electrons in the $2p$ subshell on a carbon atom occupy two different orbitals and have the same spin. These electrons can be represented as follows:

$$\underset{2p}{\uparrow \ \uparrow \ \_\_}$$

When we get to N ($Z = 7$), we have to put one electron into each of the orbitals in the $2p$ subshell, all with the same spins.

$$N\ (Z = 7) \qquad 1s^2\ 2s^2\ 2p^3 \qquad \underset{2p}{\uparrow \ \uparrow \ \uparrow}$$

Because each orbital in the $2p$ subshell now contains one electron, the next electron added to the subshell must have the opposite spin, thereby filling one of the $2p$ orbitals.

$$O\ (Z = 8) \qquad 1s^2\ 2s^2\ 2p^4 \qquad \underset{2p}{\uparrow\downarrow \ \uparrow \ \uparrow}$$

The ninth electron fills a second orbital in the subshell.

$$F\ (Z = 9) \qquad 1s^2\ 2s^2\ 2p^5 \qquad \underset{2p}{\uparrow\downarrow \ \uparrow\downarrow \ \uparrow}$$

The tenth electron completes the $2p$ subshell.

$$Ne\ (Z = 10) \qquad 1s^2\ 2s^2\ 2p^6 \qquad \underset{2p}{\uparrow\downarrow \ \uparrow\downarrow \ \uparrow\downarrow}$$

These rules are a consequence of the spin and repulsion of electrons previously discussed. Electrons tend to avoid one another and can only exist in the same orbital if their spins are paired.

Checkpoint

How many unpaired electrons are there in each of the following atoms?

(a) carbon   (b) nitrogen   (c) neon   (d) fluorine

3.17 WAVE PROPERTIES OF THE ELECTRON

We have used modern experimental evidence to develop the way electrons are distributed around the nucleus of an atom. Experiments carried out near the beginning of the twentieth century were responsible for the development of the theory which remains the underpinning of our understanding of the atomic world. Scientists became more and more perplexed as they attempted to explain results of these early experiments. It was apparent that particles as small as atoms and electrons did not conform to the same physical laws as could be applied to everyday objects.

In 1926 the Austrian physicist Erwin Schrödinger worked out a mathematical way of dealing with the problem. This new study was known as wave mechanics or quantum mechanics. Schrödinger's formulation of the way electrons interacted with the nucleus led, of course, to the same results just derived from our study of photoelectron spectroscopy, but his method relied on a mathematical treatment. He assumed that the same mathematics which applied to a light wave also could be applied to explain the interaction of an electron with the electric charge field of a nucleus. His work showed that electrons of different energies were found in different regions around the nucleus. These different regions of stability resulted from a complex interplay of the electrostatic fields generated by charged electrons and nuclei. An electron could exist only in certain parts of this wave field, and its energy was therefore quantized. Schrödinger further showed that these stable states could be characterized by (1) the distance of the electron from the nucleus, (2) the momentum of the electron, and (3) the location of these regions of stability in space. A fourth characteristic of these stable states was Pauli's principle requiring electrons in the same region of space to be spin paired. These four characteristics are described by **quantum numbers** which define the energy state of an electron of an atom.

There were certain complications introduced by the new "wave mechanics" that were beyond what we experience in ordinary life. The idea of quantization, that there is a minimum amount of energy which every energy packet must have, places restrictions on our ability to make precise measurements at the atomic level. In our everyday world we can describe the motion of a baseball or an airplane by giving their successive positions and velocities. The motion of atomic particles, however, cannot be so described.

To measure the position of an object, the object must be illuminated with light so that we can see it. The photons of illuminating light are scattered by the object and strike our eye, allowing us to measure the position of the object. However, in the atomic world the energy of a photon of light can cause electrons to change energy states. This means that the electron is now disturbed from its original state, so either its velocity or its position is no longer the same. Therefore, if we know precisely the electron's position we cannot know its velocity and vice versa. We cannot simultaneously be certain of an electron's position and velocity.

The German physicist Werner Heisenberg formulated this uncertainty in his famous principle, which states that the better the position of an electron is known, the less well its velocity is known. For an electron *not* to escape from an atom it must have a certain minimum velocity. This minimum velocity corresponds to an uncertainty in its position in the atom, which is as large as the atom itself. This means that an electron is simultaneously all around the nucleus. For this reason we often describe an electron as a cloud of electron density within the atom instead of describing it as a single particle. This rather unsettling view of an atomic system has been put into some perspective by Nobel Laureate George Gamow who, in his books on Mr. Tompkins, described how this small atomic world would appear.[1]

In one adventure, Mr. Tompkins, accompanied by a professor, is in a billiard hall in a world in which the physical laws of the atomic world apply.

He approached the table and started to watch the game. Something very queer about it! A player put a ball on the table and hit it with the cue. Watching the rolling ball, Mr. Tompkins noticed to his great surprise that the ball began to "spread out." This was the only expression he could find for the strange behavior of the ball which, moving across the green field, seemed to become more and more washed out, losing its sharp contours. It looked as if not one ball was rolling across the table but a great number of balls, all partially penetrating into each other.

"This is very unusual," murmured Mr. Tompkins.

"On the contrary," insisted the professor, "it is absolutely usual, in the sense that it's always happening to any material body. Only owing to the small value of the quantum constant and to the roughness of the ordinary methods of observation, people do not notice this indeterminacy. They arrive at the erroneous conclusion that position or velocity are always definite quantities. Actually both are always indefinite to some extent, and the better one is defined the more the other is spread out. The quantum constant just governs the relation between these two uncertainties. Look here, I am going to put definite limits on the position of this ball by putting it inside a wooden triangle."

As soon as the ball was placed in the enclosure the whole inside of the triangle became filled up with the glittering of ivory.

"You see!" said the professor, "I defined the position of the ball to the extent of the dimensions of the triangle, i.e., several inches. This results in considerable uncertainty in the velocity, and the ball is moving rapidly inside the boundary."

"Can't you stop it?" asked Mr. Tompkins.

"No—it is physically impossible. Any body in an enclosed space possesses a certain motion—we physicists call it zero-point motion. Such as, for example, the motion of electrons in any atom."

While Mr. Tompkins was watching the ball dashing to and fro in its enclosure like a tiger in a cage, something very unusual happened. The ball just "leaked out" through the wall of the triangle and the next moment was rolling toward a distant corner of the table. The strange thing was that it really did not jump over the wooden wall, but just passed through it, not rising from the table.

"Well, there you are," said Mr. Tompkins, "your 'zero-motion' has run away. Is that according to the rules?"

"Of course it is," said the professor, "in fact this is one of the most interesting consequences of quantum theory. It is impossible to hold anything inside an enclosure provided there is enough energy for running away after crossing the wall. Sooner or later the object will just 'leak through' and get away."

The inability to precisely locate an object and the possibility of a solid object passing through a solid barrier are experiences which contradict our macroscopic world. But this behavior is a consequence of the physical laws which apparently govern the atomic world.

3.18   THE SIZES OF ATOMS: METALLIC RADII

The size of an atom is an important property that influences chemical behavior. Atomic size is determined by how far out electrons extend from the nucleus. Unfortunately, the size of an isolated atom can't be measured because we can't determine the location of the electrons that surround the nucleus. We therefore estimate the size of an atom by assuming that the radius of the atom is equal to one-half the distance between nuclei of adjacent atoms in a solid. This technique is best suited to elements that are metals, which form solids composed of extended planes of atoms of that element. The results of these measurements are therefore known as **metallic radii.**

Because more than 75% of the elements are metals, metallic radii are available for most elements in the periodic table. Figure 3.22 shows the relationship between the metallic radii for elements in Groups IA and IIA. There are two general trends in the data.

- Atoms become *larger* as we go down a column of the periodic table.
- Atoms become *smaller* as we go from left to right across a row of the periodic table.

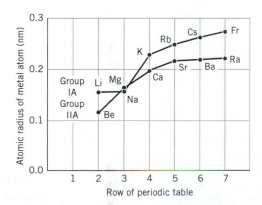

FIGURE 3.22 Metallic radii increase as we go down a column of the periodic table. With rare exception, they decrease from left to right across a row of the table.

The first trend can be rationalized. As we go down the periodic table, electrons are placed in larger and larger subshells but the core charge remains the same. When this happens, the size of the atom should increase.

The second trend is a bit surprising. We might expect atoms to become larger as we go across a row of the periodic table because each element has one more electron than the preceding element. But the additional electrons are added to the same shell. Because the number of protons in the nucleus increases as we go across a row of the table, the core charge increases and the force of attraction between the nucleus and the electrons that surround it also increases. The nucleus therefore tends to pull each electron in closer, and the atoms become smaller.

3.19   THE SIZES OF ATOMS: COVALENT RADII

The size of an atom can also be estimated by measuring the distance between adjacent atoms in compounds which are neither ionic nor metallic. These compounds are called covalent compounds. The **covalent radius** of a chlorine atom, for example, is assumed to be equal to one-half the distance between the nuclei of the atoms in a Cl_2 molecule.

The covalent radii of the main-group elements are given in Figure 3.23. The data confirm the trends observed for metallic radii. Atoms *become larger* as we go down a column of the periodic table, and they *become smaller* as we go across a row of the table.

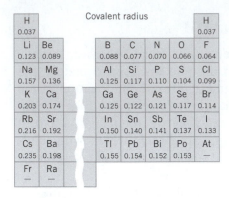

FIGURE 3.23 Covalent radii for the main-group elements. Atoms become larger as we go down a column of the periodic table and smaller as we go from left to right across a row of the table.

Table B.4 in the appendix contains covalent and metallic radii for a number of elements. The covalent radius for an element is usually a little smaller than the metallic radius. This can be explained by noting that covalent bonds tend to squeeze the atoms together, as shown in Figure 3.24.

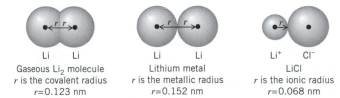

Li   Li
Gaseous Li$_2$ molecule
r is the covalent radius
r=0.123 nm

Li   Li
Lithium metal
r is the metallic radius
r=0.152 nm

Li$^+$ Cl$^-$
LiCl
r is the ionic radius
r=0.068 nm

FIGURE 3.24 The radii of atoms and ions can be determined by measuring the distance between adjacent nuclei. The covalent radius of lithium can be determined by measuring the internuclear distance in the gaseous Li$_2$ molecule. This molecule exists only at very high temperatures. Li$_2$ is covalently bonded, and the Li atoms are pulled tightly together. The metallic radius of lithium can be determined from measurements of solid lithium metal. In lithium metal each lithium atom is surrounded by other lithium atoms all immersed in a sea of electrons. The ionic radius of lithium can be determined from measurements of an ionic compound such as lithium chloride. Lithium loses an electron to chlorine to form LiCl. Lithium and chloride ions are held in place by strong coulombic forces of attraction. Measurements of the distance between Li$^+$ and Cl$^-$ nuclei must take into account that Li$^+$ ions differ in size from Cl$^-$ ions.

Checkpoint
What is the difference between covalent and metallic radii?

3.20   THE RELATIVE SIZES OF ATOMS AND THEIR IONS

In Table 3.6 and Figure 3.25 the covalent radii of neutral fluorine, chlorine, bromine, and iodine atoms are compared with the radii of their F$^-$, Cl$^-$, Br$^-$, and I$^-$ ions. In each case, the negative ion is much larger than the atom from which it has been formed. In fact, the negative ion can be more than twice as large as the neutral atom.

The only difference between an atom and its ions is the number of electrons that surround the nucleus. A neutral chlorine atom, for example, contains 17 electrons, while a Cl$^-$ ion contains 18 electrons. There is no room in the shells closest to the nucleus for an additional electron. The electron is, therefore, found in the outermost shell.

TABLE 3.6   Covalent Radii of Neutral Group VIIA Atoms and Ionic Radii of Their Negative Ions

Element	Covalent Radius (nm)	Ion	Ionic Radius (nm)
F	0.064	F⁻	0.136
Cl	0.099	Cl⁻	0.181
Br	0.1142	Br⁻	0.196
I	0.1333	I⁻	0.216

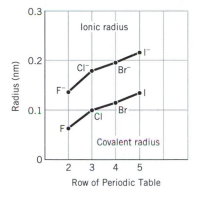

FIGURE 3.25 Comparison of the radii of F^-, Cl^-, Br^-, and I^- with the covalent radii of the corresponding neutral atoms. In each case, the negatively charged ion is larger than the neutral atom.

$$\text{Cl} \quad [\text{Ne}] \; 3s^2 \; 3p^5 \qquad\qquad \text{Cl}^- \quad [\text{Ne}] \; 3s^2 \; 3p^6$$

Because the nucleus can't hold the 18 electrons in the Cl^- ion as tightly as the 17 electrons in the neutral atom, the negative ion is significantly larger than the atom from which it forms.

Extending this line of reasoning suggests that positive ions should be smaller than the atoms from which they are formed. The electrons removed from an atom to form a positive ion are the electrons which are most loosely held, the electrons in the outer shell. The 11 protons in the nucleus of an Na^+ ion, for example, should be able to hold the 10 electrons on the ion more tightly than the 11 electrons on a neutral sodium atom. Removing an electron from the Na atom gives a sodium ion with electrons in only two shells. The Na^+ ion therefore should be much smaller than a sodium atom.

$$\text{Na} \quad 1s^2 \; 2s^2 \; 2p^6 \; 3s^1 \qquad\qquad \text{Na}^+ \quad 1s^2 \; 2s^2 \; 2p^6$$

Table 3.7 and Figure 3.26 provide data to test the hypothesis. Here the covalent radii for neutral atoms of the Group IA elements are compared with the ionic radii for the corresponding positive ions. In each case, the positive ion is much smaller than the atom from which it forms.

TABLE 3.7   Covalent Radii of Neutral Group IA Atoms and Ionic Radii of Their Positive Ions

Element	Covalent Radius (nm)	Ion	Ionic Radius (nm)
Li	0.123	Li⁺	0.068
Na	0.157	Na⁺	0.095
K	0.2025	K⁺	0.133
Rb	0.216	Rb⁺	0.148
Cs	0.235	Cs⁺	0.169

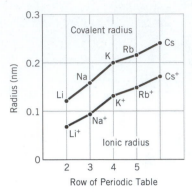

FIGURE 3.26 Comparison of the radii of Li^+, Na^+, K^+, Rb^+, and Cs^+ with the covalent radii of the corresponding neutral atoms. In each case, the positively charged ion is smaller than the neutral atom.

Exercise 3.7

Compare the sizes of neutral sodium and chlorine atoms and their Na^+ and Cl^- ions.

Solution

A neutral sodium atom is significantly larger than a neutral chlorine atom.

$$Na \quad \gg \quad Cl$$
$$(0.157 \text{ nm}) \quad (0.099 \text{ nm})$$

But an Na^+ ion is only one-half the size of a Cl^- ion.

$$Na^+ \quad \ll \quad Cl^-$$
$$(0.095 \text{ nm}) \quad (0.181 \text{ nm})$$

These particles therefore increase in size in the following order:

$$Na^+ \approx Cl \ll Na \ll Cl^-$$

Tabulated values of ionic and covalent radii are particularly useful because they can be used to estimate the distance between atoms and ions in compounds. These distances are important because properties of compounds are often dictated by how far one atom or ion is from another. For example, from the ionic radii in Tables 3.6 and 3.7 we can estimate the distance between Na^+ and Cl^- in sodium chloride to be 0.276 nm. For the distance between Na^+ and Br^- in sodium bromide we find 0.291 nm. It is no coincidence that NaCl melts at a higher temperature than NaBr. More detailed discussion of how properties depend on the sizes of the particles composing compounds is the subject of following chapters.

3.21 PATTERNS IN IONIC RADII

The ionic radii in Tables 3.6 and 3.7 confirm one of the patterns observed for both metallic and covalent radii: Atoms become larger as we go down a column of the periodic table. We can examine trends in ionic radii across a row of the periodic table by comparing data for atoms and ions that are **isoelectronic.** By definition, isoelectronic atoms or ions have the same number of electrons. Table 3.8 summarizes data on the radii of a series of isoelectronic ions and atoms of second- and third-row elements.

TABLE 3.8   Radii for Isoelectronic Second-Row and Third-Row Atoms or Ions

Atom or Ion	Radius (nm)	Electron Configuration
C^{4-}	0.260	$1s^2\,2s^2\,2p^6$
N^{3-}	0.171	$1s^2\,2s^2\,2p^6$
O^{2-}	0.140	$1s^2\,2s^2\,2p^6$
F^-	0.136	$1s^2\,2s^2\,2p^6$
Ne	0.112	$1s^2\,2s^2\,2p^6$
Na^+	0.095	$1s^2\,2s^2\,2p^6$
Mg^{2+}	0.065	$1s^2\,2s^2\,2p^6$
Al^{3+}	0.050	$1s^2\,2s^2\,2p^6$

The data in Table 3.8 can be explained if we note that these atoms or ions all have 10 electrons, but the number of protons in the nucleus increases from 6 in the C^{4-} ion to 13 in the Al^{3+} ion. As the charge on the nucleus becomes larger, the nucleus can hold a constant number of electrons more tightly. As a result, the atoms or ions become significantly smaller—in this series, by a factor of 5 from C^{4-} to Al^{3+}.

Checkpoint

Where would the radius of Si^{4+} fit into Table 3.8?

3.22   SECOND, THIRD, FOURTH, AND HIGHER IONIZATION ENERGIES

Sodium forms Na^+ ions, magnesium forms Mg^{2+} ions, and aluminum forms Al^{3+} ions. Why doesn't sodium form Na^{2+} ions, or even Na^{3+} ions? The answer can be obtained from data for the second, third, and higher ionization energies of the element.

As we have seen, more than one ionization energy can be measured for sodium, depending on whether the electron is removed from the $1s$, $2s$, $2p$, or $3s$ orbitals on the atom. The first ionization energy represents the energy it takes to remove the *outermost* electron on the atom. For sodium this would be the energy required to remove the electron from the $3s$ orbital.

The *first ionization energy* of sodium is the energy it takes to remove an electron from a neutral atom in the gas phase.

$$Na(g) + \text{energy} \longrightarrow Na^+(g) + e^-$$

The *second ionization energy* of sodium is the energy it would take to remove another electron to form an Na^{2+} ion in the gas phase.

$$Na^+(g) + \text{energy} \longrightarrow Na^{2+}(g) + e^-$$

The *third ionization energy* of sodium represents the process by which the Na^{2+} ion is converted to an Na^{3+} ion.

$$Na^{2+}(g) + \text{energy} \longrightarrow Na^{3+}(g) + e^-$$

The energy required to form an Na^{3+} ion in the gas phase is the sum of the first, second, and third ionization energies of the element.

A complete set of data for the ionization energies of the elements is given in Table B.5 in the appendix. For the moment, let's look at the first, second, third, and fourth ionization energies of sodium, magnesium, and aluminum listed in Table 3.9.

TABLE 3.9   First, Second, Third, and Fourth Ionization Energies (kJ/mol) of Gas Phase Atoms of Sodium, Magnesium, and Aluminum

	First IE	Second IE	Third IE	Fourth IE
Na	495.8	4562.4	6912	9543
Mg	737.7	1450.6	7732.6	10,540
Al	577.6	1816.6	2744.7	11,577

It doesn't take much energy to remove one electron from a sodium atom to form an Na^+ ion with a filled-shell electron configuration. Once this is done, it takes almost 10 times as much energy to break into the filled-shell configuration to remove a second electron. Because it takes more energy to remove the second electron, sodium generally forms compounds that contain Na^+ ions rather than Na^{2+} or Na^{3+} ions.

A similar pattern is observed when the ionization energies of magnesium are analyzed. The first ionization energy of magnesium is larger than that of sodium because magnesium has one more proton in its nucleus to hold on to the electrons in the 3s orbital.

$$Mg \quad [Ne] \ 3s^2$$

The second ionization energy of Mg is larger than the first because it always takes more energy to remove an electron from a positively charged ion than from a neutral atom. The third ionization energy of magnesium is enormous, however, because the Mg^{2+} ion has a filled-shell electron configuration.

The same pattern can be seen in the ionization energies of aluminum. The first ionization energy of aluminum is smaller than that of magnesium because it involves removing an electron from a 2p rather than a 2s orbital. The second ionization energy of aluminum is larger than the first, and the third ionization energy is even larger. Although it takes a considerable amount of energy to remove three electrons from an aluminum atom to form an Al^{3+} ion, the energy needed to break into the filled-shell configuration of the Al^{3+} ion is astronomical. Thus, it would be a mistake to look for an Al^{4+} ion as the product of a chemical reaction.

Exercise 3.8

Predict the group in the periodic table in which an element with the following ionization energies would most likely be found.

$$
\begin{aligned}
\text{1st IE} &= \quad 786 \text{ kJ/mol} \\
\text{2nd IE} &= \ 1{,}577 \\
\text{3rd IE} &= \ 3{,}232 \\
\text{4th IE} &= \quad 4{,}355 \\
\text{5th IE} &= 16{,}091 \\
\text{6th IE} &= 19{,}784
\end{aligned}
$$

Solution

The gradual increase in the energy needed to remove the first, second, third, and fourth electrons from this element is followed by an abrupt increase in the energy required to remove one more electron. This is consistent with an element that has four more electrons than a filled-shell configuration, and we might expect the element to be in Group IVA of the periodic table. These data are in fact the ionization energies of silicon.

$$\text{Si} \quad [\text{Ne}]\, 3s^2\, 3p^2$$

The trends in the ionization energies of the elements can be used to explain why elements on the left side of the periodic table are more likely than those on the right to form positive ions. The ionization energies of elements on the left side of the table are much smaller than those of elements on the right. Consider the first ionization energies for sodium and chlorine, for example.

$$\text{Na} \quad \text{1st IE} = 495.8 \text{ kJ/mol}$$
$$\text{Cl} \quad \text{1st IE} = 1251.1 \text{ kJ/mol}$$

Elements on the left side of the periodic table are therefore more likely to form positive ions.

Trends in ionization energies can also be used to explain why the maximum positive charge found on atoms under normal conditions for the main-group elements is equal to the group number of the element. The number of valence electrons is equal to the group number, so the maximum positive charge on an ion is also equal to the group number. Because aluminum is in Group IIIA, for example, it can lose only three electrons before it reaches a filled-shell configuration. Thus, the maximum positive charge on an aluminum ion is +3.

Checkpoint

What is the maximum charge under normal conditions for an atom from Group IVA?

3.23   AVERAGE VALENCE ELECTRON ENERGY (AVEE)[2]

Ionization energies provide a measure of how tightly the electrons are held in an isolated atom. It would be useful to have a single quantity that reflects the *average* ionization energy of the valence electrons on an atom. The quantity, which is known as the **average valence electron energy (AVEE),** can be calculated from the ionization energies for the valence electrons obtained by photoelectron spectroscopy. Because there are different numbers of electrons in the various subshells, the AVEE is calculated as a weighted average.

Consider carbon, for example, which has two electrons in the valence $2s$ subshell and two electrons in the valence $2p$ subshell. The ionization energies of electrons from the $2s$ subshell ($\text{IE}_s = 1.72$ MJ/mol) and the $2p$ subshell ($\text{IE}_p = 1.09$ MJ/mol) are given in Table 3.4. The average valence electron energy for a carbon atom can therefore be calculated as follows.

[2]L. C. Allen, *Journal of the American Chemical Society,* **111,** 9003 (1989).

$$\text{AVEE}_C = \left[\frac{(2 \times \text{IE}_s) + (2 \times \text{IE}_p)}{(2 + 2)}\right] = \left[\frac{(2 \times 1.72 \text{ MJ/mol}) + (2 \times 1.09 \text{ MJ/mol})}{4}\right]$$

$$= 1.41 \text{ MJ/mol}$$

Fluorine, on the other hand, would have an AVEE that is significantly larger.

$$\text{AVEE}_F = \left[\frac{(2 \times \text{IE}_s) + (5 \times \text{IE}_p)}{(2 + 5)}\right] = \left[\frac{(2 \times 3.88 \text{ MJ/mol}) + (5 \times 1.68 \text{ MJ/mol})}{7}\right]$$

$$= 2.31 \text{ MJ/mol}$$

The AVEE values for the main-group elements calculated from photoelectron spectroscopy data are given in Figure 3.27.

H								He
1.31								2.37
Li	Be		B	C	N	O	F	Ne
0.52	0.90		1.17	1.41	1.82	1.91	2.31	2.73
Na	Mg		Al	Si	P	S	Cl	Ar
0.50	0.74		0.92	1.13	1.39	1.35	1.59	1.85
K	Ca	Sc	Ga	Ge	As	Se	Br	
0.42	0.59	0.68	1.00	1.07	1.26	1.3	1.53	
Rb	Sr	Y	In	Sn	Sb	Te	I	
0.40	0.55	0.57	0.94	1.04	1.13	1.2	1.35	

FIGURE 3.27 Average valence electron energies (AVEE) for the main-group elements, in megajoules per mole (MJ/mol) [L. C. Allen, *Journal of the American Chemical Society*, **111,** 9003 (1989)].

With the exception of phosphorus, there is a systematic increase in AVEE from left to right across each row of the periodic table. Because the charge on the nucleus steadily increases as we go across each row, and the size of the atom gradually decreases, the valence electrons on each atom are held more tightly as we go across the row.

In addition, as the atomic number increases across a row, the energy difference between the valence subshells increases. As we proceed down a group, electrons are less strongly held and the energy difference between valence subshells decreases and AVEE decreases. Thus AVEE measures two important quantities, the attraction of an atom for its electrons and the spacing of its valence energy levels. These trends are shown in Figure 3.28, which shows the energies of the $2s$ and $2p$ electrons in B, C, N, O, and F. The difference in energy between the $2s$ and $2p$ electrons also becomes larger as shown by the increased separation of the lines in Figure 3.28.

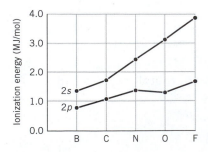

FIGURE 3.28 Ionization energies of the electrons in the $2s$ and $2p$ subshells of B, C, N, O, and F. Two trends can be observed: the ionization energies of the subshells increase (the electrons are more tightly held) as we move across the period and the difference in energy between the s and the p electrons also increases as we go across the period.

Checkpoint
Describe how Figure 3.28 would appear if relative electron energy levels were plotted instead of ionization energies (see Figure 3.19).

The metals on the left side of the periodic table have AVEE values that are relatively small compared with the nonmetals on the right side of the table. Sodium and magnesium, for example, have AVEE values of 0.50 and 0.74 MJ/mol, respectively, whereas the AVEE values for O and Cl are 1.91 and 1.59 MJ/mol. Thus sodium and magnesium are more likely to lose electrons to form positive ions than are O and Cl, and oxygen and chlorine are more likely to add electrons to form negative ions than are Na and Mg.

3.24 AVEE AND METALLICITY

Because the average valence electron energy provides a measure of how tightly an atom holds onto its valence electrons and the energy gap between valence electron subshells, it can be used to explore the dividing line between the metals and nonmetals in the periodic table. The semimetals (metalloids) that lie along the stair-step line separating the metals from the nonmetals in the periodic table all have similar AVEE values. B, Si, Ge, As, Sb, and Te all have AVEE values that lie between 1.07 and 1.26 MJ/mol, and they are the only elements in the periodic table that have values within this range (Figure 3.29).

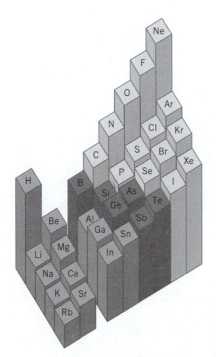

FIGURE 3.29 Three-dimensional plot of the average valence electron energies (AVEE) of the main-group elements versus position in the periodic table. The AVEE is a measure of how tightly atoms hold onto their valence electrons and the energy gap between valence subshells.

It therefore appears that the AVEE of an atom of a given element can be used to decide whether the element is a metal or nonmetal. Atoms with an AVEE value below 1.07 MJ/mol are metals, whereas those with an AVEE greater than 1.26 MJ/mol are nonmetals. Elements with AVEE values in the range of 1.07–1.26 MJ/mol have properties intermediate between those of metals and nonmetals and are known as the *semimetals,* or *metalloids.*

The power of AVEE data can be illustrated by considering the difference between the second and third versus the fourth and fifth rows of the periodic table. In the second and third rows, there is only one semimetal each (B and Si). In the fourth and fifth rows, there are two (Ge and As, and Sb and Te). If we tried to extend the pattern seen in the fourth and fifth rows of the periodic table to the second and third rows, we would predict that beryllium and aluminum would be semimetals. But the AVEE values for beryllium (0.90) and aluminum (0.92) clearly indicate that these elements should be metals. This is consistent with the observation that both elements exhibit all of the characteristic properties of a metal. They have a metallic luster; they are good conductors of heat and electricity; and they are both malleable and ductile.

In general, there are two factors that contribute to metallic behavior. Valence electrons must be easy to remove, and the energy separation between valence subshells must be small. AVEE is a convenient measure of both effects. Both are necessary to observe metallic behavior. The increasing trend for metallicity down a group also follows the AVEE values. Carbon is a nonmetal, but as we go down Group IVA through Si, Ge, Sn, and Pb metallic behavior increases. This reflects two things: the electrons are becoming easier to remove, and the valence subshells are becoming closer in energy.

Checkpoint
Why do AVEE values increase from left to right across a period and from bottom to top of a group? Is there another periodic property that follows those trends?

KEY TERMS

Amplitude

Atomic spectra

Average valence electron
 energy (AVEE)

Bohr model

Covalent radius

Electromagnetic radiation

Electron configuration

Emission spectrum

First ionization energy

Frequency

Hund's rules

Isoelectronic

Metallic character

Metallic radius

Nonmetallic character

Nucleus

Orbitals

Particles

Photoelectron
 spectroscopy
 (PES)

Photons

Quantized

Shell

Shell model

Spectrum

Spin

Subshell

Transition metals

Valence electrons

Wavelength

Waves

PROBLEMS

Structure of the Atom

1. Static electricity is due to the buildup of charge on a material. If wool is rubbed on a piece of rubber, the rubber becomes negatively charged and the wool becomes positively charged. Use the shell model for the structure of the atom to explain static electricity.

2. A chemistry text published in 1922 proposed a model of the atom based on only two subatomic particles: electrons and protons. All of the protons and some of the electrons were concentrated in the nucleus of the atom. The other electrons revolved

around the nucleus. The number of protons in the nucleus was equal to the mass of the atom. The charge on the nucleus was equal to the number of protons minus the number of electrons in the nucleus. Enough electrons were then added to the atom to neutralize the charge on the nucleus. Use this model to calculate the number of protons and electrons in a neutral fluorine atom, F, and a fluoride ion, F^-. Compare this calculation with results obtained by assuming that the nucleus is composed of protons and neutrons.

3. The charge on a single electron is $1.60217733 \times 10^{-19}$ coulomb. Calculate the charge on a mole of electrons and compare the result of your calculation with Faraday's constant, given in Table B-1 in Appendix B.

Particles and Waves

4. Describe the difference between a particle and a wave.

5. To examine the relationship between the frequency, wavelength, and speed of a wave, imagine that you are sitting at a railroad crossing, watching a train that consists of 45-foot boxcars go by. Furthermore assume it takes 1.0 second for each boxcar to pass in front of your car. Calculate the "frequency" of a boxcar and the product of the 45-foot "wavelength" times this frequency. Convert the final answer into units of miles per hour.

6. An octave in a musical scale corresponds to a change in the frequency of a note by a factor of 2. If a note with a frequency of 440 hertz is an A, then the A one octave above this note has a frequency of 880 hertz. What happens to the wavelength of the sound as the frequency increases by a factor of 2? What happens to the speed at which the sound travels to your ear?

7. The human ear is capable of hearing sound waves with frequencies between about 20 and 20,000 hertz. If the speed of sound is 340.3 meters per second at sea level, what is the wavelength in meters of the longest wave the human ear can hear?

Light and Other Forms of Electromagnetic Radiation

8. Calculate the wavelength in meters of blue-green light that has a frequency of 5.0×10^{14} cycles per second.

9. Calculate the frequency of red light that has a wavelength of 700 nanometers.

10. Which has the longer wavelength, red light or blue light?

11. Which has the larger frequency, radio waves or microwaves?

12. In a magnetic field of 2.35 tesla, ^{13}C nuclei absorb electromagnetic radiation that has a frequency of 25.147 MHz. Calculate the wavelength of this radiation. In which region of the electromagnetic spectrum does the radiation fall?

13. The meter has been defined as 1,650,763.73 wavelengths of the orange-red line of the emission spectrum of ^{86}Kr. Calculate the frequency of this radiation. In what portion of the electromagnetic spectrum does the radiation fall?

Atomic Spectra

14. Soap bubbles pick up colors because they reflect light with wavelengths equal to the thickness of the walls of the bubble. What frequency of light is reflected by a soap bubble 6 nanometers thick?

15. Methylene blue, $C_{16}H_{18}ClN_3S$, absorbs light most intensely at wavelengths of 668 and 609 nanometers. What color is the light absorbed by the dye?

16. Sodium salts give off a characteristic yellow-orange light when added to the flame of a bunsen burner. The yellow-orange color is due to two narrow bands of radiation with wavelengths of 588.9953 and 589.5923 nm. Calculate the frequencies of these emission lines.

Quantization of Energy

17. Which carries more energy, ultraviolet or infrared radiation?

18. Which carries more energy, yellow light with a wavelength of 580 nm or green light with a wavelength of 660 nm?

19. List the four lines in the visible region of emission spectrum of hydrogen in order of increasing energy.

20. Calculate the energy in joules of the radiation in the emission spectrum of the hydrogen atom that has a wavelength of 656.3 nm.

21. Calculate the energy in joules of a single particle of radiation broadcast by an amateur radio operator who transmits at a wavelength of 10 m.

22. Cl_2 molecules can dissociate to form chlorine atoms by absorbing electromagnetic radiation. It takes 243.4 kJ of energy to break the bonds in a mole of Cl_2 molecules. What is the wavelength of the radiation that has just enough energy to decompose a Cl_2 molecule to chlorine atoms? In what portion of the spectrum is this wavelength found?

First Ionization Energy

23. According to the shell model, why is the first ionization energy of Cl less than that of F?

24. Estimate the first ionization energy for Br using only the shell model.

25. If a single electron is removed from an Li atom, the resulting Li^+ cation has only two electrons, both in the $n = 1$ shell. In this respect it is very similar to an He atom. How would you expect the ionization energy of Li^+ to compare to that of an He atom? Explain your reasoning.

26. If a single electron is added to an F atom, the resulting F^- negative ion has a total of eight valence electrons in the $n = 2$ shell. In this respect it is very similar to an Ne atom. How would you expect the ionization energy of F^- to compare to that of an Ne atom? Explain your reasoning.

27. Predict the order of the first ionization energies for the atoms Br, Kr, and Rb. Explain your reasoning.

28. Write balanced chemical equations, as described in Section 3.22, for the reactions that occur when the first, second, third, and fourth ionization energies of aluminum are measured.

29. Explain why it takes energy to remove an electron from an isolated atom in the gas phase.

30. Describe the general trend in first ionization energies from left to right across the second row of the periodic table.

31. Describe the general trend in first ionization energies from top to bottom of a column of the periodic table.

32. Explain why the first ionization energy of B is smaller than that of Be and why the first ionization energy of O is smaller than that of N.

33. Explain why the first ionization energy of hydrogen is so much larger than the first ionization energy of sodium.

34. List the following elements in order of increasing first ionization energy.
    (a) Li (b) Be (c) F (d) Na

35. Which of the following elements should have the largest first ionization energy?
    (a) B   (b) C   (c) N   (d) Mg   (e) Al
36. Which of the following elements should have the smallest first ionization energy?
    (a) Mg   (b) Ca   (c) Si   (d) S   (e) Se

Photoelectron Spectroscopy

Questions 37–41 should be answered in sequence.

37. Why are two of the three peaks in the PES spectrum of neon assigned to the $n = 2$ shell rather than to the $n = 1$ shell? What is the rationale for assuming that the peak at 84.0 MJ/mol corresponds to electrons in the $n = 1$ shell?

38. Roughly sketch the photoelectron spectra for Ti and V. Give the relative intensities of the peaks.

39. What element do you think should give rise to the photoelectron spectrum shown in Figure 3P.1? Explain your reasoning.

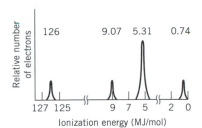

FIGURE 3P.1

40. Use the ionization energies given below for Li, Na, and Ar to predict the photoelectron spectrum of K.

Ionization Energy (MJ/mol)

Element	1s	2s	2p	3s	3p
Li	6.26	0.52			
Na	104	6.84	3.67	0.50	
Ar	309	31.5	24.1	2.82	1.52

(a) First consider the first three shells (18 electrons) of K. For these 18 electrons, indicate the relative energies of the peaks and their relative intensities.
(b) If the nineteenth electron of K is found in the $n = 4$ shell, would the ionization energy be closest to 0.42, 1.4, or 2.0 MJ/mol? Explain. (Hint: Compare to Na and Li.) Show a predicted photoelectron spectrum based on this assumption.
(c) If the nineteenth electron of K is found in the third subshell of the $n = 3$ shell, would the ionization energy be closest to 0.42, 1.4, or 2.0 MJ/mol? Explain. (Hint: Compare to Ar.) Show a predicted photoelectron spectrum based on this assumption.
(d) Given the correct photoelectron spectrum of K (Figure 3P.2), predict whether the nineteenth electron of K is found in the $n = 4$ or $n = 3$ shell. Explain your reasoning.

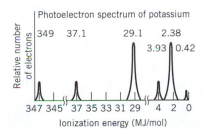

FIGURE 3P.2

41. Identify the element whose photoelectron spectrum is shown in Figure 3P.3. (Note: In Figure 3P.3, the peak which arises from the 1s electrons has been omitted.)

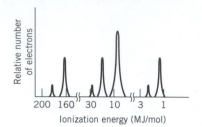

FIGURE 3P.3

Shells and Subshells of Orbitals

42. Describe what happens to the difference between the energies of subshells as the value of n becomes larger.
43. What happens to the number of subshells as the value of n becomes larger?
44. Determine the number of subshells in the $n = 3$ and $n = 4$ shells.
45. Calculate the maximum number of electrons in the $n = 1$, $n = 2$, $n = 3$, and $n = 4$ shells.
46. Calculate the maximum number of electrons that can fit into a $4d$ subshell.
47. Calculate the maximum number of unpaired electrons that can be placed in a $5d$ subshell.
48. Explain why the difference between the atomic numbers of adjacent pairs of elements in a vertical column, or group, of the periodic table is either 8, 18, or 32.

Relative Energies of Atomic Orbitals

49. Which of the following sets of subshells for yttrium is arranged in the correct order of increasing energy?
    (a) $3d, 4s, 4p, 5s, 4d$ (b) $3d, 4s, 4p, 4d, 5s$
    (c) $4s, 3d, 4p, 5s, 4d$ (d) $4s, 3d, 4p, 4d, 5s$
50. Which of the following orders of filling of orbitals is incorrect?
    (a) $3s, 4s, 5s$ (b) $5s, 5p, 5d$ (c) $5s, 4d, 5p$
    (d) $6s, 4f, 5d$ (e) $6s, 5f, 6p$
51. As atomic orbitals are filled, the $6p$ orbitals are filled immediately after which of the following orbitals?
    (a) $4f$ (b) $5d$ (c) $6s$ (d) $7s$

Electron Configurations

52. Describe the regions of the periodic table in which the s, p, d, and f subshells are filled.
53. Write the electron configurations for the elements in the third row of the periodic table.
54. Use straight lines and arrows to draw the orbital diagram for the electron configuration of the nitrogen atom.
55. Use straight lines and arrows to draw the orbital diagram for the electron configuration for nickel.

56. The electron configuration of Si is $1s^2\ 2s^2\ 2p^6\ 3s^2\ 3p^x$, where x is which of the following?
    (a) 1   (b) 2   (c) 3   (d) 4   (e) 6

57. Which of the following electron configurations for carbon satisfies Hund's rules?

    (a) $1s^2\ 2s^2\ 2p^2$ ↑ ↓ __
    $\qquad\qquad\qquad\quad$ 2p

    (b) $1s^2\ 2s^2\ 2p^2$ ↑ ↑ __
    $\qquad\qquad\qquad\quad$ 2p

    (c) $1s^2\ 2s^2\ 2p^2$ ↑ __ ↓
    $\qquad\qquad\qquad\quad$ 2p

    (d) $1s^2\ 2s^2\ 2p^2$ ↑↓ __ __
    $\qquad\qquad\qquad\quad$ 2p

58. Which of the following is the correct electron configuration for the P^{3-} ion?
    (a) [Ne]   (b) [Ne] $3s^2$ (c) [Ne] $3s^2\ 3p^3$ (d) [Ne] $3s^2\ 3p^6$

59. Which of the following is the correct electron configuration for the bromide ion, Br^-?
    (a) [Ar] $4s^2\ 4p^5$ (b) [Ar] $4s^2\ 3d^{10}\ 4p^5$ (c) [Ar] $4s^2\ 3d^{10}\ 4p^6$
    (d) [Ar] $4s^2\ 3d^{10}\ 4p^6\ 5s^1$

60. Determine the number of electrons in the third shell of a vanadium atom.

61. Determine the number of electrons in s orbitals in the Ti^{2+} ion.

62. Theoreticians predict that the element with atomic number 114 will be more stable than the elements with atomic numbers between 103 and 114. On the basis of its electron configuration, in which group of the periodic table should element 114 be placed?

63. Which is the first element to have $4d$ electrons in its electron configuration?
    (a) Ca   (b) Sc   (c) Rb   (d) Y   (e) La

64. Which of the following neutral atoms has the largest number of unpaired electrons?
    (a) Na   (b) Al   (c) Si   (d) P   (e) S

65. Which of the following ions has five unpaired electrons?
    (a) Ti^{4+} (b) Co^{2+} (c) V^{3+} (d) Fe^{3+} (e) Zn^{2+}

The Periodic Table

66. Describe some of the evidence that could be used to justify the argument that the modern periodic table is based on similarities in the chemical properties of the elements.

67. Describe some of the evidence that could be used to refute the argument that the modern periodic table groups elements with similar electron configurations.

68. Determine the row and column of the periodic table in which you would expect to find the first element to have $3d$ electrons in its electron configuration.

69. Determine the row and column of the periodic table in which you would expect to find the element that has five more electrons than the rare gas krypton.

70. Determine the group of the periodic table in which an element with the following electron configuration belongs.

$$[X] \qquad 1s^2\ 2s^2\ 2p^6\ 3s^2\ 3p^6\ 4s^2\ 3d^{10}\ 4p^6\ 5s^2\ 4d^{10}\ 5p^3$$

71. In which group of the periodic table should element 119 belong if and when it is discovered?

72. Which of the following ions do not have the electron configuration of argon?
    (a) Ga^{3+} (b) Cl^- (c) P^{3-} (d) Sc^{3+} (e) K^+

73. Which of the following ions has the electron configuration $[Ar] 3d^4$?
    (a) Ca^{2+} (b) Ti^{2+} (c) Cr^{2+} (d) Mn^{2+} (e) Fe^{2+}

Sizes of Atoms: Metallic Radii

74. Describe what happens to the sizes of the atoms as we go down a column of the periodic table. Explain.

75. Describe what happens to the sizes of the atoms as we go across a row of the periodic table from left to right. Explain.

76. At one time, the size of an atom was given in units of angstroms because the radius of a typical atom was about 1 angstrom (Å). Now they are given in a variety of units. If the radius of a gold atom is 1.442 Å, and 1 Å is equal to 10^{-8} cm, what is the radius of this atom in nanometers and in picometers?

77. Which of the following atoms has the smallest radius?
    (a) Na (b) Mg (c) Al (d) K (e) Ca

Sizes of Atoms: Covalent Radii

78. Explain why the covalent radius of an atom is smaller than the metallic radius of the atom.

79. Which of the following atoms has the largest covalent radius?
    (a) N (b) O (c) F (d) P (e) S

Sizes of Atoms: Ionic Radii

80. Explain how values of ionic radii can be used and why this information is important.

81. Describe what happens to the radius of an atom when electrons are removed to form a positive ion. Describe what happens to the radius of the atom when electrons are added to form a negative ion.

82. Predict the order of increasing ionic radius for the following ions: H^-, F^-, Cl^-, Br^-, and I^-. Compare your predictions with the data for the ions in Appendix B.4. Explain any differences between your predictions and the experiment.

Relative Sizes of Atoms and Their Ions

83. Look up the covalent radii for magnesium and sulfur atoms and the ionic radii of Mg^{2+} and S^{2-} ions in Appendix B.4. Explain why Mg^{2+} ions are smaller than S^{2-} ions even though magnesium atoms are larger than sulfur atoms.

84. Explain why the radius of a Pb^{2+} ion (0.120 nm) is very much larger than that of a Pb^{4+} ion (0.084 nm).

85. Predict the relative sizes of the Fe^{2+} and Fe^{3+} ions, which can be found in a variety of proteins, including hemoglobin, myoglobin, and the cytochromes.

Patterns in Ionic Radii

86. Sort the following atoms or ions into isoelectronic groups.
    (a) N^{3-} (b) Ar (c) F^- (d) Ne (e) P^{3-}
    (f) Ca^{2+} (g) Al^{3+} (h) Si^{4+} (i) Na^+ (j) S^{2-}
    (k) Cl^- (l) O^{2-} (m) K^+ (n) Mg^{2+}

87. Which of the following contains sets of atoms or ions that are isoelectronic?
    (a) B^{3+}, C^{4+}, H^+, He (b) Na^+, Ne, N^{3+}, O^{2-}
    (c) $Mg^{2+}, F^-, Na^+, O^{2-}$ (d) Ne, Ar, Xe, Kr
    (e) $O^{2-}, S^{2-}, Se^{2-}, Te^{2-}$

88. Predict whether the Al^{3+} or the Mg^{2+} ion is the smaller. Explain.

89. Which of the following ions has the largest radius?
    (a) Na^+ (b) Mg^{2+} (c) S^{2-} (d) Cl^- (e) Se^{2-}

90. Which of the following atoms or ions is the smallest?
    (a) Na (b) Mg (c) Na^+ (d) Mg^{2+} (e) O^{2-}

91. Which of the following ions has the smallest radius?
    (a) K^+ (b) Li^+ (c) Be^{2+} (d) O^{2-} (e) F^-

92. Which of the following isoelectronic ions is the largest?
    (a) Mn^{7+} (b) P^{3-} (c) S^{2-} (d) Sc^{3+} (e) Ti^{4+}

Second, Third, Fourth, and Higher Ionization Energies

93. Which of the following atoms or ions has the largest ionization energy?
    (a) P (b) P^+ (c) P^{2+} (d) P^{3+} (e) P^{4+}

94. Explain why the second ionization energy of sodium is so much larger than the first ionization energy of the element.

95. What is the most probable electron configuration for the element that has the following ionization energies?

$$
\begin{aligned}
\text{1st IE} &= 578 \text{ kJ/mol} \\
\text{2nd IE} &= 1{,}817 \\
\text{3rd IE} &= 2{,}745 \\
\text{4th IE} &= 11{,}577 \\
\text{5th IE} &= 14{,}831
\end{aligned}
$$

    (a) [Ne]   (b) [Ne] $3s^1$ (c) [Ne] $3s^2$
    (d) [Ne] $3s^2\,3p^1$ (e) [Ne] $3s^2\,3p^2$ (f) [Ne] $3s^2\,3p^3$

96. Which of the following ionization energies is the largest?
    (a) 1st IE of Ba   (b) 1st IE of Mg   (c) 2nd IE of Ba
    (d) 2nd IE of Mg   (e) 3rd IE of Al   (f) 3rd IE of Mg

97. Which of the following elements should have the largest second ionization energy?
    (a) Na (b) Mg (c) Al (d) Si (e) P

98. Which of the following elements should have the largest third ionization energy?
    (a) B (b) C (c) N (d) Mg (e) Al

99. List the following elements in order of increasing second ionization energy.
    (a) Li (b) Be (c) Na (d) Mg (e) Ne

100. Some elements, such as tin and lead, have more than one common ion. Use the electron configurations to predict the most likely ions of Sn and Pb.

101. Use the electron configurations to rationalize why iron forms the Fe^{2+} and Fe^{3+} ions.

AVEE

102. Why is Be more likely to form Be^{2+} than O is to form O^{2+}?

103. Use Table 3.4 to calculate the AVEE of B and F. Compare your results to the values given in Figure 3.27.

104. Without reference to Figure 3.27, arrange the following in order of increasing AVEE: (a) P, Mg, Cl   (b) S, O, Se, F   (c) K, P, O

Integrated Problems

105. The most recent estimates give values of about 10^{-10} meters for the radius of an atom and 10^{-14} meters for the radius of the nucleus of the atom. Calculate the fraction of the total volume of an atom that is essentially empty space.

106. What is the relationship between the core charges of the atoms in the third period and their atomic radii?

107. Consider the following ions/atoms: O^{2-}, F^-, Ne, Na^+, and Mg^{2+}. Order them in terms of increasing ionization energy. Also order them in terms of increasing radius. Now consider the following atoms: O, F, Ne, Na, Mg. Order them in terms of increasing ionization energy. Also order them in terms of increasing radius.

108. Chemists often describe substances from three perspectives: macroscopic, microscopic (atomic), and symbolic. Describe aluminum from these three perspectives.

109. Compare and contrast (what is the same and what is different) for the shell model of an oxygen atom obtained from first ionization energies (Table 3.3) and the quantum mechanical model developed from PES data (Table 3.4).

110. Two *hypothetical* shell models of a lithium atom are shown below.

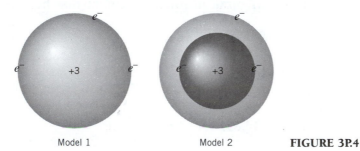

Model 1   Model 2   **FIGURE 3P.4**

(a) Use Model 1 to predict the relationship (larger, smaller, or equal) of the first ionization energy of Li compared to the first ionization energy of He. Explain your reasoning.

(b) Use Model 2 to predict the relationship (larger, smaller, or equal) of the first ionization energy of Li compared to the first ionization energy of He. Explain your reasoning.

(c) Use Table 3.3 to decide which model is most consistent with the observed ionization energy of Li. Explain your reasoning.

111. The following atoms and ions all have the same electronic structure (they are isoelectronic).

$$Ar \quad S^{2-} \quad K^+$$

Arrange them in order of increasing first ionization energy. Arrange them in order of increasing atomic radii. Explain your reasoning.

Questions 112–117 relate to the following data: An atom with an equal number of spin "up" and spin "down" electrons is known as **diamagnetic,** and the atom is repelled by a magnetic field. In this case we say that all of the electrons are "paired." If this is not the case—that is, if there are unpaired electrons—the atom is attracted to a magnetic field, and it is known as **paramagnetic.** The strength of the attraction is an experimentally measurable quantity known as the magnetic moment. The magnitude of the **magnetic moment** (measured in magnetons) is related to (but not proportional to) the number of unpaired electrons present. In other words, the larger the number of unpaired electrons, the larger the magnetic moment. Here are some experimental data collected by an investigator of this phenomenon:

TABLE I   Magnetic Moments of Several Elements

Element	Type	Magnetic Moment (magnetons)
H	Paramagnetic	1.7
He	Diamagnetic	0
B	Paramagnetic	1.7
C	Paramagnetic	2.8
N	Paramagnetic	3.9
O	Paramagnetic	2.8
Ne	Diamagnetic	0

112. Why is the situation of equal numbers of spin-up and spin-down electrons referred to as all electrons being "paired"?

113. How many unpaired electrons are in the following atoms?
     (a) C   (b) N   (c) O   (d) Ne   (e) F

114. How many "pairs" of electrons are there in a "filled" p shell?

115. On the basis of information provided in Table I, predict the results of a Stern–Gerlach experiment on each of the atoms listed.

116. An ion, X^{2+}, is known to be from the first transition metal series. The ion is paramagnetic with four unpaired electrons. What two possible elements could X be?

117. Element Z is diamagnetic. Its most common ion is Z^{2+}. Z has the next to the lowest first ionization energy in its group. The energy required to remove an electron from Z^{2+} is extremely high. Identify element Z. Give the chemical formula of the ionic compounds formed from Z^{2+} and the negative ions O^{2-} and Cl^-.

CHAPTER
3
SPECIAL TOPICS

3A.1   RULES FOR ALLOWED COMBINATIONS OF QUANTUM NUMBERS

The advantage of the quantum mechanical model of the atom developed by Erwin Schrödinger in 1926 is that it consists of mathematical equations known as *wave functions* that satisfy the requirements placed on the behavior of electrons. The disadvantage is that it is difficult to imagine a physical model of electrons as waves.

The Schrödinger model assumes that the electron is a wave and tries to describe the regions in space, or **orbitals,** where electrons are most likely to be found. Instead of trying to tell us where the electron is at any time, the Schrödinger model describes the probability that an electron can be found in a given region of space at a given time. The model no longer tells us where the electron is; it only tells us where it might be.

The Schrödinger model allows the electron to occupy three-dimensional space. It therefore requires three coordinates, or three **quantum numbers,** to describe the orbitals in which electrons can be found. The three coordinates that come from Schrödinger's wave equations are the principal (n), angular (l), and magnetic (m) quantum numbers. These quantum numbers describe the size, shape, and orientation in space of the orbitals on an atom.

The **principal quantum number (n)** describes the size of the orbital. Orbitals for which $n = 2$ are larger than those for which $n = 1$, for example. Because they have opposite electrical charges, electrons are attracted to the nucleus of the atom. Energy must therefore be absorbed to excite an electron from an orbital in which the electron is close to the nucleus ($n = 1$) into an orbital in which it is further from the nucleus ($n = 2$ or higher). The principal quantum number therefore indirectly describes the energy of an orbital.

The **angular quantum number (l)** describes the shape of the orbital. Orbitals have shapes that are best described as spherical ($l = 0$), polar ($l = 1$), or cloverleaf ($l = 2$), as shown in Figure 3A.1. They can even take on more complex shapes as the value of the angular quantum number becomes larger.

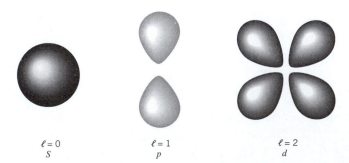

$\ell = 0$
s

$\ell = 1$
p

$\ell = 2$
d

FIGURE 3A.1 The angular quantum number specifies the shape of the orbital. When $l = 0$, the orbital is spherical. When $l = 1$, it is polar. When $l = 2$, the orbital typically has the shape of a cloverleaf.

There is only one way in which a sphere ($l = 0$) can be oriented in space. Orbitals that have polar ($l = 1$) or cloverleaf ($l = 2$) shapes, however, can point in different directions. We therefore need a third quantum number, known as the **magnetic quantum number (m),** to describe the orientation in space of a particular orbital. (It is called the *magnetic* quantum number because the effect of different orientations of orbitals was first observed in the presence of a magnetic field.)

Atomic spectra suggest that the orbitals on an atom are quantized. We therefore need a set of rules that limit the possible combinations of the n, l, and m quantum numbers.

Rules Governing Allowed Combinations of Quantum Numbers

- The three quantum numbers (n, l, and m) that describe an orbital are integers: 0, 1, 2, 3, and so on.
- The principal quantum number (n) cannot be zero. The allowed values of n are therefore 1, 2, 3, 4, and so on.
- The angular quantum number (l) can be any integer between 0 and $n - 1$. If $n = 3$, for example, l can be either 0, 1, or 2.
- The magnetic quantum number (m) can be any integer between $-l$ and $+l$. If $l = 2$, m can be either -2, -1, 0, $+1$, or $+2$.

Checkpoint

Explain why three quantum numbers are required to describe an orbital. What feature of the three-dimensional structure of an orbital is specified by each of these quantum numbers?

To show how these rules are applied, let's generate the set of allowed combinations of the n, l, and m quantum numbers when $n = 3$. According to the second rule, the angular quantum number (l) can be any integer between 0 and $n - 1$. Thus, if $n = 3$, l can be either 0, 1, or 2. The third rule limits the magnetic quantum number (m) to integers between $-l$ and $+l$. Thus, when $l = 0$, m must be 0. When $l = 1$, m can be either -1, 0, or $+1$. When $l = 2$, m can be -2, -1, 0, $+1$, or $+2$. The allowed combinations of n, l, and m for which $n = 3$ are therefore restricted to the following values.

n	l	m	
3	0	0	3s
3	1	-1	
3	1	0	3p
3	1	$+1$	
3	2	-2	
3	2	-1	
3	2	0	3d
3	2	$+1$	
3	2	$+2$	

3A.2 SHELLS AND SUBSHELLS OF ORBITALS

Orbitals that have the same value of the principal quantum number form a **shell.** Orbitals within a shell are divided into **subshells** that have the same value of the angular quantum number. Chemists describe the shell and subshell in which an orbital belongs with a two-character code such as 2*p* or 4*f*. The first character indicates the shell (*n* = 2 or *n* = 4). The second character identifies the subshell. By convention, the following lowercase letters are used to indicate different subshells.

$$
\begin{array}{ll}
s & l = 0 \\
p & l = 1 \\
d & l = 2 \\
f & l = 3
\end{array}
$$

Although there is no pattern in the first four letters (*s*, *p*, *d*, *f*), the letters progress alphabetically from that point (*g*, *h*, and so on). Some of the allowed combinations of the *n* and *l* quantum numbers are shown in Figure 3A.2.

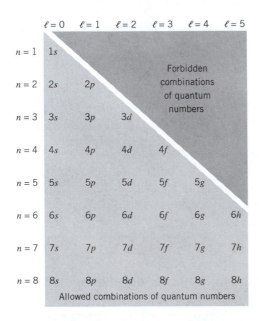

FIGURE 3A.2 Only certain combinations of the principal (*n*) and angular (*l*) quantum numbers are allowed.

The third rule limiting allowed combinations of the *n*, *l*, and *m* quantum numbers has an important consequence. It forces the number of subshells in a shell to be equal to the principal quantum number for the shell. The *n* = 3 shell, for example, contains three subshells: the 3*s*, 3*p*, and 3*d* orbitals.

Let's look at some of the possible combinations of the *n*, *l*, and *m* quantum numbers. There is only one orbital in the *n* = 1 shell because there is only one way in which a sphere can be oriented in space. The only allowed combination of quantum numbers for which *n* = 1 is the following.

$$
\begin{array}{ccc}
n & l & m \\
1 & 0 & 0 \quad 1s
\end{array}
$$

There are four orbitals in the *n* = 2 shell.

$$
\begin{array}{ccc}
n & l & m \\
2 & 0 & 0 \quad 2s \\
2 & 1 & -1 \\
2 & 1 & 0 \left.\right\} 2p \\
2 & 1 & 1
\end{array}
$$

There is only one orbital in the 2*s* subshell. But there are three orbitals in the 2*p* subshell because there are three directions in which a *p* orbital can point. One of these orbitals is oriented along the x axis, another along the y axis, and the third along the z axis of a co-ordinate system, as shown in Figure 3A.3. These orbitals are therefore known as the $2p_x$, $2p_y$, and $2p_z$ orbitals.

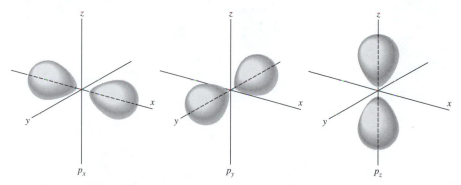

FIGURE 3A.3 When the angular quantum number (l) is 1, there are three possible values of the magnetic quantum number (m). The three values of m correspond to the three different orientations of the orbitals in space.

There are nine orbitals in the $n = 3$ shell.

n	l	m	
3	0	0	$3s$
3	1	−1	
3	1	0	$3p$
3	1	1	
3	2	−2	
3	2	−1	
3	2	0	$3d$
3	2	1	
3	2	2	

There is one orbital in the $3s$ subshell and three orbitals in the $3p$ subshell. The $n = 3$ shell, however, also includes $3d$ orbitals.

The five different orientations of orbitals in the $3d$ subshell are shown in Figure 3A.4. One of the orbitals lies in the xy plane of an xyz coordinate system and is called the $3d_{xy}$ orbital. The $3d_{xz}$ and $3d_{yz}$ orbitals have the same shape, but they lie between the axes of the coordinate system in the xz and yz planes. The fourth orbital in this subshell lies along the x and y axes and is called the $3d_{x^2-y^2}$ orbital. Most of the space occupied by the fifth orbital lies along the z axis and this orbital is called the $3d_{z^2}$ orbital.

The relationship between the number of orbitals in a shell and the principal quantum number of the shell can be understood by noting that there is one orbital in the $n = 1$ shell, four orbitals in the $n = 2$ shell, nine orbitals in the $n = 3$ shell, and so on. The number of orbitals in a shell is therefore the square of the principal quantum number: $1^2 = 1$, $2^2 = 4$, $3^2 = 9$. The relationship between the number of orbitals in a subshell and the angular quantum number for the subshell can be obtained by noting that there is one orbital in an s subshell ($l = 0$), three orbitals in a p subshell ($l = 1$), and five orbitals in a d subshell ($l = 2$). The number of orbitals in a subshell is therefore $2(l) + 1$.

All we have done so far is describe a series of orbitals in which electrons can be placed. It is important to remember that up to two electrons can be placed in an orbital. We therefore need a fourth quantum number to distinguish between these electrons. This is called the **spin quantum number** (s) because electrons behave as if they were spinning in either

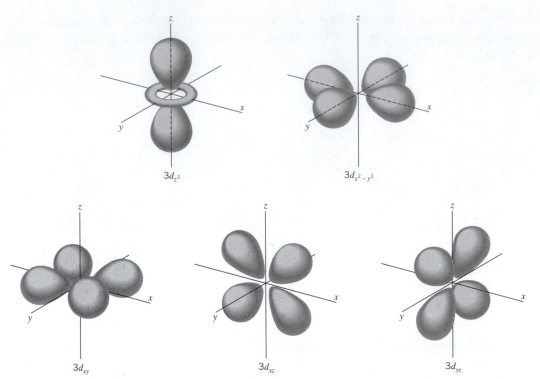

FIGURE 3A.4 When the angular quantum number (l) is 2, there are five allowed values of the magnetic quantum number (m) corresponding to the five different orientations of the orbitals in space.

a clockwise or counterclockwise fashion. One of the electrons in an orbital is arbitrarily assigned an s quantum number of $+\frac{1}{2}$, the other is assigned an s quantum number of $-\frac{1}{2}$. Thus, it takes three quantum numbers to define an orbital but four quantum numbers to identify one of the electrons that can occupy the orbital.

The allowed combinations of n, l, and m quantum numbers for the first four shells are given in Table 3A1. For each of these orbitals, there are two allowed values of the spin quantum number, s.

TABLE 3A.1   Summary of Allowed Combinations of Quantum Numbers

n	l	m	Subshell Notation	Number of Orbitals in Subshell	Number of Electrons Needed to Fill Subshell	
1	0	0	1s	1	2	total: 2
2	0	0	2s	1	2	
2	1	1, 0, −1	2p	3	6	total: 8
3	0	0	3s	1	2	
3	1	1, 0, −1	3p	3	6	
3	2	2, 1, 0, −1, −2	3d	5	10	total: 18
4	0	0	4s	1	2	
4	1	1, 0, −1	4p	3	6	
4	2	2, 1, 0, −1, −2	4d	5	10	
4	3	3, 2, 1, 0, −1, −2, −3	4f	7	14	total: 32

PROBLEMS

Quantum Numbers

3A.1. Describe the function of each of the four quantum numbers: n, l, m, and s.

3A.2. Describe the selection rules for the n, l, m, and s quantum numbers.

3A.3. Identify the quantum number that specifies each of the following.
 (a) the size of the orbital
 (b) the shape of the orbital
 (c) the way the orbital is oriented in space
 (d) the spin of the electrons that occupy an orbital

3A.4. Determine the allowed values of the angular quantum number, l, when the principal quantum number is 4. Describe the difference between orbitals that have the same principal quantum number and different angular quantum numbers.

3A.5. Determine the allowed values of the magnetic quantum number, m, when the angular quantum number is 2. Describe the difference between orbitals that have the same angular quantum number and different magnetic quantum numbers.

3A.6. Determine the allowed values of the spin quantum number, s, when $n = 5$, $l = 2$, and $m = -1$.

3A.7. Determine the number of allowed values of the magnetic quantum number when $n = 3$ and $l = 2$.

3A.8. Determine the maximum value for the angular quantum number, l, when the principal quantum number is 4.

3A.9. Which of the following is a legitimate set of n, l, m, and s quantum numbers?
 (a) $4, -2, -1, \frac{1}{2}$ (b) $4, 2, 3, \frac{1}{2}$ (c) $4, 3, 0, 1$ (d) $4, 0, 0, -\frac{1}{2}$

3A.10. Which of the following is a legitimate set of n, l, m, and s quantum numbers?
 (a) $0, 0, 0, \frac{1}{2}$ (b) $8, 4, -3, -\frac{1}{2}$ (c) $3, 3, 2, +\frac{1}{2}$ (d) $2, 1, -2, -\frac{1}{2}$
 (e) $5, 3, 3, -1$

3A.11. Calculate the maximum number of electrons that can have the quantum numbers $n = 4$ and $l = 3$.

3A.12. Calculate the number of electrons in an atom that can simultaneously possess the quantum numbers $n = 4$ and $s = +\frac{1}{2}$.

3A.13. Write the combination of quantum numbers for every electron in the $n = 1$ and $n = 2$ shells using the selection rules outlined in this chapter.

3A.14. Write the combination of quantum numbers for every electron in the $n = 1$ and $n = 2$ shells assuming that the selection rules for assigning quantum numbers are changed to the following.
 1. The principal quantum number can be any integer greater than or equal to 1.
 2. The angular quantum number can have any value between 0 and n.
 3. The magnetic quantum number can have any value between 0 and 1.
 4. The spin quantum number can have a value of either $+1$ or -1.

Shells and Subshells of Orbitals

3A.15. Describe what happens to the difference between the energies of subshells of orbitals as the value of the principal quantum number, n, becomes larger.

3A.16. Identify the symbols used to describe orbitals for which $l = 0, 1, 2,$ and 3.

3A.17. Determine the number of orbitals in the $n = 3$, $n = 4$, and $n = 5$ shells.

3A.18. Which of the following sets of n, l, m, and s quantum numbers can be used to describe an electron in a $2p$ orbital?
(a) $2, 1, 0, -\frac{1}{2}$ (b) $2, 0, 0, \frac{1}{2}$ (c) $2, 2, 1, \frac{1}{2}$ (d) $3, 2, 1, -\frac{1}{2}$
(e) $3, 1, 0, \frac{1}{2}$

3A.19. Which of the following orbitals cannot exist?
(a) $6s$ (b) $3p$ (c) $2d$ (d) $4f$ (e) $17f$

3A.20. Calculate the maximum number of electrons in the $n = 1$, $n = 2$, $n = 3$, $n = 4$, and $n = 5$ shells of orbitals.

3A.21. Calculate the maximum number of electrons that can fit into a $4d$ subshell.

3A.22. Calculate the maximum number of unpaired electrons that can be placed in a $5d$ subshell.

3A.23. Explain why the difference between the atomic numbers of pairs of elements in a vertical column, or group, of the periodic table is either 8, 18, or 32.

3A.24. Explain why the number of orbitals in a shell is always equal to n^2, where n is the principal quantum number.

3A.25. Explain why the maximum number of electrons in a shell of orbitals is equal to $2n^2$, where n is the principal quantum number.

3A.26. Explain why the number of orbitals in a subshell is always equal to $2l + 1$, where l is the angular quantum number.

Relative Energies of Atomic Orbitals

3A.27. Which pair of quantum numbers determines the energy of an electron in an orbital?
(a) n and l (b) n and m (c) n and s (d) l and m (e) l and s
(f) m and s

Electron Configurations

3A.28. Which of the following is a possible set of n, l, m, and s quantum numbers for the last electron added to form a gallium atom ($Z = 31$)?
(a) $3, 1, 0, -\frac{1}{2}$ (b) $3, 2, 1, \frac{1}{2}$ (c) $4, 0, 0, \frac{1}{2}$ (d) $4, 1, 1, \frac{1}{2}$ (e) $4, 2, 2, \frac{1}{2}$

3A.29. Which of the following is a possible set of n, l, m, and s quantum numbers for the last electron added to form an As^{3+} ion?
(a) $3, 1, -1, \frac{1}{2}$ (b) $4, 0, 0, -\frac{1}{2}$ (c) $3, 2, 0, \frac{1}{2}$ (d) $4, 1, -1, \frac{1}{2}$ (e) $5, 0, 0, \frac{1}{2}$

CHAPTER
4
THE COVALENT BOND

Ever since Dalton introduced his atomic theory in 1803, chemists have tried to understand the forces that hold atoms together in chemical compounds. The goal of this chapter is to build a model for the bonding in molecules. The way in which atoms bond is important to chemists because the properties of compounds depend on molecular structure.

4.1   VALENCE ELECTRONS

In 1902, while trying to find a way to explain the periodic table to a beginning chemistry class, G. N. Lewis discovered that the chemistry of the main-group elements shown in Figure 4.1 could be explained by assuming that atoms of these elements gain or lose electrons until they have eight electrons in the outermost shell of electrons of the atom. Eight electrons make an octet, and hence Lewis' discovery is often called the **octet rule.** The magnitude of this achievement can be appreciated by noting that this model was generated only 5 years after Thomson's discovery of the electron and 9 years *before* Rutherford proposed that the atom consisted of an infinitesimally small nucleus surrounded by a sea of electrons.

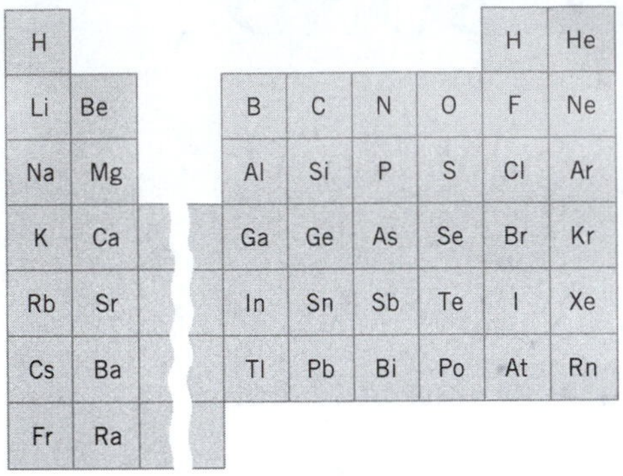

H
H He
Li Be
B C N O F Ne
Na Mg
Al Si P S Cl Ar
K Ca
Ga Ge As Se Br Kr
Rb Sr
In Sn Sb Te I Xe
Cs Ba
Tl Pb Bi Po At Rn
Fr Ra

FIGURE 4.1 The model developed by G. N. Lewis was first applied to the atoms of the main-group elements, which are found on either side of the periodic table.

The electrons in the outermost shell eventually became known as the **valence electrons.** This name reflects the fact that the number of bonds an element can form is called its *valence*. Because the number of electrons in the outermost shells in the Lewis theory controls the number of bonds the atom can form, these outermost electrons are the valence electrons.

Figure 4.2, which utilizes the PES data from Chapter 3, shows how much easier it is to remove valence electrons than core electrons as atomic number increases. When $Z = 3$ (lithium) we already begin to see a large difference in the energy to remove the 1*s* core electrons as compared to the 2*s*. The gap continues to widen, and by $Z = 10$ (neon) the 1*s* electron is extremely difficult to remove. At $Z = 18$ (argon) we see that the 2*s* and 2*p* core electrons are much more tightly held than the 3*s* and 3*p* valence electrons. There is not much difference, however, in the energy required to remove an electron from the 2*p* as compared to the 2*s* nor the 3*p* compared to the 3*s*. As Z increases the electrons in shells or subshells closer to the nucleus become increasingly buried in the atom and increasingly more difficult to remove. Figure 4.2 gives the ionization energies of both the inner shell and outer shell electrons and shows trends similar to those in Figure 3.28. Both figures show that AVEE is a measure of the energy required to remove valence electrons as well as the energy separation of the valence shells.

Ionization energy (MJ·mol⁻¹)
110
100
90
80
70
60
50
40
30
20
10
0
1s
2s
2p
3s
3p
1 3 5 7 9 11 13 15 17 19 21
H Li B Ne Na Al Ar
Z

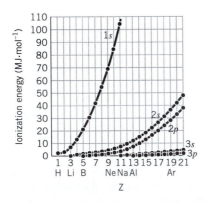

FIGURE 4.2 Photoelectron ionization energies of the atoms of the first 21 elements. The energy required to remove a 1*s* electron increases rapidly with increasing nuclear charge, *Z*. The difference in the energy required to remove an *s* electron and a *p* electron from the same shell is never large. This difference gets smaller in subsequent shells. Reprinted from R. J. Gillespie et al., *Atoms, Molecules and Reactions,* Prentice-Hall, Inc., Englewood Cliffs, New Jersey, 1994, p. 200.

The valence electrons are those electrons of an atom that can be gained or lost in a chemical reaction. The number of valence electrons of an atom can be counted by totaling up all the electrons outside the core electrons. There is an exception to this method of counting the valence electrons. Filled d or f subshells are seldom involved in a chemical reaction, and consequently the electrons in filled d or f subshells are not considered valence electrons. Consider, for example, gallium which has the following electron configuration.

$$\text{Ga} \qquad [\text{Ar}] \; 4s^2 \; 3d^{10} \; 4p^1$$

The [Ar] symbol represents all the core electrons, and since these electrons do not participate in chemical reactions they are not valence electrons. Of the 13 electrons outside the core the $3d$ subshell is filled, and so the 10 $3d$ electrons are not counted as valence electrons. This leaves two $4s$ electrons and one $4p$ electron. Gallium therefore has three valence electrons. The electron configuration of vanadium is

$$\text{V} \qquad [\text{Ar}] \; 4s^2 \; 3d^3$$

The core electrons represented by [Ar] do not count as valence electrons, but the $4s$ and $3d$ electrons do. Thus vanadium has five valence shell electrons.

This difficulty of recognizing filled d or f subshells occurs only with the transition or rare earth elements. For most atoms dealt with in this course the counting of valence electrons is not complicated. For fluorine, whose electron configuration is

$$\text{F} \qquad [\text{He}] \; 2s^2 \; 2p^5$$

all the electrons outside the [He] core are valence electrons, and fluorine thus has seven valence electrons. Figure 4.2 shows the energy gap separating the $2s$ and $2p$ electrons from the core $1s$ electrons in fluorine ($Z = 9$). Because of the energy difference between the first and second shells, only the second-shell electrons participate in bonding and are thus valence electrons. For main-group elements the number of valence electrons is equal to the group number. Fluorine in Group VIIA has seven valence electrons.

Checkpoint

How many valence electrons does an atom in Group IVA have?

Exercise 4.1

Determine the number of valence electrons in neutral atoms of the following elements.
(a) Si   (b) Mn   (c) Sb   (d) Pb

Solution

We start by writing the electron configuration for each element.

$$\begin{aligned}
\text{Si} \quad & [\text{Ne}] \; 3s^2 \; 3p^2 \\
\text{Mn} \quad & [\text{Ar}] \; 4s^2 \; 3d^5 \\
\text{Sb} \quad & [\text{Kr}] \; 5s^2 \; 4d^{10} \; 5p^3 \\
\text{Pb} \quad & [\text{Xe}] \; 6s^2 \; 4f^{14} \; 5d^{10} \; 6p^2
\end{aligned}$$

Ignoring filled d and f subshells, we conclude that neutral atoms of these elements contain the following numbers of valence electrons.
(a) Si = **4**   (b) Mn = **7**   (c) Sb = **5**   (d) Pb = **4**

4.2   THE COVALENT BOND

By 1916 Lewis realized that in addition to losing electrons there is another way atoms can combine to achieve an octet of valence electrons: They can share electrons until their valence shells contain eight electrons. Two fluorine atoms, for example, can form a stable F_2 molecule in which each atom is surrounded by eight valence electrons by sharing a pair of electrons. A pair of oxygen atoms can form an O_2 molecule in which each atom has a total of eight valence electrons by sharing two pairs of electrons.

Whenever he applied this model, Lewis noted that the atoms seem to share pairs of electrons. He also noted that most molecules contain an even number of electrons, which suggests that the electrons exist in pairs. He therefore introduced a system of notation known as **Lewis structures** in which each atom is surrounded by up to four pairs of dots corresponding to the eight possible valence electrons. The Lewis structures of F_2 and O_2 are therefore written as shown in Figure 4.3. This symbolism is still in use today. The only significant change is the use of lines to indicate bonds between atoms formed by the sharing of a pair of electrons.

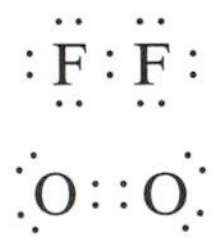

$$:\overset{..}{\underset{..}{F}}:\overset{..}{\underset{..}{F}}:$$

$$:\overset{..}{\underset{..}{O}}::\overset{..}{\underset{..}{O}}:$$

FIGURE 4.3 Lewis structures of F_2 and O_2.

The prefix *co-* is used to indicate when things are joined or equal (for example, *coexist, cooperate,* and *coordinate*). It is therefore appropriate that the term **covalent bond** be used to describe the bonds in molecules that result from the sharing of one or more pairs of electrons. As might be expected, molecules held together by covalent bonds are called **covalent molecules.**

4.3   HOW DOES THE SHARING OF ELECTRONS BOND ATOMS?

To understand how atoms can be held together by sharing a pair of electrons, let's look at the simplest covalent bond, the bond that forms when two isolated hydrogen atoms come together to form an H_2 molecule. In the Lewis model a single line is used to represent a pair of electrons in a bond as shown below.

$$H\cdot \ + \ \cdot H \longrightarrow H—H$$

Each hydrogen atom contains a proton surrounded by a spherical cloud of electron density that corresponds to a single electron. When a pair of hydrogen atoms approach one another, the electron of each atom is attracted to the proton in the nucleus of the other atom, as shown in Figure 4.4. The magnitude of the force of attraction between the particles is equal to the charge on the electron (q_e) times the charge on the proton (q_p) divided by the square of the distance between particles (r^2).

$$F = \frac{q_e \times q_p}{r^2}$$

Two forces of repulsion are also created, however, because the two negatively charged electrons repel one another, as do the two positively charged nuclei, as shown in Figure 4.4.

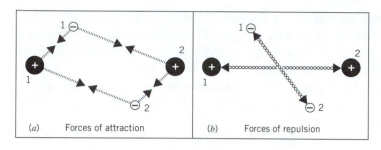

FIGURE 4.4 (*a*) Two forces of attraction act to bring a pair of hydrogen atoms together: the forces of attraction between the electron of each atom and the proton of the other atom. (*b*) Two forces of repulsion drive a pair of hydrogen atoms apart: the repulsion between the two protons and the repulsion between the two electrons.

At first glance, it might seem that the two new repulsive forces would balance the two new attractive forces. If this happened, the H_2 molecule would be no more stable than a pair of isolated hydrogen atoms. There must be a way in which the forces of attraction can be maximized, while the forces of repulsion are minimized.

The force of repulsion between the protons can be minimized if the pair of electrons are placed between the two nuclei. The distance between the electron in one atom and the nucleus of the other is now smaller than the distance between the two nuclei, as shown in Figure 4.5. As a result, the force of attraction between each electron and the nucleus of the other atom is larger than the force of repulsion between the two nuclei. This can only occur if the hydrogen atoms have electrons of *opposite spin* so that each electron may enter the domain of the other electron. The net result of pairing the electrons is to create a region in space, in which both electrons can reside, that extends over both nuclei but concentrates the electron density between the two nuclei. These two electrons are said to occupy the same bonding domain. Each hydrogen atom is considered to share two electrons and so to have two electrons in its domain.

FIGURE 4.5 If the electrons are paired and restricted to the region directly between the two nuclei, the attractive forces are larger than the repulsive forces in the hydrogen molecule. As a result, the molecule is more stable than a pair of isolated atoms.

Pairing the electrons and placing them between the two nuclei creates a system that is more stable than a pair of isolated atoms if the nuclei are close enough to share the pair of electrons, but not so close that repulsion between the nuclei becomes too large. The hydrogen atoms in an H_2 molecule are therefore held together (or bonded) by the sharing of a pair of electrons, and this bond is the strongest when the distance between the two nuclei is 0.074 nm.

Checkpoint

Explain why the distance between the nuclei of the two hydrogens in H_2^+ is 0.106 nm and that between the hydrogens in H_2 is 0.074 nm.

4.4 USING LEWIS STRUCTURES TO UNDERSTAND THE FORMATION OF BONDS

It is so easy to get caught up in the process of generating Lewis structures that we lose sight of the role they played in the development of chemistry. For at least a century before the theory was proposed, chemists had evidence that hydrogen and oxygen existed as diatomic H_2 and O_2 molecules. Lewis provided the first model that could explain why hydrogen atoms come together to form H_2, and why oxygen atoms spontaneously combine to form O_2. Although the model works well for many molecules, there are other bonding models that are superior in certain cases. The Lewis model has the advantage of simplicity and is often the first model used by chemists to describe the bonding in a molecule.

Lewis structures could also be used to understand why hydrogen and oxygen atoms combine to form a molecule with the formula H_2O. We start by noting that the hydrogen atom has only one valence electron and that the oxygen atom has six valence electrons.

$$H \qquad \mathbf{1s^1}$$
$$O \qquad 1s^2\ \mathbf{2s^2\ 2p^4}$$

We then represent neutral atoms of these elements with the following symbols.

$$H\cdot \qquad \cdot\ddot{\underset{\cdot\cdot}{O}}\cdot$$

As we bring two hydrogen atoms and an oxygen atom together, the valence electrons come under the influence of both the oxygen and hydrogen atoms. The hydrogen atoms can attract at most two electrons that must be of opposite spin into the region of space between the hydrogen and oxygen nuclei. These two electrons occupy the same bonding domain.

$$H\cdot\ +\ \cdot\ddot{\underset{\cdot\cdot}{O}}\cdot\ +\ \cdot H\ \longrightarrow\ H\!:\!\ddot{\underset{\cdot\cdot}{O}}\!:\!H$$

The result is an H_2O molecule in which each hydrogen atom shares a pair of electrons with the oxygen atom and the oxygen atom has access to an octet of valence electrons.

The Lewis structure of water is often written with a line representing each pair of electrons that is shared by two atoms.

$$H\!-\!\ddot{\underset{\cdot\cdot}{O}}\!-\!H$$

There are four regions in space, or **domains,** where electrons can be found around the oxygen atom in the water molecule. Two of the domains contain **bonding electrons,** which are used to form the covalent bonds that hold the molecule together. These domains are called **bonding domains** because they hold the pairs of electrons used for bonding. Each hydrogen atom shares one domain, a bonding domain, with oxygen. The other two domains around the oxygen atom contain pairs of electrons that are described as **nonbonding electrons.** Those domains are called **nonbonding domains.** The nonbonding electrons are in the valence shell of oxygen and are considered to belong exclusively to the oxygen atom in the molecule. Thus there are eight total electrons in the nonbonding and bonding domains of oxygen, and there are two electrons in the bonding domain of each hydrogen atom. The distance between the oxygen and hydrogen nuclei in each bond is experimentally found to be 0.096 nm.

Let's apply our technique for generating Lewis structures to carbon dioxide (CO_2). We start by determining the number of valence electrons on each atom. Carbon has four valence electrons, and each oxygen has six.

$$C \qquad [He]\ \mathbf{2s^2\ 2p^2}$$
$$O \qquad [He]\ \mathbf{2s^2\ 2p^4}$$

We can represent this information by the following symbols.

$$:\ddot{\underset{\cdot}{O}}\cdot\ +\ \cdot\dot{C}\cdot\ +\ \cdot\ddot{\underset{\cdot}{O}}:$$

We now combine one electron from each atom to form covalent bonds between the atoms.

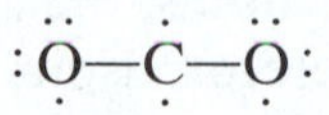

$$:\ddot{O}-\dot{C}-\ddot{O}:$$

When this is done, each oxygen atom has a total of seven valence electrons, and the carbon atom has a total of six valence electrons. Because none of the atoms has a filled valence shell, we combine another electron on each atom to form two more bonds. The result is a Lewis structure in which each atom has eight electrons in its valence shell.

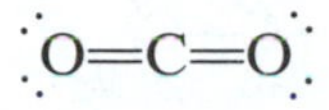

$$:\ddot{O}=C=\ddot{O}:$$

Checkpoint

In the CO_2 molecule above, where are the nonbonding electrons? How many electrons are in bonding domains?

4.5   DRAWING SKELETON STRUCTURES

The most difficult step in generating the Lewis structure of a molecule is the step in which the skeleton structure of the molecule is written. As a general rule the element with the smallest AVEE (see Figure 3.27) is more likely to be at the center of the molecule. Thus, the formulas of thionyl chloride ($SOCl_2$) and sulfuryl chloride (SO_2Cl_2) can be translated into the following skeleton structures.

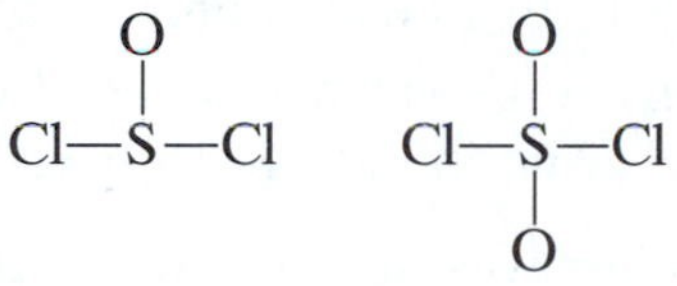

It is also useful to recognize that the formulas for complex molecules often provide hints about the skeleton structure of the molecule. Dimethyl ether, for example, is often written as CH_3OCH_3, which translates into the following skeleton structure.

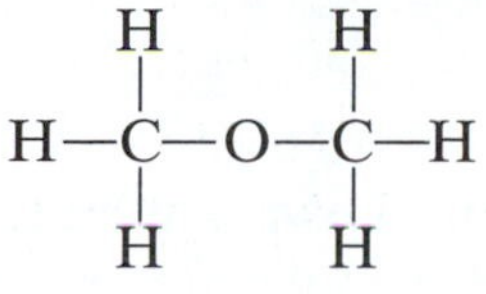

The atoms of many chemical formulas can be arranged in more than one skeleton structure. For example, the atoms in dimethyl ether shown above, C_2H_6O, can also be arranged into the following skeleton structure.

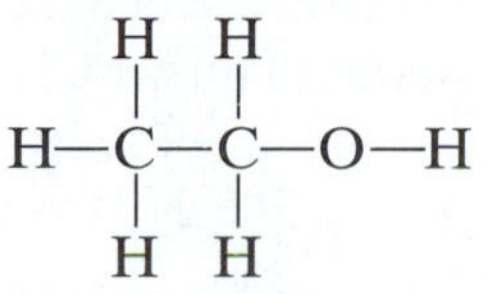

This is the skeleton structure for ethanol. As in this case, two or more possible structures may be found to be equally satisfactory. Section 4.6 details how to make certain a given structure represents a correct Lewis structure, and Section 4.12 provides another way to assist in determining which of two arrangements of atoms may be better. When two structures differ only in the arrangement of atoms, the structures are called isomers. Dimethyl ether and ethanol shown above are therefore isomers.

Finally, it is useful to recognize that many compounds that are acids contain O—H bonds. The formula for acetic acid, for example, is often written as CH_3CO_2H, because this molecule contains the following skeleton structure.

$$
\begin{array}{ccc}
\text{H} & & \text{O} \\
| & & \diagup \\
\text{H}-\text{C}-\text{C} & & \\
| & & \diagdown \\
\text{H} & & \text{O}-\text{H}
\end{array}
$$

Checkpoint

What is the difference between a skeleton structure and a Lewis structure?

4.6   A STEP-BY-STEP APPROACH TO WRITING LEWIS STRUCTURES

The method for writing Lewis structures used in Section 4.4 can be time consuming. For all but the simplest molecules, the following step-by-step process is faster.

- Write the skeleton structure of the molecule.
- Determine the number of valence electrons of the molecule.
- Use two valence electrons to form each bond in the skeleton structure.
- Try to place eight electrons in the valence shells of the atoms by distributing the remaining valence electrons as nonbonding electrons.

The first step in the process involves deciding which atoms in the molecules are connected by covalent bonds. As we have seen, the formula of the compound often provides a hint as to the skeleton structure. Consider PCl_3, for example. The formula for the molecule suggests the following skeleton structure.

$$
\begin{array}{c}
\text{Cl}-\text{P}-\text{Cl} \\
| \\
\text{Cl}
\end{array}
$$

The second step in generating the Lewis structure of a molecule involves calculating the number of valence electrons in the molecule or ion. For a neutral molecule this is the sum of the valence electrons on each atom. If the molecule carries a charge, we add one electron for each negative charge and subtract an electron for each positive charge.

Chlorine is in Group VIIA of the periodic table, which means it contains seven valence electrons. Therefore 21 valence electrons are contributed by the three chlorine atoms. A phosphorus atom has five valence electrons. Because PCl_3 has no charge, no additional electrons need to be added or subtracted. Thus, PCl_3 has a total of 26 valence electrons.

$$PCl_3 \qquad 5 + 3(7) = 26$$

The third step assumes that the skeleton structure of the molecule is held together by covalent bonds. The valence electrons are therefore divided into two categories: *bonding electrons* and *nonbonding electrons*. Because it takes two electrons to form a covalent bond, we can calculate the number of nonbonding electrons in the molecule by subtracting two electrons for each bond in the skeleton structure from the total number of valence electrons.

There are three covalent bonds in the skeleton structure for the PCl_3. As a result, 6 of the 26 valence electrons must be used as bonding electrons. This leaves 20 nonbonding electrons in the valence shell.

$$
\begin{array}{r}
26 \text{ valence electrons} \\
-6 \text{ bonding electrons} \\
\hline
20 \text{ nonbonding electrons}
\end{array}
$$

The last step in the process by which Lewis structures are generated involves using the nonbonding valence electrons to satisfy the octets (valence shells) of the atoms in the molecule. Each chlorine atom in PCl_3 already has two electrons—the electrons in the P—Cl covalent bond. Because each chlorine atom needs 6 nonbonding electrons to satisfy its octet, it takes 18 nonbonding electrons to satisfy the 3 chlorine atoms. This leaves one pair of nonbonding electrons, which can be used to fill the valence shell of the central atom.

$$
\begin{array}{c}
:\ddot{C}l\!-\!\ddot{P}\!-\!\ddot{C}l: \\
| \\
:\ddot{C}l:
\end{array}
$$

Lewis structures can also be used to describe polyatomic ions. Using the ammonium ion, NH_4^+, as an example, we begin by drawing a skeleton structure.

$$
\begin{array}{c}
H \\
| \\
H\!-\!N\!-\!H \\
| \\
H
\end{array}
$$

Nitrogen is in Group VA of the periodic table, which means that it contains five valence electrons. Each hydrogen atom has one valence electron. Thus the four hydrogens can contribute four electrons. This gives a total of nine valence electrons from the nitrogen and four hydrogens. Since this is a polyatomic cation, however, one electron is removed to give a positive charge, leaving a total of eight valence electrons. If this had been a polyatomic anion, sufficient electrons would have been added to give the overall negative charge.

Each of the four covalent bonds contains two electrons. Therefore, all eight valence electrons are used in the bonding, and no electrons remain to be distributed as nonbonding electrons.

$$
\begin{array}{r}
8 \text{ valence electrons} \\
-8 \text{ bonding electrons} \\
\hline
0 \text{ nonbonding electrons}
\end{array}
$$

This leaves us with eight electrons surrounding the nitrogen and two electrons shared with each hydrogen. Thus, each atom is surrounded by the expected number of electrons. The final structure is drawn in brackets and shows the charge on the polyatomic ion at the upper right-hand corner.

$$
\left[
\begin{array}{c}
H \\
| \\
H\!-\!N\!-\!H \\
| \\
H
\end{array}
\right]^{+}
$$

In a polyatomic ion the charge does not reside on any specific atom but is spread over the structure. For a polyatomic anion such as NO_3^-, sufficient additional electrons must be added to the valence electrons to give the overall negative charge. Thus NO_3^- has $5 + 3(6) + 1 = 24$ electrons in the Lewis structure.

Checkpoint

Draw the Lewis structures of carbon tetrachloride, CCl_4, and the hydroxide anion, OH^-.

4.7   MOLECULES THAT DON'T SEEM TO SATISFY THE OCTET RULE

Not Enough Electrons

Occasionally we encounter a molecule that doesn't seem to have enough valence electrons. When this happens, we have to remember why atoms share electrons in the first place. If we can't get a satisfactory Lewis structure by sharing a single pair of electrons, it may be possible to achieve that goal by sharing two or even three pairs of electrons.

Consider formaldehyde (H_2CO), for example, which contains 12 valence electrons.

$$H_2CO \qquad 2(1) + 4 + 6 = 12$$

The formula of the molecule suggests the following skeleton structure.

$$
\begin{array}{c}
O \\
| \\
H-C-H
\end{array}
$$

There are three covalent bonds in the skeleton structure, which means that six valence electrons must be used as bonding electrons. This leaves six nonbonding electrons. It is impossible, however, to satisfy the octets of the atoms in this molecule with only six nonbonding electrons. When the nonbonding electrons are used to satisfy the octet of the oxygen atom, the carbon atom has a total of only six valence electrons.

$$
\begin{array}{c}
:\ddot{O}: \\
| \\
H-C-H
\end{array}
$$

We therefore assume that the carbon and oxygen atoms share two pairs of electrons to make a double bond. There are now four bonds in the skeleton structure, which leaves only four nonbonding electrons. This is enough, however, to satisfy the octets of the carbon and oxygen atoms. Note that there are two single bond domains between the hydrogens and carbon and that there is one double bond domain between the carbon and oxygen atoms. Surrounding the oxygen atom there are one double bond domain and two nonbonding domains.

$$
\begin{array}{c}
\cdot\ddot{O}\cdot \\
\| \\
H-C-H
\end{array}
$$

Every once in a while, we encounter a molecule for which it is impossible to write a satisfactory Lewis structure. Consider boron trifluoride (BF_3), for example, which contains 24 valence electrons.

$$BF_3 \qquad 3 + 3(7) = 24$$

There are three covalent bonds in the skeleton structure for the molecule. Because it takes 6 electrons to form the skeleton structure, there are 18 nonbonding valence electrons. But each fluorine atom needs 6 additional electrons to satisfy its octet. Thus, all of the nonbonding electrons are used by the three fluorine atoms. As a result, we run out of electrons while the boron atom has only 6 valence electrons.

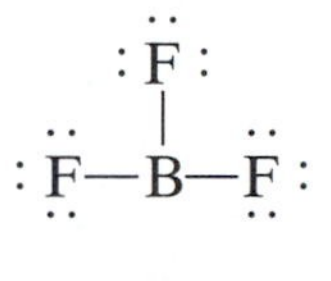

$$:\ddot{F}:$$
$$|$$
$$:\ddot{F}-B-\ddot{F}:$$

The next step would be to look for the possibility of forming a double or triple bond. However, chemists have learned that the atoms that form strong double or triple bonds are C, N, O, P, and S. Because neither boron nor fluorine belongs in that category, we have to stop with what appears to be an unsatisfactory Lewis structure.

Checkpoint

Could the following structure be written for BF_3?

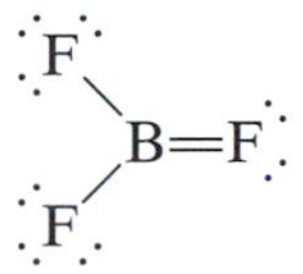

$$:\ddot{F}$$
$$\diagdown$$
$$B{=}\ddot{F}:$$
$$\diagup$$
$$:\ddot{F}$$

Too Many Electrons

It is also possible to encounter a molecule that seems to have too many valence electrons. When that happens, we expand the valence shell of the central atom. Consider the Lewis structure for sulfur tetrafluoride (SF_4), for example, which contains 34 valence electrons.

$$SF_4 \qquad 6 + 4(7) = 34$$

There are four covalent bonds in the skeleton structure for SF_4.

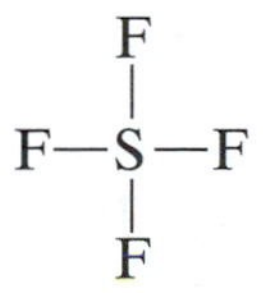

$$F$$
$$|$$
$$F-S-F$$
$$|$$
$$F$$

Because this requires using 8 valence electrons to form the covalent bonds that hold the molecule together, there are 26 nonbonding valence electrons.

Each fluorine atom needs six additional electrons to satisfy its octet. Because there are four F atoms, we need 24 nonbonding electrons for this purpose. But there are 26 nonbonding electrons in the molecule. We have satisfied the octets for all five atoms, and we still have one more pair of valence electrons. We therefore expand the valence shell of the central atom to hold more than eight electrons.

This raises an interesting question: How does the sulfur atom in SF_4 hold 10 electrons in its valence shell? The electron configuration for a neutral sulfur atom seems to suggest that only 8 electrons will fit in the valence shell of the atom because it takes 8 electrons to fill the $3s$ and $3p$ orbitals. But sulfur also has valence shell $3d$ orbitals.

$$S \qquad [Ne]\ 3s^2\ 3p^4\ 3d^0$$

Because the $3d$ orbitals on a neutral sulfur atom are all empty, one of them can be used to hold the extra pair of electrons on the sulfur atom in SF_4.

Checkpoint

Elements in the first and second rows of the periodic table do not have valence shell d orbitals. Use the fact that nitrogen and oxygen do not contain $2d$ orbitals to explain why these elements cannot expand their valence shell.

Although a very large number of molecules are composed of atoms that obey the octet rule (i.e., they have four pairs of electrons around the atom), there are exceptions. The octet rule is obeyed by C, N, O, and F in most of their compounds. Be, B, and Al often have only three electron pairs in their valence shell. Elements from periods 3 and 4 in the periodic table can accommodate six electron pairs in their valence shell. Elements of the first period (hydrogen and helium) fill their valence shell with only one electron pair.

Exercise 4.2

Write the Lewis structure for xenon tetrafluoride (XeF_4).

Solution

Xenon (Group VIIIA) has eight valence electrons, and fluorine (Group VIIA) has seven. Thus, there are 36 valence electrons in the molecule.

$$XeF_4 \qquad 8 + 4(7) = 36$$

The skeleton structure for the molecule contains four covalent bonds.

Because 8 electrons are used to form the skeleton structure, there are 28 nonbonding valence electrons. If each fluorine atom needs 6 nonbonding electrons, a total of 24 nonbonding electrons are used to complete the octets of the F atoms. This leaves 4 extra nonbonding electrons. Because the octet of each atom appears to be satisfied and we have electrons left over, we expand the valence shell of the central atom until it contains a total of 12 electrons.

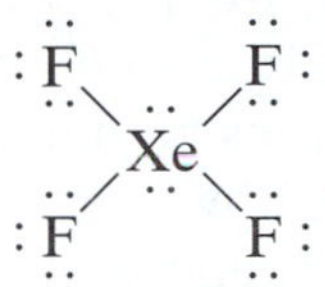

$$\ddot{:}\ddot{F} \qquad \ddot{F}\ddot{:}$$
$$\diagdown \ddots \qquad \diagup$$
$$\ddot{X}e$$
$$\diagup \qquad \ddots \diagdown$$
$$\ddot{:}\ddot{F} \qquad \qquad F\ddot{:}$$

4.8   BOND LENGTHS

The distance between the nuclei of two atoms bonded to each other is called the bond length. Bond length is a very important structural parameter which contains much information about a molecule. Consider, for example, the distance between carbon atoms in three of the compounds given in Table 4.1 and illustrated in Figure 4.6. Bond lengths are usually given in units of nanometers (nm) which is 10^{-9} meters.

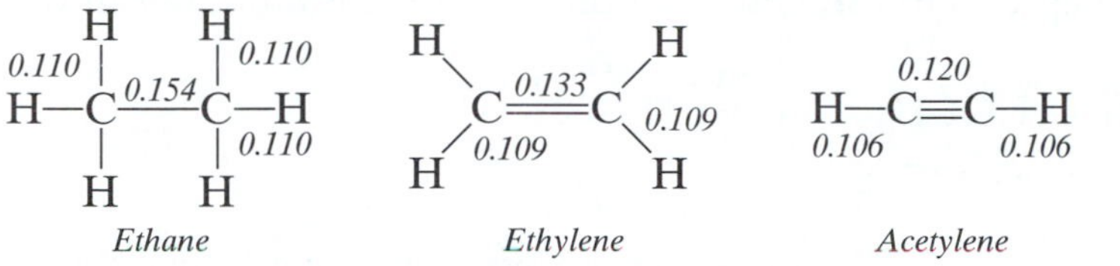

FIGURE 4.6 Bond lengths (nm) in ethane, ethylene, and acetylene. The carbon–carbon bond length gets shorter as the bond changes from a single to a double to a triple bond. The carbon–hydrogen bond length is about the same in all compounds.

TABLE 4.1   Carbon–Carbon and Carbon–Hydrogen Bond Lengths in Selected Molecules[a]

Molecule	Carbon–Carbon Bond Length (nm)	Carbon–Hydrogen Bond Length (nm)
Ethane	0.154	0.110
Graphite	0.142	
Benzene	0.139	0.110[b]
Ethylene	0.133	0.109
Acetylene	0.120	0.106[b]

[a] From Emil J. Margolis, *Bonding and Structure,* Appleton, Century, Crofts, New York, 1968, with permission from Plenum Publishing Corp.

[b] David R. Lide, Ed., CRC Handbook of Chemistry and Physics, 75 Ed., CRC Press, Boca Raton, FL, 1994.

Figure 4.6 shows the variation in the distance between carbon atoms in three different compounds. Bond lengths can be determined experimentally by making use of how radiation interacts with various molecules. The carbon–carbon bond lengths were calculated from data obtained by the absorption of light by these species.

Atoms held together by a triple bond are pulled closer to one another than when the atoms are held by double or single bonds. Thus, the bond length of the triple bond is shorter than that of a double or single bond between the same atoms. The three compounds of Figure 4.6 have carbon–carbon bond lengths that decrease from ethane to ethylene to acetylene. The Lewis structures show that the carbon–carbon bond in the three species changes from a single (ethane) to a double (ethylene) to a triple bond (acetylene). A single bond consists of two shared electrons, a double bond of four, and a triple bond of six electrons.

The carbon–hydrogen bond lengths are also shown in Figure 4.6 for ethane and ethylene. Because each H—C bond consists of a single pair of electrons, the carbon–hydrogen bond lengths are about the same within each individual molecule.

The carbon–hydrogen bond length, however, is shorter than the carbon–carbon bond length. This can be explained using the relative sizes of atoms discussed in Chapter 3. In covalently bonded compounds such as those shown here we can estimate the carbon–carbon bond length by noting that the covalent radius of a carbon atom is 0.077 nm (see Figure 3.23). Thus, two covalently bonded carbon atoms should have their nuclei separated by 0.077 + 0.077 = 0.154 nm. Similarly, the carbon–hydrogen bond length should be about 0.077 + 0.037 = 0.114 nm.

When comparing bond lengths of single, double, and triple bonds it is necessary to compare bonds between the same pair of atoms. For example, N≡O triple bonds are shorter than N=O double bonds which are shorter than N—O single bonds. However, it is not true that an N=O double bond is shorter than an N—H single bond.

Exercise 4.3

Use the covalent radii given in Chapter 3 to estimate the bond lengths for all bonds in the following compounds.

(a) H_2O (b) CH_3OH (c) CH_3OCH_3 (d) CH_3SCH_3

Solution

It is first necessary to know the structures of the molecules before bond lengths can be estimated. The Lewis structures of the molecules are

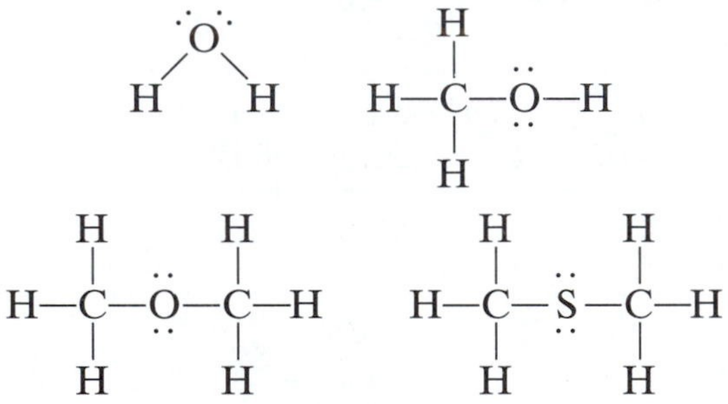

From Figure 3.23 the appropriate covalent radii are

H	0.037 nm
O	0.066 nm
C	0.077 nm
S	0.104 nm

The best estimates for the bond lengths in these molecules are

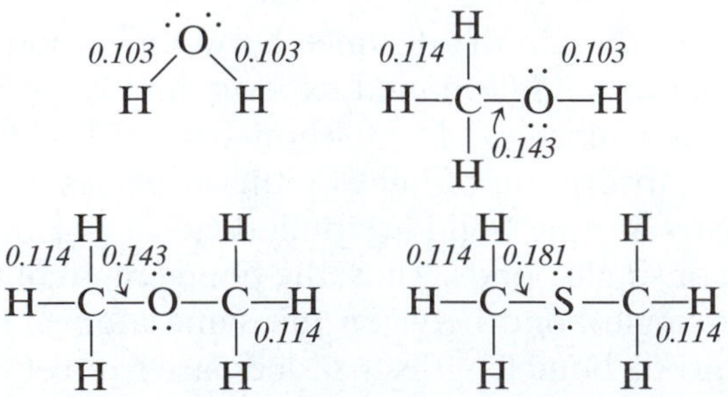

The data given in Figure 3.23 are average values taken from a variety of compounds. Therefore, the bond lengths obtained by adding these radii are only approximations, but they are quite close to the actual values. It should also be noted that the covalent radii given in Figure 3.23 apply only to single bonds. If a compound contains a double or triple bond between pairs of atoms these radii do not apply.

Checkpoint

In which compound is the nitrogen–nitrogen bond length the greatest, H_2N—NH_2 or HN=NH? Explain.

4.9 RESONANCE HYBRIDS

Two equivalent Lewis structures can be written for sulfur dioxide, as shown in Figure 4.7. The only difference between the two structures is the identity of the oxygen atom to which the double bond is formed. As a result, they must be equally satisfactory representations of the molecule. This raises an important question: Which of the Lewis structures for SO_2 is correct?

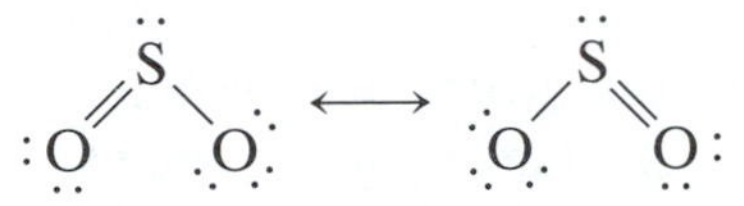

FIGURE 4.7 Two equally satisfactory Lewis structures can be written for SO_2.

Interestingly enough, neither of the structures is correct. The two Lewis structures suggest that one of the sulfur–oxygen bond lengths is shorter than the other. Every experiment that is done to probe the structure of this molecule, however, suggests that the two sulfur–oxygen bonds have identical bond lengths. Moreover, the actual bond length is intermediate between that of a single and double bond.

Often when writing Lewis structures for molecules containing double or triple bonds it is found that multiple bonds may be drawn satisfactorily in several different ways. When this occurs, as with SO_2, the best description of the structure of the molecule is a **resonance hybrid** of all the possible structures. The meaning of the term *resonance* can be best understood by an analogy. In music, the notes in a chord are often said to resonate, that is, they mix to give something that is more than the sum of its parts. In a similar sense, the two Lewis structures for the SO_2 molecule are in resonance. They mix to give a hybrid that is more than the sum of its components. The fact that SO_2 is a resonance hybrid of two Lewis structures is indicated by writing a double-headed arrow between the Lewis structures, as shown in Figure 4.7.

Isomeric structures and resonance structures are different. Isomers are structures with the same chemical formula but a different arrangement of the atoms. Resonance structures have the same arrangement of atoms but a different arrangement of electrons leading to multiple bonds in more than one position in the structures.

The relationship between the SO_2 molecule and its Lewis structures can be illustrated by an analogy. Suppose a knight of the round table returns to Camelot to describe a wondrous beast encountered during his search for the Holy Grail. He suggests that the animal looked something like a unicorn because it had a large horn in the center of its forehead. But it also looked something like a dragon because it was huge, ugly, and thick skinned. The beast is, in fact, a rhinoceros. The animal is real, but it was described as a hybrid of two mythical animals. The SO_2 molecule is also real, but we have to use two mythical Lewis structures to describe its bonding.

Exercise 4.4

Acetic acid dissociates to some extent in water to give the acetate ion, $CH_3CO_2^-$. Write two alternative Lewis structures for the acetate ion.

Solution

We can start by noting that the acetate ion contains two carbon atoms (Group IVA), three hydrogen atoms (Group IA), and two oxygen atoms (Group VIA). It also carries a negative charge, which means that $CH_3CO_2^-$ contains a total of 24 valence electrons.

$$CH_3CO_2^- \qquad 2(4) + 3(1) + 2(6) + 1 = 24$$

The skeleton structure contains six covalent bonds, which leaves 12 nonbonding electrons. Unfortunately, it takes all 12 nonbonding electrons to satisfy the octets of the oxygen atoms, which leaves no nonbonding electrons for the carbon atom bonded to the oxygens.

Because there aren't enough electrons to satisfy the octets of the atoms, we assume that there is at least one C=O double bond. There are now seven covalent bonds in the skeleton structure, which leaves only 10 nonbonding electrons. Fortunately, this is enough.

This process gives us one satisfactory Lewis structure for the acetate ion. We can get another by changing the location of the C=O double bond. As a result, the acetate ion is a resonance hybrid of the following Lewis structures.

Checkpoint

Why do the two structures of acetate ion above not represent two isomers?

Exercise 4.5

The molecule N_2O has the skeletal structure NNO. Which of the Lewis structures given below are acceptable? For structures that are not acceptable explain why.

(a) $:\ddot{N}-\ddot{N}-\ddot{O}:$ (b) $:\ddot{N}-N=\ddot{O}:$ (c) $:\ddot{N}=N=\ddot{O}:$ (d) $:N\equiv N-\ddot{O}:$

(e) $:N=N-\ddot{O}:$ (f) $:N\equiv N-\ddot{O}:$ (g) $:N=\ddot{N}-\ddot{O}:$ (h) $:\ddot{N}-N\equiv O:$

Solution

The N_2O molecule has 16 valence electrons: 5 from each nitrogen and 6 from the oxygen atom.

(a) This is not an acceptable Lewis structure because the valence shell of the central nitrogen is not filled. There are only four electrons around the central nitrogen atom.

(b) This is not an acceptable structure because there are only six electrons around the central nitrogen atom.

(c) This is an acceptable structure.

(d) This is an acceptable structure.

(e) This is not an acceptable structure because there are only six electrons around the central nitrogen atom.

(f) This is not an acceptable structure because the central nitrogen atom has more than a filled valence shell. Because nitrogen is from the second period in the periodic table it can accommodate only eight electrons.

(g) The structure drawn has the correct filled valence shells for all of the atoms, but it includes 18 electrons. Only 16 electrons are available in the valence shells of the atoms.

(h) This is an acceptable Lewis structure.

Structures (c), (d), and (h) are resonance structures of one another.

Checkpoint

Benzene, C_6H_6, has the best known resonance structure in chemistry. If the six carbon atoms are arranged in a ring, draw two resonance Lewis structures for benzene. All of the carbon–carbon bonds in benzene have the same bond length. Referring to Table 4.1, explain why the carbon–carbon bond length is less than that of ethane but greater than that of ethylene.

4.10 ELECTRONEGATIVITY

A covalent bond involves the sharing of a pair of valence electrons by two atoms. When the atoms are identical, they must share the electrons equally. There is no difference between the electron density on the two oxygen atoms in an O_2 molecule, for example. The O_2 molecule is an example of a purely covalent compound.

The same can't be said about molecules that contain different atoms. Consider the HCl molecule. If there is any difference between the relative ability of hydrogen and chlorine to draw electrons toward itself, the electrons in the H—Cl bond won't be shared equally. The electrons in the bond will be drawn closer to one atom or the other.

The relative ability of an atom to draw electrons in a bond toward itself is called the **electronegativity** (*EN*) of the atom. For many years, chemists have recognized that some atoms attract electrons in a bond better than other atoms. Thus F and O are more electronegative than Na and Mg. But there is no direct way to measure the electronegativity of an atom. Instead, properties that are assumed to depend on electronegativity are measured and then compared to one another in order to determine a relative scale of electronegativity. There are currently at least 15 electronegativity scales in use. When there is close agreement between electronegativity values of an atom in different scales, our confidence in the value is increased.

The first scale of electronegativities was created by Linus Pauling. He assigned fluorine an electronegativity of 4.0 and determined the electronegativities of the atoms of the other elements relative to fluorine. Every electronegativity scale that has been proposed since then has been adjusted so that the electronegativity of fluorine is about 4.

A new scale was proposed in 1989 that links the electronegativity of an atom to the average valence electron energy (AVEE) obtained from photoelectron spectroscopy.[1] This scale assumes that atoms which strenuously resist the loss of valence electrons are the atoms which are most likely to draw electrons in a bond toward themselves. Thus AVEE is, in fact, a generalized electronegativity, and the two terms can be used interchangeably. The AVEE data given in Figure 3.27 have been refined and used to calculate the electronegativities shown in Figure 4.8a by multiplying by a factor to give fluorine an electronegativity value of about 4. (These data also can be found in Table B.7 in the appendix and on the back inside cover.) Figure 4.8 shows two electronegativity scales. Figure 4.8a shows the scale based on AVEE, and Figure 4.8b shows the original Pauling scale. Because the AVEE scale is based on PES experimental measurements discussed in Chapter 3, we will use the AVEE electronegativity scale exclusively in this text. This scale offers the additional advantage of providing electronegativities for the noble gases. These values will be useful when discussing the compounds formed by the noble gases.

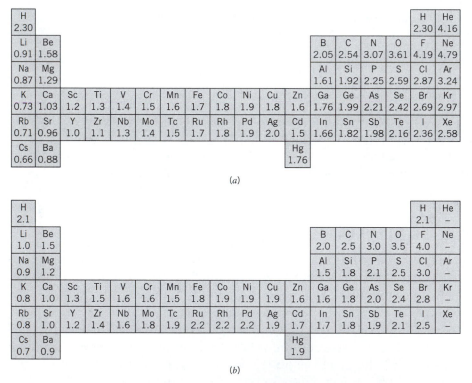

(a)

(b)

FIGURE 4.8 (*a*) Electronegativities of the elements calculated from photoelectron spectroscopy and refined AVEE. The AVEE data from Chapter 3 have been adjusted to give fluorine an electronegativity value close to 4. Reprinted from L. C. Allen and E. T. Knight, *Journal of Molecular Structure,* **261,** 313 (1992). (*b*) Electronegativities based on the Pauling scale.

[1]L. C. Allen, *Journal of the American Chemical Society,* **111,** 9003 (1989).

When the magnitude of the electronegativities of the main-group elements is added to the periodic table as a third axis, we get the results shown in Figure 4.9. There are clear patterns in the data in Figures 4.8 and 4.9.

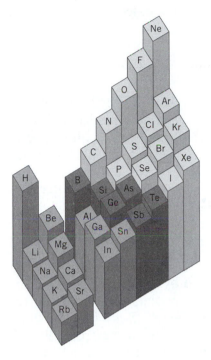

FIGURE 4.9 Three-dimensional plot of the electronegativities of the main-group elements versus position in the periodic table. Reprinted from L. C. Allen, *Journal of the American Chemical Society,* **111,** 9003 (1989).

- Electronegativity increases in a regular fashion from left to right across a row of the periodic table.
- Electronegativity decreases down a column of the periodic table.

When atoms with large differences in electronegativities combine, the electron density in the resulting bond is pulled to the more electronegative atom. Such compounds are called ionic compounds and are discussed in Chapter 5. NaCl is an example of an ionic compound. In NaCl most of the electron density in the bond is transferred from the sodium to the more electronegative atom chlorine.

In covalent molecules, such as those of this chapter, electrons will be shared between bonding atoms. If these atoms have the same electronegativity, the electrons will be shared equally and the resulting covalent bond is referred to as a **pure covalent bond.** If the covalently bonded atoms have different electronegativities, the more electronegative atom will attract more of the electron density in the bond. This type of bond is referred to as a **polar covalent bond.** One end of the bond has a partial positive charge ($\delta+$), and the other end has a partial negative charge ($\delta-$).

Checkpoint
Classify the bonds in NO and O_2 as pure covalent or polar covalent.

4.11   PARTIAL CHARGE

It would be useful to have a quantitative measure of the extent to which the charge in an individual covalent bond is located on one of the atoms that form the bond. This can be achieved by calculating the **partial charge** on the atom. The partial charge on an atom is determined by the difference between the electronegativities of the atoms that form the bond. When the difference is relatively small, the electrons are shared more or less equally. When it is relatively large, there is a significant charge separation in the bond.

To illustrate how the partial charge on an atom can be calculated, let's look at the HCl molecule.

$$H\!-\!\overset{\cdot\cdot}{\underset{\cdot\cdot}{Cl}}:$$

In Figure 4.8a the electronegativity of chlorine is given as 2.87 and that of hydrogen is 2.30. Thus the electrons of the H—Cl bond should be attracted to the chlorine slightly more than to the hydrogen. This would result in a partial charge on both atoms, with an excess of electron density on the chlorine atom. The symbol δ is used to indicate the partial charge on an atom, resulting from unequal sharing of electrons. Thus $\delta-$ above an element's symbol means that the element attracts electrons and therefore has a greater share of the electron density in the bond, and $\delta+$ indicates a lesser share of electron density.

$$^{\delta+}H\!-\!Cl^{\delta-}$$

The partial charge of an atom in a molecule is determined by comparing the electron density associated with the free atom (when the atom is not involved in a bond) to the electron density associated with the atom when bonded in a molecule. Since only the outer shell valence electrons are involved in bonding, we consider only those electrons. We begin the calculation by determining the number of valence electrons in a free atom (V). This can be determined from the group number in the periodic table. The valence electron density associated with an atom in a molecule is due to both the electrons in the covalent bond (B) and the nonbonding electrons (N) around the atom. The difference in the electronegativities of two bonded atoms has no effect on the number of nonbonding electrons. But electronegativity difference does affect the number of bonding electrons that are assigned to each atom. The calculation of partial charge therefore involves the electronegativities of the two atoms (EN_a and EN_b) that form the bond.

The partial charge on an atom engaged in a single bond is calculated by multiplying the number of bonding electrons, B_a, by the fraction of the total electronegativity of the two atoms that can be assigned to that atom.[2]

$$\delta_a = V_a - N_a - B_a\left(\frac{EN_a}{EN_a + EN_b}\right)$$

To illustrate how the formula is used, let's calculate the partial charge on the atoms in HCl.

$$\delta_{Cl} = 7 - 6 - 2\left(\frac{2.87}{2.87 + 2.30}\right) = -0.11$$

The partial charge on the chlorine is -0.11. The magnitude of the partial charge on the hydrogen atom is the same as the magnitude of the partial charge on the chlorine atom,

[2]L. C. Allen, *Journal of the American Chemical Society,* **111,** 9115 (1989).

but the sign of the charge is different. The partial charge on the hydrogen is +0.11 as shown below.

$$\delta_H = 1 - 0 - 2\left(\frac{2.30}{2.30 + 2.87}\right) = 0.11$$

The chlorine atom, being more electronegative than the hydrogen atom, attracts the electrons of the bond to itself so that the partial charge on chlorine is -0.11. This means that the chlorine atom has 11% more electron density associated with it in the HCl molecule when compared to a free chlorine atom. This molecule is therefore slightly polarized. HCl is a neutral molecule, so the sum of the partial charges on the two atoms must be zero. The electron density that is lost by the hydrogen atom must be gained by the chlorine atom. Table 4.2 lists additional partial charges on compounds formed between hydrogen and other Group VIIA atoms.

TABLE 4.2   Partial Charge on Group VIIA Atoms in Combination with Hydrogen

Molecule	δ
HF	-0.29
HCl	-0.11
HBr	-0.08

Checkpoint

In a hypothetical molecule *AB*, if the electronegativity of *B* is much greater than that of *A*, which atom, *A* or *B*, would have the negative partial charge?

4.12   FORMAL CHARGE

The actual charge on an atom in a molecule is best represented by the partial charge. However, chemists also find it useful to calculate **formal charge** because experience has shown that when several Lewis structures are possible, formal charge can help decide which structure best represents the actual bonding in the molecule. Formal charges do not represent real charges on atoms.

The first step in the calculation of formal charge involves dividing the electrons in each covalent bond equally between the atoms that form the bond. The number of valence electrons formally assigned to each atom is then compared with the number of valence electrons on a neutral atom of the element. If the atom has more valence electrons than a neutral atom, it is assigned a formal negative charge. If it has fewer valence electrons it carries a formal positive charge.

Consider the amino acid known as glycine, which is often written as $H_3N^+CH_2CO_2^-$. We can use the concept of formal charge to explain the meaning of the positive and negative signs in the Lewis structure of this molecule.

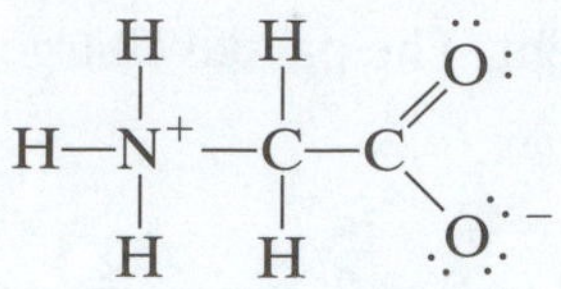

$$H-\overset{\displaystyle H}{\underset{\displaystyle H}{\overset{|}{\underset{|}{N^+}}}}-\overset{\displaystyle H}{\underset{\displaystyle H}{\overset{|}{\underset{|}{C}}}}-C\overset{\ddot{O}:}{\underset{\ddot{\underset{..}{O}}:^-}{}}$$

We start by arbitrarily dividing pairs of bonding electrons so that each atom in a bond is formally assigned one of the electrons, as shown in Figure 4.10. Once this is done, the nitrogen has four valence electrons, one fewer than a neutral nitrogen atom. The nitrogen therefore carries a formal charge of +1 in this Lewis structure. Both of the carbon atoms formally have four electrons, which is equal to the number on a neutral carbon atom. As a result, neither carbon atom carries a formal charge. The oxygen atom in the C=O double bond formally carries six electrons, which means it has no formal charge. The other oxygen atom, however, carries seven electrons, which means that it formally has a charge of −1.

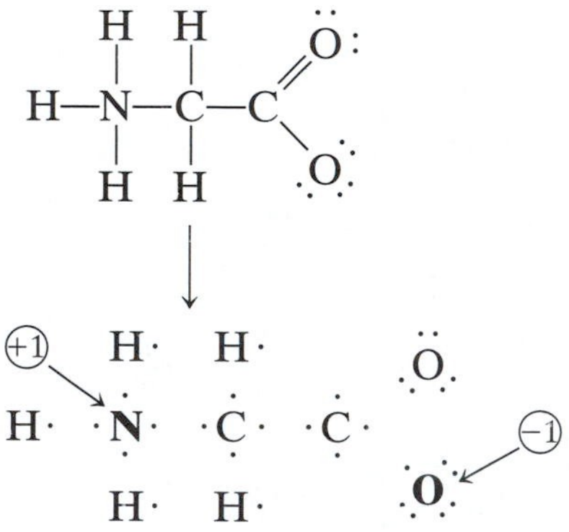

FIGURE 4.10 The first step in calculating the formal charge on the atoms in glycine involves dividing pairs of bonding electrons between the atoms in each bond. Formal charges are shown in circles to distinguish them from actual charges.

Although the glycine molecule carries no net charge, one end of the molecule carries a positive charge, and the other end carries a negative charge. As a result, it isn't surprising that chemists write the formula for this amino acid as $H_3N^+CH_2CO_2^-$.

We can summarize the process by which the formal charge on an atom is calculated by noting that we start with the number of valence electrons (V) on a neutral atom of the element. We then subtract all of the nonbonding electrons on the atom (N) and half of the bonding electrons on the atom (B). The formal charge (FC) on an atom, a, is therefore given by the following equation.

$$FC_a = V_a - N_a - \frac{B_a}{2}$$

To illustrate how this equation is used, let's calculate the formal charge on the nitrogen atom in glycine. Nitrogen has five valence electrons, there are no nonbonding electrons on the atom, and there are eight bonding electrons. As we have seen, the formal charge on the nitrogen atom is therefore +1.

$$FC_N = 5 - 0 - (8/2) = 1$$

The following example shows why chemists find formal charge useful. Three equally valid Lewis structures can be drawn for dinitrogen oxide, N_2O, a gas known as laughing gas and used as an anesthetic.

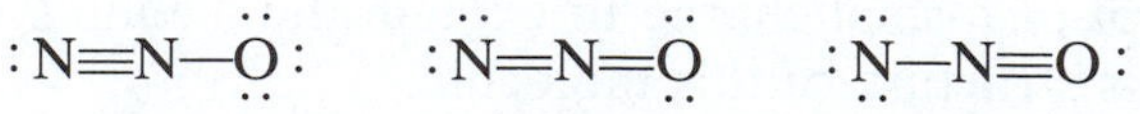

$$:N\equiv N-\ddot{\underset{..}{O}}: \qquad :\ddot{N}=N=\ddot{\underset{..}{O}} \qquad :\ddot{N}-N\equiv O:$$

$$\qquad I \qquad\qquad\qquad II \qquad\qquad\qquad III$$

Which structure best represents the actual bonding in the molecule? Are the two nitrogen atoms bonded by a single, a double, or a triple bond? We begin by using what has come to be accepted as a general rule for deciding which is the best Lewis representation of a molecule. The best structure is considered to be the one in which the atoms have the smallest formal charges and the negative formal charges are on the more electronegative atoms. The formal charge on each atom in the three Lewis structures of N_2O are shown below.

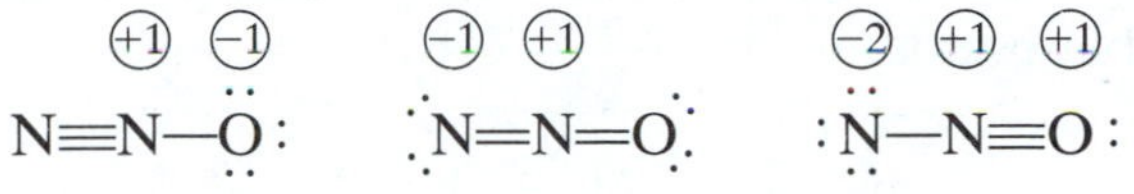

$$N\equiv N\!-\!\ddot{O}: \qquad :\ddot{N}\!=\!N\!=\!\ddot{O}: \qquad :\ddot{N}\!-\!N\equiv O:$$
$$I \qquad\qquad II \qquad\qquad III$$

Note that the sum of the formal charges must always total to the charge on the species, which in this case is zero because N_2O is an electrically neutral molecule.

Oxygen is more electronegative than nitrogen; hence, any structure that puts negative formal charge on nitrogen rather than oxygen is not good. Structure III gives nitrogen rather than oxygen a formal negative charge. In addition this structure has the largest formal charges. Structure II also puts a negative formal charge on nitrogen rather than oxygen and hence is not as good as structure I.

Are there experimental data that could be used to support structure I? Because nitrogen forms single, double, and triple bonds in the proposed structures, bond length data should be helpful. We find that the N—N bond length will be longer than N=N, which is in turn longer than the N≡N bond length. Typical nitrogen–nitrogen bond lengths are:

$$
\begin{array}{ll}
\text{N—N} & 0.146 \text{ nm} \\
\text{N=N} & 0.125 \text{ nm} \\
\text{N≡N} & 0.110 \text{ nm}
\end{array}
$$

An experimental determination of the nitrogen–nitrogen bond length in N_2O gives a result of 0.113 nm, which indicates the bond must be a triple bond. The experimental data support the formal charge calculation that suggests structure I is the best Lewis structure for N_2O.

Checkpoint

Calculate the formal charge on each atom in SO_2.

There are many instances in which the use of formal charge serves to rationalize chemical information. Nitric acid, HNO_3, is a very strong acid whereas nitrous acid, HNO_2, is not. The Lewis structures are

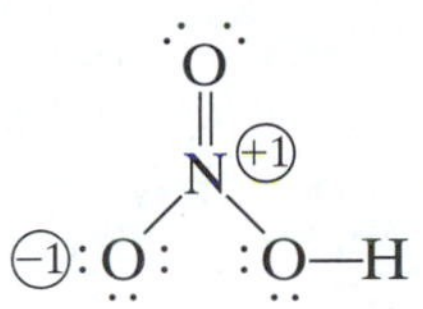

Nitric acid

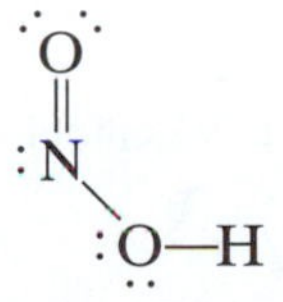

Nitrous acid

The formal charge on the nitrogen in nitric acid is +1, allowing the hydrogen atom with its positive partial charge to be easily removed. The nitrogen in nitrous acid has no formal charge and does not lose its hydrogen as easily as nitric acid. Thus nitric acid is a stronger acid than nitrous acid.

There are several schools of thought on what constitutes the best Lewis structures for molecules. The best evidence, of course, is experimental data. In the absence of conclusive data, some chemists prefer to arrange the electrons in a Lewis structure in order to minimize the formal charge on each atom. For example, the structure of SO_2 can be drawn with either one or two double bonds.

Structure I   Structure II

In structure I sulfur has eight valence electrons and a formal charge of +1. In structure II sulfur has ten valence electrons and a formal charge of zero. Without additional information, it is not possible to feel confident of the correct structure. One study of Lewis structures suggests that using additional multiple bonds and therefore expanding the octet around the central atom may not be the best representation of the structures of many molecules which have traditionally been drawn this way.[3] In this text we use expanded octets to explain the bonding in structures such as SF_4 and XeF_4 discussed in Section 4.7. We also use formal charge as a tool to select between several possible Lewis structures as discussed in this section. However, we do not expand the octet of an atom for the purpose of minimizing the formal charges in the structure. Therefore, until conclusive evidence is available, we'll use structure I to describe SO_2 rather than structure II.

Checkpoint

The acid strengths of the series of chloro-oxy acids are as follows:

$$HClO_4 > HClO_3 > HClO_2 > HClO$$

Use formal charge to rationalize this trend.

The relationship used to calculate partial charge

$$\delta_a = V_a - N_a - B_a\left(\frac{EN_a}{EN_a + EN_b}\right)$$

is related to the expression used to determine formal charge on an atom

$$FC_a = V_a - N_a - (B_a/2)$$

The difference between the two expressions is that when formal charge is calculated the electrons are assumed to be equally shared ($EN_a = EN_b$). If this relationship is used in the partial charge equation we obtain the following:

[3]L. Suidan, J. K. Badenhoop, E. D. Glendenins, and F. Weinhold, *Journal of Chemical Education*, **72**, 583 (1995).

$$\delta_a = V_a - N_a - B_a\left(\frac{EN_a}{EN_a + EN_b}\right)$$

$$\delta_a = V_a - N_a - \frac{B_a}{2} = FC_a$$

The calculated partial charge values are a better description of the actual distribution of electron density in the bond than is formal charge. Formal charge is a useful concept but is not intended to show how electrons are distributed between bonded atoms. Formal charges do not represent actual charges on atoms. The actual charge on an atom is determined in part by electronegativity differences between atoms. Note, for example, that the formal charge on both hydrogen and chlorine in HCl is zero. The electronegativity of chlorine is greater than that of hydrogen, so chlorine will attract electron density in the bond at the expense of hydrogen. In the calculation of formal charge, the electronegativities of hydrogen and chlorine are treated as if they were equal. Consequently, in formal charge calculations the bonding electrons are treated as if they were equally shared.

4.13 THE SHAPES OF MOLECULES

The shape of a molecule can play an important role in its chemistry. The changes in the three-dimensional structure of proteins that occur when an egg is heated, for example, are the primary source of the differences between a raw egg and a cooked one.

The best illustration of how sensitive biomolecules are to changes in their three-dimensional structure is provided by the chemistry of hemoglobin, a protein with a molecular weight of 65,000 amu that carries oxygen through the body. The protein contains four chains of amino acids, two α chains and two β chains.

The structure of hemoglobin is shown in Figure 4.11. Sickle-cell anemia occurs when the identity of a single amino acid among the 146 amino acids on a β chain is changed. The substitution of valine for glutamic acid at the sixth position on the chain produces a subtle change in the structure of the hemoglobin, which interferes with its ability to pick up oxygen at low pressures. The result is so severe that children who inherit this disorder from both parents seldom live past the age of 2 years.

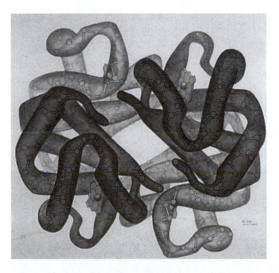

FIGURE 4.11 Change in the identity of one of the 146 amino acids on one of the chains in hemoglobin causes a large enough change in the structure of hemoglobin to interfere with its ability to carry oxygen through the blood.

Research in the 1990s

The Shapes of Molecules

Research in recent years has greatly expanded our understanding of how the shape of a molecule affects the way it binds to certain receptors on the membrane of a cell. It has long been known, for example, that cholera toxin binds to the cells that line the intestine. Once this happens, the toxin can slip through the cell membrane without destroying the cell. This initiates a series of reactions that induce the intestinal cells to secrete so much water—up to 9 gallons a day—that cholera victims die of the debilitating effects of diarrhea.

Until relatively recently, the mechanism by which cholera toxin achieved these results was unknown. In 1991 Rongguang Zhang and Edwin Westbrook at Argonne National Laboratory reported the structure of the cholera toxin [*Science,* **253,** 382–383 (1991)]. The protein contains two subunits. The B subunit, which has a molecular weight of about 55,000 amu, forms the doughnut structure shown in Figure 4.12. The A subunit is smaller, with a molecular weight of 29,000 amu. The B subunit apparently binds to the cell and then pushes the smaller A subunit through the membrane of the cell. The A subunit then acts as an enzyme, initiating the reactions that lead to the secretion of water.

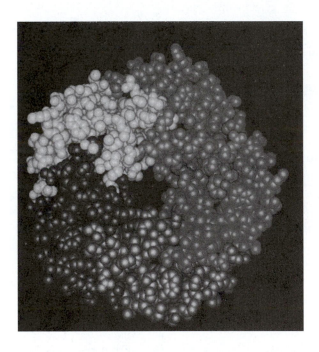

FIGURE 4.12 Poison doughnut, the B subunit of cholera toxin.

In the short term, knowledge of the structure of cholera toxin will drive research toward the discovery of vaccines that can prevent cholera. In the long term, it may provide the basis for the design of a system to deliver drugs across the cell membrane to destroy cancer cells.

Another example of how the structure of a molecule affects the way it binds to a receptor site is provided by progesterone and the synthetic analog of this steroid shown in Figure 4.13. Progesterone is a hormone that is secreted in the second half of the menstrual cycle. It binds to receptors in the endometrium, preparing the uterus for implantation of a fertilized egg. If pregnancy occurs, continued production of progesterone is essential for maintenance of the embryo, and eventually the fetus.

In 1975 Georges Teutsch initiated a project to study how small changes in the structure of progesterone might affect the ability of the molecule to bind to the five kinds of steroid

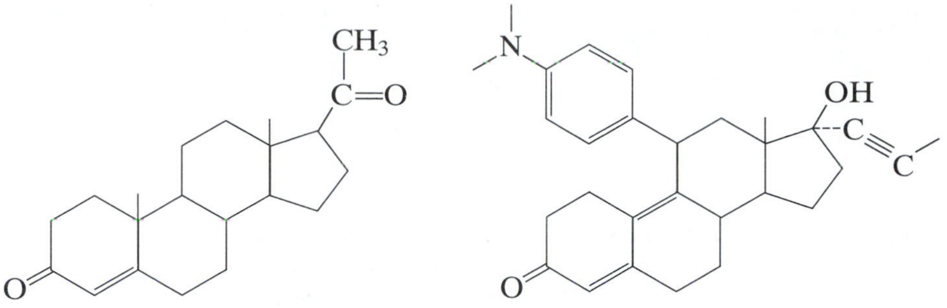

FIGURE 4.13 Structures of progesterone and a synthetic analog that is the basis of the drug known as RU-486.

receptors in a cell. Synthetic steroid hormones are divided into two categories. The *agonists* produce the same effect as the natural hormone when they bind. The *antagonists* bind to the receptor but don't switch on the activity induced when the natural hormone binds.

Teutsch and co-workers were searching for an antagonist for glucocorticoid receptors, which might increase the rate at which burns and other wounds heal. When they tested a compound originally known as RU-38486, they found that it was a powerful glucocorticoid antagonist. But it did more than that; it also bound very tightly to the progesterone receptor. As a progesterone antagonist, RU-486, as it became known, was a candidate for testing as a drug for fertility control.

In September 1988, RU-486 became available in France for the termination of pregnancy. The protocol for its use involves delivering 600 mg of the drug in a single dose, followed 36 to 48 hours later with a dose of a prostaglandin to induce the contractions necessary to expel the embryo from the uterus. When this protocol is followed, the rate of success is similar to that of surgical procedures [*New England Journal of Medicine,* **322**(10), 645–648 (1990)].

Because it binds so tightly to several kinds of hormone receptors, RU-486 has the potential to serve other functions. It could be administered with oxytocin to induce labor at the end of pregnancy, therefore reducing the number of cesarean deliveries that have to be done. It might be used to treat cancers that involve progesterone receptors, such as certain forms of breast cancer. It might also control the growth of certain noncancerous tumors that affect the brain. It might even be used for its original purpose—as a glucocorticoid antagonist. Research is sure to continue on this controversial drug and its potential uses.

4.14 PREDICTING THE SHAPES OF MOLECULES (THE ELECTRON DOMAIN MODEL)

There is no direct relationship between the formula of a compound and the shape of its molecules, as shown by the examples in Figure 4.14. The three-dimensional shapes of these molecules can be predicted from their Lewis structures, however, with a model developed in the 1960s.[4,5] When it was introduced, it was called the *valence-shell electron pair repulsion (VSEPR) model.* The model has been modified and is now referred to as the **electron domain (ED) model.**[6]

[4]R. J. Gillespie and R. S. Nyholm, *Quarterly Review of the Chemical Society,* **11,** 339 (1957).
[5]R. J. Gillespie, *Journal of Chemical Education,* **40,** 295 (1963).
[6]R. J. Gillespie, *Journal of Chemical Education,* **69,** 116 (1992).

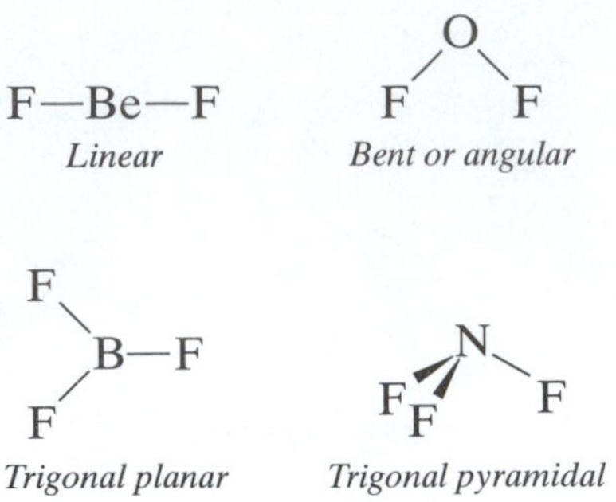

F—Be—F
Linear

O
F   F
Bent or angular

F
 \
  B—F
 /
F
Trigonal planar

N
F F F
Trigonal pyramidal

FIGURE 4.14 There is no obvious relationship between the chemical formula and the shape of a molecule. BeF_2 is linear, whereas OF_2 is bent. BF_3 is a planar molecule, whereas NF_3 is pyramidal.

It is surprising that the spatial arrangement of atoms in molecules conforms to a relatively small number of types. Angles between selected atoms tend generally to be 90°, 109°, 120°, or 180°. The limited number of ways of arranging atoms can be understood from the electron domain model.

The electrons in the valence shell of an atom in a molecule form pairs with opposite spins. Each pair occupies its own domain and is attracted to the central atom. The domains get as close to the central atom as they can but keep other domains as far away as possible. The ED model assumes that the geometry around each atom in a molecule can be predicted by keeping the domains of electron pairs separated. Thus two domains will be separated by 180°. For three domains the best arrangement is 120°, whereas four domains are 109° apart. Figure 4.15 shows how domains, treated as spheres, are arranged about a central atom in accordance with the ED model.

Two domains
180°

Three domains
120°

Four domains
109°

FIGURE 4.15 Arrangement of domains about a central atom. Two domains take up a position on opposite sides of a central atom, three are arranged at an angle of 120°, and four at 109°.

A bonding domain is shared by two atoms while a nonbonding domain belongs entirely to the valence shell of a particular atom. Therefore, a nonbonding domain tends to spread out and occupy a larger space than does a bonding domain.

We can see how the ED model is used by applying it to BeF_2. The Lewis structure of the molecule suggests that there are two pairs of bonding electrons, or two bonding domains, in the valence shell of the central atom. Note that this molecule is an exception to the octet rule because only two pairs of electrons surround the central atom. However, this Lewis structure is supported by experimental data.

$$: \ddot{F}—Be—\ddot{F} :$$

We can keep the two bonding domains as far apart as possible by arranging them on either side of the beryllium atom. The ED model therefore predicts that BeF_2 should be a **linear** molecule, with a 180° angle between the two Be—F bonds.

A more complicated example shows how the ED model can be used for molecules with multiple bonds. Consider the Lewis structure of CO_2.

$$: \ddot{O}=C=\ddot{O} :$$

There are four pairs of bonding electrons in the valence shell of the carbon atom but only two domains in which those electrons can be found. (There are two pairs of electrons in the C=O double bond on the left and two pairs in the double bond on the right.) As a result, the ED model predicts that this molecule will also have a linear geometry, with a 180° angle between the two double bonds.

It is assumed that the two electron pairs which make up the double bond occupy more space than a single bond domain. The three electron pairs of a triple bond likewise require more space than a double bond domain. The relative sizes of the spaces occupied by single, double, and triple bonds will determine how these bonding regions are distributed around the central atoms. The space occupied by a double bond, which consists of two electron pairs, is called a double bond domain. Correspondingly, the three electron pairs that form a triple bond are collectively called a triple bond domain.

There are three domains in the valence shell of the central atom in boron trifluoride (BF$_3$) where electrons can be found. The domains correspond to the three pairs of bonding electrons in the B—F bonds.

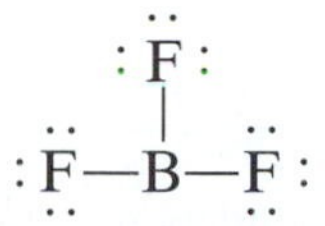

$$: \overset{\cdot\cdot}{\underset{}{F}} :$$
$$|$$
$$: \overset{\cdot\cdot}{\underset{\cdot\cdot}{F}} — B — \overset{\cdot\cdot}{\underset{\cdot\cdot}{F}} :$$

The optimum geometry to keep the three domains as far apart as possible is an equilateral triangle. The ED model therefore predicts a **trigonal planar** geometry for the BF$_3$ molecule, with a F—B—F bond angle of 120°, as shown in Figure 4.16.

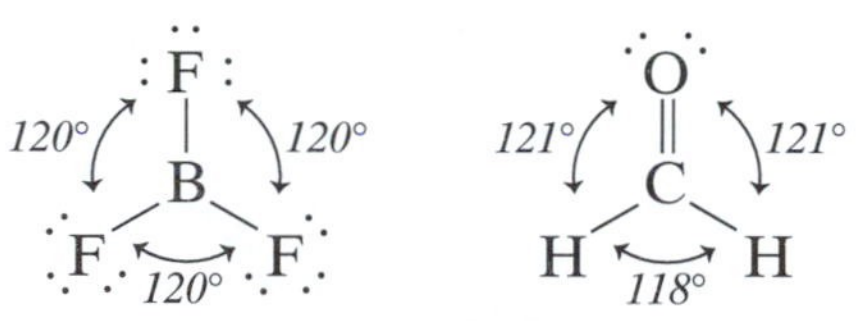

FIGURE 4.16 Three equal bonding domains surround the central B atom in BF$_3$, giving an angle of 120°. The three bonding domains around the carbon atom in formaldehyde are not the same size, and the H—C—O angle increases to 121°.

Checkpoint

The structure of benzene was discussed in the checkpoint at the end of Section 4.9. Determine the number of electron pair domains around each carbon atom.

The advantage of counting domains of electron density rather than pairs of electrons can also be illustrated by considering the geometry of formaldehyde, H$_2$CO.

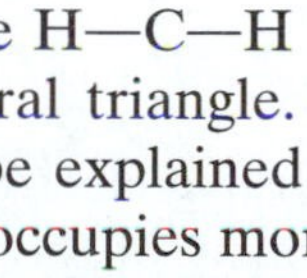

$$: \overset{\cdot\cdot}{\underset{}{O}} :$$
$$\|$$
$$H — C — H$$

The Lewis structure of the molecule suggests that there are four pairs of electrons in the valence shell of the central atom. Two pairs form a double bond domain, and there are two single bond domains. The ED model therefore predicts that this molecule would have a trigonal planar geometry, just like BF$_3$.

At first glance, we might expect the H—C—H bond angle in formaldehyde would be 120°, the internal angle in an equilateral triangle. Experiment suggests that it is a little smaller than this, only 118°. This can be explained by recalling that according to the ED model the C=O double bond domain occupies more space than the domains that contain the electrons in the C—H bonds. As a result, the H—C—O bond angles are slightly larger than 120°, and the H—C—H bond angle is slightly smaller than 120°, as shown in Figure 4.16.

BeF$_2$, CO$_2$, BF$_3$, and H$_2$O are all two-dimensional molecules, in which the atoms lie in the same plane. If we place the same restriction on methane (CH$_4$), we would get a square planar geometry, with an H—C—H angle of 90°.

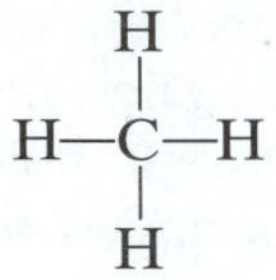

$$
\begin{array}{c}
\text{H} \\
| \\
\text{H—C—H} \\
| \\
\text{H}
\end{array}
$$

There is a more efficient way of arranging the four domains in the valence shell of the central atom, however. If the four domains are arranged toward the corners of a **tetrahedron,** the H—C—H angle increases to 109.5°, as shown in Figure 4.17.

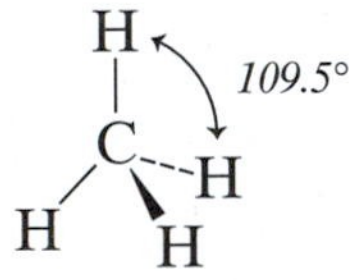

FIGURE 4.17 Four equal bonding domains around the central carbon atom in methane give bond angles of 109.5°.

PF$_5$ has five single bond domains about the phosphorus atom.

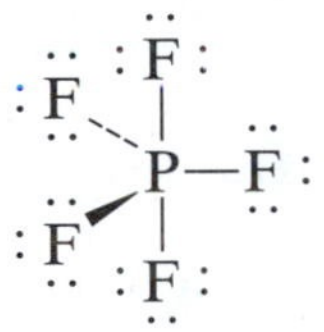

The best arrangement of the domains is toward the corners of a **trigonal bipyramid.** Three of the positions in a trigonal bipyramid are labeled equatorial because they lie along the equator of the molecule. The other two are axial because they lie along an axis perpendicular to the equatorial plane. The angle between the three equatorial positions is 120°, whereas the angle between an axial and an equatorial position is 90°.

There are six single bond domains on the central atom in SF$_6$.

The optimum geometry for the molecule would involve arranging the six domains toward the corners of an **octahedron.** The term octahedron literally means "eight sides," but it is the six corners, or vertices, that interest us. To envision the geometry of an SF$_6$ molecule, imagine four fluorine atoms lying in a plane around the sulfur atom with one fluorine atom above the plane and another below. All F—S—F angles are 90°.

Checkpoint

Predict the shape of a molecule with five electron domains around the central atom. Is your answer dependent on how many of the five domains are bonding or nonbonding domains?

4.15 THE ROLE OF NONBONDING ELECTRONS IN THE ED MODEL

All of our examples, so far, have contained only bonding electrons in the valence shell of the central atom. What happens when we apply the ED model to atoms that also contain nonbonding electrons? Consider ammonia (NH_3) and water (H_2O), for example.

$$H\!-\!\overset{\displaystyle ..}{N}\!-\!H \qquad H\!-\!\overset{\displaystyle ..}{\underset{\displaystyle ..}{O}}\!-\!H$$
$$\underset{\displaystyle H}{|}$$

In each case, there are four domains in the valence shell of the central atom where electrons can be found. The valence electrons on the central atom in both NH_3 and H_2O therefore should be distributed toward the corners of a tetrahedron, as shown in Figure 4.18. Our goal, however, isn't the prediction of the distribution of valence electrons. It is to use the distribution of electrons to predict the geometry of the molecule, which describes how the atoms are distributed in space. In previous examples, the way the valence electrons were distributed and the geometry of the molecule have been the same. Once we include nonbonding electrons, this is no longer true.

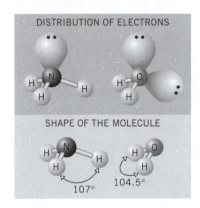

FIGURE 4.18 The ED model predicts that the valence electrons on the central atom in NH_3 and H_2O will be oriented toward the corners of a tetrahedron. The shape of the molecules, however, is determined by the positions of the atoms. Ammonia is therefore described as trigonal pyramidal, and water is described as bent, or angular.

The ED model predicts that the valence electrons on the central atoms in ammonia and water will point toward the corners of a tetrahedron. Because we can't locate the nonbonding electrons with any precision, this prediction can't be tested directly. But the results of the ED model can be used to predict the positions of the atoms in the molecules, which can be tested experimentally. If we focus on the positions of the atoms in ammonia, we predict that the NH_3 molecule should have a shape best described as **trigonal pyramidal,** with the nitrogen at the top of the pyramid. Water, on the other hand, should have a shape that can be described as **bent,** or **angular.** Both predictions have been shown to be correct, which reinforces our faith in the ED model.

Exercise 4.6

Use the Lewis structure of the PF_3 molecule shown in Figure 4.19 to predict the shape of this molecule.

Solution

The Lewis structure in Figure 4.19 suggests that there are four domains in the valence shell of the phosphorus atom where electrons can be found. There are electron pairs in the three

P—F single bonds and a pair of nonbonding electrons. The geometry for the molecule is based on arranging these domains toward the corners of a tetrahedron. The shape of the molecule is therefore *trigonal pyramidal,* like the geometry of ammonia.

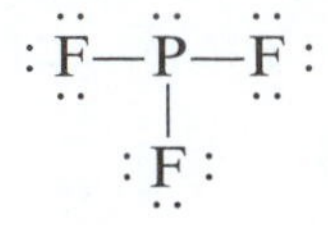

$:\ddot{F}—P—\ddot{F}:$

$:\ddot{F}:$

FIGURE 4.19 Lewis structure of the PF_3 molecule. There are one nonbonding and three bonding domains.

When we extend the ED model to molecules in which the domains are distributed toward the corners of a trigonal bipyramid, we run into the question of whether nonbonding electrons should be placed in equatorial or axial positions. Experimentally we find that nonbonding electrons usually occupy equatorial positions in a trigonal bipyramid.

To understand why, we have to recognize that nonbonding electron domains take up more space than bonding electron domains. Nonbonding domains are close to only one nucleus, and there is a considerable amount of space in which nonbonding electrons can reside and still be near the nucleus of the atom. Bonding electrons, however, must be simultaneously close to two nuclei, and only a small region of space between the nuclei satisfies this restriction.

Figures 4.20 and 4.21 can help us understand why nonbonding electrons are placed in equatorial positions in a trigonal bipyramid. If the nonbonding electron domain in SF_4 is placed in an axial position, it will be relatively close (90°) to *three* bonding pair domains. But if the nonbonding domain is placed in an equatorial position, it will be 90° away from only *two* bonding domains. As a result, an axial position is more crowded than an equatorial position in a trigonal bipyramid arrangement. Larger nonbonding domains therefore occupy the equatorial positions.

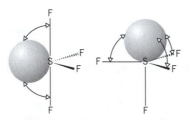

FIGURE 4.20 Crowding of bonding and nonbonding domains in SF_4 is minimized if the nonbonding domain is placed in an equatorial position, as shown in the structure on the left.

The results of applying the ED model to SF_4, ClF_3, and the I_3^- ion are shown in Figure 4.21. When the nonbonding domain on the sulfur atom in SF_4 is placed in an equatorial position, the molecule can be best described as having a **seesaw** or **teeter-totter** shape. The bonding and nonbonding domains in the valence shell of chlorine in ClF_3 can best be accommodated by placing both nonbonding domains in equatorial positions in a trigonal bipyramid. When this is done, we get a geometry that can be described as **T shaped.** The Lewis structure of the triiodide (I_3^-) ion suggests a trigonal bipyramidal distribution of valence electrons on the central atom. When the three nonbonding electron domains on the central I atom are placed in equatorial positions, we get a linear molecule.

Molecular geometries based on an octahedral distribution of electron domains are easier to predict because the corners of an octahedron are all identical.

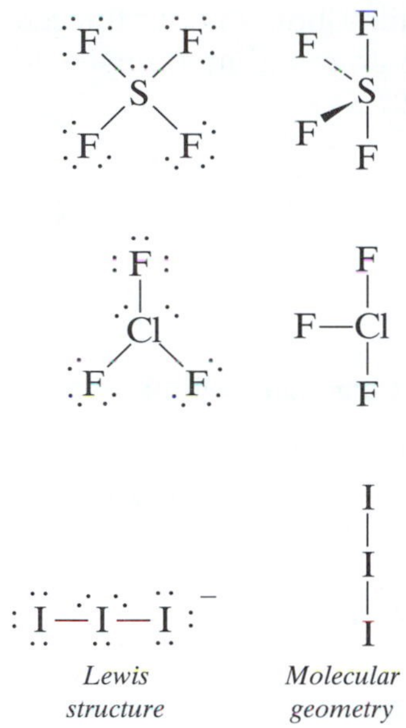

Lewis
structure

Molecular
geometry   **FIGURE 4.21** Structures of SF_4, ClF_3, and the I_3^- ion.

Exercise 4.7

Use the Lewis structures of BrF_5 and XeF_4 shown in Figure 4.22 to predict the shapes of the molecules.

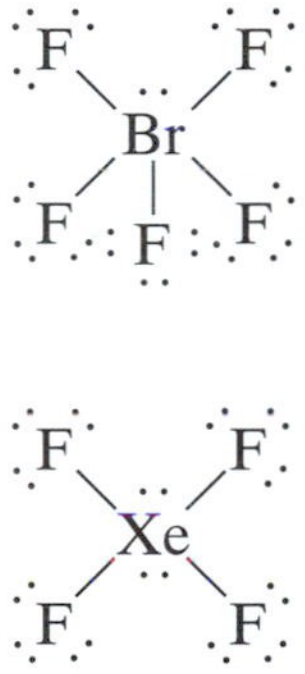

FIGURE 4.22 Lewis structures of BrF_5 and XeF_4.

Solution

The valence shell electrons on the bromine atom in BrF_5 are distributed toward the corners of an octahedron to form a molecule with a **square pyramidal** geometry.

For XeF_4 two of the six domains in the valence shell of the central atom are nonbonding domains and four are bonding domains. Because they occupy considerably more space, the nonbonding domains are kept as far apart as possible. The ED model therefore predicts a **square planar** structure.

The ED model is summarized in Table 4.3. The number of bonding and nonbonding domains around the central atom determines the molecular geometry. For example, SO_2 has three electron domains around the central atom (two bonding domains and one nonbonding domain). We can determine the geometry around the central atom by locating the number of electron domains in column one. A central atom surrounded by three electron domains will have one of the three possible molecular geometries listed for that entry in column 5. Reading across the table we find that the electrons are distributed about the central atom in a trigonal planar fashion. The two bonding and one nonbonding domains allow us to determine that the molecular geometry of SO_2 is bent.

TABLE 4.3   Relationship between Number of Electron Domains and Geometry around an Atom

Electron Domains	Bonding Domains	Nonbonding Domains	Distribution of Electrons	Molecular Geometry	Examples
2 (sp)	2	0	Linear	Linear	BeF_2, CO_2
	1	1		Linear	CO, N_2
3 (sp^2)	3	0	Trigonal planar	Trigonal-planar	BF_3, CO_3^{2-}
	2	1		Bent	O_3, SO_2
	1	2		Linear	O_2
4 (sp^3)	4	0	Tetrahedral	Tetrahedral	CH_4, SO_4^{2-}
	3	1		Trigonal-pyramidal	NH_3, H_3O^+
	2	2		Bent	H_2O, ICl_2^+
	1	3		Linear	HF, OH^-
5 (sp^3d)	5	0	Trigonal bipyramidal	Trigonal-bipyramidal	PF_5
	4	1		Seesaw	SF_4, IF_4^+
	3	2		T shaped	ClF_3
	2	3		Linear	I_3^-, XeF_2
6 (sp^3d^2)	6	0	Octahedral	Octahedral	SF_6, PF_6^-
	5	1		Square-pyramidal	BrF_5, $SbCl_5^{2-}$
	4	2		Square-planar	XeF_4, ICl_4^-

Chemists frequently refer to the molecular geometry in terms other than those which describe the shape. Two domains are arranged linearly, which is often called an sp geometry. Three electron domains are called sp^2, and four electron domains are designated sp^3. These designations are also given in Table 4.3. The geometry of the SO_2 molecule is thus described as having three electron domains or as sp^2. A more detailed description of the use of these labels is given in the appendix of this chapter.

Checkpoint
What is the shape of a molecule with three bonding domains and one nonbonding domain around the central atom?

4.16 BOND ANGLES

Just as bond lengths are important to the structure and hence the properties of a compound, so are the bond angles. To define a bond angle, at least three atoms must be located. Consider, for example, the molecule propyl alcohol.

$$
\begin{array}{ccccc}
& H_a & H_g & H_d & \\
& | & | & | & \ddot{} \\
H_b- & C_1- & C_2- & C_3- & \ddot{O}-H_f \\
& | & | & | & \ddot{} \\
& H_c & H_h & H_e &
\end{array}
$$

In the flat, two-dimensional picture of molecules usually given on the printed page, it is difficult to imagine the three-dimensional representation. There are four bonding domains about the carbon atom labeled 1. The best arrangement of those domains is a tetrahedron, so the angle from hydrogen atom a to carbon 1 to hydrogen atom b ($H_a-C_1-H_b$) is 109°. Likewise, the angles $H_a-C_1-H_c$, $H_c-C_1-C_2$, and $H_b-C_1-H_c$ are also 109°. The angle H_d-C_3-O is also 109° because the central atom, C_3, is surrounded by four bonding domains. The angle C_3-O-H_f is slightly less than 109°. The central oxygen atom is surrounded by two bonding pairs of electrons and two nonbonding pairs of electrons. Because the nonbonding pairs occupy larger domains than the bonding pairs, the bonding pairs are squeezed somewhat to form an angle slightly less than 109°.

Checkpoint

Give the following bond angles for propyl alcohol:

$$
\begin{array}{c}
C_1-C_2-C_3 \\
C_1-C_2-H_g \\
H_e-C_3-O
\end{array}
$$

The ammonia molecule contains three bonding domains and one nonbonding domain, while the water molecule contains two bonding domains and two nonbonding domains. The ED model predicts that the four electron domains in water and ammonia will be arranged in a tetrahedral geometry around the central atom. On the basis of the tetrahedral geometry shown in Figure 4.17, we might expect the bond angles in ammonia and water to be 109.5° as found in methane, CH_4. Although this is a good approximation of the bond angle, the experimentally determined values for the bond angles in ammonia and water are 107° and 104.5°, respectively. The differences can be explained by nonbonding domains taking more space than bonding domains, thus pushing the bonding domains closer together and giving a smaller than expected bond angle.

Checkpoint

The structure of benzene was discussed in the checkpoint at the end of Section 4.9. Determine the bond angles around each carbon atom.

Exercise 4.8

Give the following bond angles for acetic acid.

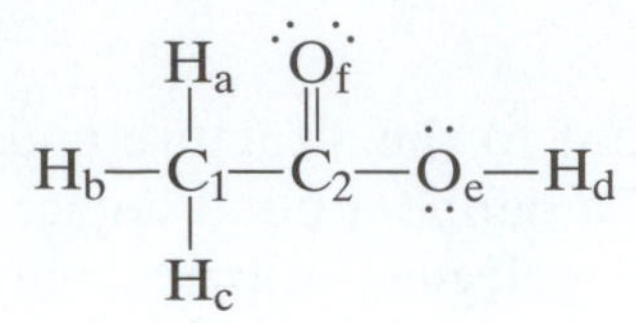

$$\begin{array}{c}
\quad\quad H_a \quad \overset{\cdot\cdot}{\underset{}{O}}_f \\
\quad\quad | \quad\quad || \\
H_b\!-\!C_1\!-\!C_2\!-\!\overset{\cdot\cdot}{\underset{\cdot\cdot}{O}}_e\!-\!H_d \\
\quad\quad | \\
\quad\quad H_c
\end{array}$$

$$
\begin{array}{ll}
H_a\!-\!C_1\!-\!H_b & C_2\!-\!O_e\!-\!H_d \\
C_1\!-\!C_2\!-\!O_f & O_f\!-\!C_2\!-\!O_e \\
C_1\!-\!C_2\!-\!O_e & H_a\!-\!C_1\!-\!C_2
\end{array}
$$

Solution

The Lewis structure of acetic acid shows that there are four bonding domains around C_1, three bonding domains around C_2, and a total of four domains (two bonding and two nonbonding) around O_e. The resulting bond angles are as follows:

$$
\begin{array}{ll}
H_a\!-\!C_1\!-\!H_b = 109^\circ & C_2\!-\!O_e\!-\!H_d = 109^\circ \\
C_1\!-\!C_2\!-\!O_f = 120^\circ & O_f\!-\!C_2\!-\!O_e = 120^\circ \\
C_1\!-\!C_2\!-\!O_e = 120^\circ & H_a\!-\!C_1\!-\!C_2 = 109^\circ
\end{array}
$$

The number of domains around the central atom of the three atoms forming the angle determines the arrangement of the atoms in space.

4.17   THE DIFFERENCE BETWEEN POLAR BONDS AND POLAR MOLECULES

The difference between the electronegativities of chlorine ($EN = 2.87$) and hydrogen ($EN = 2.30$) is sufficient that the bond in HCl has a charge separation. The partial charge on the hydrogen is positive and that on the chlorine is negative. Such a bond is said to be **polar.**

$$
\begin{array}{c}
\delta+ \leftrightarrow \delta- \\
H\!-\!Cl
\end{array}
$$

Because it contains only this one bond, the HCl molecule can also be described as polar. The polarity of a bond in a structure is represented with an arrow having a plus sign at the end (the $\delta+$ and $\delta-$ shown in the above structure could be replaced by the $\leftrightarrow$). The arrow points in the direction of the more electronegative atom, and the plus sign is next to the least electronegative atom.

The polarity of a molecule can be predicted by considering the polarities of the individual bonds, the location of nonbonding pairs, and the three-dimensional geometric shape of the molecule. The magnitude of the polarity of a bond or of a molecule is measured using a quantity known as the **dipole moment, μ.** The dipole moment for a molecule depends on two factors: (1) the magnitude of the charge and (2) the distance between the negative and positive poles of the molecule. Dipole moments are reported in units of *debye* (D). The dipole moment for HCl is small, 1.08 D (see Table 4.4). This can be understood by noting that the difference in charge in the HCl bond is relatively small ($\Delta EN = 0.57$) and that the H—Cl bond is relatively short.

TABLE 4.4   Dipole Moments of Common Compounds

Compound	Dipole Moment (D)
CH_4, methane	0
CH_3OH, methanol	1.70
CO, carbon monoxide	0.112
CO_2, carbon dioxide	0
HCl, hydrogen chloride	1.08
HI, hydrogen iodide	0.44
H_2O, water	1.85
H_2S, hydrogen sulfide	0.97

C—Cl bonds ($\Delta EN = 0.33$) are not as polar as H—Cl bonds ($\Delta EN = 0.57$), but they are significantly longer. As a result, the dipole moment for CH_3Cl is about the same as that for HCl, namely, 1.01 D. At first glance, we might expect a similar dipole moment for carbon tetrachloride (CCl_4), which contains four polar C—Cl bonds. The dipole moment of CCl_4, however, is 0. This can be rationalized by considering the structure of CCl_4 shown in Figure 4.23. The individual C—Cl bonds in this molecule are polar. However, since the dipole moments of the four C—Cl bonds are equivalent and the bonds are symmetrically arranged at the corners of a tetrahedron, the four C—Cl dipoles cancel each other. Carbon tetrachloride therefore illustrates an important point: Not all molecules that contain polar bonds have dipole moments.

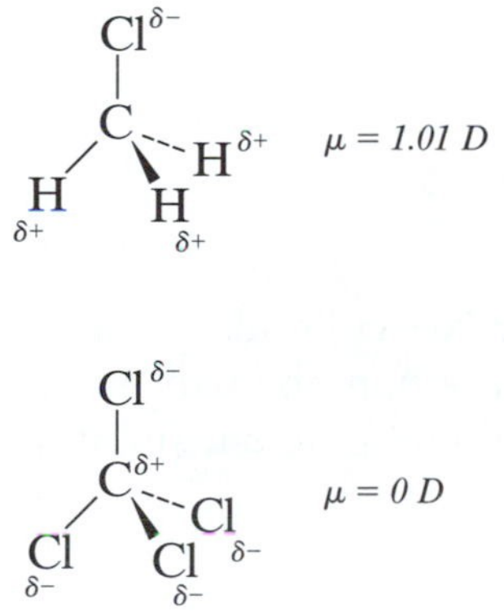

FIGURE 4.23 Although the C—Cl bonds in both CH_3Cl and CCl_4 are polar, CCl_4 has no net dipole moment.

Checkpoint

Does chloroform have a dipole moment?

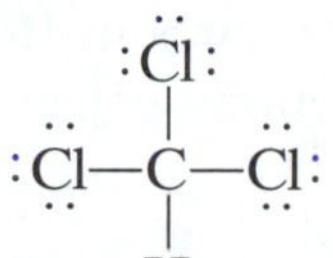

Exercise 4.9

The two-dimensional Lewis structure for dichloromethane is given below. On first inspection it would appear that the polarities of the two C—H bonds and the two C—Cl bonds should cancel one another, resulting in a nonpolar molecule with a dipole moment of zero. However, the measured dipole moment of dichloromethane is 1.60. Explain this discrepancy.

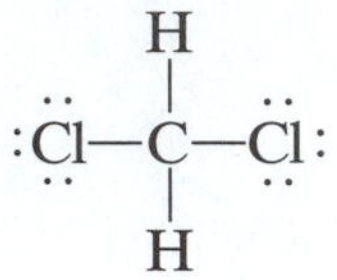

$$
\begin{array}{c}
H \\
| \\
:\ddot{C}l-C-\ddot{C}l: \\
| \\
H
\end{array}
$$

Solution

Dichloromethane is not the flat two-dimensional molecule shown in the Lewis structure but instead is tetrahedral as shown below. The dipole moments of the C—H bonds point toward the carbon atom. The dipole moments of the C—Cl bonds point away from the carbon atom. Therefore, the dipole moments of the bonds are additive to give a net dipole moment to the molecule as shown below.

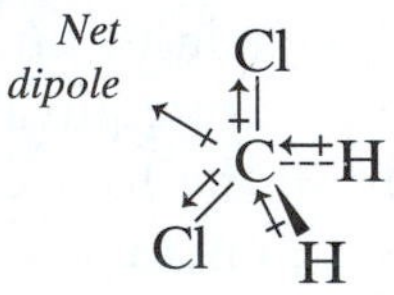

Exercise 4.10

Determine whether each of the following molecules is polar.
(a) NH_3 (b) CO_2

Solution

(a) The structure for ammonia is shown below. The dipoles of the N—H bonds are additive to give an overall dipole to the molecule. In addition the nonbonding pair makes the nitrogen in the molecule very negative, leading to a net dipole moment pointing toward the nonbonding pair.

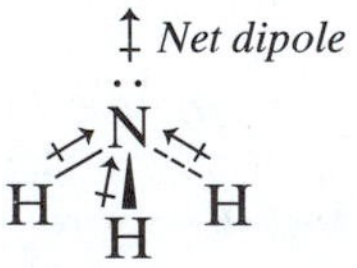

(b) The structure for carbon dioxide is shown below. Carbon dioxide is a linear molecule with an oxygen on each side of the carbon. Both C=O double bonds are polar, but their dipoles point in opposite directions, causing them to cancel one another and leading to a nonpolar molecule.

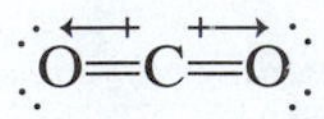

$$
:\ddot{O}=C=\ddot{O}:
$$

KEY TERMS

Angular	Formal charge	Polar compound
Bent	Ionic	Resonance hybrid
Covalent	Lewis structures	Square planar
Covalent bond	Linear	Tetrahedron
Covalent compounds	Nonbonding electrons	Trigonal planar
Dipole moment	Octahedron	Trigonal bipyramid
Domains	Octet	Trigonal pyramidal
Electron domain model	Partial charge	Valence electrons
Electronegativity	Polar bond	

PROBLEMS

Valence Electrons

1. Define the term *valence electrons*.
2. Determine the number of valence electrons in neutral atoms of the following elements.
   (a) Li   (b) C   (c) Mg   (d) Ar
3. Determine the number of valence electrons in neutral atoms of the following elements.
   (a) Fe   (b) Cu   (c) Bi   (d) I
4. Determine the number of valence electrons in the following negative ions and describe any general trends.
   (a) C^{4-} (b) N^{3-} (c) S^{2-} (d) I^-
5. Determine the number of valence electrons in the following positive ions and describe any general trends.
   (a) Na^+ (b) Mg^{2+} (c) Al^{3+} (d) Sc^{3+}
6. Explain why filled d and f subshells are ignored when the valence electrons on an atom are counted.
7. Determine the total number of valence electrons in the following molecules or ions.
   (a) BF_3 (b) CH_4 (c) NH_4^+ (d) H_2SO_4
8. Determine the total number of valence electrons in the following molecules or ions.
   (a) KrF_2 (b) SF_4 (c) SiF_6^{2-} (d) ZrF_7^{3-}
9. Which of the following molecules or ions contain the same number of valence electrons?
   (a) CO_2 (b) N_2O (c) CNO^- (d) NO_2^+
   (e) SO_2 (f) O_3 (g) NO_2^-

The Covalent Bond

10. Write electron dot symbols for the elements in the third row of the periodic table.
11. Describe how the sharing of a pair of electrons by two chlorine atoms makes a Cl_2 molecule more stable than a pair of isolated Cl atoms.
12. Describe the difference between the covalent bonds of O_2 and F_2.

Writing Lewis Structures

13. Draw skeleton structures of the following molecules.
   (a) CH_4 (b) CH_3Cl (c) H_2CO

14. Draw skeleton structures of the following ions.
    (a) NH_4^+ (b) NO_3^- (c) SO_4^{2-}

15. Write Lewis structures for the following ions or molecules.
    (a) NH_3 (b) CH_3^+ (c) H_3O^+ (d) BH_4^-

16. Write Lewis structures for the following ions.
    (a) NO_3^- (b) SO_3^{2-} (c) CO_3^{2-} (d) NO_2^+

17. Write Lewis structures for the following ions or molecules.
    (a) C_2H_6 (b) C_2H_4 (c) C_2H_2 (d) C_2^{2-}

18. Write Lewis structures for the following nitrogen-containing molecules.
    (a) N_2O (b) N_2O_3 $(ON—NO_2)$

19. Write Lewis structures for the following nitrogen-containing molecules.
    (a) ClNO   (b) $ClNO_2$ (c) NO^+ (d) NO_2^-
    (e) ONF_3

20. Explain why a satisfactory Lewis structure for N_2O_5 cannot be based on a skeleton structure that contains an N—N bond: $O_2N—NO_3$. Show how a satisfactory Lewis structure can be written if we assume that the skeleton structure is O_2NONO_2.

21. Write Lewis structures for the following ions or molecules.
    (a) O_2 (b) O_3 (c) O_2^{2-} (d) O^{2-}

22. Write Lewis structures for the following ions or molecules.
    (a) SO_2 (b) SO_3 (c) SO_3^{2-} (d) SO_4^{2-}

23. Write Lewis structures for the following ions or molecules.
    (a) XeF_2 (b) XeF_4 (c) XeF_3^+ (d) $OXeF_4$

24. A pair of NO_2 molecules can combine to form N_2O_4.

$$2\ NO_2(g) \longrightarrow N_2O_4(g)$$

Use the Lewis structures of these molecules to explain why.

25. Which of the following ions or molecules have the same electron configurations as the N_2 molecule?
    (a) CO   (b) NO   (c) CN^- (d) NO^+ (e) NO^-

Exceptions to the Lewis Octet Rule

26. Which of the following are exceptions to the Lewis octet rule?
    (a) CO_2 (b) BeF_2 (c) SF_4 (d) SO_3

27. Which of the following are exceptions to the Lewis octet rule?
    (a) BF_3 (b) H_2CO (c) XeF_4 (d) IF_3

Lewis Structures of Molecules with Double or Triple Bonds

28. Which of the following molecules does not contain a double bond?
    (a) N_2 (b) CO_2 (c) C_2H_4 (d) NO_2 (e) SO_3

29. Which of the following contain only single bonds?
    (a) CN^- (b) NO^+ (c) CO   (d) O_2^{2-} (e) Cl_2CO

30. Ingestion of oxalic acid $(H_2C_2O_4)$, which can be found in a variety of vegetables and other plants, can produce nausea, vomiting, and diarrhea. When taken in excess, oxalic acid can be toxic. Draw the Lewis structure for the molecule. (Assume that the skeleton structure can be described as HO_2CCO_2H.)

Lewis Structures and Nonbonding Electrons

31. Determine the number of nonbonding pairs of electrons on the iodine atom in the following molecules or ions.
    (a) I_2 (b) I_3^- (c) IF_3 (d) ICl_4^-

32. One reason for writing Lewis structures is to determine which ions or molecules can act as bases. Bases invariably have one or more pairs of nonbonding electrons. Which of the following ions or molecules can act as a base?
    (a) OH^- (b) O_2 (c) CO_3^{2-} (d) Br^- (e) NH_3

33. Which of the following cannot act as a base? (See the previous problem.)
    (a) H_2S (b) NH_4^+ (c) AlH_3 (d) CH_3^- (e) NH_2^-

Resonance Hybrids

34. Draw all of the possible Lewis structures for the CO_3^{2-} ion. Determine the formal charge of all of the atoms in each structure.

35. Draw the three Lewis structures that may be used to describe the SCN^- ion.

36. Which of the following Lewis structures can contribute to the description of the electron structure of N_2O? Use formal charge to rationalize your choices.

    (a) $:N\equiv O-\ddot{N}:$ (b) $\ddot{N}=N=\ddot{O}:$ (c) $:\ddot{N}-N\equiv O:$ (d) $\ddot{N}=O=\ddot{N}$

    (e) $:N\equiv N-\ddot{O}:$

37. Which of the following molecules or ions have possible resonance hybrids?
    (a) HCO_2^- (b) PH_3 (c) HCN (d) C_2H_4

38. Which of the following molecules do not have a possible resonance hybrid?
    (a) CO_2 (b) C_2H_2 (c) $CHCl_3$ (d) SO_3

Electronegativity

39. Define electronegativity and AVEE. Why can electronegativity and AVEE be used interchangeably?

40. Use Figure 3.27 and trends in covalent and metallic radii to explain the trend in electronegativity to increase from left to right across a row in the periodic table.

41. Which of the following atoms is the most electronegative?
    (a) S   (b) As   (c) P   (d) Se
    (e) Cl   (f) Br

42. Which of the following series of atoms are arranged in order of decreasing electronegativity?
    (a) $C > Si > P > As > Se$ (b) $O > P > Al > Mg > K$
    (c) $Na > Li > B > N > F$ (d) $K > Mg > Be > O > N$
    (e) $Li > Be > B > C > N$

Partial Charge

43. Calculate the partial charge on the fluorine atom in the following molecules:
    (a) F_2 (b) HF (c) ClF

44. Calculate the partial charge on both atoms in the following molecules: IBr, ICl.

45. Calculate the partial charges on the two atoms in BF. What are the formal charges on B and F?

46. If ordinary purified table salt, NaCl, is heated to 801°C the solid melts to produce Na^+ and Cl^-. Continued heating causes evaporation of the molten salt. Some of the species that are found in the gas phase are sodium chloride molecules, $NaCl(g)$. Similar behavior is encountered for other solid salts when heated to sufficiently high temperatures. Calculate the partial charges on both atoms in the following gas phase molecules: LiCl, NaCl, KCl, RbCl, CsCl. Is the trend in partial charge what you would have expected on the basis of electronegativities?

Formal Charge

47. Calculate the formal charge on the sulfur atom in the following molecules or ions.
    (a) SO_2 (b) SO_3 (c) SO_3^{2-}

48. Calculate the formal charge on the bromine atom in the following molecules.
    (a) HBr (b) Br_2 (c) HOBr (d) BrF_5

49. Calculate the formal charge on the nitrogen atom in the following ions, molecules, or compounds.
    (a) NH_3 (b) NH_4^+ (c) N_2H_4 (d) NH_2^-

50. Calculate the formal charge on the nitrogen atom in the following molecules or ions.
    (a) N_2O (b) N_2O_3 (c) N_2O_5

51. Calculate the formal charge on the boron and nitrogen atoms in BF_3, NH_3, and the F_3B—NH_3 molecule formed when the two gases combine.

52. Calculate the formal charge on the two different sulfur atoms in the thiosulfate ion, $S_2O_3^{2-}$. Assume a skeleton structure that could be described as S—SO_3.

53. Draw two correct Lewis structures for ClNO and use formal charge to decide which is the best representation. What experimental evidence could be used to support your choice?

54. Draw two correct Lewis structures for PO_4^{3-} and decide on the basis of formal charge which is the best. What experimental evidence would help support your choice of structure?

55. Which would you expect to be more stable: ozone (O_3) or oxygen (O_2)? Use formal charge to support your answer.

56. Below are two skeleton structures for nitrous acid, HNO_2. Which is the best arrangement of atoms, on the basis of formal charge?

$$H—O—N—O \qquad \begin{array}{c} H—N—O \\ | \\ O \end{array}$$

Predicting the Shapes of Molecules

57. Predict the geometry around the central atom in the following molecules or ions.
    (a) PH_3 (b) GaH_3 (c) ICl_3 (d) XeF_3^+

58. Predict the geometry around the central atom in the following molecules or ions.
    (a) PO_4^{3-} (b) SO_4^{2-} (c) XeO_4 (d) MnO_4^-

59. The same elements often form compounds with very different shapes. Predict the geometry around the central atom in the following molecules or ions.
    (a) SnF_2 (b) SnF_3^- (c) SnF_4 (d) SnF_6^{2-}

60. Sulfur reacts with fluorine to form a pair of neutral molecules—SF_4 and SF_6—that in turn form positive and negative ions. Predict the geometry around the sulfur atom in each of the following.
    (a) SF_3^+ (b) SF_4 (c) SF_5^- (d) SF_6

61. Iodine and fluorine combine to form interhalogen compounds that can exist as either neutral molecules, positive ions, or negative ions. Predict the geometry around the iodine atom in each of the following.
(a) IF_2^- (b) IF_3 (c) IF_4^+ (d) IF_4^-

62. Predict the geometry around the central atom in each of the following oxides of nitrogen.
(a) N_2O (b) NO_2^- (c) NO_3^-

63. Predict the geometry around the central atom in the following molecules.
(a) $Hg(CH_3)_2$ (b) $Pb(CH_2CH_3)_4$

64. Which of the following compounds is best described as T shaped?
(a) XeF_3^+ (b) NO_3^- (c) NH_3 (d) ClO_3^- (e) SF_4

65. Which of the following molecules or ions have the same shape or geometry?
(a) NH_2^- and H_2O (b) NH_2^- and BeH_2 (c) H_2O and BeH_2
(d) NH_2^-, H_2O, and BeH_2

66. Which of the following molecules or ions have the same shape or geometry?
(a) SF_4 and CH_4 (b) CO_2 and H_2O (c) CO_2 and BeH_2
(d) N_2O and NO_2 (e) PCl_4^+ and PCl_4^-

67. Which of the following molecules are best described as bent, or angular?
(a) H_2S (b) CO_2 (c) $ClNO$ (d) NH_2^- (e) O_3

68. Which of the following molecules are planar?
(a) SO_3 (b) SO_3^{2-} (c) NO_3^- (d) PF_3 (e) BH_3

69. Which of the following molecules are tetrahedral?
(a) SiF_4 (b) CH_4 (c) NF_4^+ (d) BF_4^- (e) TeF_4

70. Which of the following molecules are linear?
(a) C_2H_2 (b) CO_2 (c) NO_2^- (d) NO_2^+ (e) H_2O

71. Which of the following elements would form a linear compound with the formula XO_2?
(a) Se (b) P (c) C (d) O (e) F

72. Explain why the nonbonding electrons occupy equatorial positions in ClF_3, not axial positions.

Polar Bonds versus Polar Molecules

73. Explain why CH_3Cl is a polar molecule but CCl_4 is not.

74. Use the Lewis structure of CO_2 to explain why the molecule has no dipole moment.

75. Which of the following molecules should be polar?
(a) CH_3OH (b) H_2O (c) CH_3OCH_3 (d) CH_3CO_2H

76. Explain why formaldehyde (H_2CO) is a polar molecule.

77. Explain why both thionyl chloride ($SOCl_2$) and sulfuryl chloride (SO_2Cl_2) are polar molecules.

Integrated Problems

78. Polymerization is the combining together of individual units to produce a long chain. Phosphate detergent additives and many biologically important molecules consist of chains of individual PO_3^- units.

Draw the Lewis structure of PO_3^-. Use formal charges to explain why individual PO_3^- units aggregate with themselves to produce long chains. Would you expect SO_3 to polymerize?

79. (a) The partial charge in molecules with single bonds containing more than two atoms can be estimated if each bond is taken into account, although the procedure is not strictly correct. For example, in H_2O the partial charge on O is estimated by considering the two O—H bonds:

$$\delta_O = 6 - 4 - 2\left(\frac{EN_O}{EN_H + EN_O}\right) - 2\left(\frac{EN_O}{EN_H + EN_O}\right) = -0.44$$

Thus, for water the partial charges are

$$
\begin{array}{c}
-0.44 \\
:O: \\
H \quad H \\
+0.22 \quad +0.22
\end{array}
$$

A more exact calculation gives a charge on oxygen that is less negative. Can you suggest why the simple calculation might have been anticipated to be in error?   (b) For small molecules, however, these estimates of partial charge are acceptable. Calculate the partial charges on all the atoms in HOF and F_2O. Do the values seem reasonable? Why?

80. Chemical formulas usually give a good idea of the skeleton structure of a compound, but some formulas are misleading. The formulas of common acids are often written as follows: HNO_3, H_2SO_4, H_3PO_4, $HClO_4$. Write Lewis structures for those four acids, assuming that their skeleton structures are more appropriately described by the formulas $HONO_2$, $(HO)_2SO_2$, $(HO)_3PO$, and $HOClO_3$.

81. Write Lewis structures for the following compounds and then describe why the compounds are unusual.
    (a) NO   (b) NO_2 (c) ClO_2 (d) ClO_3

82. Which element forms a compound with the following Lewis structure?

$$:\!O\!=\!X\!-\!\ddot{O}:$$

    (a) Al   (b) Si   (c) P   (d) S   (e) Cl

83. Which element forms a compound with the following Lewis structure?

$$:\!O\!=\!X\!-\!\ddot{C}l:$$

    (a) Be   (b) B   (c) C   (d) N   (e) O   (f) F

84. Which element forms a polyatomic ion with the formula XF_6^{2-} that has no nonbonding electrons in the valence shell of the central atom?
    (a) N   (b) C   (c) Si   (d) S   (e) P

85. In which group of the periodic table does element X belong if there are two pairs of nonbonding electrons in the valence shell of the central atom in the XF_4^- ion?

86. In which group of the periodic table does element X belong if the distribution of electrons in the valence shell of the central atom in the XCl_4^- ion is octahedral?

87. In which group does element X belong if the shape of the XF_2^- ion is linear?

88. The charged species OCN^- is well known in chemistry.
    (a) Write all possible Lewis structures for the species.
    (b) Which of the structures is best? Explain.
    (c) Typical experimental carbon to oxygen bond lengths are as follows:

$$\begin{array}{ccc} C-O & C=O & C\equiv O \\ 150 \text{ pm} & 133 \text{ pm} & 120 \text{ pm} \end{array}$$

What type of carbon/oxygen bond (single, double, triple, or somewhere in between) do you expect for OCN^-? Predict the carbon/oxygen bond length for OCN^-. Explain your answers. In addition to bond length, what other piece of experimental evidence would support your conclusion concerning the type of carbon/oxygen bond?
    (d) What is the O—C—N bond angle? Explain your answer.
    (e) Would you expect OCN^- to have a dipole moment? Explain.

89. Draw the Lewis structures and select the one species of each of the following pairs that has the largest dipole moment. Show your work and explain your answer. The central atom is underlined.
    (a) $\underline{C}S_2$, $\underline{S}O_2$
    (b) HC̲N, NC̲CN (C_2N_2)
    (c) G̲eF$_4$, S̲eF$_4$

90. The molecule thymine, one of four DNA bases, is known to have the following structure:

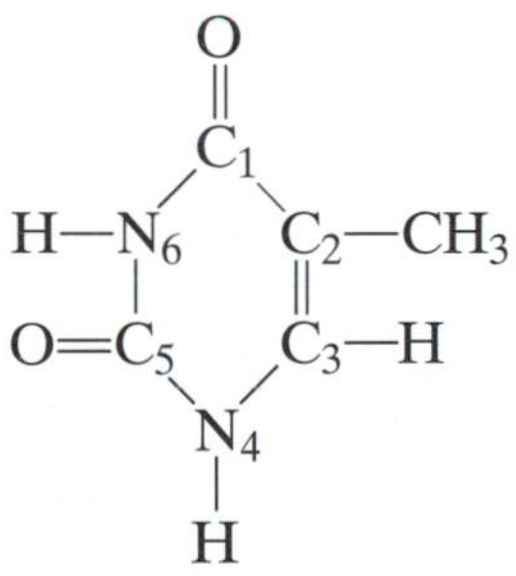

    (a) Add the missing electrons to the structure.
    (b) Give the bond angles around each numbered atom.

91. In Section 4.12 we discussed two possible structures for SO_2 and stated that without experimental evidence we could not distinguish between the two. Structure I contains one single bond and one double bond while structure II contains two double bonds. Explain why measurements of bond length cannot be used to distinguish between the two structures.

92. Each of the following Lewis structures is incorrect. Explain what is wrong with each one and give the correct Lewis structure.
    (a) 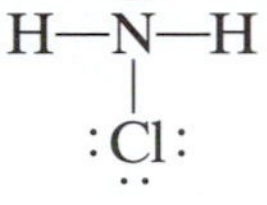 H—N—H
          |
        :Cl̈:

    (b) 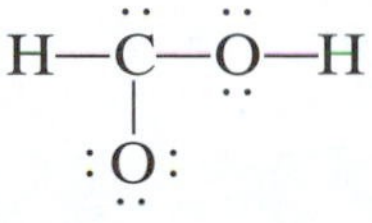 H—C̈—Ö—H
          |
        :Ö:

(c)
$$\left[\begin{array}{c} \ddot{:}\underset{}{\overset{\cdots}{Cl}}\ddot{:} \\ | \\ :\underset{\cdot\cdot}{\overset{\cdot\cdot}{Cl}} - I - \underset{\cdot\cdot}{\overset{\cdot\cdot}{Cl}}: \\ | \\ :\underset{\cdot\cdot}{\overset{\cdot\cdot}{Cl}}: \end{array} \right]^{-}$$

(d) $:\ddot{O} - H - \ddot{\underset{\cdot\cdot}{Cl}}:$

(e) $H = C = \ddot{N}:$

C H A P T E R
4
SPECIAL TOPICS

4A.1   THE SHAPES OF ORBITALS

Chemists describe the shell and subshell in which an orbital belongs with a two-character code, such as $2p$ or $4f$. The first character indicates the shell ($n = 2$ or $n = 4$). The second character identifies the subshell. This raises an interesting question: What is the difference between the s, p, and d orbitals that we use so often in writing the electron configuration of an element?

Consider the $3s$, $3p$, and $3d$ orbitals, for example. These orbitals are in the same shell, which means they have similar sizes. They differ, however, in shape. The s orbital has a spherical shape, the p orbital has a shape that can best be described as polar, and the typical d orbital has a cloverleaf shape, as shown in Figure 4A.1.

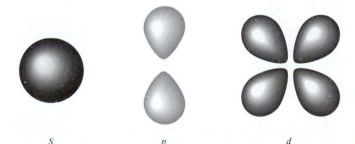

s p d **FIGURE 4A.1** Shapes of s, p, and d orbitals.

There is only one $3s$ orbital in an atom because there is only one way of orienting a sphere in space. There are three $3p$ orbitals because there are three directions in which a polar orbital can point. One of the $3p$ orbitals lies along the x axis, another along the y axis, and the third along the z axis of a coordinate system, as shown in Figure 4A.2. The orbitals are therefore known as the $3p_x$, $3p_y$, and $3p_z$ orbitals.

The five orbitals in the $3d$ subshell have the shapes shown in Figure 4A.3.

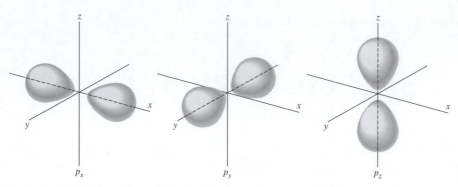

FIGURE 4A.2 Orientation of the *p* orbitals.

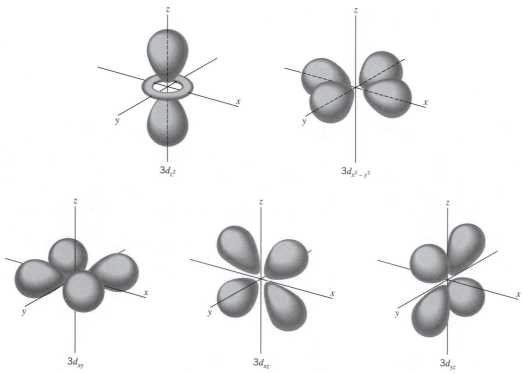

FIGURE 4A.3 Orientation of the *d* orbitals.

4A.2 VALENCE BOND THEORY

The octet rule and Lewis structures give us a simple view of covalent bonding, but they don't tell us why covalent bonds are formed or how electrons are shared between atoms. There are two models that provide a deeper understanding of the covalent bond: molecular orbital theory (see Section 4A.5) and valence bond theory.

Valence bond theory assumes that valence atomic orbitals on adjacent atoms can interact to form a new orbital that lies between the atoms. A covalent bond is formed when a pair of electrons with opposite spins is added to the new orbital. Consider, for example, the formation of an H_2 molecule from a pair of hydrogen atoms that each has an unpaired electron in an 1*s* orbital.

A similar approach can be used to explain the covalent bond in HF. Each atom has an orbital that has an unpaired electron.

H $\underset{1s^1}{\uparrow}$

F $\underset{1s^2}{\uparrow\downarrow}$ $\underset{2s^2}{\uparrow\downarrow}$ $\underset{2p^5}{\underline{\uparrow\downarrow \quad \uparrow\downarrow \quad \uparrow}}$

But now the bond is formed by the interaction between a $1s$ orbital on one atom and a $2p$ orbital on the other.

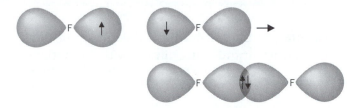

The valence bond model for the F_2 molecule is based on the head-to-head overlap between half-filled $2p$ orbitals on adjacent atoms.

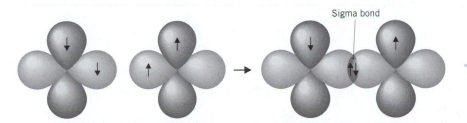

The covalent bonds in H_2, HF, and F_2 are called **sigma (σ) bonds** because they look like an s orbital when viewed along the bond. **Pi (π) bonds** are formed when orbitals interact to form a new orbital that looks like a p orbital when viewed along the bond.

The best way to differentiate between sigma and pi bonds is to consider the bonding in the O_2 molecule. The electron configuration of the oxygen atom tells us that there are two unpaired electrons on each atom.

O $\underset{1s^2}{\uparrow\downarrow}$ $\underset{2s^2}{\uparrow\downarrow}$ $\underset{2p^4}{\underline{\uparrow\downarrow \quad \uparrow \quad \uparrow}}$

Let's assume that one of the unpaired electrons is in the $2p_z$ orbital that points toward the neighboring atom. The interaction between the $2p_z$ orbitals on adjacent atoms leads to a sigma bond, just as it did in the F_2 molecule.

But this leaves us with an unpaired electron on each atom in one of the $2p$ orbitals perpendicular to the axis along which the sigma bond forms. The edge-on interaction between the half-filled orbitals leads to the formation of a bond that looks like a p orbital when viewed along the bond axis. In other words, it leads to a *pi bond*.

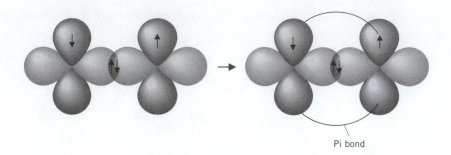

Pi bond

The combination of the sigma and pi bonds between the oxygen atoms suggests that the O_2 molecule is held together by a double bond, which is consistent with the Lewis structure for oxygen.

$$:\ddot{O}::\ddot{O}:$$

The analysis also suggests, however, that the bond between the atoms is not necessarily twice as strong as an O—O single bond.

Checkpoint
Draw the Lewis structure and the valence bond description of the bonding in N_2. Comment on any similarities or differences in the two descriptions.

4A.3   HYBRID ATOMIC ORBITALS

The simple version of the valence bond theory introduced in the previous section gives a satisfactory picture of the bonding in simple diatomic molecules such as H_2, HF, F_2, O_2, and N_2. But it gives less satisfactory results when applied to slightly more complex molecules, such as CH_4 and H_2O.

The electron configuration of carbon has only two unpaired electrons, and therefore carbon should form only two bonds.

$$C \quad \underset{1s^2}{\uparrow\downarrow} \quad \underset{2s^2}{\uparrow\downarrow} \quad \underset{2p^2}{\uparrow \quad \uparrow \quad \underline{}}$$

The fact that carbon atoms form four bonds can be explained by noting that it takes relatively little energy to excite one of the electrons from the filled $2s$ orbital into one of the empty $2p$ orbitals to give four orbitals that each contain an unpaired electron.

$$C^* \quad \underset{1s^2}{\uparrow\downarrow} \quad \underset{2s^1}{\uparrow} \quad \underset{2p^3}{\uparrow \quad \uparrow \quad \uparrow}$$

The energy invested in moving the electron is more than repaid by the energy released when two more bonds are formed.

A more serious problem arises when the valence bond theory introduced in the previous section is used to predict the geometry around an atom in a molecule. Consider water, for example. Combining a pair of hydrogen atoms with unpaired electrons in a $1s$ orbital with the two unpaired electrons on an oxygen atom ($2p$ orbitals) gives the correct number of bonds.

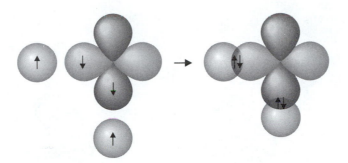

In this case, however, we predict that the H—O—H bond angle would be 90° because that is the angle between the $2p$ orbitals on the oxygen atom. The experimental angle is 105°.

The problem was solved by assuming that the valence atomic orbitals on an individual atom can be combined to form **hybrid atomic orbitals.** The geometry of a linear BeF_2 molecule can be explained, for example, by mixing the $2s$ orbital on the beryllium atom with one of the $2p$ orbitals to form a set of sp hybrid orbitals that point in opposite directions, as shown in Figure 4A.4. One of the valence electrons on the beryllium atom is then placed in each of the orbitals, and the orbitals are allowed to overlap with half-filled $2p$ orbitals on a pair of fluorine atoms to form a linear BeF_2 molecule.

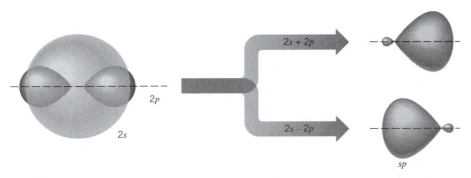

FIGURE 4A.4 The sp hybrid orbitals used by the beryllium atom in the linear BeF_2 molecule are formed by combining the wave functions for the $2s$ and $2p$ orbitals on that atom. When the wave functions are added we get an sp orbital that points in one direction. When one of the wave functions is subtracted from the other we get an sp orbital that points in the opposite direction.

The geometry of trigonal planar molecules such as BF_3 can be explained by mixing a $2s$ orbital with a pair of $2p$ orbitals on the central atom to form three sp^2 hybrid orbitals that point toward the corners of an equilateral triangle. Molecules whose geometries are based on a tetrahedron, such as CH_4 and H_2O, can be understood by mixing a $2s$ orbital with all three $2p$ orbitals to obtain a set of four sp^3 orbitals that are oriented toward the corners of a tetrahedron.

The hybrid atomic orbital model can be extended to molecules whose shapes are based on trigonal bipyramidal or octahedral distributions of electrons by including valence shell d orbitals. When one of the $3d$ orbitals is mixed with the $3s$ and the three $3p$ orbitals on an atom, the resulting sp^3d hybrid orbitals point toward the corners of a trigonal bipyramid. When two of the $3d$ orbitals are mixed with the $3s$ and $3p$ orbitals, the result is a set of six sp^3d^2 hybrid orbitals that point toward the corners of an octahedron.

The geometries of the five different sets of hybrid atomic orbitals (sp, sp^2, sp^3, sp^3d, and sp^3d^2) are shown in Figure 4A.5.

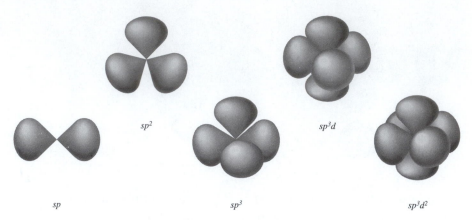

FIGURE 4A.5 Shapes of the sp, sp^2, sp^3, sp^3d, and sp^3d^2 hybrid orbitals.

The relationship between hybridization and the distribution of electrons in the valence shell of an atom is summarized in Table 4A.1.

TABLE 4A.1   Relationship between the Distribution of Electron Domains on an Atom and the Hybridization of an Atom

Electron Domains	Distribution of Electron Domains	Hybridization	Examples
2	Linear	sp	BeF_2, CO_2, N_2
3	Trigonal planar	sp^2	BF_3, CO_3^{2-}, O_3
4	Tetrahedral	sp^3	CH_4, SO_4^{2-}, NH_3, H_2O
5	Trigonal bipyramidal	sp^3d	PF_5, ClF_3
6	Octahedral	sp^3d^2	SF_6, XeF_4

Exercise 4A.1

Determine the hybridization of the central atom in O_2, H_3O^+, $TeCl_4$, and ICl_4^- from the following Lewis structures

(a) O_2 (b) H_3O^+ (c) $TeCl_4$ (d) ICl_4^-

Solution

(a) The Lewis structure for the O_2 molecule suggests that the distribution of electron domains around each oxygen atom should be trigonal planar. The hybridization of the oxygen atoms in the O_2 molecule is therefore assumed to be sp^2.

(b) The Lewis structure suggests a tetrahedral distribution of electron domains around the oxygen atom in the ion. The oxygen atom in H_3O^+ is therefore assumed to be sp^3 hybridized.

(c) The Lewis structure suggests a trigonal bipyramidal distribution of domains in the valence shell of the tellurium atom. The tellurium atom therefore forms an sp^3d hybrid.

(d) The Lewis structure suggests an octahedral distribution of domains in the valence shell of the iodine atom. The iodine atom is therefore sp^3d^2 hybridized.

4A.4   MOLECULES WITH DOUBLE AND TRIPLE BONDS

The hybrid atomic orbital model can also be used to explain the formation of double and triple bonds. Let's consider the bonding in formaldehyde (H_2CO), for example, which has the following Lewis structure.

$$
\begin{array}{c}
\ddot{\text{O}} \\
\| \\
\text{C} \\
\diagup \quad \diagdown \\
\text{H} \qquad \text{H}
\end{array}
$$

There are three places where electrons can be found in the valence shell of both the carbon and oxygen atoms in the molecule. As a result, the electron domain theory predicts that the valence electrons on the atom will be oriented toward the corners of an equilateral triangle. The carbon atom is therefore assumed to be sp^2 hybridized.

When we create a set of sp^2 hybrid orbitals, we combine the $2s$ and two of the $2p$ orbitals on the atom. One of the valence electrons on the carbon atom is placed in each of the three sp^2 hybrid orbitals. The fourth valence electron is placed in the $2p$ orbital that wasn't used during hybridization.

There are six valence electrons on a neutral oxygen atom. A pair of electrons is placed in each of two of the sp^2 hybrid orbitals. One electron is then placed in the sp^2 hybrid orbital that points toward the carbon atom, and another is placed in the unhybridized $2p$ orbital.

The two C—H bonds are formed when the unpaired electron in one of the sp^2 hybrid orbitals on carbon interacts with a $1s$ electron on a hydrogen atom, as shown in Figure 4A.6. A C—O bond is formed when the electron in the third sp^2 hybrid orbital on carbon interacts with the unpaired electron in the sp^2 hybrid orbital on the oxygen atom. These bonds are sigma bonds.

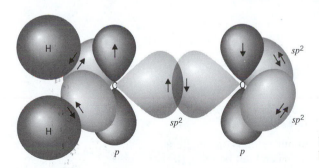

FIGURE 4A.6 Hybrid atomic orbital picture of bonding in formaldehyde.

The electron in the unhybridized $2p$ orbital on the carbon atom then interacts with the electron in the unhybridized $2p$ orbital on the oxygen atom to form a second covalent bond between these atoms. This is called a pi bond.

When double bonds were first introduced, we noted that they occur most often in compounds that contain C, N, O, P, or S. There are two reasons for this. First, double bonds by their very nature are covalent bonds. They are therefore most likely to be found among the elements that form covalent compounds. Second, the interaction between p orbitals to form a π bond requires that the atoms come relatively close together, and thus π bonds tend to be the strongest for atoms that are relatively small.

4A.5   MOLECULAR ORBITAL THEORY

Bonding theory based on atomic orbitals focuses on the bonds formed between valence electrons on atoms, and is called *valence bond* theory.

The valence bond model of SO_2 can't adequately explain the fact that the molecule contains two equivalent bonds with a bond order between that of an S—O single bond and an S=O double bond. The best it can do is suggest that SO_2 is a mixture, or hybrid, of the two Lewis structures that can be written for the molecule.

This problem, and many others, can be overcome by using a more sophisticated model of bonding based on **molecular orbitals.** Molecular orbital theory is more powerful than valence bond theory because the orbitals reflect the geometry of the molecule to which they are applied. But this power carries a significant cost in terms of the ease with which the model can be visualized. We can understand the basics of molecular orbital theory by constructing the molecular orbitals for the simplest possible molecules: homonuclear diatomic molecules such as H_2, O_2, and N_2.

Molecular orbitals are obtained by combining the atomic orbitals on the atoms to give orbitals that are characteristic of the molecule, not the individual atoms. Consider the H_2 molecule, for example. One of the molecular orbitals in the H_2 molecule is constructed by adding the mathematical functions for the two $1s$ atomic orbitals that come together to form the molecule. Another orbital is formed by subtracting one of these functions from the other, as shown in Figure 4A.7.

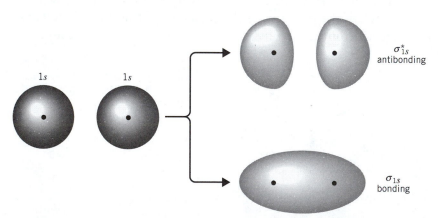

FIGURE 4A.7 The interaction between a pair of $1s$ atomic orbitals on neighboring hydrogen atoms leads to the formation of bonding (σ) and antibonding (σ^*) molecular orbitals.

One of the orbitals is called a **bonding molecular orbital** because electrons in the orbital spend most of their time in the region directly between the two nuclei. It is called a *sigma* (σ) molecular orbital because it looks like an s orbital when viewed along the H—H bond. Electrons placed in the other orbital spend most of their time away from the region between the two nuclei. That orbital is therefore an **antibonding,** or *sigma star* (σ^*), molecular orbital.

The σ bonding molecular orbital concentrates electrons in the region directly between the two nuclei. Placing an electron in that orbital therefore stabilizes the H_2 molecule. Since the σ^* antibonding molecular orbital forces the electron to spend most of its time away from the area between the nuclei, placing an electron in that orbital makes the molecule less stable.

Electrons are added to molecular orbitals, one at a time, starting with the lowest energy molecular orbital. The two electrons associated with a pair of hydrogen atoms are placed in the lowest energy, or σ bonding, molecular orbital shown in Figure 4A.8. When this is done, the energy of an H_2 molecule is lower than that of a pair of isolated atoms. The H_2 molecule is therefore more stable than a pair of isolated atoms.

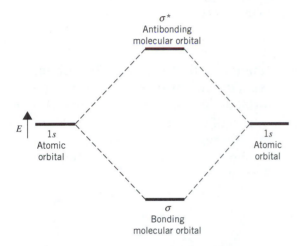

FIGURE 4A.8 Relative energies of the $1s$ atomic orbitals on a pair of isolated hydrogen atoms and the σ and σ^* molecular orbitals they form when they interact.

The molecular orbital model can be used to explain why He_2 molecules don't exist. Combining a pair of helium atoms with $1s^2$ electron configurations would produce a molecule with pairs of electrons in both the σ bonding and the σ^* antibonding molecular orbitals. The total energy of an He_2 molecule would be essentially the same as the energy of a pair of isolated helium atoms, and there would be nothing to hold the He_2 molecule together.

The fact that a diatomic He_2 molecule is neither more nor less stable than a pair of isolated helium atoms illustrates an important principle: The core orbitals on an atom make no contribution to the stability of the molecules that contain the atom. The only molecular orbitals that are important are those formed when valence shell orbitals are combined. The molecular orbital diagram for an O_2 molecule would therefore ignore the $1s$ electrons on both oxygen atoms and concentrate on the interactions between the $2s$ and $2p$ valence orbitals.

The $2s$ orbitals on one oxygen atom combine with the $2s$ orbitals on another to form a σ_{2s} bonding and a σ_{2s}^* antibonding molecular orbital, just like the σ_{1s} and σ_{1s}^* orbitals formed from the $1s$ atomic orbitals. If we arbitrarily define the z axis of our coordinate system as the axis between the two atoms in an O_2 molecule, the $2p_z$ orbitals on the atoms in the molecule point directly toward each other. When those orbitals interact, they meet head-on to form a σ_{2p} bonding and a σ_{2p}^* antibonding molecular orbital, as shown in Figure 4A.9.

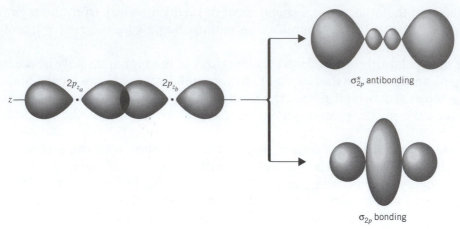

FIGURE 4A.9 When a pair of $2p$ orbitals are combined so that they meet head-on, bonding (σ_{2p}) and antibonding (σ_{2p}*) molecular orbitals are formed. These are called σ orbitals because they concentrate the electrons along the axis between the two nuclei, just like the σ_{1s} and σ_{2s} orbitals.

The $2p_x$ and $2p_y$ orbitals on one oxygen atom interact with the $2p_x$ and $2p_y$ orbitals on the other atom to form molecular orbitals that have a different shape, as shown in Figure 4A.10. These molecular orbitals are called *pi* (π) orbitals. Whereas σ and σ* orbitals concentrate the electrons along the axis on which the nuclei of the atoms lie, π and π* orbitals concentrate the electrons above and below the axis. Because there are two $2p$ atomic orbitals on each atom that meet edge-on we get two π orbitals and two π* orbitals. These are called the π_x and π_y and the π_x* and π_y* molecular orbitals.

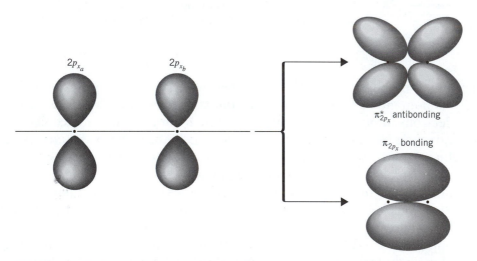

FIGURE 4A.10 When a pair of $2p_x$ orbitals are combined so that they meet edge-on, bonding (π_x) and antibonding (π_x*) molecular orbitals are formed. A similar set of bonding (π_y) and antibonding (π_y*) orbitals is formed when a pair of $2p_y$ atomic orbitals are combined.

The interaction of four valence atomic orbitals on one atom with a set of four atomic orbitals on another atom therefore leads to the formation of a total of eight molecular orbitals: σ_{2s}, σ_{2s}*, σ_{2p}, σ_{2p},* π_x, π_y, π_x*, and π_y*.

There is a difference between the energies of the $2s$ and $2p$ orbitals on an atom. As a result, the σ_{2s} and σ_{2s}* orbitals both lie at lower energies than the σ_{2p}, σ_{2p}*, π_x, π_y, π_x*, and π_y*. To sort out the relative energies of the six molecular orbitals formed when the $2p$

atomic orbitals on a pair of atoms are combined, we need to understand the relationship between the strength of the interaction between a pair of orbitals and the relative energies of the molecular orbitals they form.

Because the $2p_z$ orbitals meet head-on, the interaction between them is stronger than the interactions between the $2p_x$ or $2p_y$ orbitals, which meet edge-on. As a result, the σ_{2p} orbital lies at a lower energy than the π_x and π_y orbitals, and the $\sigma_{2p}{}^*$ orbital lies at higher energy than the $\pi_x{}^*$ and $\pi_y{}^*$ orbitals, as shown in Figure 4A.11.

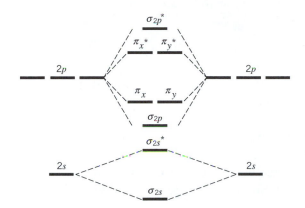

FIGURE 4A.11 The interaction between $2p_z$ atomic orbitals that meet head-on is stronger than the interaction between $2p_x$ or $2p_y$ orbitals that meet edge-on. As a result, the difference between the energies of the σ_{2p} and $\sigma_{2p}{}^*$ orbitals is larger than the difference between the π_x and $\pi_x{}^*$ or π_y and $\pi_y{}^*$ orbitals.

Unfortunately an interaction is missing from the model. It is possible for the $2s$ orbital on one atom to interact with the $2p_z$ orbital on the other. This interaction introduces a slight change in the relative energies of the molecular orbitals, to give the diagram shown in Figure 4A.12. Experiments have shown that O_2 and F_2 are best described by the model in Figure 4A.11, but B_2, C_2, and N_2 are best described by the model shown in Figure 4A.12.

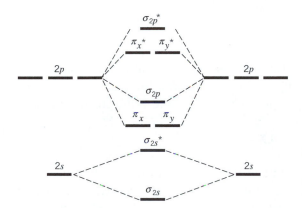

FIGURE 4A.12 When hybridization is added to the molecular orbital theory, there is a slight change in the energies of the orbitals formed by overlap of the $2p$ atomic orbitals on neighboring atoms.

Exercise 4A.2

Construct a molecular orbital diagram for the O_2 molecule.

Solution

There are six valence electrons on a neutral oxygen atom and therefore 12 valence electrons in an O_2 molecule. These electrons are added to the diagram in Figure 4A.11, one at a time, starting with the lowest energy molecular orbital.

Because Hund's rules apply to the filling of molecular orbitals, molecular orbital theory predicts that there should be two unpaired electrons on this molecule—one electron each in the $\pi_x{}^*$ and $\pi_y{}^*$ orbitals.

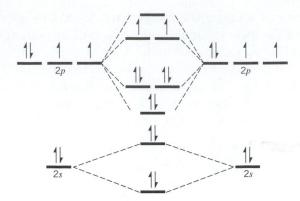

When writing the electron configuration of an atom, we usually list the orbitals in the order in which they fill.

$$Pb \qquad [Xe]\ 6s^2\ 4f^{14}\ 5d^{10}\ 6p^2$$

We can write the electron configuration of a molecule by doing the same thing. Concentrating only on the valence orbitals, we write the electron configuration of O_2 as follows.

$$O_2 \qquad \sigma_{2s}^2\ \sigma_{2s}^{*2}\ \sigma_{2p}^2\ \pi_x^2\ \pi_y^2\ \pi_x^{*1}\ \pi_y^{*1}$$

The number of bonds between a pair of atoms is called the **bond order.** Bond orders can be calculated from Lewis structures, which are the heart of the valence bond model. Oxygen, for example, has a bond order of two.

$$\ddot{O}{=}\ddot{O}$$

When there is more than one Lewis structure for a molecule, the bond order is an average of these structures. The bond order in sulfur dioxide, for example, is 1.5—the average of an S—O single bond in one Lewis structure and an S=O double bond in the other.

$$:\!\ddot{O}{-}\ddot{S}{=}\ddot{O}: \longleftrightarrow :\!\ddot{O}{=}\ddot{S}{-}\ddot{O}:$$

In molecular orbital theory, we calculate bond orders by assuming that two electrons in a bonding molecular orbital contribute one net bond and that two electrons in an antibonding molecular orbital cancel the effect of one bond. We can calculate the bond order in the O_2 molecule by noting that there are eight valence electrons in bonding molecular orbitals and four valence electrons in antibonding molecular orbitals in the electron configuration of the molecule given in Exercise 4A.2. Thus, the bond order is two.

$$\text{Bond order} = \frac{\text{bonding electrons} - \text{antibonding electrons}}{2} = \frac{8 - 4}{2} = 2$$

Although the Lewis structure and molecular orbital models of oxygen yield the same bond order, there is an important difference between the models. The electrons in the Lewis structure are all paired, but there are two unpaired electrons in the molecular orbital description of the molecule. As a result, we can test the predictions of the theories by studying the effect of a magnetic field on oxygen.

Atoms or molecules in which the electrons are paired are **diamagnetic,** that is, they are repelled by both poles of a magnet. Those that have one or more unpaired electrons are **paramagnetic,** or attracted to a magnetic field. Liquid oxygen is attracted to a magnetic field and can actually bridge the gap between the poles of a horseshoe magnet. The molecular orbital model of O_2 is therefore superior to the valence bond model which cannot explain the paramagnetism of oxygen.

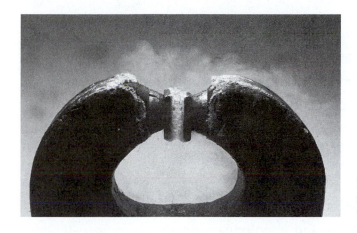

Liquid O_2 is so strongly attracted to a magnetic field that it will bridge the gap between the poles of a horseshoe magnet.

Exercise 4A.3

Use the molecular orbital diagram in Figure 4A.12 to calculate the bond order in nitrogen oxide (NO). Compare the results of this calculation with the bond order obtained from the Lewis structure.

Solution

There are 11 valence electrons in the NO molecule, and there are two possible Lewis structures, depending on whether we locate the unpaired electron on the nitrogen atom or on the oxygen atom.

$$\ddot{N}{=}\ddot{O}{\cdot} \longleftrightarrow {\cdot}\ddot{N}{=}\ddot{O}$$

Both Lewis structures contain an N=O double bond, however, so the bond order in the valence bond model for NO is 2.

To calculate the bond order in the molecular orbital model of NO, we have to predict the electron configuration of the molecule by adding electrons, one at a time, to the molecular orbitals in Figure 4A.12.

$$\text{NO} \qquad \sigma_{2s}^{2}\ \sigma_{2s}^{*2}\ \pi_{x}^{2}\ \pi_{y}^{2}\ \sigma_{2p}^{2}\ \pi_{x}^{*1}$$

There are eight valence electrons in bonding molecular orbitals and three valence electrons in antibonding molecular orbitals in the compound. The bond order in the molecular orbital model of NO is therefore 2.5.

$$\text{Bond order} = \frac{\text{bonding electrons} - \text{antibonding electrons}}{2} = \frac{8 - 3}{2} = 2.5$$

The strength of the covalent bond in NO has been shown by experiment to be significantly stronger than a typical N=O double bond, in agreement with the predictions of molecular orbital theory for the molecule.

PROBLEMS

Hybrid Atomic Orbitals

4A.1. Determine the hybridization of the central atom in the following molecules or ions.
(a) NH_3 (b) NH_4^+ (c) NO   (d) NO_2 (e) NO_2^+

4A.2. Determine the hybridization of the central atom in the following molecules or ions.
(a) CH_4 (b) H_2CO (c) HCO_2^-

4A.3. Determine the hybridization of the central atom in the following molecules or ions.
(a) SF_4 (b) BrO_3^- (c) XeF_3^+ (d) Cl_2CO

A Hybrid Atomic Orbital Description of Multiple Bonds

4A.4. Write the Lewis structures for molecules of ethylene, C_2H_4, and acetylene, C_2H_2. Use hybrid atomic orbitals to describe the bonding in the compounds.

4A.5. Write the Lewis structures for carbon monoxide and carbon dioxide. Use hybrid atomic orbitals to describe the bonding in the compounds.

Molecular Orbital Theory for a Diatomic Molecule

4A.6. Describe the difference between σ and π molecular orbitals.

4A.7. Describe the difference between bonding and antibonding molecular orbitals.

4A.8. Explain why the difference between the energies of the σ_{2p} bonding and σ_{2p}^* antibonding orbitals is larger than the difference between the π_x or π_y bonding and π_x^* or π_y^* antibonding orbitals.

4A.9. Describe the molecular orbitals formed by the overlap of the following atomic orbitals. (Assume that the bond lies along the z axis of the coordinate system.)
(a) $2s + 2s$ (b) $2p_x + 2p_y$ (c) $2p_y + 2p_y$ (d) $2p_z + 2p_z$ (e) $2s + 2p_z$

4A.10. Write the electron configuration for the following diatomic molecules and calculate the bond order in each molecule.
(a) H_2 (b) C_2 (c) N_2 (d) O_2 (e) F_2

4A.11. Use molecular orbital theory to predict whether the H_2^+, H_2^-, and H_2^{2-} ions should be more stable or less stable than a neutral H_2 molecule.

4A.12. Use molecular orbital theory to explain why the oxygen–oxygen bond is stronger in the O_2 molecule than in the O_2^{2-} ion.

4A.13. Use molecular orbital theory to predict whether the bond order in the superoxide ion, O_2^-, should be higher or lower than the bond order in a neutral O_2 molecule.

4A.14. Use molecular orbital theory to predict whether the peroxide ion, O_2^{2-}, should be paramagnetic.

4A.15. Write the electron configuration for the following diatomic molecules. Calculate the bond order in each molecule.
(a) HF   (b) CO   (c) CN^- (d) ClO^- (e) NO^+

4A.16. Classify the following molecules as paramagnetic or diamagnetic.
(a) HF   (b) CO   (c) CN^- (d) NO   (e) NO^+

CHAPTER
5
IONIC AND METALLIC BONDS

5.1   THE ACTIVE METALS

As we have seen, there are no abrupt changes in the physical properties of the elements as we go across a row of the periodic table or down a column. As a result, the change from metal to nonmetal must be gradual. Instead of arbitrarily dividing elements into metals and nonmetals, it is better to describe some elements as being more metallic and other elements as being more nonmetallic. Recall the trends in average valence electron energy (AVEE) discussed at the end of Chapter 3 that allowed us to differentiate metals from nonmetals. Metals have low AVEE values which indicates that valence electrons are relatively easy to remove and that valence subshells are close in energy. Nonmetals have high AVEE values which means that valence electrons are more difficult to remove and that the subshells are widely spaced in energy. Elements become less metallic and more nonmetallic as we go across a row of the periodic table from left to right.

Metallic character decreases $\longrightarrow$

Na Mg Al Si P S Cl Ar

Nonmetallic character increases $\longrightarrow$

Because AVEE values decrease as we go down a column of the Periodic Table, the metallic character of an element increases.

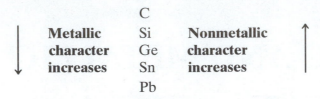

Checkpoint
Arrange the following metals in increasing order of metallic character: K, Ca, and Rb.

The primary difference between the various metals in the periodic table is the ease with which they undergo chemical reactions. The elements toward the bottom left corner of the periodic table, which have the most metallic character, are the metals that are the most **active** in the sense of being the most **reactive** (see Figure 5.1). Lithium, sodium, and potassium all react with water, for example. As we go down the column the elements react more vigorously because they become more active as they become more metallic.

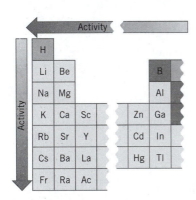

FIGURE 5.1 The main-group metals in the bottom left-hand corner of the periodic table are the most active metals.

Metals are often divided into four classes on the basis of their activity, as shown in Table 5.1. The most active metals are so reactive that they readily combine with the O_2 and H_2O in the atmosphere and must therefore be stored under an inert liquid, such as mineral oil. Those metals are found exclusively in Groups IA and IIA of the periodic table.

TABLE 5.1   Common Metals Divided into Classes on the Basis of Activity

Class I Metals: The Active Metals
    Li, Na, K, Rb, Cs   (Group IA)
    Ca, Sr, Ba   (Group IIA)

Class II Metals: The Less Active Metals
            Mg, Al, Zn, Mn

Class III Metals: The Structural Metals
            Cr, Fe, Sn, Pb, Cu

Class IV Metals: The Coinage Metals
            Ag, Au, Pt

Metals in the second class are slightly less active than Class I metals. They don't react with water at room temperature, but they react rapidly with acids. The less active metals can be used for a variety of purposes. Aluminum, for example, is used for aluminum foil and beverage cans. It is important to protect these metals from exposure to acids, however, so aluminum cans are lined with a plastic coating to prevent contact with the acidic soft drinks or fruit juices they contain.

Aluminum metal is more "active" than iron. So much energy is given off when aluminum metal reacts with Fe_2O_3 to give iron metal and Al_2O_3 that this reaction is called the "thermite" reaction.

The third class contains metals such as iron, tin, and lead that react only with strong acids. Some cereals that are described as "iron-enriched" contain tiny pieces of iron metal, not Fe^{3+} salts, as the source of the iron they provide because the iron metal dissolves in the strongly acidic stomach fluid to form Fe^{3+} ions.

Metals in the fourth class are so unreactive they are essentially inert at room temperature. These metals are ideal for making jewelry and coins because they do not react with the majority of the substances with which they come into daily contact. As a result, they are often called the coinage metals.

5.2 MAIN-GROUP METALS AND THEIR IONS

Group IA: The Alkali Metals

We begin our discussion of the formation of ions by examining three separate areas of the periodic table: main-group metals, main-group nonmetals, and transition metals. The **main-group elements** in the periodic table are made up of the elements in Groups IA through VIIA. Chapter 3 introduced you to electron configuration and its relationship to the periodic chart. We will use electron configuration to explain the formation of ions. This will allow us to use the periodic chart as a tool to predict the most stable ion(s) formed by the atoms of an element.

The metals in Group IA include lithium (Li), sodium (Na), potassium (K), rubidium (Rb), cesium (Cs), and francium (Fr). These elements are called the **alkali metals** because they form compounds, such as NaOH, that were once known as *alkalies*. Sodium and potassium are relatively common elements. In fact, they are among the eight most abundant elements in the earth's crust, as shown in Table 5.2. The other alkali metals are much less abundant. Discussions of the chemistry of alkali metals therefore focus on sodium and potassium.

TABLE 5.2   Percent by Weight of the Most Common Elements in the Earth's Crust

Element	Percent by Weight
O	46.60
Si	27.72
Al	8.13
Fe	5.00
Ca	3.63
Na	**2.83**
K	**2.59**
Mg	2.09

A sodium atom has the electron configuration

$$1s^2\, 2s^2\, 2p^6\, 3s^1 \qquad \text{or} \qquad [\text{Ne}]\, 3s^1$$

When forming an ion a sodium atom will lose its $3s$ electron to form a sodium ion with an electron configuration of $1s^2\, 2s^2\, 2p^6$. This is the same electron configuration as Ne. A neon atom and a sodium ion are said to be **isoelectronic** because they have the same electron configuration. Only one electron is lost from sodium during the formation of its ion because the electrons in the second shell are closer to the nucleus and therefore more strongly attracted to the nucleus than the $3s$ electron. The electron configurations of the alkali metals are characterized by a single valence electron.

Li	$[\text{He}]\, 2s^1$	Rb	$[\text{Kr}]\, 5s^1$
Na	$[\text{Ne}]\, 3s^1$	Cs	$[\text{Xe}]\, 6s^1$
K	$[\text{Ar}]\, 4s^1$	Fr	$[\text{Rn}]\, 7s^1$

Chapter 3 discusses how the electron configuration of an atom is related to the atom's position in the periodic table. Group IA elements all have the same valence shell electron configuration (xs^1) and are also characterized by unusually small AVEE values. This means that it is easier to remove an electron from these elements than from almost any other elements in the periodic table. As a result, the chemistry of these elements is dominated by their tendency to lose an electron to form positively charged ions (Li^+, Na^+, K^+). The periodic chart can therefore be used to predict a $+1$ charge on the ions formed by Group IA elements.

Group IIA: The Alkaline Earth Metals

The elements in Group IIA (Be, Mg, Ca, Sr, Ba, and Ra) are all metals, and all but Be and Mg are active metals. These elements are often called the **alkaline earth metals.** The term *alkaline* reflects the fact that many compounds of these metals can neutralize acids. The term *earth* was historically used to describe the fact that many of those compounds are insoluble in water.

Most of the chemistry of the alkaline earth metals (Group IIA) can be predicted from the behavior of the alkali metals (Group IA). Once again, these elements are characterized by relatively small average valence electron energies. As a result, the chemistry of these elements is also dominated by their tendency to form positive ions. In this case, however, each element has two valence electrons, which means that they tend to form M^{2+} ions.

$$\text{Mg} \quad [\text{Ne}]\, 3s^2 \qquad\qquad \text{Mg}^{2+} \quad [\text{Ne}]$$

The alkaline earth metals in Group IIA are significantly less reactive than the alkali metals in Group IA. They are also harder and have higher melting points than the Group IA metals.

Checkpoint

Why do atoms of Group IIA tend to lose two electrons while those of Group IA lose only one?

Group IIIA Metals

Group IIIA contains a semimetal (B) and four metals (Al, Ga, In, and Tl). Discussions of the chemistry of the Group IIIA metals often focus on aluminum because gallium, indium, and thallium are very scarce and therefore of limited interest. Aluminum, on the other hand, is the third most abundant element in the earth's crust (see Table 5.2).

The average valence electron energy for aluminum is only slightly larger than that of the metals in Groups IA and IIA. Thus, the chemistry of aluminum is dominated by the tendency of its atoms to lose electrons to form positive ions. Because the electron configuration of aluminum contains three valence electrons, this metal forms the Al^{3+} ion in many if not most of its compounds.

$$\text{Al} \quad [\text{Ne}]\, 3s^2\, 3p^1 \qquad\qquad \text{Al}^{3+} \quad [\text{Ne}]$$

5.3 MAIN-GROUP NONMETALS AND THEIR IONS

In Chapter 3, we noted that the average valence electron energy of the elements increases as we go from left to right across a row of the periodic table. As we go across a row we therefore reach the point at which it takes too much energy to remove enough electrons from a neutral atom to form an ion with a noble gas electron configuration. The atoms of the main-group elements toward the right-hand side of the periodic table therefore tend to gain electrons to form negatively charged ions with one of the noble gas electron configurations. Ionic compounds, or salts, that contain both positive and negative ions are therefore often formed when main-group elements at opposite sides of the periodic table combine.

The alkali metals react with the nonmetals in Group VIIA (F_2, Cl_2, Br_2, I_2, and At_2) to form ionic compounds called salts. Chlorine, for example, reacts with sodium metal to produce sodium chloride (table salt).

$$2\,\text{Na}(s) + \text{Cl}_2(g) \longrightarrow 2\,\text{NaCl}(s)$$

Because they form salts with so many metals, the elements in Group VIIA are known as the **halogens.** The name comes from the Greek word for salt (*hals*) and the Greek word meaning "to produce" (*gennan*). The salts formed by the halogens are called **halides.** These salts include **fluorides** (LiF), **chlorides** (NaCl), **bromides** (KBr), and **iodides** (NaI). The halide ions all carry a charge of -1. The halogens need only one more electron to achieve a filled-shell electron configuration.

$$\text{Cl} \quad [\text{Ne}] \, 3s^2 \, 3p^5 \qquad\qquad \text{Cl}^- \quad [\text{Ne}] \, 3s^2 \, 3p^6 = [\text{Ar}]$$

The halogens exist in their free elemental forms as diatomic molecules (F_2, Cl_2, Br_2, I_2, and At_2). However, no halogen is found in nature in its elemental form. When halogens are found as part of ionic compounds they have a -1 charge. Fluoride ions are found in minerals such as fluorite (CaF_2) and cryolite (Na_3AlF_6). Chloride ions are found in rock salt ($NaCl$), in oceans (which are $\sim 2\%$ Cl^- by weight), and in lakes that have a high salt content, such as the Great Salt Lake in Utah. Both bromide and iodide ions are found at low concentrations in the oceans, as well as in brine wells.

Hydrogen is the only element in the periodic table that is listed in more than one group. In most periodic tables it can be found among the metals in Group IA and the nonmetals in Group VIIA. This can be understood by looking at the position of hydrogen in the three-dimensional version of the periodic table shown in Figure 4.9. Hydrogen is almost exactly in the middle of the elements in terms of its electronegativity. When it reacts with an element that is less electronegative, it tends to pick up one electron, like the halogens, to form an ion with the electron configuration of the noble gas helium.

$$\text{H} \quad 1s^1 \qquad\qquad \text{H}^- \quad 1s^2 = [\text{He}]$$

Compounds that contain the H^- ion are known as **hydrides.** Thus, potassium reacts with hydrogen to form potassium hydride.

$$2 \, \text{K}(s) + \text{H}_2(g) \longrightarrow 2 \, \text{KH}(s)$$

The elements in Group VIA often react with metals to form compounds in which they contain a negatively charged ion with a -2 charge. Thus, oxygen forms **oxides** that contain the O^{2-} ion.

$$\text{O} \quad [\text{He}] \, 2s^2 \, 2p^4 \qquad\qquad \text{O}^{2-} \quad [\text{He}] \, 2s^2 \, 2p^6 = [\text{Ne}]$$

Oxygen is the most abundant element on the planet. The earth's crust is 46.6% oxygen by weight, the oceans are 86% oxygen, and the atmosphere is 21% oxygen. Oxygen combines with many other elements to form compounds. In the earth's crust oxygen is present primarily as a -2 ion in ionic compounds. In the oceans oxygen is found in the covalently bonded water molecule. Oxygen exists in its elemental form as a diatomic molecule (O_2) in the atmosphere.

Like oxygen, sulfur will tend to form a -2 ion.

$$\text{S} \quad [\text{Ne}] \, 3s^2 \, 3p^4 \qquad\qquad \text{S}^{2-} \quad [\text{Ne}] \, 3s^2 \, 3p^6 = [\text{Ar}]$$

Sulfur also forms compounds with most of the other elements in the periodic chart. Elemental sulfur usually consists of cyclic S_8 molecules in which each atom fills its valence shell by forming single bonds to two neighboring atoms. Because sulfur forms unusually strong S—S single bonds, it is capable of forming cyclic molecules that contain 6, 7, 8, 10, and 12 sulfur atoms.

Although N_2 is virtually inert to chemical reactions at room temperature, the most active metals react with nitrogen to form **nitrides,** such as lithium nitride, Li_3N. The nitride ion carries a charge of -3 because it takes three electrons to transform the electron configuration of nitrogen to a filled-shell electron configuration.

$$N \quad 1s^2\, 2s^2\, 2p^3 \qquad\qquad N^{3-} \quad 1s^2\, 2s^2\, 2p^6 = [Ne]$$

Nitrogen is an essential component of the proteins, nucleic acids, vitamins, and hormones that make life possible. However, most plants and animals cannot use elemental nitrogen from the atmosphere. Animals pick up the nitrogen they need from nitrogen compounds in the plants or other animals in their diet. Plants require nitrogen compounds in the soil. A sufficient supply of nitrogen compounds in the soil is maintained by the process of nitrogen fixation carried out by blue-green algae and bacteria in which elemental nitrogen is converted to ammonia, NH_3.

5.4 TRANSITION METALS AND THEIR IONS

The transition metals that lie between the main-group elements on either side of the periodic table have the ability to form more than one ion. Most transition metals form an M^{2+} ion because they all have two valence electrons in an s subshell. Iron, for example, can form a $+2$ ion by losing its 4s electrons.

$$Fe \quad\longrightarrow\quad Fe^{2+} \quad + 2\,e^-$$
$$[Ar]\, 4s^2\, 3d^6 \qquad\quad [Ar]\, 3d^6$$

Iron can also form a $+3$ ion by losing one of the $3d$ electrons to form an electron configuration in which the d orbitals are half-filled, that is, with one electron in each of the five orbitals in the subshell.

$$Fe \quad\longrightarrow\quad Fe^{3+} \quad + 3\,e^-$$
$$[Ar]\, 4s^2\, 3d^6 \qquad\quad [Ar]\, 3d^5$$

Transition metals are like main-group metals in many ways. They look like metals, they are malleable and ductile, they conduct heat and electricity, and they form positive ions.

Exercise 5.1

Predict the most stable ion and the corresponding electron configuration for the following atoms.
(a) Ca   (b) P

Solution

(a) Calcium will lose two electrons from its valence shell to achieve the electron configuration of argon.

$$Ca^{2+} \quad 1s^2\, 2s^2\, 2p^6\, 3s^2\, 3p^6 = [Ar]$$

(b) Phosphorus will gain three electrons in order to achieve the electron configuration of argon.

$$P^{3-} \quad 1s^2\, 2s^2\, 2p^6\, 3s^2\, 3p^6 = [Ar]$$

Exercise 5.2

The most stable ions of titanium are Ti^{2+} and Ti^{4+}. Predict the electron configuration of the two ions.

Solution

Titanium has the following electron configuration:

$$Ti \qquad [Ar]\ 4s^2\ 3d^2$$

The element forms Ti^{2+} ions by the loss of the two electrons in the valence $4s$ orbital.

$$Ti^{2+} \qquad [Ar]\ 3d^2$$

Ti^{4+} ions are formed by the loss of the two valence electrons from the $4s$ subshell plus the loss of the two $3d$ electrons to give the noble gas electron configuration of argon.

$$Ti^{4+} \qquad [Ar]$$

5.5   PREDICTING THE PRODUCTS OF REACTIONS THAT PRODUCE IONIC COMPOUNDS

The products of many reactions between main-group metals and other elements can be predicted from the electron configurations of the elements. Consider the reaction between sodium and chlorine to form sodium chloride, for example.

$$2\ Na(s) + Cl_2(g) \longrightarrow 2\ NaCl(s)$$

The net effect of the reaction is to transfer an electron from a neutral sodium atom to a neutral chlorine atom to form Na^+ and Cl^- ions that have filled-shell configurations.

$$Na\cdot + \cdot \overset{..}{\underset{..}{Cl}}: \longrightarrow [Na]^+[:\overset{..}{\underset{..}{Cl}}:]^-$$

Sodium is so reactive it is tempting to fall into the trap of thinking that "a sodium atom wants to lose an electron to achieve a filled-shell configuration." Nothing could be further from the truth. As we saw in Chapter 3, it takes a considerable amount of energy to remove an electron from any neutral atom. It takes less energy, however, to remove an electron from a sodium atom than almost any other element. But only one electron is readily removable. Once the $3s^1$ electron is lost, and we have an Na^+ ion with a filled-shell electron configuration, it takes an enormous amount of energy to remove a second electron. As a result, sodium forms the Na^+ ion in virtually all of its chemical reactions.

We can often predict the product of reactions that produce ionic compounds by using average valence electron energies (AVEE) and electron configurations of the elements involved in the reaction. Consider the reaction between potassium metal and hydrogen gas, for example.

$$K(s) + H_2(g) \longrightarrow$$

Potassium and hydrogen have similar electron configurations.

$$K \quad [Ar]\, 4s^1 \qquad\qquad H \quad 1s^1$$

To form an ionic compound an electron has to be transferred from one element to the other. We can decide which element should lose an electron by comparing the AVEE (or electronegativity) for potassium (0.42 MJ/mol) with that for hydrogen (1.31 MJ/mol). The data suggest that each potassium atom consumed in the reaction will lose an electron to a hydrogen atom to form K^+ and H^- ions.

$$K \cdot + \cdot H \longrightarrow [K]^+[H\!:]^-$$

The balanced equation for the reaction is written as follows.

$$2\, K(s) + H_2(g) \longrightarrow 2\, KH(s)$$

Exercise 5.3

Lithium metal is used in high energy, lightweight batteries. The batteries are sealed from the atmosphere in order to prevent the lithium metal from reacting with the oxygen in the air. Write a balanced equation for the following reaction.

$$Li(s) + O_2(s) \longrightarrow$$

Solution

We start by using the electron configurations of the elements to predict the ions that form in this reaction.

$$Li \quad [He]\, 2s^1 \qquad\qquad O \quad [He]\, 2s^2\, 2p^4$$

Everything we learned about the structure of the atom in Chapter 3 suggests that it should be easier to remove an electron from lithium than from oxygen. We can confirm this by comparing the average valence electron energies for lithium (0.52 MJ/mol) and oxygen (1.91 MJ/mol). We therefore expect that lithium will lose electrons and oxygen will gain electrons to form the Li^+ and O^{2-} ions.

$$Li^+ \quad [He] \qquad\qquad O^{2-} \quad [He]\, 2s^2\, 2p^6 = [Ne]$$

We then combine the ions to predict the product of the reaction—Li_2O—and write a balanced equation for the reaction.

$$4\, Li(s) + O_2(g) \longrightarrow 2\, Li_2O(s)$$

5.6   OXIDES, PEROXIDES, AND SUPEROXIDES

The method just used to predict the products of reactions of the main-group metals is remarkably powerful. Exceptions to its predictions arise, however, when very active metals react with oxygen, which is one of the most reactive nonmetals.

Lithium is well behaved. It reacts with O_2 to form an **oxide,** as predicted in Exercise 5.3.

$$4\,Li(s) + O_2(g) \longrightarrow 2\,Li_2O(s)$$

Sodium, however, reacts with O_2 under normal conditions to form a compound that contains twice as much oxygen.

$$2\,Na(s) + O_2(g) \longrightarrow Na_2O_2(s)$$

Compounds that are unusually rich in oxygen are called **peroxides.** The prefix *per-* means "above normal" or "excessive." Na_2O_2 is a peroxide because it contains more than the usual amount of oxygen.

The difference between oxides and peroxides can be understood by remembering the basic assumption behind the chemistry of the alkali metals: These elements form compounds in which the metal is present as a +1 ion. Oxides, such as Li_2O, contain the O^{2-} ion.

$$[Li]^+ [\,:\!\ddot{O}\!:\,]^{2-} [Li]^+$$

Peroxides, such as Na_2O_2, contain the O_2^{2-} polyatomic ion.

$$[Na]^+ [\,:\!\ddot{O}\!-\!\ddot{O}\!:\,]^{2-} [Na]^+$$

The formation of sodium peroxide can be explained by assuming that sodium is so reactive that the metal is consumed before each O_2 molecule can combine with enough sodium to form Na_2O. This explanation is supported by the fact that sodium reacts with O_2 in the presence of a large excess of the metal, or in the presence of a limited amount of O_2, to form the oxide expected when the reaction goes to completion.

$$4\,Na(s) + O_2(g) \longrightarrow 2\,Na_2O(s)$$

It is also consistent with the fact that the very active alkali metals—potassium, rubidium, and cesium—react so rapidly with oxygen they form **superoxides,** in which the alkali metal reacts with O_2 in a 1:1 mole ratio. Superoxides contain the polyatomic ion O_2^-.

$$K(s) + O_2(g) \longrightarrow KO_2(s)$$

Potassium superoxide forms on the surface of potassium metal, even when the metal is stored under an inert solvent. As a result, old pieces of potassium metal are potentially dangerous. When someone tries to cut the metal, the pressure of the knife pushing down on the area where the superoxide touches the metal can induce the following reaction.

$$KO_2(s) + 3\,K(s) \longrightarrow 2\,K_2O(s)$$

Because potassium oxide is more stable than potassium superoxide, the reaction gives off enough energy to boil potassium metal off the surface, which reacts explosively with the oxygen and water vapor in the atmosphere.

The reactions between the alkali metals and oxygen raise an important point. We can predict the product of any reaction between a main-group metal and a main-group nonmetal from the AVEE values for the elements and their electron configurations. But we always have to check our predictions against the reality of an experiment because other factors, such as the rate at which one of the reactants is consumed, may affect the formula of the substance actually produced in the reaction.

The alkaline earth metals also react with oxygen to form oxides and peroxides. Because the activity of the alkaline earth metals increases as we go down the column, magnesium forms the oxide (MgO), whereas barium forms the peroxide (BaO$_2$).

Checkpoint

What is the formula of the peroxide of strontium, Sr?

5.7 THE IONIC BOND

Ionic compounds are held together by the force of attraction between ions of opposite charge. The magnitude of the **electrostatic** or **coulombic** attraction between the particles depends on the product of the charge on the two ions and the square of the distance between the ions.

$$F = \frac{q_+ \times q_-}{r^2}$$

The bond that forms when the ions are brought together is called, logically enough, an **ionic bond.** In the formation of an ionic bond between a metal and nonmetal, there is a transfer of electron density from the metal to the nonmetal to form the ionic bond. This can be described using Lewis structures, which show the valence shell electrons.

$$\text{K} \cdot + \cdot \ddot{\underset{\cdot\cdot}{\text{I}}} : \longrightarrow [\text{K}]^+ [: \ddot{\underset{\cdot\cdot}{\text{I}}} :]^-$$

$$2\,\text{Li} \cdot + \cdot \ddot{\underset{\cdot}{\text{S}}} : \longrightarrow [\text{Li}]^+ [: \ddot{\underset{\cdot\cdot}{\text{S}}} :]^{2-} [\text{Li}]^+$$

There is a fundamental difference between the covalent compounds discussed in the previous chapter and the ionic compounds in this chapter. Covalent compounds usually exist in the form of molecules that contain a limited number of atoms. Even compounds as complex as glucose ($C_6H_{12}O_6$) and vitamin B$_{12}$ ($C_{63}H_{88}N_{14}O_{14}PCo$) consist of distinct molecules that contain a well-defined number of atoms.

The same can't be said of ionic compounds. They exist as huge, three-dimensional networks of ions of opposite charge that are held together by ionic bonds, as shown in Figure 5.2. Each Na$^+$ ion in the structure is surrounded by six Cl$^-$ ions, and each Cl$^-$ ion is surrounded by six Na$^+$ ions. The simplest whole number ratio of sodium ions to chloride ions is therefore 1:1, so the formula of the compound is written as NaCl.

NaCl(s)

FIGURE 5.2 Three-dimensional network structure of NaCl. The smaller ions are Na$^+$ and the larger ions are Cl$^-$.

5.8   STRUCTURES OF IONIC COMPOUNDS

When we want to describe the structure of a covalent compound, we draw a picture of its molecules. When we want to envision the structure of an ionic compound, we draw a picture of the simplest repeating unit, or **unit cell,** in the three-dimensional network of its ions. By describing the size, shape, and contents of the unit cell—and the way the repeating units stack to form a three-dimensional solid—we can unambiguously describe the structure of the compound. Figure 5.3 shows the unit cells for several common ionic compounds. Because a unit cell is a part of a larger three-dimensional structure, ions at the corners, edges, and faces are shared by neighboring unit cells. Only when this is taken into consideration does the ionic formula match the ratio of ions in the unit cell.

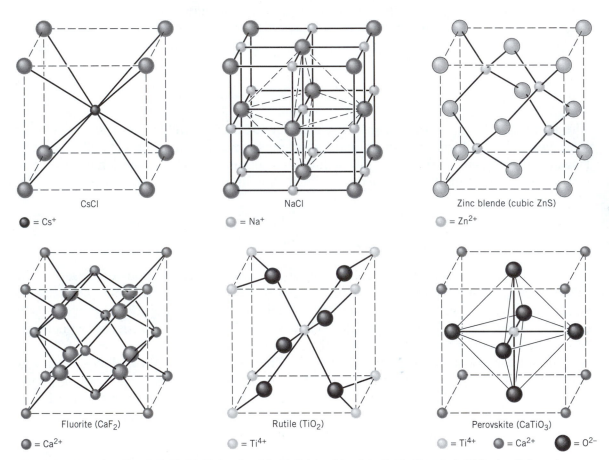

CsCl   NaCl   Zinc blende (cubic ZnS)

● = Cs⁺   ● = Na⁺   ● = Zn²⁺

Fluorite (CaF₂)   Rutile (TiO₂)   Perovskite (CaTiO₃)

● = Ca²⁺   ● = Ti⁴⁺   ● = Ti⁴⁺   ● = Ca²⁺   ● = O²⁻

FIGURE 5.3 Unit cells of CsCl, NaCl (rock salt), ZnS (zinc blende), CaF₂ (fluorite), TiO₂ (rutile), and CaTiO₃ (perovskite).

In 1850 Auguste Bravais showed that every crystal, no matter how complex its structure, could be classified as one of 14 unit cells shown in Figure 5.4 that meet the following criteria.

- The unit cell is the simplest repeating unit in the crystal.
- Opposite faces of a unit cell are parallel.
- The edge of the unit cell connects equivalent points.

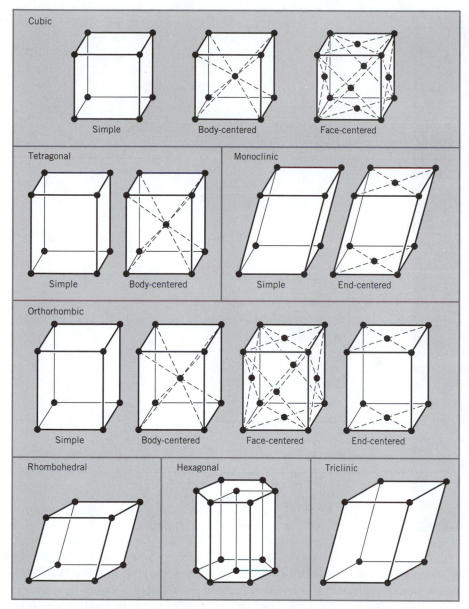

FIGURE 5.4 The 14 Bravais unit cells.

This chapter focuses on only one classification of unit cell, the cubic unit cell. There are three types of cubic unit cells—simple cubic, body-centered cubic, and face-centered cubic—shown at the top of Figure 5.4 and in Figure 5.5. Cubic unit cells are important for two reasons. First, a number of metals, ionic solids, and intermetallic compounds crystallize in cubic unit cells. Second, these unit cells are relatively easy to visualize because for a particular substance the cell-edge lengths are all the same and the cell angles are all 90°.

Each unit cell contains a number of **lattice points.** A particle can be found at each lattice point in the crystal. This particle can be an atom, an ion, or a molecule.

By convention, unit cells are defined so that the cell edges always connect equivalent points. Therefore, an identical particle must be found at each of the eight corners of a cubic unit cell. The unit cell of NaCl shown in Figure 5.3, for example, contains an equivalent Cl^- ion at each of the eight corners of the unit cell.

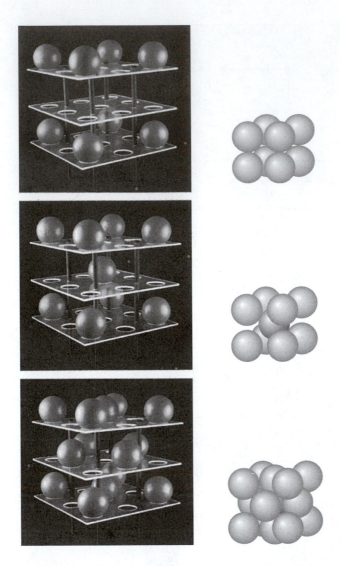

FIGURE 5.5 Models of simple cubic (top), body-centered cubic (center), and face-centered cubic unit cells (bottom).

The **simple cubic** unit cell, as its name implies, is the simplest of all unit cells. It consists of a minimum of eight equivalent particles at the eight corners of a cube.

Other types of cubic unit cells are derived from the simple cubic cell. Particles can be present on the edges or faces of the unit cell, or within the body of the unit cell. But the unit cell must contain equivalent particles at each corner.

The **body-centered cubic** unit cell also has eight identical particles on the eight corners of the unit cell. However, there is a ninth identical particle in the center of the body of the unit cell. If the particle at the center of the unit cell differs from the ones that define the eight corners, the crystal is classified as simple cubic.

The **face-centered cubic** unit cell starts with identical particles on the eight corners of the cube. But the structure also contains the same particles in the centers of the six faces of the unit cell, for a total of 14 identical lattice points.

Checkpoint

Cubic cells exist as part of a network of cells in ionic crystals. The atoms at the corners of a unit cell do not belong exclusively to one unit cell but instead are shared by eight unit cells. Sketch a diagram to demonstrate that an atom at the corner of a simple cubic cell is actually part of eight simple cubic cells.

5.9 METALLIC BONDS

The covalent bond that binds one chlorine atom to another to form a Cl_2 molecule has one thing in common with the ionic bond between the Na^+ and Cl^- ions in NaCl. In both cases, the electrons in the bond are **localized.** Either the electrons reside on one of the atoms or ions that form the bond or they are shared by a pair of atoms.

Metal atoms don't have enough valence electrons to reach a filled-shell configuration by sharing electrons with their neighbors. The valence electrons on a metal atom are therefore shared with many neighboring atoms, not just one. In effect, these valence electrons are **delocalized** over a number of metal atoms.

In order for electrons to be delocalized it is necessary that the valence subshell levels not be separated by a large energy gap. Because the delocalized valence electrons aren't tightly bound to individual atoms, they are free to move through the metal. A useful picture of the structure of metals therefore envisions the metal atoms as positive ions locked in a crystal lattice surrounded by a sea of valence electrons that move among the ions, as shown in Figure 5.6. The force of attraction between the positive metal ions and the sea of mobile negative electrons forms a **metallic bond** that holds the particles together.

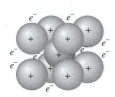 

FIGURE 5.6 The force of attraction between the positively charged metal ions and the surrounding sea of electrons in a metal is called a metallic bond. The electrons are not bound to individual atoms but may move throughout the metal.

Metals exist as extended three-dimensional arrays of atoms which pack so that each atom can touch as many neighboring atoms as possible. When a metal is heated, or beaten with a hammer, the planes of atoms that form the structure can slip past one another, which explains why metals are malleable and ductile. Each atom in the structure has a limited number of loosely held valence electrons that it can share with all of its neighbors. This is only possible when the atoms are relatively close together, which means that the metal must be a solid under normal conditions. Because the atoms pack so tightly, metals can easily transfer kinetic energy from one atom to another. As a result, they are good conductors of heat.

As we have seen, atoms become larger as we go toward the bottom left-hand corner of the periodic table. Here we find atoms that have relatively small AVEE values. Therefore, elements in that region of the periodic table have metallic character. This means the electrons are not tightly held and the valence subshells are close in energy.

Because the valence electrons in a metal are delocalized instead of being strictly associated with a single atom, they are free to move from one atom to another. The energy levels in isolated atoms differ from the energy levels of atoms that are bonded to one another. As a result, in a metal there is even a smaller difference in energy between the *s, p,* and *d* subshells within the valence shell of bonded metal atoms as compared to isolated atoms. This allows electrons to move between the subshells of an atom or adjacent atoms. The electrons that freely move through the subshells are said to form *conduction bands*. If we draw the metal into a thin wire, and connect the wire to a source of an electric current, electrons that enter the wire displace electrons that were already present on the atoms closest to the source of the current. Electrons flow through the conduction band until they eventually displace electrons from the other side of the wire. Metals are therefore good conductors of electricity.

As for ionic compounds, the chemical formulas of metallic compounds (Na, Cu, Fe, $CuAl_2$) represent the smallest whole number ratio of the atoms present in the compound. Metals in their elemental form (Na, Cu, Fe) do not consist of individual atoms but instead are composed of a three-dimensional network of positive metal ions in a sea of electrons. Because a pure metal in its elemental state, such as iron, consists only of the element iron, we write the chemical formula as Fe. There are also metallic compounds made up of more than one element. An example is $CuAl_2$, an intermetallic compound with a fixed composition. Pure aluminum metal is too weak to be used as a structural metal in cars or airplanes. However, the addition of microcrystals of $CuAl_2$ to aluminum strengthens the aluminum metal by interfering with the way planes of atoms slip past each other. The result is a metal that is both harder and stronger than pure aluminum.

Checkpoint

How does the bond in $CuAl_2$ differ from the bond in NaCl?

5.10   THE RELATIONSHIP AMONG IONIC, COVALENT, AND METALLIC BONDS

There are enormous differences among the physical properties of sodium chloride, sodium metal, and chlorine, as shown in Table 5.3. The differences in properties result from the structural differences arising from the ionic bonds in NaCl, the metallic bonds in Na metal, and the covalent bonds in Cl_2 molecules. Chemists classify compounds as ionic, metallic, or covalent on the basis of macroscopic and physical properties. Although physical properties alone are not necessarily indicative of bond type, physical properties can be explained and predicted using bond type, Lewis structures, and geometry. The classification of a compound is closely related to the type of bonding between its atoms. Thus compounds with primarily metallic bonding are called metallic compounds and those with predominantly ionic bonding are called ionic compounds.

TABLE 5.3   Some Physical Properties of NaCl, Na, and Cl_2

	NaCl	Na	Cl_2
Phase at room temperature	Solid	Solid	Gas
Density (g/cm^3)	2.2	0.97	0.0032
Melting point	801°C	97.81°C	−100.98°C
Boiling point	1413°C	882.9°C	−34.6°C

As we have seen, each Na^+ ion in NaCl is surrounded by six Cl^- ions, and vice versa. Removing an ion from the compound therefore involves breaking at least six bonds. Ionic compounds such as NaCl tend to have high melting points and boiling points. Ionic compounds are therefore invariably solids at room temperature.

Each Na atom in sodium metal is bound to as many neighboring atoms as possible; this plus strong attractive forces means that the metal is a solid at room temperature. Sodium metal has a very high boiling point because it takes a great deal of energy to disrupt all of the bonds necessary to remove individual Na atoms from the metal.

$Cl_2(g)$ consists of molecules in which one chlorine atom is tightly bound to another chlorine atom to form a covalent bond. The covalent bonds within the molecules are at least as strong as ionic bonds, but we observe in Table 5.3 that the melting and boiling points of chlorine are much lower than those for ionic compounds such as sodium chloride. The low melting and boiling points of chlorine indicate that the strong covalent chlorine–chlorine bonds are not broken during melting or boiling. The chlorine molecules remain as intact Cl_2 molecules during a change in state. It is necessary only to break relatively weak forces of attraction that hold one chlorine molecule to another in order for chlorine to melt or boil.

The difference between the physical properties of NaCl and Cl_2 is so large that it is easy to believe that the bond between two atoms is either ionic or covalent. G. N. Lewis recognized that this isn't true. In the paper in which he first described bonds based on the sharing of electrons, Lewis argued that words such as *ionic* and *covalent* referred to the extremes at either end of a continuous spectrum of bonding.

To see how he came to that conclusion, let's compare what happens when covalent and ionic bonds form. When two chlorine atoms come together to form a covalent bond, each atom contributes one electron to form a pair of electrons shared equally by the two chlorine atoms. The electrons are shared equally because both chlorine atoms have the same electronegativity.

$$:\overset{..}{\underset{..}{Cl}}\cdot\ +\ \cdot\overset{..}{\underset{..}{Cl}}: \longrightarrow\ :\overset{..}{\underset{..}{Cl}}-\overset{..}{\underset{..}{Cl}}:$$

If the electronegativities of the atoms in a compound are about the same and the elements are from the right-hand side of the periodic chart, the atoms share electrons, and the substance is considered to be **covalent.** Examples of covalent compounds are methane (CH_4) and nitrogen dioxide (NO_2).

$$
\begin{array}{ll}
\quad\quad CH_4 & \quad\quad NO_2 \\
C\quad EN = 2.54 & O\quad EN = 3.61 \\
\underline{H\quad EN = 2.30} & \underline{N\quad EN = 3.07} \\
\quad \Delta EN = 0.24 & \quad \Delta EN = 0.54
\end{array}
$$

The compounds, which consist of discrete molecules, have relatively low melting points (MP) and boiling points (BP), and they are both gases at room temperature. These are characteristic properties of low molecular weight covalent compounds.

$$
\begin{array}{lcc}
 & CH_4 & NO_2 \\
MP & -182.5°C & -163.6°C \\
BP & -161.5°C & -151.8°C
\end{array}
$$

If the electronegativities of the atoms in a substance are about the same and if the atoms come from the left-hand side of the periodic chart, the substance will be **metallic.** Sodium, although made up of only one element, is one example of a metallic substance because the sodium atoms are held together by metallic bonds. The properties of sodium given in Table 5.3 correspond to those of metallic substances. The compound CdLi, with $\Delta EN = 0.6$, is also classified as metallic.

When a sodium atom combines with a chlorine atom to form an ionic bond, each atom contributes one electron to form a pair of electrons, but that pair of electrons spends most of its time on the more electronegative chlorine.

$$Na\cdot\ +\ \cdot\overset{..}{\underset{..}{Cl}}: \longrightarrow\ [Na]^+[:\overset{..}{\underset{..}{Cl}}:]^-$$

When the difference between the electronegativities of the elements in a compound is relatively large, the compound is best classified as **ionic.** Large differences in electronegativity are normally associated with bonding between metals from the left side of the periodic chart and nonmetals from the right side. NaCl and SrF_2 are good examples of ionic compounds. In each case, the electronegativity of the nonmetal is at least two units larger than that of the metal.

NaCl		*SrF₂*	
Cl	$EN = 2.87$	F	$EN = 4.19$
Na	$EN = 0.87$	Sr	$EN = 0.96$
	$\Delta EN = 2.00$		$\Delta EN = 3.23$

We can therefore assume a net transfer of electrons from the metal to the nonmetal to form positive and negative ions. The ions attract each other electrostatically to form a three-dimensional network. We write the following Lewis structures for these compounds.

$$NaCl \qquad [Na]^+ [:\ddot{Cl}:]^-$$

$$SrF_2 \qquad [:\ddot{F}:]^- [Sr]^{2+} [:\ddot{F}:]^-$$

These compounds have high melting points and boiling points, which are characteristic properties of ionic compounds.

	NaCl	*SrF₂*
MP	801°C	1473°C
BP	1413°C	2489°C

When ionic compounds dissolve in water, they break apart to form the ions that compose them. The resulting ions are free to move throughout the solution, and therefore the aqueous solution can conduct an electric current.

$$NaCl(s) \xrightarrow{H_2O} Na^+(aq) + Cl^-(aq)$$

Inevitably, there must be compounds that fall between the covalent and ionic extremes. For such compounds, the difference between the electronegativities of the elements is large enough to be significant, but not large enough to classify the compound as ionic. Consider water, for example.

H₂O	
O	$EN = 3.61$
H	$EN = 2.30$
	$\Delta EN = 1.31$

Water is neither purely ionic nor purely covalent. It doesn't contain positive and negative ions, as indicated by the first Lewis structure in Figure 5.7. But the electrons are not shared equally, as indicated by the second Lewis structure in Figure 5.7. Water is best described as a **polar covalent compound.** One end, or pole, of the molecule has a partial positive charge ($\delta+$), and the other end has a partial negative charge ($\delta-$).

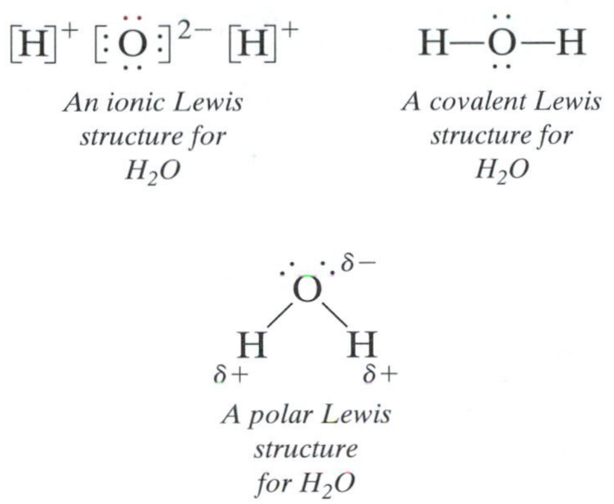

$[H]^+ \ [:\overset{\cdot\cdot}{\underset{\cdot\cdot}{O}}:]^{2-} \ [H]^+$

*An ionic Lewis
structure for
H_2O*

$H\!-\!\overset{\cdot\cdot}{\underset{\cdot\cdot}{O}}\!-\!H$

*A covalent Lewis
structure for
H_2O*

$\overset{\cdot\cdot \ \ \cdot\cdot \ \delta-}{O}$

H   H
$\delta+$ $\delta+$

*A polar Lewis
structure
for H_2O*

FIGURE 5.7 Water is neither purely ionic nor purely covalent; it is a polar covalent compound.

Ionic and covalent bonds differ in the extent to which a pair of electrons is shared by the atoms that form the bond. When one of the atoms is much better at drawing electrons toward itself than the other, the bond is *ionic*. When the atoms are approximately equal in their ability to draw electrons toward themselves, the atoms share the pair of electrons more or less equally, and the bond is *covalent*. Unfortunately, the terms *ionic* and *covalent* describe the extremes of a continuum of bonding. There is some covalent character in even the most ionic compounds. The chemistry of magnesium oxide, for example, can be understood if we assume that MgO contains Mg^{2+} and O^{2-} ions. But no compounds are 100% ionic. There is experimental evidence, for example, that the true charge (partial charge) on the magnesium and oxygen atoms in MgO is +1.5 and −1.5, respectively.

Checkpoint

What is the difference in electronegativity between aluminum and chlorine? Using only the difference in electronegativity, would you expect $AlCl_3$ to exhibit primarily ionic or covalent bonding?

Electronegativity is a powerful concept that summarizes the tendency of an atom of an element to gain, lose, or share electrons when it combines with another atom. The electronegativity difference between atoms was successful in predicting the bond type in the above examples. However, electronegativity difference (ΔEN) alone is not sufficient to categorize the bonding between atoms as primarily ionic, covalent, or metallic. For example, BF_3 ($\Delta EN = 2.14$) and SiF_4 ($\Delta EN = 2.27$) have electronegativity differences that lead us to expect the compounds to behave as if they were ionic, but both compounds are covalent. They are both gases at room temperature, and their boiling points are −99.9°C and −86°C, respectively. Thus, electronegativity difference alone cannot be relied on to predict bond type or compound classification.

The problem of understanding and predicting bond type is complicated by the fact that the three types of bonding are closely related. This is because all bonds are based on the force of attraction between particles (electrons, nuclei, and ions) of opposite charge, and the bonds all result from the transfer or sharing of electrons. Many bonds have characteristics of two or more bond types.

One approach to the problem was taken in the 1970s by William L. Jolly, who constructed the diagram shown in Figure 5.8.[1] In the 1990s, Leland Allen, William Jensen, and Gordon Sproul returned to a triangular representation to describe the bonding in systems

[1] W. L. Jolly, *The Principles of Inorganic Chemistry*, McGraw-Hill, New York, 1974, p. 187.

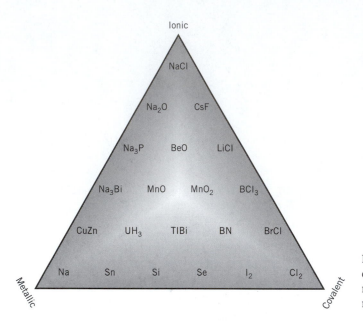

FIGURE 5.8 The three different forms of chemical bonds—ionic, covalent, and metallic—form a two-dimensional plane, not a linear continuum.

that don't fit the extremes of ionic, covalent, or metallic bonding.[2,3,4] Whereas Jolly's triangle was descriptive, the more recent bond-type triangles have predictive power. Figure 5.9 shows the basic structure of a bond-type triangle constructed for the elements between lithium and fluorine. In a bond-type triangle difference in electronegativity between two atoms is plotted on the vertical axis, and the average electronegativity of the atoms within a compound is plotted along the base of the triangle (the horizontal axis).

In both triangles, the vertex on the left represents pure metallic bonding between identical metal atoms. The vertex on the right corresponds to a pure covalent bond between identical nonmetal atoms. Compounds near the top of the triangle represent ideal ionic bonds, such as LiF, in which the transfer of electrons is essentially complete.

Compounds near the three vertices are predominantly metallic, covalent, or ionic, respectively. Those that lie toward the middle of the diagram are the most difficult to describe in terms of any one of the three categories of chemical bonds. Compounds in the middle area have properties intermediate between the three types of bonding. Boron separates the second-row metals from nonmetals in the periodic table. The two dark lines that intersect at boron in Figure 5.9 divide the triangle into metallic, ionic, and covalent regions. *The principal reason for categorizing substances by type of bonding is to more accurately predict the properties of the compound.*

The position of a compound within the bond-type triangle can help visualize the degree of ionic, covalent, or metallic character in the compound. Along the right side of this diagram, for example, the interface between covalent and ionic compounds is encountered. At this interface, substances have some of the characteristics of both ionic and covalent compounds. Going down the left side of the triangle we find the interface between ionic and metallic compounds. Some compounds that may appear at first glance to be ionic may in fact be metallic. As the triangle is ascended from the base to the ionic vertex there is a steady increase in ionic character.

[2]L. C. Allen, *Journal of the American Chemical Society,* **114,** 1510 (1992).
[3]W. B. Jensen, *Bulletin of the History of Chemistry,* **13–14,** 47 (1992–1993).
[4]G. Sproul, *Journal of Physical Chemistry,* **98,** 13221 (1994).

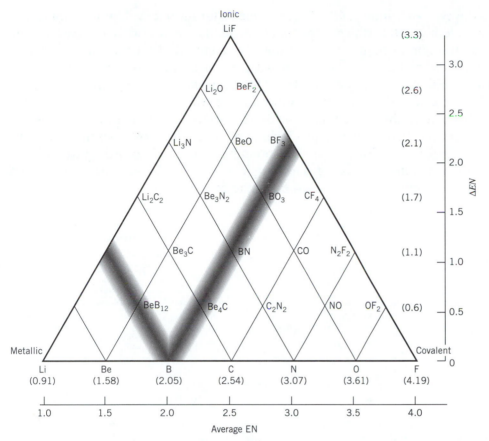

FIGURE 5.9 Bond type triangle for the second-row elements. Electronegativity difference is plotted on the vertical axis, and the average electronegativity is plotted along the base. The bond type varies from ionic at the top of the triangle to metallic at the bottom left to covalent at the bottom right. A compound may be classified according to bond type by locating it within a bond type triangle. The dark lines separate the triangle into three regions: metallic, ionic, and covalent.

5.11   BOND TYPE TRIANGLES

Bond type triangles are constructed using electronegativities. Electronegativities describe how electrons are divided between atoms in a bond. The atoms under consideration are arranged from left to right along the base of the triangle in order of increasing electronegativity (EN), as shown in Figure 5.9. The atoms along the base are then sequentially combined with one another to form the possible **binary compounds** (compounds composed of two elements).

The vertical location, y axis, of each of the compounds is determined by the difference between the electronegativities of the two elements (ΔEN). The horizontal location, x axis, of each compound is determined from the average of the electronegativities of the two elements in the compound. Lithium fluoride has the largest electronegativity difference of any of the elements listed in Figure 5.9; it is the most ionic compound on the triangle. It is therefore placed at the top of the triangle at a position determined by its average EN. Other compounds are placed in the triangle by calculating their ΔEN and the average EN and plotting them at the appropriate positions. For example, the compound BN is located above the base of the triangle midway between boron and nitrogen. The electronegativity at this point on the base of the triangle (2.56) is the average of the electronegativity of boron (2.05) and the electronegativity of nitrogen (3.07). ΔEN, which gives the vertical distance from the base, is 1.02. The location of compounds in the triangle gives immediate information about the type of bonding. CO, for example, is far more covalent than ionic, but perhaps not as covalent as NO, which appears closer to the right corner of the triangle.

The bond type triangle in Figure 5.9 does not give compound formulas for combinations of lithium with either boron or beryllium. At present no known binary compounds containing B or Be combined with Li are known. However, the possible properties of such compounds could be predicted from their position on the triangle.

An enlarged triangle is shown in Figure 5.10. This triangle is based on the previous bond type triangle but has been expanded to include many of the most common elements and their binary compounds. In Figure 5.10 compounds are plotted according to their average electronegativities along the base of the triangle. Electronegativity differences again are plotted increasing from the base to the apex of the triangle.

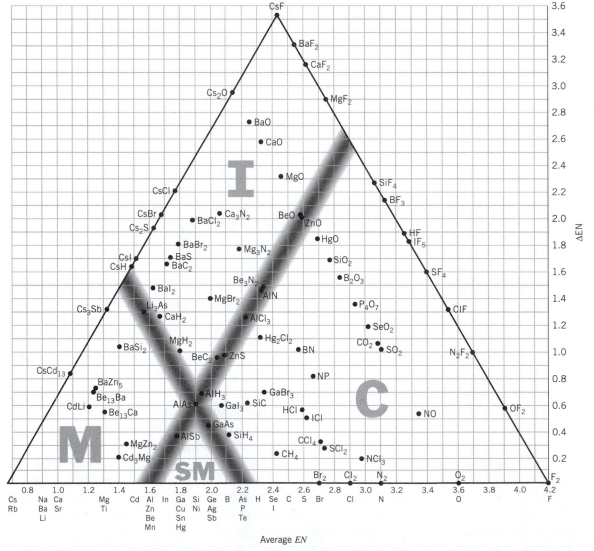

FIGURE 5.10 Bond type triangle for elements with electronegativities ranging from Cs to F. The lines separating the regions into ionic (I), metallic (M), covalent (C), and semimetallic compounds (SM) may be determined empirically and correspond approximately to lines parallel to the triangle sides drawn through Al and Te.

By plotting more compounds on the triangle we can make a better estimate of the position of the interfaces between the three types of bonding. Sproul[4] has classified more than 300 binary compounds by bond type using this division.

The separation of metals from nonmetals on the periodic chart is given by a diagonal stair-step line. In Chapter 3 we noted that the AVEE provides a rationale for the position of the diagonal line. Tellurium and arsenic, which have about the same electronegativities (AVEE values), define the right-hand limit of metallic behavior in the periodic chart, whereas silicon and aluminum represent the left-hand limits of nonmetallic behavior. In Figure 5.10 the same elements that separate metals from nonmetals in the periodic chart are used to separate the triangle into different types of bonding. Heavy lines parallel to the sides of the triangle separating elements above Te and below Al have been drawn in Figure 5.10.

This expanded representation of the bond type triangle not only allows classification of a variety of binary compounds but also makes clear that changing atom combinations can dramatically alter the physical properties of a substance. Along the right-hand side of the triangle, for example, compounds containing fluorine pass from covalent to ionic depending on electronegativity differences. On the left-hand side, compounds of Cs can be seen to change from metallic to ionic bonding as the electronegativity difference increases.

The area spanned by tellurium and aluminum is called the metalloid or semimetal region.[5] In this region of the triangle the changeover from metallic to covalent bonding is occurring, and these compounds have properties that are intermediate between the two bonding types. Many compounds that fall in this region have electrical conductivities between conductors and insulators and are called semiconductors. Other compounds that fall outside this region but are close to the ionic–covalent interface are also semiconductors.

Checkpoint

In the previous checkpoint you were asked to use only ΔEN to describe the type of bonding in $AlCl_3$. Now use the bond type triangle to describe the bonding of $AlCl_3$.

As noted in Section 5.10, in the compounds BF_3 and SiF_4, the difference between the electronegativities of a pair of elements is not sufficient to categorize the compounds the elements form. The bond type triangle shows us why the value of ΔEN does not provide enough information to categorize a compound. Both the average electronegativities of the elements in the compounds and the electronegativity differences must be taken into consideration. The bond type triangle in Figure 5.10 shows both BF_3 and SiF_4 within the covalent region of the triangle.

Predominantly covalent substances are formed when atoms that have relatively large electronegativities and small ΔEN values combine. Metallic substances are formed when we combine elements that have small electronegativities and a small ΔEN. Ionic compounds occur, as we might expect, when elements with very different electronegativities are combined (a large ΔEN).

The bond type triangle allows separation of the interatomic bonds of compounds into three contributing factors. Thus, each compound on the triangle is described in terms of the relative contributions of the three bonding types to the overall bond.

As we move from left to right on the triangle the bonding electrons change from delocalization in metallic compounds to localized shared electrons in covalent compounds. The average electronegativity increases from left to right on the bond type triangle and thus is a measure of the degree of covalent character of the bond. As we move from the bottom to the top of the triangle the transfer of bonding electrons becomes more complete.

[5]L. C. Allen, *Journal of the American Chemical Society,* **114,** 1510 (1992).

The difference in electronegativity increases from the bottom to the top of the triangle and thus is a measure of the ionic character of a bond.

The compounds CdLi, AlAs, SiC, NO, and OF_2 have about the same electronegativity difference between atoms ($\Delta EN \approx 0.6$) and therefore lie along a narrow band parallel to the base of the triangle in Figure 5.10. The bond type triangle shows that these compounds have very different bonding characteristics and therefore very different properties. CdLi is a metallic compound, AlAs lies on the metallic–covalent interface, SiC is just beyond the metallic–covalent interface, and NO and OF_2 are predominantly covalent compounds. Along this horizontal line the covalent character of the bonding increases while the ionic character of the bonding remains about the same. On the other hand, if a series of compounds such as $GaBr_3$, Hg_2Cl_2, and CaO, all with average electronegativities of approximately 2.3, is followed from the base to the apex of the bond type triangle, the covalent character of the bond remains the same while the ionic character increases.

As the interfaces between the types of bonding are approached, classification into the various categories becomes more difficult. Compounds lying close to the interfaces have properties intermediate between the bonding classifications separated by the dividing line. However, the compounds that fall into the intermediate region exhibit some of the most interesting and exciting properties and are areas of current chemical research. These include semiconductors, high-temperature superconductors, and ceramics for automobile engines.

Exercise 5.4

Use electronegativities from Table B.7 and the bond type triangle in Figure 5.10 to describe the bonding in the following compounds.
(a) $AlBr_3$ (b) $SnCl_4$ (c) CaS   (d) InNa

Solution

(a) $EN_{Al} = 1.61$ $EN_{Br} = 2.69$ Avg. $EN = 2.15$ $\Delta EN = 1.08$
Find the average electronegativity between Al and Br (2.15) on the base of the triangle and then move up the triangle to the difference in electronegativity (1.08). This point falls on the interface line between ionic and covalent bonding. Therefore, $AlBr_3$ would be expected to have characteristics of both ionic and covalent bonding.

(b) $EN_{Sn} = 1.82$ $EN_{Cl} = 2.87$ Avg. $EN = 2.35$ $\Delta EN = 1.05$
This places $SnCl_4$ in the area for predominately covalent bonding.

(c) $EN_{Ca} = 1.03$ $EN_{S} = 2.59$ Avg. $EN = 1.81$ $\Delta EN = 1.56$
CaS has primarily ionic bonding.

(d) $EN_{In} = 1.66$ $EN_{Na} = 0.87$ Avg. $EN = 1.27$ $\Delta EN = 0.79$
InNa has principally metallic bonding.

5.12   LIMITATIONS OF BOND TYPE TRIANGLES

Bond type triangles are sometimes unable to distinguish between the bonding in compounds formed by different stoichiometric amounts of the same elements. $TiCl_2$ and MnO, for example, have many of the properties of ionic compounds and are predicted by the bond type triangle to have significant ionic character. They are both solids at room temperature, and they have very high melting points, as expected for ionic compounds.

$$TiCl_2 \qquad\qquad MnO$$
$$MP = 1035°C \qquad MP = 1785°C$$

$TiCl_4$ and Mn_2O_7, on the other hand, are both liquids at room temperature, with melting points below 0°C and relatively low boiling points, as might be expected for covalent compounds.

$$TiCl_4 \qquad\qquad Mn_2O_7$$
$$MP = -24.1°C \qquad MP = -20°C$$
$$BP = 136.4°C \qquad BP = 25°C$$

Bond type triangles cannot predict these differences in covalent and ionic character. A more sophisticated model of bonding is required to explain the differences in covalent and ionic character of the compounds. However, that is beyond the scope of this text.

5.13   OXIDATION NUMBERS

The bond type triangle can be used to make a number of predictions that will play an important role in our discussion of the structure of solids in Chapter 9. For now, it helps us understand why it is so difficult to decide whether to talk about ZnS as if it contained Zn^{2+} and S^{2-} ions or polar covalent bonds in which zinc and sulfur atoms share a pair of electrons almost, but not quite, equally.

Because it is sometimes difficult to distinguish an ionic from a covalent compound, chemists have developed the concept of **oxidation number** or **oxidation state** to allow covalent and ionic compounds to be described and treated by the same model. Oxidation numbers treat both ionic and covalent compounds (except atomic molecular species such as O_2, Cl_2, P_4, and S_8) as if they were ionic.

It doesn't matter whether the compound actually contains ions. The oxidation number is the charge an atom would have *if the compound were ionic*. Both the strontium atom in SrF_2 and the carbon atom in CO, for example, are assigned oxidation numbers of +2 or are described as being in the +2 oxidation state, even though one of the compounds is fairly ionic and the other is predominantly covalent. As we will see, oxidation numbers are nothing more than a bookkeeping method for keeping track of the flow of electrons in a chemical reaction.

For the active metals in Groups IA and IIA, oxidation numbers give a good description of the charge that the metal has in its compounds. The main-group metals in Groups IIIA and IVA, however, form compounds that have a significant amount of covalent character. Although we assign an oxidation number of +3 to aluminum and -1 to bromine, it is misleading to assume that aluminum bromide contains Al^{3+} and Br^- ions. It actually exists as Al_2Br_6 molecules.

The problem becomes even more severe when we turn to the chemistry of the transition metals. MnO, for example, is ionic enough to be considered a salt that contains Mn^{2+} and O^{2-} ions. Mn_2O_7, on the other hand, is a covalent compound that boils at room temperature. It is therefore more useful to think about Mn_2O_7 as if it contained manganese in a +7 oxidation state, not Mn^{7+} ions.

For many years, chemists have used a set of general rules to determine the oxidation number of an element.

- The oxidation number is 0 in any neutral substance that contains atoms of only one element. Aluminum foil, iron metal, and the H_2, O_2, O_3, P_4, and S_8 molecules all contain atoms that have an oxidation number of 0.
- The oxidation number is equal to the charge on the ion for ions that contain only a single atom. The oxidation number of the Na^+ ion, for example, is +1, whereas the oxidation number for the Cl^- ion is −1.
- The oxidation number of hydrogen is +1 when it is combined with a *more electronegative element*. Hydrogen is therefore in the +1 oxidation state in CH_4, NH_3, H_2O, and HCl.
- The oxidation number of hydrogen is −1 when it is combined with a *less electronegative element*. Hydrogen is therefore in the −1 oxidation state in LiH, NaH, CaH_2, and $LiAlH_4$.
- The elements in Groups IA and IIA form compounds in which the metal atoms have oxidation numbers of +1 and +2, respectively.
- Oxygen usually has an oxidation number of −2. Exceptions include molecules and polyatomic ions that contain O—O bonds, such as O_2, O_3, H_2O_2, and the O_2^{2-} ion.
- Elements in Group VIIA have an oxidation number of −1 when the atom is bonded to a less electronegative element.
- The sum of the oxidation numbers of the atoms in a neutral substance is zero.

$$H_2O \qquad (2\ hydrogens)(+1) + (1\ oxygen)(-2) = 0$$

- The sum of the oxidation numbers in a polyatomic ion is equal to the charge on the ion.

$$OH^- \qquad (1\ oxygen)(-2) + (1\ hydrogen)(+1) = -1$$

- The least electronegative element is assigned a positive oxidation state. Sulfur carries a positive oxidation state in SO_2, for example, because it is less electronegative than oxygen.

$$SO_2 \qquad (1\ sulfur)(+4) + (2\ oxygens)(-2) = 0$$

Figure 5.11 shows the common oxidation numbers for many of the elements in the periodic table. There are several clear patterns in the data.

- Elements in the same group often have the same oxidation numbers.
- The largest, or most positive, oxidation number of an atom is often equal to the group number of the element (Group VIIIB atoms are an exception). The largest oxidation number for phosphorus, in Group VA, for example, is +5.
- The smallest, or most negative, oxidation number of a nonmetal often can be found by subtracting 8 from the group number. The most negative oxidation state of phosphorus, for example, is $5 - 8 = -3$.

For compounds that contain transition metal ions, such as the Fe^{2+} and Fe^{3+} ions, these rules provide the easiest way of determining oxidation numbers. In the next section an alternate method of determining oxidation numbers is introduced.

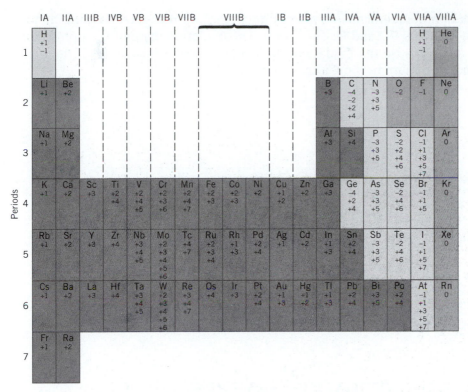

FIGURE 5.11 Oxidation numbers of the common elements. All elements have an oxidation state of zero in neutral substances that contain atoms of only one element. Different shades are used to distinguish between elements that form only positive and only negative oxidation states and those that exhibit both positive and negative oxidation states.

Exercise 5.5

Assign the oxidation numbers of the atoms in the following compounds.

(a) Aluminum oxide, also known as alumina, is used in pigments, ceramics, and abrasives: Al_2O_3.

(b) The noble gases are characterized by their extreme chemical inertness. For many years they were believed to be incapable of combining with other elements to form compounds. Xenon tetrafluoride, XeF_4, one of the first noble gas compounds, was discovered in 1962.

(c) Potassium dichromate, $K_2Cr_2O_7$, is one of the principal components in the Breathalyzer test used to determine whether someone has been drinking alcohol.

Solution

(a) The sum of the oxidation numbers in Al_2O_3 must be zero because the compound is neutral. If we assume that oxygen is in the -2 oxidation state, the oxidation state of the aluminum must be $+3$.

$$Al_2O_3 \quad 2(+3) + 3(-2) = 0$$

(b) Because the oxidation number of the fluorine is -1, the xenon atom must be present in the $+4$ oxidation state.

$$XeF_4 \quad (+4) + 4(-1) = 0$$

(c) Assigning oxidation numbers in $K_2Cr_2O_7$ is simplified if we recognize that it is an ionic compound that contains K^+ and $Cr_2O_7^{2-}$ ions. The oxidation state of the potassium is $+1$ in the K^+ ion. Because the oxidation number of oxygen is usually -2, the oxidation state of the chromium in the $Cr_2O_7^{2-}$ ion must be $+6$.

$$Cr_2O_7^{2-} \qquad 2(+6) + 7(-2) = -2$$

Exercise 5.6

Arrange the following compounds in order of increasing oxidation state for the carbon atom.

(a) CO, carbon monoxide   (b) CO_2, carbon dioxide   (c) H_2CO, formaldehyde
(d) CH_3OH, methanol   (e) CH_4, methane

Solution

$$CH_4 < CH_3OH < H_2CO < CO < CO_2$$
$$\;\;-4 \qquad\;\; -2 \qquad\qquad\; 0 \qquad +2 \qquad +4$$

5.14   CALCULATING OXIDATION NUMBERS

In addition to using the rules from the previous section to determine the oxidation numbers of atoms, oxidation numbers can be calculated using a relation based on the following equation (which was used to calculate partial charge in Chapter 4):

$$\delta_a = V_a - N_a - B_a \left(\frac{EN_a}{EN_a + EN_b} \right)$$

where δ_a is the partial charge on an atom in a molecule, a; V_a is the number of valence electrons in a neutral atom of element a; N_a is the number of nonbonding electrons on the atom; B_a is the number of bonding electrons on the atom; and EN is the electronegativity. The symbol b represents the atom to which atom a is bonded.

Formal charge, FC, was calculated in Chapter 4 by assuming that all bonds behave as covalent bonds in which the electrons are equally shared between two bonded atoms. In other words, the atoms were treated as if $EN_a = EN_b$. When calculating B_a, the number of bonding electrons, the bonding electrons are divided equally between the atoms that form covalent bonds. Since it is assumed that EN_a and EN_b are equal, the last term in the equation becomes $EN_a/(EN_a + EN_b) = 1/2$. This results in the following equation for formal charge.

$$FC_a = V_a - N_a - B_a/2$$

When we calculate the partial charge on each atom in a bond, we try to estimate what fraction of the electrons in the bond should be assigned to each atom. When we calculate formal charge we divide the bonding electrons equally between the atoms that form covalent bonds. When we calculate *oxidation numbers* we treat each bond as if it were an ionic bond and the electrons in the bond were transferred to the more electronegative element in the bond. In other words, we treat the atoms as if $EN_a \gg EN_b$. After the electrons in a bond are transferred to the more electronegative atom, the number of nonbonding, N, and bonding, B, electrons in the partial charge equation can be combined into a single term, Y. Y represents the number of electrons assigned to an atom in the Lewis structure *after the electrons in each covalent bond have been assigned to the more electronegative element.*

Therefore, for the calculation of the oxidation number we need only two pieces of information from the partial charge equation: (1) the number of valence electrons of the neutral atom (V) and (2) the number of electrons assigned to the atom in the Lewis structure of the molecule (Y) after electrons of a bond are assigned to the more electronegative atom. We then calculate the oxidation number (OX) of an atom with the following simplified equation.

$$OX = V - N - B$$
$$OX = V - Y$$

The relationship among the three ways of describing the charge associated with atoms within a compound are summarized below. It should be remembered that partial charge is the best description of the actual distribution of electron density in a molecule. Formal charge and oxidation numbers are invented bookkeeping methods. Formal charge, which treats all bonds as if they were covalent bonds, is particularly useful in determining which of several possible Lewis structures is more probable. We will find that oxidation numbers, which treat all bonds as ionic bonds, are useful in describing oxidation–reduction reactions.

$$\delta_a = V_a - N_a - B_a \left(\frac{EN_a}{EN_a + EN_b} \right)$$

$$EN_a = EN_b$$
$$FC_a = V_a - N_a - (B_a/2)$$

$$EN_a \gg EN_b$$
$$OX_a = V_a - N_a - B_a$$
$$OX_a = V_a - Y_a$$

The above equation for determining oxidation numbers is particularly useful in molecules in which an element has more than one oxidation state, such as in organic molecules (molecules that contain primarily C and H). Consider, for example, ethanol, the alcohol in "alcoholic" beverages. Ethanol has the following Lewis structure.

$$
\begin{array}{ccc}
\text{H} & \text{H} & \\
| & | & \ddots \\
\text{H--C--C--O--H} & & \\
| & | & \ddots \\
\text{H} & \text{H} &
\end{array}
$$

We start by recognizing that O ($EN = 3.61$) is more electronegative than C ($EN = 2.54$), which is more electronegative than H ($EN = 2.30$). We therefore arbitrarily assign the electrons in each covalent bond to the more electronegative atom. When a covalent bond exists between a pair of equivalent atoms, the electrons in the bond are divided equally between the atoms. In the case of the bond between the two carbon atoms, one electron is assigned to each carbon since both atoms have the same electronegativity.

$$
\begin{array}{ccccc}
 & \text{H} & & \text{H} & \\
 & & \ddots & & \ddots \\
\text{H} & :\overset{..}{\underset{.}{C}}_a \cdot & \cdot \overset{..}{\underset{.}{C}}_b & :\overset{..}{\underset{..}{O}}: & \text{H} \\
 & \text{H} & & \text{H} &
\end{array}
$$

Oxygen appears in Group VIA of the periodic table and has six valence electrons. Oxygen is assigned eight electrons in the above structure. The oxidation number of the oxygen atom is therefore −2.

$$OX_{oxygen} = V - Y = 6 - 8 = -2$$

Hydrogen, which is in Group IA, has no valence electrons once the electrons in the bonds are assigned to the more electronegative oxygen or carbon atoms. The oxidation number of the hydrogen atom is therefore +1.

$$OX_{hydrogen} = 1 - 0 = +1$$

The two carbon atoms in the structure of ethanol have different oxidation numbers. Carbon is in Group IVA and has four valence electrons. C_a is assigned seven electrons in the above structure, giving it an oxidation number of −3.

$$OX_{carbon\ a} = 4 - 7 = -3$$

C_b has only five electrons assigned to it in the above structure. Thus it will have an oxidation number of −1.

$$OX_{carbon\ b} = 4 - 5 = -1$$

Note that the oxidation numbers sum to zero for the neutral CH_3CH_2OH molecule.

Exercise 5.7

Acetic acid, $C_2H_4O_2$, is the compound that gives vinegar its sour taste. The structure of acetic acid is given below. Use the method described in this section to determine the oxidation numbers for all atoms in acetic acid.

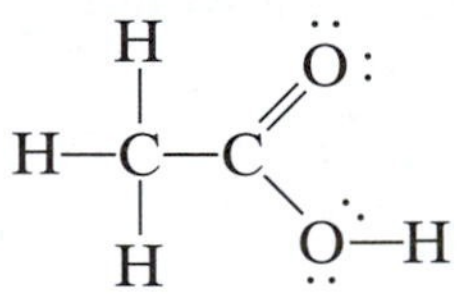

Solution

First the electrons in each bond must be assigned to the more electronegative element.

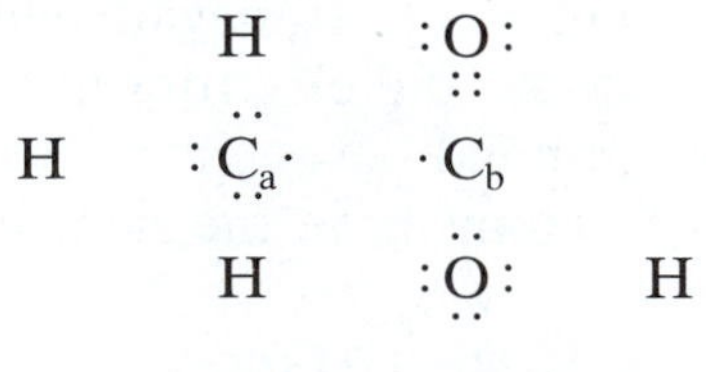

Note that all four electrons in the carbon–oxygen double bond are assigned to oxygen. In the carbon–carbon bond one electron is assigned to each carbon since they have equal electronegativities.

Oxygen	$OX = 6 - 8 = -2$
Hydrogen	$OX = 1 - 0 = +1$
Carbon$_a$	$OX = 4 - 7 = -3$
Carbon$_b$	$OX = 4 - 1 = +3$

The sum of the oxidation numbers for all atoms is zero. Had the molecule had a charge, the sum of the oxidation numbers would total to that charge.

Checkpoint
The partial charge gives the best description of the charge on atoms in a compound. What do formal charge and oxidation numbers represent?

5.15 OXIDATION–REDUCTION REACTIONS

One of the most common and important types of chemical reactions is an **oxidation–reduction** reaction. These reactions are characterized by a transfer of electrons or electron density from one species to another. The major utility of oxidation numbers is to provide a way of following electrons during an oxidation–reduction reaction. Consider the reaction that occurs when magnesium metal burns in the presence of oxygen.

$$2\ Mg(s) + O_2(g) \longrightarrow 2\ MgO(s)$$

The term *oxidation* was originally used to describe reactions, such as this, in which an element combines with oxygen.

$$2\ Mg(s) + O_2(g) \longrightarrow 2\ MgO(s)$$
$$\underset{oxidation}{\underline{\hspace{3cm}}}$$

The term *reduction* comes from a Latin stem meaning "to lead back." Anything that leads back to magnesium metal therefore involves reduction. The reaction between magnesium oxide and carbon at 2000°C to form magnesium metal and carbon monoxide is an example of the reduction of magnesium oxide to magnesium metal.

$$MgO(s) + C(s) \longrightarrow Mg(s) + CO(g)$$
$$\underset{reduction}{\underline{\hspace{3cm}}}$$

The oxidation reaction can be described in terms of the transfer of electrons from magnesium to oxygen. From this perspective, the reaction between magnesium and oxygen is written as follows.

$$2\ Mg + O_2 \longrightarrow 2\ [Mg]^{2+}[O]^{2-}$$

In the course of the reaction, each magnesium atom loses two electrons to form a Mg^{2+} ion.

$$Mg \longrightarrow Mg^{2+} + 2\ e^-$$

In addition, each O_2 molecule gains four electrons to form a pair of O^{2-} ions.

$$O_2 + 4\,e^- \longrightarrow 2\,O^{2-}$$

Because electrons are neither created nor destroyed in a chemical reaction, oxidation and reduction are linked. The four electrons gained by oxygen require that four electrons are lost by magnesium, causing two Mg atoms to form two Mg^{2+} ions. The number of electrons gained during reduction must equal the number of electrons lost during oxidation. It is impossible to have oxidation without the reduction, as shown in Figure 5.12.

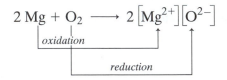

FIGURE 5.12 Oxidation cannot occur in the absence of reduction.

Oxidation numbers are particularly useful in describing the oxidation–reduction reactions of covalent molecules in which there is only a shift of electron density rather than a transfer of full electrons. Consider the following reaction, for example.

$$CO(g) + H_2O(g) \longrightarrow CO_2(g) + H_2(g)$$

The total number of electrons in the valence shell of each atom remains the same in the reaction.

$$:C\equiv O: + H-\overset{..}{\underset{..}{O}}-H \longrightarrow :O=C=O: + H-H$$

What changes in the reaction is the oxidation state of the atoms. The oxidation state of carbon increases from +2 to +4, while the oxidation state of the hydrogen decreases from +1 to 0.

$$\underset{+2}{CO} + \underset{+1}{H_2O} \longrightarrow \underset{+4}{CO_2} + \underset{0}{H_2}$$

Oxidation and reduction are therefore best defined as follows.

Oxidation occurs when the oxidation number of an atom becomes more positive.
Reduction occurs when the oxidation number of an atom becomes more negative.

Thus in the reaction of CO with H_2O the oxidation state of carbon increases and CO is said to be oxidized. The oxidation state of the hydrogen atoms in water decreases and water is said to be reduced.

What can we deduce from the fact that the oxidation state of the nitrogen atom is +5 in nitric acid? Because nitrogen is found in Group VA of the periodic table, an oxidation state of +5 is the highest oxidation state to which a nitrogen atom can be oxidized. We can therefore conclude that the nitrogen atom in nitric acid can only be reduced. It cannot be oxidized. Nitric acid is therefore likely to react with substances that can reduce the nitrogen atom to a lower oxidation state. It reacts with copper metal, for example, to form Cu^{2+} ions and either NO or NO_2, depending on the concentration of the acid.

Checkpoint

Are the following reactions oxidation–reduction reactions?

$$H_3PO_4(aq) + NH_3(aq) \longrightarrow NH_4H_2PO_4(aq)$$

$$4\ NH_3(g) + 5\ O_2(g) \longrightarrow 4\ NO(g) + 6\ H_2O(g)$$

$$\underset{\displaystyle O}{\overset{\displaystyle\ }{CH_3OH(l) + CO(g) \longrightarrow CH_3\overset{\displaystyle O}{\overset{\displaystyle \|}{C}}\text{—}OH(l)}}$$

5.16 NOMENCLATURE

Long before chemists knew the formulas for chemical compounds, they developed a system of **nomenclature** (from the Latin words *nomen,* "name," and *calare,* "to call") that gave each compound a unique name. Today we often use chemical formulas, such as NaCl, $C_{12}H_{22}O_{11}$, and $Co(NO)_6(ClO_4)_3$, to describe chemical compounds. But we still need unique names that unambiguously identify each compound.

Common Names

Some compounds have been known for so long that a systematic nomenclature cannot compete with well-established common names. Examples of compounds for which common names are used include water (H_2O), ammonia (NH_3), and methane (CH_4).

Naming Ionic Compounds, or Salts

The names of ionic compounds are written by listing the name of the positive ion followed by the name of the negative ion.

NaCl	sodium chloride
$(NH_4)_2SO_4$	ammonium sulfate
Fe_2O_3	iron(III) oxide
$NaHCO_3$	sodium hydrogen carbonate
$Al(ClO_4)_3$	aluminum perchlorate

We therefore need a series of rules that allow us to name unambiguously the positive and negative ions before we can name ionic compounds.

Naming Positive Ions

Positive ions that consist of a single atom carry the name of the element from which they are formed.

Na^+	sodium	Zn^{2+}	zinc
Ca^{2+}	calcium	H^+	hydrogen
K^+	potassium	Sr^{2+}	strontium

As we have seen, some metals—particularly the transition metals—form positive ions in more than one oxidation state. One of the earliest methods of distinguishing between the ions used the suffixes *-ic* and *-ous* added to the Latin name of the element to represent the higher and lower oxidation states, respectively.

Fe^{2+}	ferrous	Fe^{3+}	ferric
Sn^{2+}	stannous	Sn^{4+}	stannic
Cu^+	cuprous	Cu^{2+}	cupric

Chemists now use a simpler method, in which the charge on the ion is indicated by a Roman numeral in parentheses immediately after the name of the element.

Fe^{2+} iron(II) Fe^{3+} iron(III)
Sn^{2+} tin(II) Sn^{4+} tin(IV)
Cu^+ copper(I) Cu^{2+} copper(II)

There are only a limited number of polyatomic positive ions. These ions often have common names ending with the suffix -onium.

H_3O^+ hydronium NH_4^+ ammonium

Naming Negative Ions

Negative ions that consist of a single atom are named by adding the suffix -ide to the stem of the name of the element.

F^- fluoride O^{2-} oxide
Cl^- chloride S^{2-} sulfide
H^- hydride P^{3-} phosphide

Rules for Naming Polyatomic Negative Ions

The names of the common polyatomic negative ions are given in Table 1.5. At first glance, the nomenclature of these ions seems hopeless. There are several rules, however, that can bring some order to this apparent chaos.

- The name of the ion usually ends in -ite or -ate.
- There are several polyatomic ions in which an atom combines with oxygen in at least two different stoichiometric amounts.
- The -ate ending indicates the polyatomic ion in which the central atom has the higher oxidation state (i.e., the central atom bonds with the larger number of oxygens). The NO_3^- ion, for example, with three oxygens attached to the central nitrogen atom, is the nitrate ion.
- The -ite ending indicates a lower oxidation state for the central atom (i.e., there are fewer oxygen atoms attached to the central atom). Thus, the NO_2^- ion, in which two oxygen atoms are attached to the central nitrogen atom, is the nitrite ion.
- When more than two polyatomic ions exist in which a different number of oxygens are attached to the same central atom, the prefix per- (as in hyper-) and the prefix hypo- are used to indicate the very largest and very smallest number of oxygens. Consider the ClO^-, ClO_2^-, ClO_3^-, and ClO_4^- ions, in which there are 1, 2, 3, and 4 oxygen atoms. The ions with an intermediate number of oxygens (2 and 3) are known as the chlorite (ClO_2^-) and chlorate (ClO_3^-) ions, respectively. The ion with the largest number of oxygens (4) is the perchlorate (ClO_4^-) ion, and the ion with the smallest number of oxygens is the hypochlorite (ClO^-) ion.

There are a handful of exceptions to these generalizations. The names of the hydroxide (OH^-), cyanide (CN^-), and peroxide (O_2^{2-}) ions, for example, have the -ide ending because they were once thought to be monatomic ions.

Exercise 5.8

Name the following ionic compounds.

(a) $FePO_4$ is used in the automobile industry as an anticorrosion film that improves the adherence of paint to metal.

(b) $SrCO_3$ is used as an X-ray absorber in the glass faceplate of color television tubes.

(c) $Ca(ClO)_2$ is used in bleaching and sanitizing applications. It is the solid bleach in Clorox 2.

Solution

(a) Iron ions can exist with two charges, Fe^{2+} and Fe^{3+}. Since phosphate has a -3 charge, the compound $FePO_4$ contains the Fe^{3+} ion. The compound is therefore known as iron(III) phosphate.

(b) Because strontium always has a $+2$ charge in its compounds, $SrCO_3$ is known as strontium carbonate.

(c) The compound $Ca(ClO)_2$ contains Ca^{2+} and ClO^- ions. Because the metal forms only $+2$ ions, the compound is known as calcium hypochlorite.

Naming Simple Covalent Compounds

One of the most difficult tasks faced in determining the name of a compound is deciding whether to name it as an ionic or covalent compound. A bond-type triangle can be used to determine the predominant type of bonding in binary compounds. As a general rule, nonmetals bonded together are usually classified as covalent compounds.

Chemists write formulas in which the least electronegative element is written first, followed by the more electronegative element(s). The suffix *-ide* is then added to the stem of the name of the more electronegative atom.

HCl   hydrogen chloride
NO   nitrogen oxide

The number of atoms of an element in simple covalent compounds is indicated by adding one of the following Greek prefixes to the name of the element.

1	mono-	6	hexa-
2	di-	7	hepta-
3	tri-	8	octa-
4	tetra-	9	nona-
5	penta-	10	deca-

The prefix *mono-* is seldom used because it is redundant. The principal exception to the rule is carbon monoxide (CO).

Exercise 5.9

Name the following compounds.

(a) NO_2 is a component of a series of reactions responsible for Los Angeles smog.

(b) SF_4 is a highly reactive toxic colorless gas.

Solution

(a) Nitrogen dioxide   (b) Sulfur tetrafluoride

Naming Acids

Simple covalent compounds that contain hydrogen, such as HCl, HBr, and HCN, often dissolve in water to produce acids. The solutions are named by adding the prefix *hydro-* to the name of the compound and then replacing the suffix -ide with -ic. For example, hydrogen chloride (HCl) dissolves in water to form hydrochloric acid, hydrogen bromide (HBr) forms hydrobromic acid, and hydrogen cyanide (HCN) forms hydrocyanic acid.

Many of the oxygen-rich polyatomic negative ions in Table 1.5 form acids that are named by replacing the suffix -ate with -ic and the suffix -ite with -ous.

$CH_3CO_2^-$	acetate	CH_3CO_2H	acetic acid
CO_3^{2-}	carbonate	H_2CO_3	carbonic acid
BO_3^{3-}	borate	H_3BO_3	boric acid
NO_3^-	nitrate	HNO_3	nitric acid
NO_2^-	nitrite	HNO_2	nitrous acid
SO_4^{2-}	sulfate	H_2SO_4	sulfuric acid
SO_3^{2-}	sulfite	H_2SO_3	sulfurous acid
ClO_4^-	perchlorate	$HClO_4$	perchloric acid
ClO_3^-	chlorate	$HClO_3$	chloric acid
ClO_2^-	chlorite	$HClO_2$	chlorous acid
ClO^-	hypochlorite	$HClO$	hypochlorous acid
PO_4^{3-}	phosphate	H_3PO_4	phosphoric acid
MnO_4^-	permanganate	$HMnO_4$	permanganic acid
CrO_4^{2-}	chromate	H_2CrO_4	chromic acid

Salts containing anions with acidic hydrogens can be named by indicating the presence of the acidic hydrogen as follows.

$NaHCO_3$ sodium hydrogen carbonate (also known as sodium bicarbonate)
$NaHSO_3$ sodium hydrogen sulfite (also known as sodium bisulfite)
KH_2PO_4 potassium dihydrogen phosphate

Exercise 5.10

Name the following compounds.
(a) $NaClO_3$ (b) $Al_2(SO_4)_3$ (c) P_4S_3 (d) SCl_4

Solution

The key to naming the compounds is recognizing that the first two are ionic compounds and the last two are covalent compounds.
(a) sodium chlorate   (b) aluminum sulfate   (c) tetraphosphorus trisulfide
(d) sulfur tetrachloride

KEY TERMS

Active metal	Coulombic	Halide
Alkali metal	Covalent bond	Halogen
Body-centered cubic	Delocalized	Hydride
Bond type triangle	Electronegativity	Iodide
Bromide	Electrostatic	Ionic bond
Chloride	Face-centered cubic	Lattice point
Conduction band	Fluoride	Localized

Main-group element	Oxidation number	Reduction
Metallic bond	Oxide	Simple cubic
Nitride	Phosphide	Sulfide
Nomenclature	Polar covalent	Transition metal
Oxidation	Reactive	Unit cell

PROBLEMS

The Active (Reactive) Metals

1. List the elements in the third row of the periodic table in order of decreasing metallic character. List the elements in order of increasing nonmetallic character. Explain the similarities between the trends. Identify each element as either a metal, nonmetal, or semimetal.

2. List the elements in Group VA of the periodic table in terms of increasing metallic character. Identify each element in the group as a metal, nonmetal, or semimetal.

3. Which of the following sets of elements is arranged in order of increasing nonmetallic character?
    (a) Sr < Al < Ga < N   (b) K < Mg < Rb < Si   (c) Ge < P < As < N
    (d) Al < B < N < F

4. Describe the general trends in the activity, or reactivity, of the main-group metals.

5. What do we mean when we say that sodium is more metallic than lithium?

6. For each of the following pairs of metals, which is more active?
    (a) Mg or Ca   (b) Na or Mg   (c) K or Mg   (d) Mg or Al

7. Which metal in each of the following pairs would you expect to react more rapidly with water?
    (a) Na or K   (b) Na or Mg   (c) Mg or Ca   (d) Ca or Al

Group IA: The Alkali Metals

8. Define the terms *alkali metal, halide, hydride, sulfide, nitride, phosphide, oxide, peroxide,* and *superoxide.* Give at least one example of each.

9. Write the formulas for the bromide, hydride, sulfide, nitride, phosphide, oxide, and peroxide of rubidium.

10. Describe the difference between the hydrogen atoms in metal hydrides such as LiH and nonmetal hydrides such as CH_4 and H_2O.

11. White phosphorus bursts spontaneously into flame in the presence of air and is therefore stored under water. Cesium metal can also burst into flame in the presence of air, but it can't be stored under water. Why not?

12. Cesium metal is so active that it reacts explosively with cold water and even with ice at temperatures as low as $-116°C$. Explain why cesium is more active than sodium.

Group IIA: The Alkaline Earth Metals

13. Write the formulas for the fluoride, hydride, sulfide, nitride, phosphide, oxide, and peroxide of barium.

14. Explain why calcium reacts with water at room temperature, but magnesium only reacts with steam at high temperatures.

15. Explain why BaO_2 must be a peroxide, $[Ba^{2+}][O_2^{2-}]$, and not an oxide $[Ba^{4+}][O^{2-}]_2$.

Ion Formation

16. Just as there is a difference between an orange and an orange peel, there is a difference between Fe metal and Fe^{3+} ions. If you were told that you have a shortage of iron, calcium, or zinc in your diet, which would you add to your diet, pieces of the metal or salts of their ions?

17. Calculate the charge on the positive ions formed by the following elements.
    (a) Mg (b) Al (c) Si (d) Cs (e) Ba

18. Calculate the charge on the negative ions formed by the following elements.
    (a) C (b) P (c) S (d) I

Identifying Main-Group Metals from the Products of Their Reactions

19. Which one of the following elements is most likely to form an oxide with the formula XO and also a hydride with the formula XH_2?
    (a) Na (b) Mg (c) Al (d) Si (e) P

20. A main-group metal reacts with hydrogen and oxygen to form compounds with the formulas XH_4 and XO_2. In which group of the periodic table does the element belong?

21. In which column of the periodic table do we find main-group metals that form sulfides with the formula M_2S_3 and react with acid to form M^{3+} ions and H_2 gas?

Predicting the Products of Reactions between Main-Group Metals and Nonmetal Elements

22. Explain why sodium reacts with chlorine to form NaCl and not $NaCl_2$ or $NaCl_3$.

23. Explain why magnesium reacts with chlorine to form $MgCl_2$ and not $MgCl_3$.

24. Describe why lithium reacts with nitrogen to give Li_3N and not a compound with another formula.

25. Describe the most important factors in determining the formula of an ionic compound such as NaCl or $MgCl_2$.

26. Predict the product of the reaction between aluminum and nitrogen.

27. Predict the product of the reaction between strontium metal and phosphorus.

28. The light meters in automatic cameras are based on the sensitivity of gallium arsenide to light. Predict the formula of gallium arsenide.

29. Write balanced equations for the reactions of sodium metal with each of the following elements.
    (a) F_2 (b) O_2 (c) H_2 (d) S_8 (e) P_4

30. Write balanced equations for the reactions of calcium metal with each of the following elements.
    (a) H_2 (b) O_2 (c) S_8 (d) F_2 (e) N_2 (f) P_4

31. Predict the product of the reactions of F_2 with each of the following metals.
    (a) Zn (b) Al (c) Sn (d) Mg (e) Bi

Unit Cells

32. Define the term *unit cell.* Describe the common properties of all unit cells.

33. Describe the difference between simple cubic, body-centered cubic, and face-centered cubic unit cells.

34. Sodium hydride crystallizes in a face-centered cubic unit cell of H^- ions with Na^+ ions at the center of the unit cell and in the center of each edge of the unit cell. How many Na^+ ions does each H^- ion touch? How many H^- ions does each Na^+ ion touch?

Ionic, Covalent, and Metallic Bonding

35. New chemistry students sometimes assume that the difference between ionic and covalent compounds is black and white. Explain why chemists believe that ionic and covalent are the two extremes of a continuum of differences in the bonding between two atoms.

36. Predict whether the following compounds are ionic or covalent by using both the general rule that metals combine with nonmetals to form ionic compounds and a bond type triangle.
    (a) OF_2 (b) CS_2 (c) MgO (d) ZnS

37. Predict whether the following compounds are ionic or covalent by using both the general rule that metals combine with nonmetals to form ionic compounds and a bond type triangle.
    (a) IF_3 (b) $SiCl_4$ (c) BF_3 (d) Na_2S

38. Which of the following elements forms bonds with fluorine that are the most covalent?
    (a) P   (b) Ca   (c) Al   (d) O   (e) Se

39. Which one of the following elements forms the most covalent bond with oxygen?
    (a) Sr   (b) In   (c) Sb   (d) Te   (e) Se

40. Which of the following compounds are best described as polar covalent?
    (a) CO   (b) H_2O (c) BeF_2 (d) $MgBr_2$ (e) AlI_3 (f) ZnS (g) CdLi

41. Which of the following compounds are best described as polar covalent?
    (a) CaH_2 (b) BrF_3 (c) NF_3 (d) $SiCl_4$ (e) AsH_3 (f) $MgZn_2$

42. Describe what happens to the physical properties of a metal when it combines with another metal to form an alloy.

43. Describe what happens to the physical properties of a nonmetal when it combines with another nonmetal to form a covalent compound.

44. Describe what happens to the physical properties of a metal and a nonmetal when they combine to form an ionic compound, or salt.

45. Describe the differences between the physical properties of covalent compounds such as CO_2, ionic compounds such as NaCl, and metallic compounds such as GaAs.

46. Which of the following compounds should be ionic?
    (a) ZnS   (b) $AlCl_3$ (c) SnF_2 (d) BH_3 (e) H_2S

47. Which of the following compounds should be covalent?
    (a) CH_4 (b) CO_2 (c) $SrCl_2$ (d) NaH   (e) SF_4

48. Which of the following pairs of elements should combine to give ionic compounds?
    (a) $Mg + O_2$ (b) $S_8 + F_2$ (c) Na + Hg   (d) $K + I_2$

49. Which of the following pairs of elements should combine to give metallic compounds?
    (a) $N_2 + O_2$ (b) $Cl_2 + F_2$ (c) $Cl_2 + Cs$ (d) $S_8 + Na$ (e) Cu + Sn

50. Many covalent compounds have very distinctive odors, such as the H_2S given off by rotten eggs, and the CH_3CO_2H in vinegar. Ionic compounds such as NaCl and Al_2O_3 have no odor. What is the difference between the physical properties of ionic and covalent compounds that is responsible for this characteristic?

Bond Type Triangles

51. Construct a bond type triangle for the elements of the third period, Na through Cl. Electronegativities are given in Figure 4.8. First draw a horizontal base line starting with the electronegativity of Na and ending with Cl. Place all other elements of the period according to their electronegativities along the line. Locate the midpoint along the base line between Na and Cl. At a distance above the midpoint corresponding to the ΔEN for Na and Cl, draw the two equal sides of the triangle. Determine the chemical formula of the binary compounds formed by as many of the pairs of elements as you can. Plot the compounds on the triangle using the average electronegativity and the difference in electronegativity, ΔEN, for the elements in each binary compound. How would you classify each compound?

52. Use Figure 5.10 to determine whether there is an electronegativity difference, ΔEN, above which a compound can always be predicted to be ionic.

53. Use Figure 5.10 to determine whether there is an electronegativity difference, ΔEN, below which a binary compound can always be classified as covalent.

54. For each of the following characteristics select three binary compounds from Figure 5.10 which exhibit the characteristic.
    (a) ΔEN is relatively the same for the three compounds, but the bonding type changes from metallic to semimetallic to covalent.
    (b) ΔEN ranges from about 3 to about 1, but all compounds are ionic.
    (c) The average EN of the atoms in the compounds stays the same, but the ionic character increases.
    (d) The ΔEN of the atoms in the compounds stays the same, but the covalent character increases.
    (e) ΔEN is about the same, but the bonding type changes from metallic to ionic to covalent.

55. Classify the following compounds according to bonding type:
    (a) HgS (b) GaSb (c) Li_3N (d) NaBr (e) $SnBr_4$ (f) Na_3P (g) InP
    (h) InN (i) TeO_2

56. For each of the following properties list a binary solid compound that exhibits the property.
    (a) conducts electricity in the solid state
    (b) is an insulator
    (c) is a semiconductor
    (d) has a very high melting point
    (e) has a low melting point
    (f) electrons are shared unequally between the atoms composing the compound

57. Why are two variables, ΔEN and average EN, required to classify a compound according to bonding type?

58. Make a plot of ΔEN versus the minimum electronegativity for the compounds of Figure 5.9. Add a few of the compounds given in Figure 5.10 to the plot and interpret the results.

59. Make a plot of the maximum electronegativity versus the minimum electronegativity for the elements of Figure 5.9. Add a few of the compounds given in Figure 5.10 to the plot and interpret the results.

Oxidation Numbers

60. An area of active research interest in recent years has involved compounds such as $Re_2Cl_8{}^{2-}$, $Cr_2Cl_9{}^{3-}$, and $Mo_2Cl_8{}^{4-}$ that contain bonds between metal atoms. Calculate the oxidation number of the metal atom in each compound.

61. Determine the oxidation state of phosphorus in the following compounds.
    (a) K_3P (b) $AlPO_4$ (c) $PO_3{}^{3-}$ (d) P_2Cl_4

62. Calculate the oxidation number of the aluminum atom in the following compounds.
    (a) $LiAlH_4$ (b) $Al(H_2O)_6{}^{3+}$ (c) $Al(OH)_4{}^-$

63. The active ingredient in Rolaids antacid tablets has the formula $NaAl(OH)_2CO_3$. Calculate the oxidation state of the aluminum atom in the compound.

64. Which of the following compounds contain hydrogen in a negative oxidation state?
    (a) H_2S (b) H_2O (c) NH_3 (d) H_3PO_4 (e) $LiAlH_4$ (f) HF (g) CaH_2
    (h) CH_4

65. Calculate the oxidation number of the chlorine atom in the following compounds.
    (a) Cl_2 (b) Cl^- (c) ClO^- (d) $ClO_2{}^-$ (e) $ClO_3{}^-$ (f) $ClO_4{}^-$

66. Calculate the oxidation number of the iodine atom in the following compounds. Group together any compounds in which iodine has the same oxidation number.
    (a) HI (b) KI (c) I_2 (d) HOI (e) KIO_3 (f) I_2O_5 (g) KIO_4 (h) H_5IO_6

67. The noble gases rarely form compounds. Therefore, their most common oxidation number is zero. However, Xe does form a limited number of compounds. Calculate the oxidation number of the xenon atom in the following compounds and describe any trends in the oxidation states.
    (a) XeF_2 (b) XeF_4 (c) $XeOF_2$ (d) XeF_6 (e) $XeOF_4$ (f) XeO_3
    (g) XeO_4 (h) $XeO_6{}^{4-}$

68. Calculate the oxidation number of barium in BaO_2. Does your answer make sense? If it doesn't, determine the consequence of assuming that barium is present in its usual oxidation state in the compound.

69. Carbon can have any oxidation number between -4 and $+4$. Calculate the oxidation number of carbon in the following compounds. Group compounds with the same oxidation number and describe any trends.
    (a) CCl_4 (b) $COCl_2$ (c) CO (d) CO_2 (e) CS_2 (f) CH_3Li (g) CH_4
    (h) H_2CO (i) Na_2CO_3 (j) HCO_2H

70. Sulfur can have any oxidation number between $+6$ and -2. Calculate the oxidation number of sulfur in the following compounds. Group compounds with the same oxidation number and describe any trends.
    (a) S_8 (b) H_2S (c) ZnS (d) SF_4 (e) SF_6 (f) SO_2 (g) SO_3 (h) $SO_3{}^{2-}$
    (i) $SO_4{}^{2-}$ (j) H_2SO_3 (k) H_2SO_4

71. Calculate the oxidation number of manganese in the following compounds. Group compounds with the same oxidation number and describe any trends.
    (a) MnO (b) Mn_2O_3 (c) MnO_2 (d) MnO_3 (e) Mn_2O_7 (f) $Mn(OH)_2$
    (g) $Mn(OH)_3$ (h) H_2MnO_4 (i) $HMnO_4$ (j) $CaMnO_3$ (k) $MnSO_4$

72. Calculate the oxidation number of titanium in the following compounds. Group compounds with the same oxidation number and describe any trends.
    (a) TiO (b) TiO_2 (c) Ti_2O_3 (d) Ti_3O_7 (e) $TiCl_3$ (f) $TiCl_4$
    (g) K_2TiO_3 (h) H_2TiCl_6 (i) $Ti(SO_4)_2$

73. Prussian blue is a pigment with the formula $Fe_4[Fe(CN)_6]_3$. If the compound contains the $Fe(CN)_6^{4-}$ ion, what is the oxidation state of the other four iron atoms? Turnbull's blue is a pigment with the formula $Fe_3[Fe(CN)_6]_2$. If the compound contains the $Fe(CN)_6^{3-}$ ion, what is the oxidation state of the other three iron atoms?

74. Determine the oxidation state of each atom in the following organic molecules.

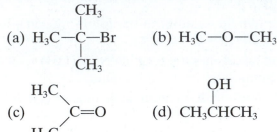

75. Determine the oxidation number of each atom in the following organic molecules.

$$(a)\ \overset{O}{\overset{\|}{HCOH}}$$
Formic acid

$$(b)\ \overset{O}{\overset{\|}{HCH}}$$
Formaldehyde

$$(c)\ CH_3\overset{O}{\overset{\|}{C}}OH$$
Acetic acid

$$(d)\ CH_3\overset{O}{\overset{\|}{CH}}$$
Acetaldehyde

76. Determine the oxidation number of each atom in the following organic molecules.

$$(a)\ CH_3\overset{O}{\overset{\|}{C}}Cl$$

(b) CH_3NH_2

$$(c)\ CH_3\overset{O}{\overset{\|}{C}}OCH_2CH_3$$

(d) CH_3CH_3

(e) $CH_2{=}CH{-}CH_3$ (f) $HC{\equiv}CH$

Oxidation–Reduction Reactions

77. Decide whether each of the following reactions involves oxidation–reduction. If it does, identify what is oxidized and what is reduced.
    (a) $CO_2(g) + H_2O(l) \rightarrow H_2CO_3(aq)$
    (b) $Fe_2O_3(s) + 3\ CO(g) \rightarrow 2\ Fe(s) + 3\ CO_2(g)$
    (c) $CO_2(g) + H_2(g) \rightarrow CO(g) + H_2O(g)$
    (d) $CO(g) + 2\ H_2(g) \rightarrow CH_3OH(l)$

78. Decide whether each of the following reactions involves oxidation–reduction. If it does, identify what is oxidized and what is reduced.
    (a) $Mg(s) + 2\ HCl(g) \rightarrow MgCl_2(s) + H_2(g)$
    (b) $I_2(s) + 3\ Cl_2(g) \rightarrow 2\ ICl_3(s)$
    (c) $NaOH(aq) + HCl(aq) \rightarrow NaCl(aq) + H_2O(aq)$
    (d) $2\ Na(s) + 2\ H_2O(l) \rightarrow 2\ NaOH(aq) + H_2(g)$

79. Decide whether each of the following reactions involves oxidation–reduction. If it does, identify what is oxidized and what is reduced.
    (a) $P_4O_{10}(g) + 10C(s) \rightarrow 10CO(g) + P_4(g)$
    (b) $P_4(s) + 5\ O_2(g) \rightarrow P_4O_{10}(s)$

80. What is the change in oxidation state of C and S in the following reaction? Which is oxidized and which is reduced?

$$4\,C(s) + S_8(l) \longrightarrow 4\,CS_2(l)$$

81. The Ag_2S that forms when silver tarnishes can be removed by polishing the silver with a source of cyanide ion or by wrapping it in aluminum foil and immersing it in salt water.

$$Ag_2S(s) + 4\,CN^-(aq) \longrightarrow 2\,Ag(CN)_2^-(aq) + S^{2-}(aq)$$
$$3\,Ag_2S(s) + 2\,Al(s) \longrightarrow 6\,Ag(s) + Al_2S_3(s)$$

Which of the reactions involves oxidation–reduction?

Nomenclature

82. Describe what is wrong with the common names for the following compounds and write a better name for each compound.
    (a) phosphorus pentoxide (P_2O_5)   (b) iron oxide (Fe_2O_3)
    (c) chlorine monoxide (Cl_2O)   (d) copper bromide ($CuBr_2$)
83. Explain why calcium bromide is a satisfactory name for $CaBr_2$, but $FeBr_2$ must be called iron(II) bromide.
84. Write the formulas for the following compounds.
    (a) tetraphosphorus trisulfide   (b) silicon dioxide
    (c) carbon disulfide   (d) carbon tetrachloride
    (e) phosphorus pentafluoride
85. Write the formulas for the following compounds.
    (a) silicon tetrafluoride   (b) sulfur hexafluoride
    (c) oxygen difluoride   (d) dichlorine heptoxide
    (e) chlorine trifluoride
86. Write the formulas for the following compounds.
    (a) tin(II) chloride   (b) mercury(II) nitrate
    (c) tin(IV) sulfide   (d) chromium(III) oxide
    (e) iron(II) phosphide
87. Write the formulas for the following compounds.
    (a) beryllium fluoride   (b) magnesium nitride
    (c) calcium carbide   (d) barium peroxide
    (e) potassium carbonate
88. Write the formulas for the following compounds.
    (a) cobalt(III) nitrate   (b) iron(III) sulfate
    (c) gold(III) chloride   (d) manganese(IV) oxide
    (e) tungsten(VI) chloride
89. Name the following compounds.
    (a) KNO_3 (b) Li_2CO_3 (c) $BaSO_4$ (d) PbI_2
90. Name the following compounds.
    (a) $AlCl_3$ (b) Na_3N (c) Ca_3P_2 (d) Li_2S (e) MgO
91. Name the following compounds.
    (a) NH_4OH (b) H_2O_2 (c) $Mg(OH)_2$ (d) $Ca(ClO)_2$ (e) $NaCN$

92. Name the following compounds.
    (a) Sb_2S_3 (b) $SnCl_2$ (c) SF_4 (d) $SrBr_2$ (e) $SiCl_4$
93. Write the formulas of the following common acids.
    (a) acetic acid   (b) hydrochloric acid
    (c) sulfuric acid   (d) phosphoric acid
    (e) nitric acid
94. Write the formulas of the following less common acids.
    (a) carbonic acid   (b) hydrocyanic acid
    (c) boric acid   (d) phosphorous acid
    (e) nitrous acid
95. If sodium carbonate is Na_2CO_3 and sodium hydrogen carbonate is $NaHCO_3$, what are the formulas for sodium sulfite and sodium hydrogen sulfite?
96. The prefix *thio-* describes compounds in which sulfur replaces oxygen, for example, cyanate (OCN^-) and thiocyanate (SCN^-). If SO_4^{2-} is the sulfate ion, what is the formula for the thiosulfate ion?
97. Name the compound in each of the following minerals.
    (a) fluorite (CaF_2)   (b) galena (PbS)
    (c) quartz (SiO_2)   (d) rutile (TiO_2)
    (e) hematite (Fe_2O_3)
98. Name the compound in each of the following minerals.
    (a) calcite ($CaCO_3$)   (b) barite ($BaSO_4$)

Integrated Problems

99. The atoms composing the compounds SeO_2, CaH_2, and Cs_3Sb all have about the same difference in electronegativity, ΔEN. Yet only one of the compounds conducts electricity in the solid state, only one has a high melting point and dissolves in water to give a solution that conducts electricity, and only one has a relatively low melting point. Identify each of the compounds and explain your identification.
100. The common oxidation numbers for iron are +2 and +3. Iron reacts with $O_2(g)$ and $Cl_2(g)$ to form two different oxides and two different chlorides. What would be the formula of the four compounds formed? Classify the compounds according to bond type. Explain your answer.
101. Compare and contrast (what is the same and what is different) the distribution of electrons in covalent, ionic, and metallic bonds.
102. Partial charge, formal charge, and oxidation number are all used to describe the arrangement of electrons or charge associated with atoms within a compound. Determine the formal charge, partial charge, and oxidation number on both atoms in IBr. Give a short description for each of the three calculations describing what information the calculated value gives you. Which of the three calculations most accurately describes the distribution of charge between I and Br? What is the utility of the two calculations that are not the best description of the distribution of charge?
103. Describe what happens to the *distribution of electrons* in a bond between two elements as you move from left to right on a bond type triangle. Describe what happens to the *distribution of electrons* in a bond between two elements as you move from the bottom to the top of a bond type triangle.
104. Describe the types of bonding in $Ba(NO_3)_2$. What is different about the bonding in barium nitrate as compared to other compounds described in this chapter? Draw a Lewis structure for the compound.

C H A P T E R

6

GASES

The Nobel Prize winning physicist Richard Feynman once asked, "If, in some cataclysm, all of scientific knowledge were to be destroyed, and only one sentence passed on to the next generations of creatures, what statement would contain the most information in the fewest words?"[1] Feynman then gave his answer to the question, "I believe it is the atomic hypothesis that all things are made of atoms—little particles that move about in perpetual motion, attracting each other when they are a little distance apart, but repelling upon being squeezed into one another." This is, as Feynman observed, a statement of enormous power. So far we have developed the idea that all things are made of atoms, but we have

---

[1]*The Feynman Lectures on Physics,* Addison-Wesley Publishing Company, Reading, Massachusetts, 1963.

not yet discussed the other parts of Feynman's statement, that atoms move about in perpetual motion and that particles attract and repel each other.

Atoms are never at rest. Even atoms which are bonded together to form a molecule move relative to one another. The sodium and chloride ions of sodium chloride jiggle even though they are locked into an ordered three-dimensional structure. Gaseous molecules, such as those of oxygen in the atmosphere, speed rapidly about the room at several hundred miles per hour. What is the origin of this perpetual motion?

6.1   TEMPERATURE

We don't need to be told that the temperature is 95°F (35°C) on a summer afternoon to know it is hot. Nor do we need to be told that it is −15°F (−26°C) on a clear winter night to realize it is cold. For most purposes, we can rely on our senses to distinguish between hot and cold.

But there are times when our senses can be misled. Imagine you are getting out of bed in the middle of a cold winter night and stepping onto a "cold" floor and then onto a small throw rug. Your senses tell you that the rug is warmer than the floor. Unfortunately, your senses are wrong. The floor is not colder than the rug. Both objects are just as warm (or just as cold) as the air in the room. When you reach into the freezer, metal ice-cube trays feel colder than plastic trays. But this can't be true; all the trays in the freezer are equally cold. Then why do they feel differently?

Just as some materials conduct electricity better than others, some materials conduct heat better than others. Metals are good conductors of heat. The metal atoms are held in place by a sea of electrons, which are free to move throughout the metal. As a metal is heated its atoms vibrate more rapidly and the electrons transport heat from one part of the metal to another. Materials whose electrons are more tightly bound do not conduct heat as well as or in the same way as metals. Covalent and ionic compounds conduct heat by passing the heat along through vibrating strings of atoms and ions instead of through electrons. Different substances in contact with one another transfer heat only when the atoms in one substance acquire sufficient heat to cause them to vibrate so wildly that they collide with neighboring atoms in the other substance, causing its atoms to vibrate more. Good thermal conductors feel cooler at room temperature than poorer thermal conductors at the same temperature because good conductors transfer more heat away from our hand, making our hand feel cool.

Because our senses can be misled, it is useful to have a reliable measure of the degree to which an object is either hot or cold. This quantity is called the **temperature** of an object.

6.2   TEMPERATURE AS A PROPERTY OF MATTER

Temperature is nothing more than a quantitative measure of the degree to which an object is either "hot" or "cold." Temperature can be measured on either relative or absolute scales (Figure 6.1). The Celsius (°C) and Fahrenheit (°F) scales measure *relative* temperatures. These scales compare the temperature of a system with arbitrary standards such as boiling water and an ice–water bath. The Kelvin (K) scale measures *absolute* temperatures. An object that has a temperature of 0°C has a temperature of 32°F and a temperature of 273 K.

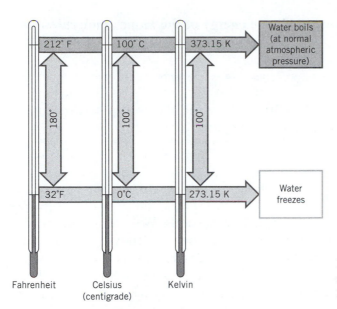

FIGURE 6.1 The three common temperature scales. Note that water boils at 212°F, 100°C, or 373.15 K and freezes at 32°F, 0°C, or 273.15 K.

The Kelvin scale of temperature has an interesting interpretation. At a very low temperature, called absolute zero on the Kelvin scale (−273.15°C), the units that compose a substance (atoms, molecules, or ions) are as close together as they can get, and any motion of the units is at a minimum. What happens when a substance at absolute zero is warmed? The atoms, molecules, or ions that compose it begin to move—to vibrate or jiggle more vigorously. The warmer we make the substance, the more its components jostle about. A familiar example is the melting of ice.

Ice melts when it is heated because the average motion of the water molecules increases with temperature. Molecules can move in three ways: (1) vibration, (2) rotation, and (3) translation (Figure 6.2). Water molecules *vibrate* when H—O bonds are stretched or bent. *Rotation* involves the motion of a molecule around its center of gravity. *Translation* literally means change from one place to another and describes the motion of molecules through space.

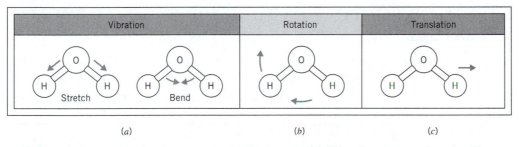

FIGURE 6.2 A water molecule can move in three ways. (*a*) Vibration occurs when O—H bonds stretch or bend. (*b*) Rotation involves movement around the center of gravity of the molecule. (*c*) Translation occurs when a water molecule moves through space.

As the system becomes warmer, the motion of the water molecules eventually becomes too large to allow the molecules to be locked into the rigid structure of ice. At this point, the solid melts to form a liquid. Eventually, the kinetic energy of the molecules becomes so large and their movement so rapid that the liquid boils to form a gas in which each particle moves more or less randomly through space. The energy associated with the motion of molecules is called kinetic energy. This leads us to a new definition of temperature:

Temperature is a measure of the average kinetic energy of the atoms, molecules, or ions that compose a substance.

Checkpoint

If liquid water and gaseous water are both at 100°C, how will the average kinetic energy of the water molecules in the liquid and gas compare?

6.3   THE STATES OF MATTER

Scientists divide matter into three **states:** gases, liquids, and solids. We shall begin our discussion of the states of matter with gases. There are two reasons for studying gases before liquids and solids. First, the behavior of gases is easier to describe, because most of the properties of gases don't depend on the identity of the gas. We can therefore develop a model for a gas without worrying about whether the gas is O_2, N_2, H_2, or a mixture of gases. Second, a relatively simple yet powerful model known as the *kinetic molecular theory* is available that explains most of the behavior of gases.

The term *gas* comes from the Greek word for chaos because gases consist of a chaotic collection of particles in constant, random motion. In the course of our discussion of gases we will provide the basis for answering the following questions.

- Why does popcorn "pop" when we heat it?
- Why does a hot-air balloon rise when the air in the balloon is heated?
- Why does a balloon filled with helium rise? Why does a balloon filled with CO_2 sink?
- At 25°C and 1 atmosphere of pressure, which weighs more: dry air or air that is saturated with water vapor?
- Is the volume of the O_2 in the room in which you are sitting the same as the volume of the N_2? Is the pressure of the O_2 the same as the pressure of the N_2?

6.4   ELEMENTS OR COMPOUNDS THAT ARE GASES AT ROOM TEMPERATURE

Before examining the chemical and physical properties of gases, it may be useful to ask: What kinds of elements or compounds are gases at room temperature? To help answer the question, a list of some common compounds that are gases at room temperature is given in Table 6.1.

There are several patterns in the data in Table 6.1.

- Common gases at room temperature include both elements (such as H_2 and O_2) and compounds (such as CO_2 and NH_3).
- Elements that are gases at room temperature are all *nonmetals* (such as He, Ar, N_2, and O_2).
- Compounds that are gases at room temperature are all *covalent compounds* (such as CO_2, SO_2, and NH_3) that contain two or more nonmetals.
- With rare exception, the gases have relatively small atomic or molecular weights.

As a general rule, elements and compounds that consist of relatively light, covalent molecules are most likely to be gases at room temperature.

TABLE 6.1   Common Gases at Room Temperature

Element or Compound	Molecular Weight (g/mol)
H_2 (hydrogen)	2.02
He (helium)	4.00
CH_4 (methane)	16.04
NH_3 (ammonia)	17.03
Ne (neon)	20.18
HCN (hydrogen cyanide)	27.03
CO (carbon monoxide)	28.01
N_2 (nitrogen)	28.01
NO (nitrogen oxide)	30.01
C_2H_6 (ethane)	30.07
O_2 (oxygen)	32.00
PH_3 (phosphine)	34.00
H_2S (hydrogen sulfide)	34.08
HCl (hydrogen chloride)	36.46
F_2 (fluorine)	38.00
Ar (argon)	39.95
CO_2 (carbon dioxide)	44.01
N_2O (dinitrogen oxide)	44.01
C_3H_8 (propane)	44.10
NO_2 (nitrogen dioxide)	46.01
O_3 (ozone)	48.00
C_4H_{10} (butane)	58.12
SO_2 (sulfur dioxide)	64.07
BF_3 (boron trifluoride)	67.81
Cl_2 (chlorine)	70.91
Kr (krypton)	83.80
CF_2Cl_2 (dichlorodifluoromethane)	120.92
SF_6 (sulfur hexafluoride)	146.05
Xe (xenon)	131.29

6.5   THE PROPERTIES OF GASES

Gases have three characteristic properties: (1) they are easy to compress, (2) they expand to fill their containers, and (3) they occupy far more space than equivalent masses of liquids or solids under normal atmospheric conditions.

Compressibility

The internal combustion engine found in most cars provides a good example of the ease with which gases can be compressed. In a typical four-stroke engine, the piston is first pulled out of the cylinder to create a partial vacuum, which draws a mixture of gasoline vapor and air into the cylinder (Figure 6.3). The piston is then pushed into the cylinder, compressing the gasoline–air mixture to a fraction of the original volume.

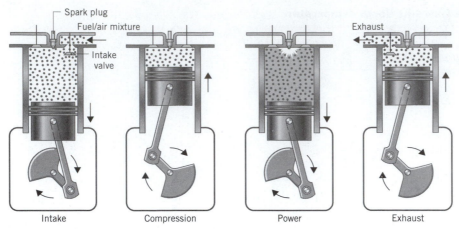

FIGURE 6.3 The operation of a four-stroke engine can be divided into four cycles: intake, compression, power, and exhaust stages.

The ratio of the volume of the gas in the cylinder after the first stroke, to its volume after the second stroke, is the *compression ratio* of the engine. Modern cars run at compression ratios of about 9:1, which means the gasoline–air mixture in the cylinder is compressed by a factor of 9 in the second stroke. After the gasoline–air mixture is compressed, the spark plug at the top of the cylinder fires, and the resulting explosion pushes the piston out of the cylinder in the third stroke. Finally, the piston is pushed back into the cylinder in the fourth stroke, clearing out the exhaust gases.

Liquids are much harder to compress than gases. They are so difficult to compress that the hydraulic brake systems used in most cars operate on the principle that there is essentially no change in the volume of the brake fluid when pressure is applied to the liquid. Most solids are even harder to compress. The only exceptions belong to a unique class of compounds that includes natural and synthetic rubber. Most rubber balls that seem easy to compress, such as a racquetball, are filled with air, which is compressed when the ball is squeezed.

Expandability

Anyone who has walked into a kitchen where bread is baking has experienced the fact that gases expand to fill their containers, as the air in the kitchen becomes filled with wonderful odors. Unfortunately the same thing happens when someone breaks open a rotten egg and the characteristic odor of hydrogen sulfide (H_2S) rapidly diffuses through the room. Because gases expand to fill their containers, the volume of a gas must be equal to the volume of its container.

Volumes of Gases versus Volumes of Liquids or Solids

The difference between the volume of a gas and the volume of the liquid or solid from which it forms can be illustrated with the following examples. One gram of liquid oxygen at its boiling point ($-183°C$) has a volume of 0.894 mL. The same amount of O_2 gas at $-183°C$ and atmospheric pressure has a volume of over 200 mL. Similar results are obtained when the volumes of solids and gases are compared. One gram of solid CO_2 at $-79°C$ has a volume of 0.641 mL. At $-79°C$ and atmospheric pressure, the same amount of CO_2 gas has a volume of over 300 mL.

The consequences of the enormous change in volume that occurs when a liquid or solid is transformed into a gas are frequently used to do work. The steam engine, which brought about the industrial revolution, is based on the fact that water boils to form a gas (steam) which has a much larger volume. The gas therefore escapes from the container in which it has been generated, and the escaping steam can be made to do work. The same principle is at work when dynamite is used to blast rocks. In 1867 the Swedish chemist Alfred Nobel discovered that the highly dangerous liquid explosive known as nitroglycerin could be absorbed onto clay or sawdust to produce a solid that was much more stable and therefore safer to use. When dynamite is detonated, the nitroglycerin decomposes to produce a mixture of CO_2, H_2O, N_2, and O_2 gases.

$$4\ C_3H_5N_3O_9(l) \longrightarrow 12\ CO_2(g) + 10\ H_2O(g) + 6\ N_2(g) + O_2(g)$$

Because 29 mol of gas is produced for every 4 mol of liquid that decomposes, and each mole of gas occupies a volume a few hundred times larger than a mole of liquid, the reaction produces a shock wave that destroys anything in its vicinity.

Popcorn "pops" because of the enormous difference between the volume of the liquids inside the kernel and the volume of the gases these liquids produce when they boil.

The same phenomenon occurs on a much smaller scale when we pop popcorn. When kernels of popcorn are heated in oil, the liquids inside the kernel turn into gases. The pressure that builds up inside the kernel is enormous and eventually causes the kernel to explode.

6.6 PRESSURE VERSUS FORCE

The volume of a gas is one of its characteristic properties. Another characteristic property is the **pressure** the gas exerts on its surroundings. Many of us got our first exposure to the pressure of a gas when we rode to the neighborhood gas station to check the pressure in our bicycle tires. Depending on the kind of bicycle we had, we added air to the tires until the pressure gauge read between 30 and 70 pounds per square inch (70 lb/in^2) or psi. Two important properties of pressure can be gleaned from this example.

- The pressure of a gas becomes greater as more gas is added to the container (as long as the volume is constant and the temperature does not change).

- Pressure is measured in units (such as lb/in^2) that describe the **force** exerted by the gas divided by the **area** over which the force is distributed.[2]

The first conclusion can be summarized in the following relationship, where P is the pressure of the gas and n is the number of moles of gas in the container.

$$P \propto n \ (T \text{ and } V \text{ constant})$$

The symbol $\propto$ means "proportional to," and the relationship is read as "pressure is proportional to the number of moles of gas." Because the pressure increases as gas is added to a container with a fixed volume at a given temperature, P is directly proportional to n.

The second conclusion describes the relationship between pressure and force. Pressure is defined as the force exerted on an object divided by the area over which the force is distributed.

$$\text{Pressure} = \frac{\text{force}}{\text{area}}$$

The difference between pressure and force can be illustrated with an analogy based on a 10-penny nail, a hammer, and a piece of wood (Figure 6.4). By resting the nail on its point and hitting the head with the hammer, we can drive the nail into the wood. But what happens if we turn the nail over and rest the head of the nail against the wood? If we hit the nail with the same force, we can't get the nail to penetrate into the wood.

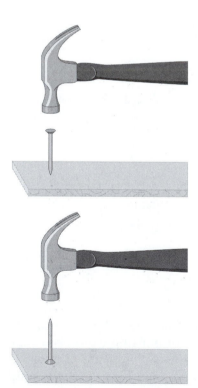

FIGURE 6.4 The *force* exerted by a hammer hitting a nail is the same regardless of whether the hammer hits the nail on the head or on the point. But the *pressure* exerted on the wood is very different.

---

[2]The SI unit for pressure is the pascal (Pa), which is defined as a force of 1 newton averaged over an area of $1 \ m^2$.

When we hit the nail on the head, the force of the blow is applied to the very small area of the wood in contact with the point of the nail, and the nail slips easily into the wood. But when we turn the nail over, and hit it on the point, the same force is distributed over a much larger area. The force is now distributed over the surface of the wood that touches any part of the nail head. As a result, the pressure applied to the wood is much smaller and the nail just bounces off the wood.

Checkpoint

Is the pressure exerted on the floor by a person wearing a spike heel greater than, less than, or equal to the pressure exerted when the same person wears a flat sandal? Why?

Exercise 6.1

(a) Calculate the pressure exerted by the shoes of a 200-lb man wearing size 10 shoes, if each shoe makes contact with an area of the floor that is 20 in^2.

(b) Calculate the pressure exerted by each heel of a 100-lb woman in high heels, if the area beneath the heel of each shoe is 0.25 in^2.

Solution

(a) The pressure can be calculated by dividing the force by the area over which it is distributed. Because one-half of the weight of the man is applied to each shoe, the pressure in this case is 5 lb/in^2.

$$\text{Pressure} = \frac{\text{force}}{\text{area}} = \frac{100 \text{ lb}}{20 \text{ in}^2} = 5.0 \text{ lb/in}^2$$

(b) We can assume that about one-fourth of the weight of the woman is applied to each heel, if her weight is evenly divided between the heel and sole of each shoe.

$$\text{Pressure} = \frac{\text{force}}{\text{area}} = \frac{25 \text{ lb}}{0.25 \text{ in}^2} = 1.0 \times 10^2 \text{ lb/in}^2$$

The pressure exerted by the heels of the 100-lb woman is 20 times greater than the pressure exerted by the man, even though he weighs twice as much.

6.7 ATMOSPHERIC PRESSURE

What would happen if we bent a long piece of glass tubing into the shape of the letter U and then carefully filled one arm of the U-tube with water and the other arm with ethyl alcohol? Most people expect the height of the columns of liquid in the two arms of the tube to be the same. Experimentally, we find the results shown in Figure 6.5. A 100-cm column of water balances a 127-cm column of ethyl alcohol, regardless of the diameter of the glass tubing.

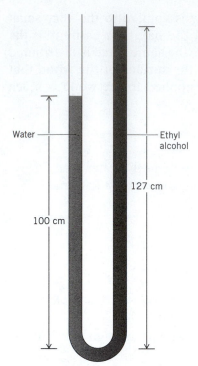

Water

Ethyl
alcohol

127 cm

100 cm

FIGURE 6.5 Because of the difference between the densities of water and ethyl alcohol, the weight of a column of water 100 cm long balances the weight of a column of ethyl alcohol 127 cm long in the other arm of a U-tube.

We can explain this observation by comparing the densities of water (1.00 g/cm^3) and ethyl alcohol (0.789 g/cm^3). A column of water 100 cm tall exerts a pressure proportional to 100 g/cm^2.[3]

$$100 \text{ cm} \times \frac{1.00 \text{ g}}{1 \text{ cm}^3} = 100 \text{ g/cm}^2$$

A column of ethyl alcohol 127 cm tall exerts the same pressure.

$$127 \text{ cm} \times \frac{0.789 \text{ g}}{1 \text{ cm}^3} = 100 \text{ g/cm}^2$$

Because the pressure of the water pushing down on one arm of the U-tube is equal to the pressure of the alcohol pushing down on the other arm of the tube, the system is in balance. This demonstration provides the basis for understanding how a mercury barometer can be used to measure the pressure of the atmosphere.

The Discovery of the Barometer

In the early 1600s, Galileo argued that suction pumps were able to draw water from a well because of the "force of vacuum" inside the pump. After Galileo's death, the Italian mathematician and physicist Evangelista Torricelli (1608–1647) proposed another explanation. He suggested that the air in the atmosphere has weight and that the force of the atmosphere pushing down on the surface of the water drives the water into the suction pump when it is evacuated.

In 1646 Torricelli described an experiment in which a glass tube about 1 meter long was sealed at one end, filled with mercury, and then inverted into a dish filled with mercury, as

---

[3]Pressure is defined as force/area. The force is the mass times the acceleration of gravity (ma). Thus the actual pressure is 100g/cm^2 × 980 cm/sec^2.

shown in Figure 6.6. Some, but not all, of the mercury drained out of the glass tube into the dish. Torricelli explained this by assuming that mercury drains from the glass tube until the pressure of the column of mercury pushing down on the *inside* of the tube exactly balances the pressure of the atmosphere pushing down on the surface of the liquid *outside* the tube.

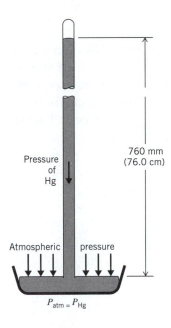

760 mm
(76.0 cm)

Pressure
of
Hg

Atmospheric   pressure

$P_{atm} = P_{Hg}$

FIGURE 6.6 On a sunny day, at sea level, the pressure of a 760-mm-high column of mercury inside a glass tube balances the pressure of the atmosphere pushing down on the pool of mercury that surrounds the tube. The pressure of the atmosphere is therefore said to be equivalent to 760 mmHg.

Torricelli predicted that the height of the mercury column would change from day to day as the pressure of the atmosphere changed. Today, his apparatus is known as a *barometer*, from the Greek word *baros,* meaning "weight," because it literally measures the weight of the atmosphere. Repeated experiments showed that the average pressure of the atmosphere at sea level is equal to the pressure of a column of mercury 760 mm tall. Thus, a standard unit of pressure known as the *atmosphere* was defined as follows.

$$1 \text{ atm} = 760 \text{ mmHg} \qquad (\text{exactly})$$

To recognize Torricelli's contributions, some scientists describe pressure in units of torr, defined as follows.

$$1 \text{ torr} = 1 \text{ mmHg}$$

Exercise 6.2

Calculate the pressure in units of atmospheres on a day when a barometer gives a measurement of 745.8 mmHg.

Solution

The conversion between mmHg and atmospheres is based on the following definition.

$$1 \text{ atm} = 760.0 \text{ mmHg}$$

Using that equality to generate an appropriate unit factor gives the following result.

$$745.8 \text{ mmHg} \times \frac{1 \text{ atm}}{760.0 \text{ mmHg}} = 0.9813 \text{ atm}$$

Although chemists still work with pressures in units of atm or mmHg, neither unit is accepted in the SI system (see Appendix page A-3). The SI unit of pressure is the pascal (Pa). The relationship between one standard atmosphere pressure and the pascal is given by the following.

$$1 \text{ atm} = 101{,}325 \text{ Pa} = 101.325 \text{ kPa} = 0.101325 \text{ MPa} = 14.7 \text{ lb/in}^2$$

The pressure of the atmosphere can be demonstrated by connecting a 1-gallon can to a vacuum pump. Normally the pressure inside the can balances the pressure of the atmosphere pushing on the outside of the can. When the vacuum pump is turned on, however, the can rapidly collapses as it is evacuated. The surface area of a 1-gallon can is about 250 in^2. At 14.7 lb/in^2, this corresponds to a total force over the surface of the can of about 3700 lb. For the sake of comparison, it might be noted that each of the 18 wheels of a 70,000-lb truck carries only about 3900 lb.

The pressure of the atmosphere can be demonstrated by connecting an empty paint-thinner can to a vacuum pump. Within seconds of turning on the vacuum pump, the can collapses.

The Difference between Pressure of a Gas and Pressure Resulting from Weight

There is an important difference between the pressure of a gas and the other examples of pressure discussed in this section. The pressure exerted by a 70,000-pound truck is directional. The truck exerts all of its pressure on the surface beneath its wheels. In contrast, gas pressure is the same in all directions. To demonstrate this, we can fill a glass cylinder with water and rest a glass plate on top of the cylinder. When we turn the cylinder over, the plate doesn't fall to the floor because the pressure of the air outside the cylinder pushing up on the bottom of the plate is larger than the pressure exerted by the water in the cylinder pushing down on the plate. It would take a column of water 33.9 feet tall to produce as much pressure as the gas in the atmosphere.

6.8 BOYLE'S LAW

Torricelli's work with the mercury barometer caught the eye of the British scientist Robert Boyle. Boyle's most famous experiments were done in a J-tube apparatus similar to the one shown in Figure 6.7. By adding mercury to the open end of the tube, Boyle was able to trap a small volume of air in the sealed end.

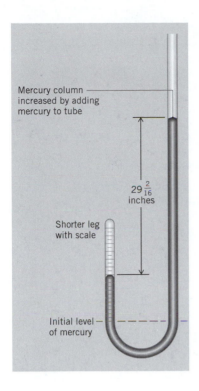

Mercury column increased by adding mercury to tube

$29\frac{2}{16}$ inches

Shorter leg with scale

Initial level of mercury

FIGURE 6.7 Boyle's law is based on data obtained with a J-tube apparatus. The temperature and number of moles of gas are constant. The volume is in arbitrary units and the height is in inches.

Boyle was interested in a phenomenon he called the "spring of air." His experiments were based on the fact that gases are *elastic*. (They return to their original size and shape after being stretched or squeezed.) When he added more mercury to the open end of the J-tube, Boyle noticed that the air in the sealed end was compressed into a smaller volume.

Table 6.2 contains some of the experimental data he reported in his book, *New Experiments Physico-Mechanicall, Touching the Spring of Air, and its Effects. . .* , published in 1662. The first column in Table 6.2 lists the volume of the gas in the sealed end of the J-tube, in arbitrary units. The second column is the difference in heights of the mercury in the sealed and open arms of the J-tube, to the nearest ±1/16 inch. The third column is the product of the volume of the gas (V) and the pressure (P) calculated from Boyle's data. The number of moles of gas and temperature were held constant for the experiments.

The product of the pressure times the volume for any measurement in the table is equal to the product of the pressure times the volume for any other measurement.

$$P_1V_1 = P_2V_2$$

This expression, or its equivalent,

$$P \propto 1/V \qquad (T \text{ and } n \text{ constant})$$

is now known as **Boyle's law.**

Checkpoint

Does the diameter of the tube used in Boyle's experiments affect the results obtained?

TABLE 6.2   Boyle's Data on the Dependence of the Volume of a Gas on the Pressure of the Gas

Volume	Pressure	$P \times V$
48	$29\frac{2}{16}$	1.4×10^3
46	$30\frac{9}{16}$	1.4×10^3
44	$31\frac{15}{16}$	1.4×10^3
42	$33\frac{8}{16}$	1.4×10^3
40	$35\frac{5}{16}$	1.4×10^3
38	37	1.4×10^3
36	$39\frac{4}{16}$	1.4×10^3
34	$41\frac{10}{16}$	1.4×10^3
32	$44\frac{3}{16}$	1.4×10^3
30	$47\frac{1}{16}$	1.4×10^3
⋮	⋮	⋮

6.9   AMONTONS' LAW

Toward the end of the 1600s, the French physicist Guillaume Amontons built a thermometer based on the fact that the pressure of a gas is directly proportional to its temperature. Thus, the relationship between the pressure and the temperature of a gas is known as **Amontons' law.**

$$P \propto T \qquad (V \text{ and } n \text{ constant})$$

Amontons' law explains why car manufacturers recommend adjusting the pressure of your tires before you start on a trip. The flexing of the tire as you drive inevitably raises the temperature of the air in the tire. When this happens, the pressure of the gas inside the tires increases.

Amontons' law can be demonstrated with the apparatus shown in Figure 6.8. Data obtained with the apparatus at various temperatures are given in Table 6.3.

The relationship between the temperature and pressure of a gas provided the first definition for absolute zero on the temperature scale. In 1779 Joseph Lambert defined **absolute zero** as the temperature at which the pressure of a gas becomes zero. When the data in Table 6.3 are plotted, the pressure of a gas approaches zero when the temperature is about $-270°C$, as shown in Figure 6.9. When more accurate measurements are made, the pressure of a gas extrapolates to zero when the temperature is $-273.15°C$. Absolute zero on the Celsius scale is therefore $-273.15°C$.

TABLE 6.3   The Dependence of the Pressure of a Gas on Its Temperature

Temperature (°C)	Pressure (lb/in²)
100	18.1
74	16.7
24	14.5
0	13.2
−47	10.8

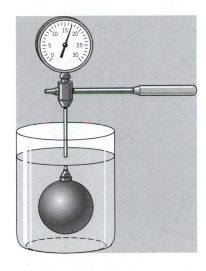

FIGURE 6.8 The apparatus for demonstrating Amontons' law consists of a pressure gauge connected to a metal sphere of constant volume, which is immersed in water at different temperatures.

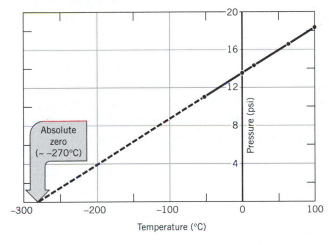

FIGURE 6.9 When data obtained with the Amontons' law apparatus in Figure 6.8 are extrapolated, the pressure of a gas approaches zero when the temperature of the gas is approximately −270°C.

The relationship between the temperature and pressure data in Table 6.3 can be simplified by converting the temperatures from the Celsius to the Kelvin scale.

$$T_K = T_{°C} + 273.15$$

When this is done, a plot of the temperature versus the pressure of a gas gives a straight line that passes through the origin. Any two points along the line therefore fit the following equation.

$$\frac{P_1}{P_2} = \frac{T_1}{T_2}$$

It is important to remember that this equation is only valid if the temperatures are converted from the Celsius to the Kelvin scale before calculations are done, and the volume of gas and number of moles of gas must be constant.

Checkpoint

If Boyle's and Amontons' laws are combined, what is the resulting relationship?

6.10   CHARLES' LAW

On June 5, 1783, Joseph and Etienne Montgolfier used a fire to inflate a spherical balloon about 30 feet in diameter that traveled about $1\frac{1}{2}$ miles before it came back to earth. News of the remarkable achievement spread throughout France, and Jacques-Alexandre-César Charles immediately tried to duplicate the performance. As a result of his work with balloons, Charles noticed that the volume of a gas is directly proportional to its temperature when the pressure and number of moles of gas are fixed.

$$V \propto T \qquad (P \text{ and } n \text{ constant})$$

This relationship between the temperature and volume of a gas, which became known as **Charles' law,** can be used to explain how a hot-air balloon works. Ever since the third century B.C., it has been known that an object floats when it weighs less than the fluid it displaces. If a gas expands when heated, then a given weight of hot air occupies a larger volume than the same weight of cold air. Hot air is therefore less dense than cold air. Once the air in a balloon gets hot enough, the net weight of the balloon plus the hot air is less than the weight of an equivalent volume of cold air, and the balloon starts to rise. When the gas in the balloon is allowed to cool, the balloon returns to the ground.

Charles' law can be demonstrated with the apparatus shown in Figure 6.10. A 30-mL syringe and a thermometer are inserted through a rubber stopper into a flask that has been cooled to 0°C. The ice bath is then removed and the flask is immersed in a warm-water bath. The gas in the flask expands as it warms, slowly pushing the piston out of the syringe. The total volume of the gas in the system is equal to the volume of the flask plus the volume of the syringe. Table 6.4 contains typical data obtained with this apparatus.

TABLE 6.4   Dependence of the Volume of a Gas on Its Temperature

Temperature (°C)	Volume (mL)
0	107.9
5	109.7
10	111.7
15	113.6
20	115.5
25	117.5
30	119.4
35	121.3
40	123.2

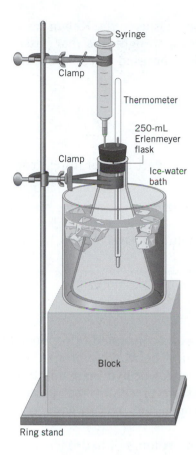

Syringe

Clamp

Thermometer

250-mL
Erlenmeyer
flask

Clamp

Ice-water
bath

Block

Ring stand

FIGURE 6.10 Charles' law can be demonstrated with a simple apparatus. When the flask is removed from the ice bath and placed in a warm
water bath, the gas in the flask expands, slowly pushing up on the piston of the syringe.

The data in Table 6.4 are plotted in Figure 6.11. The graph provides us with another
way of defining absolute zero on the temperature scale. It is the temperature at which the
volume of a gas becomes zero when the plot of volume versus temperature for a gas is extrapolated. As expected, the value of absolute zero obtained by extrapolating the data in
Table 6.4 is essentially the same as the value obtained from the graph of pressure versus
temperature in the preceding section. Absolute zero can therefore be more accurately defined as the temperature at which the pressure and the volume of a gas extrapolate to zero.

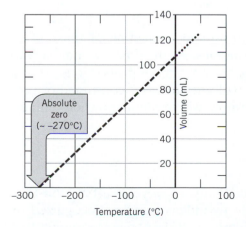

Absolute
zero
(~ −270°C)

Volume (mL)

−300   −200   −100   0   100

Temperature (°C)

FIGURE 6.11 When data obtained with the Charles' law apparatus in Figure 6.10 are extrapolated, the volume of the gas approaches zero when the temperature of the gas is approximately −270°C.

When the temperatures in Table 6.4 are converted from the Celsius to the Kelvin scale, a plot of the volume versus the temperature of a gas becomes a straight line that passes through the origin. Any two points along the line can therefore be used to construct the following equation, which is known as Charles' law.

$$\frac{V_1}{V_2} = \frac{T_1}{T_2}$$

Before you use the equation, it is important to remember to convert temperatures from °C to K.

Checkpoint

Could Charles' law have been predicted from Boyle's and Amontons' laws, or is it independent of those laws?

6.11   GAY-LUSSAC'S LAW

Joseph Louis Gay-Lussac (1778–1850) studied the volume of gases consumed or produced in a chemical reaction because he was interested in the reaction between hydrogen and oxygen to form water. He argued that measurements of the *weights* of hydrogen and oxygen consumed in the reaction could be influenced by the moisture present in the reaction flask, but the moisture would not affect the *volumes* of hydrogen and oxygen gases consumed in the reaction.

Much to his surprise, Gay-Lussac found that 199.89 parts by volume of hydrogen were consumed for every 100 parts by volume of oxygen. Thus, hydrogen and oxygen seemed to combine in a simple 2:1 ratio by volume.

$$\text{hydrogen} + \text{oxygen} \longrightarrow \text{water}$$
$$\underset{2\ volumes}{} \quad \underset{1\ volume}{}$$

Gay-Lussac found similar whole number ratios for the reactions between other pairs of gases. He also obtained similar results when he analyzed the volumes of gases given off when compounds decomposed. Ammonia, for example, decomposed to give three times as much hydrogen by volume as nitrogen.

$$\text{ammonia} \longrightarrow \text{nitrogen} + \text{hydrogen}$$
$$\underset{1\ volume}{} \quad \underset{3\ volumes}{}$$

On December 31, 1808, Gay-Lussac announced his results at a meeting of the Societé Philomatique in Paris in terms of a **law of combining volumes.** At the time, he summarized the law as follows: Gases combine among themselves in very simple proportions. Today, Gay-Lussac's law is stated as follows: The ratio of the volumes of gases consumed or produced in a chemical reaction is equal to the ratio of simple whole numbers when temperature and pressure are constant.

6.12   AVOGADRO'S HYPOTHESIS

Gay-Lussac's law of combining volumes was announced only a few years after John Dalton proposed his atomic theory. The link between the two ideas was first recognized by the Italian physicist Amadeo Avogadro three years later, in 1811. Avogadro argued that

Gay-Lussac's law of combining volumes could be explained by assuming that equal volumes of different gases collected under similar conditions contain the same number of particles.

HCl and NH_3 combine in a 1:1 ratio by volume, for example, because one molecule of HCl is consumed for every molecule of NH_3 in the reaction and equal volumes of the gases contain the same number of molecules.

$$NH_3(g) + HCl(g) \longrightarrow NH_4Cl(s)$$

Anyone who has blown up a balloon can accept the notion that the volume of a gas is proportional to the number of particles in the gas.

$$V \propto n \qquad (T \text{ and } P \text{ constant})$$

The more air you add to a balloon, the bigger it gets. Unfortunately this example does not test **Avogadro's hypothesis,** that equal volumes of *different gases* contain the same number of particles. The best way to prove the validity of the hypothesis is to measure the number of molecules in a given volume of different gases, which can be done with the apparatus shown in Figure 6.12. The temperature and pressure are held constant.

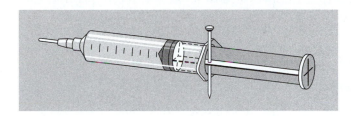

FIGURE 6.12 Apparatus used to demonstrate Avogadro's hypothesis.

A small hole is drilled through the plunger of a 50-mL plastic syringe. The plunger is pushed into the syringe and the syringe is sealed with a syringe cap. The plunger is then pulled out of the syringe until the volume reads 50 mL, and a nail is inserted through the hole in the plunger so that the plunger is not drawn back into the barrel of the syringe. The "empty" syringe is then weighed, the syringe is filled with 50 mL of a gas, and the syringe is reweighed. The difference between the measurements is the mass of 50 mL of the gas.

The results of experiments with six gases are given in Table 6.5. The number of molecules in a 50-mL sample of any one of the gases can be calculated from the mass of the sample, the molecular weight of the gas, and the number of molecules in a mole. Consider the following calculation of the number of H_2 molecules in 50 mL of hydrogen gas, for example.

$$0.005 \text{ g } H_2 \times \frac{1 \text{ mol } H_2}{2.02 \text{ g } H_2} \times \frac{6.02 \times 10^{23} \text{ molecules}}{1 \text{ mol}} = 1 \times 10^{21} \text{ } H_2 \text{ molecules}$$

The last column in Table 6.5 summarizes the results obtained when the calculation is repeated for each gas. The number of significant figures in the answer changes from one calculation to the next. But the number of molecules in each sample is the same, within experimental error. We therefore conclude that equal volumes of different gases collected under the same conditions of temperature and pressure do in fact contain the same number of particles.

TABLE 6.5   Experimental Data for the Mass of 50-mL Samples of Different Gases (T and P are constant)

Compound	Mass of 50 mL Gas (g)	Molecular Weight of Gas (g/mol)	Number of Molecules in 50 mL Gas
H_2	0.005	2.02	1×10^{21}
N_2	0.055	28.01	1.2×10^{21}
O_2	0.061	32.00	1.1×10^{21}
CO_2	0.088	44.01	1.2×10^{21}
C_4H_{10}	0.111	58.12	1.15×10^{21}
CCl_2F_2	0.228	120.91	1.14×10^{21}

6.13   THE IDEAL GAS EQUATION

So far, gases have been described in terms of four variables: pressure (P), volume (V), temperature (T), and the number of moles of gas (n). In the course of this chapter, five relationships between pairs of these variables have been discussed. In each case, two of the variables were allowed to change while the other two were held constant. The discussion of the bicycle tire, for example, showed that the pressure of a gas is directly proportional to the moles of gas when the temperature and volume of the gas are held constant.

$$P \propto n \qquad (T \text{ and } V \text{ constant})$$

Other relationships between pairs of variables include the following.

Boyle's law:	$P \propto 1/V$	(T and n constant)
Amontons' law:	$P \propto T$	(V and n constant)
Charles' law:	$V \propto T$	(P and n constant)
Avogadro's hypothesis:	$V \propto n$	(P and T constant)

Each of the relationships is a special case of a more general relationship known as the **ideal gas equation.**

$$PV = nRT$$

In this equation, R is a proportionality constant known as the **ideal gas constant** and T is the absolute temperature. The value of R depends on the units used to express the four variables P, V, n, and T. By convention, most chemists use the following set of units.

P: atmospheres T: kelvins
V: liters n: moles

According to the ideal gas law, the product of pressure times volume of an ideal gas divided by the product of amount of gas times absolute temperature is a constant.

$$\frac{PV}{nT} = R$$

We can calculate the value of R from experimental data similar to those presented in Table 6.4. In the experiments the number of moles of gas and the pressure are held constant. For exactly 1 mol of gas at 1.0000 atm and 0°C (standard temperature and pressure, **STP**) the volume is found to be 22.414 L. Therefore, R can be calculated using the following equation:

$$\frac{(1.0000 \text{ atm})(22.414 \text{ L})}{(1.0000 \text{ mol})(273.15 \text{ K})} = \textbf{0.082057 L} \cdot \textbf{atm/mol} \cdot \textbf{K}$$

For most ideal gas calculations, four significant figures are sufficient. The standard value of the ideal gas constant is therefore 0.08206 L · atm/mol · K.

6.14   IDEAL GAS CALCULATIONS: PART I

The ideal gas equation can be used to predict the value of any one of the variables that describes a gas from known values of the other three.

Exercise 6.3

Many gases are available for use in the laboratory in compressed gas cylinders in which they are stored at high pressures. Calculate the mass of O_2 that could be stored at 21°C and 170 atm in a cylinder with a volume of 60.0 L.

Solution

For this problem we know three of the four variables in the ideal gas equation. We know the pressure (170 atm), volume (60.0 L), and temperature (21°C) of the gas, as shown in Figure 6.13. This suggests that the ideal gas equation will play an important role in solving the problem.

$$PV = nRT$$

60.0 L
170 atm
21°C (294 K)

FIGURE 6.13 Compressed gas cylinder.

Before we can use the equation, we have to convert the temperature from °C to K.

$$T_K = T_{°C} + 273 = 294 \text{ K}$$

We can then substitute the known information into the ideal gas equation:

$$(170 \text{ atm})(60.0 \text{ L}) = (n)(0.08206 \text{ L} \cdot \text{atm/mol} \cdot \text{K})(294 \text{ K})$$

and then solve the equation for the number of moles of gas in the container.

$$n = \frac{(170 \text{ atm})(60.0 \text{ L})}{(0.08206 \text{ L} \cdot \text{atm/mol} \cdot \text{K})(294 \text{ K})} = 423 \text{ mol}$$

We then use the mass of a mole of O_2 to calculate the number of grams of oxygen that could be stored in the cylinder.

$$423 \ \text{mol} \ O_2 \times \frac{32.00 \text{ g } O_2}{1 \ \text{mol} \ O_2} = \mathbf{1.35 \times 10^4 \text{ g } O_2}$$

According to the calculation, 13.5 kilograms of O_2 can be stored in a 60.0-L compressed gas cylinder at 21°C and 170 atm.

The key to solving ideal gas problems often involves recognizing what is known and deciding how to use the information.

Exercise 6.4

Calculate the mass of the air in a hot-air balloon that has a volume of 4.00×10^5 L when the temperature of the gas is 30°C and the pressure is 748 mmHg. Assume that the air in the atmosphere is a mixture of roughly 78% N_2 (28.0 g/mol), 21% O_2 (32.0 g/mol), and 1% Ar (39.9 g/mol) and therefore has an average molecular weight of about 29.0 g/mol.

Solution

We know the pressure (748 mmHg), volume (4.00×10^5 L), and temperature (30°C) of the gas, as shown in Figure 6.14. We might therefore consider using the ideal gas equation to calculate the number of moles of gas in the balloon.

$$PV = nRT$$

4.00 × 10⁵ L
748 mmHg
(0.984 atm)
30°C
(303 K)

FIGURE 6.14 Hot-air balloon.

Before we can do that, however, we have to convert the pressure to units of atmospheres,

$$748 \ \text{mmHg} \times \frac{1 \text{ atm}}{760 \ \text{mmHg}} = 0.984 \text{ atm}$$

and the temperature to kelvins.

$$T_K = T_{°C} + 273 = 303 \text{ K}$$

We can then substitute the information into the ideal gas equation and solve for the variable we don't know.

$$n = \frac{PV}{RT} = \frac{(0.984 \text{ atm})(4.00 \times 10^5 \text{ L})}{(0.08206 \text{ L} \cdot \text{atm/mol} \cdot \text{K})(303 \text{ K})} = 1.58 \times 10^4 \text{ mol}$$

We then go back and reread the problem to see whether we are any closer to its solution.

The problem asks for the mass of the air in the balloon, which can be calculated from the number of moles of gas and the average molecular weight of air.

$$1.58 \times 10^4 \text{ mol} \times \frac{29.0 \text{ g}}{1 \text{ mol}} = \textbf{4.58} \times \textbf{10}^5 \textbf{ g}$$

The balloon therefore contains just over 1000 pounds of air.

The ideal gas equation can even be applied to problems that don't seem to ask for one of the variables in this equation.

Exercise 6.5

Calculate the molecular weight of butane if 0.5813 g of the gas fills a 250.0-mL flask at a temperature of 24.4°C and a pressure of 742.6 mmHg.

Solution

We know something about the pressure (742.6 mmHg), volume (250.0 mL), and temperature (24.4°C) of the gas, as shown in Figure 6.15. We might therefore start by calculating the number of moles of gas in the sample. To do this, we need to convert the pressure, volume, and temperature into appropriate units.

250.0 mL
24.4°C
(297.6 K)

0.5813 g butane
742.6 mmHg
(0.9771 atm)   **FIGURE 6.15** Stoppered round-bottom flask containing butane.

$$742.6 \text{ mmHg} \times \frac{1 \text{ atm}}{760.0 \text{ mmHg}} = 0.9771 \text{ atm}$$

$$250.0 \text{ mL} \times \frac{1 \text{ L}}{1000 \text{ mL}} = 0.2500 \text{ L}$$

$$T_K = T_{°C} + 273.15 = 297.6 \text{ K}$$

Solving the ideal gas equation for the number of moles of gas and substituting the known values of the pressure, volume, and temperature into the equation give the following result.

$$n = \frac{PV}{RT} = \frac{(0.9771 \text{ atm})(0.2500 \text{ L})}{(0.08206 \text{ L} \cdot \text{atm/mol} \cdot \text{K})(297.6 \text{ K})} = \textbf{0.01000 mol}$$

The problem asks for the molecular weight of the gas. At this point, we know the mass of the gas in the sample and the number of moles of gas in the sample. We can therefore calculate the molecular weight of butane by dividing the mass of the sample by the number of moles of gas in the sample.

$$\frac{0.5813 \text{ g}}{0.01000 \text{ mol}} = \textbf{58.13 g/mol}$$

The molecular weight of butane is therefore 58.13 g/mol.

The ideal gas equation can even be used to solve problems that don't seem to contain enough information.

Exercise 6.6

Calculate the density in grams per liter of O_2 gas at 0°C and 1.00 atm.

Solution

This time we have information about only two of the variables in the ideal gas equation.

$$P = 1.00 \text{ atm} \qquad T = 0°C$$

What can we calculate from this information? One way to answer that question is to rearrange the ideal gas equation, putting the knowns on one side and the unknowns on the other.

$$\frac{n}{V} = \frac{P}{RT}$$

We now have enough information to calculate the number of moles of gas per liter.

$$\frac{n}{V} = \frac{P}{RT} = \frac{(1.00 \text{ atm})}{(0.08206 \text{ L} \cdot \text{atm/mol} \cdot \text{K})(273 \text{ K})} = \textbf{0.0446 mol } O_2\textbf{/L}$$

The problem asks for the density of the gas in grams per liter. Because we know the number of moles of O_2 per liter, we can use the molar mass of O_2 to calculate the number of grams per liter.

$$\frac{0.0446 \text{ mol } O_2}{1 \text{ L}} \times \frac{32.00 \text{ g } O_2}{1 \text{ mol } O_2} = \textbf{1.43 g } O_2\textbf{/L}$$

The density of O_2 gas at 0°C and 1.00 atm is therefore 1.43 g/L.

Exercise 6.7

Explain why balloons filled with helium at room temperature (21°C) and 1.00 atm float, whereas balloons filled with CO_2 sink.

Solution

Avogadro's hypothesis suggests that equal volumes of different gases at the same temperature and pressure contain the same number of gas particles. But this hypothesis is nothing more than a special case of the ideal gas law, which predicts that the number of moles of gas per liter at room temperature ($21°C$) and 1 atm does not depend on the identity of the gas.

$$\frac{n}{V} = \frac{P}{RT} = \frac{(1.00 \text{ atm})}{(0.08206 \text{ L} \cdot \text{atm/mol} \cdot \text{K})(294 \text{ K})} = 0.0414 \text{ mol/L}$$

The number of grams per liter, however, depends on the mass of a mole of each gas. Helium is less dense than air, whereas CO_2 is more dense than air.

$$d_{He} = \frac{0.0414 \text{ mol}}{1 \text{ L}} \times \frac{4.00 \text{ g He}}{1 \text{ mol}} = 0.166 \text{ g He/L}$$

$$d_{air} = \frac{0.0414 \text{ mol}}{1 \text{ L}} \times \frac{29.0 \text{ g air}}{1 \text{ mol}} = 1.20 \text{ g air/L}$$

$$d_{CO_2} = \frac{0.0414 \text{ mol}}{1 \text{ L}} \times \frac{44.0 \text{ g CO}_2}{1 \text{ mol}} = 1.82 \text{ g CO}_2\text{/L}$$

A balloon filled with helium therefore weighs significantly less than the air it displaces, and it floats. A balloon filled with CO_2, on the other hand, weighs more than the air it displaces, so it sinks.

Checkpoint

Use the difference between the average molecular weights of air (29.0 g/mol) and water (18.0 g/mol) to explain why 1 L of dry air weighs more than 1 L of air that has been saturated with water vapor.

6.15   IDEAL GAS CALCULATIONS: PART II

Gas law problems often ask you to predict what happens when one or more changes are made in the variables that describe the gas. The most powerful approach to such problems is based on the fact that the ideal gas constant is, in fact, a constant.

We start by solving the ideal gas equation for the ideal gas constant.

$$R = \frac{PV}{nT}$$

Thus, the ratio PV/nT under any set of conditions must be equal to the ratio under any other set of conditions.

$$\frac{P_1 V_1}{n_1 T_1} = \frac{P_2 V_2}{n_2 T_2}$$

We then substitute the known values of pressure, temperature, volume, and amount of gas into the equation and solve for the appropriate unknown.

Exercise 6.8

A sample of NH_3 gas fills a 27.0-L container at $-15°C$ and 2.58 atm. Calculate the volume of the gas at 21°C and 751 mmHg.

Solution

We can start by listing what we know and what we don't know about the system.

Initial conditions   *Final conditions*

$P_1 = 2.58$ atm $P_2 = 751$ mmHg

$V_1 = 27.0$ L $V_2 = ?$

$T_1 = -15°C$ $T_2 = 21°C$

$n_1 = ?$ $n_2 = ?$

We convert the data to a consistent set of units for use in the ideal gas equation.

Initial conditions   *Final conditions*

$P_1 = 2.58$ atm $P_2 = 0.988$ atm

$V_1 = 27.0$ L $V_2 = ?$

$T_1 = 258$ K $T_2 = 294$ K

$n_1 = ?$ $n_2 = ?$

We can then substitute the information into the following equation.

$$\frac{P_1 V_1}{n_1 T_1} = \frac{P_2 V_2}{n_2 T_2}$$

When this is done we get one equation with three unknowns: n_1, n_2, and V_2.

$$\frac{(2.58 \text{ atm})(27.0 \text{ L})}{(n_1)(258 \text{ K})} = \frac{(0.988 \text{ atm})(V_2)}{(n_2)(294 \text{ K})}$$

It is impossible to solve one equation with three unknowns. But a careful reading of the problem suggests that the number of moles of NH_3 is the same before and after the temperature and pressure change.

$$n_1 = n_2$$

Substituting this equality into the unknown equation gives the following.

$$\frac{(2.58 \text{ atm})(27.0 \text{ L})}{(n_1)(258 \text{ K})} = \frac{(0.988 \text{ atm})(V_2)}{(n_1)(294 \text{ K})}$$

Multiplying both sides of the equation by n_1 gives one equation with one unknown.

$$\frac{(2.58 \text{ atm})(27.0 \text{ L})}{(258 \text{ K})} = \frac{(0.988 \text{ atm})(V_2)}{(294 \text{ K})}$$

We therefore solve for the only unknown left in the equation.

$$V_2 = \frac{(2.58 \text{ atm})(27.0 \text{ L})(294 \text{ K})}{(258 \text{ K})(0.988 \text{ atm})} = \textbf{80.3 L}$$

According to the calculation, the volume of the gas increases from 27.0 L at $-15°C$ and 2.58 atm to 80.3 L at 21°C and 751 mmHg.

6.16 DALTON'S LAW OF PARTIAL PRESSURES

The *CRC Handbook of Chemistry and Physics* describes the atmosphere as being 78.084% N_2, 20.946% O_2, 0.934% Ar, and 0.033% CO_2 by volume after the water vapor has been removed. What image does this description evoke in your mind? Do you believe that only 20.946% of the room in which you are sitting contains O_2? Or do you believe that the atmosphere in your room is a more or less homogeneous mixture of the gases?

In Section 6.5 we argued that gases expand to fill their containers. The volume of O_2 in your room is therefore the same as the volume of N_2. Both gases expand to fill the room.

What about the pressure of the different gases in your room? Is the pressure of the O_2 in the atmosphere the same as the pressure of the N_2? We can answer the question by rearranging the ideal gas equation as follows.

$$P = n \times \frac{RT}{V}$$

According to the equation, the pressure of a gas is proportional to the number of moles of gas, if the temperature and volume are held constant. Because the temperature and volume of the O_2 and N_2 in the atmosphere are the same, the pressure of each gas is proportional to the number of the moles of the gas. Because there is more N_2 in the atmosphere than O_2, the contribution to the total pressure of the atmosphere from N_2 is larger than the contribution from O_2.

John Dalton was the first to recognize that the total pressure of a mixture of gases is the sum of the contributions of the individual components of the mixture. The part of the total pressure of a mixture that results from one component is called the **partial pressure** of that component. **Dalton's law of partial pressures** states that the total pressure (P_T) of a mixture of gases is the sum of the partial pressures of the various components.

$$P_T = P_1 + P_2 + P_3 + \ldots$$

Exercise 6.9

Calculate the total pressure of a mixture that contains 1.00 g of H_2 and 1.00 g of He in a 5.00-L container at 21°C.

Solution

We start, as always, by listing what we know about the problem.

$$V = 5.00 \text{ L} \qquad T = 21°C$$

We also know the masses of H_2 and He in the flask, however, so we can calculate the number of moles of each gas in the container.

$$1.00 \text{ g } H_2 \times \frac{1 \text{ mol } H_2}{2.016 \text{ g } H_2} = 0.496 \text{ mol } H_2$$

$$1.00 \text{ g He} \times \frac{1 \text{ mol He}}{4.003 \text{ g He}} = 0.250 \text{ mol He}$$

Because we know the volume, the temperature, and the number of moles of each component of the mixture, we can calculate the partial pressures of H_2 and He. We start by rearranging the ideal gas equation as follows.

$$P = \frac{nRT}{V}$$

We then use that equation to calculate the partial pressure of each gas.

$$P_{hydrogen} = \frac{(0.496 \text{ mol } H_2)(0.08206 \text{ L} \cdot \text{atm/mol} \cdot \text{K})(294 \text{ K})}{(5.00 \text{ L})} = \textbf{2.39 atm}$$

$$P_{helium} = \frac{(0.250 \text{ mol } He)(0.08206 \text{ L} \cdot \text{atm/mol} \cdot \text{K})(294 \text{ K})}{(5.00 \text{ L})} = \textbf{1.21 atm}$$

The total pressure in the mixture is the sum of the partial pressures of the two components.

$$P_T = P_{hydrogen} + P_{helium} = \textbf{3.60 atm}$$

Dalton derived the law of partial pressures from his work on the amount of water vapor that could be absorbed by air at different temperatures. It is therefore fitting that this law is used most often to correct for the amount of water vapor picked up when a gas is collected by displacing water. Suppose, for example, that we want to collect a sample of O_2 prepared by heating potassium chlorate until it decomposes.

$$2 \text{ KClO}_3(s) \longrightarrow 2 \text{ KCl}(s) + 3 \text{ O}_2(g)$$

The gas given off in the reaction can be collected by filling a flask with water, inverting the flask in a trough, and then letting the gas bubble into the flask.

Because some of the water in the flask will evaporate during the experiment, the gas that collects in the flask is going to be a mixture of O_2 and water vapor. The total pressure of the gas is the sum of the partial pressures of the two components.

$$P_T = P_{oxygen} + P_{water}$$

The total pressure of the mixture must be equal to atmospheric pressure. (If it were any greater, the gas would push water out of the container. If it were any less, water would be forced into the container.) If we had some way to estimate the partial pressure of the water in the system, we could therefore calculate the partial pressure of the oxygen gas.

The partial pressure of the gas that collects in a closed container above a liquid is known as the **vapor pressure** of the liquid. Table B.3 in the appendix gives the vapor pressure of water at various temperatures. If we know the temperature at which a gas is collected by displacing water, and we assume that the gas is saturated with water vapor at that temperature, we can calculate the partial pressure of the gas by subtracting the vapor pressure of water from the total pressure of the mixture of gases collected in the experiment.

Exercise 6.10

Calculate the number of grams of O_2 that can be collected by displacing water from a 250-mL bottle at 21°C and 746.2 mmHg.

Solution

It is often useful to work a problem backward. In this case, our goal is the mass of O_2 in the flask. To reach this goal we need to calculate the number of moles of O_2 in the sample. To do this, we need to know the pressure, volume, and temperature of the O_2. We already know some of the information, as shown in Figure 6.16. We know the volume of the O_2 because a gas expands to fill its container. We also know the temperature of the O_2. But we don't know the pressure of the O_2.

$$V_{\text{oxygen}} = 250 \text{ mL} \qquad T_{\text{oxygen}} = 21°C \qquad P_{\text{oxygen}} = ?$$

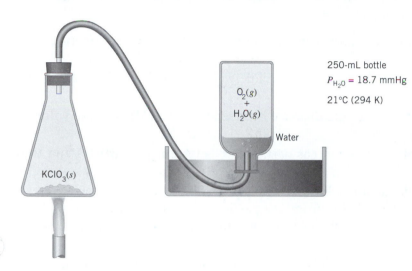

250-mL bottle
$P_{\text{H}_2\text{O}} = 18.7$ mmHg

21°C (294 K)

$O_2(g)$
+
$H_2O(g)$

Water

KCIO$_3$(s)

FIGURE 6.16 Gases, such as O_2, that are not very soluble in water can be collected by displacing water from a container.

All we know is the *total pressure* of the oxygen plus the water vapor that collects in the bottle.

$$P_T = P_{\text{oxygen}} + P_{\text{water}} = 746.2 \text{ mmHg}$$

Table B.3 gives the vapor pressure of water at 21°C as 18.7 mmHg. If we assume that the gas collected in the experiment is saturated with water vapor, we can write the following.

$$P_T = P_{\text{oxygen}} + P_{\text{water}}$$
$$746.2 \text{ mmHg} = P_{\text{oxygen}} + 18.7 \text{ mmHg}$$

In other words, the partial pressure of O_2 must be less than the total pressure in the bottle by 18.7 mmHg.

$$P_{\text{oxygen}} = 746.2 \text{ mmHg} - 18.7 \text{ mmHg} = 727.5 \text{ mmHg}$$

We now know the pressure (727.5 mmHg), volume (250 mL), and temperature (21°C) of the O_2 collected in the experiment. Once we convert the measurements to appropriate units, we can use the ideal gas equation to calculate the number of moles of O_2 collected.

$$n = \frac{PV}{RT} = \frac{(0.9572 \text{ atm})(0.250 \text{ L})}{(0.08206 \text{ L} \cdot \text{atm/mol} \cdot \text{K})(294 \text{ K})} = 0.00992 \text{ mol } O_2$$

We now have enough information to calculate the number of grams of O_2 collected.

$$0.00992 \; \cancel{\text{mol}} \, O_2 \times \frac{32.00 \text{ g } O_2}{1 \; \cancel{\text{mol}}} = 0.317 \text{ g } O_2$$

6.17 THE KINETIC MOLECULAR THEORY

The experimental observations about the behavior of gases discussed so far can be explained with a theoretical model known as the **kinetic molecular theory.** This theory is based on the following postulates, or assumptions.

1. Gases are composed of a large number of particles that behave like hard, spherical objects in a state of constant, random motion.
2. The particles move in a straight line until they collide with another particle or the walls of the container.
3. The particles are much smaller than the distance between particles. Most of the volume of a gas is therefore empty space.
4. There is no force of attraction between gas particles or between the particles and the walls of the container.
5. Collisions between gas particles or collisions with the walls of the container are perfectly elastic. Energy can be transferred from one particle to another during a collision, but the total kinetic energy of the particles after the collision is the same as it was before the collision.
6. The average kinetic energy of a collection of gas particles depends on the temperature of the gas and nothing else.

The assumptions of the kinetic molecular theory can be illustrated with the apparatus shown in Figure 6.17, which consists of a glass plate surrounded by walls mounted on a vibrating motor. A handful of steel ball bearings is placed on top of the glass plate to represent the gas particles.

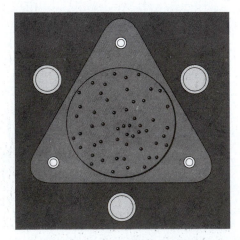

FIGURE 6.17 The six postulates of the kinetic molecular theory can be demonstrated with a molecular dynamics simulator.

When the motor is turned on the glass plate vibrates, which makes the ball bearings move in a constant, random fashion (postulate 1). Each ball moves in a straight line until it collides with another ball or with the walls of the container (postulate 2). Although collisions are frequent, the average distance between the ball bearings is much larger than the diameter of the balls (postulate 3). There is no force of attraction between the individual ball bearings nor between the ball bearings and the walls of the container (postulate 4).

The collisions that occur in the apparatus are very different from those that occur when a rubber ball is dropped on the floor. Collisions between the rubber ball and the floor are *inelastic,* as shown in Figure 6.18. A portion of the energy of the ball is lost each time it hits the floor, until it eventually rolls to a stop. In our apparatus, however, the collisions are perfectly *elastic.* The balls have just as much kinetic energy—on the average—after a collision as before (postulate 5).

FIGURE 6.18 Most collisions are inelastic. Some energy is transferred each time a ball collides with the floor, for example.

Any object in motion has a **kinetic energy** that is defined as one-half of the product of its mass times its velocity squared.

$$KE = \tfrac{1}{2}mv^2$$

At any time, some of the ball bearings on the apparatus are moving faster than others, but the system can be described by an *average kinetic energy.* When we increase the "temperature" of the system by increasing the voltage to the motors, we find that the average kinetic energy of the ball bearings increases (postulate 6).

Checkpoint

The kinetic theory assumes that the temperature of a gas on the macroscopic scale is directly proportional to the average kinetic energy of the particles on the atomic scale. Use this assumption to explain what happens to the particles of a gas when the temperature of a gas increases.

We have described gaseous systems in terms of an "average" kinetic energy. The use of the term "average" implies that not all particles are moving with the same kinetic energy. Some particles must be moving with less than the average kinetic energy and some more rapidly than the average. In other words, there is a distribution of kinetic energies among the particles. Collisions of rapidly moving particles with slower ones always result in a transfer of energy from the rapid particle to the slow particle. About 10^{27} collisions per second take place in a volume of 1 milliliter of gas at 1 atm pressure and room temperature. This assures that there is continuous exchange of both kinetic energy and velocities between the particles of a gas. At any given instant a few particles are moving very slowly, a few are moving very rapidly, and the bulk of the particles have velocities and hence kinetic energies somewhere between the two extremes.

Figure 6.19 shows a plot of kinetic energy along the *x* axis versus the relative number of particles which possess a particular kinetic energy along the *y* axis. When temperature

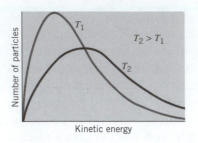

FIGURE 6.19 The kinetic molecular theory states that the average kinetic energy of a gas is proportional to the temperature of the gas, and nothing else. At any given temperature, however, some of the gas particles are moving faster than others.

increases, the number of particles having a greater kinetic energy also increases, as indicated in Figure 6.19 by the shift in the distribution curve to the right. The distribution of kinetic energies has significant influence on a number of properties of interest to chemists. For instance, as we'll see later, liquids may be treated by the kinetic model, and the evaporation of liquids to form gases can be described in terms similar to those used to describe gases. The effect of temperature on chemical and physical properties may be rationalized in terms of particles having a distribution of kinetic energies.

6.18   HOW THE KINETIC MOLECULAR THEORY EXPLAINS THE GAS LAWS

The kinetic molecular theory can be used to explain each of the experimentally determined gas laws. The pressure of a gas results from collisions between the gas particles and the walls of the container. Each time a gas particle hits the wall, it exerts a force on the wall. The frequency of collisions with the walls and the kinetic energy of the gas determine the magnitude of the force. Because pressure is force per unit area, a change in force or surface area results in a change in pressure.

The Link between P and n (T and V constant)

If the temperature is held constant, the average kinetic energy of the gas particles remains unchanged. If volume is held constant the surface area remains the same. Any increase in the number of gas particles in the container increases the frequency of collisions with the walls and therefore the pressure of the gas.

Amontons' Law ($P \propto T$; n and V constant)

The last postulate of the kinetic molecular theory states that the average kinetic energy of a gas particle depends only on the temperature of the gas. Thus, the average kinetic energy of the gas particles increases as the gas becomes warmer. Because the mass of the particles is constant, their kinetic energy can only increase if the average velocity of the particles increases. The faster the particles are moving when they hit the wall, the greater the force they exert on the wall and the greater the frequency of collisions with the wall. Because the volume is held constant, the surface area is constant. Since the force per collision and the number of collisions become larger as the temperature increases, the pressure of the gas must increase as well.

Boyle's Law ($P \propto 1/V$; T and n constant)

Gases can be compressed because most of the volume of a gas is empty space. If we compress a gas without changing its temperature, the average kinetic energy of the gas particles stays the same. There is no change in the speed with which the particles move, but the surface area of the container is smaller. Thus, the particles travel from one end of the container to the other in a shorter time. This means that they hit the walls more often. Any increase in the frequency of collisions with the walls must lead to an increase in the pressure of the gas. Thus, the pressure of a gas becomes greater as the volume of the gas becomes smaller.

Charles' Law ($V \propto T$; P and n constant)

The average kinetic energy of the particles in a gas is proportional to the temperature of the gas. Because the mass of the particles is constant, the particles must move faster as the gas becomes warmer. If they move faster, the particles will exert a greater force on the container each time they hit the walls, which leads to an increase in the pressure of the gas. If the walls of the container are flexible, they will expand until the pressure of the gas once again balances the pressure of the atmosphere. The volume of the gas in a flexible container therefore increases as the temperature of the gas increases.

Avogadro's Hypothesis ($V \propto n$; T and P constant)

As the number of gas particles increases, the frequency of collisions with the walls of the container must increase. This, in turn, leads to an increase in the pressure of the gas. Flexible containers, such as a balloon, will expand until the pressure of the gas inside the balloon once again balances the pressure of the gas outside. Thus, the volume of the gas is proportional to the number of gas particles.

Dalton's Law of Partial Pressures ($P_T = P_1 + P_2 + P_3 + \ldots$; V and T constant)

Imagine what would happen if six ball bearings of a different size were added to the ball bearings all ready in the apparatus in Figure 6.17. The total pressure would increase because there would be more collisions with the walls of the container. But the pressure resulting from the collisions between the original ball bearings and the walls of the container would remain the same. There is so much empty space in the container that each type of ball bearing hits the walls of the container as often in the mixture as it would if there was only one kind of ball bearing on the glass plate. The total number of collisions with the wall in the mixture is therefore equal to the sum of the collisions that would occur when each size of ball bearing is present by itself. In other words, the total pressure of a mixture of gases is equal to the sum of the partial pressures of the individual gases.

KEY TERMS

Absolute zero

Amontons' law

Avogadro's hypothesis

Boyle's law

Charles' law

Compressibility

Dalton's law of partial
  pressure

Expandability

Force

Gay-Lussac's law

Ideal gas constant

Ideal gas law

Kinetic energy

Kinetic molecular
  theory

Law of combining
  volumes

Partial pressure

Pressure

States

STP

Vapor pressure

PROBLEMS

Gases at Room Temperature

1. Which of the following elements and compounds are most likely to be gases at room temperature?
   (a) Ar (b) CO (c) CH_4 (d) $C_{10}H_{22}$ (e) Cl_2 (f) Fe_2O_3 (g) Na
   (h) NaCl (i) Pt (j) S_8

Properties of Gases

2. Describe in detail how you would measure the following properties of a gas. In each case, describe the equipment you would use and the steps you would have to take to make the measurements.
   (a) volume (b) mass (c) pressure (d) density

3. What are the three characteristic properties of gases that distinguish gases from solids from liquids?

4. Imagine two identical flasks at the same temperature. One contains 2 g of H_2 and the other contains 28 g of N_2. Which of the following properties are the same for the two flasks?
   (a) pressure (b) average kinetic energy (c) density
   (d) the number of molecules per container

5. Explain why the difference between the radius of a helium atom and that of a xenon atom has no effect on the volume of a mole of the gases.

6. Use the fact that the pressure of a gas is the same in all directions to predict what would happen to the weight of an empty metal cylinder when it is filled with helium gas. Will it increase, decrease, or remain the same?

Pressure versus Force

7. Calculate the force on the bottom of the column of mercury in a barometer when the atmospheric pressure is 1.00 atm. Assume that the diameter of the tube is 1.00 cm and the density of mercury is 13.6 g/cm^3. Calculate the pressure in pounds per square inch and newtons per square meter, Pa.

Atmospheric Pressure

8. What is the pressure in units of atmospheres when a barometer reads 745.8 mmHg? What is the pressure in units of pascals?

9. One atmosphere pressure will support a column of mercury 760 millimeters tall in a barometer with a tube 1.00 centimeter in diameter. What would be the height of the column of mercury if the diameter of the tube were twice as large?

10. Atmospheric pressure is announced during weather reports in the United States in units of inches of mercury. How many inches of mercury would exert a pressure of 1.00 atm? In Canada, atmospheric pressure is reported in units of pascals. What is 1.00 atm in pascals?

11. If the directions that come with your car tell you to inflate the tires to 200 kPa pressure and you have a tire pressure gauge calibrated in pounds per square inch (psi), what pressure in psi will you use?

12. The vapor pressure of the mercury gas that collects at the top of a barometer is 2×10^{-3} mmHg. Calculate the vapor pressure of the gas in atmospheres.

13. Calculate the force exerted on the earth by the atmosphere if atmospheric pressure is 14.7 lb/in^2 and the surface area of the planet is 5.1×10^8 km^2.

Boyle's Law

14. A 425-mL sample of O_2 gas was collected at 742.3 mmHg. What would be the pressure in mmHg if the gas were allowed to expand to 975 mL at constant temperature?

15. What would happen to the volume of a balloon filled with 0.357 L of H_2 gas collected at 741.3 mmHg if the atmospheric pressure increased to 758.1 mmHg?

16. What is the volume of a scuba tank if it takes 2000 L of air collected at 1 atm to fill the tank to a pressure of 150 atm?

17. Calculate the volume of a balloon that could be filled at 1.00 atm with the helium in a 2.50-L compressed gas cylinder in which the pressure is 200 atm at 25°C.

Amontons' Law

18. A butane lighter fluid can is filled with 5.00 atm of C_4H_{10} gas at 21°C. Calculate the pressure in the can when it is stored in a warehouse on a hot summer day when the temperature reaches 38°C.

19. An automobile tire was inflated to a pressure of 32 lb/in^2 at 21°C. At what temperature would the pressure reach 60 psi?

20. At 25°C, four-fifths of the pressure of the atmosphere is due to N_2 and one-fifth is due to O_2. At 100°C, what fraction of the pressure is due to N_2?

Charles' Law

21. Calculate the percent change in the volume of a toy balloon when the gas inside is heated from 22°C to 75°C in a hot water bath.

22. A sample of O_2 gas with a volume of 0.357 liter was collected at 21°C. Calculate the volume of the gas when it is cooled to 0°C if the pressure remains constant.

23. Explain why a balloon filled with O_2 at room temperature collapses when cooled in liquid nitrogen, whereas a balloon filled with He does not.

Gay-Lussac's Law

24. Calculate the ratio of the volumes of sulfur dioxide and oxygen produced when sulfuric acid decomposes. Temperature and pressure are constant.

$$2\ H_2SO_4(aq) \longrightarrow 2\ SO_2(g) + O_2(g) + 2\ H_2O(l)$$

25. Calculate the volume of H_2 and N_2 gas formed when 1.38 L of NH_3 decomposes at a constant temperature and pressure.

$$2\ NH_3(g) \longrightarrow N_2(g) + 3\ H_2(g)$$

26. Ammonia burns in the presence of oxygen to form nitrogen oxide and water.

$$4\ NH_3(g) + 5\ O_2(g) \longrightarrow 4\ NO(g) + 6\ H_2O(g)$$

What volume of NO can be prepared when 15.0 L of ammonia reacts with excess oxygen, if all measurements are made at the same temperature and pressure?

27. Acetylene burns in oxygen to form CO_2 and H_2O.

$$2\ C_2H_2(g) + 5\ O_2(g) \longrightarrow 4\ CO_2(g) + 2\ H_2O(g)$$

Calculate the total volume of the products formed when 15.0 L of C_2H_2 burns in the presence of 15.0 L of O_2, if all measurements are made at the same temperature and pressure.

28. Methane reacts with steam to form hydrogen and carbon monoxide.

$$CH_4(g) + H_2O(g) \longrightarrow CO(g) + 3\ H_2(g)$$

It can also react with steam to form carbon dioxide.

$$CH_4(g) + 2\ H_2O(g) \longrightarrow CO_2(g) + 4\ H_2(g)$$

What are the products of a reaction if 1.50 L of methane is found by experiment to react with 1.50 L of water vapor?

Avogadro's Hypothesis

29. Which weighs more, dry air at 25°C and 1 atm, or air at that temperature and pressure that is saturated with water vapor? (Assume that the average molecular weight of air is 29.0 g/mol.)
30. Which of the following samples would have the largest volume at 25°C and 750 mmHg?
    (a) 100 g CO_2 (b) 100 g CH_4 (c) 100 g NO   (d) 100 g SO_2
31. Nitrous oxide decomposes to form nitrogen and oxygen. Use Avogadro's hypothesis to determine the formula for nitrous oxide if 2.36 L of the compound decomposes to form 2.36 L of N_2 and 1.18 L of O_2.
32. Why does a balloon filled with neon rise? Does a balloon filled with NH_3 sink? (Assume that the average molecular weight of air is 29.0 g/mol.)

The Ideal Gas Equation

33. Predict the shape of the following graphs for an ideal gas (assume all other variables are held constant).
    (a) pressure versus volume   (b) pressure versus temperature
    (c) volume versus temperature   (d) kinetic energy versus temperature
    (e) pressure versus the number of moles of gas
34. Which of the following graphs could not give a straight line for an ideal gas?
    (a) V versus T (b) T versus P (c) P versus $1/V$
    (d) n versus $1/T$ (e) n versus $1/P$
35. Which of the following statements is always true for an ideal gas?
    (a) If the temperature and volume of a gas both increase at constant pressure, the amount of gas must also increase.
    (b) If the pressure increases and the temperature decreases for a constant amount of gas, the volume must decrease.
    (c) If the volume and the amount of gas both decrease at constant temperature, the pressure must decrease.
36. Calculate the value of the ideal gas constant in units of mL · psi/mol · K if 1.00 mol of an ideal gas at 0°C occupies a volume of 22,400 mL at 14.7 lb/in^2 pressure.

Ideal Gas Calculations: Part I

37. Nitrogen gas sells for roughly 50 cents per 100 cubic feet at 0°C and 1 atm. What is the price per gram of nitrogen?
38. Calculate the pressure in atmospheres of 80 g of CO_2 in a 30-L container at 23°C.

39. Calculate the temperature at which 1.5 g of O_2 has a pressure of 740 mmHg in a 1-L container.

40. Calculate the number of kilograms of O_2 gas that can be stored in a compressed gas cylinder with a volume of 40 L when the cylinder is filled at 150 atm and 21°C.

41. Calculate the pressure in an evacuated 250-mL container at 0°C when the O_2 in 1.00 cm^3 of liquid oxygen evaporates. Assume that liquid oxygen has a density of 1.118 g/cm^3.

42. Assume that a 1.00-L flask was evacuated, 5.0 g of liquid ammonia was added to the flask, and the flask was sealed with a cork. When the NH_3 in the flask warms to 21°C, will the cork be blown out of the mouth of the flask?

Ideal Gas Calculations: Part II

43. What is the volume of the gas in a balloon at −195°C if the balloon has been filled to a volume of 5.0 L at 25°C?

44. CO_2 gas with a volume of 25.0 L was collected at 25°C and 0.982 atm. Calculate the pressure of the gas if it is compressed to a volume of 0.150 L and heated to 350°C.

45. Calculate the pressure of 4.80 g of ozone, O_3, in a 2.45-L flask at 25°C. Assume that the ozone decomposes completely to molecular oxygen.

$$2 O_3(g) \longrightarrow 3 O_2(g)$$

Calculate the pressure inside the flask once the reaction is complete.

46. One mole of an ideal gas occupies a volume of 22.4 L at 0°C and 1.00 atm. What would be the volume of a mole of an ideal gas at 22°C and 748.8 mmHg?

47. Assume that 10.0 L of O_2 gas was collected at 120°C and 749.3 mmHg. Calculate the volume of the gas when it is cooled to 0°C and stored in a container at 1.00 atm.

48. Assume that 5.0 L of CO_2 gas was collected at 25°C and 2.5 atm. At what temperature would the gas have to be stored to fill a 10.0-L flask at 0.978 atm?

49. Assume that two 10-L samples of O_2 collected at 120°C and 749.3 mmHg are combined and stored in a 1.25-L flask at 27°C. Calculate the pressure of the gas.

Ideal Gas Calculations Involving the Density of a Gas

50. Calculate the density of CH_2Cl_2 in the gas phase at 40° and 1.00 atm. Compare this with the density of liquid CH_2Cl_2 (1.336 g/cm^3).

51. Calculate the density of methane gas, CH_4, in kilograms per cubic meter at 25°C and 956 mmHg.

52. Calculate the density of helium at 0°C and 1.00 atm and compare this with the density of air (1.29 g/cm^3) at 0°C and 1.00 atm. Explain why 1.00 ft^3 of helium can lift a weight of 0.076 lb under these conditions.

53. Calculate the ratio of the densities of H_2 and O_2 at 0°C and 100°C.

54. Which of the noble gases in Group VIIIA of the periodic table has a density of 3.7493 g/L at 0°C and 1.00 atm?

55. Calculate the average molecular weight of air assuming a sample of air weighs 1.700 times as much as an equivalent volume of ammonia, NH_3.

Dalton's Law of Partial Pressures ($P_T = P_1 + P_2 + P_3 + \ldots$)

56. Calculate the partial pressure of propane in a mixture that contains equal weights of propane (C_3H_8) and butane (C_4H_{10}) at 20°C and 746 mmHg.

57. Calculate the partial pressure of helium in a 1.00-L flask that contains equal numbers of moles of N_2, O_2, and He at a total pressure of 7.5 atm.

58. Calculate the total pressure in a 10.0-L flask at 27°C of a sample of gas that contains 6.0 g of H_2, 15.2 g of N_2, and 16.8 g of He.

59. A 1-L flask is filled with carbon monoxide at 27°C until the pressure is 0.200 atm. Calculate the total pressure after 0.450 g of carbon dioxide has been added to the flask.

60. A few milliliters of water was added to a 1-L flask at 25°C and the pressure of the water that evaporated was found to be 23.8 mmHg at that temperature. What would be the pressure of the water if the experiment were repeated in a 2-L flask?

61. Calculate the volume of the hydrogen obtained when the water vapor is removed from 289 mL of H_2 gas collected by displacing water from a flask at 15°C and 0.988 atm.

The Kinetic Molecular Theory

62. What would happen to a balloon if the collisions between gas molecules were not perfectly elastic?

63. What would happen to a balloon if the gas molecules were in a state of constant motion, but the motion was not random?

64. Explain why the pressure of a gas is evidence for the assumption that gas particles are in a state of constant random motion.

65. Use the kinetic molecular theory to explain why the pressure of a gas is proportional to the number of gas particles and the temperature of the gas but inversely proportional to the volume of the gas.

66. Which of the following statements explains why a hot-air balloon rises when the air in the balloon is heated?
    (a) The average kinetic energy of the molecules increases, and the collisions between the molecules and the walls of the balloon make it rise.
    (b) The pressure of the gas inside the balloon increases, pushing up on the balloon.
    (c) The gas expands, forcing some of it to escape from the bottom of the balloon, and the decrease in the density of the gas lifts the balloon.
    (d) The balloon expands, causing it to rise.
    (e) The hot air rising inside the balloon produces enough force to lift the balloon.

Combined Stoichiometry and Gas Law Problems

67. Equal volumes of oxygen and an unknown gas at the same temperature and pressure weigh 3.00 g and 7.50 g, respectively. Which of the following is the unknown gas?
    (a) CO   (b) CO_2 (c) NO   (d) NO_2 (e) SO_2 (f) SO_3

68. What is the molecular weight of acetone if 0.520 g of acetone occupies a volume of 275.5 mL at 100°C and 756 mmHg?

69. Boron forms a number of compounds with hydrogen, including B_2H_6, B_4H_{10}, B_5H_9, B_5H_{11}, and B_6H_{10}. For which compound would a 1.00-g sample occupy a volume of 390 cm^3 at 25°C and 0.993 atm?

70. Cyclopropane is an anesthetic that is 85.63% carbon and 14.37% hydrogen by mass. What is the molecular formula of the compound if 0.45 L of cyclopropane reacts with excess oxygen at 120°C and 0.72 atm to form 1.35 L of carbon dioxide and 1.35 L of water vapor?

71. Calculate the molecular formula of diazomethane, assuming the compound is 28.6% C, 4.8% H, and 66.6% N by mass and the density of the gas is 1.72 g/L at 25°C and 1.00 atm.

72. What are the molecular formulas for phosphine, PH_x, and diphosphine, P_2H_y, if the densities of the gases are 1.517 and 2.944 g/L, respectively, at 0°C and 1.00 atm?

73. Calculate the weight of magnesium that would be needed to generate 500 mL of hydrogen gas at 0°C and 1.00 atm.

$$Mg(s) + 2\ HCl(aq) \longrightarrow Mg^{2+}(aq) + 2\ Cl^-(aq) + H_2(g)$$

74. Calculate the formula of the oxide formed when 10.0 g of chromium metal reacts with 6.98 L of O_2 at 20°C and 0.994 atm.

75. Calculate the volume of CO_2 gas measured at 756 mmHg and 23°C given off when 150 kg of limestone is heated until it decomposes.

$$CaCO_3(s) \longrightarrow CaO(s) + CO_2(g)$$

76. Calculate the volume of O_2 that would have to be inhaled at 20°C and 1.00 atm to consume 1.00 kg of fat, $C_{57}H_{110}O_6$.

$$2\ C_{57}H_{110}O_6(s) + 163\ O_2(g) \longrightarrow 114\ CO_2(g) + 110\ H_2O(l)$$

77. Calculate the volume of CO_2 gas collected at 23°C and 0.991 atm that can be prepared by reacting 10.0 g of calcium carbonate with excess acid.

$$CaCO_3(aq) + 2\ H^+(aq) \longrightarrow Ca^{2+}(aq) + CO_2(g) + H_2O(l)$$

78. Determine the identity of an unknown metal if 1.00 g of the metal reacts with excess acid according to the following equation to produce 374 mL of H_2 gas at 25°C and 1.00 atm.

$$M(s) + 2\ H^+(aq) \longrightarrow M^{2+}(aq) + H_2(g)$$

Integrated Problems

79. Imagine two identical flasks at the same temperature. One contains 2 grams of H_2 and the other contains 28 grams of N_2. Which of the following properties are the same for the two flasks?
    (a) pressure (b) average kinetic energy (c) density
    (d) number of molecules per container (e) weight of the container

80. At 25°C, four-fifths of the pressure of the atmosphere is due to N_2 and one-fifth is due to O_2. What fraction of the pressure is due to N_2 at 100°C?

81. Which of the noble gases in Group VIIIA of the periodic table has a density of 5.86 grams per liter at 0°C and 1.00 atm?

82. Two flasks with the same volume and temperature are connected by a valve. One gram of hydrogen is added to one flask, and 1 gram of oxygen is added to the other.
    (a) In which flask is the average kinetic energy of the gas molecules the greatest?
    (b) In which flask are the gas molecules moving the most rapidly?

83. When a meterologist reports precipitation in the weather forecast this is usually associated with a low pressure system. Explain why precipitation would be associated with a low atmospheric pressure. (Hint: See problem 29.)

84. If equal weights of O_2 and N_2 are placed in identical containers at the same temperature, which of the following statements is true?
    (a) Both flasks contain the same number of molecules.

(b) The pressure in the flask that contains the N_2 will be greater than the pressure in the flask that contains the O_2.

(c) There will be more molecules in the flask that contains O_2 than the flask that contains N_2.

(d) This question cannot be answered unless we know the weights of O_2 and N_2 in the flask.

(e) None of the above are correct.

85. N_2H_4 decomposes at a fixed temperature in a closed container to form $N_2(g)$ and $H_2(g)$. If the reaction goes to completion, the final pressure will be:

(a) the same as the initial pressure

(b) twice the initial pressure

(c) three times the initial pressure

(d) one-half of the initial pressure

(e) one-third of the initial pressure

86. For the apparatus diagramed below, what will be the final partial pressures of O_2 and N_2 after the stopcock is opened? The temperature of both flasks is the same and does not change after opening the stopcock. What will be the final pressure in the apparatus?

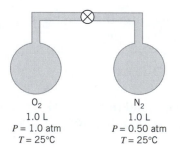

O_2	N_2
1.0 L	1.0 L
$P = 1.0$ atm	$P = 0.50$ atm
$T = 25°C$	$T = 25°C$

87. N_2 gas is in a 2.0-L container at 298 K and 1 atm. Explain what will happen to both the kinetic energy and the frequency of collisions of the nitrogen molecules for each of the following conditions.

(a) The number of moles of N_2 is halved.

(b) The temperature is changed to 15°C.

(c) The temperature is changed to 1500°C.

(d) The volume is decreased to 1 L.

CHAPTER
6
SPECIAL TOPICS

6A.1 Graham's Laws of Diffusion and Effusion
6A.2 The Kinetic Molecular Theory and Graham's Laws
6A.3 Deviations from Ideal Gas Law Behavior: The van der Waals Equation
6A.4 Analysis of the van der Waals Constants

6A.1 GRAHAM'S LAWS OF DIFFUSION AND EFFUSION

A few of the physical properties of gases depend on the identity of the gas. One of these physical properties can be seen when the movement of gases is studied.

In 1829 Thomas Graham used an apparatus similar to the one shown in Figure 6A.1 to study the **diffusion** of gases—the rate at which two gases mix. The apparatus consists of a glass tube sealed at one end with plaster that has holes large enough to allow a gas to enter or leave the tube. When the tube is filled with H_2 gas, the level of water in the tube slowly rises because the H_2 molecules inside the tube escape through the holes in the plaster more rapidly than the molecules in air can enter the tube. By studying the rate at which the water level in the apparatus changed, Graham was able to obtain data on the rate at which different gases mixed with air.

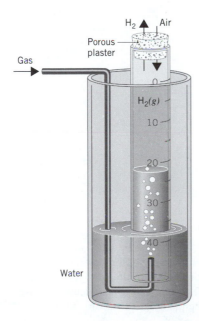

FIGURE 6A.1 The rate of diffusion, or mixing of a gas with air, can be studied with the apparatus shown. If the gas escapes from the tube faster than air enters the tube, the amount of water in the tube will increase. If air enters the tube faster than the gas escapes, water will be displaced from the tube.

Graham found that the rates at which gases diffuse is inversely proportional to the square root of their densities.

$$\text{Rate}_{\text{diffusion}} \propto \frac{1}{\sqrt{\text{density}}}$$

This relationship eventually became known as **Graham's law of diffusion.**

To understand the importance of this discovery we have to remember that equal volumes of different gases at the same temperature and pressure contain the same number of particles. As a result, the number of moles of gas per liter at a given temperature and pressure is constant, which means that the density of a gas is directly proportional to its molecular weight (MW). Graham's law of diffusion can therefore also be written as follows.

$$\text{Rate}_{\text{diffusion}} \propto \frac{1}{\sqrt{MW}}$$

Similar results were obtained when Graham studied the rate of **effusion** of a gas, which is the rate at which the gas escapes through a pinhole into a vacuum. The rate of effusion of a gas is also inversely proportional to the square root of either the density or the molecular weight of the gas.

$$\text{Rate}_{\text{effusion}} \propto \frac{1}{\sqrt{\text{density}}} \propto \frac{1}{\sqrt{MW}}$$

Graham's law of effusion can be demonstrated with the apparatus in Figure 6A.2. A thick-walled filter flask is evacuated with a vacuum pump. A syringe is filled with 25 mL of gas, and the time required for the gas to escape through the syringe needle into the evacuated filter flask is measured with a stop watch. The experimental data in Table 6A.1 were obtained by using a special needle with a very small (0.015-cm) hole through which the gas could escape.

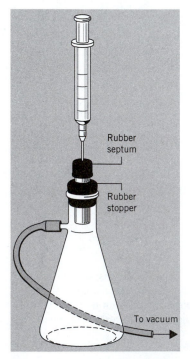

Rubber septum

Rubber stopper

To vacuum

FIGURE 6A.2 The rate of effusion of a gas can be demonstrated with the apparatus shown. The time required for the gas in the syringe to escape into an evacuated flask is measured. The faster the gas molecules effuse, the less time it takes for a given volume of the gas to escape into the flask.

TABLE 6A.1   Time Required for 25-mL Samples of Different Gases to Escape through a 0.015-cm Hole into a Vacuum

Compound	Time (s)	Molecular Weight (g/mol)
H_2	5.1	2.02
He	7.2	4.00
NH_3	14.2	17.0
Air	18.2	29.0
O_2	19.2	32.0
CO_2	22.5	44.0
SO_2	27.4	64.1

As we can see when the data are graphed in Figure 6A.3, the *time* required for 25-mL samples of different gases to escape into a vacuum is proportional to the square root of the molecular weight of the gas. The *rate* at which the gases effuse is therefore inversely proportional to the square root of the molecular weight. Graham's observations about the rate at which gases diffuse (mix) or effuse (escape through a pinhole) suggest that relatively light gas particles, such as H_2 molecules and He atoms, move faster than relatively heavy gas particles, such as CO_2 and SO_2 molecules.

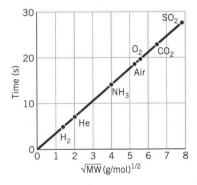

FIGURE 6A.3 Graph of the time required for 25-mL samples of different gases to escape into an evacuated flask versus the square root of the molecular weight of the gas. Relatively heavy molecules move more slowly, and it takes more time for the gas to escape.

6A.2   THE KINETIC MOLECULAR THEORY AND GRAHAM'S LAWS

The kinetic molecular theory can be used to explain the results Graham obtained when he studied the diffusion and effusion of gases. The last postulate of the kinetic theory states that the temperature of a system is proportional to the average kinetic energy of its particles and nothing else. In other words, the temperature of a system increases if and only if there is an increase in the average kinetic energy of its particles.

Two gases at the same temperature, such as H_2 and O_2, must have the same average kinetic energy. This can be represented by the following equation.

$$\tfrac{1}{2} m_{H_2} v_{H_2}^{\,2} = \tfrac{1}{2} m_{O_2} v_{O_2}^{\,2}$$

The equation can be simplified by multiplying both sides by two.

$$m_{H_2} v_{H_2}^{\,2} = m_{O_2} v_{O_2}^{\,2}$$

It can then be rearranged to give the following.

$$\frac{v_{H_2}^2}{v_{O_2}^2} = \frac{m_{O_2}}{m_{H_2}}$$

Taking the square root of both sides of the equation gives a relationship between the ratio of the velocities at which the two gases move and the square root of the ratio of their molecular weights.

$$\frac{v_{H_2}}{v_{O_2}} = \sqrt{\frac{m_{O_2}}{m_{H_2}}}$$

This equation is a modified form of Graham's law. It suggests that the velocity (or rate) at which gas molecules move is inversely proportional to the square root of their molecular weights.

Exercise 6A.1

Calculate the average velocity of an H_2 molecule at $0°C$ if the average velocity of an O_2 molecule at that temperature is 500 m/s.

Solution

The relative velocities of the H_2 and O_2 molecules at a given temperature are described by the following equation.

$$\frac{v_{H_2}}{v_{O_2}} = \sqrt{\frac{m_{O_2}}{m_{H_2}}}$$

Substituting the molecular weights of H_2 and O_2 and the average velocity of an O_2 molecule into the equation gives the following result.

$$\frac{v_{H_2}}{500 \text{ m/s}} = \sqrt{\frac{32.0 \text{ g/mol}}{2.02 \text{ g/mol}}}$$

Solving the equation for the average velocity of an H_2 molecule gives a value of 1.99×10^3 m/s or about 4.5×10^3 mi/h.

Because it is easy to make mistakes when setting up a ratio problem such as this, it is important to check the answer to see whether it makes sense. Graham's law suggests that light molecules move faster on the average than heavy molecules. In this case, the answer makes sense because H_2 molecules are much lighter than O_2 molecules and they should therefore travel much faster.

6A.3   DEVIATIONS FROM IDEAL GAS LAW BEHAVIOR: THE VAN DER WAALS EQUATION

The behavior of real gases usually agrees with the predictions of the ideal gas equation to within ±5% at normal temperatures and pressures. At low temperatures or high pressures, real gases deviate significantly from ideal gas behavior. In 1873, while searching for a way to link the behavior of liquids and gases, the Dutch physicist Johannes van der Waals developed an explanation for these deviations and an equation that was able to fit the behavior of real gases over a much wider range of pressures.

van der Waals realized that two of the assumptions of the kinetic molecular theory were questionable. The kinetic theory assumes that gas particles occupy a negligible fraction of the total volume of the gas. It also assumes that the force of attraction between gas molecules is zero.

The first assumption works at pressures close to 1 atm. But it becomes increasingly less valid as the gas is compressed. Imagine for the moment that the atoms or molecules in a gas were all clustered in one corner of a cylinder, as shown in Figure 6A.4. At normal pressures, the volume occupied by the particles is a negligibly small fraction of the total volume of the gas. But at high pressures, this is no longer true. As a result, real gases are not as compressible at high pressures as an ideal gas. The volume of a real gas is therefore larger than expected from the ideal gas equation at high pressures.

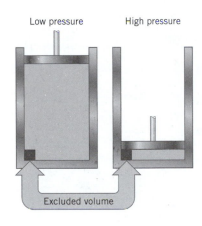

Low pressure   High pressure

Excluded volume

FIGURE 6A.4 The volume actually occupied by the particles in a gas is relatively small at low pressures, but it can be a significant fraction of the total volume at high pressures. For O_2, for example, the gas molecules occupy 0.13% of the total volume at 1.00 atm but 17% of the volume at 100 atm.

van der Waals recognized that we can correct for the fact that the volume of a real gas is too large at high pressures by *subtracting* a term from the volume of the real gas before we substitute it into the ideal gas equation. He therefore introduced a constant (*b*) into the ideal gas equation that was related to the volume actually occupied by a mole of gas particles. Because the volume of the gas particles depends on the number of moles of gas in the container, the term that is subtracted from the real volume of the gas is equal to the number of moles of gas times *b*.

$$P(V - nb) = nRT$$

When the pressure is relatively low, and the volume is reasonably large, the *nb* term is too small to make any difference in the calculation. But at high pressures, when the volume of the gas is small, the *nb* term corrects for the fact that the volume of a real gas is larger than expected from the ideal gas equation.

The assumption that there is no force of attraction between gas particles can't be true. If it were, gases would never condense to form liquids. In reality, there is a small force of attraction between gas molecules that tends to hold the molecules together. This force of attraction has two consequences: (1) gases condense to form liquids at low temperatures, and (2) the pressure of a real gas is sometimes lower than expected for an ideal gas.

To correct for the fact that the pressure of a real gas is lower than expected from the ideal gas equation, van der Waals *added* a term to the pressure in the equation. The term contained a second constant (*a*) and has the form an^2/V^2. The complete **van der Waals equation** is therefore written as follows.

$$\left(P + \frac{an^2}{V^2}\right)(V - nb) = nRT$$

The van der Waals equation is something of a mixed blessing. It provides a much better fit with the behavior of a real gas than the ideal gas equation. But it does this at the cost of a loss in generality. The ideal gas equation is equally valid for any gas, whereas the van der Waals equation contains a pair of constants (*a* and *b*) that change from gas to gas. Values of the van der Waals constants for gases are given in Table 6A.2.

TABLE 6A.2   van der Waals Constants for Various Gases

Compound	$a\ (L^2 \cdot atm/mol^2)$	$b\ (L/mol)$
He	0.03412	0.02370
Ne	0.2107	0.01709
H_2	0.2444	0.02661
Ar	1.345	0.03219
O_2	1.360	0.03803
N_2	1.390	0.03913
CO	1.485	0.03985
CH_4	2.253	0.04278
CO_2	3.592	0.04267
NH_3	4.170	0.03707

The ideal gas equation predicts that a plot of *PV* versus *P* for a gas at constant *n* and *T* would be a horizontal line because *PV* should be constant. Experimental data for *PV* versus *P* for H_2 and N_2 gas at 0°C and CO_2 at 40°C are given in Figure 6A.5. As the pressure increases, the product of pressure times volume for N_2 and CO_2 first falls below the line expected from the ideal gas equation and then rises above the line.

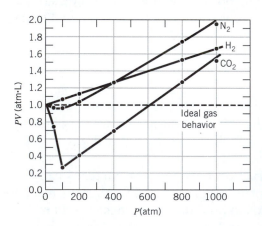

FIGURE 6A.5 Plot of the product of pressure times volume for samples of H_2, N_2, and CO_2 gases versus the pressure of the gases, at constant temperature.

This behavior can be understood by comparing the results of calculations using the ideal gas equation and the van der Waals equation for 1.00 mol of CO_2 at 0°C in containers of different volumes. Let's start with a 22.4-L container. According to the ideal gas equation, the pressure of the gas should be 1.00 atm.

$$P = \frac{nRT}{V} = \frac{(1.00 \text{ mol})(0.08206 \text{ L} \cdot \text{atm/mol} \cdot \text{K})(273 \text{ K})}{(22.4 \text{ L})} = \textbf{1.00 atm}$$

Substituting what we know about CO_2 into the van der Waals equation gives a much more complex equation.

$$\left(P + \frac{an^2}{V^2}\right)(V - nb) = nRT$$

$$\left[\frac{P + (3.592 \text{ L}^2 \cdot \text{atm/mol}^2)(1.00 \text{ mol})^2}{(22.4 \text{ L})^2}\right][22.4 \text{ L} - (1.00 \text{ mol})(0.04267 \text{ L/mol})]$$

$$= (1.00 \text{ mol})(0.08206 \text{ L} \cdot \text{atm/mol} \cdot \text{K})(273 \text{ K})$$

The equation can be solved, however, for the pressure of the gas.

$$\textbf{P = 0.995 atm}$$

At normal temperatures and pressures, the ideal gas and van der Waals equations give essentially the same results.

Let's now repeat the calculation, assuming that the gas is compressed so that it fills a container which has a volume of only 0.200 L. According to the ideal gas equation, the pressure would have to increase to 112 atm to compress 1.00 mol of CO_2 at 0°C to a volume of 0.200 L.

$$P = \frac{nRT}{V} = \frac{(1.00 \text{ mol})(0.08206 \text{ L} \cdot \text{atm/mol} \cdot \text{K})(273 \text{ K})}{(0.200 \text{ L})} = \textbf{112 atm}$$

The van der Waals equation, however, predicts that the pressure will only have to increase to 52.6 atm to achieve the same results.

$$\left(P + \frac{an^2}{V^2}\right)(V - nb) = nRT$$

$$\left[\frac{P + (3.592 \text{ L}^2 \cdot \text{atm/mol}^2)(1.00 \text{ mol})^2}{(0.200 \text{ L})^2}\right][0.200 \text{ L} - (1.00 \text{ mol})(0.04267 \text{ L/mol})]$$

$$= (1.00 \text{ mol})(0.08206 \text{ L} \cdot \text{atm/mol} \cdot \text{K})(273 \text{ K})$$

$$\textbf{P = 52.6 atm}$$

As the pressure of CO_2 increases, the van der Waals equation initially gives pressures that are *lower* than the ideal gas equation, as shown in Figure 6A.5, because of the strong force of attraction between CO_2 molecules.

Let's now compress the gas even further, raising the pressure until the volume of the gas is only 0.0500 L. The ideal gas equation predicts that the pressure would have to increase to 448 atm to compress 1.00 mol of CO_2 at 0°C to a volume of 0.0500 L.

$$P = \frac{nRT}{V} = \frac{(1.00 \text{ mol})(0.08206 \text{ L} \cdot \text{atm/mol} \cdot \text{K})(273 \text{ K})}{(0.0500 \text{ L})} = \textbf{448 atm}$$

The van der Waals equation predicts that the pressure will have to reach 1.62×10^3 atm to achieve the same results.

$$\left(P + \frac{an^2}{V^2}\right)(V - nb) = nRT$$

$$\left[\frac{P + (3.592 \; L^2 \cdot \text{atm/mol}^2)(1.00 \; \text{mol}^2)}{(0.0500 \; L)^2}\right][0.0500 \; L - (1.00 \; \text{mol})(0.04267 \; L/\text{mol})]$$

$$= (1.00 \; \text{mol})(0.08206 \; L \cdot \text{atm/mol} \cdot K)(273 \; K)$$

$$P = 1.62 \times 10^3 \text{ atm}$$

The van der Waals equation gives results that are *larger* than the ideal gas equation at very high pressures, as shown in Figure 6A.5, because of the volume occupied by the CO_2 molecules.

6A.4   ANALYSIS OF THE VAN DER WAALS CONSTANTS

The van der Waals equation contains two constants, a and b, that are characteristic properties of a particular gas. The first of the constants corrects for the force of attraction between gas particles. Compounds for which the force of attraction between particles is strong have large values for a. If you think about what happens when a liquid boils, you may expect compounds with large values of a to have higher boiling points. (As the force of attraction between gas particles becomes stronger, we have to go to higher temperatures before we can break the forces of attraction between the molecules in the liquid to form a gas.) It isn't surprising to find a correlation between the value of the a constant in the van der Waals equation and the boiling points of a number of simple compounds, as shown in Figure 6A.6. Gases with very small values of a, such as H_2 and He, must be cooled to almost absolute zero before they condense to form a liquid.

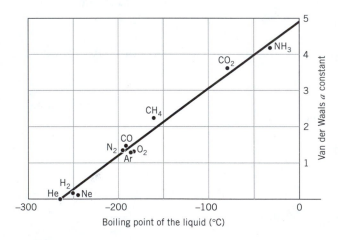

FIGURE 6A.6 The boiling point of a liquid is an indirect measure of the force of attraction between its molecules. Thus, it isn't surprising to find a correlation between the value of the van der Waals constant a, which measures the force of attraction between gas particles, and the boiling point of a number of simple compounds that are gases at room temperature.

The other van der Waals constant, b, is a rough measure of the size of a gas particle. As the volume of the container becomes smaller and smaller, the volume occupied by the molecules becomes more significant.

PROBLEMS

Graham's Laws of Diffusion and Effusion

6A-1. List the following gases in order of increasing rate of diffusion.
   (a) Ar   (b) Cl_2 (c) CF_2Cl_2 (d) SO_2 (e) SF_6

6A-2. Bromine vapor is roughly five times as dense as oxygen gas. Calculate the relative rates at which $Br_2(g)$ and $O_2(g)$ diffuse.

6A-3. Two flasks with the same volume are connected by a valve. One gram of hydrogen is added to one flask, and 1 g of oxygen is added to the other. What happens to the weight of the gas in the flask filled with hydrogen when the valve is opened?

6A-4. What happens to the relative amounts of N_2, O_2, Ar, CO_2, and He in air as air diffuses from one flask to another through a pinhole?

6A-5. N_2O and NO are often known by the common names nitrous oxide and nitric oxide. Associate the correct formula with the appropriate common name if nitric oxide diffuses through a pinhole 1.21 times as fast as nitrous oxide.

6A-6. If it takes 6.5 s for 25.0 cm^3 of helium gas to effuse through a pinhole into a vacuum, how long would it take for 25.0 cm^3 of CH_4 to escape under the same conditions?

6A-7. Calculate the molecular weight of an unknown gas if it takes 60.0 s for 250 cm^3 of the gas to escape through a pinhole in a flask into a vacuum and if it takes 84.9 s for the same volume of oxygen to escape under identical conditions.

6A-8. A lecture hall has 50 rows of seats. If laughing gas (N_2O) is released from the front of the room at the same time hydrogen cyanide (HCN) is released from the back of the room, in which row (counting from the front) will students first begin to die laughing? (Assume Graham's law of diffusion is valid.)

6A-9. The atomic weight of radon was first estimated by comparing its rate of diffusion with that of mercury vapor. What is the atomic weight of radon if mercury vapor diffuses 1.082 times as fast?

Derivations from Ideal Gas Behavior

6A-10. Predict whether the force of attraction between particles makes the volume of a real gas larger or smaller than an ideal gas.

6A-11. Predict whether the fact that the volume of gas particles is not zero makes the volume of a real gas larger or smaller than an ideal gas.

6A-12. Identify the term in the van der Waals equation used to explain why gases become cooler when they are allowed to expand rapidly.

6A-13. Describe the conditions under which significant deviations from ideal gas behavior are observed.

6A-14. Calculate the fraction of empty space in CO_2 gas, assuming 1 L of the gas at 0°C and 1.00 atm can be compressed until it changes to a liquid with a volume of 1.26 cm^3.

6A-15. The following data were obtained in a study of the pressure and volume of a sample of acetylene.

| P (atm): | 1 | 45.8 | 84.2 | 110.5 | 176.0 | 282.2 | 398.7 |
| V (L): | 1 | 0.01705 | 0.00474 | 0.00411 | 0.00365 | 0.00333 | 0.00313 |

Calculate the product of pressure times volume for each measurement. Plot PV versus P and explain the shape of the curve.

6A-16. Use the van der Waals constants for helium, neon, and argon to calculate the relative sizes of the atoms of these gases.

6A-17. Calculate the pressure of 1.00 mol of O_2 at 0°C in 1.0-L, 0.10-L, and 0.010-L containers using both the ideal gas equation and the van der Waals equation.

CHAPTER
7
MAKING AND BREAKING OF BONDS

7.1 ENERGY

We need energy to run our automobiles, power our batteries, heat our homes, and fuel our bodies. What do we mean when we say "energy," and what exactly is the origin of this energy?

It has often been stated that the United States runs on petroleum. Petroleum is oil, oil that is trapped underground. It is from petroleum that gasoline is made, and in fact petroleum is our society's main source of energy. From petroleum it is possible to separate out various chemical compounds. Some of these compounds are called hydrocarbons because they are composed of only carbon and hydrogen atoms. Table 7.1 lists some of the hydrocarbons with their names.

When the hydrocarbons are mixed with oxygen and ignited in the cylinder of an automobile by a spark, a chemical reaction occurs. For example, with the hydrocarbon octane the following reaction takes place:

$$2\ C_8H_{18}(g) + 25\ O_2(g) \longrightarrow 16\ CO_2(g) + 18\ H_2O(g)$$

TABLE 7.1   Common Hydrocarbons Found in Petroleum

Name	Structure
Butane	H H H H | | | | H—C—C—C—C—H | | | | H H H H
Octane	H H H H H H H H | | | | | | | | H—C—C—C—C—C—C—C—C—H | | | | | | | | H H H H H H H H
Isooctane (2,2,4-trimethyl pentane)	H CH₃ H CH₃ H | | | | | H—C—C—C—C—C—H | | | | | H CH₃ H H H
Isopentane (2-methyl butane)	H H H H | | | | H—C—C—C—C—H | | | | H CH₃ H H

As a result of the reaction the hot product gases expand, the pistons move, and the automobile's wheels turn. The burning of hydrocarbons in the furnace supplies the heat for our homes. The food we ingest is metabolized in the body by series of complex reactions. The reaction of sucrose, a sugar, with oxygen in the body can be summarized by the following chemical equation.

$$C_{12}H_{22}O_{11}(s) + 12\ O_2(g) \longrightarrow 12\ CO_2(g) + 11\ H_2O(l)$$

The consequences of metabolism are that the body is provided with the energy to do work and to maintain an appropriate temperature. A flashlight battery supplies electrons, which pass through a wire in a bulb and produce light. The electrons are driven through the wire as a result of a chemical reaction that takes place in the battery. All of the processes just mentioned occur because the energy needed to move pistons, drive electrons through wires, and heat our bodies and homes can be derived from energy-producing chemical reactions.

What is the ultimate source of the energy produced in those reactions? How much energy is available from such reactions? How can the reactions be used to obtain the maximum amount of energy from a given process? The answers to those questions form a major part of the remainder of the book.

Because this is a course in chemistry, the examples given above all relate to chemical processes. There are other kinds of energy, however, and some understanding of the different ways energy may be produced is essential to a more complete understanding of chemical energy.

First, what *is* energy? Energy may be classified as either kinetic or potential energy. **Kinetic energy** is the energy of motion. Molecules in translational or rotational motion, as discussed in Chapter 6, possess kinetic energy. **Potential energy** is the energy of position. A box lifted up a ladder is at a higher potential energy than a box left on the ground because the position of the box on the ladder is higher in the earth's gravitational field than a box on the ground. A vibrating molecule has both kinetic and potential energy. As the atoms in a molecule move relative to one another, the energy is part kinetic because of the motion and part potential because of the changing distance between atoms. Many substances we encounter daily are made up of molecules that are attracted to one another. On heating most substances expand. The energy supplied by heat not only causes the atoms of the molecules to move more vigorously but also causes the molecules to move away from each other. Thus, both potential and kinetic energy change during heat expansion because adding energy increases motion and the increased motion leads to increased distance between the atoms and molecules.

Energy may occur in several different forms and be transferred from one object to another. When a billiard ball is struck by a cue stick, the mechanical energy of the motion of the cue stick is transformed into the kinetic energy of the billiard ball.

In Chapter 3 we described the technique of photoelectron spectroscopy (PES). In PES, photons are used to transfer the electromagnetic energy contained in electromagnetic waves to an electron, causing the electron to be ejected from an atom. The resulting kinetic energy of the electron is measured and used to calculate the energy required to remove the electron from the atom.

Energy can be converted from one form to another. An example of the conversion of one form of energy into another occurs in the engine of an automobile. The car's battery converts chemical energy into electrical energy which is then converted into mechanical energy. The spectrum of hydrogen studied in Chapter 3 showed that electrons which move from one energy level to a level closer to the nucleus give up energy, which is emitted as electromagnetic radiation.

No matter what form energy takes or how it is transferred, the total energy before a process takes place and the total energy after the process is completed is the same, that is, energy is conserved. Conservation of energy means that energy cannot be created nor destroyed. In an elastic collision between a moving cue ball and a stationary billiard ball, the billiard ball moves away with increased kinetic energy. The cue ball must have lost energy and therefore must have slowed down after the collision. In photoelectron spectroscopy (Chapter 3), when a photon strikes an atom, causing an electron to be ejected with some amount of kinetic energy, we use the following equation to calculate the amount of energy required to remove the electron from the atom.

$$h\nu = \text{IE} + \text{KE}$$

The equation is based on conservation of energy. The amount of kinetic energy of the ejected electron plus the energy required to dislodge the electron from the atom are assumed to be equal to the total energy of the photon.

This chapter is concerned with the energy transfers and conversions associated with chemical reactions. These processes are part of the area of study known as **thermodynamics.** Chemical energy is the energy associated with the force of attraction between electrons and nuclei in molecules and metals and between oppositely charged ions in ionic compounds. In other words, chemical energy is the energy that is due to chemical bonds.

Energy changes that occur during a chemical reaction are due to the making and breaking of chemical bonds.[1] The amount of energy associated with a chemical reaction is directly related to the strengths of the chemical bonds that are broken and formed during a chemical reaction. The relationship between energy and the strengths of chemical bonds can be illustrated by the following hypothetical construction of molecules.

---

[1]There are also attractive forces that may be formed or overcome that can produce energy changes, as discussed further later in this chapter and in Chapter 8.

Imagine a collection of gaseous atoms consisting of two moles of carbon atoms, six moles of hydrogen atoms, and one mole of oxygen atoms. The atoms can be used to make two structures that are consistent with the rules developed for Lewis structures, one of which is dimethyl ether.

$$2\ C(g) + 6\ H(g) + O(g)$$

$$
\begin{array}{ccccc}
 & H & & H & \\
 & | & & | & \\
H- & C- & \ddot{O}- & C- & H \\
 & | & & | & \\
 & H & & H &
\end{array}
$$

Dimethyl ether

If these atoms are brought together to make 1 mol of gaseous dimethyl ether, 3151 kJ of energy is released. What is the origin of the energy? The energy is produced by the making of six carbon–hydrogen bonds and two carbon–oxygen bonds (a total of eight bonds). *Energy is released during the formation of chemical bonds.*

The starting atoms can be arranged in a different combination to form a different compound, ethyl alcohol.

$$2\ C(g) + 6\ H(g) + O(g)$$

$$
\begin{array}{cccc}
 & H & H & \\
 & | & | & \\
H- & C- & C- & \ddot{O}-H \\
 & | & | & \\
 & H & H &
\end{array}
$$

Ethyl alcohol

The compound contains five carbon–hydrogen bonds, one carbon–carbon bond, one carbon–oxygen bond, and one oxygen–hydrogen bond (a total of eight bonds). The formation of 1 mol of gaseous ethyl alcohol from its gaseous atoms releases 3204 kJ of energy. Thus, more energy is released in the formation of ethyl alcohol than in the formation of dimethyl ether. The bonds in ethyl alcohol are therefore stronger.

The effect of bond strength on the energy released during a chemical reaction can be further illustrated with a reaction more practical than the formation of compounds from their gaseous atoms. Consider the reactions of ethyl alcohol and dimethyl ether with oxygen, that is, the combustion of the two compounds.

$$CH_3CH_2OH(g) + 3\ O_2(g) \longrightarrow 2\ CO_2(g) + 3\ H_2O(g)$$
Ethyl alcohol

$$CH_3-O-CH_3(g) + 3\ O_2(g) \longrightarrow 2\ CO_2(g) + 3\ H_2O(g)$$
Dimethyl ether

One mole of each compound reacts with 3 mol of oxygen and produces 2 mol carbon dioxide and 3 mol water. The combustion of 1 mol of ethyl alcohol releases 1275 kJ of energy, whereas the combustion of 1 mol of dimethyl ether releases 1327 kJ of energy. The same products are formed in both reactions, but different bonds must be broken in the two reactant molecules. *Breaking chemical bonds requires an input of energy.* We have seen in the above discussion that the chemical bonds in ethyl alcohol are stronger than those in dimethyl ether. Therefore, more energy must be expended to break the bonds in ethyl alcohol. This explains why less energy results from the combustion of ethyl alcohol as compared to dimethyl ether. It also shows that, molecule per molecule, dimethyl ether would be a better energy producer than ethyl alcohol.

Checkpoint

In the following reaction are bonds broken or are they made?

$$O(g) + O(g) \longrightarrow O_2(g)$$

Will this reaction release energy or require an input of energy?

7.2   HEAT

Heat is energy in transit. Heat is one way in which energy can be transferred from one object to another. The transfer of heat is usually associated with a change in temperature. Although heat and temperature are related to one another, they are not the same thing. Temperature, as discussed in Chapter 6, is a measure of the "hotness" or "coldness" of an object and may be measured using the Fahrenheit, Celsius, or Kelvin scales. Because heat is defined as the transfer of energy, it must be measured in the same units as energy (joules). Although heat is associated with energy transfer, it is incorrect to consider a system or object as *containing* heat energy. A system that contains one of the forms of energy discussed in the previous section can transfer some of that energy to another object by means of heat.

If a hot brick is placed in contact with a cold brick, energy will be transferred as heat from the hot brick to the cold one. The hot brick will get cooler and the cool brick will get warmer. Eventually the two bricks will come to the same temperature. The hot brick has lost energy and the cold brick has gained energy.

7.3   HEAT AND THE KINETIC MOLECULAR THEORY

Chemists often divide the universe into a system and its surroundings. The **system** is that small portion of the universe in which we are interested. It may consist of the water in a beaker or a gas trapped in a piston and cylinder, as shown in Figure 7.1. The **surroundings** are everything else—in other words, the rest of the universe.

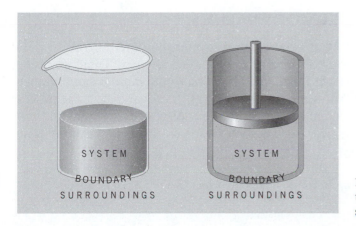

FIGURE 7.1 In the kinetic theory heat is transferred across the boundary between a system and its surroundings.

The system and its surroundings are separated by a **boundary.** The boundary can be as real as the glass in a beaker or the walls of a balloon. It can also be imaginary, such as a line 200 nm from the surface of a metal that arbitrarily divides the air close to the metal surface from the rest of the atmosphere. The boundary can be rigid or it can be elastic. In the kinetic theory, heat is transferred across the boundary between a system and its surroundings.

If a hot brick is placed in contact with a cold brick, kinetic energy will be transferred from the hot brick to the cold one. The particles composing the hot brick are jiggling about more rapidly than those of the cold brick. The motion of the particles of the hot brick transfers some energy to the cold brick. Eventually the particles in the two bricks come to the same average energy; in other words, the temperatures of the two bricks become the same.

The amount of energy needed to increase the temperature of a system (consisting of a single state of matter)[2] by a given amount depends on the nature of the system. Different materials are composed of different atoms, molecules, or ions arranged in different ways. Energy supplied to a solid can increase the motion (rotation, vibration, and motion about a lattice point) of the units composing a solid substance. Energy supplied to a liquid can increase the rotation, vibration, and translational motion of the molecules in the liquid. When energy is transferred to a gas, the speeds of the molecules in their random motion, as well as their rotation and vibration, increase. The increased motions can cause the units composing the system to move a little farther apart from one another. Because the units (atoms, molecules, or ions) are attracted to one another, moving them farther apart increases their potential energy (much like stretching a rubber band increases its potential energy). Thus, the transfer of energy by means of heat can increase the kinetic and potential energy of the system. The kinetic theory of heat can be summarized as follows. *Heat, when it enters a system, produces an increase in the average motion with which the particles of the system move.*[3]

The idea that atoms, molecules, and ions are in continuous random motion plays a very important role in much of the chemistry that is to follow. In addition to providing a means of understanding temperature and heat, the kinetic theory allows many properties of liquids, solids, and gases and the factors that influence how fast a chemical reaction occurs to be understood.

Checkpoint

Use the kinetic theory to explain what happens to the particles of a gas in a balloon when heat enters the balloon from its surroundings.

7.4   THE FIRST LAW OF THERMODYNAMICS

The first law of thermodynamics is a law of experience. It cannot be derived from general principles, but no experiment has yet been devised that contradicts the law. Thus, we accept this natural law as being of universal validity. *The first law of thermodynamics states that energy is conserved.* In other words, in any process, the total energy before and after the process has been carried out is the same. By total energy we mean the energy of everything that may conceivably be altered as a result of the process. The first law also tells us that energy is additive. That is, the total energy of a system is the sum of the energies of its parts.

---

[2]Changes between different states of matter are discussed in more detail in Chapter 8. Those changes involve a change in potential energy.

[3]A system as considered here consists of a single state of matter.

Energy is neither created nor destroyed. The energy absorbed by the system (the ice cubes) in this example is exactly equal to the energy lost by its surroundings (the tea).

Consider the following experiment. An ideal gas is contained in a piston and cylinder as shown in Figure 7.2. The piston is secured by stops, which prevent its movement. Then a hot brick is brought into contact with the apparatus. The average kinetic energy of the particles composing the brick is higher than that of the molecules of gas in the container; that is, the brick is hotter than the gas in the piston–cylinder apparatus. The motion of the jostling brick particles is transferred to the wall of the container and, through collisions with the wall, to the molecules of the gas. The temperature of the gas increases, the thermal motion of the gas molecules increases, collisions of the gas molecules with the walls of the container are more frequent and more forceful, and the pressure in the container increases. The brick has lost some energy, all of which has been gained by the gas and container.

FIGURE 7.2 Piston and cylinder apparatus containing an ideal gas. The piston is held in place by stops.

If the brick and container are considered to be the surroundings, the gas may be taken to be the system. The first law of thermodynamics then assures us that the total energy change, ΔE_{total}, may be written as

$$\Delta E_{total} = \Delta E_{sys} + \Delta E_{surr} = 0$$

The subscripts "sys" and "surr" stand for the system and its surroundings, respectively. The additive nature of energy allows the energy changes to be expressed as a sum of all of the individual energy changes that are part of a process, and the conservation condition assures that ΔE_{total} must be zero.

Checkpoint
A hot brick is thrown into cold water. According to the first law of thermodynamics, could the brick get hotter and the water colder?

In our piston experiment all the energy that entered the system went directly to the increased energy of the molecules of gas. The piston did not move despite the increased pressure in the container because the piston was held in place. The energy of the surroundings and that of the gas could and did change. We may summarize the energy changes in the following way. Before the brick was brought into contact with the container the total energy was

$$(E_{total})_{before} = (E_{surr})_{before} + (E_{gas})_{before}$$

and after the brick exchanged energy with the gas and container the total energy was

$$(E_{total})_{after} = (E_{surr})_{after} + (E_{gas})_{after}$$

The change in energy can be related to the energy before and after the process as follows:

$$\Delta E_{total} = (E_{total})_{after} - (E_{total})_{before} = 0$$
$$\Delta E_{total} = (E_{surr} + E_{gas})_{after} - (E_{surr} + E_{gas})_{before}$$
$$= [(E_{surr})_{after} - (E_{surr})_{before}] + [(E_{gas})_{after} - (E_{gas})_{before}]$$
$$\Delta E_{total} = \Delta E_{surr} + \Delta E_{gas} = 0$$

or

$$-\Delta E_{surr} = \Delta E_{gas}$$

The energy lost by the surroundings was gained by the gas. The sign convention for energy values is that energy gained by a system is considered positive and energy lost from a system is negative.

Now imagine a second experiment in which the process is the same but the stops on the piston are removed as shown in Figure 7.3. As before, we must take account of everything that may have its energy changed as a result of the process. With the stops removed, the piston can move upward. In order for the piston to move, some energy will have to be used to lift the weight of the piston. When the piston has moved up because of the increased pressure in the container, it will be higher in the earth's gravitational field and thus have increased energy. We'll call this energy change ΔE_{piston}. The change in energy is

$$\Delta E_{total} = \Delta E_{surr} + \Delta E_{sys}$$

Here the system consists of both the gas and the piston.

FIGURE 7.3 Piston and cylinder apparatus with an ideal gas but with the stops removed so that the piston can move.

$$\Delta E_{\text{total}} = \Delta E_{\text{surr}} + (\Delta E_{\text{gas}} + \Delta E_{\text{piston}})$$
$$\Delta E_{\text{total}} = \Delta E_{\text{surr}} + \Delta E_{\text{gas}} + \Delta E_{\text{piston}} = 0$$

The energy of the piston is increased by $mg\,\Delta h$, where m is the mass of the piston, g is the acceleration due to gravity, and Δh is the height to which the piston has been raised. The raising of a weight requires work. Chemists often write the energy change of the piston as

$$\Delta E_{\text{piston}} = mg\,\Delta h = \text{work} = w$$

Thus the first law becomes

$$\Delta E_{\text{total}} = \Delta E_{\text{surr}} + (\Delta E_{\text{gas}} + w) = 0$$
$$-\Delta E_{\text{surr}} = \Delta E_{\text{gas}} + w$$

In this case the energy lost by the surroundings appears in the increased temperature of the gas and the increased height of the piston. Not all of the energy lost by the surroundings has gone into increasing the temperature of the gas molecules as was the case in the first experiment. In the second experiment some of the energy went into increasing the temperature of the gaseous molecules, but some also went into the work of lifting the piston. If the same amount of energy is supplied to the gas in both experiments the final temperature of the gas in the first experiment will be higher than in the second.

The difference in the two processes is that, in the first case, the container was kept at a constant volume while in the second experiment the piston moved to expand the volume of gas. The pressure remains constant in the second experiment because the piston moves to maintain a balance between the pressure produced by the gas in the container and the pressure pushing down on the gas from the piston.

Experiments done in the laboratory are usually done in open flasks and not in piston–cylinder types of containers. An expanding system in the laboratory, however, does push against something if not a piston. What is this something that must be pushed back if the system is to expand? It is the atmosphere. The weight pushing down on the system under investigation is thus the weight of the atmosphere. What happens when a system contracts? The atmosphere slips a little closer to the earth, and the system can actually gain energy as a result of the process. In either case, for processes carried out under conditions of constant pressure, the first law is written

$$\Delta E_{\text{total}} = \Delta E_{\text{surr}} + (\Delta E_{\text{sys}} + w) = 0$$

A new term is defined called the **enthalpy (H)** which is a measure both of the energy change of the system plus any work done to or by the system. For example, an expanding gas raising a weight represents work done by the gas.

$$(\Delta E_{sys} + w) = \Delta H_{sys}$$

Checkpoint

If the same quantity of heat is transferred from the brick to the gas in a fixed volume piston cylinder (Figure 7.2) and in a movable piston cylinder (Figure 7.3), in which container will the temperature of the gas increase the most?

In the two experiments just described energy was supplied to the system from the surroundings. Now consider another process that is of considerable interest to a chemist.

The same assembly is used as previously described except the hot brick is removed and the ideal gas is replaced with a mixture of ethane gas and oxygen gas (Figure 7.4). The following reaction between ethane and oxygen can be induced to occur.

$$2 \, C_2H_6(g) + 7 \, O_2(g) \longrightarrow 4 \, CO_2(g) + 6 \, H_2O(g)$$

FIGURE 7.4 The ideal gas in the piston–cylinder apparatus has been replaced with a mixture of ethane and oxygen. There are no stops on the piston, and the source of external heat, the brick, has been removed.

The temperature in the container increases not as a result of externally supplied energy but because of the energy produced by the breaking and making of chemical bonds in the reaction. The piston is also raised and work has been done. The heat for the process came not from an external thermal source but from a chemical process. Heat was evolved by the chemical reaction because the sum of the bond strengths in the products is greater than the sum of the reactant bond strengths. Part of the evolved heat went into raising the piston and part into heating the container and gases. Under conditions of constant pressure, the heat is referred to as the enthalpy change produced by the chemical reaction.

The first law of thermodynamics summarizes the changes that take place in the experiment.

$$\Delta E_{total} = \Delta E_{sys} + \Delta E_{surr} + \Delta E_{piston}$$

Let us take the system to be the gaseous mixture in the cylinder. The system is heated because of the energy released in the chemical reaction; the product gases expand and raise the piston. Thus we identify ΔE_{sys} with ΔE_{rxn}, and ΔE_{piston} is the work done. The container is the surroundings. Then the first law becomes

$$\Delta E_{total} = \Delta E_{rxn} + \Delta E_{surr} + w$$

The enthalpy change for the reaction, ΔH_{rxn}, measures the heat of the reaction, ΔE_{rxn}, and the work done in lifting the piston.

$$\Delta E_{total} = \Delta H_{rxn} + \Delta E_{surr}$$

For any process $\Delta E_{total} = 0$ and $-\Delta H_{rxn} = \Delta E_{surr}$. ΔE_{surr} is seen to be determined by how much heat is exchanged with the container.

Table 7.2 gives the reactions of several hydrocarbons with oxygen along with the heat produced at constant pressure. The heat produced is therefore the enthalpy change for the reactions. Mixtures of hydrocarbon gases burned in a furnace with oxygen from the atmosphere are commonly used to heat homes.

TABLE 7.2   Enthalpy Changes for Several Common Hydrocarbons (298 K, 1 atm)

Hydrocarbon	Name	Structure	Heat Released Per Mole of Hydrocarbon Reacted with Oxygen (kJ)
$CH_4(g)$	Methane	H—C—H (with H above and below)	890
$C_2H_6(g)$	Ethane	H—C—C—H	1560
$C_3H_8(g)$	Propane	H—C—C—C—H	2222
$C_4H_{10}(g)$	Butane	H—C—C—C—C—H	2877
$C_5H_{12}(g)$	Pentane	H—C—C—C—C—C—H	3540

Exercise 7.1

We previously saw that breaking bonds requires an input of energy and that heat is released during formation of chemical bonds. What does this tell us about the relative strengths of the bonds in the products and reactants in the combustion reactions of Table 7.2?

Solution

The combustion reactions all release heat. Therefore, the sum of the bond strengths must be greater in the products than in the reactants. If this were not the case more heat would be required to break the bonds in the reactants than was returned on formation of the products, and this would result in a net input of heat.

7.5 STATE FUNCTIONS

Every system can be described in terms of certain measurable properties. A gas, for example, can be described in terms of the number of moles of particles it contains, its temperature, its pressure, its volume, its mass, or its density. Those properties describe the **state** of the system at a particular moment in time. Some of the properties depend on the size of the sample, such as mass and volume, and are therefore examples of **extensive properties** of the system. Others, such as temperature and density, are **intensive properties** that don't depend on the size of the sample being studied.

Properties can also be classified on the basis of whether they are **state functions.** By definition, one of the properties of a system is a state function if it depends only on the state of the system, not on the path used to get to that state.

Consider the temperature of a liquid. The fact that the temperature is 75.1°C at some moment in time doesn't tell us anything about the history of the system. It doesn't tell us how often the liquid has been heated or cooled before we take the measurement. Temperature is therefore a state function because it only reflects the state of the system at the moment at which it is measured. *Energy also is a state function,* that is, energy depends only on the state of the system, not on the path used to get to that state.

An analogy can illustrate that energy is a state function. Suppose we have a crate on the first floor of a New York skyscraper and we wish to take the crate to the tenth floor. To do that we will have to expend someone's or something's energy. The energy of the crate will change because the crate's potential energy is greater on the tenth floor than on the first by an amount that depends on the mass of the crate, the acceleration of gravity, and the difference in height between the tenth and first floors. The only variable factor influencing the energy of the crate is height. If the crate were transported to the fifteenth floor the energy of the crate would be greater than it was on the tenth or first floor. How the crate gets from one floor to the other does not change the energy of the crate. The energy change undergone by the crate depends only on the initial floor and the final floor. This is the meaning of a state function. Only initial and final conditions matter. Suppose the crate were transported to the tenth floor from the first and then returned to the first floor. What would be the energy change of the crate? The initial and final conditions are the same, so the energy of the crate has not changed.

Checkpoint

Suppose a crate is dropped from the tenth floor of a building. When at rest on the tenth floor what kind of energy does the crate have? During the fall what kind of energy does the crate have? What happens to the energy acquired by the falling crate when it comes to rest on the ground?

7.6 THE ENTHALPY OF A SYSTEM

All chemical reactions, no matter how simple or complex, have one thing in common. They all involve the breaking and reforming of bonds between atoms or ions. At some step, for example, the reaction between sodium metal and chlorine gas must involve breaking bonds between metal atoms in sodium metal and between chlorine atoms in Cl_2 gas. And the reaction must eventually involve the formation of the ionic bonds that hold the Na^+ and Cl^- ions together in NaCl.

$$2 \, Na(s) + Cl_2(g) \longrightarrow 2 \, NaCl(s)$$

Because the making and breaking of bonds plays such a central role in chemistry, a way to determine the energy consumed or produced in a chemical reaction is of extreme importance to chemists. The first law of thermodynamics provides us with a way to monitor changes in the energy of the system that accompanies a chemical reaction.

$$\Delta E_{total} = \Delta E_{sys} + \Delta E_{surr} = 0$$

All we have to do is find a way to measure the heat given off or absorbed by a chemical reaction. If we can achieve that goal, the heat given off or consumed in the reaction will be exactly equal to the change in the energy of the system that occurs during the reaction.

Experiments involving measurement of heat changes can be done in a **calorimeter,** shown in Figure 7.5. Because the volume of the container in which the reaction is run cannot change, the heat given off or absorbed during the reaction goes entirely to produce change in the surroundings. By monitoring the change in temperature of a water bath that surrounds the reaction container, the energy change of the surroundings that results from the making and breaking of chemical bonds can be calculated. Because $\Delta E_{surr} = -\Delta E_{sys}$, ΔE_{sys} can be found.

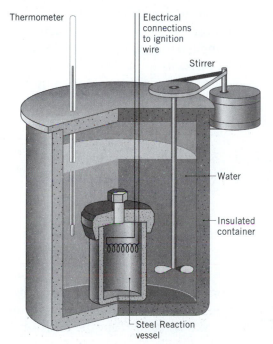

Thermometer

Electrical connections to ignition wire

Stirrer

Water

Insulated container

Steel Reaction vessel

FIGURE 7.5 Bomb calorimeter. Because the volume of the system is constant, no work of expansion can be done. As a result, $\Delta E_{surr} = -\Delta E_{sys}$.

Chemists, however, usually carry out reactions in containers such as beakers or flasks that are open to the atmosphere. These reactions occur under conditions of *constant pressure*. When the volume of a system can change, not all of the heat supplied to or taken from the system goes into producing a temperature change. To adjust for the difference in conditions of constant volume and constant pressure, chemists use the *enthalpy* of the system. Enthalpy, like energy, is a state function. The change in enthalpy that accompanies a chemical reaction at constant pressure is exactly equal to the heat given off or absorbed by the reaction.[4]

Experimentally determined calorimetric data are used to determine enthalpy values for various compounds. The data are compiled for use in calculating enthalpy changes for chemical reactions. Thus it is possible to use such data as given in Appendix B.14 to calculate the enthalpy change for many processes without having to experimentally measure the heat change directly. This compilation is a rather remarkable achievement of the branch of chemistry called thermochemistry. It is possible, for example, to use nothing more than the data in Appendix B.14 not only to calculate the economics of one fuel compared to another based on energy production but also to decide whether a given chemical reaction will occur and, as Chapter 13 will show, to determine what the best conditions are for maximizing the yield of a chemical reaction.

Calorimeters are routinely used to determine the caloric content of food. The food in question may be a sugar, such as sucrose, which is table sugar. Sucrose is reacted with oxygen in the calorimeter, and the heat released from the reaction is measured by determining how much the temperature of the surrounding water goes up.

$$C_{12}H_{22}O_{11}(s) + 12\ O_2(g) \longrightarrow 12\ CO_2(g) + 11\ H_2O(l)$$

In the reaction 5645 kJ of heat is released by the combustion of 1 mol of sucrose. The reaction is the same as the overall reaction that takes place in the body if sucrose is eaten. Thus an equivalent amount of energy, 5645 kJ, would be supplied to or stored by the body. This energy is generally referred to in terms of Calories (with a capital C) rather than kilojoules when talking about nutritional value. One Calorie (Cal) is equivalent to 4.184 kJ. Thus the caloric value of 1 mol of sucrose is about 1349 food Cal.[5] One mole of sucrose is 342 g, so that if one teaspoon (about 5 g) of sucrose is ingested it will provide about 20 Cal. If that energy is not expended by work or exercise it remains in the body. Slow walking expends about 150 Cal per hour, so an 8-minute walk will use up the Calories from a teaspoon of sugar.

Checkpoint

A tablespoon of sugar can provide several Calories of energy. What does this tell us about how the number of bonds and the strengths of these bonds in the reactants compare to those of the products?

7.7   ENTHALPIES OF REACTION

Chemical reactions are divided into two classes on the basis of whether they give off or absorb heat from their surroundings. **Exothermic** reactions give off heat to the surroundings. **Endothermic** reactions absorb heat from the surroundings. Reactions that give off heat are therefore *exothermic*. Reactions that absorb, or take in, heat from the surroundings are therefore *endothermic*.

---

[4]At constant pressure, the change in the enthalpy that occurs during a chemical reaction (ΔH) is equal to the change in energy of the system (ΔE) plus the product of the pressure times the change that occurs in the volume of the system during the reaction. For reactions in which the number of moles of gaseous products and reactants are equal, $\Delta H = \Delta E$.

[5]A Calorie (with a capital C) is equal to 1000 calories (with a small c).

The reaction between NH_4SCN and $Ba(OH)_2$ is an example of a spontaneous endothermic reaction. This reaction absorbs so much heat from its surroundings that it can freeze a beaker to a wooden board if the outside of the beaker is moistened.

The heat given off or absorbed in a chemical reaction at constant pressure is known as the **enthalpy of reaction.** When a reaction gives off heat to its surroundings, the energy of the system decreases. As a result, the enthalpy of the system decreases. Exothermic reactions are therefore characterized by negative values of ΔH.

Exothermic reactions: ΔH is negative   ($\Delta H < 0$)

Endothermic reactions, on the other hand, take in heat from their surroundings. As a result, the enthalpy of the system increases. Endothermic reactions are therefore characterized by positive values of ΔH.

Endothermic reactions: ΔH is positive   ($\Delta H > 0$)

An example of an exothermic reaction occurs when a balloon filled with hydrogen gas is ignited in the presence of oxygen. The reaction is accompanied by a large ball of fire and loud boom. The following chemical equation describes the reaction.

$$\boxed{2\ H_2(g) + O_2(g)}$$

$$\Delta H = -483.64\ kJ/mol_{rxn}$$

$$\boxed{2\ H_2O(g)}$$

When 2 mol of hydrogen reacts with 1 mol of oxygen to produce 2 mol of water, 483.64 kJ of heat is released. Changes in enthalpy are written in units of kilojoules per mole of reaction (kJ/mol_{rxn}), where "mol_{rxn}" represents the balanced chemical equation as a whole unit. To have meaning, a change in enthalpy must be associated with a specific chemical equation. Although the term "mol_{rxn}" is used to describe the reaction, it does not necessarily mean that there is only 1 mol of reactant or product. In the above chemical equation there are 3 mol of reactants (2 mol H_2 and 1 mol O_2) and 2 mol of product, but the reaction as a whole is referred to as one unit or one mole of chemical reaction.

If the chemical equation is changed there must be an accompanying change made in the enthalpy. For example, if the coefficients in the above chemical equation are all doubled, the change in enthalpy will also be doubled. The coefficients can be thought of as giving the number of moles of a substance per mole of reaction, for example, 4 mol H_2/mol_{rxn}.

$$\boxed{4\ H_2(g) + 2\ O_2(g)}$$

$$\Delta H = -967.28\ kJ/mol_{rxn}$$

$$\boxed{4\ H_2O(g)}$$

Because twice as many moles of reactants undergo the reaction, twice the amount of energy is released. The term "mol_{rxn}" in the enthalpy change of -967.28 kJ/mol_{rxn} refers now to the second chemical equation.

$$4 H_2(g) + 2 O_2(g) \longrightarrow 4 H_2O(g)$$

The reaction of hydrogen and oxygen to produce water is exothermic and therefore releases heat. If the reaction is considered in the reverse direction, an input of heat would be required.

$$\boxed{2 H_2(g) + O_2(g)}$$

$\Delta H = +483.64 \ kJ/mol_{rxn}$

$$\boxed{2 H_2O(g)}$$

The decomposition reaction has a positive ΔH and is therefore endothermic. Whenever a chemical equation is reversed, the magnitude of the change in enthalpy will remain constant but the sign will change.

Checkpoint

What is the sign of the enthalpy change for the combustion reactions in Table 7.2?

The enthalpy change at 298 K for the reaction

$$N_2(g) + 3 H_2(g) \longrightarrow 2 NH_3(g)$$

can be determined to be -92.2 kJ/mol_{rxn}. That enthalpy change applies to the reaction in which 1 mol of nitrogen reacts with 3 mol of hydrogen to produce 2 mol of ammonia. What will be the enthalpy change for the reaction if 0.200 mol of nitrogen reacts? We note that stoichiometry requires that if 0.200 mol of nitrogen reacts, 0.600 mol of hydrogen must be consumed and 0.400 mol of ammonia must be produced. We see from the equation that when 1 mol of nitrogen reacts, 92.2 kJ of heat is produced. This is equivalent to stating that the enthalpy change of the reaction is -92.2 kJ/mol N_2 reacted. If only 0.200 mol of N_2 reacts, the change in enthalpy is given by

$$(0.200 \text{ mol } N_2 \text{ reacted})\left(\frac{-92.2 \text{ kJ}}{\text{mol } N_2 \text{ reacted}}\right) = -18.4 \text{ kJ}$$

We could do the same calculation for hydrogen, for which -92.2 kJ of heat is produced for every 3 mol of hydrogen reacted:

$$(0.600 \text{ mol } H_2 \text{ reacted})\left(\frac{-92.2 \text{ kJ}}{3 \text{ mol } H_2 \text{ reacted}}\right) = -18.4 \text{ kJ}$$

Similarly, for NH_3 we could write:

$$(0.400 \text{ mol } NH_3 \text{ produced})\left(\frac{-92.2 \text{ kJ}}{2 \text{ mol } NH_3 \text{ produced}}\right) = -18.4 \text{ kJ}$$

Exercise 7.2

Pentaborane, B_5H_9, was once considered as a potential rocket fuel. B_5H_9 reacts with excess oxygen according to the following equation:

$$2\ B_5H_9(g) + 12\ O_2(g) \longrightarrow 5\ B_2O_3(g) + 9\ H_2O(g)$$

At 298 K the enthalpy change for the reaction is -8686.6 kJ/mol$_{rxn}$. Calculate the change in enthalpy when 0.600 mol of pentaborane is consumed.

Solution

As the reaction is written 8686.6 kJ of heat is produced from the combustion of 2 mol of B_5H_9. Therefore,

$$(0.600 \text{ mol } B_5H_9 \text{ reacted})\left(\frac{-8686.6 \text{ kJ}}{2 \text{ mol } B_5H_9 \text{ reacted}}\right) = -2.61 \times 10^3 \text{ kJ}$$

Note also that, if 0.600 mol of B_5H_9 is consumed, 3.60 mol of O_2 must be supplied and 1.50 mol of B_2O_3 and 2.70 mol of H_2O are produced. It does not matter which chemical species is chosen for the calculation of ΔH because stoichiometric constraints relate the number of moles of all species. We could use the product water to calculate the change in enthalpy:

$$(2.70 \text{ mol } H_2O \text{ produced})\left(\frac{-8686.6 \text{ kJ}}{9 \text{ mol } H_2O \text{ produced}}\right) = -2.61 \times 10^3 \text{ kJ}$$

Exercise 7.3

The reaction between hydrogen and oxygen is similar to the combustion reactions in Table 7.2. It has been suggested that hydrogen-powered cars could help solve the pollution of our atmosphere. Could the reaction

$$2\ H_2O(g) \longrightarrow 2\ H_2(g) + O_2(g)$$

be used to power an automobile? Could the reverse reaction be used?

$$2\ H_2(g) + O_2(g) \longrightarrow 2\ H_2O(g)$$

Explain the origin of the heat released or absorbed in the reactions.

Solution

The reaction $2\ H_2O(g) \rightarrow 2\ H_2(g) + O_2(g)$ is endothermic and could not be used to run an engine. The reverse reaction $2\ H_2(g) + O_2(g) \rightarrow 2\ H_2O(g)$ is exothermic and could supply energy for running a car. The origin of the heat is in the making and breaking of bonds. The Lewis structures of the products and reactants are as follows:

$$2\ H{-}H + \ddot{\underset{\cdot\cdot}{O}}{=}\ddot{\underset{\cdot\cdot}{O}} \longrightarrow 2\ \overset{\cdot\cdot}{\underset{H\ \ \ \ H}{O}}$$

Two H—H bonds and one oxygen–oxygen double bond must be broken to form two $H_2O(g)$ molecules, each containing two H—O bonds. Because the reaction is exothermic, the bond strengths of the water must be stronger than those broken in H_2 and O_2. The reverse reaction is endothermic because more enthalpy is required to break the bonds in water than is gained by the formation of the bonds in H_2 and O_2.

7.8   ENTHALPY AS A STATE FUNCTION

Both the energy and the enthalpy of a system are state functions. They depend only on the state of the system at any moment, not its history. To examine the consequences of this fact, let's consider the following reaction.

$$H_2(g) + Cl_2(g) \longrightarrow 2\ HCl(g)$$

Since enthalpy is a state function we can visualize the reaction as occurring by two simple processes. First, we break the bonds in the starting materials to form hydrogen and chlorine atoms in the gas phase.

$$\boxed{2\ H(g) + 2\ Cl(g)}$$

$$\boxed{H_2(g) + Cl_2(g)}$$

Then the atoms are recombined to form the product of the reaction.

$$\boxed{2\ H(g) + 2\ Cl(g)}$$

$$\boxed{2\ HCl(g)}$$

It doesn't matter whether the reaction actually occurs by these steps or not. Because enthalpy is a state function, ΔH for the hypothetical reaction will be equal to ΔH for the reaction whatever the pathway by which the starting materials are transformed into the products.

To break the covalent bonds in a mole of Cl_2 molecules to form 2 mol of chlorine atoms, 243.4 kJ of enthalpy is required.

$$Cl_2(g) \longrightarrow 2\ Cl(g) \qquad \Delta H = 243.4\ \text{kJ/mol}_{rxn}$$

It takes 435.3 kJ of enthalpy to break apart a mole of H_2 molecules to form 2 mol of hydrogen atoms.

$$H_2(g) \longrightarrow 2\ H(g) \qquad \Delta H = 435.3\ \text{kJ/mol}_{rxn}$$

We therefore have to invest a total of 678.7 kJ of energy in the system to transform a mole of H_2 molecules and a mole of Cl_2 molecules into two moles of H atoms and two moles of Cl atoms.

Bond breaking

$$Cl_2(g) \longrightarrow 2\ Cl(g) \qquad \Delta H = 243.4\ \text{kJ/mol}_{rxn}$$

$$H_2(g) \longrightarrow 2\ H(g) \qquad \Delta H = 435.3\ \text{kJ/mol}_{rxn}$$

$$H_2(g) + Cl_2(g) \longrightarrow 2\ H(g) + 2\ Cl(g) \qquad \Delta H = 678.7\ \text{kJ/mol}_{rxn}$$

We now turn to the process by which the atoms recombine to form HCl molecules. The bond between hydrogen and chlorine atoms is relatively strong. We get back 431.6 kJ for each mole of H—Cl bonds that are formed. Because we have 2 mol of hydrogen atoms and 2 mol of chlorine atoms, we get 2 mol of HCl molecules. Thus, the bond making process gives off a total of 863.2 kJ.

Bond making $2\ H(g) + 2\ Cl(g) \longrightarrow 2\ HCl(g)$ $\Delta H = -863.2\ \text{kJ/mol}_{rxn}$

The overall change in the enthalpy of the system that occurs during the reaction can be calculated by combining ΔH for the two hypothetical steps in the reaction.

$$\begin{array}{ll} 678.7\ \text{kJ/mol}_{rxn} & \text{Bond breaking} \\ -863.2\ \text{kJ/mol}_{rxn} & \text{Bond making} \\ \hline -184.5\ \text{kJ/mol}_{rxn} & \end{array}$$

According to the calculation, the overall reaction is exothermic (ΔH is negative) by a total of -184.5 kJ when 1 mol of H_2 reacts with a mole of Cl_2 to form 2 mol of HCl.

$$H_2(g) + Cl_2(g) \longrightarrow 2\ HCl(g) \qquad \Delta H = -184.5\ \text{kJ/mol}_{rxn}$$

Does the reaction proceed through the hypothetical steps? It doesn't matter whether it does or it doesn't. At the heart of reaction thermodynamics is the concept that the enthalpy of a system is a state function. As a result, *the value of ΔH for a reaction doesn't depend on the path used to convert the starting materials into the products of the reaction.* It only depends on the initial and final conditions—the reactants and products of the reaction.

7.9   STANDARD-STATE ENTHALPIES OF REACTION

The heat given off or absorbed by a chemical reaction depends on the conditions of the reaction. Three factors are important: (1) the amounts of the starting materials and products involved in the reaction, (2) the temperature at which the reaction is run, and (3) the pressure of any gases involved in the reaction. The reaction in which methane is burned can be used to illustrate why the reaction conditions must be specified.

Assume that we start with a mixture of CH_4 and O_2 at 25°C in a container just large enough so that the pressure of each gas is 1 atm.

$$CH_4(g) + 2\ O_2(g) \longrightarrow CO_2(g) + 2\ H_2O(g)$$

Under those conditions, the reaction gives off 802.4 kJ/mol of CH_4 consumed. However, if we start with the reactants at 1000°C and 1 atm pressure and produce products at 1000°C and 1 atm, the reaction gives off only 792.4 kJ/mol_{rxn}. This illustrates the importance when reporting thermodynamic data of specifying the conditions under which a reaction occurs.

Thermodynamic data are often measured at 25°C (298 K). Measurements taken at other temperatures are identified by adding a subscript specifying the temperature in kelvins.

The data collected for the combustion of methane at 1000°C, for example, would be reported as follows: $\Delta H_{1273} = -792.4$ kJ/mol$_{rxn}$.

The effect of pressure and amount of materials on the heat given off or absorbed in a chemical reaction is controlled by defining a set of standard conditions for thermodynamic experiments. By definition, the **standard state** for thermodynamic measurements involving gases fulfills the following requirement.

- The pressure of any gas involved in the reaction is 1 atm.

Enthalpy measurements done under the standard-state condition are indicated by adding a superscript ° to the symbol for enthalpy. The *standard-state* enthalpy of reaction for the combustion of natural gas at 25°C, for example, would be reported as follows:
$$\Delta H° = -802.4 \text{ kJ/mol}_{rxn}.$$

Checkpoint
What does the symbol $\Delta H°_{373}$ mean?

7.10   CALCULATING ENTHALPIES OF REACTION

The origin of the heat change that accompanies chemical reactions can be more clearly seen by examining the bonding in the products and reactants. Consider the following reaction:

$$CO(g) + H_2O(g) \longrightarrow CO_2(g) + H_2(g)$$

We can use Lewis structures to visualize the reaction.

$$:C\equiv O: + H-\overset{..}{\underset{..}{O}}-H \longrightarrow \overset{..}{\underset{.}{O}}=C=\overset{..}{\underset{.}{O}} + H-H$$

Once again, we can imagine a process for converting reactants into products. We break all of the bonds in the starting materials to form isolated atoms in the gas phase. Note that a carbon–oxygen triple bond and two oxygen–hydrogen single bonds are broken.

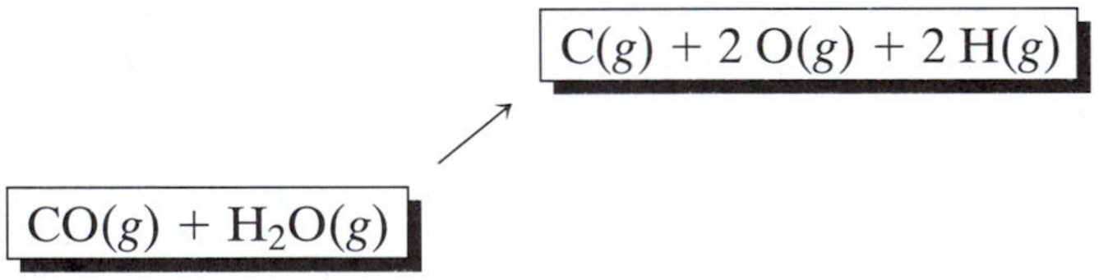

$$C(g) + 2\,O(g) + 2\,H(g)$$

$$CO(g) + H_2O(g)$$

We then allow the atoms to recombine to form the products of the reaction.

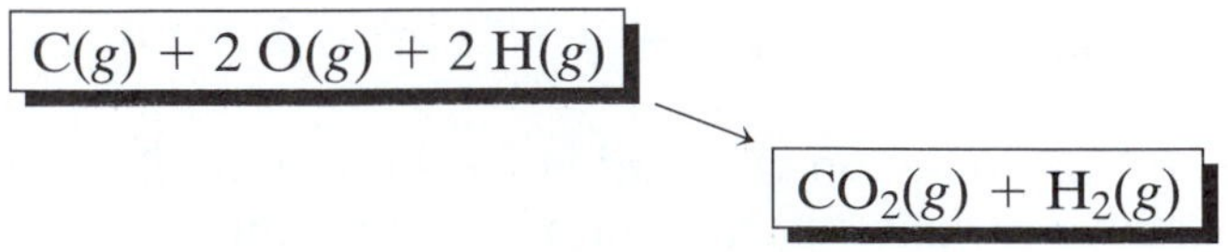

$$C(g) + 2\,O(g) + 2\,H(g)$$

$$CO_2(g) + H_2(g)$$

Experimentally we find that it takes 1076.4 kJ/mol$_{rxn}$ to break apart CO molecules to form C and O atoms. It takes 926.3 kJ/mol$_{rxn}$ to break the bonds in H_2O molecules to form H and O atoms. Thus, the bond breaking process takes a total of 2002.7 kJ.

Bond breaking

$$CO(g) \longrightarrow C(g) + O(g) \qquad \Delta H° = 1076.4 \text{ kJ/mol}_{rxn}$$
$$H_2O(g) \longrightarrow 2 H(g) + O(g) \qquad \Delta H° = 926.3 \text{ kJ/mol}_{rxn}$$
$$\overline{CO(g) + H_2O(g) \longrightarrow C(g) + 2 H(g) + 2 O(g) \qquad \Delta H° = 2002.7 \text{ kJ/mol}_{rxn}}$$

When the isolated atoms in the gas phase are recombined to form new bonds, we see that two carbon–oxygen double bonds are formed, giving off a total of 1608.5 kJ per mole of CO_2. A hydrogen–hydrogen single bond is also created. We have already found that we get 435.3 kJ per mole of H_2 molecules formed when hydrogen atoms combine. We therefore get a total of 2043.8 kJ back when the C, H, and O atoms combine to form one mole of CO_2 and H_2.

Bond making

$$C(g) + 2 O(g) \longrightarrow CO_2(g) \qquad \Delta H° = -1608.5 \text{ kJ/mol}_{rxn}$$
$$2 H(g) \longrightarrow H_2(g) \qquad \Delta H° = -435.3 \text{ kJ/mol}_{rxn}$$
$$\overline{C(g) + 2 H(g) + 2 O(g) \longrightarrow CO_2(g) + H_2(g) \qquad \Delta H° = -2043.8 \text{ kJ/mol}_{rxn}}$$

When we combine the change in the enthalpy of the system for the two hypothetical steps we conclude that the overall enthalpy of reaction is relatively small, compared with the enthalpy change associated with either bond breaking or bond making.

$$\begin{array}{ll} 2002.7 \text{ kJ/mol}_{rxn} & \text{Bond breaking} \\ \underline{-2043.8 \text{ kJ/mol}_{rxn}} & \text{Bond making} \\ -41.1 \text{ kJ/mol}_{rxn} & \end{array}$$

The reaction is exothermic, and it will give off heat because the sum of bond strengths in the products is larger than that in the reactants. But the amount of heat given off by the reaction is not large.

7.11 ENTHALPIES OF ATOM COMBINATION

Chemists are often interested in knowing whether under constant atmospheric pressure a reaction gives off or absorbs heat—and how much heat is given off or absorbed. The heat released or absorbed at a constant pressure is equivalent to the enthalpy change and can be determined experimentally in the laboratory or calculated as was done in the previous section. It is important to remember that even the calculations are based on experimentally measured values.

To apply the technique used in the previous section to other reactions we need a set of data that allows us to predict how much heat will be absorbed when we transform the starting materials into their isolated atoms in the gas phase and how much heat will be given off when the atoms recombine to give the products of the reaction. There are two ways the data could be compiled. Consider, for example, the compound ammonia, NH_3. We could choose to give the enthalpy change for the reaction

$$NH_3(g) \longrightarrow N(g) + 3 H(g)$$

for which the enthalpy required to break ammonia apart into its atoms is $+1171.76$ kJ/mol$_{rxn}$. When the reaction is written in that way the enthalpy change is called the **enthalpy of atomization** because the reaction refers to the breaking apart of the compound into its atoms.

Another way to arrange the data in a table is based on the reverse of the reaction just shown.

$$N(g) + 3 H(g) \longrightarrow NH_3(g)$$

In this case ammonia is formed from its atoms, and the enthalpy change for the reaction is -1171.76 kJ/mol$_{rxn}$. Such enthalpy changes are called **enthalpies of atom combination.** It makes no difference how we tabulate the data as long as we specify to which process, atomization or atom combination, the enthalpies refer. By a long-standing convention chemists prefer to use the formation reaction, and that is how data are compiled in Appendix B.14. If we look under $NH_3(g)$ in Table B.14 we will find the enthalpy of atom combination (ac) listed as -1171.76, which refers to the formation reaction

$$N(g) + 3 H(g) \longrightarrow NH_3(g)$$

at one atmosphere pressure and 25°C. By definition the symbol ΔH°_{ac} represents the enthalpy change under the standard conditions when 1 mol of the compound listed in the table is formed from its gaseous atoms.

When we start with a substance that is a gas at 25°C and 1 atm, the enthalpy of atom combination involves nothing more than forming the bonds within that substance. The value of ΔH°_{ac} for methane, for example, reflects the heat that is released when the four C—H bonds in a mole of CH_4 molecules are formed from the gaseous atoms of carbon and hydrogen.

$$C(g) + 4 H(g) \longrightarrow CH_4(g) \qquad \Delta H^{\circ}_{ac} = -1662.09 \text{ kJ/mol}_{rxn}$$

Checkpoint
What are the values of ΔH°_{ac} for N, and H? Explain your answers. What is the value of ΔH°_{ac} of NH_3? Explain your answer.

Relatively few reactions involve only starting materials and products that exist as gases at room temperature and atmospheric pressure. The enthalpy of atom combination for liquid methanol (CH_3OH), for example, includes two terms in its measurement. First is the heat released when we convert the carbon, hydrogen, and oxygen atoms into methanol in the gas phase.

$$C(g) + 4 H(g) + O(g) \longrightarrow CH_3OH(g) \qquad \Delta H^{\circ}_{ac} = -2037.11 \text{ kJ/mol}_{rxn}$$

The origin of the heat is the formation of the bonds of gaseous methanol. The covalent bonds, formed by the sharing of electrons between atoms, are called intramolecular bonds because they are within the molecule itself.

The second part of the enthalpy of atom combination is how much heat is released when the gas is condensed to form liquid methanol. The latter quantity may be directly determined by calorimetric measurements.

$$CH_3OH(g) \longrightarrow CH_3OH(l) \qquad \Delta H° = -38.00 \text{ kJ/mol}_{rxn}$$

The release of heat during a change of state is not due to formation of new intramolecular bonds. The covalent bonds in liquid and gaseous methanol are essentially the same. What then is responsible for the enthalpy change when gaseous methanol is condensed to the liquid state?

In the gaseous state the methanol molecules are far apart and touch only when they collide. In the liquid state the molecules of methanol are in constant contact with one another. Molecules contain electrons distributed between atoms according to the electronegativities of the atoms. The partial charges developed on the atoms can give a charge separation that produces an electric dipole moment. The dipoles are significant only when molecules are close together, and they serve to attract molecules to one another. Similar types of attractive forces exist between ions or other particles that make up a liquid or solid substance. Such forces are called *intermolecular forces* when they exist between molecules and *interionic forces* when they occur between ions. When a gas is condensed its particles come close enough that they are attracted to one another, and as a consequence heat is released. Thus, the enthalpies of atom combination for liquids and solids give directly the sum of the contributing bond enthalpies and any additional interactions by intermolecular or interionic forces.

The enthalpy of atom combination for gaseous methanol from its gaseous atoms is $-2037.11 \text{ kJ/mol}_{rxn}$, whereas the enthalpy of atom combination for liquid methanol from its gaseous atoms is $-2075.11 \text{ kJ/mol}_{rxn}$.

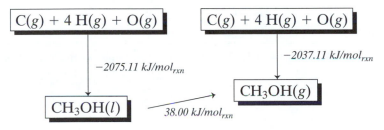

The difference in the two values (38.00 kJ/mol$_{rxn}$) represents the energy released when a mole of gaseous methanol condenses to form liquid methanol or the enthalpy which must be absorbed to evaporate one mole of liquid methanol to form gaseous methanol. The change in enthalpy is associated with forming or overcoming intermolecular forces.

Enthalpy of atom combination data for more than 200 substances are given in the table of standard-state enthalpies, free energies, and entropies of atom combination found in Appendix B.14. The first column of Table B.14 lists the chemical formulas and physical states of a variety of substances. The second column lists the enthalpies of atom combination associated with forming all of the bonds in 1 mol of the substance from individual gaseous atoms. In addition, Table 7.3 gives the enthalpies of atom combination data of several gas phase organic compounds. The data can be used to determine the change in enthalpy associated with any chemical reaction if data for all of the reactants and products are available in the table.

TABLE 7.3   Enthalpies of Atom Combination for Several Gaseous Organic Compounds[a]

Name	Formula	ΔH°_{ac}, kJ/mol_{rxn} at 298.15 K
Methane	CH_4	−1662.09
Ethane	$C_2H_6(g)$	−2823.94
Propane	$C_3H_8(g)$	−3992.9
n-Butane	$C_4H_{10}(g)$	−5169.38
Isobutane	$C_4H_{10}(g)$	−5177.75
n-Pentane	$C_5H_{12}(g)$	−6337.9
n-Hexane	$C_6H_{14}(g)$	−7509.1
Methanol	$CH_3OH(g)$	−2037.11
Ethanol	$CH_3CH_2OH(g)$	−3223.53
n-Propanol	$CH_3CH_2CH_2OH(g)$	−4394.2
n-Butanol	$CH_3CH_2CH_2CH_2OH(g)$	−5564.5
n-Pentanol	$CH_3CH_2CH_2CH_2CH_2OH(g)$	−6735.9
Dimethyl ether	$CH_3{-}O{-}CH_3(g)$	−3171.3
Ethylmethyl ether	$CH_3{-}O{-}CH_2CH_3(g)$	−4354.6
Diethyl ether	$CH_3CH_2{-}O{-}CH_2CH_3(g)$	−5541.4
Dipropyl ether	$CH_3CH_2CH_2{-}O{-}CH_2CH_2CH_3(g)$	−7883.1
Ethylene	$CH_2{=}CH_2(g)$	−2251.70
Propene	$CH_3CH{=}CH_2(g)$	−3432.6
1-Butene	$CH_2{=}CHCH_2CH_3(g)$	−4604.9
1-Pentene	$CH_2{=}CHCH_2CH_2CH_3(g)$	−5777.4
1-Hexene	$CH_2{=}CHCH_2CH_2CH_2CH_3(g)$	−6947.7
1,3-Butadiene	$CH_2{=}CHCH{=}CH_2(g)$	−4058.9
Benzene	$C_6H_6(g)$	−5523.07
Toluene	$C_6H_5CH_3(g)$	−6690.0
Ethylbenzene	$C_6H_5CH_2CH_3(g)$	−7870.3
1,3,5-Trimethylbenzene	$C_6H_3(CH_3)_3(g)$	−9067.2

[a]From J.D. Cox and G. Pilcher, *Thermochemistry of Organic and Organometallic Compunds,* Academic Press, New York, 1970.

To illustrate how the data are used, let's consider the reaction to synthesize ammonia from nitrogen and hydrogen.

$$N_2(g) + 3\ H_2(g) \longrightarrow 2\ NH_3(g)$$

Because enthalpy is a state function, we can divide the reaction into a sequence of bond breaking and bond making. Heat would have to be absorbed to break the N≡N triple bonds in a mole of N_2 and the H—H single bonds in three moles of H_2 to form isolated nitrogen and hydrogen atoms in the gas phase.

$$\boxed{2\ N(g) + 6\ H(g)}$$

ΔH is positive (+) ↗

$$\boxed{N_2(g) + 3\ H_2(g)}$$

Heat will be given off when the atoms come together to form NH_3 molecules in the gas phase.

$$\boxed{2\,N(g) + 6\,H(g)}$$

ΔH *is negative* $(-)$

$$\boxed{2\,NH_3(g)}$$

The primary reason for introducing the sign convention is to remind us that endothermic reactions, which absorb heat from the surroundings, are represented by *positive* values of ΔH. Exothermic reactions, which give off heat, are described by *negative* values of ΔH. Note that if we want the enthalpy change for breaking the bonds in a compound, we must reverse the sign on the enthalpy given in Appendix B.14.

It always takes energy to break the bonds in a molecule to form isolated atoms in the gas phase. When we calculate the heat that must be absorbed to break the bonds for this hypothetical reaction, we must therefore use positive values for the enthalpy for each of the starting materials consumed in the step. Heat is always given off when atoms recombine to form molecules. Negative enthalpies are therefore used to calculate the heat given off when the products of the reaction are formed.

The energetics of the bond breaking in the synthesis of ammonia can be calculated from the enthalpies tabulated in Table B.14 for H_2 ($\Delta H^\circ_{ac} = -435.30$ kJ/mol$_{rxn}$) and N_2 ($\Delta H^\circ_{ac} = -945.41$ kJ/mol$_{rxn}$). In the course of this hypothetical step in the reaction, 1 mol of N_2 and 3 mol of H_2 are transformed into hydrogen and nitrogen atoms in the gas phase. The enthalpy of reaction for breaking the nitrogen–nitrogen triple bonds in a mole of N_2 is therefore equal to the enthalpy required to break all the bonds in the molecule. At the same time, 3 mol of H_2 is converted into 6 mol of hydrogen atoms in the gas phase. The enthalpy of reaction for breaking the H—H single bonds in 3 mol of H_2 is equal to three times the enthalpy required for one mole of H_2 molecules.

Bond breaking			
	$N_2(g) \longrightarrow 2\,N(g)$	$\Delta H^\circ =$	945.41 kJ/mol$_{rxn}$
	$3\,H_2(g) \longrightarrow 6\,H(g)$	$\Delta H^\circ =$	1305.9 kJ/mol$_{rxn}$
	$N_2(g) + 3\,H_2(g) \longrightarrow 2\,N(g) + 6\,H(g)$	$\Delta H^\circ =$	2251.3 kJ/mol$_{rxn}$

The heat given off during bond formation can be calculated from the enthalpy of atom combination of NH_3 ($\Delta H^\circ_{ac} = -1171.76$ kJ/mol$_{rxn}$ in Table B.14). When the isolated nitrogen and hydrogen atoms come together to form NH_3 molecules, 2 mol of NH_3 is formed. The enthalpy of reaction for the formation of 2 moles of NH_3 is therefore twice the enthalpy of atom combination of NH_3.

Bond making $\qquad 2\,N(g) + 6\,H(g) \longrightarrow 2\,NH_3(g) \qquad \Delta H^\circ_{ac} = -2343.52$ kJ/mol$_{rxn}$

Note that bond breaking is endothermic (ΔH is positive) and that heat is released when the atoms recombine to form bonds, which means that bond making is exothermic (ΔH is negative).

All we have to do to calculate the overall enthalpy of reaction is add the results of our calculations for the two hypothetical steps in the reaction.

$$
\begin{array}{r}
2251.3 \ \text{kJ/mol}_{rxn} \\
-2343.52 \ \text{kJ/mol}_{rxn} \\
\hline
-92.2 \ \text{kJ/mol}_{rxn}
\end{array}
$$

The overall reaction is therefore exothermic.

It is important to remember when doing enthalpy calculations in this way that very few chemical reactions actually occur by first breaking all the bonds in the starting materials to form isolated atoms in the gas phase, followed by recombination of the atoms to form the products of the reaction. But, for our purposes, that doesn't matter. If all we want to know is whether the reaction gives off or absorbs heat, we can *assume* that it occurs by atomization and recombination steps. Because enthalpy is a state function, ΔH for a reaction doesn't depend on the path used to convert the starting materials into the products.

Checkpoint

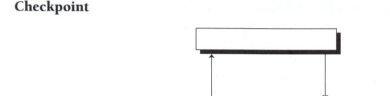

For the purposes of calculating enthalpy changes a chemical reaction can be diagrammed as shown above. If the total bond strengths in the products is less than that in the reactants, label the boxes above as *reactants, products,* and *gaseous atoms.* Draw and label a similar diagram if the total bond strengths in the products is greater than that in the reactants. Also label the arrows that represent the *energy to break bonds, the energy to form bonds,* and the *enthalpy change* for the overall reaction.

Exercise 7.4

Charcoal is mainly carbon. Use enthalpy of atom combination data from Appendix B.14 to predict whether the reaction between charcoal and oxygen to form carbon dioxide is endothermic or exothermic and to calculate the amount of heat given off or absorbed in the reaction below.

$$C(s) + O_2(g) \longrightarrow CO_2(g)$$

Solution

We can imagine the process by using the following diagram:

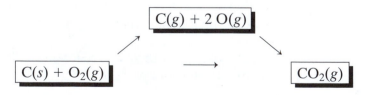

When solid carbon is formed from its isolated carbon atoms in the gas phase, a significant release of heat occurs as shown in the enthalpy of atom combination data.

$$C(g) \longrightarrow C(s) \qquad \Delta H°_{ac} = -716.68 \text{ kJ/mol}_{rxn}$$

Likewise, atom combination data show that when the O$=$O double bond in an O_2 molecule is formed, 498.34 kJ/mol$_{rxn}$ of heat is released.

$$2\ O(g) \longrightarrow O_2(g) \qquad \Delta H^\circ_{ac} = -498.34\ \text{kJ/mol}_{rxn}$$

Therefore, the hypothetical step in which we break all the bonds in the starting materials requires a large input of heat (enthalpy).

Bond breaking

$$
\begin{array}{ll}
C(s) \longrightarrow C(g) & \Delta H^\circ = \quad 716.68\ \text{kJ/mol}_{rxn} \\
O_2(g) \longrightarrow 2\ O(g) & \Delta H^\circ = \quad 498.34\ \text{kJ/mol}_{rxn} \\
\hline
C(s) + O_2(g) \longrightarrow C(g) + 2\ O(g) & \Delta H^\circ = 1215.02\ \text{kJ/mol}_{rxn}
\end{array}
$$

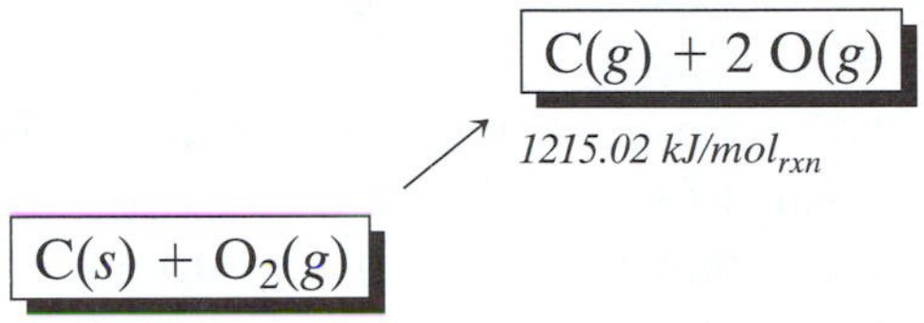

$$\boxed{C(g) + 2\ O(g)}$$

1215.02 kJ/mol$_{rxn}$

$$\boxed{C(s) + O_2(g)}$$

But the C$=$O double bonds in a CO_2 molecule are unusually strong ($\Delta H^\circ_{ac} = -1608.53$ kJ/mol). Thus, a great deal of energy is given off when the isolated atoms come together to form CO_2 molecules.

Bond making $$C(g) + 2\ O(g) \longrightarrow CO_2(g) \qquad \Delta H^\circ_{ac} = -1608.53\ \text{kJ/mol}_{rxn}$$

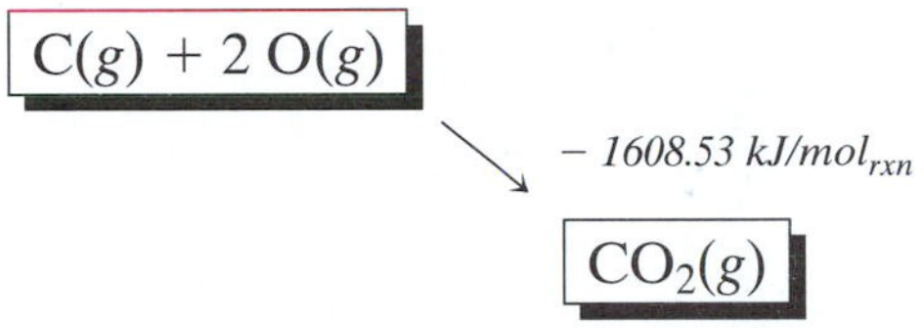

$$\boxed{C(g) + 2\ O(g)}$$

$-$ 1608.53 kJ/mol$_{rxn}$

$$\boxed{CO_2(g)}$$

The overall reaction is therefore strongly exothermic.

$$
\begin{array}{r}
1215.02\ \text{kJ/mol}_{rxn} \\
-1608.53\ \text{kJ/mol}_{rxn} \\
\hline
-393.51\ \text{kJ/mol}_{rxn}
\end{array}
$$

The heat produced by the reaction is used in backyard grills to cook hamburgers.

Exercise 7.5

The reason that charcoal can be used in a grill is that the carbon–oxygen double bonds in CO_2 are so strong that the reaction proceeds with an evolution of heat. Carbon (charcoal) can also react with oxygen to produce carbon monoxide (CO). Is the reaction endothermic or exothermic?

Solution

The reaction is

$$2\ C(s) + O_2(g) \longrightarrow 2\ CO(g)$$

The enthalpy change is

$$2\,C(g) + 2\,O(g)$$

$1931.7\ kJ/mol_{rxn}$ $-2152.8\ kJ/mol_{rxn}$

$$2\,C(s) + O_2(g) \qquad\longrightarrow\qquad 2\,CO(g)$$

$$\Delta H° = 1931.7 - 2152.8 = -221.1\ kJ/mol_{rxn}$$

The reaction is exothermic but not as exothermic as

$$C(s) + O_2(g) \longrightarrow CO_2(g) \qquad \Delta H° = -393.51\ kJ/mol_{rxn}$$

The reason is that the carbon to oxygen triple bond in CO is not as strong as the two carbon–oxygen double bonds in CO_2. The reaction of carbon with oxygen can and does produce carbon monoxide particularly if inadequate supplies of oxygen are available for conversion to CO_2. Carbon monoxide can combine with the iron in human blood and is toxic. To cut down on the production of CO from automobiles, catalytic converters are used to convert CO to CO_2 by the reaction

$$2\,CO(g) + O_2(g) \longrightarrow 2\,CO_2(g)$$

Checkpoint
Is this reaction endothermic or exothermic?

$$2\,CO(g) + O_2(g) \longrightarrow 2\,CO_2(g)$$

7.12 USING ENTHALPIES OF ATOM COMBINATION TO PROBE CHEMICAL REACTIONS

Enthalpies of atom combination can be used in ways other than to calculate the enthalpy change for a reaction. Values of these enthalpies measure the total bond strengths for gaseous compounds. They provide a direct means of comparing the strengths of the bonds holding a compound together.

In 1833 Jöns Jacob Berzelius suggested that compounds with the same formula but different structures should be called **isomers** (literally, "equal parts"). The following compounds are isomers because they both contain the same number of carbon, hydrogen, and oxygen atoms.

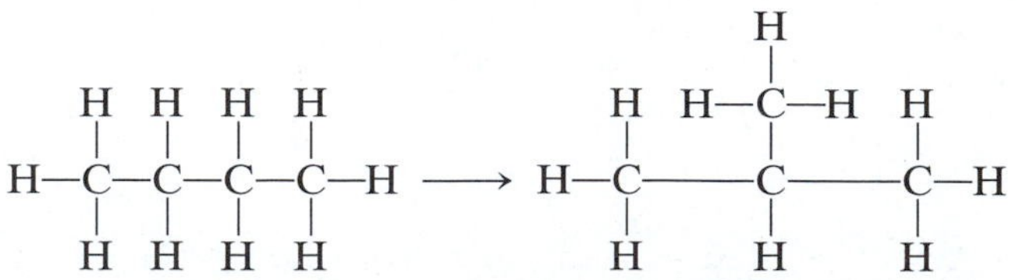

Butane   *Isobutane*

Butane is known as a straight-chain hydrocarbon, because it contains one continuous chain of C—C bonds. Isobutane, on the other hand, is an example of a branched hydrocarbon.

Because they are isomers, the compounds must have the same molecular formula: C_4H_{10}. But there is an even greater similarity between the two compounds. Each compound contains 3 C—C bonds and 10 C—H bonds. Are the bonds all of the same strength? Enthalpy of atom combination data from Table 7.3 shows us that butane ($\Delta H^\circ_{ac} = -5169.38$ kJ/mol$_{rxn}$) has a slightly less negative enthalpy of atom combination than isobutane ($\Delta H^\circ_{ac} = -5177.75$ kJ/mol$_{rxn}$). If we form the compounds from their atoms, more heat will be released when the atoms combine to form isobutane. Thus, isobutane(g) has stronger bonds than butane(g).

We can also calculate the enthalpy change for the reaction given above.

Bond breaking in butane	ΔH° =	5169.38 kJ/mol$_{rxn}$
Bond making in isobutane	ΔH°_{ac} =	-5177.75 kJ/mol$_{rxn}$
	ΔH° =	-8.37 kJ/mol$_{rxn}$

We see that the overall enthalpy change for the conversion of butane(g) into isobutane(g) is negative (-8.37 kJ/mol$_{rxn}$), which means that the bonds formed in the product, isobutane, are stronger than those broken in the reactant, butane. This information tells us that it may be possible to convert straight-chain hydrocarbons to branched chains. Experimental data confirm this. At high temperatures (500–600°C) and high pressures (25–50 atm) straight-chain hydrocarbons isomerize to form branched hydrocarbons. The reaction can be run under more moderate conditions in the presence of a catalyst, such as a mixture of silica (SiO_2) and alumina (Al_2O_3). Reactions such as this play a vital role in the refining of gasoline, because branched alkanes burn more evenly than straight-chain alkanes and are therefore less likely to cause an engine to "knock."

Although we tend to describe butane as a straight-chain hydrocarbon, it is important to remember that the ED model would predict that the geometry around each carbon atom in the molecule would be tetrahedral. The shapes of the butane and isobutane molecules are shown in Figure 7.6.

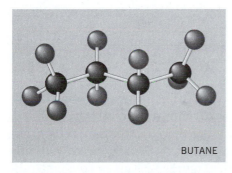
BUTANE

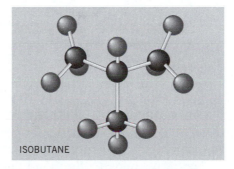
ISOBUTANE

FIGURE 7.6 Two isomers with the formula C_4H_{10} are possible: butane and isobutane. Butane is a straight-chain hydrocarbon, and isobutane is a branched hydrocarbon. All carbon atoms in both hydrocarbons have 109° bond angles with hydrogen.

Exercise 7.6

Fumaric acid and maleic acid are isomers that differ only in the orientation of substituents around the C=C double bond in the compounds.

$$\begin{array}{cc} \text{HO}_2\text{C} \diagdown \diagup \text{CO}_2\text{H} & \text{HO}_2\text{C} \diagdown \diagup \text{H} \\ \text{C}\!\!=\!\!\text{C} & \text{C}\!\!=\!\!\text{C} \\ \text{H} \diagup \diagdown \text{H} & \text{H} \diagup \diagdown \text{CO}_2\text{H} \\ \textit{Maleic acid} & \textit{Fumaric acid} \end{array}$$

In maleic acid, the two hydrogen atoms and the two —CO_2H groups are on the same side of a plane passing through the two carbons of the C=C double bond. In fumaric acid, they are on opposite sides of the bond. Use the enthalpies of atom combination of fumaric acid ($\Delta H_{ac}^{\circ} = -5545.03$ kJ/mol$_{rxn}$) and maleic acid ($\Delta H_{ac}^{\circ} = -5524.53$ kJ/mol$_{rxn}$) to predict whether we should be able to convert fumaric acid to maleic acid, or maleic acid to fumaric acid.

Solution

More heat is released when fumaric acid is formed from its atoms than when maleic acid forms. Thus fumaric acid has *stronger bonds* than maleic acid. This suggests that it may be possible to convert maleic acid to fumaric acid by some chemical process. Again we can also calculate the enthalpy change for the conversion of maleic acid to fumaric acid:

Bond breaking	5524.53 kJ/mol$_{rxn}$
Bond making	-5545.03 kJ/mol$_{rxn}$
	-20.50 kJ/mol$_{rxn}$

Experimentally we can convert maleic acid to fumaric acid by heating the maleic acid in the presence of a strong acid for about 30 minutes.

Checkpoint

Compare the enthalpies of atom combination of $CH_3CH_2OH(g)$ and $CH_3CH_2OH(l)$. What can you conclude from the difference in the two enthalpies?

7.13 BOND LENGTH AND THE ENTHALPY OF ATOM COMBINATION

In Chapters 3 and 4 we noted that the distance between neighboring atoms, the bond length, could often be adequately estimated by taking the periodic trends in size into account. For example, we expect a carbon–oxygen single bond to be longer than a carbon–hydrogen single bond because the oxygen covalent radius is larger than that of hydrogen. We can now relate the strengths of certain bonds to the bond length. Figure 3.23 gives covalent radii from which the bond lengths for the series of halogen acids given in Table 7.4 can be estimated. The enthalpies of atom combination taken from Appendix B.14 are also given in Table 7.4.

The strongest hydrogen–halogen bond, H—F, has the shortest bond length. The other halogen compounds with hydrogen follow the trend of increasing bond length with decreasing bond strength. In general, if the atoms composing a bond are similar, with the same types of bonds being formed, the atom pair with the largest internuclear distance will have the weakest bond strength.

TABLE 7.4   Bond Lengths and Enthalpies of Atom Combination for the Hydrogen Halides

Molecule	Bond Length (nm)	$\Delta H^\circ_{ac}(kJ/mol_{rxn})$
HF(g)	0.101	-567.7
HCl(g)	0.136	-431.64
HBr(g)	0.151	-365.93
HI(g)	0.170	-298.01

Checkpoint

Which molecule, ClF(g) or BrF(g), has the weakest bond?

Exercise 7.7

The bond length in NO(g) is about the same as the bond length in CO(g), but the enthalpy of atom combination for CO(g) is almost double that for NO(g). How can this be explained?

Solution

On the basis of atomic covalent radii we would expect the bond length of a single bond in NO to be smaller than that of a single bond in CO. That the bond lengths are about the same suggests that different types of bonding may be present. The very much larger enthalpy of bond formation confirms this. Therefore, a close look at the structures of the two compounds is required. The Lewis structures are

$$:\!N\!=\!O\!: \qquad :C\!\equiv\!O: $$

The structures make clear that in CO the bond is a triple bond, whereas in NO the bond is a relatively weaker double bond.

KEY TERMS

Boundary	Exothermic	Kinetic theory of heat
Calorimeter	Extensive property	Standard state
Endothermic	First law of	State
Enthalpy	thermodynamics	State function
Enthalpy of atom	Heat	Surroundings
combination	Intensive property	System
Equation of state	Isomer	Temperature

PROBLEMS

Heat and Temperature

1. Nonchemists often confuse heat and temperature. Describe how you would explain the difference to them.

2. What physical properties on both the atomic and macroscopic scale change when a balloon filled with helium is heated? Use the kinetic theory of heat to explain each of the changes.

3. Define the terms *system, surroundings,* and *boundary.* Give three examples of a system separated from its surroundings by a boundary, either real or imaginary.

4. Nonchemists often believe that things that are hot contain a lot of heat. Use the thermodynamics concepts of system, surroundings, and boundaries to explain why they are wrong.

The First Law of Thermodynamics

5. What property of an ideal gas is directly proportional to the energy of the gas? Describe how changes in the energy of more complex systems can be detected.

6. The first law of thermodynamics is often described as saying that "energy is conserved." Describe why it is incorrect to assume that the first law suggests that the "energy of a *system* is conserved."

7. Describe what happens to the energy of a system when the system does work on its surroundings. What happens to the energy of the system when it loses heat to its surroundings?

8. Give examples of both a system doing work on its surroundings and a system losing heat to its surroundings. Describe what happens to the temperature of the system in each case.

9. Give examples of both a system having work done on it by its surroundings and a system gaining heat from its surroundings. What happens to the temperature of the system in each case?

10. Explain why the first law of thermodynamics is often described as suggesting that there is no such thing as a free lunch.

11. Give a verbal definition of the first law of thermodynamics.

12. Describe what happens to the energy of the system when an exothermic reaction is run under conditions of constant volume.

Enthalpy versus Internal Energy

13. Describe the difference between ΔE and ΔH for a chemical reaction.

14. Under what conditions is the heat given off or absorbed by a reaction equal to the change in the energy of the system? Under what conditions is it equal to the enthalpy of reaction?

15. Explain why there is only one value of $\Delta H°$ for a reaction at 25°C, but many values of ΔH.

State Functions

16. Give examples of at least five physical properties that are state functions.

17. Which of the following descriptions of a trip are state functions?
    (a) work done   (b) energy expended
    (c) cost   (d) distance traveled
    (e) tire wear   (f) gasoline consumed
    (g) change in location of the car
    (h) elevation change (i) latitude change
    (j) longitude change

18. Which of the following are state functions?
    (a) temperature (b) energy (c) enthalpy
    (d) pressure   (e) volume (f) heat
    (g) work

Enthalpies of Reaction

19. Why is separating atoms from molecules an endothermic process?

20. What does the sign of ΔH tell you about the net flow of energy in a given chemical reaction?

21. Why is the energy released when a bond is formed precisely the same as the amount of energy needed to break the bond?

22. Predict whether each of the following reactions would be exothermic or endothermic.
    (a) $CO(g) \rightarrow C(g) + O(g)$
    (b) $2 H(g) + O(g) \rightarrow H_2O(g)$
    (c) $Na^+(g) + Cl^-(g) \rightarrow NaCl(s)$
    What is the sign of ΔH in each of the reactions?

23. What is the value of ΔH for the overall process of separating 1 mol of CH_4 into its constituent atoms and then reforming 1 mol of CH_4?

24. Does the amount of enthalpy released when a molecule is formed from its gaseous atoms depend on the amount of substance formed? For example, how much enthalpy is released when 2 mol of $CH_4(g)$ are formed as opposed to 1 mol?

25. If the sum of the enthalpies of atom combination for all of the reactants is more negative than that for the products, will the value of ΔH be positive or negative?

26. Predict without using tables which of the following reactions would be endothermic.
    (a) $H_2(g) \rightarrow 2 H(g)$
    (b) $H_2O(g) \rightarrow H_2O(l)$

27. Predict without using tables which of the following reactions would be endothermic.
    (a) $2 C_8H_{18}(g) + 25 O_2(g) \rightarrow 16 CO_2(g) + 18 H_2O(g)$
    (b) $Na^+(g) + Cl^-(g) \rightarrow Na^+(aq) + Cl^-(aq)$
    (c) $Na^+(g) + e^- \rightarrow Na(g)$

Standard-State Enthalpies of Reaction

28. Oxyacetylene torches are fueled by the combustion of acetylene, C_2H_2.

$$2 C_2H_2(g) + 5 O_2(g) \longrightarrow 4 CO_2(g) + 2 H_2O(g)$$

If the enthalpy change for the reaction is -2511.14 kJ/mol$_{rxn}$ how much heat can be produced by the reaction of
    (a) 2 mol of C_2H_2
    (b) 1 mol of C_2H_2
    (c) 0.500 mol of C_2H_2
    (d) 0.2000 mol of C_2H_2
    (e) 10 g of C_2H_2

29. If the enthalpy change for the following reaction

$$C(s) + H_2O(g) \longrightarrow CO(g) + H_2(g)$$

is 131.29 kJ/mol$_{rxn}$, how much heat will be absorbed by the reaction of

(a) 1 mol of $H_2O(g)$
(b) 2 mol of $H_2O(g)$
(c) 0.0300 mol of $H_2O(g)$
(d) 0.0500 mol of $C(s)$

30. How much heat is released when 1 mol of nitrogen reacts with 2 mol of O_2 to give 2 mol of $NO_2(g)$, if $\Delta H°$ for the reaction is 33.2 kJ/mol$_{rxn}$?

$$N_2(g) + 2\ O_2(g) \longrightarrow 2\ NO_2(g)$$

31. Calculate the standard-state molar enthalpy of reaction for the following reaction, if 1.00 g of magnesium gives off 46.22 kJ of heat when it reacts with excess fluorine.

$$Mg(s) + F_2(g) \longrightarrow MgF_2(s)$$

32. Calculate $\Delta H°$ for the following reaction, assuming that 1.00 g of hydrogen gives off 4.65 kJ of heat when it reacts with 1.00 g of calcium.

$$Ca(s) + H_2(g) \longrightarrow CaH_2(s)$$

Enthalpies of Atom Combination

33. Use the enthalpy of atom combination data in Appendix B.14 to determine whether heat is given off or absorbed when limestone is converted to lime and carbon dioxide.

$$CaCO_3(s) \longrightarrow CaO(s) + CO_2(g)$$

34. Calculate $\Delta H°$ for the following reaction from the enthalpy of atom combination data in Appendix B.14.

$$CO(g) + NH_3(g) \longrightarrow HCN(g) + H_2O(g)$$

35. Phosphine (PH_3) is a foul-smelling gas, which often burns on contact with air. Use the enthalpy of atom combination data in Appendix B.14 to calculate $\Delta H°$ for the reaction, to obtain an estimate of the amount of energy given off when the compound burns.

$$PH_3(g) + 2\ O_2(g) \longrightarrow H_3PO_4(s)$$

36. Carbon disulfide (CS_2) is a useful, but flammable, solvent. Calculate $\Delta H°$ for the following reaction from the enthalpy of atom combination data in Appendix B.14.

$$CS_2(l) + 3\ O_2(g) \longrightarrow CO_2(g) + 2\ SO_2(g)$$

37. The disposable lighters that so many smokers carry use butane as a fuel. Calculate $\Delta H°$ for the combustion of butane from the enthalpy of atom combination data in Appendix B.14 and Table 7.3.

$$2\ C_4H_{10}(g) + 13\ O_2(g) \longrightarrow 8\ CO_2(g) + 10\ H_2O(g)$$

38. The first step in the synthesis of nitric acid involves burning ammonia. Calculate $\Delta H°$ for the following reaction from the enthalpy of atom combination data in Appendix B.14.

$$4\ NH_3(g) + 5\ O_2(g) \longrightarrow 4\ NO(g) + 6\ H_2O(g)$$

39. Lavoisier believed that all acids contained oxygen because so many compounds he studied that contained oxygen form acids when they dissolve in water. Calculate $\Delta H°$ for the reaction between tetraphosphorus decaoxide and water to form phosphoric acid from the enthalpy of atom combination data in Appendix B.14.

$$P_4O_{10}(s) + 6\ H_2O(l) \longrightarrow 4\ H_3PO_4(aq)$$

40. Calculate $\Delta H°$ for the decomposition of hydrogen peroxide:

$$2\ H_2O_2(aq) \longrightarrow 2\ H_2O(l) + O_2(g)$$

41. Calculate $\Delta H°$ for the thermite reaction.

$$Fe_2O_3(s) + 2\ Al(s) \longrightarrow 2\ Fe(s) + Al_2O_3(s)$$

42. Calculate $\Delta H°$ for the reaction of Al with Cr_2O_3 and compare to Problem 7-41 to predict which reaction will liberate the most heat per mole of Al consumed.

$$Cr_2O_3(s) + 2\ Al(s) \longrightarrow 2\ Cr(s) + Al_2O_3(s)$$

43. The first step in extracting iron ore from pyrite, FeS_2, involves roasting the ore in the presence of oxygen to form iron(III) oxide and sulfur dioxide.

$$4\ FeS_2(s) + 11\ O_2(g) \longrightarrow 2\ Fe_2O_3(s) + 8\ SO_2(g)$$

Calculate $\Delta H°$ for the reaction.

44. In which of the following reactions is the sum of the bond strengths greater in the products than in the reactants?
    (a) $CH_3OH(l) \rightarrow HCHO(g) + H_2(g)$
    (b) $2\ CH_3OH(l) \rightarrow 2\ CH_4(g) + O_2(g)$
    (c) $CH_3OH(l) \rightarrow CO(g) + 2\ H_2(g)$

45. Which compound, P_4 or P_2, has the strongest average P—P bonds?

46. Use the enthalpy of atom combination data given in Table 7.3 and Appendix B.14 to calculate the enthalpy changes for the combustion reactions of each of the hydrocarbons listed in Table 7.2. In all cases the hydrocarbons are gaseous, and the products of the reaction are $CO_2(g)$ and $H_2O(l)$.

Integrated Problems

47. Use the enthalpy of combustion for methane, given below, to estimate the energy released when 100 cubic feet of natural gas is burned.

$$CH_4(g) + 2\ O_2(g) \longrightarrow CO_2(g) + 2\ H_2O(g) \qquad \Delta H° = -802.36\ \text{kJ/mol}_{\text{rxn}}$$

48. Which do you predict has the stronger bond, C—H or C—Cl? Calculate the average C—H bond enthalpy in CH_4 from $\Delta H°_{ac}$. Calculate the average C—Cl bond enthalpy in CCl_4 from $\Delta H°_{ac}$. Compare the two bond enthalpies. Is this the result you predicted?

49. If the enthalpy change for breaking all the bonds in the reactants is greater than the enthalpy change for making the bonds of the products, what is the sign of ΔH?

50. If the enthalpy change for breaking all the bonds in the reactants is less than the enthalpy change for making the bonds of the products, what is the sign for ΔH?

51. A measure of the forces which operate between molecules in a liquid can be obtained by comparing the enthalpy required to separate the molecules to the gaseous phase. In which of the following liquids are the intermolecular forces strongest?
    (a) CH_3COOH (b) CH_3CH_2OH (c) C_6H_6 (d) CCl_4

52. Determine the average C—H bond enthalpy for the following compounds (see Table 7.3).
    (a) CH_4 (b) C_2H_6 (c) C_3H_8 (d) C_4H_{10}, n-butane

53. The enthalpy change per mole of hydrocarbon combusted with oxygen is given in Table 7.2. Calculate the amount of heat released per mole of covalent bonds broken in each hydrocarbon listed. Is there a relation between molecular structure and the heat released? If so, what is it?

54. Isomers are compounds that have the same number and kinds of atoms but have a different arrangement of the atoms. The enthalpies of atom combination for several pairs of gaseous isomers are given below. For each pair decide which has the strongest bonds.
    (a) $CH_3CH_2CH_2OH$ and $CH_3\overset{|}{C}HOH$

                                          CH_3
          -4394.2 kJ/mol$_{rxn}$ -4410.7 kJ/mol$_{rxn}$
    (b) $CH_2{=}CHCH_2CH_3$ and $CH_3CH{=}CHCH_3$
          -4604.9 kJ/mol$_{rxn}$ -4611.9 kJ/mol$_{rxn}$
    (c) $H_2C{=}CHCH{=}CHCH_3$ and $H_2C{=}CHCH_2CH{=}CH_2$
          -5243.3 kJ/mol$_{rxn}$ -5213.5 kJ/mol$_{rxn}$

55. When hydrocarbons are bonded only by single bonds, they are said to be saturated; pentane, $CH_3CH_2CH_2CH_2CH_3$, is an example. If carbon–carbon double or triple bonds are present, the compound is said to be unsaturated. Unsaturated hydrocarbons are generally better for human nutrition, hence the claims made by manufacturers to have reduced saturated fats in foods such as margarine. An unsaturated hydrocarbon can be saturated by adding hydrogen across the double bond.

$$CH_3CH{=}CHCH_3(g) + H_2(g) \longrightarrow CH_3CH_2CH_2CH_3(g)$$

Is the sum of the bond strengths greater in the products or the reactants in the above reaction? Draw Lewis structures for the reactants and products and give all bond angles. Which species are planar? The C=C bond is rigid and therefore the —CH$_3$ groups attached to those carbon atoms can appear either on the same side (in which case the compound is called *cis*-2-butene) or on opposite sides (in which case the compound is named *trans*-2-butene).

cis-2-Butene
$\Delta H°_{ac} = -4611.86$ kJ/mol$_{rxn}$

trans-2-Butene
$\Delta H°_{ac} = -4616.58$ kJ/mol$_{rxn}$

Given the enthalpies of atom combination above and data from Appendix B.14, suggest a way that a chemical reaction could be used to differentiate between the two forms of 2-butene.

56. Both dimethyl ether CH_3—O—CH_3 and ethyl alcohol CH_3CH_2OH have been suggested as possible fuels. When reacted with oxygen, O_2, both compounds yield $CO_2(g)$ and $H_2O(g)$. The reactions are called combustion reactions.
    (a) Write balanced chemical equations that describe the combustion reaction between dimethyl ether(g) and O_2. Write a second reaction for the combustion reaction between ethyl alcohol(g) and O_2.

(b) Calculate ΔH for both reactions. (See Table 7.3.) Why is the heat different from that given in section 7.1?

(c) Which is the better fuel? In other words, which releases the most heat on combustion with O_2?

(d) Which molecule, dimethyl ether or ethyl alcohol, has the stronger bonds? Explain.

57. For the following reaction

$$SiBr_4(g) + 2\ Cl_2(g) \longrightarrow SiCl_4(g) + 2\ Br_2(g)$$

(a) Calculate $\Delta H°$ for the reaction. [$\Delta H°_{ac}$ for $SiBr_4(g)$ is -1272 kJ/mol$_{rxn}$.]

(b) Calculate the average Si—Br, Si—Cl, Cl—Cl, and Br—Br bond enthalpies.

(c) Which do you expect to be the stronger bond, Si—Br or Si—Cl? Explain. Which do you expect to be the stronger bond, Br—Br or Cl—Cl? Explain. Do your predictions agree with the calculations in part (b)?

(d) Is the reaction endothermic or exothermic? Explain the sign of $\Delta H°$.

58. When carbon is burned in air the following reaction takes place and releases heat.

$$C(s) + O_2(g) \longrightarrow CO_2(g)$$

Which of the following is responsible for the heat produced?

(a) breaking oxygen–oxygen bonds

(b) making carbon–oxygen bonds

(c) breaking carbon–carbon bonds

(d) both (a) and (c) are correct

(e) (a), (b), and (c) are correct

(f) none of the above are correct

59. In which molecule would you expect the nitrogen–nitrogen bond strengths to be the greatest? Explain.

(a) H_2N—NH_2

(b) F_2N—NF_2

(c) HN=NH

(d) N≡N

60. Determine the average bond strengths in N_2, H_2, and NH_3. Would you expect the following reaction to be exothermic or endothermic? Use the average bond strengths to support your answer.

$$N_2(g) + 3\ H_2(g) \longrightarrow 2\ NH_3(g)$$

61. For the reaction

$$2\ H_2(g) + O_2(g) \longrightarrow 2\ H_2O(g)$$

determine the average bond strengths for hydrogen–hydrogen, oxygen–oxygen, and oxygen–hydrogen bonds. Do you expect the reaction to be exothermic or endothermic? Explain.

62. In terms of the bonds made and the bonds broken, explain why the following reaction is exothermic.

$$Si(s) + 2\ H_2(g) \longrightarrow SiH_4(g)$$

Use enthalpies of atom combination (Table B.14) to support your answer.

C H A P T E R
7
SPECIAL TOPICS

7A.1   Hess's Law

7A.2   Enthalpies of Formation

7A.1   HESS'S LAW

As we have seen, the enthalpy of a system is a state function. As a result, the value of ΔH for a reaction doesn't depend on the path used to go from one of the states to the other.

In 1840, German Henri Hess, professor of chemistry at the University and Artillery School in St. Petersburg, Russia, came to the same conclusion on the basis of experiment and proposed a general rule known as **Hess's law,** which states that the enthalpy of reaction (ΔH) is the same regardless of whether a reaction occurs in one step or in several steps. Thus, as we have seen, we can calculate the enthalpy of reaction by adding the enthalpies associated with a series of hypothetical steps into which the reaction can be broken.

Exercise 7A.1

The heat given off under standard-state conditions when water is formed from its elements as both a liquid and a gas has been measured.

$$H_2(g) + \tfrac{1}{2} O_2(g) \longrightarrow H_2O(l) \qquad \Delta H° = -285.83 \text{ kJ/mol}_{rxn}$$
$$H_2(g) + \tfrac{1}{2} O_2(g) \longrightarrow H_2O(g) \qquad \Delta H° = -241.82 \text{ kJ/mol}_{rxn}$$

Use the data and Hess's law to calculate $\Delta H°$ for the following reaction.

$$H_2O(l) \longrightarrow H_2O(g)$$

Solution

The key to the problem is finding a way to combine the two reactions for which experimental data are known to give the reaction for which $\Delta H°$ is unknown. We can do this by reversing the direction in which the first reaction is written and then adding it to the equation for the second reaction. We assume that 1 mol of water is decomposed into its elements in the first reaction and a mole of water vapor is formed from its elements in the second reaction.

$$
\begin{array}{lr}
H_2O(l) \longrightarrow H_2(g) + \tfrac{1}{2} O_2(g) & \Delta H° = 285.83 \text{ kJ/mol}_{rxn} \\
\underline{H_2(g) + \tfrac{1}{2} O_2(g) \longrightarrow H_2O(g)} & \underline{\Delta H° = -241.82 \text{ kJ/mol}_{rxn}} \\
H_2O(l) \longrightarrow H_2O(g) & \Delta H° = 44.01 \text{ kJ/mol}_{rxn}
\end{array}
$$

Note that when we reversed the direction in which the first reaction was written we had to change the sign of $\Delta H°$ for the reaction. $H_2(g)$ and $\frac{1}{2} O_2(g)$ cancel out and do not appear in the final chemical equation because they are on *both* sides of the chemical equations added together to obtain the final chemical equation.

There was just enough information in Exercise 7A.1 to solve the problem. The next exercise forces us to choose, from a wealth of information, the reactions that will be combined.

Exercise 7A.2

Before pipelines were built to deliver natural gas, individual towns and cities contained plants that produced a fuel known as "town gas" by passing steam over red-hot charcoal.

$$C(s) + H_2O(g) \longrightarrow CO(g) + H_2(g)$$

Calculate $\Delta H°$ for the reaction from the following information.

$$C(s) + \tfrac{1}{2} O_2(g) \longrightarrow CO(g) \qquad \Delta H° = -110.53 \text{ kJ/mol}_{rxn}$$
$$C(s) + O_2(g) \longrightarrow CO_2(g) \qquad \Delta H° = -393.51 \text{ kJ/mol}_{rxn}$$
$$CO(g) + \tfrac{1}{2} O_2(g) \longrightarrow CO_2(g) \qquad \Delta H° = -282.98 \text{ kJ/mol}_{rxn}$$
$$H_2(g) + \tfrac{1}{2} O_2(g) \longrightarrow H_2O(g) \qquad \Delta H° = -241.82 \text{ kJ/mol}_{rxn}$$

Solution

In this calculation we recognize that the desired equation can be found by using the first reaction to generate $CO(g)$ from $C(s)$ and the reverse of the fourth reaction to generate $H_2(g)$ from $H_2O(g)$.

$$
\begin{array}{ll}
C(s) + \tfrac{1}{2} O_2(g) \longrightarrow CO(g) & \Delta H° = -110.53 \text{ kJ/mol}_{rxn} \\
\underline{H_2O(g) \longrightarrow H_2(g) + \tfrac{1}{2}O_2(g)} & \underline{\Delta H° = 241.82 \text{ kJ/mol}_{rxn}} \\
C(s) + H_2O(g) \longrightarrow CO(g) + H_2(g) & \Delta H° = 131.29 \text{ kJ/mol}_{rxn}
\end{array}
$$

7A.2   ENTHALPIES OF FORMATION

Hess's law suggests that we can save a great deal of work measuring enthalpies of reaction by using a little imagination in choosing the reactions for which measurements are made. The question is, What is the best set of reactions to study so that we get the greatest benefit from the smallest number of experiments?

Exercise 7A.2 predicted $\Delta H°$ for the reaction

$$C(s) + H_2O(g) \longrightarrow CO(g) + H_2(g)$$

by combining enthalpy of reaction measurements for the following reactions.

$$C(s) + \tfrac{1}{2} O_2(g) \longrightarrow CO(g)$$
$$H_2(g) + \tfrac{1}{2} O_2(g) \longrightarrow H_2O(g)$$

The reactions used to solve Exercise 7A.2 have one thing in common. Each leads to the formation of a compound from the elements in their most thermodynamically stable form. The enthalpy of reaction for each of the reactions is therefore the **enthalpy of formation** of the compound, ΔH_f°. By definition, ΔH_f° is the enthalpy associated with the reaction that forms 1 mol of a compound from its elements in their most thermodynamically stable states. Enthalpies of formation are similar to enthalpies of atom combination in that both represent the formation of a compound from its elements. However, the values differ because enthalpies of formation represent formation from the elements in their *most thermodynamically stable states* as compared to enthalpies of atom combination in which compounds are formed from *gaseous atoms*. Enthalpies of formation are frequently used to calculate ΔH for chemical reactions.

Exercise 7A.3

Which of the following equations describes a reaction for which ΔH° is equal to the enthalpy of formation of a compound, ΔH°_f?

(a) $Mg(s) + \frac{1}{2} O_2(g) \rightarrow MgO(s)$
(b) $MgO(s) + CO_2(g) \rightarrow MgCO_3(s)$
(c) $Mg(s) + C(s) + \frac{3}{2} O_2(g) \rightarrow MgCO_3(s)$
(d) $Mg(g) + O(g) \rightarrow MgO(s)$

Solution

Equations (a) and (c) describe enthalpy of formation reactions. Those reactions result in the formation of a mole of compound from the most thermodynamically stable form of its elements. Equation (b) can't be an enthalpy of formation reaction because the product of the reaction is not formed from its elements. Equation (d) is an enthalpy of atom combination reaction not an enthalpy of formation reaction.

Hess's law can be used to calculate the enthalpy of reaction for a chemical reaction from the enthalpies of formation of the reactants and products of the reaction. Since enthalpies of formation describe the change in enthalpy for the production of *one mole* of a substance from its most thermodynamically stable elements, the ΔH_f° values are often written in terms of *kilojoules per mole of product*, similar to the manner in which enthalpies of atom combination were written as kilojoules per mole of reaction.

Exercise 7A.4

Use Hess's law to calculate ΔH° for the reaction

$$MgO(s) + CO_2(g) \longrightarrow MgCO_3(s)$$

from the following enthalpy of formation data.

$$Mg(s) + \frac{1}{2} O_2(g) \longrightarrow MgO(s) \qquad \Delta H_f^\circ = -601.70 \text{ kJ/mol MgO}$$
$$C(s) + O_2(g) \longrightarrow CO_2(g) \qquad \Delta H_f^\circ = -393.51 \text{ kJ/mol CO}_2$$
$$Mg(s) + C(s) + \frac{3}{2} O_2(g) \longrightarrow MgCO_3(s) \qquad \Delta H_f^\circ = -1095.8 \text{ kJ/mol MgCO}_3$$

Solution

The reaction in which we are interested converts two reactants (MgO and CO_2) into a single product ($MgCO_3$). We might therefore start by reversing the direction in which we write the enthalpy of formation reactions for MgO and CO_2, thereby decomposing these substances into their elements in their most thermodynamically stable form. We can then add the enthalpy of formation reaction for $MgCO_3$, thereby forming the product from its elements in their most stable form.

$$
\begin{aligned}
MgO(s) &\longrightarrow Mg(s) + \tfrac{1}{2}O_2(g) & \Delta H^\circ &= 601.70 \text{ kJ/mol}_{rxn} \\
CO_2(g) &\longrightarrow C(s) + O_2(g) & \Delta H^\circ &= 393.51 \text{ kJ/mol}_{rxn} \\
Mg(s) + C(s) + \tfrac{3}{2}O_2(g) &\longrightarrow MgCO_3(s) & \Delta H^\circ &= -1095.8 \text{ kJ/mol}_{rxn} \\
\hline
MgO(s) + CO_2(g) &\longrightarrow MgCO_3(s) & \Delta H^\circ &= -100.6 \text{ kJ/mol}_{rxn}
\end{aligned}
$$

Adding the three equations gives the desired unknown reaction. ΔH° for the reaction is therefore the sum of the enthalpies of the three hypothetical steps.

No matter how complex the reaction, the procedure used in Exercise 7A.4 works. All we have to do as the reaction becomes more complex is add more intermediate steps. This approach works because enthalpy is a state function. Thus, ΔH° is the same regardless of the path used to get from the starting materials to the products of the reaction. Instead of running the reaction in a single step,

$$MgO(s) + CO_2(g) \longrightarrow MgCO_3(s)$$

we can split it into two steps. In the first step, the starting materials are converted to the elements from which they form in their most thermodynamically stable states.

$$MgO(s) + CO_2(g) \longrightarrow Mg(s) + C(s) + \tfrac{3}{2}O_2(g)$$

In the second step, the elements combine to form the products of the reaction.

$$Mg(s) + C(s) + \tfrac{3}{2}O_2(g) \longrightarrow MgCO_3(s)$$

Standard-state enthalpy of formation data for a variety of elements and compounds can be found in Table 7A.1.

One more point needs to be understood before Table 7A.1 can be used effectively. By definition, the enthalpy of formation of any element in its most thermodynamically stable form is zero. Under standard-state conditions, the most thermodynamically stable form of oxygen, for example, is the diatomic molecule in the gas phase: $O_2(g)$. By definition, the enthalpy of formation of this substance is equal to the enthalpy associated with the reaction in which it is formed from its elements in their most thermodynamically stable form. For O_2 molecules in the gas phase, ΔH_f° is therefore equal to the heat given off or absorbed in the following reaction.

$$O_2(g) \longrightarrow O_2(g)$$

Because the initial and final states of the reaction are identical, no heat can be given off or absorbed, so ΔH_f° for $O_2(g)$ is zero.

We are now ready to use standard-state enthalpy of formation data to predict enthalpies of reaction.

TABLE 7A.1   Standard-State Enthalpies of Formation

Substance	ΔH_f° (kJ/mol$_{rxn}$)	Substance	ΔH_f° (kJ/mol$_{rxn}$)	Substance	ΔH_f° (kJ/mol$_{rxn}$)
Aluminum		**Bismuth**		**Carbon, continued**	
$Al(s)$	0	$Bi(s)$	0	$COCl_2(g)$	−218.8
$Al(g)$	326.4	$Bi(g)$	207.1	$CH_4(g)$	−74.81
$Al^{3+}(aq)$	−531	$Bi_2O_3(s)$	−573.88	$HCHO(g)$	−108.57
$Al_2O_3(s)$	−1675.7	$BiCl_3(s)$	−379.1	$H_2CO_3(aq)$	−699.65
$AlCl_3(s)$	−704.2	$BiCl_3(g)$	−265.7	$HCO_3^-(aq)$	−691.99
$AlF_3(s)$	−1504.1	$Bi_2S_3(s)$	−143.1	$CO_3^{2-}(aq)$	−677.14
$Al_2(SO_4)_3(s)$	−3440.84			$CH_3OH(l)$	−238.66
		Boron		$CH_3OH(g)$	−200.66
Antimony		$B(s)$	0	$CCl_4(l)$	−135.44
$Sb(s)$	0	$B(g)$	562.7	$CCl_4(g)$	−102.9
$Sb_4(g)$	205.0	$B_2O_3(s)$	−1272.77	$CHCl_3(l)$	−134.47
$Sb_2O_5(s)$	−971.9	$B_2H_6(g)$	35.6	$CHCl_3(g)$	−103.14
$Sb_4O_6(s)$	−1440.6	$B_5H_9(l)$	42.68	$CH_2Cl_2(l)$	−121.46
$SbH_3(g)$	145.105	$B_5H_9(g)$	73.2	$CH_2Cl_2(g)$	−92.47
$SbCl_3(s)$	−382.17	$B_{10}H_{14}(s)$	−45.2	$CH_3Cl(g)$	−80.84
$SbCl_3(g)$	−313.8	$H_3BO_3(s)$	−1094.33	$CS_2(l)$	89.70
$SbCl_5(l)$	−440.2	$BF_3(g)$	−1137.00	$CS_2(g)$	117.36
$SbCl_5(g)$	−394.3	$BCl_3(l)$	−427.2	$HCN(g)$	135.1
$Sb_2S_3(s)$	−174.9	$B_3N_3H_6(l)$	−541.00	$CH_3NO_2(l)$	−113.09
		$B_3N_3H_6(g)$	−511.75	$C_2H_2(g)$	226.73
Arsenic				$C_2H_4(g)$	52.26
$As(s)$	0	**Bromine**		$C_2H_6(g)$	−84.68
$As_4(g)$	143.9	$Br_2(l)$	0	$CH_3CHO(l)$	−192.30
$As_2O_5(s)$	−924.87	$Br_2(g)$	30.907	$CH_3CO_2H(l)$	−484.5
$As_4O_6(s)$	−1313.94	$Br(g)$	111.884	$CH_3CO_2H(g)$	−432.25
$AsH_3(g)$	66.44	$HBr(g)$	−36.40	$CH_3CO_2H(aq)$	−485.76
$AsF_3(g)$	−785.76	$HBr(aq)$	−121.55	$CH_3CO_2^-(aq)$	−486.01
$AsCl_3(g)$	−261.5	$BrF(g)$	−93.85	$CH_3CH_2OH(l)$	−277.69
$AsBr_3(g)$	−130	$BrF_3(g)$	−255.60	$CH_3CH_2OH(g)$	−235.10
$As_2S_3(s)$	−169	$BrF_5(g)$	−428.9	$CH_3CH_2OH(aq)$	−288.3
				$C_6H_6(l)$	49.028
Barium		**Calcium**		$C_6H_6(g)$	82.927
$Ba(s)$	0	$Ca(s)$	0		
$Ba(g)$	180	$Ca(g)$	178.2	**Chlorine**	
$Ba^{2+}(aq)$	−537.64	$Ca^{2+}(aq)$	−542.83	$Cl_2(g)$	0
$BaO(s)$	−553.5	$CaO(s)$	−635.09	$Cl(g)$	121.679
$Ba(OH)_2$	−3342.2	$Ca(OH)_2(s)$	−986.09	$Cl^-(aq)$	−167.159
·8 $H_2O(s)$		$CaCl_2(s)$	−795.8	$ClO_2(g)$	102.5
$BaCl_2(s)$	−858.6	$CaSO_4(s)$	−1434.11	$Cl_2O(g)$	80.3
$BaCl_2(aq)$	−871.95	$CaSO_4·2 H_2O(s)$	−2022.63	$Cl_2O_7(l)$	238
$BaSO_4(s)$	−1473.2	$Ca(NO_3)_2(s)$	−938.39	$HCl(g)$	−92.307
$Ba(NO_3)_2(s)$	−992.07	$CaCO_3(s)$	−1206.92	$HCl(aq)$	−167.159
$Ba(NO_3)_2(aq)$	−952.36	$Ca_3(PO_4)_2(s)$	−4120.8	$ClF(g)$	−54.48
Beryllium		**Carbon**		**Chromium**	
$Be(s)$	0	$C(s, graphite)$	0	$Cr(s)$	0
$Be(g)$	324.3	$C(s, diamond)$	1.895	$Cr(g)$	396.6
$Be^{2+}(aq)$	−382.8	$C(g)$	716.682	$CrO_3(s)$	−589.5
$BeO(s)$	−609.6	$CO(g)$	−110.525	$CrO_4^{2-}(aq)$	−881.15
$BeCl_2(s)$	−490.4	$CO_2(g)$	−393.509	$Cr_2O_3(s)$	−1139.7

TABLE 7A.1   Standard-State Enthalpies of Formation (continued)

Substance	ΔH_f° (kJ/mol$_{rxn}$)	Substance	ΔH_f° (kJ/mol$_{rxn}$)	Substance	ΔH_f° (kJ/mol$_{rxn}$)
Chromium, continued		**Iodine, continued**		**Magnesium, continued**	
$Cr_2O_7^{2-}(aq)$	−1490.3	$IF_5(g)$	−822.49	$Mg^{2+}(aq)$	−466.85
$(NH_4)_2Cr_2O_7(s)$	−1806.7	$IF_7(g)$	−943.9	$MgO(s)$	−601.70
$PbCrO_4(s)$	−930.9	$ICl(g)$	17.78	$MgH_2(s)$	−75.3
		$IBr(g)$	40.84	$Mg(OH)_2(s)$	−924.54
Cobalt				$MgCl_2(s)$	−641.32
$Co(s)$	0	**Iron**		$MgCO_3(s)$	−1095.8
$Co(g)$	424.7	$Fe(s)$	0	$MgSO_4(s)$	−1284.9
$Co^{2+}(aq)$	−58.2	$Fe(g)$	416.3		
$Co^{3+}(aq)$	92	$Fe^{2+}(aq)$	−89.1	**Manganese**	
$CoO(s)$	−237.94	$Fe^{3+}(aq)$	−48.5	$Mn(s)$	0
$Co_3O_4(s)$	−891	$Fe_2O_3(s)$	−824.2	$Mn(g)$	280.7
$Co(NH_3)_6^{3+}(aq)$	−584.9	$Fe_3O_4(s)$	−1118.4	$Mn^{2+}(aq)$	−220.75
		$Fe(OH)_2(s)$	−569.0	$MnO(s)$	−385.22
Copper		$Fe(OH)_3(s)$	−823.0	$MnO_2(s)$	−520.03
$Cu(s)$	0	$FeCl_3(s)$	−399.49	$Mn_2O_3(s)$	−959.0
$Cu(g)$	338.32	$FeS_2(s)$	−178.2	$Mn_3O_4(s)$	−1387.8
$Cu^+(aq)$	71.67	$Fe(CO)_5(l)$	−774.0	$KMnO_4(s)$	−837.2
$Cu^{2+}(aq)$	64.77	$Fe(CO)_5(g)$	−733.9	MnS	−214.2
$CuO(s)$	−157.3				
$Cu_2O(s)$	−168.6	**Lead**		**Mercury**	
$CuCl_2(s)$	−220.1	$Pb(s)$	0	$Hg(l)$	0
$CuS(s)$	−53.1	$Pb(g)$	195.0	$Hg(g)$	61.317
$Cu_2S(s)$	−79.5	$Pb^{2+}(aq)$	−1.7	$Hg^{2+}(aq)$	171.1
$CuSO_4(s)$	−771.36	$PbO(s)$	−217.32	$HgO(s)$	−90.83
$Cu(NH_3)_4^{2+}(aq)$	−348.5	$PbO_2(s)$	−277.4	$HgCl_2(s)$	−224.3
		$PbCl_2(s)$	−359.41	$Hg_2Cl_2(s)$	−265.22
Fluorine		$PbCl_4(l)$	−329.3	$HgS(s)$	−58.2
$F_2(g)$	0	$PbS(s)$	−100.4		
$F(g)$	78.99	$PbSO_4(s)$	−919.94	**Nitrogen**	
$F^-(aq)$	−332.63	$Pb(NO_3)_2(s)$	−451.9	$N_2(g)$	0
$HF(g)$	−271.1	$PbCO_3(s)$	−699.1	$N(g)$	472.704
$HF(aq)$	−320.08			$NO(g)$	90.25
		Lithium		$NO_2(g)$	33.18
Hydrogen		$Li(s)$	0	$N_2O(g)$	82.05
$H_2(g)$	0	$Li(g)$	159.37	$N_2O_3(g)$	83.72
$H(g)$	217.65	$Li^+(aq)$	−278.49	$N_2O_4(g)$	9.16
$H^+(aq)$	0	$LiH(s)$	−90.54	$N_2O_5(g)$	11.35
$OH^-(aq)$	−229.994	$LiOH(s)$	−484.93	$NO_3^-(aq)$	−205.0
$H_2O(l)$	−285.830	$LiF(s)$	−615.97	$NOCl(g)$	51.71
$H_2O(g)$	−241.818	$LiCl(s)$	−408.61	$NO_2Cl(g)$	12.6
$H_2O_2(l)$	−187.78	$LiBr(s)$	−351.23	$HNO_2(aq)$	−119.2
$H_2O_2(aq)$	−191.17	$LiI(s)$	−270.41	$HNO_3(g)$	−135.06
		$LiAlH_4(s)$	−116.3	$HNO_3(aq)$	−207.36
Iodine		$LiBH_4(s)$	190.8	$NH_3(g)$	−46.11
$I_2(s)$	0			$NH_3(aq)$	−80.29
$I_2(g)$	62.438	**Magnesium**		$NH_4^+(aq)$	−132.51
$I(g)$	106.838	$Mg(s)$	0	$NH_4NO_3(s)$	−365.56
$HI(g)$	26.48	$Mg(g)$	147.70	$NH_4NO_3(aq)$	−339.87
$IF(g)$	−95.65			$NH_4Cl(s)$	−314.43

TABLE 7A.1   Standard-State Enthalpies of Formation (continued)

Substance	ΔH_f° (kJ/mol$_{rxn}$)	Substance	ΔH_f° (kJ/mol$_{rxn}$)	Substance	ΔH_f° (kJ/mol$_{rxn}$)
Nitrogen, continued		**Silicon, continued**		**Sulfur, continued**	
$N_2H_4(l)$	50.63	$SiCl_4(l)$	−687.0	$SO_3(g)$	−395.72
$N_2H_4(g)$	95.40	$SiCl_4(g)$	−657.01	$SO_4^{2-}(aq)$	−909.27
$HN_3(g)$	294.1			$SOCl_2(g)$	−212.5
		Silver		$SO_2Cl_2(g)$	−364.0
Oxygen		$Ag(s)$	0	$H_2S(g)$	−20.63
$O_2(g)$	0	$Ag(g)$	284.55	$H_2SO_3(aq)$	−608.81
$O(g)$	249.170	$Ag^+(aq)$	105.579	$H_2SO_4(aq)$	−909.27
$O_3(g)$	142.7	$Ag(NH_3)_2{}^+(aq)$	−111.29	$SF_4(g)$	−774.9
		$Ag_2O(s)$	−31.05	$SF_6(g)$	−1209
Phosphorus		$AgCl(s)$	−127.068	$SCN^-(aq)$	76.44
P(white)	0	$AgBr(s)$	−100.37		
$P_4(g)$	58.91	$AgI(s)$	−61.84	**Tin**	
$P_2(g)$	144.3			$Sn(s)$	0
$P(g)$	314.64	**Sodium**		$Sn(g)$	302.1
$PH_3(g)$	5.4	$Na(s)$	0	$SnO(s)$	−285.8
$P_4O_6(s)$	−1640.1	$Na(g)$	107.32	$SnO_2(s)$	−580.7
$P_4O_{10}(s)$	−2984.0	$Na^+(aq)$	−240.13	$SnCl_2(s)$	−325.1
$PO_4^{3-}(aq)$	−1277.4	$NaH(s)$	−56.275	$SnCl_4(l)$	−511.3
$PF_3(g)$	−918.8	$NaOH(s)$	−425.609	$SnCl_4(g)$	−471.5
$PF_5(g)$	−1595.8	$NaOH(aq)$	−470.114		
$PCl_3(l)$	−319.7	$NaCl(s)$	−411.153	**Titanium**	
$PCl_3(g)$	−287.0	$NaCl(g)$	−176.65	$Ti(s)$	0
$PCl_5(g)$	−374.9	$NaCl(aq)$	−407.27	$Ti(g)$	469.9
$H_3PO_4(s)$	−1279.0	$NaNO_3(s)$	−467.85	$TiO(s)$	−519.7
$H_3PO_4(aq)$	−1277.4	$Na_3PO_4(s)$	−1917.40	$TiO_2(s, \text{rutile})$	−944.8
		$Na_2SO_3(s)$	−1123.0	$TiCl_4(l)$	−804.2
Potassium		$Na_2SO_4(s)$	−1387.08	$TiCl_4(g)$	−763.2
$K(s)$	0	$Na_2CO_3(s)$	−1130.68		
$K(g)$	89.24	$NaHCO_3(s)$	−950.81	**Tungsten**	
$K^+(aq)$	−252.38	$NaC_2H_3O_2$	−708.81	$W(s)$	0
$KOH(s)$	−424.764	$Na_2CrO_4(s)$	−1342.2	$W(g)$	849.4
$KCl(s)$	−436.747	$Na_2Cr_2O_7(s)$	−1978.6	$WO_3(s)$	−842.87
$KNO_3(s)$	−494.63				
$K_2Cr_2O_7(s)$	−2061.5	**Sulfur**		**Zinc**	
$KMnO_4(s)$	−837.2	$S_8(s)$	0	$Zn(s)$	0
		$S_8(g)$	102.30	$Zn(g)$	130.729
Silicon		$S(g)$	278.805	$Zn^{2+}(aq)$	−153.89
$Si(s)$	0	$S^{2-}(aq)$	33.1	$ZnO(s)$	−348.28
$Si(g)$	455.6	$SO_2(g)$	−296.830	$ZnCl_2(s)$	−415.05
$SiO_2(g)$	−910.94	$SO_3(s)$	−454.51	$ZnS(s)$	−205.98
$SiH_4(g)$	34.3	$SO_3(l)$	−441.04	$ZnSO_4(s)$	−982.8
$SiF_4(g)$	−1614.94				

Exercise 7A.5

In Exercise 7.2 we calculated the enthalpy change associated with the combustion of 0.600 mol of pentaborane, B_5H_9. Use the enthalpy of formation data below to calculate the heat given off when a mole of B_5H_9 reacts with excess oxygen according to the following equation.

$$2\ B_5H_9(g) + 12\ O_2(g) \longrightarrow 5\ B_2O_3(s) + 9\ H_2O(g)$$

Solution

Table 7A.1 contains the following data for the reactants and products of the reaction.

Compound	ΔH_f° (kJ/mol$_{rxn}$)
$B_5H_9(g)$	73.2
$B_2O_3(s)$	−1272.77
$O_2(g)$	0
$H_2O(g)$	−241.82

We start by translating the data into a series of equations in which we transform each of the starting materials into its elements and then allow the elements to recombine to form the products of the reaction.

$$
\begin{aligned}
2\ B_5H_9(g) &\longrightarrow 10\ B(s) + 9\ H_2(g) & 2 \times -73.2\ \text{kJ/mol}_{rxn} &= -146.4\ \text{kJ/mol}_{rxn}\\
12\ O_2(g) &\longrightarrow 12\ O_2(g) & 12 \times 0\ \text{kJ/mol}_{rxn} &= 0\ \text{kJ/mol}_{rxn}\\
10\ B(s) + 7\tfrac{1}{2}\ O_2(g) &\longrightarrow 5\ B_2O_3(s) & 5 \times -1272.77\ \text{kJ/mol}_{rxn} &= -6363.85\ \text{kJ/mol}_{rxn}\\
9\ H_2(g) + 4\tfrac{1}{2}\ O_2(g) &\longrightarrow 9\ H_2O(g) & 9 \times -241.82\ \text{kJ/mol}_{rxn} &= -2176.38\ \text{kJ/mol}_{rxn}\\
& & \Delta H^{\circ} &= -8686.6\ \text{kJ/mol}_{rxn}
\end{aligned}
$$

According to the balanced equation, this is the enthalpy change when *two* moles of B_5H_9 is consumed. ΔH° for the reaction is therefore −4343.3 kJ/mol B_5H_9. This is five times the molar enthalpy of reaction for the combustion of CH_4. On a per gram basis, it is about 40% larger than the energy released when methane burns. It isn't surprising that B_5H_9 was once studied as a potential rocket fuel.

PROBLEMS

Hess's Law

7A-1. Explain how Hess's law is a direct consequence of the fact that the enthalpy of a system is a state function.

7A-2. Use the following data to calculate ΔH° for the conversion of graphite into diamond.

$$
\begin{aligned}
C(s, \text{graphite}) + O_2(g) &\longrightarrow CO_2(g) & \Delta H^{\circ} &= -393.51\ \text{kJ/mol}_{rxn}\\
C(s, \text{diamond}) + O_2(g) &\longrightarrow CO_2(g) & \Delta H^{\circ} &= -395.41\ \text{kJ/mol}_{rxn}
\end{aligned}
$$

7A-3. Use the following data

$$
\begin{aligned}
H_2(g) + \tfrac{1}{2}\ O_2(g) &\longrightarrow H_2O(l) & \Delta H^{\circ} &= -285.83\ \text{kJ/mol}_{rxn}\\
H_2(g) + O_2(g) &\longrightarrow H_2O_2(aq) & \Delta H^{\circ} &= -191.17\ \text{kJ/mol}_{rxn}
\end{aligned}
$$

to calculate $\Delta H°$ for the decomposition of hydrogen peroxide.

$$2\ H_2O_2(aq) \longrightarrow 2\ H_2O(l) + O_2(g)$$

7A-4. In the presence of a spark, nitrogen and oxygen react to form nitrogen oxide.

$$N_2(g) + O_2(g) \longrightarrow 2\ NO(g)$$

Calculate $\Delta H°$ for the reaction from the following data.

$$\tfrac{1}{2}\ N_2(g) + O_2(g) \longrightarrow NO_2(g) \qquad \Delta H° = 33.2\ \text{kJ/mol}_{rxn}$$
$$NO(g) + \tfrac{1}{2}\ O_2(g) \longrightarrow NO_2(g) \qquad \Delta H° = -57.1\ \text{kJ/mol}_{rxn}$$

7A-5. Enthalpy of reaction data can be combined to determine $\Delta H°$ for reactions that are difficult, if not impossible, to study directly. Nitrogen and oxygen, for example, do not react directly to form dinitrogen pentoxide.

$$2\ N_2(g) + 5\ O_2(g) \longrightarrow 2\ N_2O_5(g)$$

Use the following data to determine $\Delta H°$ for the hypothetical reaction in which nitrogen and oxygen combine to form N_2O_5.

$$N_2(g) + 3\ O_2(g) + H_2(g) \longrightarrow 2\ HNO_3(aq) \qquad \Delta H° = -414.7\ \text{kJ/mol}_{rxn}$$
$$N_2O_5(g) + H_2O(l) \longrightarrow 2\ HNO_3(aq) \qquad \Delta H° = -140.24\ \text{kJ/mol}_{rxn}$$
$$2\ H_2(g) + O_2(g) \longrightarrow 2\ H_2O(l) \qquad \Delta H° = -571.7\ \text{kJ/mol}_{rxn}$$

7A-6. Use the following data

$$3\ C(s) + 4\ H_2(g) \longrightarrow C_3H_8(g) \qquad \Delta H° = -103.85\ \text{kJ/mol}_{rxn}$$
$$C(s) + O_2(g) \longrightarrow CO_2(g) \qquad \Delta H° = -393.51\ \text{kJ/mol}_{rxn}$$
$$H_2(g) + \tfrac{1}{2}\ O_2(g) \longrightarrow H_2O(g) \qquad \Delta H° = -241.82\ \text{kJ/mol}_{rxn}$$

to calculate the heat of combustion of propane, C_3H_8.

$$C_3H_8(g) + 5\ O_2(g) \longrightarrow 3\ CO_2(g) + 4\ H_2O(g)$$

7A-7. Use the following data

$$C_4H_9OH(l) + 6\ O_2(g) \longrightarrow 4\ CO_2(g) + 5\ H_2O(g) \qquad \Delta H° = -2456.1\ \text{kJ/mol}_{rxn}$$
$$(C_2H_5)_2O(l) + 6\ O_2(g) \longrightarrow 4\ CO_2(g) + 5\ H_2O(g) \qquad \Delta H° = -2510.0\ \text{kJ/mol}_{rxn}$$

to calculate $\Delta H°$ for the following reaction.

$$(C_2H_5)_2O(l) \longrightarrow C_4H_9OH(l)$$

Enthalpies of Formation

7A-8. For which of the following substances is $\Delta H_f°$ equal to zero?
(a) $P_4(s)$ (b) $H_2O(g)$ (c) $H_2O(l)$ (d) $O_3(g)$ (e) $Cl(g)$ (f) $F_2(g)$
(g) $Na(g)$

7A-9. Use the enthalpy of formation data in Table 7A.1 to calculate the enthalpy change for the following reaction. Compare the results of the calculation to that of Problem 7-33.

$$CaCO_3(s) \longrightarrow CaO(s) + CO_2(g)$$

7A-10. Calculate $\Delta H°$ for the following reaction from the enthalpy of formation data in Table 7A.1. Compare to Problem 7-34.

$$CO(g) + NH_3(g) \longrightarrow HCN(g) + H_2O(g)$$

7A-11. Use the enthalpy of formation data in Table 7A.1 to calculate $\Delta H°$ for the combustion of PH_3. Compare the answer to Problem 7-35.

$$PH_3(g) + 2\,O_2(g) \longrightarrow H_3PO_4(s)$$

7A-12. Carbon disulfide (CS_2) is a useful, but flammable, solvent. Calculate $\Delta H°$ for the following reaction from the enthalpy of formation data in Table 7A.1.

$$CS_2(l) + 3\,O_2(g) \longrightarrow CO_2(g) + 2\,SO_2(g)$$

7A-13. Calculate $\Delta H°$ for the combustion of butane from the enthalpy of formation data in Table 7A.1. The enthalpy of formation of butane is -126.2 kJ/mol$_{rxn}$.

$$2\,C_4H_{10}(g) + 13\,O_2(g) \longrightarrow 8\,CO_2(g) + 10\,H_2O(g)$$

7A-14. The first step in the synthesis of nitric acid involves burning ammonia. Calculate $\Delta H°$ for the following reaction from the enthalpy of formation data in Table 7A.1.

$$4\,NH_3(g) + 5\,O_2(g) \longrightarrow 4\,NO(g) + 6\,H_2O(g)$$

7A-15. Calculate $\Delta H°$ for the following from the enthalpy of formation data in Table 7A.1. Compare the answer to Problem 7-39.

$$P_4O_{10}(s) + 6\,H_2O(l) \longrightarrow 4\,H_3PO_4(aq)$$

7A-16. Small quantities of oxygen can be prepared in the laboratory by heating potassium chlorate ($KClO_3$) until it decomposes. Calculate $\Delta H°$ for the following reaction from the enthalpy of formation data in Table 7A.1. The enthalpy of formation of $KClO_3(s)$ is -391.2 kJ/mol.

$$2\,KClO_3(s) \longrightarrow 2\,KCl(s) + 3\,O_2(g)$$

7A-17. Use enthalpies of formation to predict which of the following reactions gives off the most heat per mole of aluminum consumed.

$$Fe_2O_3(s) + 2\,Al(s) \longrightarrow 2\,Fe(s) + Al_2O_3(s)$$
$$Cr_2O_3(s) + 2\,Al(s) \longrightarrow 2\,Cr(s) + Al_2O_3(s)$$

8
LIQUIDS AND SOLUTIONS

8.1   THE STRUCTURE OF GASES, LIQUIDS, AND SOLIDS

Liquids, solids, and gases constitute three phases of matter. Most of the substances that are liquids at room temperature and pressure are composed of molecules. Ionic and metallic compounds are generally solids at room temperature and pressure. Although ionic and metallic substances can be melted to form the liquid state, we shall confine our discussion in this chapter to molecular liquids.

The kinetic molecular theory explains the characteristic properties of gases by assuming that gas particles are in a state of constant, random motion and that the diameter of the particles is very small compared with the distance between particles (Figure 8.1c). Because most of the volume of a gas is empty space, the simplest analogy might compare the particles of a gas to fruit flies in a jar.

Many of the properties of solids have been captured in the way the term *solid* is used in English. It describes something that holds its shape, such as a solidly constructed house. It implies continuity; there is a definite position for each particle. It implies the absence of empty space, as in a solid chocolate Easter bunny. Finally, it describes things that occupy three dimensions, as in solid geometry. A solid might be compared to a brick wall, in which the individual bricks form a regular structure and the amount of empty space is kept to a minimum (Figure 8.1a).

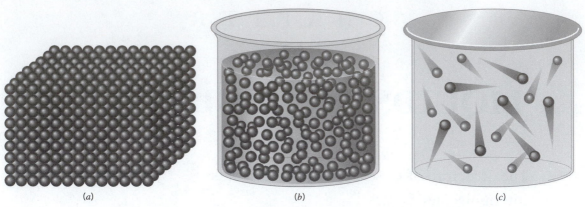

FIGURE 8.1 Particles in a solid (*a*) are packed tightly in a regular pattern. The particles in a liquid (*b*) do not pack as tightly as they do in a solid. The structure of liquids also contains small, molecule-size holes that enable the liquid to flow so that it can conform to the shape of its container. Gas particles (*c*) are in random motion.

Liquids have properties between the extremes of gases and solids. Like gases, they flow to conform to the shape of their containers (Figure 8.1*b*). Like solids, they can't expand to fill their containers, and they are very difficult to compress. The structure of a liquid might be compared to a bag full of marbles being shaken vigorously, back and forth.

The difference in the structures of gases, liquids, and solids might best be understood by comparing the densities of substances in the three phases. As shown by the data in Table 8.1, typical solids are about 20% more dense than the corresponding liquid, whereas the liquid is much more dense than the gas. The models shown in Figure 8.1 are consistent with the data.

TABLE 8.1   Densities of Solid, Liquid, and Gaseous Forms of Three Elements

	Solid (g/cm³)	Liquid (g/cm³)	Gas (g/cm³)
Ar	1.65	1.40	0.001784
N_2	1.026	0.8081	0.001251
O_2	1.426	1.149	0.001429

Because water is the only substance that we routinely encounter as a solid, a liquid, and a gas, it may be useful to consider what happens to water as we change the temperature. At low temperatures, water is a solid in which the individual molecules are locked into a rigid structure. As we raise the temperature, the average kinetic energy of the molecules increases, due to an increase in the motion with which these molecules move about their lattice positions in the solid.

To understand the effect of molecular motion, we need to differentiate between intramolecular bonds and intermolecular forces, as shown in Figure 8.2. The covalent bonds between the hydrogen and oxygen atoms in a water molecule are called **intramolecular bonds.**[1] The attractive forces between the neighboring water molecules are called **intermolecular forces.**[2]

---

[1] The prefix *intra-* comes from a Latin stem meaning "within or inside." Intramural sports, for example, match teams from within the same institution.

[2] The prefix *inter-* comes from a Latin stem meaning "between." It is used in words such as *interact, intermediate,* and *international.*

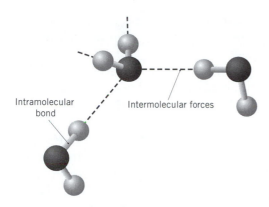

FIGURE 8.2 The covalent bonds between the hydrogen and oxygen atoms in a water molecule are called intramolecular bonds. The attractive forces between water molecules are called intermolecular forces.

The *intramolecular* bonds that hold the atoms in H_2O molecules together are much stronger than the *intermolecular* forces between water molecules. It takes 463 kJ to break the H—O bonds in a mole of water molecules but only about 50 kJ to break the intermolecular forces that hold a mole of water molecules to one another in the solid state.

Increased motion disrupts the attractive intermolecular forces between water molecules. As solid water (ice) becomes warmer, the kinetic energy of the water molecules eventually becomes too large to allow the molecules to be locked into the rigid structure of ice. At this point, the solid melts to form a liquid, in which the intermolecular forces between molecules now are sufficiently weakened that the molecules may move through the entire volume of the liquid. As the temperature continues to increase, the kinetic energy of the water molecules becomes so large, and they move so rapidly, that most of the attractive intermolecular forces are overcome, and the liquid boils to form a gas in which each particle moves more or less randomly through space. At no point is the increased motion sufficient to overcome the strengths of the covalent bonds. Therefore, water exists as a molecular species in the gaseous state up to very high temperatures.

At low temperatures molecules have less kinetic energy to overcome the attractive forces and hence approach each other more closely. At higher temperatures the molecules possess sufficient energy to allow for a larger separation. The distance between molecules is one factor that determines the strength of intermolecular forces. Thus how closely molecules may approach one another is an important factor in determining how strongly molecules attract each other. It follows then that, in addition to temperature, the size and shape of molecules become significant for the strength of intermolecular attractive forces.

We now have a means of rationalizing why a substance forms a solid, liquid, or gas at room temperature. The difference between the three phases of matter is based on a competition between the strength of intermolecular forces and the kinetic energy of the system. When the force of attraction between the particles is relatively weak, the substance is likely to be a gas at room temperature. When the force of attraction is strong, the substance is more likely to be a solid. As might be expected, a substance is a liquid at room temperature when the intermolecular forces are neither too strong nor too weak. For a given intermolecular force strength, the higher the temperature, the more likely the substance is to be a gas.

8.2 INTERMOLECULAR FORCES

The kinetic theory of gases assumes that there is no force of attraction between the particles in a gas. If the assumption were correct, gases would never condense to form liquids and solids at low temperatures. In 1873 the Dutch physicist Johannes van der Waals derived an equation that not only included the force of attraction between gas particles but also corrected for the fact that the volume of the particles becomes a significant fraction of the total volume of the gas at high pressures.

The van der Waals equation described in the appendix of Chapter 6 gives a better description of the experimental data for real gases than can be obtained with the ideal gas equation. But that was not van der Waals' goal. He was trying to develop a model that would explain the behavior of liquids by including terms that reflected the size of the atoms or molecules in the liquid and the strength of the forces between the atoms or molecules. The weak intermolecular forces in liquids and solids are therefore often called **van der Waals forces.** Intermolecular forces can be divided into four categories: (1) dipole–dipole, (2) dipole–induced dipole, (3) induced dipole–induced dipole, and (4) hydrogen bonding. Therefore, depending on the structure, a given molecular substance may exhibit one, two, three, or all four of the intermolecular forces. Because the forces of attraction between particles vary greatly from one substance to another it is not possible to describe the properties of liquids or solids with just one equation as we did with gases in Chapter 6.

Dipole–Dipole Forces

Many molecules are held together by intramolecular bonds that fall between the extremes of pure ionic and pure covalent bonds. The difference between the electronegativities of the atoms in the molecules is large enough that the electrons are not shared equally, and yet small enough that the electrons aren't drawn exclusively to one of the atoms to form positive and negative ions. The bonds in the molecules are said to be polar because they have positive and negative ends, or poles. When polar bonds and nonbonding pairs of electrons are distributed in a molecule such that there is a separation of charge within the molecule, the molecule will be **polar** with a positive and a negative pole.

The acetone molecules that compose nail polish remover have a dipole moment (2.88 D) because the methyl groups (CH_3) have a slight positive charge and the oxygen atom has a slight negative charge. Because of the force of attraction between the positive end of one molecule and the negative end of another, there is a dipole–dipole force of attraction between adjacent acetone molecules, as shown in Figure 8.3.

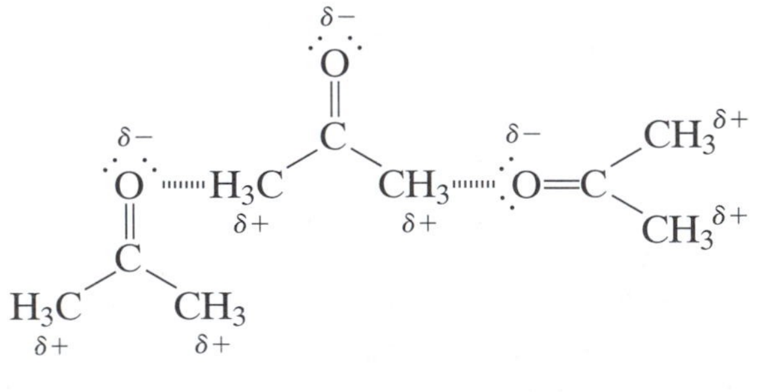

FIGURE 8.3 A dipole–dipole force exists between adjacent acetone molecules. The methyl groups (CH_3) of acetone have a positive partial charge and the oxygen atom a negative partial charge, thus producing an overall dipole moment.

The dipole–dipole interaction in acetone is relatively weak; it requires only a small amount of energy to pull acetone molecules apart from one another. In contrast, the covalent bonds (C—H, C—C, and C=O) in acetone are much stronger. The strength of the dipole–dipole attraction depends on the magnitude of the dipole moment and on how closely the molecules may approach one another. The closer the approach, the greater are the intermolecular attractive forces. If molecules approach too closely, however, a repulsive force occurs.

Dipole–Induced Dipole Forces

What would happen if we mixed acetone with carbon tetrachloride, which has no dipole moment? In carbon tetrachloride the individual carbon–chlorine bonds are polar due to the difference in electronegativity between carbon and chlorine atoms. Because the molecule has a symmetric tetrahedral molecular geometry the polarity of the four bonds cancel one another to yield a uniform charge distribution with a dipole moment of zero. However, the electrons in a molecule are not static but are in constant motion. When a carbon tetrachloride molecule comes close to a polar acetone molecule, the electrons in carbon tetrachloride can shift to one side of the molecule to produce a very small dipole moment, as shown in Figure 8.4.

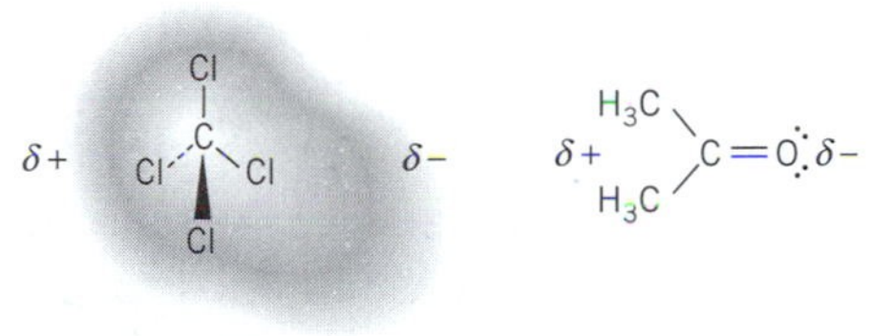

FIGURE 8.4 When a carbon tetrachloride molecule comes close to a polar acetone molecule, the distribution of electrons around carbon tetrachloride is distorted. The shaded area around the carbon tetrachloride molecule represents the electron density. The electron density is attracted toward the partial positive charge of the acetone. A small dipole moment is induced in the carbon tetrachloride molecule, which allows a weak dipole–induced dipole force of attraction.

By distorting the distribution of electrons in carbon tetrachloride the polar acetone molecule induces a small dipole moment in the CCl_4 molecule, which creates a dipole–induced dipole force of attraction between the acetone and carbon tetrachloride molecules. Dipole–induced dipole interactions become stronger the closer the molecules approach one another and weaken rapidly as the molecules move apart.

Induced Dipole–Induced Dipole Forces

Bromine has no dipole and therefore has neither dipole–dipole nor dipole–induced dipole forces yet bromine, Br_2, is a liquid at room temperature. A liquid can exist only if there are forces of attraction between adjacent molecules sufficient to hold the molecules together. The electrons in bromine are in constant motion. There is some probability that for an instant in time there may be more electron density on one side of the molecule than on the other, causing a temporary dipole. The temporary dipole can induce a temporary dipole in an adjacent bromine molecule, as shown in Figure 8.5. Such fluctuations in electron density occur constantly, creating temporary induced dipole–induced dipole forces of attraction—also known as **dispersion forces** or **London forces**—between pairs of molecules or atoms throughout a liquid. All molecules experience some degree of induced dipole–induced dipole interaction. This is the only type of intermolecular force present in substances, such as Br_2, that are made up of nonpolar molecules.

FIGURE 8.5 Fluctuations in electron density occur around the nuclei of neighboring bromine molecules, creating induced dipole–induced dipole forces of attraction.

Because intermolecular forces must be overcome to melt a solid or to boil a liquid, melting points and boiling points can serve as a measure of the relative strengths of the intermolecular forces which hold the molecules together. Table 8.2 shows the melting points and boiling points of chlorine, bromine, and iodine. All three molecules are nonpolar and have only dispersion forces, but they have significantly different melting and boiling points.

As molecular weight increases so do the melting and boiling points. At room temperature and pressure chlorine is a gas, bromine a liquid, and iodine a solid. The data in Table 8.2 indicate that the dispersion forces in iodine are much stronger than the dispersion forces in bromine, which are stronger than the dispersion forces in chlorine.

TABLE 8.2   Melting Points and Boiling Points of Three Molecules That Have Only Dispersion Forces

Compound	Molecular Weight (g/mol)	Melting Point (°C)	Boiling Point (°C)
Cl_2	70.91	−101.0	−34.6
Br_2	159.81	−7.2	58.8
I_2	253.81	113.5	184.4

The magnitude of an induced dipole moment depends on how easily electrons may be moved, which is what chemists call polarizability. The polarizability depends on the number of electrons, how tightly they are held, and the shape of the molecule. For similar species dispersion force interactions increase with increasing molecular weight. We see an increase in molecular weight as we move from Cl_2 to Br_2 to I_2, which indicates an increase in the strength of the dispersion forces between adjacent molecules. The increase can be explained by the increased distance of the valence electrons from the nucleus as we move down the periodic table from Cl to Br to I. As the distance from the nucleus increases the electrons are held less tightly and can more easily be pulled to one side of the molecule. In general, the more electrons, the more polarizable is an atom, molecule, or ion, and the greater can be the induced dipole.

The shape of a molecule is also a factor in determining the magnitude of dispersion forces. This factor can be demonstrated using three molecules with equal molecular weights. Such molecules are composed of the same atoms but have different structures and are called isomers. The data in Figure 8.6 show how the shape of a molecule influences the boiling point of a compound. One of the compounds, neopentane, has very symmetrical molecules, each with four identical CH_3 groups arranged in a tetrahedral pattern around a central carbon atom. The molecules shown in Figure 8.6 all have the same molecular weight and have only dispersion interactions between molecules. The more symmetrical neopentane does not have as much surface area in contact with its neighboring molecule as the straight-chain n-pentane molecule. Thus, because the surface area in contact is smaller for neopentane, the dispersion forces are weaker and the boiling point is lower than for n-pentane. Isopentane, being a branched molecule, falls in between neopentane and n-pentane in how closely it can approach its neighbors, and consequently isopentane has a boiling point in between that of neopentane and n-pentane.

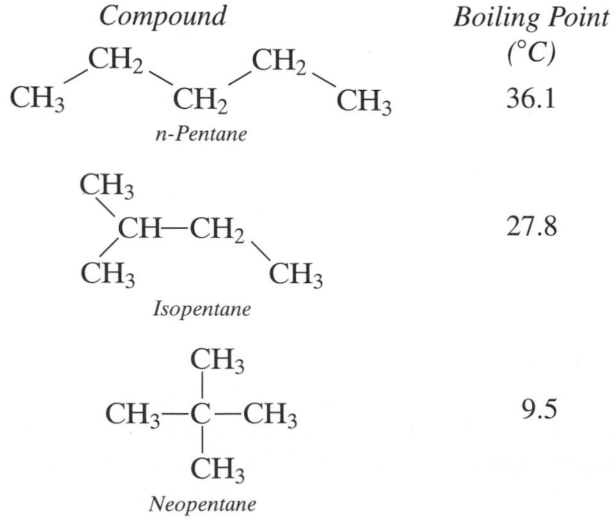

FIGURE 8.6 Boiling points for the three isomers with the formula C_5H_{12}.

The relationship between the molecular weight of a compound and its boiling point is shown in Figure 8.7. The compounds in Figure 8.7 all have the same generic formula, C_nH_{2n+2}, and are all straight-chain hydrocarbons. The only differences between the compounds are their size and molecular weights. As shown in Figure 8.7, the relationship between the molecular weights of the compounds and their boiling points is not a straight line, but it is a remarkably smooth curve.

The molecules shown in Figure 8.7 increase in length from 1 to 12 carbons. Because the molecules are nonpolar, the only intermolecular interactions between them are due to dispersion interactions. With each additional increase in the length of the chain of carbon atoms, an additional contribution to the dispersion interactions is made. Increased molecular weight corresponds to additional polarizable electrons and hence to increased intermolecular interactions. For similar compounds, dispersion interactions increase directly with increased molecular weight and length of the molecule. Hence, the attractive forces holding the molecules together as a liquid increase. This leads to an increase in the boiling point. Thus, dispersion forces become more significant as a species becomes larger.

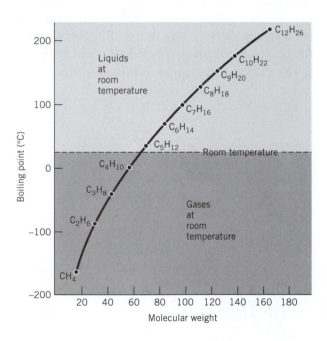

FIGURE 8.7 There is a gradual increase in the boiling points of compounds with the generic formula C_nH_{2n+2} as the molecules in the series become heavier.

Checkpoint

Arrange the noble gas atoms in order of increasing dispersion interactions.

Hydrogen Bonding

The intermolecular force called **hydrogen bonding** is actually a type of dipole–dipole interaction. Hydrogen bonds are separated from other examples of van der Waals forces because they are unusually strong: 15–25 kJ/mol. However, it should be noted that even though the name of the interaction is hydrogen *bonding*, a hydrogen bond is not a covalent, ionic, or metallic bond but instead is an intermolecular force between adjacent molecules.

Molecules that can form hydrogen bonds have relatively polar H—X bonds, such as NH_3, H_2O, and HF. The hydrogen bond is created when a hydrogen atom forms a bridge between two very electronegative atoms. The hydrogen is covalently bonded to one of the atoms and hydrogen bonded to the other. The H—X bond must be polar to create the partial positive charge on the hydrogen atom that allows the bridging interactions to exist. As the X atom in the H—X bond becomes more electronegative, hydrogen bonding between molecules becomes more important. Hydrogen bonding is most significant when the hydrogen atom is bonded to N, O, or F atoms. An illustration of hydrogen bonding is shown as the dashed lines between hydrogen and oxygen atoms on adjacent water molecules in Figure 8.2.

Checkpoint

Explain why hydrogen bonds between NH_3 molecules are weaker than those between H_2O molecules.

As previously discussed, melting and boiling points can be used to compare the strengths of intermolecular forces. Figure 8.8 shows the relationship between the melting points and boiling points of the hydrides of elements in Groups IVA, VIA, and VIIA. The boiling points of the hydrides of Group IVA (CH_4, SiH_4, GeH_4, and SnH_4) increase in a somewhat linear fashion with molecular weight as would be predicted from our discussion of dispersion forces. However, the melting points of H_2O and HF do not follow the expected trends for the hydrides of elements in Groups VIA and VIIA. The unusually high melting and boiling points of H_2O and HF are due to the strong hydrogen bonds formed between water molecules and between HF molecules. The effect of hydrogen bonds in water is discussed in more detail in Section 8.9.

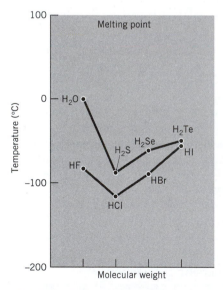 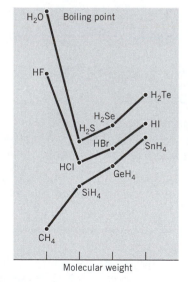

FIGURE 8.8 Plot of the melting points and boiling points of hydrides of elements in Groups IVA, VIA, and VIIA. The melting points and boiling points of HF and H_2O are anomalously large because of the strength of the hydrogen bonds between molecules in those compounds.

8.3   RELATIVE STRENGTHS OF INTERMOLECULAR FORCES

We can now classify the forces of attraction between particles in terms of four categories: dipole–dipole, dipole–induced dipole, induced dipole–induced dipole or dispersion forces, and hydrogen bonds. Consider the classification of the compounds in Table 8.3, for example.

TABLE 8.3   Intermolecular Forces

Molecule	Structure	Dipole–Dipole	Dipole–Induced Dipole	Dispersion Forces	Hydrogen Bonding
Propyl alcohol	$CH_3CH_2CH_2OH$	Yes	Yes	Yes	Yes
Diethyl ether	$CH_3CH_2OCH_2CH_3$	Yes	Yes	Yes	No
Ethyl fluoride	CH_3CH_2F	Yes	Yes	Yes	No
Tin tetrachloride	$SnCl_4$	No	No	Yes	No
Acetic acid	$CH_3\overset{\displaystyle O}{\overset{\displaystyle \|}{C}}OH$	Yes	Yes	Yes	Yes

We note that dispersion forces are present in all the compounds. In fact dispersion forces must always be present in all molecular substances because these forces depend on induced dipoles resulting from electronic motion. All compounds contain electrons and therefore all compounds will have dispersion forces. The other forces, dipole–dipole, dipole–induced dipole, and hydrogen bonding, depend on the structure of the molecule, and their presence can be deduced only if a suitable structure such as the Lewis structure for the compound is known.

Table 8.4 shows the relationship between three classes of organic compounds, alkanes, aldehydes, and carboxylic acids. There are two trends that we can examine in Table 8.4. The first is the relationship between boiling point and molecular weight. For each class of compounds as molecular weight increases there is a corresponding increase in boiling point ($BP_{butanal} < BP_{pentanal} < BP_{hexanal}$). Figure 8.7 shows the same trend graphically for the alkanes. When comparing similar molecular compounds of significantly different molecular weights, the compounds with the highest molecular weight will generally have the highest boiling point and, therefore, the strongest intermolecular forces. This relationship is due to the dispersion force interactions that increase with increasing molecular weight.

The second trend that can be observed in Table 8.4 is shown by comparing the boiling points of molecules from different categories that have similar molecular weights. Heptane, hexanal, and pentanoic acid have similar molecular weights (100.2, 100.2, and 102.1 g/mol, respectively), but their respective boiling points are 98.4, 128, and 186°C. To understand the difference in boiling points we must examine the structure of the molecules and determine the types of intermolecular forces between molecules.

TABLE 8.4   Boiling Points of Three Classes of Organic Compounds

Alkane	MW (g/mol)	BP (°C)	Aldehyde	MW (g/mol)	BP (°C)	Carboxylic Acid	MW (g/mol)	BP (°C)
Butane $CH_3(CH_2)_2CH_3$	58.1	−0.5	Butanal $CH_3(CH_2)_2CHO$	72.1	75.7	Butanoic acid $CH_3(CH_2)_2COOH$	88.1	164
Pentane $CH_3(CH_2)_3CH_3$	72.2	36.1	Pentanal $CH_3(CH_2)_3CHO$	86.1	103	Pentanoic acid $CH_3(CH_2)_3COOH$	102.1	186
Hexane $CH_3(CH_2)_4CH_3$	86.2	69.0	Hexanal $CH_3(CH_2)_4CHO$	100.2	128	Hexanoic acid $CH_3(CH_2)_4COOH$	116.2	205
Heptane $CH_3(CH_2)_5CH_3$	100.2	98.4	Heptanal $CH_3(CH_2)_5CHO$	114.2	153	Heptanoic acid $CH_3(CH_2)_5COOH$	130.2	223
Octane $CH_3(CH_2)_6CH_3$	114.2	126	Octanal $CH_3(CH_2)_6CHO$	128.2	171	Octanoic acid $CH_3(CH_2)_6COOH$	144.2	239

The structures of the three compounds are given in Figure 8.9. Alkanes such as heptane are composed of only carbon and hydrogen. Carbon and hydrogen have relatively similar electronegativities so that the covalent bonds are not very polar. In addition the symmetric structure of alkanes leads to a uniform distribution of charge within the molecule and results in a molecule with a very small dipole moment. The addition of electronegative oxygen to form aldehydes and carboxylic acids causes compounds such as hexanal and pentanoic acid to have a charge separation and therefore a significant dipole moment. Both aldehydes and carboxylic acids are polar and can have dipole–dipole interactions. This causes aldehydes and carboxylic acids to have higher boiling points than alkanes of similar molecular weights.

The boiling point of pentanoic acid, however, is considerably higher than that of hexanal. This indicates that the intermolecular forces attracting pentanoic acid molecules to one another are stronger than the forces found in hexanal. The structure of pentanoic acid in Figure 8.9 shows that a hydrogen atom is attached to a very electronegative oxygen atom. This allows for the formation of hydrogen bonds between this hydrogen atom on one molecule and an oxygen atom on an adjacent molecule. Therefore, pentanoic acid has dispersion, dipole–induced dipole, dipole–dipole, and hydrogen bonding intermolecular forces, giving it a high boiling point.

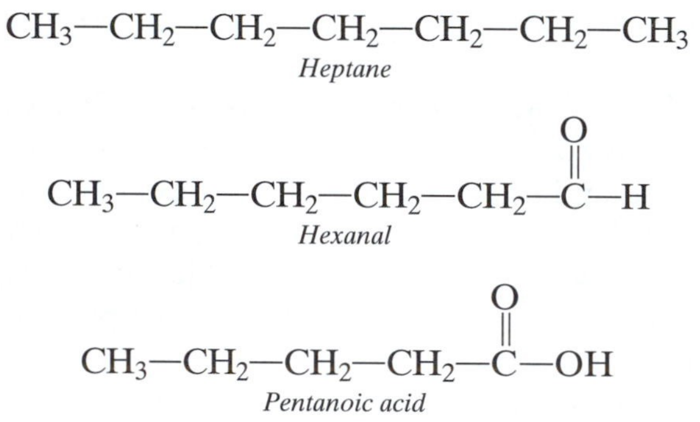

$$CH_3—CH_2—CH_2—CH_2—CH_2—CH_2—CH_3$$
Heptane

$$CH_3—CH_2—CH_2—CH_2—CH_2—\overset{\overset{\textstyle O}{\|}}{C}—H$$
Hexanal

$$CH_3—CH_2—CH_2—CH_2—\overset{\overset{\textstyle O}{\|}}{C}—OH$$
Pentanoic acid

FIGURE 8.9 Structures of heptane, hexanal, and pentanoic acid.

Exercise 8.1

The following three compounds have similar molecular weights. Place them in order of increasing boiling point.

Formaldehyde: $H_2C{=}O$ Methanol: $CH_3—OH$ Ethane: CH_3CH_3

Solution

Since the compounds all have similar molecular weights they would be expected to have similar dispersion forces. The structures indicate that the following intermolecular forces would be present.

Formaldehyde: Formaldehyde is polar and therefore would have dispersion, dipole–dipole, and dipole–induced dipole interactions.

Methanol: Methanol is polar and also has a hydrogen atom covalently bonded to an electronegative oxygen that is capable of forming hydrogen bonds with neighboring molecules. Therefore, it would be expected to have dispersion, dipole–dipole, dipole–induced dipole, and hydrogen bonding interactions.

Ethane: Ethane is nonpolar and would have only dispersion forces.

The order of boiling points would be ethane < formaldehyde < methanol.

Checkpoint

Arrange the following compounds in order of increasing boiling point.

(a) carbon tetrachloride (CCl_4), acetone (C_3H_6O), and tetrabromobutane ($C_4H_6Br_4$)

Place the following compounds in order of decreasing boiling point.

(b) octanoic acid ($C_8H_{16}O_2$), decane ($C_{10}H_{22}$), and nonanal ($C_9H_{18}O$)

Intermolecular forces play a major role in determining the properties of liquids. In the following sections we will discuss these properties.

8.4   THE KINETIC THEORY OF LIQUIDS

The kinetic theory of gases discussed in Chapter 6 can be extended to liquids. Liquids and solids are more complex, but, like gases, the particles in a liquid are in constant motion. The *average kinetic energy* of the particles in a liquid is also the same as the average kinetic energy of the particles in a gas at the same temperature. We must include the term *average* in the statement because there is an enormous range of kinetic energies possessed by molecules at a given temperature. Not all molecules at the same temperature have the same kinetic energy.

The force of attraction between the particles in a liquid, however, is large enough to hold the particles relatively close together. As a result, collisions between particles in a liquid occur much more frequently than in a gas. In a typical gas, a particle will collide with its neighbors about 10^9 times per second, whereas in a liquid the frequency of collisions is 10^{13} times per second.

In a liquid the particles undergo almost constant collisions with their neighbors to form clusters of particles that stick together for a moment, and then come apart. The force of attraction between the particles in a liquid can be estimated by measuring the amount of heat required to transform a given quantity of a liquid into the corresponding gas (or "vapor") at a given temperature. The result of the measurement is known as the **enthalpy of vaporization** of the liquid, ΔH°_{vap}. As might be expected, the enthalpy of vaporization depends on the strength of the forces of attraction that hold the particles together in the liquid. Thus, ΔH°_{vap} is a good measure of the strength of the intermolecular forces in a liquid. The larger the force of attraction between the particles, the larger the enthalpy of vaporization.

Chapter 7 described enthalpies of atom combination as the energy released due to the forces of attraction that result when gaseous atoms combine to form a compound. The forces of attraction consist of both intramolecular bonds and intermolecular or interionic forces. Therefore, enthalpies of atom combination can be used as a direct means of comparing intermolecular forces between molecules. For example, the enthalpies of atom combination for liquid and gaseous carbon tetrachloride are -1338.84 and -1306.3 kJ/mol_{rxn}, respectively. The covalent intramolecular bonds in the liquid and the gaseous CCl_4 are substantially the same (covalent bonds between C and Cl). Therefore, any difference between the two enthalpies must be due to intermolecular forces. The more negative enthalpy of atom combination for liquid CCl_4 shows that the intermolecular forces are stronger in the liquid than in the gaseous CCl_4. As we have also seen in Chapter 7 the enthalpy change at 298K for the reaction $CCl_4(l) \rightarrow CCl_4(g)$ may be calculated from the data.

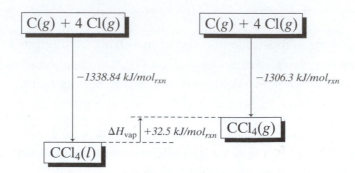

Remember that the above diagram is used only to illustrate how the change in enthalpy can be calculated from the enthalpies of atom combination found in Appendix B.14. It is *not* meant to imply that the CCl_4 phase change from liquid to gas involves the breaking of bonds to form atoms. No C—Cl bonds are broken during the phase change.

The enthalpy change, which is also the enthalpy of vaporization, is 32.5 kJ/mol$_{rxn}$. The enthalpy change is positive, indicating that heat must be put into $CCl_4(l)$ to change it to the gaseous state; that is, intermolecular forces that attract the CCl_4 molecules in the liquid state must be overcome.

A similar approach may be taken to determine the relative magnitude of intermolecular forces in solids. In this case the reaction involves converting a solid to a liquid, and the enthalpy change that accompanies the process is called the **enthalpy of fusion,** ΔH°_{fus}. The major difference is that for a liquid to gas phase change essentially all of the intermolecular forces in the liquid must be broken when forming the gas because there is very little interaction of particles in the gas phase. During melting, however, not all intermolecular forces in the solid phase must be broken to form the liquid. Intermolecular interactions are still important in the liquid phase.

Checkpoint

The enthalpy of fusion for water is 6 kJ/mol$_{rxn}$ at 0°C.

$$H_2O(s) \rightarrow H_2O(l)$$

Which phase change, liquid to gas or solid to liquid, involves breaking the greatest intermolecular forces? Which state, solid, liquid, or gas, has the strongest intermolecular forces?

The kinetic theory of liquids can explain the effect of changes in the temperature of a liquid on many of its characteristic properties. Consider the density of a liquid, for example. The density of a substance is determined by the mass and shape of its particles and how close the particles are together. As the temperature of a liquid increases, the average kinetic energy of its particles increases. The increased thermal motion will cause the particles to move further apart. As a result, the density of a liquid usually decreases with increasing temperature.

8.5   THE VAPOR PRESSURE OF A LIQUID

When chemists discuss a liquid to gas phase change, the gaseous state is often described as a vapor. The term vapor is usually reserved to describe the gaseous form of a substance that exists as a liquid at room temperature and room pressure. The terms vapor and gaseous describe the same state of matter. The phase change from a liquid to a gas is referred to as either *boiling* or *evaporation* depending on the conditions under which the phase change occurs. The phase change from gas to liquid is referred to as **condensation.**

A liquid doesn't have to be heated to its boiling point before it can become a gas. Water, for example, evaporates from an open container at room temperature (20°C), even though the boiling point of water is 100°C. According to the kinetic molecular theory the *average kinetic energy* of particles in the liquid, solid, and gas states depends on the temperature of the substance. However, not all molecules have the same energy at a given temperature. Instead, they possess a range of kinetic energies.

Figure 8.10 describes the fraction of molecules in the liquid phase that have specific kinetic energies at two different temperatures. As temperature increases from T_1 to T_2 the shape of the curve changes, and there is an overall increase in the average kinetic energy of the molecules. Even at temperatures well below the boiling point of the liquid, some of the particles are moving fast enough to escape from the liquid. The initial energy value of the shaded portion of the graph represents the minimum energy a molecule must possess in order to escape into the vapor state. Any kinetic energy greater than that value will be sufficient to allow the molecule to escape the liquid state and move into the gaseous state. At temperature T_2 a larger number of molecules have sufficient energy to move into the vapor state than at temperature T_1.

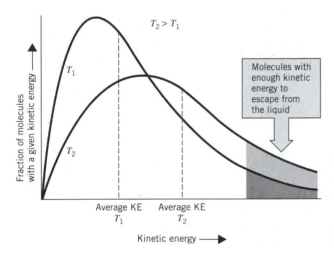

FIGURE 8.10 At a given temperature some of the particles in a liquid have enough energy to form a gas. As the temperature increases, the fraction of the molecules moving fast enough to escape from the liquid increases. As a result, the vapor pressure of the liquid also increases.

When a liquid evaporates, the average kinetic energy of the molecules composing the liquid decreases. As a result, the liquid becomes cooler. It therefore absorbs energy from its surroundings until it returns to thermal equilibrium. But as soon as that happens, some of the molecules in the liquid state acquire enough energy to escape from the liquid. In an open container, the process continues until all of the liquid evaporates.

Figure 8.11 describes what happens to a liquid placed in a closed container that is maintained at a constant temperature with respect to time. Water is used here as the example, but other liquids would behave in the same way. There is initially a vacuum in the container with a pressure of 0 atm (no water has yet evaporated). As time passes some of the molecules begin to escape from the surface of the liquid to form a gas. At time t_1, the total pressure has increased due to the pressure exerted by the water vapor. The pressure exerted by the water vapor is called the partial pressure of the water vapor. At time t_1, some of the molecules in the gas state are condensing into the liquid state.

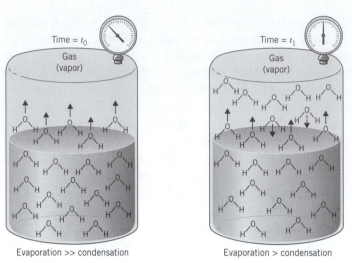
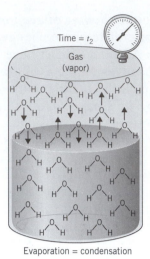

FIGURE 8.11 Change in the pressure exerted by a vapor with respect to time. The vapor pressure of the liquid is the equilibrium partial pressure of the gas (or vapor) that collects above the liquid in a closed container at a given temperature.

At time t_2, the total pressure has continued to increase due to the evaporation of additional water. The rate at which the liquid evaporates to form a gas, however, has become equal to the rate at which the gas condenses to form the liquid (illustrated by an equal number of molecules entering the liquid state as enter the gas state). At this point, the system is said to be in **equilibrium** (from Latin, "a state of balance"). The space above the liquid is saturated with water vapor. The number of water molecules in the vapor phase remains constant, and hence the pressure in the container will no longer change as long as the temperature is constant. The pressure due to the water vapor in the closed container at equilibrium is called the **vapor pressure.** The vapor pressure is the maximum pressure that can be exerted by a vapor at a given temperature. Individual molecules may continue to move between the liquid and gas state, but at equilibrium the total number of molecules in the liquid and gas state remain constant and therefore the pressure remains constant. Note, however, that the number of molecules in the liquid and solid state do not have to be equal. What is equal is the number of molecules of gas entering the liquid state and the number of molecules of liquid entering the gas state.

Exercise 8.2

Suppose instead of a vacuum in the closed container of Figure 8.11, the initial pressure in the container was 1 atm. The 1 atm of pressure can be assumed to be due to dry air composed primarily of nitrogen and oxygen. Describe what happens to the pressure with respect to time.

Solution

The pressure at t_0 would be 1 atm prior to evaporation. As time progresses to t_1 the water evaporates, and the total pressure in the container becomes equal to 1 atm plus the partial pressure of the water vapor. As equilibrium is established at t_2 the total pressure in the container will be 1 atm plus the partial pressure due to water vapor. Because the system is at equilibrium at t_2 the partial pressure due to water vapor at that time is the vapor pressure of water at the given temperature.

We have used two terms that are quite similar but are very important to differentiate. In the above discussion we used the term *pressure exerted by the vapor* to describe the partial pressure exerted by the water vapor in the closed container. The *pressure exerted by the vapor* changed with respect to time. We then used the term *vapor pressure* to describe the pressure exerted by a vapor under the very special condition of equilibrium with its liquid. *Vapor pressure* is a constant for a given liquid as long as the temperature is constant.

Checkpoint

A liquid is placed in an evacuated container (initial pressure equals 0) and allowed to reach equilibrium with its vapor. What will happen to the pressure exerted by the vapor if the volume of the container is decreased without a change in temperature? Describe what must happen on the molecular level.

The kinetic theory suggests that the vapor pressure of a liquid depends on its temperature. As can be seen in Figure 8.10, the fraction of the molecules that have enough energy to escape from a liquid increases with the temperature of the liquid. The vapor pressure of a liquid is determined by the strength of the intermolecular forces that hold the molecules of the liquid together. The stronger the attraction between molecules, the smaller is the tendency for the molecules to escape from the liquid into the gas phase, and hence the lower is the vapor pressure. Any increase in the temperature of the liquid, however, will increase the kinetic energy of its molecules. The increased motion of the molecules offsets the intermolecular attractive forces. Thus, the vapor pressure of the liquid will increase with increasing temperature.

The vapor pressures of water at temperatures from 0°C to 50°C are given in Appendix B, Table B.3. Figure 8.12 shows that the relationship between vapor pressure and temperature is not linear. The vapor pressure of water increases more rapidly than the temperature of the system.

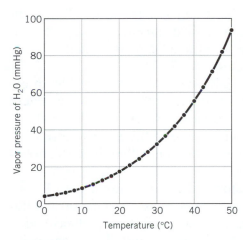

FIGURE 8.12 Plot of the vapor pressure of water versus temperature.

Below the surface of a liquid, the force of **cohesion** (literally, "sticking together") between particles is the same in all directions, as shown in Figure 8.13. Particles on the surface of the liquid feel a net force attracting them back toward the body of the liquid. As a result, the liquid takes on the shape that has the smallest possible surface area. The force that controls the shape of the liquid is called the **surface tension.** The stronger the force of attraction between the particles in the liquid, the larger is the surface tension. As the temperature increases, the increased motion of the particles in a liquid partially overcomes the attractions between the particles. As a result, the surface tension of the liquid decreases with increasing temperature.

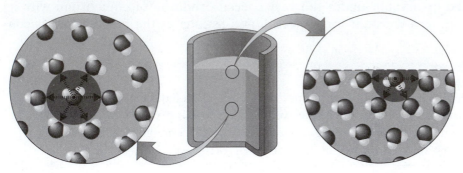

FIGURE 8.13 Molecules in a liquid feel a force of cohesion that pulls them into the body of the liquid. For molecules in the interior of the liquid, the forces are exerted from all directions, resulting in a net force of zero. However, molecules at the surface are attracted only by molecules to the side or below, resulting in a net force in the downward direction.

8.6   MELTING POINT AND FREEZING POINT

Does water always become hotter when it is heated? Think about what happens when you heat a pot of water. At first, the water gets hotter, but eventually it starts to boil. From that moment on, the temperature of the water remains the same (100°C at 1 atm), regardless of how much heat is supplied to the water, until all of the liquid is boiled away.

Imagine another experiment in which a thermometer is immersed in a snowbank on a day when the temperature finally gets above 0°C. The snow gradually melts as it gains heat from the air above it. But the temperature of the snow remains at 0°C until the last snow melts.

Apparently some heat can enter a system without changing its temperature. This is encountered whenever there is a change in the state of matter. Heat can enter or leave a sample without any detectable change in its temperature when a solid melts, when a liquid freezes or boils, or when a gas condenses to form a liquid.

Figure 8.14 shows the results obtained when ice initially at −100°C is heated in an expandable but closed container at 1 atm pressure. Initially, the heat that enters the system is used to increase the temperature of the ice from −100°C to 0°C. At 0°C the heat entering the system is used to melt the ice, and there is no change in the temperature until all of the ice is gone. The amount of heat required to melt the ice is called the **enthalpy of fusion.**

Once the ice melts, the temperature of the water slowly increases from 0°C to 100°C. But once the water starts to boil, the heat that enters the sample is used to convert the liquid to a gas, and the temperature of the sample remains constant until the liquid has boiled away. The amount of heat required to vaporize a liquid is called the **enthalpy of vaporization.** As heat continues to be added, the steam in the closed container increases in temperature, causing the container to expand even further.

At the boiling point of a liquid the energy input from heating overcomes the intermolecular forces that attract the molecules to one another. Since the energy is used to overcome the intermolecular forces, there is no increase in the kinetic energy of the molecules and hence no change in the temperature of the system.

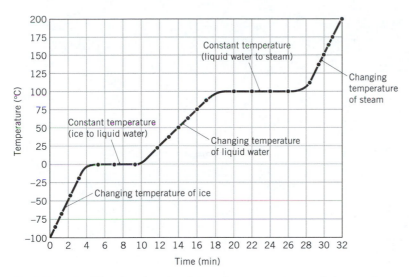

FIGURE 8.14 Plot showing the change in temperature and changes in phase as ice is heated to form liquid water and then steam.

Pure, crystalline solids have a characteristic **melting point,** which is the temperature at which the solid melts to become a liquid. The melting point of solid oxygen, for example, is −218.4°C at 1 atm pressure. Liquids have a characteristic temperature at which they turn into solids, known as the **freezing point.**

It is difficult, if not impossible, to heat a solid above its melting point because the heat that enters the solid at its melting point is used to convert the solid to a liquid. It is possible, however, to cool some liquids to temperatures below their freezing points without forming a solid. When this is done, the liquid is said to be *supercooled.*

Because it is difficult to heat solids to temperatures above their melting points, and because pure solids melt over a very small temperature range, melting points are often used to help identify compounds. We can distinguish between the three sugars known as *glucose* (MP = 150°C), *fructose* (MP = 103–105°C), and *sucrose* (MP = 185–186°C), for example, by determining the melting point of a small sample.

Measurements of the melting point of a solid can also provide information about the purity of the substance. Pure, crystalline solids melt over a very narrow range of temperatures, whereas mixtures melt over a broad temperature range. Mixtures also tend to melt at temperatures below the melting points of the pure solids. A common example is provided by the addition of salt to ice, which lowers the melting point of the ice.

8.7 BOILING POINT

Boiling of molecular liquids occurs when intermolecular forces that hold molecules in the liquid phase to one another are broken and the molecules move into the gaseous state. Figure 8.15 shows a diagram illustrating the interactions that must be overcome to boil water. Note that no covalent bonds are broken during the boiling process. When a liquid is heated, it eventually reaches a temperature at which the vapor pressure is large enough that bubbles form inside the body of the liquid. This temperature is called the **boiling point.** Once the liquid starts to boil, the temperature remains constant until all of the liquid has been converted to a gas.

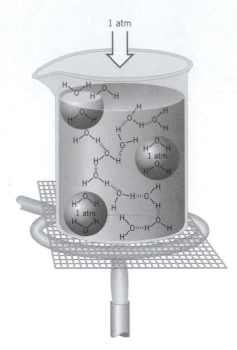

FIGURE 8.15 Molecular scale diagram showing water boiling.

Normal boiling point is defined as the temperature at which a liquid will boil at 1 atm pressure. The normal boiling point of water is 100°C. However, liquids can boil at many different temperatures depending on the pressure exerted on the liquid. If you try to cook an egg in boiling water while camping in the Rocky Mountains at an elevation of 10,000 feet, you will find that it takes longer for the egg to cook because water boils at only 90°C at that elevation.

Before microwave ovens became popular, pressure cookers were used to decrease the amount of time it took to cook food. In a typical pressure cooker, water can remain a liquid at temperatures as high as 120°C, and food cooks in as little as one-third the normal time.

To explain why water boils at 90°C in the mountains and at 120°C in a pressure cooker, even though the normal boiling point of water is 100°C, we have to understand why a liquid boils.

A liquid boils when the pressure of the vapor escaping from the liquid is equal to the pressure exerted on the liquid by its surroundings.

We recognize that a liquid is boiling by the formation of bubbles in the liquid. The bubbles are balls of vapor formed by molecules that have acquired sufficient energy to enter the vapor phase. The vapor in the bubble is in equilibrium with the liquid. Therefore, the pressure exerted by the vapor within the bubble is equal to the vapor pressure of the liquid at the temperature of the liquid. If the external pressure, usually the atmospheric pressure, is greater than the vapor pressure developed in the gaseous pockets, the pockets will be crushed and no visible evidence of boiling will be seen. If, however, the vapor pressure in the bubbles is equal to that of the external pressure, they will not collapse and will be seen to rise to the surface.

This means that we can cause a liquid to boil in two ways: by increasing the temperature of the liquid or by decreasing the pressure exerted on the liquid. As the temperature of a liquid is increased there is a corresponding increase in the vapor pressure, as shown for water in Figure 8.16. When the vapor pressure has increased such that it equals the pressure exerted on the surface of the liquid, boiling will occur.

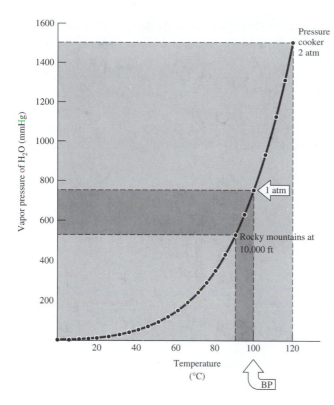

FIGURE 8.16 A liquid boils when its vapor pressure is equal to the pressure exerted on the liquid by the surroundings. The normal boiling point of water is 100°C. In the mountains, atmospheric pressure is less than 1 atm, and water boils at a temperature below 100°C. In a pressure cooker at 2 atm, water does not boil until the temperature reaches 120°C.

The normal boiling point of water is 100°C because this is the temperature at which the vapor pressure of water is 760 mmHg, or 1 atm. At 10,000 feet above sea level, the pressure of the atmosphere is only 526 mmHg. At that elevation, water boils when its vapor pressure is 526 mmHg. Figure 8.16 shows that water needs to be heated to only 90°C for its vapor pressure to reach 526 mmHg.

Pressure cookers are equipped with a valve that lets gas escape when the pressure inside the pot exceeds some fixed value. This valve is often set at 15 pounds/in^2 (psi), which combined with the usual prevailing atmospheric pressure of approximately 15 psi, means that the water vapor inside the pot must reach a pressure of 2 atm before it can escape. Because water doesn't reach a vapor pressure of 2 atm until the temperature is 120°C, it boils in the pressurized container at 120°C.

A second way to cause a liquid to boil is to reduce the pressure exerted on the surface of the liquid. The vapor pressure of water is roughly 20 mmHg at room temperature. We can therefore make water boil at room temperature by reducing the pressure in its container to less than 20 mmHg.

Anything that influences the vapor pressure of a liquid also has an effect on the boiling point. Thus, for substances composed of molecules with similar shapes and masses, the liquid with the strongest intermolecular forces of attraction will have the highest boiling point.

8.8 SPECIFIC HEAT

Heat can be used to cause a change in temperature. Thermal energy (heat) can be absorbed to increase the kinetic energy of the particles composing a solid, liquid, or gas. The energy absorbed can also cause the molecules to move farther apart, thus weakening the intermolecular interactions. In this case, the energy absorbed has increased the potential and kinetic energy. The amount of heat required to raise the temperature of equal amounts of different substances by one degree depends on the strengths of the intermolecular interactions and the intramolecular bonds.

The results of experiments done by Joseph Black between 1759 and 1762 showed what happens when liquids at different temperatures are mixed. When equal volumes of water at 100°F and 150°F were mixed, Black found that the temperature of the mixture was the average of the two samples (125°F), as shown in Figure 8.17a. He concluded that the amount of heat lost by the sample that was at 150°F was equal to the amount of heat absorbed by the sample that was at 100°F.

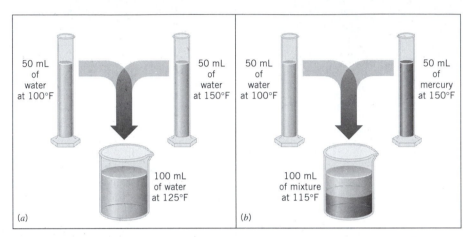

FIGURE 8.17 (*a*) The temperature of a mixture of equal volumes of water is the average of the temperatures before the samples are combined. (*b*) When equal volumes of water and mercury are mixed, the temperature of the mixture is much closer to the temperature of the water before mixing.

When water at 100°F was mixed with an equal volume of mercury at 150°F, however, the temperature of the liquids after mixing was only 115°F (Figure 8.17*b*). The temperature of the mercury fell by 35°F, but the temperature of the water increased by only 15°F. Black assumed that the heat lost by the mercury as it cooled down was equal to the heat gained by the water as it became warmer. He therefore concluded that the temperature of the water changed by a smaller amount because water has a larger "capacity for heat." In other words, it takes more heat to produce a given change in the temperature of water than it does to produce the same change in the temperature of an equivalent volume of mercury. Subsequent experiments have shown that it is not only the capacity for heat but also the mass that determines the temperature change of a substance when it is heated.

Because water has a larger capacity for heat than does mercury, it takes more heat to raise the temperature of a given mass of liquid water by one degree than it does to raise the temperature of the same mass of mercury by one degree. Any measurement of the capacity for heat must take into account not only the mass of the sample being heated and the change in the temperature observed, but also the identity of the substance being heated.

By convention, the heat needed to raise the temperature of *one gram* of substance by *one degree Celsius* is known as the **specific heat.** The units of specific heat were originally cal/g·°C. Because one degree on the Celsius scale is equal to one kelvin, specific heats can also be described in units of cal/g·K.

As might be expected, it is also possible to describe the heat required to raise the temperature of *one mole* of a substance by one degree. When this is done, we get the **molar heat capacity** (usually called heat capacity) in units of either cal/mol·°C or cal/mol·K. Molar heat capacities are generally used by chemists because they compare equal numbers of particles.

When the SI system (Appendix A) was introduced, the approved unit of heat became the *joule,* which was defined by the following equality.

$$4.184 \text{ J} = 1 \text{ cal}$$

The approved unit of temperature in this system is the kelvin. In the SI system, the units of specific heat are J/g·K, and the units of molar heat capacity are J/mol·K.

Absorption of heat by a substance increases the kinetic energy of the particles that compose the substance. Absorption of equal amounts of heat by equal moles of substances will increase the temperature of the substance with the lower molar heat capacity more than that of a substance with a higher molar heat capacity. Thus, substances with high molar heat capacities require more thermal energy to give their particles increased motion than do those of low molar heat capacity. Table 8.5 lists specific heats and molar heat capacities of a variety of substances. Note in particular the high specific heat and molar heat capacity of liquid water.

TABLE 8.5 Specific Heats and Molar Heat Capacities of Common Substances

Substance	Specific Heat (J/g·K)	Molar Heat Capacity (J/mol·K)
$Al(s)$	0.901	24.3
$C(s)$	0.7197	8.644
$Cu(s)$	0.3844	24.43
$H_2O(l)$	4.184	75.38
$Fe(s)$	0.449	25.1
$Hg(l)$	0.1395	27.98
$O_2(g)$	0.9172	29.35
$N_2(g)$	1.040	29.12
$NaCl(s)$	0.8641	50.50

Exercise 8.3

If 10 moles each of aluminum, copper, and iron absorb equivalent amounts of heat, which one of the metals will experience the largest increase in temperature?

Solution

Since metals with an equal number of moles are being compared, molar heat capacity with units based on the number of moles (J/mol·K) rather than specific heat with units based on mass (J/g·K) must be used. The molar heat capacities of aluminum, copper, and iron are given in Table 8.5. The molar heat capacity of aluminum is smaller than the molar heat capacities of copper and iron. Therefore, aluminum will experience the largest increase in temperature.

Checkpoint

Table 8.5 lists iron as having a smaller specific heat than aluminum but a larger molar heat capacity. Why does the relationship differ when specific heats and molar heat capacities are compared?

8.9   HYDROGEN BONDING AND THE ANOMALOUS PROPERTIES OF WATER

We are so familiar with the properties of water that it is difficult to appreciate the extent to which its behavior is unusual.

- Most solids expand when they melt. Water expands when it *freezes*.
- Most solids are more dense than the corresponding liquids. Ice (0.917 g/cm^3) is not as dense as water; therefore, ice floats on liquid water.
- Water has a melting point at least 100°C higher than expected on the basis of the melting points of H_2S, H_2Se, and H_2Te (see Figure 8.8).
- Water has a boiling point almost 200°C higher than expected from the boiling points of H_2S, H_2Se, and H_2Te.
- Water has the largest surface tension of any common liquid except liquid mercury.
- Water is an excellent solvent. It can dissolve compounds, such as NaCl, that are insoluble or only slightly soluble in other liquids.
- Liquid water has an unusually high specific heat. It takes more heat to raise the temperature of 1 g of water by 1°C than any other common liquid.

The anomalous properties all result from the strong intermolecular forces in water. In Section 4.17 we concluded that water is best described as a polar molecule in which there is a partial separation of charge to give positive and negative poles. The force of attraction between a positively charged hydrogen atom on one water molecule and the negatively charged oxygen atom on another gives rise to the intermolecular force called hydrogen bonding, as shown in Figure 8.18. The hydrogen bonds in water are particularly important because of the dominant role that water plays in the chemistry of living systems.

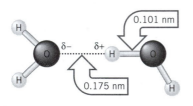

FIGURE 8.18 The primary attraction between neighboring water molecules is known as *hydrogen bonding*. The polarity of water molecules creates unusually strong intermolecular forces.

The hydrogen bonds between water molecules in ice produce the open structure shown in Figure 8.19. When ice melts, some of the hydrogen bonds are broken, and the structure collapses to form a liquid that is about 10% denser. This unusual property of water has several important consequences. The expansion of water when it freezes is responsible for the cracking of concrete, which forms potholes in streets and highways. But it also means that ice floats on top of rivers and streams.

As discussed in Section 8.2 water has a much higher boiling point than would be predicted from its molecular weight. Figure 8.8 shows a steady increase in boiling point in the series CH_4, SiH_4, GeH_4, and SnH_4. However, in the other two series shown in the figure, the boiling points of H_2O and HF are inconsistently large because of the strong hydrogen bonds between molecules in liquid H_2O and HF. If this doesn't seem important, try to imagine what life would be like if water boiled at -80°C.

FIGURE 8.19 Structure of ice. Note that in each of the hydrogen bonds the hydrogen atoms are closer to one of the oxygen atoms than the other.

The unusually large specific heat of water is also related to the strength of the hydrogen bonds between water molecules. Anything that increases the motion of water molecules, and therefore the temperature of water, must interfere with the hydrogen bonds between the molecules. The fact that it takes so much energy to overcome hydrogen bonds means that water can store enormous amounts of thermal energy. Although the water in lakes and rivers gets warmer in the summer and cooler in the winter, the large specific heat of water limits the range of temperatures; otherwise, extremes of temperature would threaten the life that flourishes in those environments. The specific heat of water is also responsible for the ability of oceans to act as a thermal reservoir that moderates the swings in temperature which occur from winter to summer.

8.10 SOLUTIONS: LIKE DISSOLVES LIKE

The initial portion of this chapter has dealt primarily with the structure and properties of pure liquids. The final sections of the chapter deal with solutions. Recall from Chapter 2 that solutions can involve solvents and solutes composed of many states of matter. The discussion of solutions in this chapter, however, is restricted to solvents that are liquids.

The photograph that accompanies this section illustrates what happens when we add a pair of solutes to a pair of solvents.

Solutes: I_2 and $KMnO_4$
Solvents: H_2O and CCl_4

Water and carbon tetrachloride form two separate liquid phases in a separatory funnel (center); $KMnO_4$ dissolves in the water layer on the top of the separatory funnel to form an intensely colored solution (left); I_2 dissolves in the CCl_4 layer on the bottom of the separatory funnel to form an intensely colored solution (right).

The solutes have two things in common. They are both solids, and they both have an intense violet or purple color. The solvents are both colorless liquids, liquids that do not mix with one another.

The difference between the solutes is easy to understand. Iodine consists of individual I_2 molecules held together by relatively weak intermolecular forces. Potassium permanganate consists of K^+ and MnO_4^- ions held together by the strong force of attraction between ions of opposite charge. It is therefore much easier to separate the I_2 molecules in solid iodine than it is to separate $KMnO_4$ into its constituent ions.

There is also a significant difference between the solvents, CCl_4 and H_2O. The difference between the electronegativities of the carbon and chlorine atoms in CCl_4 is so small ($\Delta EN = 0.33$) that there is relatively little ionic character in the C—Cl bonds. Even if there were some separation of charge in the bonds, the CCl_4 molecule wouldn't be polar because it has a symmetrical shape in which the four chlorine atoms point toward the corners of a tetrahedron, as shown in Figure 8.20. CCl_4 is therefore best described as a **nonpolar solvent.**

FIGURE 8.20 Although there is some separation of charge within the individual bonds in CCl_4, the symmetrical shape of the CCl_4 molecule ensures that there is no net dipole moment. CCl_4 is therefore *nonpolar.*

The difference between the electronegativities of the hydrogen and oxygen atoms in water is much larger ($\Delta EN = 1.31$), and the H—O bonds in the molecule are therefore polar. If the H_2O molecule were linear, the polarity of the two O—H bonds would cancel, and the molecule would have no net dipole moment. Water molecules, however, have a bent or angular shape. As a result, they have distinct positive and negative poles, and water is a polar molecule, as shown in Figure 8.21. Water is therefore classified as a **polar solvent.**

FIGURE 8.21 Because water molecules are bent, or angular, they have distinct negative and positive poles. H_2O is therefore an example of a *polar molecule.*

Because the solvents do not mix, when water and carbon tetrachloride are added to a flask, two separate liquid phases are clearly visible. We can use the densities of CCl_4 (1.594 g/cm^3) and H_2O (1.0 g/cm^3) to decide which phase is water and which is carbon tetrachloride. The denser CCl_4 settles to the bottom of the flask.

When a few crystals of iodine are added and the contents of the flask are shaken, the I_2 dissolves in the CCl_4 layer to form a violet solution. The water layer becomes a very light brown, which suggests that very little I_2 dissolves in water.

When the experiment is repeated with potassium permanganate, the water layer picks up the characteristic dark purple color of the MnO_4^- ion, and the CCl_4 layer remains colorless. This suggests that $KMnO_4$ dissolves in water but not in carbon tetrachloride. The results of the experiment are summarized in Table 8.6. Table 8.6 raises two important questions. Why does $KMnO_4$ dissolve in water, but not in carbon tetrachloride? Why does I_2 dissolve in carbon tetrachloride, but to only a very small extent in water?

TABLE 8.6   Solubilities of I_2 and $KMnO_4$ in CCl_4 and Water

	H_2O	CCl_4
I_2	Slightly soluble	Very soluble
$KMnO_4$	Very soluble	Insoluble

It takes a lot of energy to separate the K^+ and MnO_4^- ions in potassium permanganate. But the ions can form strong attractive interactions with neighboring water molecules, as shown in Figure 8.22. The energy released from the formation of the ion–dipole interactions compensates for the energy that has to be invested to separate individual ions in the $KMnO_4$ crystal. No such forces are possible between the K^+ or MnO_4^- ions and the nonpolar CCl_4 molecules. As a result, $KMnO_4$ does not dissolve in CCl_4.

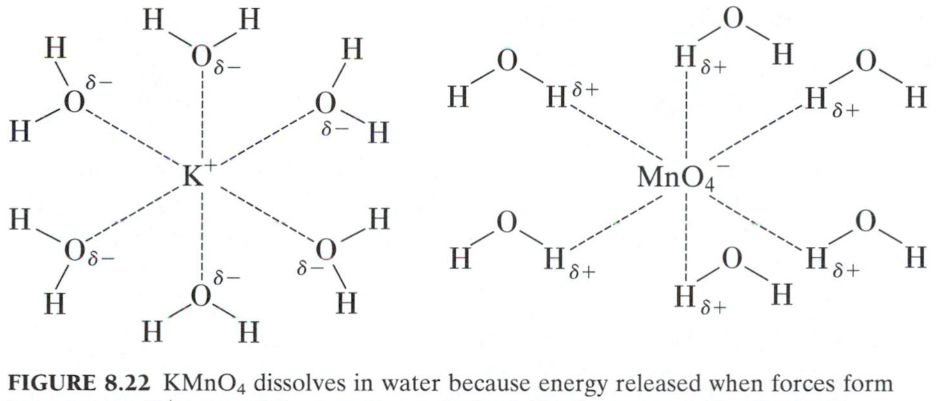

FIGURE 8.22 $KMnO_4$ dissolves in water because energy released when forces form between the K^+ ion and the negative end of neighboring water molecules and between the MnO_4^- ion and the positive end of the solvent molecules compensates for the energy it takes to separate the K^+ and MnO_4^- ions.

The I_2 molecules in iodine and the CCl_4 molecules in carbon tetrachloride are both held together by weak intermolecular forces. The intermolecular forces must be broken in order to separate solute and solvent molecules from one another. Figure 8.23 shows that similar intermolecular forces are formed between I_2 and CCl_4 molecules when the I_2 is dissolved in CCl_4. I_2 therefore readily dissolves in CCl_4 because the intermolecular forces that are broken in the solute and solvent are very similar to the intermolecular forces that are formed between the solute and solvent molecules. The molecules in water are held together by hydrogen bonds that are stronger than most intermolecular forces. No interaction between I_2 and H_2O molecules is strong enough to compensate for the hydrogen bonds between water molecules that have to be broken to dissolve iodine in water, so relatively little I_2 dissolves in H_2O.

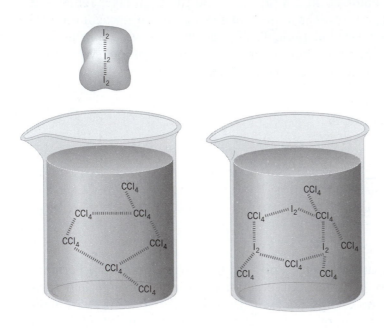

FIGURE 8.23 I_2 dissolves in CCl_4 because the intermolecular forces in I_2 and in CCl_4 are similar to the intermolecular forces between CCl_4 and I_2.

We can summarize the results of the experiment by noting that *nonpolar solutes* (such as I_2) dissolve in *nonpolar solvents* (such as CCl_4), whereas many ionic solutes (such as $KMnO_4$) dissolve in *polar solvents* (such as H_2O).

Exercise 8.4

Elemental phosphorus is often stored under water because it doesn't dissolve in water. Elemental phosphorus is very soluble in carbon disulfide, however. Use the structure of the P_4 molecule shown in Figure 8.24 to explain why P_4 is soluble in CS_2 but not in water.

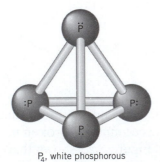

P_4, white phosphorous

FIGURE 8.24 Pure elemental phosphorus is a white, waxy solid that consists of tetrahedral P_4 molecules in which the P—P—P bond angle is only 60°.

Solution

The P_4 molecule is a perfect example of a nonpolar solute. It is therefore more likely to be soluble in nonpolar solvents than in polar solvents such as water.

The Lewis structure of CS_2 suggests that the molecule is linear.

$$\ddot{S}=C=\ddot{S}$$

Thus, even if there is some separation of charge in the C=S double bond, the molecule would have no net dipole moment, because of its symmetry. The electronegativities of carbon ($EN = 2.54$) and sulfur ($EN = 2.59$), however, suggest that the C=S double bonds are almost perfectly covalent. CS_2 is therefore a nonpolar solvent, which readily dissolves nonpolar P_4. As a general rule *like dissolves like.*

Exercise 8.5

The iodide ion reacts with iodine in aqueous solution to form I_3^-, or triiodide ion.

$$I^-(aq) + I_2(aq) \longrightarrow I_3^-(aq)$$

What would happen if CCl_4 were added to an aqueous solution that contained a mixture of KI, I_2, and KI_3?

Solution

Two layers would form with CCl_4 on the bottom and the aqueous layer on top. KI and KI_3 are both ionic compounds. One contains K^+ and I^- ions; the other contains K^+ and I_3^- ions. The ionic compounds are more soluble in polar solvents, such as water, than in nonpolar solvents, such as CCl_4. KI and KI_3 would therefore remain in the aqueous solution. I_2 is a nonpolar molecule, which is more soluble in a nonpolar solvent, such as carbon tetrachloride. Some I_2 would therefore leave the aqueous layer and enter the CCl_4 layer, where it would exhibit the characteristic violet color of solutions of molecular iodine.

8.11 WHY DO SOME SOLIDS DISSOLVE IN WATER?

When asked to describe what happens when a teaspoon of sugar is stirred into a cup of coffee, people often answer, "The sugar initially settles to the bottom of the cup. When the coffee is stirred, it dissolves to produce a sweeter cup of coffee." When asked to extend the description to the molecular level, they hesitate. If they are chemistry students, like yourself, they often ask to be reminded of the formula for sugar. When told that the sugar used in cooking is *sucrose,* $C_{12}H_{22}O_{11}$, they write equations such as the following.

$$C_{12}H_{22}O_{11}(s) \xrightarrow{H_2O} C_{12}H_{22}O_{11}(aq)$$

Although there is nothing wrong with the equation, it doesn't explain what happens at the molecular level when sugar dissolves. Nor does it provide any hints about why sugar dissolves but other solids do not. To understand what happens at the molecular level when a solid dissolves, we have to refer back to the discussion of molecular and ionic solids in Chapter 5.

The sugar we use to sweeten coffee or tea is a *molecular solid,* in which the individual molecules are held together by hydrogen bond intermolecular forces. When sugar dissolves in water, the hydrogen bond intermolecular forces between the individual sucrose molecules are broken, and the individual $C_{12}H_{22}O_{11}$ molecules are released into solution.

It takes energy to break the forces between the $C_{12}H_{22}O_{11}$ molecules in sucrose. It also takes energy to break the hydrogen bonds in water that must be disrupted to insert one of the sucrose molecules into the solution. Sugar dissolves in water because the slightly polar sucrose molecules form hydrogen bond intermolecular forces with the polar water molecules. The intermolecular forces that form between the solute and the solvent help compensate for the energy needed to disrupt the structure of both the pure solute and the solvent. In the case of sugar and water, the process works so well that up to 1800 g of sucrose can dissolve in a liter of water (see Figure 8.25).

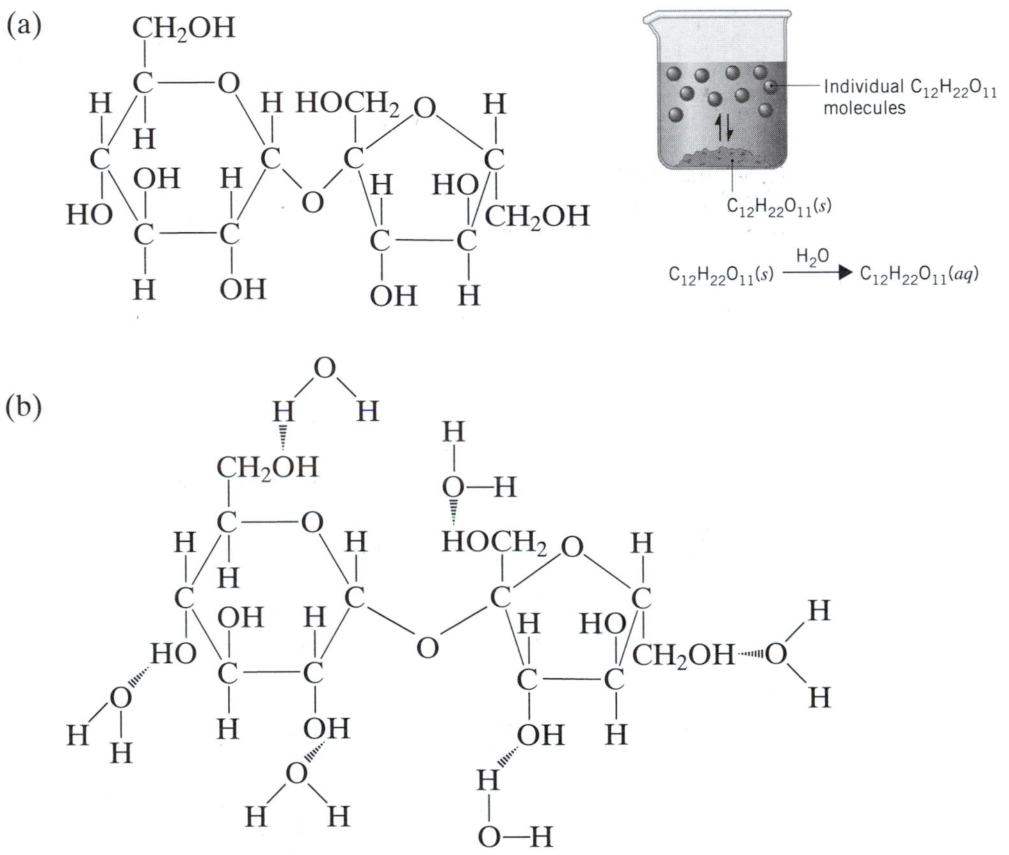

FIGURE 8.25 (*a*) Structure of sucrose. (*b*) Hydrogen bonds are broken in both sucrose and water and then formed between sucrose and water in the solution.

Ionic solids (or salts) contain positive and negative ions that are held together by the strong force of attraction between particles with opposite charges. When an ionic solid dissolves in water, the ions that form the solid are released into solution, where they are attracted to the polar solvent molecules (see Figure 8.26). As a result, we can generally assume that salts dissociate into their ions when they dissolve in water. Ionic compounds dissolve in water if the energy given off when the ions interact with water molecules partially compensates for the energy needed to break the ionic bonds in the solid and for the energy required to separate the water molecules so that the ions can be inserted into the solution.

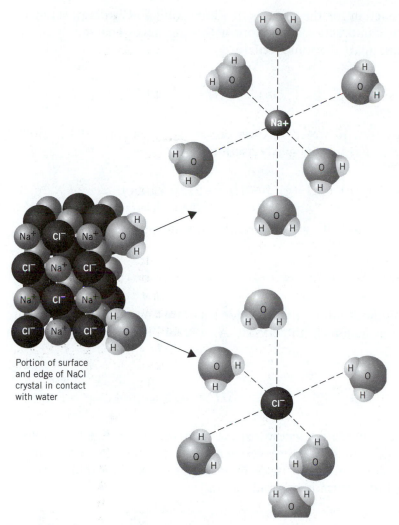

FIGURE 8.26 Interaction of water and ionic compounds. The electrostatic interaction between the polar water molecules and the ions is a dipole–ion force. Reprinted with permission from L. J. Malone, *Basic Concepts of Chemistry*, John Wiley & Sons, Inc., New York, 1994, p. 351.

It takes an enormous amount of heat to break apart an ionic crystal. For example, it takes 2033 kJ/mol to transform solid $BaCl_2$ into Ba^{2+} and 2 Cl^- ions in the gas phase.

$$BaCl_2(s) \longrightarrow Ba^{2+}(g) + 2\ Cl^-(g) \qquad \Delta H° = 2033 \text{ kJ/mol}_{rxn}$$

It takes so much energy to separate the Ba^{2+} and 2 Cl^- ions in $BaCl_2$ that we might not expect the compound to dissolve in water. The force of attraction between Ba^{2+} and 2 Cl^- ions with water molecules is so large, however, that 2047 kJ/mol$_{rxn}$ of energy is released when the Ba^{2+} and 2 Cl^- ions interact with water molecules.

$$Ba^{2+}(g) + 2\ Cl^-(g) \xrightarrow{H_2O} Ba^{2+}(aq) + 2\ Cl^-(aq) \qquad \Delta H° = -2047 \text{ kJ/mol}_{rxn}$$

Checkpoint

Calculate the enthalpy change for the dissolution of $BaCl_2(s)$ in water to form $Ba^{2+}(aq)$ and 2 $Cl^-(aq)$ from the data in Appendix B.14.

The overall enthalpy of reaction for the process in which solid $BaCl_2$ dissolves in water is exothermic. Therefore, the interaction of the ions with water more than compensates for the energy needed to break apart the ionic structure.

$$BaCl_2(s) \xrightarrow{H_2O} Ba^{2+}(aq) + 2\ Cl^-(aq) \qquad \Delta H° = -14\ kJ/mol_{rxn}$$

Silver chloride is very insoluble in water. As was previously seen for $BaCl_2$, a lot of heat is required to separate $AgCl(s)$ into its gas phase ions.

$$AgCl(s) \longrightarrow Ag^+(g) + Cl^-(g) \qquad \Delta H° = 915.7\ kJ/mol_{rxn}$$

The heat released when the Ag^+ and Cl^- ions interact with water is also large.

$$Ag^+(g) + Cl^-(g) \xrightarrow{H_2O} Ag^+(aq) + Cl^-(aq) \qquad \Delta H° = -850.2\ kJ/mol_{rxn}$$

But the heat released when the ions interact with water isn't large enough to compensate for the heat needed to separate the ions in the crystal. As a result, the overall enthalpy of reaction is very unfavorable.

$$AgCl(s) \xrightarrow{H_2O} Ag^+(aq) + Cl^-(aq) \qquad \Delta H° = 65.5\ kJ/mol_{rxn}$$

With a positive $\Delta H°$ of this magnitude we would expect relatively little AgCl to dissolve in water. In fact, less than 0.002 g of AgCl dissolves in a liter of water at room temperature. The solubility of silver chloride in water is so small that AgCl is often said to be "insoluble" in water, even though that term is misleading.

Checkpoint

Calculate $\Delta H°$ for the reaction below using the data from Table B.14.

$$AgCl(s) \xrightarrow{H_2O} Ag^+(aq) + Cl^-(aq)$$

8.12 SOLUBILITY EQUILIBRIA

Discussions of solubility equilibria are based on the following assumption: *When they dissolve, solids often break apart to give primarily the molecules or ions from which they are formed.* Most molecular solutes dissolve to give individual molecules

$$C_{12}H_{22}O_{11}(s) \xrightarrow{H_2O} C_{12}H_{22}O_{11}(aq)$$

and ionic solids dissociate to give solutions of the positive and negative ions they contain.

$$NaCl(s) \xrightarrow{H_2O} Na^+(aq) + Cl^-(aq)$$

Solutes such as NaCl that completely break up into ions when they dissolve are called **strong electrolytes.** Solutes such as sucrose that do not break up into ions when they dissolve are called **nonelectrolytes.** There are some solutes that partially break up into ions and partially exist as soluble molecules in solution; they are called **weak electrolytes.**

We can detect the presence of Na^+ and Cl^- ions in an aqueous solution with the conductivity apparatus shown in Figure 8.27. The apparatus consists of a lightbulb connected to a pair of metal wires that can be immersed in a beaker of water. The circuit in the conductivity apparatus isn't complete. In order for the lightbulb to glow when the apparatus is plugged into an electrical outlet, there must be a way for electrical charge to flow through the solution from one of the metal wires to the other.

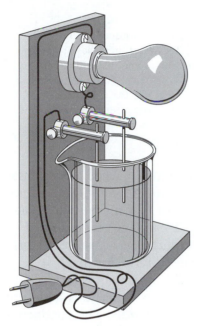

FIGURE 8.27 A conductivity apparatus can be used to demonstrate the difference between aqueous solutions of ionic and covalent solids.

When an ionic solid is dissolved in the beaker of water, the resulting positive and negative ions are free to move through the aqueous solution. The net result is a flow of electric charge through the solution that completes the circuit and lets the lightbulb glow. As might be expected, the brightness of the bulb is proportional to the concentration of the ions in the solution. Slightly soluble salts such as calcium sulfate make the lightbulb glow dimly. When the wires are immersed in a solution of a very soluble salt, such as NaCl, the lightbulb glows brightly.

The conductivity apparatus in Figure 8.27 gives us only qualitative information about the relative concentrations of the ions in different solutions. It is possible to build a more sophisticated instrument that gives quantitative measurements of the conductivity of a solution, which is directly proportional to the concentration of the ions in the solution. Figure 8.28 shows what happens to the conductivity of water as we gradually add infinitesimally small amounts of AgCl to water and wait for the solid to dissolve before taking measurements.

The system conducts a very small electric current even before any AgCl is added because of small quantities of H_3O^+ and OH^- ions in water. The solution becomes a slightly better conductor when AgCl is added because some of the salt dissolves to give Ag^+ and Cl^- ions, which can carry an electric current through the solution. The conductivity continues to increase as more AgCl per is added, until about 0.002 g of the salt has dissolved per liter of solution.

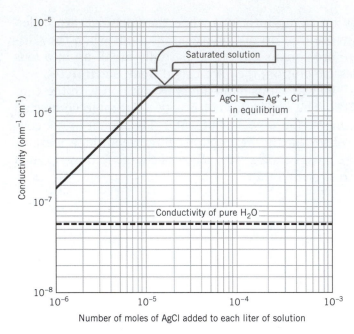

FIGURE 8.28 The conductivity of a solution of AgCl in water increases at first as the AgCl dissolves and dissociates into Ag^+ and Cl^- ions. Once the solution has become saturated with AgCl, the conductivity remains the same no matter how much solid is added.

The fact that the conductivity doesn't increase after the solution has reached a concentration of 0.002 g AgCl/liter tells us that there is a limit on the solubility of the salt in water. Once the solution reaches that limit, no more AgCl dissolves, regardless of how much solid we add to the system. This is exactly what we would expect if the solubility of AgCl is controlled by an equilibrium. Once the solution reaches equilibrium, the rate at which AgCl dissolves to form Ag^+ and Cl^- ions is equal to the rate at which the ions recombine to form AgCl.

Evidence to support that conclusion comes from Figure 8.29, which shows what happens to the conductivity of water when a large excess of solid silver chloride is added to the water. When the salt is first added, it dissolves and dissociates rapidly. The conductivity of the solution therefore increases rapidly at first.

$$AgCl(s) \xrightarrow[\text{dissociate}]{\text{dissolve}} Ag^+(aq) + Cl^-(aq)$$

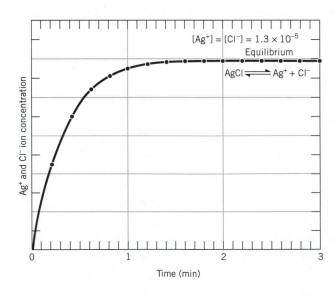

FIGURE 8.29 Concentration of Ag^+ and Cl^- ions versus time as solid AgCl is added to water.

The concentrations of the ions soon become large enough that the reverse reaction starts to compete with the forward reaction, which leads to a decrease in the rate at which Ag^+ and Cl^- ions enter the solution. The reverse reaction is the formation of insoluble silver chloride. Reactions in which soluble species form an insoluble product are called **precipitation reactions.**

$$Ag^+(aq) + Cl^-(aq) \xrightarrow[\text{precipitate}]{\text{associate}} AgCl(s)$$

Eventually, the Ag^+ and Cl^- ion concentrations become large enough that the rate at which precipitation occurs exactly balances the rate at which AgCl dissolves. Once that happens, there is no change in the concentration of the ions with time, and the reaction is at equilibrium. When the system reaches equilibrium it is called a **saturated solution** because it contains the maximum concentration of ions that can exist in equilibrium with the solid salt at a given temperature. The amount of salt that must be added to a given volume of solvent to form a saturated solution is called the **solubility** of the salt. Chemists use double arrows to indicate a reaction at equilibrium.

$$AgCl(s) \xrightleftharpoons{H_2O} Ag^+(aq) + Cl^-(aq)$$

8.13 SOLUBILITY RULES

There are two primary factors that can be used to predict whether a molecular solute will be soluble in a water: similar molecular structure and similar intermolecular forces of attraction. However, predicting whether an ionic compound will be soluble in water is not so straightforward. There are a number of patterns in the data obtained from measuring the solubilities of different ionic compounds. The patterns form the basis for the rules outlined in Table 8.7, which can guide predictions of whether a given ionic compound will dissolve in water. The rules are based on the following definitions of the terms *soluble, insoluble,* and *slightly soluble* (see Figure 8.30).

TABLE 8.7   Solubility Rules for Ionic Compounds in Water

Soluble Salts

The Na^+, K^+, and NH_4^+ ions form *soluble salts*. Thus, NaCl, KNO_3, and $(NH_4)_2CO_3$ are *soluble salts.*

The nitrate ion (NO_3^-) forms *soluble salts*. Thus, $Cu(NO_3)_2$ and $Fe(NO_3)_3$ are soluble.

The chloride (Cl^-), bromide (Br^-), and iodide (I^-) ions usually form *soluble salts*. Exceptions include salts of the Pb^{2+}, Hg_2^{2+}, Ag^+, and Cu^+ ions. $CuBr_2$ is soluble, but CuBr is not.

The sulfate ion (SO_4^{2-}) usually forms *soluble salts*. Exceptions include $BaSO_4$, $SrSO_4$, and $PbSO_4$, which are insoluble, and Ag_2SO_4, $CaSO_4$, and Hg_2SO_4, which are slightly soluble.

Insoluble Salts

Sulfides (S^{2-}) are usually *insoluble*. Exceptions include Na_2S, K_2S, $(NH_4)_2S$, MgS, CaS, SrS, and BaS.

Oxides (O^{2-}) are usually *insoluble*. Exceptions include Na_2O, K_2O, SrO, and BaO, which are soluble, and CaO, which is slightly soluble.

Hydroxides (OH^-) are usually *insoluble*. Exceptions include NaOH, KOH, $Sr(OH)_2$, and $Ba(OH)_2$, which are soluble, and $Ca(OH)_2$, which is slightly soluble.

Chromates (CrO_4^{2-}), phosphates (PO_4^{3-}), and carbonates (CO_3^{2-}) are usually *insoluble*. Exceptions include salts of the Na^+, K^+, and NH_4^+ ions, such as Na_2CrO_4, K_3PO_4, and $(NH_4)_2CO_3$.

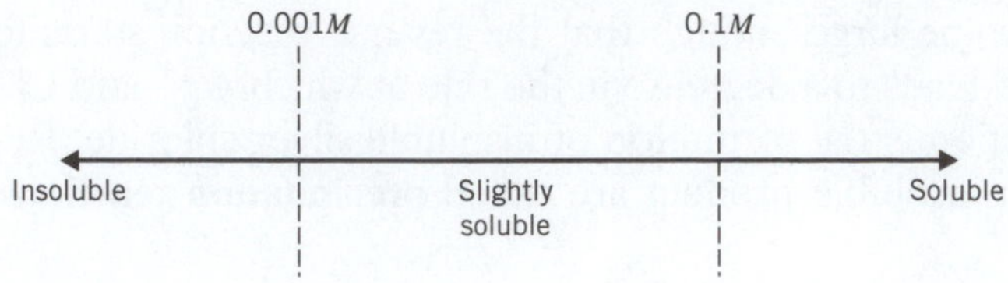

0.001*M*   0.1*M*

Insoluble ————————————————————————— Soluble

Slightly
soluble

FIGURE 8.30 Solubilities of ionic compounds in water cover a wide range that is divided into the categories insoluble, slightly soluble, and soluble.

- A salt is defined as being soluble if it dissolves in water to give a solution with a concentration of at least 0.1 mol/L at room temperature.
- A salt is defined as being insoluble if the concentration of an aqueous solution is less than 0.001 *M* at room temperature.
- Slightly soluble salts give solutions that fall between the extremes.

The definitions are used to categorize salts with similar solubilities. The solubilities of salts cover a wide range of concentrations. Salts with very low solubilities that are categorized as insoluble will dissolve to some small extent. Soluble salts will have a limit to their solubility.

8.14 HYDROPHILIC AND HYDROPHOBIC MOLECULES

Hydrocarbons, such as the alkanes discussed in Section 8.3, are compounds that contain only carbon and hydrogen. Because of the molecular geometry and the small difference between the electronegativities of carbon and hydrogen ($\Delta EN = 0.24$), hydrocarbons are nonpolar. As a result, they don't dissolve in polar solvents such as water. Hydrocarbons are therefore described as insoluble in water.

When one of the hydrogen atoms in a hydrocarbon is replaced with an —OH group, the compound is known as an **alcohol,** as shown in Figure 8.31. Because alcohols have a structure that contains an —OH group, as in water, alcohols have properties between the extremes of hydrocarbons and water. When the hydrocarbon chain is short, the alcohol is soluble in water. Methanol (CH_3OH) and ethanol (CH_3CH_2OH) are infinitely soluble in water, for example. There is no limit on the amount of those alcohols that can dissolve in a given quantity of water. The alcohol in beer, wine, and hard liquors is ethanol, and mixtures of ethanol and water can have any concentration between the extremes of pure alcohol (200 proof) and pure water (0 proof).

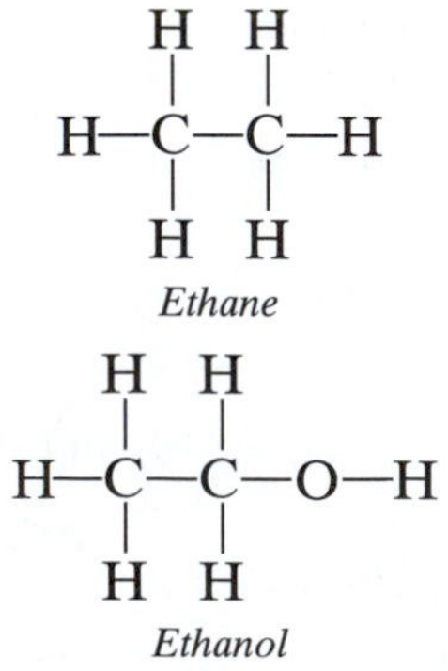

Ethane

Ethanol

FIGURE 8.31 When a hydrogen in a hydrocarbon such as ethane is replaced by an —OH group, the resulting compound is called an alcohol.

As the hydrocarbon chain becomes longer, the alcohol becomes less soluble in water, as shown in Table 8.8. One end of the longer alcohol molecules has so much nonpolar character it is called **hydrophobic** (literally, "water hating"), as shown in Figure 8.32. The other end contains an —OH group that can form hydrogen bonds to neighboring water molecules and is therefore said to be **hydrophilic** (literally, "water loving"). As the hydrocarbon chain becomes longer, the hydrophobic character of the molecule increases, and the solubility of the alcohol in water gradually decreases until it becomes essentially insoluble in water.

TABLE 8.8   Solubilities of Alcohols in Water

Formula	Name	Solubility in Water (g/100 g)
CH_3OH	Methanol	Infinitely soluble
CH_3CH_2OH	Ethanol	Infinitely soluble
$CH_3(CH_2)_2OH$	Propanol	Infinitely soluble
$CH_3(CH_2)_3OH$	Butanol	9
$CH_3(CH_2)_4OH$	Pentanol	2.7
$CH_3(CH_2)_5OH$	Hexanol	0.6
$CH_3(CH_2)_6OH$	Heptanol	0.18
$CH_3(CH_2)_7OH$	Octanol	0.054
$CH_3(CH_2)_9OH$	Decanol	Insoluble in water

Hydrophilic
head

$$CH_3CH_2CH_2CH_2CH_2CH_2CH_2CH_2CH_2CH_2OH$$

Hydrophobic
tail

FIGURE 8.32 One end of the decanol molecule is nonpolar, and therefore *hydrophobic;* the other end is polar, and therefore *hydrophilic.*

People encountering the terms *hydrophilic* and *hydrophobic* for the first time sometimes have difficulty remembering which stands for water hating and which stands for water loving. If you remember that Hamlet's girlfriend was named Ophelia (not Ophobia), you can remember that the prefix *philo-* is commonly used to describe love—for example, in *philanthropist, philharmonic, philosopher*—and that *phobia* means dislike.

Checkpoint

Amino acids are classified as either hydrophilic or hydrophobic on the basis of their side chains. Use the structure of the side chains of the following amino acids to justify the classifications shown below.

	Amino Acid	Side Chain
Hydrophobic	Alanine	$—CH_3$
	Cysteine	$—CH_2CH_2SCH_3$
Hydrophilic	Lysine	$—CH_2CH_2CH_2CH_2NH_3^+$
	Serine	$—CH_2OH$

Table 8.9 shows that the ionic compound NaCl is relatively soluble in water. Water, being polar, is able to cluster around the positively and negatively charged ions formed when NaCl dissolves, as shown in Figure 8.26. As the solvent becomes more hydrocarbonlike, the solubility of NaCl decreases because the longer chain hydrocarbon solvents do not interact as strongly with Na^+ and Cl^-.

TABLE 8.9   Solubility of Sodium Chloride in Water and in Alcohols

Formula of Solvent	Solvent Name	Solubility of NaCl (g/100 g solvent)
H_2O	Water	35.92
CH_3OH	Methanol	1.40
CH_3CH_2OH	Ethanol	0.065
$CH_3(CH_2)_2OH$	Propanol	0.012
$CH_3(CH_2)_3OH$	Butanol	0.005
$CH_3(CH_2)_4OH$	Pentanol	0.0018

8.15   SOAPS, DETERGENTS, AND DRY-CLEANING AGENTS

The chemistry behind the manufacture of soap hasn't changed since it was made from animal fat and the ash from wood fires almost 5000 years ago. Solid animal fats (such as the tallow obtained during the butchering of sheep and cattle) and liquid plant oils (such as palm oil and coconut oil) are still heated in the presence of a strong base to form a soft, waxy material that enhances the ability of water to wash away the grease and oil that forms on our bodies and our clothes.

Animal fats and plant oils contain compounds known as *fatty acids*. Fatty acids, such as stearic acid (see Figure 8.33), have small, polar, hydrophilic heads attached to long, nonpolar, hydrophobic tails. Fatty acids are seldom found by themselves in nature. They are usually bound to molecules of glycerol ($HOCH_2CHOHCH_2OH$) to form triglycerides, such as the one shown in Figure 8.34. The triglycerides break down in the presence of a strong base to form the Na^+ or K^+ salt of the fatty acid, as shown in Figure 8.35. The reaction is called *saponification*, which literally means "the making of soap."

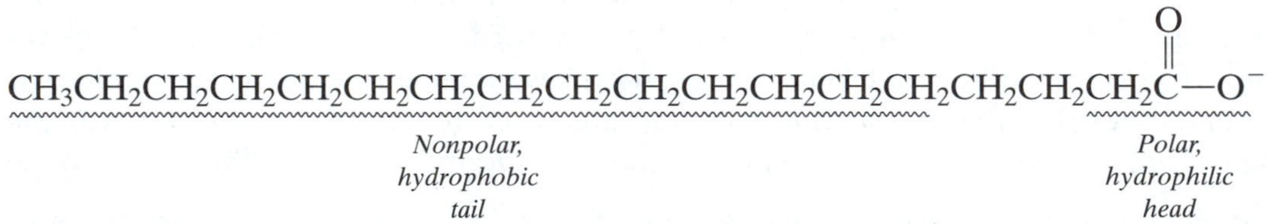

$$CH_3CH_2CH_2CH_2CH_2CH_2CH_2CH_2CH_2CH_2CH_2CH_2CH_2CH_2CH_2CH_2CH_2\overset{\displaystyle O}{\overset{\|}{C}}-O^-$$

Nonpolar, hydrophobic tail   Polar, hydrophilic head

FIGURE 8.33 The hydrocarbon chain on one end of a fatty acid molecule is nonpolar and hydrophobic, whereas the $-CO_2H$ group on the other end of the molecule is polar and hydrophilic.

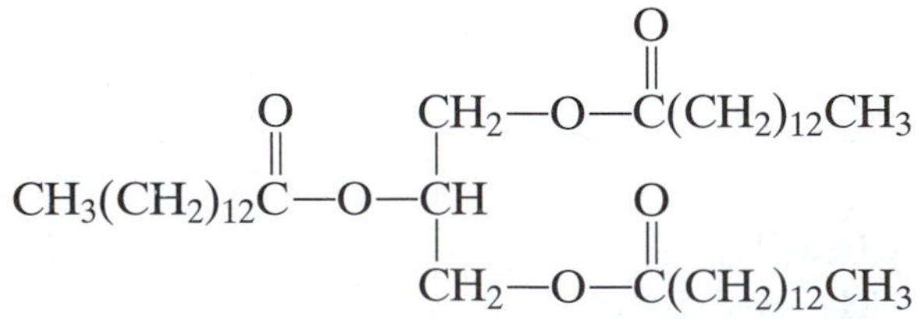

FIGURE 8.34 Structure of the triglyceride known as *trimyristin*, which can be isolated in high yield from nutmeg.

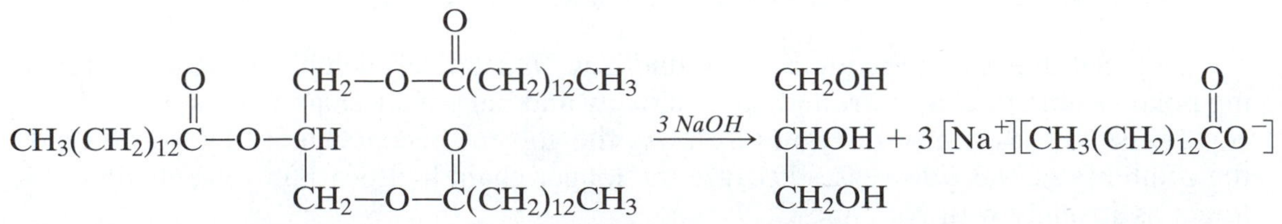

FIGURE 8.35 Saponification of the trimyristin extracted from nutmeg.

The cleaning action of soap results from the fact that soap molecules are *surfactants*—they tend to concentrate on the surface of water. They cling to the surface because they try to orient their polar CO_2^- heads toward water molecules and their nonpolar $CH_3CH_2CH_2$ . . . tails away from neighboring water molecules.

Water can't wash the soil out of clothes by itself because the soil particles that cling to textile fibers are covered by a layer of nonpolar grease or oil molecules, which repels water. The nonpolar tails of the soap molecules on the surface of water dissolve in the grease or oil that surrounds a soil particle, as shown in Figure 8.36. The soap molecules therefore disperse, or *emulsify*, the soil particles, which makes it possible to wash the particles out of the clothes.

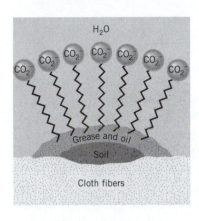

FIGURE 8.36 Soap molecules disperse, or emulsify, soil particles coated with a layer of nonpolar grease or oil molecules.

Most soaps are more dense than water. They can be made to float, however, by incorporating air into the soap during its manufacture. Most soaps are also opaque; they absorb rather than transmit light. Translucent soaps can be made by adding alcohol, sugar, and glycerol, which slow down the growth of soap crystals while the soap solidifies. Liquid soaps are made by replacing the sodium salts of the fatty acids with the more soluble K^+ or NH_4^+ salts.

In the 1950s, more than 90% of the cleaning agents sold in the United States were soaps. Today soap represents less than 20% of the market for cleaning agents. The primary reason for the decline in the popularity of soap is the reaction between soap and "hard" water. The most abundant positive ions in tap water are Na^+, Ca^{2+}, and Mg^{2+}. Water that is particularly rich in Ca^{2+}, Mg^{2+}, or Fe^{3+} ions is said to be hard. Hard water interferes with the action of soap because the ions combine with soap molecules to form insoluble precipitates that have no cleaning power. The precipitates not only decrease the concentration of the soap molecules in solution, they actually bind soil particles to clothing, leaving a dull, gray film.

One way around the problem is to "soften" the water by replacing the Ca^{2+} and Mg^{2+} ions with Na^+ ions. Many water softeners are filled with a resin that contains $—SO_3^-$ ions attached to a polymer, as shown in Figure 8.37. The resin is treated with NaCl until each $—SO_3^-$ ion picks up an Na^+ ion. When hard water flows over the resin, Ca^{2+} and Mg^{2+} ions bind to the $—SO_3^-$ ions on the polymer chain and Na^+ ions are released into the solution. Periodically, the resin becomes saturated with Ca^{2+} and Mg^{2+} ions. When that happens, the resin has to be regenerated by being washed with a concentrated solution of NaCl.

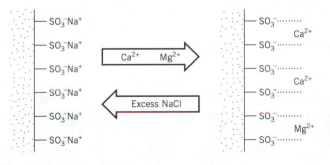

FIGURE 8.37 When a water softener is "charged," it is washed with a concentrated NaCl solution until all of the $—SO_3^-$ ions pick up an Na^+ ion. The softener then picks up Ca^{2+} and Mg^{2+} from hard water, replacing those ions with Na^+ ions.

There is another way around the problem of hard water. Instead of removing Ca^{2+} and Mg^{2+} ions from water, we can find a cleaning agent that doesn't form insoluble salts with those ions. Synthetic detergents are an example of such cleaning agents. Detergent molecules consist of long, hydrophobic hydrocarbon tails attached to polar, hydrophilic —SO_3^- or —OSO_3^- heads, as shown in Figure 8.38. By themselves, detergents don't have the cleaning power of soap. "Builders" are therefore added to synthetic detergents to increase their strength. The builders are often salts of highly charged ions, such as the triphosphate ion ($P_3O_{10}^{5-}$).

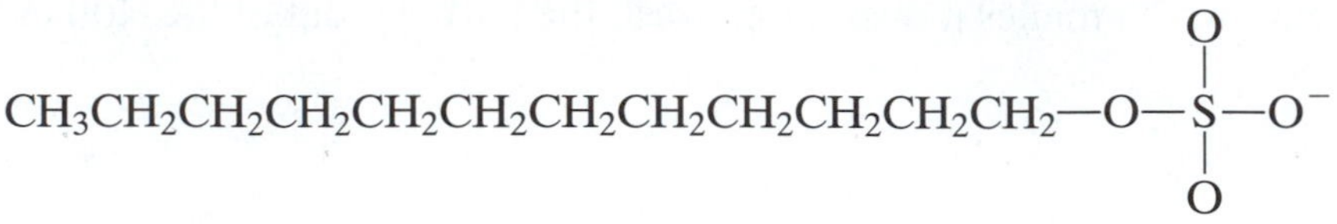

$$CH_3CH_2CH_2CH_2CH_2CH_2CH_2CH_2CH_2CH_2CH_2CH_2-O-\underset{\underset{O}{|}}{\overset{\overset{O}{|}}{S}}-O^-$$

FIGURE 8.38 Structure of one of the components of a synthetic detergent.

Cloth fibers swell when they are washed in water. This leads to changes in the dimensions of the cloth that can cause wrinkles—which are local distortions in the structure of the fiber—or even more serious damage, such as shrinking. The problems can be avoided by "dry cleaning," which uses a nonpolar solvent that does not adhere to, or wet, the cloth fibers. The nonpolar solvents used in dry cleaning dissolve the nonpolar grease or oil layer that coats soil particles, freeing the soil particles to be removed by detergents added to the solvent or by the tumbling action inside the machine. Dry cleaning has the added advantage that it can remove oily soil at lower temperatures than soap or detergent dissolved in water, so it is safer for delicate fabrics.

When dry cleaning was first introduced in the United States between 1910 and 1920, the solvent was a mixture of hydrocarbons isolated from petroleum when gasoline was refined. Over the years, those flammable hydrocarbon solvents have been replaced by halogenated hydrocarbons, such as trichloroethane (Cl_3C-CH_3), trichloroethylene ($Cl_2C=CHCl$), and perchloroethylene ($Cl_2C=CCl_2$).

KEY TERMS

Alcohol	Heat capacity	Nonpolar solvent
Boiling point	Hydrocarbons	Polar
Cohesion	Hydrogen bonds	Polar solvent
Dipole moment	Hydrophilic	Solubility
Dispersion forces	Hydrophobic	Specific heat
Electrolyte	Immiscible	Strong electrolyte
Enthalpy of fusion	Intermolecular forces	van der Waals forces
Enthalpy of vaporization	Intramolecular bonds	Vapor pressure
Equilibrium	Melting point	Weak electrolyte
Freezing point	Nonelectrolyte	

PROBLEMS

The Structure of Gases, Liquids, and Solids

1. Describe the differences in the properties of gases, liquids, and solids on the atomic scale. Explain how the differences give rise to the observed differences in the macroscopic properties of the three states of matter.

2. Describe the difference between *intermolecular* forces and *intramolecular* bonds, giving examples of each. Which are stronger?

3. Describe the differences between the four forms of van der Waals forces.

4. Propose an explanation for the fact that induced dipole–induced dipole forces increase as the number of electrons on an atom increases.

5. Describe how assuming that there are holes in the structure of a liquid explains the ease with which liquids flow.

6. Explain why liquids are usually less dense than the corresponding solid.

7. Calculate the number of molecules in 1.00 cm^3 of O_2 in the solid, liquid, and gaseous phases. Use the data to check the assumption that 99.9% of the volume of a gas is empty space (see Table 8.1).

8. Explain why salts (such as NaCl) are solids at room temperature.

Relative Intermolecular Forces

9. Predict the order in which the boiling points of the following compounds should increase. Explain your reasoning.
   (a) NH_3 (b) PH_3 (c) AsH_3 (d) SbH_3

10. List the types of intermolecular forces that would be associated with each of the following molecules
   (a) SO_2 (b) CH_3OH (c) ICl_3 (d) SF_4

11. Which compound would you expect to have the highest boiling point? Explain your reasoning.
   (a) methane, CH_4 (b) chloromethane, CH_3Cl
   (c) dichloromethane, CH_2Cl_2 (d) chloroform, $CHCl_3$
   (e) carbon tetrachloride, CCl_4

12. Explain why the boiling points of hydrocarbons that have the generic formula C_nH_{2n+2} increase with molecular weight.

13. Explain why propane (C_3H_8) is a gas but pentane (C_5H_{12}) is a liquid at room temperature.

14. Explain why methane (CH_4) is a liquid only over a very narrow range of temperatures.

15. Explain why *n*-pentane is a liquid over a much larger range of temperatures than neopentane.

Vapor Pressure

16. Explain why it is important to specify the temperature at which the vapor pressure of a liquid is measured.

17. Explain why water eventually evaporates from an open container at room temperature ($\sim 20°C$), even though it normally boils at 100°C.

18. One postulate of the kinetic theory of gases suggests that the temperature of a gas is directly proportional to the *average kinetic energy* of the particles in the gas. Why is the term "average" used?

19. Explain why the vapor pressure of water becomes greater as the temperature of the water increases.

20. Explain why a cloth soaked in water feels cool when placed on your forehead.

21. What would happen to the vapor pressure of liquid bromine, Br_2, at 20°C if the liquid was transferred from a narrow 10-mL graduated cylinder into a wide petri dish or crystallizing dish? Would it increase, decrease, or remain the same? What would happen

to the vapor pressure in a closed container if more liquid were added to the container? Would it increase, decrease, or remain the same?

22. Explain what it means to say that the liquid and vapor in a closed container are in *equilibrium*.

23. Explain why each of the following increases the rate at which water evaporates from an open container.
    (a) increasing the temperature of the water
    (b) increasing the surface area of the water
    (c) blowing air over the surface of the water
    (d) decreasing the atmospheric pressure on the water

24. Why does snow change directly to a gas on a very dry day?

25. Why does a pressure cooker cook food more rapidly than an open cooker?

Melting Point, Freezing Point, and Boiling Point

26. Explain why a "3-minute egg" cooked while camping in the Rocky Mountains does not taste as good as it does when cooked while camping near the Great Lakes.

27. Explain why it takes more time to boil food in Denver than in Salt Lake City.

28. Explain why water boils when the pressure on the system is reduced.

29. At what temperature does water boil when the pressure is 50 mmHg? Use Figure 8.12.

30. What pressure has to be achieved before water can boil at 20°C? Use Figure 8.12.

31. Increasing the temperature of a liquid will do which of the following?
    (a) increase the boiling point
    (b) increase the melting point
    (c) increase the vapor pressure
    (d) increase the amount of heat required to boil a mole of the liquid
    (e) all of the above

32. Liquid air is composed primarily of liquid oxygen (BP = −183°C) and liquid nitrogen (BP = −196°C). It can be separated into its component gases by increasing the temperature until one of the gases boils off. Which gas boils off first?

33. Predict which of the following liquids should have the lowest boiling point from their vapor pressures (VP) at 0°C.
    (a) acetone, VP = 67 mmHg   (b) benzene, VP = 24.5 mmHg
    (c) ether, VP = 183 mmHg   (d) methyl alcohol, VP = 30 mmHg
    (e) water, VP = 4.6 mmHg

34. According to the data in Table 8.4, butane (C_4H_{10}) should be a gas at room temperature (BP = −0.5°C). Use this to explain why you can hear a gas escape when a can of butane lighter fluid is opened with the nozzle pointed up. If you shake one of the cans, however, you can hear a liquid bounce against either end of the can. Furthermore, when you open the can with the nozzle pointed down, you can see a liquid escape. Explain how butane is stored as a liquid in cans at room temperature.

Surface Tension

35. The force of cohesion between mercury atoms is much larger than the force of cohesion between water molecules. Conversely, the attractive force between water molecules and glass is much larger than the attractive force between mercury atoms and glass. Use these observations to explain why mercury forms small drops when it spills on glass rather than a single large puddle such as water.

36. Describe how the surface tension of water can be used to explain the fact that a steel sewing needle floats on the surface of water.

37. If a steel sewing needle floats on the surface of water, why does a steel ball bearing sink to the bottom?

38. Explain why a drop of water seems to spread out on the surface of a car that has not been waxed for many years but beads up on the surface of a freshly waxed car.

39. Explain the advantages and disadvantages in a plant of having leaves that have a large surface area. Explain why plants coat the surfaces of leaves with wax. Explain why plants that grow in arid climates seldom have leaves as broad as those found on maple trees.

Hydrogen Bonding and the Anomalous Properties of Water

40. Explain why the boiling point and melting point of water are much higher than you would expect from the boiling points and melting points of H_2S, H_2Se, and H_2Te.

41. Explain why hydrogen bonding is very strong in HF and H_2O. Explain why hydrogen bonding is much weaker in HCl and H_2S.

42. What makes a compound a good hydrogen bond donor? What makes a compound a good hydrogen bond acceptor?

43. Explain why the strength of hydrogen bonds decreases in the following order: HF $> H_2O > NH_3$.

44. Explain why water has an unusually large heat capacity.

Solutions: Like Dissolves Like

45. Use a drawing to describe what happens when I_2 molecules dissolve in CCl_4 and when $KMnO_4$ dissolves in water.

46. One way of screening potential anesthetics involves testing whether the compound dissolves in olive oil, because all common anesthetics, including nitrous oxide (N_2O), cyclopropane (C_3H_6), and halothane (C_2HF_3ClBr), are soluble in olive oil. What property do the compounds have in common?

47. Carboxylic acids with the general formula $CH_3(CH_2)_nCO_2H$ have a nonpolar CH_3CH_2 . . . tail and a polar . . . CO_2H head. What effect does increasing the value of n have on the solubility of carboxylic acids in polar solvents, such as water? What is the effect on their solubility in nonpolar solvents, such as CCl_4?

48. Which of the following compounds would be the most soluble in a nonpolar solvent, such as CCl_4?
(a) H_2O (b) CH_3OH (c) $CH_3CH_2CH_2OH$ (d) $CH_3CH_2CH_2CH_2CH_2OH$
(e) $CH_3CH_2CH_2CH_2CH_2CH_2CH_2OH$

49. Potassium iodide reacts with iodine in aqueous solution to form the triiodide ion, I_3^-.

$$KI(aq) + I_2(aq) \longrightarrow KI_3(aq)$$

What would happen if we added CCl_4 to the reaction mixture?
(a) The KI and KI_3 would dissolve in the CCl_4 layer.
(b) The I_2 would dissolve in the CCl_4 layer.
(c) Both KI and I_2, but not KI_3, would dissolve in the CCl_4 layer.
(d) Neither KI, KI_3, nor I_2 would dissolve in the CCl_4 layer.
(e) No distinct CCl_4 layer would form because CCl_4 is soluble in water.

50. Phosphorus pentachloride can react with itself in a reversible reaction to form a salt that contains the PCl_4^+ and PCl_6^- ions.

$$2\ PCl_5 \rightleftharpoons PCl_4^+ + PCl_6^-$$

The extent to which the reaction occurs depends on the solvent in which it is run. Predict whether a nonpolar solvent, such as CCl_4, favors the products or the reactants of the reaction. Predict what would happen to the reaction if we used a polar solvent, such as acetonitrile (CH_3CN).

Why Do Some Solids Dissolve in Water?

51. Explain why some molecular solids are soluble in water but others are not.
52. Explain why $BaCl_2$ is soluble in water but AgCl is not.
53. Use an example to explain what is meant by the general rule that salts dissociate when they dissolve in water.

Solubility Equilibria

54. Explain why the lightbulb in the conductivity apparatus in Figure 8.27 glows more brightly when the wires are immersed in a solution of NaCl than when the wires are immersed in tap water.
55. Explain why the addition of a few small crystals of silver chloride makes water a slightly better conductor of electricity. Explain why the conductivity gradually increases as more AgCl is added, until it eventually reaches a maximum. Describe what is happening in the solution when its conductivity reaches the maximum.

Solubility Rules

56. Which of the following salts are *insoluble* in water?
    (a) $Ba(NO_3)_2$ (b) $BaCl_2$ (c) $BaCO_3$ (d) BaS   (e) $BaC_2H_3O_2$
57. Which of the following salts are *insoluble* in water?
    (a) $(NH_4)_2SO_4$ (b) K_2CrO_4 (c) Na_2S (d) $Pb(NO_3)_2$ (e) $Cr(OH)_3$
58. Which of the following salts are *soluble* in water?
    (a) PbS   (b) PbO   (c) $PbCrO_4$ (d) $PbCO_3$ (e) $Pb(NO_3)_2$

Integrated Problems

59. Young children often believe that water decomposes into its elements when it boils. How would you explain to a bright middle-school student the difference between the bonds that are broken when NaCl or diamond boils and the forces broken when water boils? How would you explain to the same student the difference between the boiling points of NaCl (BP = 1465°C), diamond (BP = 4827°C), and water (BP = 100°C)?
60. Because they are isomers, ethanol (CH_3CH_2OH) and dimethyl ether (CH_3OCH_3) have the same molecular weight. Explain why ethanol (BP = 78.5°C) has a much higher boiling point than dimethyl ether (BP = −23.6°C).
61. The following compounds have the same molecular weight. Explain why one of the compounds has a higher boiling point than the other.

$$CH_3NCH_3 \qquad\qquad CH_3CH_2NH$$
$$\vert \qquad\qquad\qquad\qquad\quad \vert$$
$$CH_3 \qquad\qquad\qquad\quad CH_3$$

BP = 3°C   *BP = 35°C*

62. You like a really hot cup of tea. Would you prefer to live in Denver or in Salt Lake City if hot tea is your only consideration? Why?

63. A thermometer is taped to the outside of an open flask filled with water. The water is heated to boiling and allowed to continue to boil until the water is gone. Sketch a rough plot of how the temperature would change with respect to time for water initially at 25°C. Suppose the same setup is used except that no heat is supplied to the open flask which is initially at 25°C. If you come back some time later all of the water will have evaporated. How do you think the temperature recorded by the thermometer would have changed with respect to time? What is the difference between the phase change of water at 100°C and at 25°C?

64. Which illustration most correctly describes the boiling of water at 100°C?

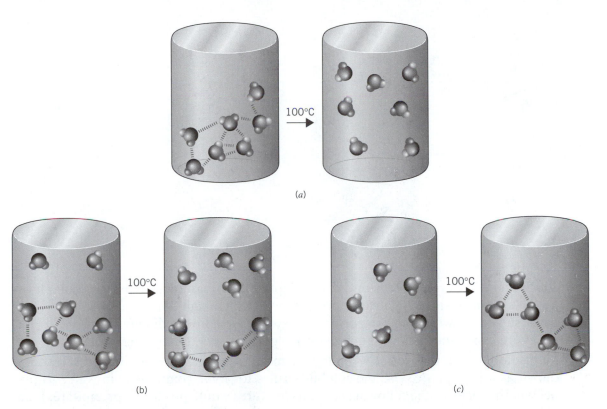

65. For each pair of the following substances pick the one with the higest melting point and explain your reasoning.

(a)

$$\begin{array}{ccc} & \text{H} \quad \text{H} & \text{H} \quad \text{H} \\ & | \quad\quad | & | \quad\quad | \\ \text{H}-\text{C}-\text{O}-\text{C}-\text{H}, & \text{H}-\text{C}-\text{C}-\text{O}-\text{H} \\ & | \quad\quad | & | \quad\quad | \\ & \text{H} \quad \text{H} & \text{H} \quad \text{H} \end{array}$$

(b) CCl_4, CBr_4
(c) LiCl, CsCl
(d) CCl_4, NaCl

You have available two solvents, H_2O and hexane (C_6H_{14}). Select one compound from above that should be soluble in water and explain why. Select one compound from above that should be soluble in hexane and explain why.

66. Which of the following solutions contains the largest number of Na^+ ions, 500 mL of a 1.5 M solution of Na_2CO_3 or 1.00 L of a 0.75 M solution of NaCl?

67. Compare and contrast (what is the same and what is different) boiling and evaporation on both the macroscopic and molecular scales.

68. Compare the enthalpy change associated with boiling one mole of water (converting one mole of water into steam at 100°C) to evaporating one mole of water (converting one mole of water to water vapor at room temperature).

69. The diagram below shows liquid methanol, CH₃OH, being prepared to be poured into a beaker of water. Draw a figure that illustrates the solution that will result. Clearly show the intermolecular forces. Describe the intermolecular forces that are broken in the solute and solvent and formed in the solution. Do you expect methanol to be very soluble in water?

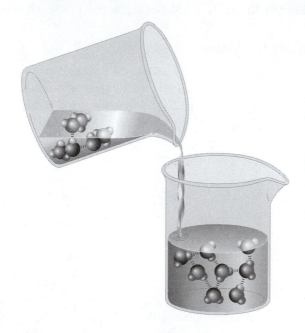

70. Which of the following statements is true? Explain your reasoning on the molecular level.
    (a) A substance with a high vapor pressure will have a high boiling point.
    (b) A substance with a high vapor pressure will have a low boiling point.
    (c) There can be no relationship between vapor pressure and boiling point.

71. An insoluble solid that forms when two or more soluble species are mixed together is known as a precipitate. Use the solubility rules for ionic compounds in water to predict whether each of the following aqueous mixtures would produce a precipitate. Identify the precipitate in cases where one is formed. (Hint: Begin by determining what compounds might be formed by mixing the solutions together.)
    (a) $FeCl_3$ and KOH
    (b) Na_2CO_3 and $(NH_4)_2SO_4$
    (c) $Pb(NO_3)_2$ and NaCl

72. Liquid benzene, C_6H_6, has a vapor pressure of 325 mmHg at 80°C. If 1.00 g of benzene is placed in an evacuated 1.00-L flask at 80°C, determine the mass of benzene that will evaporate and the final pressure in the container. Determine the same variables if the experiment is repeated using a 500-mL evacuated container. Explain what is different about the two experiments.

73. Use the data from Table 8.4 to answer the following questions.
    (a) 1-Pentanol is an alcohol with the molecular formula $CH_3CH_2CH_2CH_2CH_2OH$ and a molecular mass of 88.2 g/mol. How would you expect the boiling point of pentanol to be related to the boiling points of butanoic acid, pentanal, and hexane? Explain your reasoning.
    (b) Predict the boiling point for nonanal, $CH_3(CH_2)_8$ OH. Explain your reasoning.

74. A sealed container equipped with a movable piston contains a liquid in equilibrium with its vapor. A pressure gauge is attached to the container to measure the pressure of the vapor. Explain what is happening on the molecular level for each of the following experimental observations.

   (a) A valve on the container is opened and some of the gas is allowed to escape. The pressure is observed to immediately drop when the valve is opened. When the valve is closed, however, the pressure is observed to increase and then stabilize. (The volume of the container and the temperature remain constant.)

   (b) The movable piston on the container is depressed, causing the volume to decrease. The pressure increases as the piston moves in but then stabilizes at a constant pressure even as the piston continues to be depressed. (The temperature of the container remains constant.)

75. Octane, dibutyl ether, and 1-octanol have the following structures.

$$CH_3CH_2CH_2CH_2CH_2CH_2CH_2CH_3$$
Octane

$$CH_3CH_2CH_2CH_2-O-CH_2CH_2CH_2CH_3$$
Dibutyl ether

$$CH_3CH_2CH_2CH_2CH_2CH_2CH_2CH_2OH$$
1-Octanol

   (a) Arrange the three compounds in order of increasing vapor pressure.

   (b) Arrange the three compounds in order of decreasing boiling point.

   (c) The molar heat capacity of octane is 188 J/mol·K and that of dibutyl ether is 205 J/mol·K. Account for the difference in molar heat capacity.

   (d) If one mole of octane at 50°C is mixed with one mole of dibutyl ether at 20°C will the final temperature be 35°C? Will one compound lose the same amount of heat as the other gains? Will the temperature of octane decrease or increase? Will the temperature of dibutyl ether decrease or increase? What will be the relationship between the temperature increase or decrease of octane and dibutyl ether?

76. Which of the following properties would you expect to generally increase as temperature increases: vapor pressure, surface tension, and molar heat capacity? Explain.

77. The molar heat capacity of liquid water at room temperature is about 75 J/mol·K. What would you expect the molar heat capacity of gaseous water to be at room temperature? Explain.

78. The specific heat and the molar heat capacity of gaseous H_2S are less than that of gaseous H_2O at room temperature. How can you explain the difference?

79. Two molecular compounds have the same molecular weight, but one boils at 195°C and the other at 142°C. What factors can account for the difference in boiling point? Which compound has the lowest vapor pressure? Do either or both of the compounds have a dipole moment? How would you expect the molar heat capacities to compare?

CHAPTER
8
SPECIAL TOPICS

8A.1   COLLIGATIVE PROPERTIES

Dissolving a solute in a solvent results in a solution with physical properties that are different from the pure solute or pure solvent. In this section we look at some of the physical properties of solutions. **Colligative properties** are physical properties of solutions that depend on the number of solute particles in a solution but not on the identity of the solute particles. This means that two different solutions composed of different solutes but having the same solvent and the same concentrations of solute particles would exhibit the same colligative properties.

To begin our discussion of colligative properties we must introduce a new unit for measuring concentration, **mole fraction.** The ratio of the amount of solute to solvent in a solution can be described in terms of the mole fraction of the solute or the solvent in a solution. By definition, the mole fraction of any component of a solution is the number of moles of that component divided by the total number of moles of solute and solvent. The symbol for mole fraction is the Greek letter chi, χ. The mole fraction of the *solute* is defined as the number of moles of solute divided by the total number of moles of solute and solvent.

$$\chi_{solute} = \frac{\text{moles of solute}}{\text{moles of solute} + \text{moles of solvent}}$$

Conversely, the mole fraction of the *solvent* is the number of moles of solvent divided by the total number of moles of solute and solvent.

$$\chi_{solvent} = \frac{\text{moles of solvent}}{\text{moles of solute} + \text{moles of solvent}}$$

In a solution that contains a single solute dissolved in a solvent, the sum of the mole fractions of solute and solvent must be equal to 1.

$$\chi_{solute} + \chi_{solvent} = 1$$

Exercise 8A.1

A saturated solution of hydrogen sulfide in water can be prepared by bubbling H_2S gas into water until no more dissolves. Calculate the mole fraction of both H_2S and H_2O in the solution if 0.385 gram of H_2S gas dissolves in 100 grams of water at 20°C and 1 atm.

Solution

The number of moles of solute in the solution can be obtained.

$$0.385 \text{ g } H_2S \times \frac{1 \text{ mol } H_2S}{34.08 \text{ g } H_2S} = 0.0113 \text{ mol } H_2S$$

To determine the mole fraction of the solute and solvent we also need to know the number of moles of water in the solution.

$$100 \text{ g } H_2O \times \frac{1 \text{ mol } H_2O}{18.02 \text{ g } H_2O} = 5.55 \text{ mol } H_2O$$

The mole fraction of the solute is the number of moles of H_2S divided by the total number of moles of both H_2S and H_2O.

$$\chi_{\text{solute}} = \frac{0.0113 \text{ mol } H_2S}{0.0113 \text{ mol } H_2S + 5.55 \text{ mol } H_2O} = 0.00203$$

The mole fraction of the solvent is the number of moles of H_2O divided by the moles of both H_2S and H_2O.

$$\chi_{\text{solvent}} = \frac{5.55 \text{ mol } H_2O}{0.0113 \text{ mol } H_2S + 5.55 \text{ mol } H_2O} = 0.998$$

Note that the sum of the mole fractions of the two components of the solution is 1.

$$\chi_{\text{solute}} + \chi_{\text{solvent}} = 0.00203 + 0.998 = 1.000$$

We saw in Chapter 6 that the ideal gas law was valid only for ideal gases. In the same way the equations that we'll use to describe the colligative properties of solutions are valid only for **ideal solutions.** An ideal solution is one in which the forces that hold the solute particles together are equal to the forces that hold the solvent particles together and are also equal to the forces of attraction between the solute and solvent particles in the solution. Therefore, the forces of attraction that must be broken in the pure solute and pure solvent are equal to the forces formed between the solute and solvent in the solution. This results in a change in enthalpy of zero for the solution process (forces broken = forces formed). This is not the case for most real solutions. As a result the equations that we develop to describe colligative properties will give good approximations for real-world solutions but will not be exact.

8A.2   DEPRESSION OF THE PARTIAL PRESSURE OF A SOLVENT

One of the properties of liquids that we discussed in Chapter 8 was vapor pressure, the pressure of a vapor in equilibrium with its liquid. When a solute is added to a liquid solvent there is a decrease in the pressure exerted by the vapor of the solvent above the solution.[3] We'll define $P°$ as the vapor pressure of the pure liquid—the solvent—and P as the pressure exerted by the vapor of the solvent over a solution.

[3]To explain the observed decrease in vapor pressure requires the introduction of a new term, entropy. A solution has a higher entropy, or is more disordered, than the pure solvent. Entropy is discussed more fully in Chapter 13.

$$P° = \text{vapor pressure of the pure liquid, or solvent}$$
$$P = \text{partial pressure of the solvent in a solution}$$
$$P < P°$$

Partial pressure	*Vapor pressure*
of the solvent	*above the*
above a solution	*pure solvent*

Between 1887 and 1888, Francois-Marie Raoult showed that the pressure of the solvent of a solution is equal to the mole fraction of the solvent times the vapor pressure of the pure solvent. This equation is known as **Raoult's law.**

$$P = \chi_{\text{solvent}}P°$$

Partial pressure	*Vapor pressure*
of the solvent	*above the*
above a solution	*pure solvent*

When the solvent is pure, and the mole fraction of the solvent is equal to 1, P is equal to $P°$. As the mole fraction of the solvent becomes smaller, the partial pressure of the solvent over the solution also becomes lower.

Let's assume, for the moment, that the solvent is the only component of the solution that is volatile enough to have a measurable vapor pressure. This would be true, for example, of a solution of table salt, NaCl, dissolved in water. The partial pressure of the solution will be equal to the pressure produced by the solvent escaping from the solution. Raoult's law states that the difference between the vapor pressure of the pure solvent and the partial pressure over the solution increases as the mole fraction of the solvent decreases.

Note that the vapor pressure of a solvent is not a colligative property. Only the *change in the vapor pressure* that occurs when a solute is added to the solvent can be included among the colligative properties of a solution.

When a solute is added to a pure solvent, the change in the partial pressure of the solvent above the solution is the difference between the vapor pressure of the pure solvent and the partial pressure exerted by the solvent above the solution.

$$\Delta P = P° - P$$

Substituting Raoult's law into the above equation gives the following result.

$$\Delta P = P° - \chi_{\text{solvent}}P° = (1 - \chi_{\text{solvent}})P°$$

The equation can be simplified by remembering the relationship between the mole fraction of the solute and the mole fraction of the solvent.

$$\chi_{\text{solute}} + \chi_{\text{solvent}} = 1$$

Substituting the relationship into the equation that defines ΔP gives another form of Raoult's law.

$$\Delta P = \chi_{\text{solute}}P°$$

This equation reminds us that, for an ideal solution, as more solute is dissolved in the solvent, the change in pressure, ΔP, increases.

One of the consequences of vapor pressure depression of solvents by solutes can be seen in Figure 8A.1. Figure 8A.1 shows a beaker containing pure solvent and a second beaker

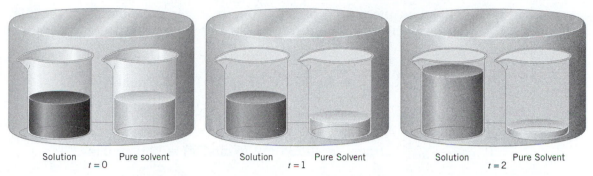

Solution   Pure solvent $t = 0$ Solution   Pure Solvent $t = 1$ Solution   Pure Solvent $t = 2$

FIGURE 8A.1 As evaporation takes place the sealed container fills with the vapor of the solvent. Due to the difference in the partial pressures of the solvent over the pure solvent and over the solution, the beaker containing the pure solvent evaporates while vapor condenses into the beaker containing the solution.

containing a solution with a nonvolatile solute. Both beakers are inside a sealed container. The pure solvent and the solvent in the solution begin to evaporate, attempting to establish equilibrium between the liquid and the vapor phase. However, the vapor pressure associated with the pure solvent will be greater than the pressure exerted by the solvent in the solution. The pressure of the vapor of the pure solvent will exceed the pressure of the solvent in the solution and will lead to condensation of the vapor at the surface of the solution. As time passes the solvent will evaporate from the beaker of pure solvent and will condense in the beaker containing the solution. In general, a liquid will move from an area where the partial pressure of its vapor is high to an area where the partial pressure of its vapor is low.

Solutions that contain a *volatile* solute will have not only a partial pressure associated with the solvent but also one associated with the solute.

Exercise 8A.2

A solution is prepared by mixing 500 mL of ethanol (C_2H_6O) and 500 mL of water at 25°C. The vapor pressures of pure water and pure ethanol at that temperature are 23.76 and 59.76 mmHg, respectively. The densities of water and ethanol at 25°C are 0.9971 and 0.786 g/mL, respectively. Determine the partial pressure due to each component of the solution and the total pressure.

Solution

We begin by determining the number of moles of each of the components of the solution.

$$500 \text{ mL H}_2\text{O} \times \frac{0.9971 \text{ g H}_2\text{O}}{1 \text{ mL H}_2\text{O}} \times \frac{1 \text{ mol H}_2\text{O}}{18.02 \text{ g H}_2\text{O}} = 27.7 \text{ mol H}_2\text{O}$$

$$500 \text{ mL C}_2\text{H}_6\text{O} \times \frac{0.786 \text{ g C}_2\text{H}_6\text{O}}{1 \text{ mL C}_2\text{H}_6\text{O}} \times \frac{1 \text{ mol C}_2\text{H}_6\text{O}}{46.07 \text{ g C}_2\text{H}_6\text{O}} = 8.53 \text{ mol C}_2\text{H}_6\text{O}$$

We then calculate the mole fractions of the two components of the solution.

$$\chi_{\text{water}} = \frac{27.7 \text{ mol}}{27.7 \text{ mol} + 8.53 \text{ mol}} = 0.765$$

$$\chi_{\text{ethanol}} = \frac{8.53 \text{ mol}}{27.7 \text{ mol} + 8.53 \text{ mol}} = 0.235$$

According to Raoult's law, the partial pressure due to water escaping from the solution is equal to the product of the mole fraction of water and the vapor pressure of pure water.

$$P_{water} = \chi_{water}P^\circ_{water} = (0.765)(23.76 \text{ mmHg}) = 18.2 \text{ mmHg}$$

The partial pressure of ethanol is found in the same fashion.

$$P_{ethanol} = \chi_{ethanol}P^\circ_{ethanol} = (0.235)(59.76 \text{ mmHg}) = 14.0 \text{ mmHg}$$

The total pressure of the gases escaping from the solution is the sum of the partial pressures of the two gases.

$$P_T = P_{water} + P_{ethanol} = 32.2 \text{ mmHg}$$

Although the solution is a 50:50 mixture by volume, slightly more than three-quarters of the particles in the solution are water molecules. As a result, the total pressure of the solution more closely resembles the vapor pressure of pure water than it does pure ethanol. In addition, the change in partial pressure associated with the ethanol is much greater than the change for water.

8A.3   BOILING POINT ELEVATION

Chapter 8 described how a pure liquid could be boiled by heating the liquid to increase its vapor pressure until the vapor pressure became equal to the pressure pushing down on the surface of the liquid. If a nonvolatile solute is added to the pure liquid, the resulting pressure due to the vapor of the solvent above the solution will be reduced. This means that it will be necessary to increase the temperature even more in order to increase the pressure of the solvent's vapor until it becomes equal to the pressure pushing down on the surface of the solution. Therefore, the boiling point of a solution will be greater than the boiling point of the pure solvent. This is known as boiling point elevation.

Because changes in the boiling point of the solvent (ΔT_{BP}) that occur when a solute is added result from changes in the partial pressure of the solvent, the magnitude of the change in the boiling point is also proportional to the mole fraction of the solute.

In very dilute solutions, the mole fraction of the solute is proportional to the molality of the solution. Molality, m, is defined as the number of moles of solute per kilogram of solvent.

$$m = \frac{\text{mol of solute}}{\text{kg of solvent}}$$

Molality is similar to molarity except the denominator is *kilograms* of *solvent* rather than *liters of solution*. Molality has an important advantage over molarity. The molarity of an aqueous solution changes with temperature, because the density of water is sensitive to temperature. The molality of a solution does not change with temperature because molality is defined in terms of the mass of the solvent, not its volume.

The equation that describes the magnitude of the boiling point elevation that occurs when a solute is added to a solvent can be written as follows.

$$\Delta T_{BP} = k_b m$$

Here, ΔT_{BP} is the **boiling point elevation,** that is, the change in boiling point that occurs when a solute dissolves in the solvent, and k_b is a proportionality constant known as the *molal boiling point elevation constant* for the solvent. Molal boiling point elevation constants for selected compounds are given in Table 8A.1.

TABLE 8A.1   Freezing Point Depression and Boiling Point Elevation Constants

Compound	Freezing Point (°C)	k_f (°C/m)
Water	0	1.853
Acetic acid	16.66	3.90
Benzene	5.53	5.12
p-Xylene	13.26	4.3
Naphthalene	80.29	6.94
Cyclohexane	6.54	20.0
Carbon tetrachloride	−22.95	29.8
Camphor	179.8	40

Compound	Boiling Point (°C)	k_b (°C/m)
Water	100	0.515
Ethyl ether	34.55	2.02
Carbon disulfide	46.23	2.35
Benzene	80.10	2.53
Carbon tetrachloride	76.75	5.03
Camphor	207.42	5.95

Because colligative properties are dependent only on the number of particles in solution, they can be used in the laboratory for the determination of molecular weights.

Exercise 8A.3

How do we know that the most common form of sulfur consists of S_8 molecules? Calculate the molecular weight of sulfur if 35.5 grams of sulfur dissolves in 100 grams of CS_2 to produce a solution that has a boiling point of 49.48°C.

Solution

The relationship between the boiling point of the solution and the molecular weight of sulfur is not immediately obvious. We therefore start by asking, What do we know about the problem? We might start by drawing a figure, such as Figure 8A.2, that helps us organize the information in the problem.

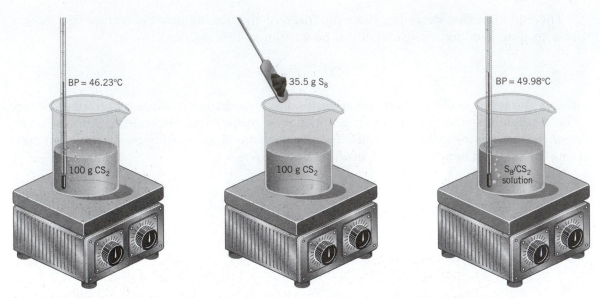

FIGURE 8A.2 Diagram for Exercise 8A.3, where 35.5 g of S_8 is dissolved in 100 g CS_2.

We know the boiling point of the solution, so we might start by looking up the boiling point of the pure solvent in order to calculate the change in the boiling point that occurs when the sulfur is dissolved in CS_2.

$$\Delta T_{BP} = 49.48°C - 46.23°C = 3.25°C$$

We also know that the change in the boiling point is proportional to the molality of the solution.

$$\Delta T_{BP} = k_b m$$

Since we know the change in the boiling point (ΔT_{BP}) and because the boiling point elevation constant for the solvent (k_b) can be found in a table, we might decide to calculate the molality of the solution at this point.

$$m = \frac{\Delta T_{BP}}{k_b} = \frac{3.25°C}{2.35°C/m} = 1.38 \ m$$

In the search for the solution to a problem, it is useful periodically to consider what we have achieved so far. At this point, we know the molality of the solution and the mass of the solvent used to prepare the solution. We can therefore calculate the number of moles of sulfur present in the carbon disulfide solution.

$$\frac{1.38 \ \text{mol sulfur}}{1000 \ \text{g } CS_2} \times 100.0 \ \text{g } CS_2 = 0.138 \ \text{mol sulfur}$$

We now know the number of moles of sulfur in the solution and the mass of the sulfur. We can therefore calculate the number of grams per mole of sulfur.

$$\frac{35.5 \ \text{g}}{0.138 \ \text{mol}} = 257 \ \text{g/mol}$$

From the periodic table we know the atomic mass of sulfur is 32.07 g/mol of sulfur atoms. Dividing this into the mass of a mole of sulfur molecules tells us that a molecule of sulfur must contain 8 sulfur atoms.

$$\frac{257 \text{ g/mol}}{32.07 \text{ g/mol}} = 8.01 \text{ sulfur atoms per molecule}$$

8A.4 FREEZING POINT DEPRESSION

It is observed that the addition of a solute to a solvent will lower the freezing point of the solution below that of the pure solvent.

An example of the utility of freezing point depression is the addition of salt to ice to decrease the temperature of the melting ice when making homemade ice cream. The salting of highways to prevent the formation of ice during the winter is another example.

An equation, similar to the boiling point elevation equation, can be written to describe what happens to the freezing point (or melting point) of a solvent when a solute is added to the solvent.

$$\Delta T_{FP} = -k_f m$$

In the equation, ΔT_{FP} is the **freezing point depression,** that is, the change in freezing point that occurs when the solute dissolves in the solvent, and k_f is the *molal freezing point depression constant* for the solvent. A negative sign is used in the equation to indicate that the freezing point of the solvent decreases when a solute is added.

Exercise 8A.4

Determine the molecular weight of acetic acid if a solution containing 30.0 grams of acetic acid in 1000 g of water freezes at $-0.93°C$. Do the results agree with the assumption that acetic acid has the formula CH_3CO_2H?

Solution

The freezing point depression for the solution is equal to the difference between the freezing point of the solution ($-0.93°C$) and the freezing point of pure water ($0°C$).

$$\Delta T_{FP} = -0.93°C - 0.0°C = -0.93°C$$

We now turn to the equation that defines the relationship between freezing point depression and the molality of the solution.

$$\Delta T_{FP} = -k_f m$$

Because we know the change in the freezing point, and because we can find the freezing point depression constant for water in Table 8A.1, we have enough information to calculate the molality of the solution.

$$m = -\frac{\Delta T_{FP}}{k_f} = \frac{0.93°C}{1.853°C/m} = 0.50 \text{ } m$$

At this point, we might return to the statement of the problem, to see if we are making any progress toward an answer. According to the calculation, there are 0.50 moles of acetic acid per kilogram of solvent. The problem stated that there were 30.0 grams of acetic acid per 1000 g of solvent. Because we simultaneously know the number of grams and the number of moles of acetic acid in the sample, we can calculate the molecular weight of acetic acid.

$$\frac{30.0 \text{ g}}{0.50 \text{ mol}} = 60 \text{ g/mol}$$

The results of the experiment are in good agreement with the molecular weight (60.05 g/mol) expected if the formula for acetic acid is CH_3CO_2H.

Exercise 8A.5

Explain why a 0.100 m solution of HCl dissolved in benzene has a freezing point depression of 0.512°C, whereas a 0.100 m solution of HCl in water has a freezing point depression of 0.369°C.

Solution

We can predict the change in the freezing point that should occur in the solutions from the freezing point depression constant for the solvent and the molality of the solution. For benzene, the results of the calculation agree with the experimental value.

$$\Delta T_{FP} = -k_f m = -(5.12°C/m)(0.100 \; m) = -0.512°C$$

For water, however, the calculation gives a predicted value for the freezing point depression that is half of the observed value.

$$\Delta T_{FP} = -k_f m = -(1.853°C/m)(0.100 \; m) = -0.185°C$$

To explain the results, it is important to remember that colligative properties depend on the relative number of solute particles in a solution, not their identity. If the acid dissociates (breaks up into its ions) to an appreciable extent, the solution will contain more solute particles than we might expect from its molality.

If HCl dissociates completely in water, the total concentration of solute particles (H^+ and Cl^- ions) in the solution will be twice as large as the molality of the solution. The freezing point depression for the solution therefore will be twice as large as the change that would be observed if HCl did not dissociate.

$$HCl(g) \xrightarrow{H_2O} H^+(aq) + Cl^-(aq)$$

If we assume that 0.100 m HCl dissociates to form H^+ and Cl^- ions in water, the freezing point depression for the solution should be −0.371°C, which is slightly larger than what is observed experimentally.

$$\Delta T_{FP} = -k_f m = -(1.853°C/m)(2 \times 0.100 \; m) = -0.371°C$$

This exercise suggests that HCl does not dissociate into ions when it dissolves in benzene, but dilute solutions of HCl dissociate more or less quantitatively in water.

In 1884 Jacobus Henricus van't Hoff introduced another term into the freezing point depression and boiling point elevation expressions to explain the colligative properties of solutions of compounds that dissociate when they dissolve in water.

$$\Delta T_{FP} = -k_f(i)m$$

Substituting the experimental value for the freezing point depression of a 0.100 m HCl solution into the equation gives a value for the i term of 1.99. If HCl did not dissociate in water, i would be 1. If it dissociates completely, i would be 2. The experimental value of 1.99 suggests that about 99% of the HCl molecules dissociate in the solution.

PROBLEMS

Vapor Pressure Depression

8A-1. Predict what will happen to the rate at which water evaporates from an open flask when salt is dissolved in the water, and explain why the rate of evaporation changes.

8A-2. If you place a beaker of pure water (I) and a beaker of a saturated solution of sugar in water (II) in a sealed bell jar, the level of water in beaker I will slowly decrease, and the level of the sugar solution in beaker II will slowly increase. Explain why.

8A-3. Explain why the vapor pressure of a liquid at a particular temperature is not a colligative property but the change in the partial pressure of the liquid when a solute is added is a colligative property.

Boiling Point Elevation and Freezing Point Depression

8A-4. Explain how the decrease in the vapor pressure of a solvent that occurs when a solute is added to the solvent leads to an increase in the solvent's boiling point.

8A-5. What is the change in the vapor pressure of a solvent that occurs when a solute is added?

8A-6. Predict the shape of a plot of the freezing point of a solution versus the molality of the solution.

8A-7. Explain why salt is added to the ice that surrounds the container in which ice cream is made.

8A-8. Explain why many cities and states spread salt on icy highways.

8A-9. Compare the values of k_f and k_b for water, benzene, carbon tetrachloride, and camphor. Explain why measurements of molecular weight based on freezing point depression might be more accurate than those based on boiling point elevation.

Colligative Properties: Calculations

8A-10. What is the approximate freezing point of a saturated solution of caffeine ($C_8H_{10}O_2N_4 \cdot H_2O$) in water if it takes 45.6 grams of water to dissolve 1.00 gram of caffeine?

8A-11. A 0.100 m solution of sulfuric acid in water freezes at $-0.371°C$. Which of the following statements is consistent with this observation?
(a) H_2SO_4 does not dissociate in water.
(b) H_2SO_4 dissociates into H^+ and HSO_4^- ions in water.

(c) H_2SO_4 dissociates in water to form two H^+ ions and one SO_4^{2-} ion.

(d) H_2SO_4 associates in water to form $(H_2SO_4)_2$ molecules.

8A-12. The "Tip of the Week" in a local newspaper suggested using a fertilizer such as ammonium nitrate or ammonium sulfate instead of salt to melt snow and ice on sidewalks, because salt can damage lawns. Which of the following compounds would give the largest freezing point depression when 100 grams is dissolved in 1 kilogram of water?

(a) NaCl   (b) NH_4NO_3 (c) $(NH_4)_2SO_4$

8A-13. *p*-Dichlorobenzene (PDCB) is replacing naphthalene as the active ingredient in mothballs. Calculate the value of k_f for camphor if a 0.260 *m* solution of PDCB in camphor decreases the freezing point of camphor by 9.8°C.

8A-14. If an aqueous solution boils at 100.50°C, at what temperature does it freeze?

8A-15. What is the boiling point of a solution of 10.0 grams of P_4 in 25.0 g of carbon disulfide?

8A-16. We usually assume that salts such as KCl dissociate completely when they dissolve in water.

$$KCl(s) \xrightarrow{H_2O} K^+(aq) + Cl^-(aq)$$

Estimate the percentage of the KCl that actually dissociates in water if the freezing point of a 0.100 *m* solution of the salt in water is $-0.345°C$.

8A-17. Calculate the freezing point of a 0.100 *m* solution of acetic acid in water if the CH_3CO_2H molecules are 1.33% ionized in the solution.

CHAPTER
9
SOLIDS

9.1   SOLIDS

Solids can be divided into three categories on the basis of the way that the particles pack together. **Crystalline solids** are three-dimensional analogs of a brick wall. They have a regular structure in which particles pack in a repeating pattern, row to row and layer to layer, from one edge of the solid to the other. **Amorphous solids** (literally, "solids without form") have a random structure with little if any long-range order, for example, glass. Many solids, such as aluminum and steel, have a structure that falls between the two extremes. Such **polycrystalline solids** are aggregates of large numbers of small crystals or grains in which the structure is regular, but the crystals or grains are arranged in a random fashion.

Solids can also be classified on the basis of the forces that hold the particles together. This approach categorizes solids as either molecular, network covalent, ionic, or metallic. As shown in Chapter 5, a bond type triangle may be used to show how electronegativity and electronegativity differences can be used to determine whether the type of bonding between atoms is ionic, metallic, or covalent. The triangle shown in Figure 9.1 illustrates that the bonding in elemental cesium is metallic, cesium fluoride contains ionic bonds, and elemental fluorine is covalently bonded. Ionic and covalent bonds are often imagined as if

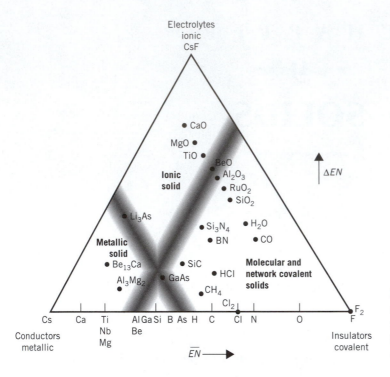

FIGURE 9.1 Bond type triangle for selected compounds. Such triangles may be used to classify solids as molecular, network covalent, ionic, or metallic.

they were opposite ends of a two-dimensional model of bonding in which compounds that contain polar bonds fall somewhere between the two extremes.

Ionic . . . polar . . . covalent

In reality, there are three kinds of bonds between adjacent atoms: ionic, covalent, and metallic. Nonmetals combine to form elements and compounds that contain primarily covalent bonds, such as F_2, HCl, and CH_4. Metals combine with nonmetals to form salts, such as CsF and CaO, which are held together by predominately ionic bonds. The force of attraction between atoms in metals, such as copper, or between the atoms in alloys, such as brass and bronze, or in intermolecular compounds, such as Li_3As, are mainly metallic bonds. The type of bonding in a compound can be used to classify the solid; thus, if a compound can be located on a bond type triangle the properties of the solid can be predicted.

9.2   MOLECULAR AND NETWORK COVALENT SOLIDS

Molecular Solids

The iodine (I_2) used to make the antiseptic known as tincture of iodine, the cane sugar ($C_{12}H_{22}O_{11}$) found in a sugar bowl, and the polyethylene used to make garbage bags all have one thing in common. They are all examples of compounds that are **molecular solids** at room temperature. Water and bromine are liquids that form molecular solids when cooled slightly; H_2O freezes at $0°C$ and Br_2 freezes at $-7°C$. Many substances that are gases at room temperature will form molecular solids when cooled far enough; F_2, at the extreme right of the bond type triangle in Figure 9.1, freezes to form a molecular solid at $-220°C$.

Molecular solids contain both *intramolecular* bonds and *intermolecular* forces. The atoms within the individual molecules are held together by relatively strong intramolecular covalent bonds. Therefore, molecular solids are composed of molecules that contain covalent bonding and are located in the covalent region of a bond type triangle. The molecules that contain covalent bonds are then held together by much weaker intermolecular forces. Because the forces between molecules are relatively weak, molecular solids are often soft substances, with low melting points.

Dry ice, or solid carbon dioxide, is a perfect example of a molecular solid. The van der Waals forces holding the CO_2 molecules together are weak enough that at $-78°C$ dry ice **sublimes**—it goes directly from the solid to the gas phase.

Strong intramolecular bond

$$\delta+ \quad \delta-$$
$$O=C=O$$

←——— *Weak intermolecular forces*

$$O=C=O$$
$$\delta+ \quad \delta-$$

Changes in the strength of the van der Waals forces that hold molecular solids together can have important consequences for the properties of the solid. Polyethylene $(-CH_2-CH_2-)_n$ is a soft plastic that melts at relatively low temperatures. Replacing one of the hydrogens on every other carbon atom with a chlorine atom produces a plastic known as poly(vinyl chloride), or PVC, which is hard enough to be used to make the plastic pipes that are slowly replacing metal pipes for plumbing.

$$
\begin{array}{cccccc}
\text{H} & \text{H} & \text{H} & \text{Cl} & \text{H} & \text{H} \\
| & | & | & | & | & | \\
-\text{C} & -\text{C} & -\text{C} & -\text{C} & -\text{C} & -\text{C}- \\
| & | & | & | & | & | \\
\text{H} & \text{Cl} & \text{H} & \text{H} & \text{H} & \text{Cl}
\end{array}
$$

Poly(vinyl chloride) or PVC

Much of the strength of PVC can be attributed to the van der Waals force of attraction between the chains of $(-CH_2-CHCl-)_n$ molecules that form the solid. From poly(vinyl chloride) $(-CH_2-CHCl-)_n$ and poly(vinylidene chloride) $(-CH_2-CCl_2-)_n$, a substance is formed that is sold under the trade name Saran. The same increase in the force of attraction between chains that makes PVC harder than polyethylene gives a thin film of Saran a tendency to be attracted to itself. Saran wrap therefore clings to itself, whereas the polyethylene in sandwich bags does not.

The family of substances known as the *halogens* (F_2, Cl_2, Br_2, and I_2) can provide a basis for understanding the effect of differences in the strengths of intermolecular forces on the properties of a molecular solid. Consider chlorine, for example, which exists as diatomic Cl_2 molecules in the gas phase at room temperature. When the gas is cooled, the average kinetic energy of the Cl_2 molecules decreases. As the motion of the molecules decreases, the force of attraction between the molecules becomes large enough to hold the molecules together, and the gas condenses to form a liquid. Further cooling transforms the liquid into a molecular solid as shown by its position on the bond type triangle (Figure 9.1).

The unit on which the solid is built is still the diatomic Cl_2 molecule. The covalent bond holding one chlorine atom to another in Cl_2 is relatively strong, about 243 kJ/mol. The intermolecular forces that hold one Cl_2 molecule to another are much smaller, only about 18 kJ/mol. Thus, the intermolecular forces responsible for the formation of the solid are very different from the intramolecular bonds within the molecule.

Because the Cl_2 molecule has no dipole moment, the intermolecular forces that hold Cl_2 molecules together result solely from induced dipole–induced dipole or dispersion forces. The dispersion forces are nondirectional, and the molecules pack in the solid in the geometry that allows them to come as close together as possible.

Covalent molecules with a dipole moment, such as HCl and HBr, also form a molecular solid structure (Figure 9.1) that allows the molecules to pack as tightly as possible. Polar molecules, however, also have a directional component to the intermolecular forces, namely, dipole–dipole interactions. This force controls the orientation of the HCl and HBr molecules as they pack, so that the negative end of one dipole is oriented toward the positive end of the other.

The chief distinction between the solid and liquid phases for covalent molecules is the regular pattern of packing in the solid versus the randomness of the structure in the liquid. Consider ice, for example. The individual H_2O molecules in the molecular solid are held together by a combination of dipole, dispersion, and hydrogen bond forces. Two parameters can be used to estimate the relative strength of the resulting intermolecular forces, the melting point of the compound, and the enthalpy of fusion, ΔH_{fus}. The melting point, as we saw in Section 8.6, is the temperature at which the solid melts at atmospheric pressure. The enthalpy of fusion is the heat required to melt the substance, in units of kilojoules per mole for the reaction as written. The enthalpy of fusion of H_2O is relatively small, only 6.00 kJ/mol_{rxn} ($H_2O(s) \rightarrow H_2O(l)$). This is only a small fraction of the strength of the hydrogen bonds between water molecules because melting the solid doesn't break all of the hydrogen bonds between the water molecules, only some of them. To break all of the hydrogen bonds we have to boil water; the enthalpy of vaporization of H_2O is 40.88 kJ/mol_{rxn} ($H_2O(l) \rightarrow H_2O(g)$) at the boiling point.

Melting points and enthalpies of fusion are convenient measures of the strengths of the intermolecular interactions that hold molecular solids together. Table 9.1 gives the melting points and the enthalpies of fusion of the halogens. The only forces that hold the crystals together are dispersion forces. Because dispersion forces depend on the number of electrons, as the size of the halogen atoms increases the dispersion force interactions should also increase. This is reflected in the increase in both the melting point and the enthalpy of fusion with increasing molecular weight.

TABLE 9.1   Melting Points and Enthalpies of Fusion of the Halogens

Halogen	Molecular Weight (g/mol)	MP (°C)	ΔH_{fus} (kJ/mol_{rxn})
F_2	38	−219.6	0.51
Cl_2	71	−101	6.41
Br_2	160	−7.2	10.8
I_2	254	112.9	15.3

The effect of adding dipole–dipole and hydrogen bond interactions to the intermolecular forces that hold molecules together can be seen in the data in Table 9.2. Dispersion and dipole forces exist in all three compounds. Two of the compounds, H_2O and CH_3OH, also form hydrogen bonds. As the number of hydrogen atoms that can form hydrogen bonds increases from zero in CH_3OCH_3 to one per molecule in CH_3OH and then two per molecule in H_2O, there is a significant increase in the melting point. The enthalpy of fusion is determined by the intermolecular attractive forces. The number of electrons, and hence the polarizability, of the compounds decreases from CH_3OCH_3 to CH_3OH to H_2O, and it might be expected that water would have the smallest enthalpy of fusion. The fact that water has the highest enthalpy of fusion shows the magnitude of the influence of the hydrogen bonding.

TABLE 9.2 Melting Points and Enthalpies of Fusion

Compound	Molecular Weight (g/mol)	MP (°C)	ΔH_{fus} (kJ/mol$_{rxn}$)
CH_3OCH_3	46	−141.5	4.94
CH_3OH	32	−97.9	3.18
H_2O	18	0	6.00

Network Covalent Solids

Network covalent solids include substances, such as diamond, whose crystals can be viewed as a single giant molecule made up of an almost endless number of covalent bonds. Each carbon atom in diamond is covalently bound to four other carbon atoms oriented toward the corners of a tetrahedron, as shown in Figure 9.2. Because all of the bonds in the structure are equally strong, diamond is the hardest natural substance, and it melts at 3550°C. Quartz is a network covalent solid composed of SiO_2 and is located in the covalent region of Figure 9.1. Network covalent solids are often very hard, and they are notoriously difficult to melt. Both molecular solids and network covalent solids are located in the covalent region of a bond type triangle. A bond type triangle can therefore not be used to distinguish between these two types of solids.

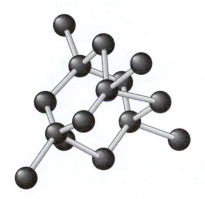

FIGURE 9.2 A perfect diamond is a single molecule in which each carbon atom is tightly bound to four neighboring carbon atoms arranged toward the corners of a tetrahedron.

Checkpoint

Describe the differences on the atomic and macroscopic scales between molecular solids and network covalent solids.

9.3 IONIC SOLIDS

As discussed in Section 5.7, **ionic solids** are salts, such as NaCl, that are held together in an extended three-dimensional network of ions by the strong force of attraction between ions of opposite charge (see Figure 9.3).

$$F = \frac{q_1 \times q_2}{r^2}$$

Because the force of attraction depends inversely on the square of the distance between the positive and negative charges, the strength of an ionic bond depends inversely on the

FIGURE 9.3 Ionic compounds are made up of a three-dimensional network of positive and negative ions.

sizes of the ions that form the solid. When the ions are large, the bond is relatively weak. But the ionic bond is still strong enough to ensure that salts have relatively high melting points and boiling points. Sodium chloride, for example, melts at 801°C and boils at 1413°C.

Solids retain their shape, are difficult to compress, and are denser than liquids and gases. These characteristic properties suggest that solids contain particles which are packed as tightly as possible. Ionic compounds form solids in which the force of attraction between the ions of opposite charge is maximized by keeping the ions as close together as possible. Ionic solids are located in the ionic region of a bond type triangle as shown in Figure 9.1.

Checkpoint

Describe the differences on the atomic and macroscopic scales between molecular solids and ionic solids.

Some understanding of the strength of the bonding in an ionic compound can be obtained by considering the enthalpy required to completely disrupt the structure of an ionic solid to form isolated ions in the gas phase.

$$NaCl(s) \longrightarrow Na^+(g) + Cl^-(g)$$

The process can be visualized by first transforming the salt into sodium and chlorine atoms in the gas phase. This requires the input of 640 kJ/mol$_{rxn}$, which is the enthalpy of atomization of sodium chloride. That value can be determined by reversing the sign of the enthalpy of atom combination given in Appendix B.14.

$$\boxed{Na(g) + Cl(g)}$$
$$\boxed{NaCl(s)} \quad \Delta H_a = 640\ kJ/mol_{rxn}$$

An electron is then removed from sodium, which requires an input of energy corresponding to the first ionization energy of that element.

$$\boxed{Na^+(g) + Cl(g)}$$
$$\boxed{Na(g) + Cl(g)} \quad \Delta H_{IE} = 496\ kJ/mol_{rxn}$$
$$\boxed{NaCl(s)} \quad \Delta H_a = 640\ kJ/mol_{rxn}$$

An electron is then added to a neutral chlorine atom to form a Cl$^-$ ion in the gas phase. The energy associated with the step is called the **electron affinity** of the element. It is usually, but not always, exothermic because most neutral atoms will give off heat when they accept an extra electron.

We can now complete the thermodynamic cycle by bringing the Na^+ and Cl^- ions in the gas phase together to form the solid NaCl. Because the force of attraction between the ions is relatively large, this is a strongly exothermic step.

Lattice energy is the energy required to break an ionic compound into its gaseous ions. Thus, the lattice energy for NaCl is $+787$ kJ/mol$_{rxn}$.

The lattice energies of a series of compounds formed by combining one of the alkali metals with a halogen are given in Table 9.3. Note that the lattice energies of the compounds decrease as the size of the ions increases because of an increase in the distance between the centers of the positive and negative charges on the ions. It therefore takes less energy to break one of the solids apart as the ions become larger, or less energy is given off when one of the compounds is formed from the corresponding positive and negative ions in the gas phase.

TABLE 9.3   Lattice Energies of Alkali Metal Halides (kJ/mol$_{rxn}$)

	F^-	Cl^-	Br^-	I^-
Li^+	1046	861	818	762
Na^+	923	787	747	704
K^+	821	718	682	649
Rb^+	785	689	660	630
Cs^+	740	659	631	604

The lattice energies for ionic compounds formed when one of the alkaline earth metals combines with oxygen to form an oxide (MgO, CaO) shows a trend similar to the halides given in Table 9.3. The lattice energy for MgO, however, is about five times as large as the lattice energy for NaCl. Part of the difference can be explained by noting that MgO formally contains ions with charges of $+2$ and -2. Thus, the product of the charge on the positive and negative ions is four times larger in MgO than it is in NaCl, which contains $+1$ and -1 ions. The remainder of the difference results from the fact that the Mg^{2+} ion is smaller than the Na^+ ion, and the O^{2-} ion is smaller than the Cl^- ion.

Checkpoint
Which has the larger lattice energy, $MgCl_2$ or MgF_2?

9.4   METALLIC SOLIDS

Molecular, ionic, and network covalent solids all have one thing in common. With only rare exceptions, the electrons in the solids are *localized*. They either reside on one of the atoms or ions, or they are shared by a pair of atoms or a small group of atoms.

As we saw in Section 5.9, metal atoms don't acquire enough electrons to fill their valence shells by sharing electrons with their immediate neighbors. Electrons in the valence shell are therefore shared by many atoms, instead of just two. In effect, the valence electrons are *delocalized* over many metal atoms. Because the electrons aren't tightly bound to individual atoms, they are free to migrate through the metal. As a result, metals are good conductors of electricity and heat. Electrons that enter the metal at one edge can displace other electrons to give rise to a net flow of electrons through the metal.

The bonds that hold metals together are very different from ionic and covalent bonds and therefore are placed in a category of their own: **metallic bonds.** Metallic bonding occurs when ΔEN and the average EN of the atoms are low (i.e., on the lower left corner of Figure 9.1). In a metal, the metal atoms form bonds with many neighboring atoms. Thus metals are usually solids in which each atom is surrounded by as many neighboring atoms as possible. Lithium, for example, crystallizes in a structure in which each atom touches eight nearest neighbors. The distance between the nuclei of the atoms is 0.303 nm.

Lithium has three electrons: $1s^2 2s^1$. There is a significant difference between the ease with which an electron can be removed from the $1s$ and $2s$ orbitals on a lithium atom, however. According to the data in Table 3.4, it takes 520 kJ/mol to remove an electron from the $2s$ orbital on lithium, but 6260 kJ/mol to remove one of the electrons from the $1s$ orbital. The core electrons in the $1s$ orbitals on a lithium atom are bound so tightly to the nucleus of the atom that they are unaffected by other atoms. Thus, there is only one valence electron to be shared per lithium atom in the metal, and the electron must be shared with all of the neighboring atoms.

In the gas phase, lithium can form a diatomic Li_2 molecule that is held together by the sharing of a pair of electrons by the two lithium nuclei.[1] The distance between the lithium atoms in the Li_2 molecule is 0.267 nm, which is considerably smaller than the distance between lithium atoms in the metal. This suggests that the covalent bond in the Li_2 molecule is significantly stronger than the metallic bonds in lithium metal. However, there are more bonds per lithium atom in the metal. As a result, the enthalpy of atomization $Li(s) \rightarrow Li(g)$ ($-\Delta H_{ac}$ from Appendix B.14) for lithium metal is 159 kJ/mol$_{rxn}$, whereas the bond holding the two atoms together in Li_2 is only 57 kJ/mol$_{rxn}$.

Those elements shown on the periodic chart as metals, that is, those elements whose AVEE values (see Chapter 3) lies to the left of the metalloid dividing line, have a common characteristic. Their electron configurations are such that the number of outermost electrons is smaller than the number of vacancies which could accommodate them. A sodium atom, for example, has one outermost electron, a $3s$ electron. One other electron could be accommodated by the $3s$ subshell, or six additional electrons could occupy the empty $3p$ subshell. Figure 4.2 and the AVEE values show that as the periodic chart is descended, the energy difference between consecutive shells is decreasing, that is, the s, p, and d subshells are becoming very close in energy. Electron configurations as developed in Chapter 3 apply to isolated atoms. Whenever an atom enters into combination with another atom or group of atoms the energy levels are different from those in the isolated atom. Because of these slight perturbations, the small difference in energy between the s, p, and d subshells becomes even smaller, and the energies of the valence subshells of metallic atoms coalesce. Then electrons may move easily to all available subshells, and the elec-

---

[1]Although dilithium molecules can exist in the gas phase, the famous "dilithium crystals" that fuel the Starship *Enterprise* exist only in the imagination of Gene Roddenberry.

trons are said to be delocalized. This means that the electrons are not confined to the space between nuclei of atoms. Bonding becomes nondirectional, and the atoms pack together as tightly as possible. Thus, a substance held together by metallic bonds can be considered to be the atomic cores (called cations) in a sea of electrons.

As a group in the periodic table is descended the sizes of the atoms increase, making removal of outer shell electrons easier, and the energy gaps between subshells decrease. Those two factors account for the characteristics of metallic bonding and explain why metallic behavior increases as we move from top right to bottom left in the periodic table.

Checkpoint

The AVEE value or electronegativity of an atom is made up of two important contributions. What are they, and why are they important for understanding metallic behavior?

9.5 PHYSICAL PROPERTIES THAT RESULT FROM THE STRUCTURE OF METALS

Metals have certain characteristic physical properties.

- They have a metallic shine, or luster.
- They are usually solids at room temperature.
- They are *malleable* (from the Latin word for "hammer"): they can be hammered, pounded, or pressed into different shapes.
- They are *ductile:* they can be drawn into thin sheets or wires without breaking.
- They conduct heat and electricity.

The structures of metals can be used to explain their characteristic properties.

Imagine that you place one hand on the window of a car that has been sitting in the hot summer's sun and your other hand on the chrome-plated door handle. If the glass isn't tinted, it will feel cooler than the door handle. Most of the light that strikes the glass passes through. A portion of the energy carried by the light that hits the door handle, however, is absorbed by the metal and turned into thermal energy.

Light is absorbed when the energy of the radiation is equal to the energy needed to excite an electron to a higher energy state or when the energy can be used to move an electron through the solid. Because electrons are delocalized in metals—and therefore free to move through the solid—metals absorb light easily. Other solids, such as glass, don't have electrons that can move through the solid, so they can't absorb light the way metals do. Such solids are colorless and can only be colored by adding an impurity in which the energy associated with exciting an electron from one orbital to another falls in the visible portion of the spectrum. Glass is usually colored by adding a small quantity of one of the transition metals. Cobalt produces a blue color, chromium makes the glass appear green, and traces of gold give a deep red color.

As you look into the car window, you might see your reflection. But the chrome-plated surface of the door handle has a characteristic metallic luster because the metal reflects (literally, "throws back") a significant fraction of the light that hits its surface. Silver is better than any other metal at reflecting light: roughly 88% of the light that hits the surface of a silver mirror is reflected.

Why are metals solid? Some nonmetals, such as hydrogen and oxygen, are gases at room temperature because the atoms of these elements form molecules that are held together by weak intermolecular forces. Metal cations are held closely together by strong metallic bonds in a three-dimensional network and are therefore solids at room temperature (except mercury).

Metals are malleable and ductile because they pack in body-centered cubic, hexagonal closest packed, or cubic closest packed structures. In theory, changing the shape of the metal is simply a matter of applying a force that makes the atoms in one of the planes slide past the atoms in an adjacent plane, as shown in Figure 9.4. In practice, it is easier to do this when the metal is hot. The layers of atomic cores in metals can slip easily over one another because there are no directional forces tending to keep them in locked positions, and thus metals will be ductile and malleable.

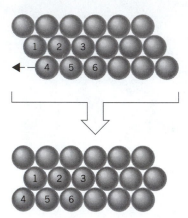

FIGURE 9.4 Metals are malleable and ductile because planes of atoms can slip past one another to reach equivalent positions.

Why are metals good conductors of heat and electricity? As we have already seen, the delocalization of valence electrons in a metal allows the solid to conduct an electric current. Metals conduct heat by the movement of electrons. Because the electrons are relatively free in metals, the electrons can quickly transport heat throughout the metal.

Checkpoint

Describe the differences on the atomic and macroscopic scales between metallic solids and ionic solids.

Table 9.4 summarizes the structure and physical properties of the solids discussed in this chapter.

9.6 SEMIMETALS

The chemist's classification of substances according to the type of bonding—metallic, covalent, or ionic—is not based on a readily measured property. However, there are characteristics that are easily measured and that also serve as a useful way of categorizing substances. Bond type triangles take advantage of the correspondence between the bonding classification and observable properties.

Metals conduct electricity as solids and in the liquid state. Ionic substances produce ions on dissolution in water and on melting and therefore conduct electricity in aqueous solution or in the liquid state but not in the solid. Thus solutions of ionic salts in water or molten ionic salts are called **electrolytes.** Because in covalent substances the electrons are localized (i.e., tightly held between nuclei), covalent compounds are poor conductors of electricity and are often insulators. The labels electrolytes, conductors, and insulators appearing over the vertices of the triangle in Figure 9.1 are observable properties that are closely related to the bonding type. The farther the distance from the vertices of the triangle, the more mixed the properties of materials become.

TABLE 9.4   Classifications and Properties of Solids

Classification of Solid	Primary Type of Bonding	Force Holding Solid Together	Physical Properties	Examples
Molecular	Covalent	Intermolecular forces	Low melting, electrical insulators	H_2O, Cl_2, HCl, CO_2
Network covalent	Covalent	Covalent bonds	Very high melting, electrical insulators, very hard	Diamond (C), quartz (SiO_2)
Ionic	Ionic	Ionic bonds	High melting points, electrical conductor in the molten and aqueous state	NaCl, CaO, LiF, $BaCl_2$
Metallic	Metallic	Metallic bonds	Range of melting points, conductors of heat and electricity, lustrous, malleable, ductile	Na, Fe, Al, $CuAl_2$, $BaZn_5$

In general, elements that have low AVEE values (electronegativities) have small energy gaps in their valence subshells and tend to form delocalized (metallic) bonds. Elements that have large AVEE values (electronegativities) have a large energy separation of their valence subshells. Therefore, delocalized bonding is not as favored as localized bonding, and these elements tend to form covalent bonds and are insulators. Semimetals fall in between the extremes of delocalized (metallic) and localized (covalent or ionic) bonding.

A dividing line on the periodic chart separates metals from nonmetals. The elements along the dividing line are called semimetals or metalloids and have properties between those of metals and nonmetals. On the bond type triangle (Figure 9.1) the region bounded by Al and As is the semimetal region (see also Figure 5.10). Because semimetals come between conductors of electricity and insulators, semimetals can be made to conduct or insulate. Hence, the name semiconductors is often applied to these substances.

9.7   THE SEARCH FOR NEW MATERIALS

Materials scientists seek to understand the physical and chemical characteristics of solids and their properties. Often this understanding requires knowledge of chemistry, physics, engineering, and biology. Materials chemistry is concerned with the relationship between structure and the properties and performance of a material. Examples of these materials include piezoelectric crystals, crystals that deform when an electric field is applied. Such materials are used in loudspeakers, pressure gauges, and buzzers. Floppy disks and hard drives for computers use other new materials that have come from materials science research. Catalytic converters, sunglasses, and superconductivity all have resulted from an understanding of the atomic structure of solids.

It has been at least 5000 years since the early Egyptians first mixed sand (SiO_2) with the earthy residue left behind when the Nile flooded each year—which contained a mixture of $CaCO_3$, Na_2CO_3, NaCl, and CuO—and then heated the mixture to form **glass.** Glass is a fascinating material. Its structure, in some ways, more closely resembles a liquid than a solid. It has therefore been defined as a material analogous to the liquid state that forms as the result of a reversible change in viscosity as it cools, but which has achieved such a high viscosity that, for all practical purposes, it is rigid. Alternatively, it can be defined as the product of a fusion—or melting process—that has cooled to a rigid condition without forming crystals.

The difference between the structures of sand and glass is one of long-range order. Sand forms a regular crystal. There is an alternating array of Si—O bonds, arranged in a tetrahedral geometry around each silicon atom, to form a regular, three-dimensional solid.

$$
\begin{array}{ccccc}
 & | & | & | & \\
 & O & O & O & \\
 & | & | & | & \\
-O- & Si-O- & Si-O- & Si & -O- \\
 & | & | & | & \\
 & O & O & O & \\
 & | & | & | & \\
-O- & Si-O- & Si-O- & Si & -O- \\
 & | & | & | & \\
 & O & O & O & \\
 & | & | & | & \\
-O- & Si-O- & Si-O- & Si & -O- \\
 & | & | & | & \\
 & O & O & O & \\
 & | & | & | & \\
\end{array}
$$

Each of the oxygen atoms in sand acts as a bridge between two silicon atoms. Because the Si—O bonds are more or less covalent, the result is a network covalent solid—a crystal that can be considered to be a single giant molecule.

When sand is heated and various impurities are incorporated into the molten liquid that forms, the regular structure of the solid can't re-form. On the average, one in three of the oxygen atoms no longer serves as a bridge between silicon atoms; it is now a terminal oxygen atom. The result is a material that doesn't have the long-range order of a perfect crystal, or even the short-range order of a polycrystalline material. There are enough bonds within the material to make it rigid but not enough to make it crystalline.

Glass is just one example of a family of materials known as **ceramics.** By definition, ceramics are materials that are hard, strong, and light, although they are often brittle. The term *ceramic* comes from a Greek word meaning "burnt stuff." Because they are often made in high temperature reactions, most ceramics are heat resistant. They are also excellent insulators of both heat and electricity.

Ceramics are usually made from such common minerals as clay and sand under normal conditions and fired at high heat. They are inorganic solids that fall into the nonmetallic classification. As such, the bonding in ceramics can be covalent network or ionic. Ceramics have different properties depending on their atomic structure.

Another basic component for building ceramics that has been used for at least 5000 years is clay, which is essentially a hydrated compound of aluminum and silicon with the empirical formula $H_2Al_2Si_2O_9$. When wet, clay is plastic and can be shaped as desired. The shape is retained when the clay is heated in a kiln. Bricks that are made from clay have a porous structure and fracture easily. If the minerals feldspar ($KAlSi_3O_8$) and quartz (SiO_2) are added to clay before it is fired, a smooth material, known as porcelain, is formed, which has a glasslike surface and is therefore less porous.

In 1891 Edward Acheson synthesized a new class of ceramics by reacting silicon dioxide with an excess of carbon in an electric furnace at 2300 K.

$$SiO_2(s) + 3\ C(s) \longrightarrow SiC(s) + 2\ CO(g)$$

Both the structure and properties of silicon carbide are analogous to diamond. Both materials are inert to chemical reactions, except at very high temperatures; both have very high melting points; and both are among the hardest substances known. Shortly after he synthesized silicon carbide, Acheson founded the Carborundum Company to market the material. Then, as now, materials in this class are most commonly used as abrasives.

Most of the ceramics produced each year are used as glass, porcelain, floor tiles, bricks, clay sewage pipes, concrete, cement, or one of the abrasives such as SiC. Since World War II, however, there has been a growing interest in the electrical and magnetic properties of certain ceramics. Ceramics can be made that vary by as much as a factor of 10^{19} in their ability to conduct an electric current. Some ceramics, such as CrO_2, conduct electricity as well as a metal. Others, such as SiC located near the covalent–semimetal interface in Figure 9.1, are semiconductors, like the semimetals silicon and germanium. Still others, such as glass and porcelain, are insulators. It is even possible to produce ceramics that have pronounced magnetic field effects yet don't conduct electricity. These ceramics play an important role in the development of memory circuits and permanent magnets.

The properties of solids depend on several factors. One of the first considerations, however, is whether the materials are likely to be ionic, metallic, or covalent solids. There are also intermediate possibilities, such as semimetals, semiconductors, and semielectrolytes. Many of the characteristics of solids can be anticipated on the basis of their position in a bond type triangle.

Suppose that you wanted to prepare a ceramic that was crystalline, heat resistant, and a good insulator. The best place to search for such ceramics would be toward the bottom and center of the appropriate bond type triangle, such as the one shown in Figure 9.1. Compounds that lie too close to the right corner would tend to form molecules that lack the long-range order needed to form a ceramic. Compounds that lie close to the left corner would be more likely to form metallic bonds that would make the material a conductor. Compounds that lie close to the top of the triangle would have the long-range order and insulating properties that are desired, but such solids are often too brittle to form useful materials.

Both silicon carbide (SiC) and boron nitride (BN) make ceramic materials. Characteristics of the two ceramics are that they are strong but brittle. Both are poor conductors of electricity, and both are very hard materials. BN is, in fact, comparable in hardness to diamond.

Suppose that you were interested in producing a new solid with electrical conductivities between that of metals and insulators, that is, a new semiconductor. Once again, you would look toward the bottom of the triangle, but this time you might shift the focus slightly to the left of center. GaAs, for example, lies on the border of the semimetal region and toward the covalent (insulator) region and is a semiconductor (Figure 9.1). The element silicon is a semiconductor and has been commonly used in the manufacture of silicon chips for integrated circuits. GaAs is a new semiconducting material that has certain advantages over silicon. GaP is also promising for use as a semiconductor.

Suppose the desired characteristics of a material are that the material be crystalline, stand up to high temperatures, and not conduct electricity or heat. What elements might compose such a material? These qualifications are met by atoms that have electronegativities that place them in the upper middle regions of a bond type triangle. Such combinations might be aluminum and oxygen or magnesium and oxygen.

The bond type triangle shown in Figure 9.1 can help us understand the remarkable differences in the properties of solids that might not seem dissimilar from the positions of their elements in the periodic table. Consider BeO and CO, for example. BeO is also known as beryllia; it melts at 2250°C, is very hard, and is a ceramic. Carbon monoxide falls in the covalent region of the triangle and forms a molecular solid at temperatures below −200°C.

The electronegativity difference between Be and O is larger than that for C and O, and the average electronegativity is smaller for BeO than for CO. These two conditions place BeO in a very different region of the triangle.

Compounds such as Li_3As are located in the metallic region. As we move away from the metallic region, atom combinations produce compounds that become increasingly insulating. Ceramic materials such as SiC, BN, and BeO are found near or along the interface between ionic and covalent areas. Semiconductors are located along the semimetal–covalent boundary, reflecting the changeover from conducting toward insulating materials.

Checkpoint
What types of solids are B_4C and MoC? Suggest an application for each of the compounds.

Research in the 1990s

The Search for High Temperature Superconductors

When an electric current is passed through any material at room temperature, some of the energy of the electrons is dissipated in the form of heat. In metals resistance to an electric current decreases as the metal is cooled. In 1911, Heike Onnes found that when mercury was cooled to temperatures below 4.1 K, its resistance fell to zero. Above that temperature, mercury was a conductor of electricity. Below the transition point, it became a **superconductor.** By 1913, Onnes found that tin and lead also became superconductors at temperatures below 4 K.

Onnes recognized the potential of superconductivity for constructing magnets with unusually strong magnetic fields. Standard electromagnets are made by winding a coil of insulated copper wire around an iron alloy core. As the current passes through the copper wire, a magnetic field is created. The field induces an alignment of electrons in the iron alloy core, which in turn produces a magnetic field in the core that is up to 1000 times larger than the field produced by the copper wire. There is an upper limit to the strength of the field that iron alloy magnets can produce, however. The magnets "saturate" at a magnetic field above about 2 Tesla, which is 40,000 times larger than the earth's magnetic field.

Onnes believed that superconducting magnets could be produced that would achieve much higher fields. Unfortunately, none of the superconducting metals he studied were able to carry enough electric current. It took 50 years before alloys of niobium and tantalum were discovered that could carry the current needed to produce high-field magnets.

Commercial spectrometers that used superconducting magnets made from niobium–tantalum alloys became available toward the end of the 1960s. The primary disadvantage of the instruments was the fact that the alloy has to be cooled to the temperature of liquid helium (4.2 K) before it becomes a superconductor. The cost of maintaining one of the instruments could be decreased by as much as a factor of 1000 if it could operate at liquid nitrogen temperatures (77 K).

The search for "high temperature" superconductors is an important object lesson in the proper role of theory and experiment. At first glance, we might expect ReO_3 and RuO_2 to be insulators, like other metal oxides. In practice, those oxides conduct electricity the way a metal would. In 1964, it was found that other metal oxides, such as NbO and TiO, conduct electricity so well that they become superconductors when cooled to extremely low temperatures (1 K). This is also unexpected, based on their position in a bond type triangle.

A major step in the evolution of high temperature superconductors occurred in 1986, when Alex Müller and Georg Bednorz at the IBM Research Laboratory in Zurich discovered that certain ceramic materials that contained lanthanum, barium, copper, and oxygen became superconductors when cooled to temperatures below 35 K. Their results contained two elements of surprise. First, ceramics—such as the plates on which we eat dinner—were considered to be insulators, not conductors. Second, the transition temperature for superconductivity in the new material was higher than that for any known metal or metal alloy.

Within a few years, a family of superconducting ceramics had been discovered that were all based on compounds of copper and oxygen. Müller and Bednorz worked with ceramics that were derivatives of a compound with the formula La_2CuO_4. If forced to assign oxidation states to the compound, most chemists would write it as $[La^{3+}]_2[Cu^{2+}][O^{2-}]_4$. The parent compound is an insulator. When some of the lanthanum atoms are replaced with barium atoms, however, a nonstoichiometric superconductor with the formula $La_{2-x}Ba_xCuO_4$ is obtained.

Applying the concept of oxidation states to the compound, we are formally replacing an La^{3+} ion by a Ba^{2+} ion each time a barium atom is incorporated into the structure. If the net charge on the compound is going to stay the same, the oxidation state of the copper atom must increase. Each time an La^{3+} ion is replaced with a Ba^{2+} ion, a Cu^{2+} ion formally becomes a Cu^{3+} ion. Müller and Bednorz found that when enough barium had been incorporated to raise the average oxidation state of the copper to +2.2, the compound became a superconductor at low temperatures.

The electron that is formally removed from the copper atom is apparently delocalized and therefore capable of moving through the solid when it is cooled to low temperatures. Similar results can be obtained by incorporating either strontium or calcium into La_2CuO_4. $La_{1.8}Sr_{0.2}CuO_4$ has the highest transition temperature of any member of the family: 40 K.

Superconductivity has also been observed with a family of compounds known as 1-2-3 superconductors. The first member of the family was discovered in 1987, when $YBa_2Cu_3O_7$ was found to be a superconductor when cooled to 95 K—above the temperature of liquid nitrogen. (The common name of the superconductors is based on the fact that there are three metals in a 1:2:3 ratio.) The compound also contains copper in a fractional oxidation state. The yttrium atom can be assumed to exist in the +3 oxidation state. Thus one Y^{3+} and two Ba^{2+} ions contribute a charge of +7 toward balancing the charge of −14 on the seven oxygens. The remaining charge of +7 has to be distributed over the three copper atoms, for an average oxidation state of +2.33.

The most common demonstration of high temperature superconductors is based on the Meissner effect. In 1933, the German physicists W. Meissner and R. Ochsenfeld found that superconductors repel an external magnetic field. On the macroscopic scale, the superconductor seems to repel the magnet that produced the magnetic field. As a result, when one of the high temperature superconductors is cooled with liquid nitrogen, it levitates off the surface of a magnet.

9.8   THE STRUCTURE OF METALS AND OTHER MONATOMIC SOLIDS

We can describe the structure of pure metals by assuming that the atoms of the metals are identical perfect spheres. The same model can be used to describe the structure of the solid noble gases (He, Ne, Ar, Kr, Xe) at low temperatures. These substances all crystallize in one of four basic structures, known as simple cubic (SC), body-centered cubic (BCC), hexagonal closest packed (HCP), and cubic closest packed (CCP).

Solids are very difficult to compress because the amount of space between particles in a solid is at a minimum. As a rule, therefore, we can conclude that the most probable structure for a solid is the structure that makes the most efficient use of space. When a solid crystallizes, the particles that form the solid pack as tightly as possible. To illustrate the principle, let's try to imagine the best way to pack spheres, such as Ping-Pong balls, into an empty box.

One approach involves carefully packing the Ping-Pong balls to form a square-packed plane of spheres, as shown in Figure 9.5. A second plane of spheres can be stacked directly on top of the first. The result is a regular structure in which the simplest repeating unit is a cube of eight spheres, as shown in Figure 9.6. The structure is called **simple cubic packing.** Each sphere in the structure touches four identical spheres in the same plane. It also touches one sphere in the plane above and one in the plane below. Each sphere is therefore said to have a **coordination number** of 6. If the spheres represent atoms, each atom in the structure can form bonds to its six nearest neighbors.

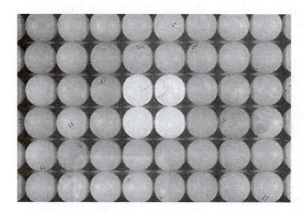

FIGURE 9.5 A square-packed plane of spheres.

FIGURE 9.6 A simple cubic packing of spheres.

One way to decide whether the simple cubic structure is an efficient way of packing spheres is to ask, What happens when we shake the box? Do the Ping-Pong balls stay in the same positions, or do they settle into a different structure? It is fairly easy to show that a simple cubic structure is not an efficient way of using space. Only 52% of the available space is actually occupied by the spheres in a simple cubic structure. The rest is empty space. Because the structure is inefficient, only one element—polonium—crystallizes in a simple cubic structure.

This raises an interesting question: How can we use space more efficiently? Another approach starts by separating the spheres to form a square-packed plane in which the spheres do not quite touch each other, as shown in Figure 9.7. The spheres in the second plane pack above the holes in the first plane, as shown in Figure 9.8. Spheres in the third plane pack above holes in the second plane. Spheres in the fourth plane pack above holes in the third plane, and so on. The result is a structure in which the odd-numbered planes of atoms are identical and the even-numbered planes are identical. The *ABABABAB* . . . repeating structure of square-packed planes is known as **body-centered cubic packing.**

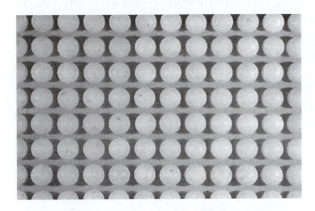

FIGURE 9.7 A square-packed plane in which the spheres do not quite touch.

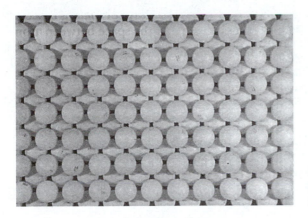

FIGURE 9.8 The spheres in the second plane of a body-centered cubic structure pack above the holes in the plane shown in Figure 9.7.

The structure is called *body-centered cubic* because each sphere touches four spheres in the plane above and four more in the plane below, arranged toward the corners of a cube. Thus, the repeating unit in the structure is a cube of eight spheres with a ninth identical sphere in the center of the body—in other words, a body-centered cube, as shown in Figure 9.9. The coordination number in the structure is 8.

FIGURE 9.9 Body-centered cubic structure. All spheres represent identical atoms.

Body-centered cubic packing is a more efficient way of using space than simple cubic packing, as 68% of the space in the structure is filled. Body-centered cubic packing is an important structure for metals. All of the metals in Group IA (Li, Na, K, Rb, Cs, and Fr), barium in Group IIA, and a number of the early transition metals (such as V, Cr, Mo, W, and Fe) pack in a body-centered cubic structure.

Two structures pack spheres so efficiently they are called **closest packed structures.** Both start by packing the spheres in planes in which each sphere touches six others oriented toward the corners of a hexagon, as shown in Figure 9.10. A second plane is then formed by packing spheres above the triangular holes in the first plane, as shown in Figure 9.11.

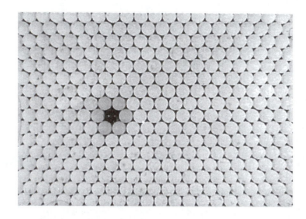

FIGURE 9.10 A closest packed plane in which each sphere touches six others oriented toward the corners of a hexagon.

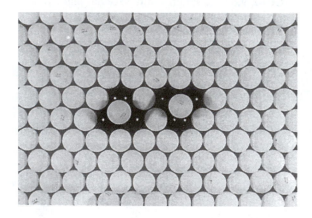

FIGURE 9.11 Atoms in the second plane of closest packed structures pack above the triangular holes in the first plane shown in Figure 9.10.

What about the next plane of spheres? The spheres in the third plane could pack directly *above the spheres* in the first plane to form an *ABABABAB* . . . repeating structure. Because such a structure is composed of alternating planes of hexagonal closest packed spheres, it is called a **hexagonal closest packed** structure. Each sphere touches three spheres in the plane above, three spheres in the plane below, and six spheres in the same plane, as shown in Figure 9.12. Thus, the coordination number in a hexagonal closest packed structure is 12, and 74% of the space in a hexagonal closest packed structure is filled. No more efficient way of packing spheres is known, and the hexagonal closest packed structure is important for such metals as Be, Co, Mg, and Zn, as well as the rare gas He at low temperatures.

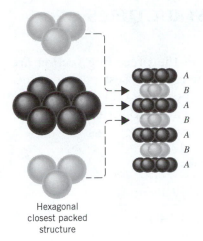

Hexagonal
closest packed
structure

FIGURE 9.12 Each atom in a hexagonal closest packed structure touches six atoms in the same plane, three in the plane above, and three in the plane below. The result is an *ABABAB* . . . repeating pattern of closest packed planes. All spheres represent identical atoms.

There is another way of stacking hexagonal closest packed planes of spheres. The atoms in the third plane can be packed *above the holes* in the first plane that have not been used to form the second plane. The fourth hexagonal closest packed plane of atoms then packs directly above the first. The net result is an *ABCABCABC* . . . structure, which is called **cubic closest packed.** Each sphere in the structure touches six others in the same plane, three in the plane above, and three in the plane below, as shown in Figure 9.13. Thus, the coordination number is still 12.

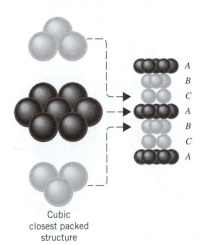

Cubic
closest packed
structure

FIGURE 9.13 Each atom in a cubic closest packed structure also touches six atoms in the same plane, three in the plane above, and three in the plane below. But the atoms in the top plane are rotated by 180° relative to the bottom plane. The planes of atoms therefore form an *ABCAB-CABC* . . . repeating pattern. All spheres represent identical atoms.

The difference between hexagonal and cubic closest packed structures can be understood by comparing Figures 9.12 and 9.13. In the hexagonal closest packed structure, the atoms in the first and third planes lie directly above each other. In the cubic closest packed structure, the atoms in those planes are oriented in different directions.

The cubic closest packed structure is just as efficient as the hexagonal closest packed structure. (Both use 74% of the available space.) Many metals, including Ag, Al, Au, Ca, Cu, Ni, Pb, and Pt, crystallize in a cubic closest packed structure. All the rare gases except helium behave in the same manner when cooled to temperatures low enough to allow solidification.

9.9   COORDINATION NUMBERS AND THE STRUCTURES OF METALS

The coordination numbers of the four structures described in the preceding section are summarized in Table 9.5. Some metals pack in hexagonal or cubic closest packed structures. Not only do those structures use space as efficiently as possible, they also have the largest possible coordination numbers, which allows each metal atom to form bonds to the largest number of neighboring metal atoms.

TABLE 9.5   Coordination Numbers for Common Crystal Structures

Structure	Coordination Number	Stacking Pattern
Simple cubic	6	*AAAAAAAA* . . .
Body-centered cubic	8	*ABABABAB* . . .
Hexagonal closest packed	12	*ABABABAB* . . .
Cubic closest packed	12	*ABCABCABC* . . .

It is less obvious why one-third of the metals pack in a body-centered cubic structure, in which the coordination number is only 8. The popularity of the body-centered cubic structure can be understood by referring to Figure 9.14. The coordination number for body-centered cubic structures given in Table 9.5 counts only the atoms that actually touch a given atom in the structure. Figure 9.14 shows that each atom also *almost touches* four neighbors in the same plane, a fifth neighbor two planes above, and a sixth two planes below. The distance from each atom to the nuclei of the nearby atoms is only 15% larger than the distance to the nuclei of the atoms that it actually touches. Each atom in a body-centered cubic structure therefore interacts with 14 other atoms—eight strong interactions to the atoms that it touches and six weaker interactions to the atoms it almost touches.

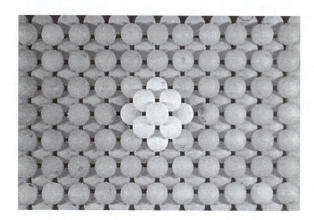

FIGURE 9.14 Each atom in a body-centered cubic structure touches four atoms in the plane above and four in the plane below. In addition, each atom *almost* touches six more atoms.

This makes it easier to understand why a metal might prefer the body-centered cubic structure to the hexagonal or cubic closest packed structure. Each metal atom in the closest packed structures interacts with 12 neighboring atoms. In the body-centered cubic structure, each atom interacts with 14 neighboring atoms.

9.10   UNIT CELLS: THE SIMPLEST REPEATING UNIT IN A CRYSTAL

So far, our description of solids has focused on the way the particles pack to fill space. Another way of describing the structures of solids was introduced in Section 5.8. This approach assumes that crystals are three-dimensional analogs of a piece of wallpaper. Wallpaper has a regular repeating design that extends from one edge to the other. Crystals have a similar repeating design, but in this case the design extends in three dimensions from one edge of the solid to the other.

We can unambiguously describe a piece of wallpaper by specifying the size, shape, and contents of the simplest repeating unit in the design. We can describe a three-dimensional crystal by specifying the size, shape, and contents of the simplest repeating unit and the way the repeating units stack to form the crystal. The simplest repeating unit in a crystal is called a **unit cell,** which is defined in terms of **lattice points**—the points in space about which the particles are free to vibrate in a crystal.

This section focuses on the three unit cells shown in Figure 9.15: simple cubic, body-centered cubic, and face-centered cubic. These unit cells are important for two reasons. First, a number of metals, ionic solids, and intermetallic compounds crystallize in cubic unit cells. Second, these unit cells have identical edge lengths for a given cubic cell and the cell angles are all 90°. Ionic solids have structures similar to those discussed in Section 9.8. However, ionic solids are composed of two or more different ions. We can describe the structure of ionic solids by assuming that the ions are perfect spheres but of different sizes (see Figure 9.3).

FIGURE 9.15 Models of simple-cubic (left), body-centered cubic (center), and face-centered cubic unit cells (right).

The **simple cubic unit cell** is the simplest repeating unit in a simple cubic structure. Each corner of the unit cell is defined by a lattice point at which an identical particle can be found. By convention, the edge of a unit cell always connects equivalent points. Each of the eight corners of the unit cell therefore must contain an identical particle. Other particles can be present on the edges or faces of the unit cell, or within the body of the unit cell. But the minimum that must be present for the unit cell to be classified as simple cubic is eight equivalent particles on the eight corners.

The **body-centered cubic unit cell** is the simplest repeating unit in a body-centered cubic structure. Once again, there are eight identical particles on the eight corners of the unit cell. In this case, however, there is a ninth identical particle in the center of the body of the unit cell.

Checkpoint

Iron metal and cesium chloride have similar structures. The simplest repeating unit in iron is a cube of eight iron atoms with a ninth iron atom in the center of the body of the cube. The simplest repeating unit in CsCl is a cube of Cl^- ions with a Cs^+ ion in the center of the body. Explain why one of the structures is classified as a body-centered cubic unit cell and the other as a simple cubic unit cell.

The **face-centered cubic unit cell** also starts with identical particles on the eight corners of the cube. But the structure also contains the same particles in the centers of the six faces of the unit cell, for a total of 14 identical lattice points. The face-centered cubic unit cell is the simplest repeating unit in a cubic closest packed structure. In fact, the presence of face-centered cubic unit cells in the structure explains why the structure is known as *cubic* closest packed.

9.11   MEASURING THE DISTANCE BETWEEN PARTICLES IN A UNIT CELL

Nickel was identified in Section 9.8 as one of the metals that crystallizes in a cubic closest packed structure. When we consider that a nickel atom has a mass of only 9.75×10^{-23} g and a radius of only 1.24×10^{-10} m, it is a remarkable achievement for us to be able to describe the structure of the metal. The obvious question is, How do we know that nickel packs in a cubic closest packed structure?

The only way to determine the structure of matter on an atomic scale is to use a probe that is even smaller. As we have seen, one of the most useful probes for studying matter on the atomic scale is electromagnetic radiation. In 1912, Max von Laue found that X rays which struck the surface of a crystal were diffracted into patterns that resembled the patterns produced when light passes through a very narrow slit. Shortly thereafter, William Lawrence Bragg, who was just completing an undergraduate degree in physics at Cambridge University, explained von Laue's results. Bragg argued that X rays were reflected from planes of atoms near the surface of the crystal, as shown in Figure 9.16. He then concluded that the only way the X rays could stay in phase was if some integer (n) times the wavelength of the radiation (λ) was twice the distance (d) between adjacent planes of atoms times the sine of the angle θ.

$$n\lambda = 2d \sin \theta$$

This relationship, which became known as the **Bragg equation,** allows us to calculate the distance between planes of atoms in a crystal from the pattern of diffraction of X rays of known wavelength.

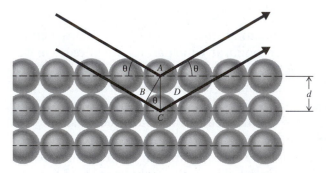

FIGURE 9.16 Diffraction of X rays by the first and second planes in a crystal.

The pattern by which X rays are diffracted by nickel metal suggests that the metal packs in a cubic unit cell with a distance between planes of atoms of 0.3524 nm. Thus, the length of the edge of a unit cell in the crystal must be 0.3524 nm. Knowing that nickel crystallizes in a cubic unit cell is not enough. We still have to decide whether it is a simple cubic, body-centered cubic, or face-centered cubic unit cell. As we'll see in the next section, we can do this by measuring the density of the metal.

Checkpoint
Why does the diffraction of X rays by solids show that solids consist of an ordered arrangement of particles?

9.12   DETERMINING THE UNIT CELL OF A CRYSTAL

Atoms on the corners, edges, and faces of a unit cell are shared by more than one unit cell, as shown in Figure 9.17. An atom on a face is shared by two unit cells, so only half of the

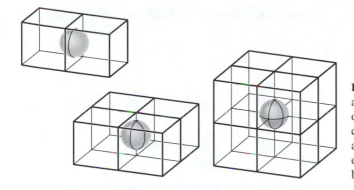

FIGURE 9.17 Because an atom on the face of a unit cell is shared by two unit cells, only one-half of the atom belongs to each of the cells. For similar reasons, one-quarter of an atom on the edge of a unit cell, and one-eighth of an atom on the corner of a unit cell, belong to the unit cell.

atom belongs to each of the cells. An atom on an edge is shared by four unit cells, and an atom on a corner is shared by eight unit cells. Thus, only one-quarter of an atom on an edge and one-eighth of an atom on a corner can be assigned to each of the unit cells that share the atoms.

If nickel crystallized in a simple cubic unit cell, there would be a nickel atom on each of the eight corners of the cell. Because only one-eighth of each atom could be attributed to a given unit cell, each unit cell in a simple cubic structure would have one net nickel atom.

Simple-cubic structure:

8 corners × 1/8 = 1 net atom/unit cell

If nickel formed a body-centered cubic structure, there would be two atoms per unit cell, because the nickel atom in the center of the body wouldn't be shared with any other unit cells.

Body-centered cubic structure:

(8 corners × 1/8) + 1 body = 2 net atoms/unit cell

If nickel crystallized in a face-centered cubic structure, the six atoms on the faces of the unit cell would contribute three net nickel atoms, for a total of four atoms per unit cell.

Face-centered cubic structure:

(8 corners × 1/8) + (6 faces × 1/2) = 4 net atoms/unit cell

Because they have different numbers of atoms in a unit cell, each of the structures would have a significantly different density. Let's therefore calculate the theoretical density for nickel on the basis of each structure and the unit cell edge length for nickel given in the previous section: 0.3524 nm. To do this, we need to know the volume of a unit cell in cubic centimeters and the weight of a single nickel atom. The type of cubic cell associated with nickel can now be found by calculating the theoretical density of nickel for each cell. The calculated density that matches the experimental density will give us the correct unit cell for nickel.

The volume (V) of the unit cell is equal to the cell edge length (a) cubed.

$$V = a^3 = (0.3524 \text{ nm})^3 = 0.04376 \text{ nm}^3$$

Because there are 10^9 nm in a meter and 100 cm in a meter, there must be 10^7 nm in a cm.

$$\frac{10^9 \text{ nm}}{1 \text{ m}} \times \frac{1 \text{ m}}{100 \text{ cm}} = 10^7 \text{ nm/cm}$$

Converting the volume of the unit cell to cubic centimeters gives the following result.

$$4.376 \times 10^{-2} \text{ nm}^3 \times \frac{(1 \text{ cm})^3}{(10^7 \text{ nm})^3} = 4.376 \times 10^{-23} \text{ cm}^3$$

The weight of a single nickel atom can be calculated from the atomic weight of the metal and Avogadro's constant.

$$\frac{58.69 \text{ g Ni}}{1 \text{ mol}} \times \frac{1 \text{ mol}}{6.022 \times 10^{23} \text{ atoms}} = 9.746 \times 10^{-23} \text{ g/atom}$$

Now that the volume is known, the theoretical density that nickel would have in each of the three unit cells can be determined.

Simple cubic structure:

$$\frac{9.746 \times 10^{-23} \text{ g/unit cell}}{4.376 \times 10^{-23} \text{ cm}^3\text{/unit cell}} = 2.227 \text{ g/cm}^3$$

The density of nickel, if it crystallized in a simple-cubic structure, would therefore be 2.227 g/cm^3.

Body-centered cubic structure:

$$\frac{2(9.746 \times 10^{-23} \text{ g/unit cell})}{4.376 \times 10^{-23} \text{ cm}^3\text{/unit cell}} = 4.454 \text{ g/cm}^3$$

Because there would be twice as many nickel atoms per unit cell if nickel crystallized in a body-centered cubic structure, the density of nickel in this structure would be twice as large as for the simple cubic.

Face-centered cubic structure:

$$\frac{4(9.746 \times 10^{-23} \text{ g/unit cell})}{4.376 \times 10^{-23} \text{ cm}^3\text{/unit cell}} = 8.909 \text{ g/cm}^3$$

There would be four nickel atoms per unit cell in a face-centered cubic structure, and the density of nickel in this structure would be four times as large as for the simple cubic.

The experimental value for the density of nickel is 8.90 g/cm^3. The obvious conclusion is that nickel crystallizes in a face-centered cubic unit cell and therefore has a cubic closest packed structure.

Checkpoint

What three factors account for the density of a metal?

9.13   CALCULATING THE SIZE OF AN ATOM OR ION

Estimates of the radii of most metal atoms can be found in Appendix B.4. Where do the data come from? How do we know, for example, that the metallic radius of a nickel atom is 0.1246 nm? The starting point for calculating the radius uses the results of the previous two sections. We now know that the nickel crystallizes in a cubic unit cell with a cell edge length of 0.3524 nm, and we know that the unit cell for the crystal is face-centered cubic. One of the faces of a face-centered cubic unit cell is shown in Figure 9.18.

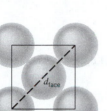

FIGURE 9.18 The diagonal across the face of a face-centered cubic unit cell is equal to four times the radius of the atoms that form the cell.

According to Figure 9.18, the diagonal across the face of the unit cell is equal to four times the metallic radius of a nickel atom.

$$d_{face} = 4\,r_{Ni}$$

The Pythagorean theorem states that the long edge of a right triangle is equal to the sum of the squares of the other sides. The diagonal across the face of the unit cell is therefore related to the unit cell edge length by the following equation.

$$d_{face}^2 = a^2 + a^2$$

Taking the square root of both sides gives the following result.

$$d_{face} = a\sqrt{2}$$

Because the diagonal across the face is four times the metallic radius of a nickel atom, the following substitution can be made.

$$4r_{Ni} = a\sqrt{2}$$

Thus, the metallic radius of a nickel atom is 0.1246 nm.

$$r_{Ni} = \frac{a\sqrt{2}}{4} = \frac{0.3524 \text{ nm} \times \sqrt{2}}{4} = 0.1246 \text{ nm}$$

A similar approach can be taken for estimating the size of an ion. Consider cesium chloride, for example, which crystallizes in a simple cubic unit cell of Cl^- ions with a Cs^+ ion in the center of the body of the cell, as shown in Figure 9.19. The Cs^+ ion in the center of the unit cell must touch the Cl^- ions at the corners. The diagonal across the body of the CsCl unit cell is equal to the sum of the radii of two Cl^- ions and two times the radius of a Cs^+ ion.

$$d_{body} = 2\,r_{Cs^+} + 2\,r_{Cl^-}$$

The three-dimensional equivalent of the Pythagorean theorem suggests that the square of the diagonal across the body of a cube is the sum of the squares of the three sides.

$$d_{body}^2 = a^2 + a^2 + a^2$$

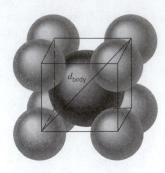

FIGURE 9.19 The diagonal across the body of the CsCl unit cell is equal to twice the sum of the radii of the Cs^+ and Cl^- ions.

Taking the square root of both sides of the equation gives the following result.

$$d_{body} = a\sqrt{3}$$

If the cell edge length in CsCl is experimentally determined to be 0.4123 nm, the diagonal across the body in the unit cell is 0.7141 nm.

$$d_{body} = a\sqrt{3} = 0.4123 \text{ nm} \times \sqrt{3} = 0.7141$$

The sum of the ionic radii of Cs^+ and Cl^- ions is half that distance, or 0.3571 nm.

$$r_{Cs^+} + r_{Cl^-} = \frac{d_{body}}{2} = \frac{0.7141 \text{ nm}}{2} = 0.3571 \text{ nm}$$

If we had an estimate of the size of either the Cs^+ or Cl^- ion, we could use the results of our calculation to estimate the size of the other ion. By combining the analysis of many ionic compounds, it is possible to create a set of consistent data for the size of the ions that form the crystals. Some of the data were reported in Section 3.20; a more complete set can be found in Appendix B.4. The small discrepancy between the sum of the ionic radii of the Cs^+ ion (0.169 nm) and the Cl^- ion (0.181 nm) reported in tables of ionic radii and the results of the calculation for CsCl reflect the fact that ionic radii seem to vary slightly from one crystal to another.

Checkpoint

For the three cubic structures studied, which has the simplest relationship between the radii of the ions and the length of the edge of the unit cell?

KEY TERMS

Amorphous solid	Crystalline solid	Metallic solid
Body-centered cubic packing	Cubic closest packing	Molecular solid
	Electron affinity	Network covalent solid
Body-centered cubic unit cell	Face-centered cubic unit cell	Polycrystalline solid
	Hexagonal closest packing	Semiconductor
Bragg equation	Ionic solid	Simple cubic packing
Ceramic	Lattice energy	Simple cubic unit cell
Closest packed structure	Lattice points	Superconductor
Coordination number	Metallic bond	Unit cell

PROBLEMS

Molecular, Covalent, Ionic, and Metallic Solids

1. Which of the following compounds should conduct an electric current when dissolved in water?
(a) $MgCl_2$ (b) CO_2 (c) CH_3OH (d) KNO_3 (e) Ca_3P_2

2. Why do ionic compounds conduct electricity better in the liquid state than the solid state?

3. One of the simplest ways of distinguishing between two covalent compounds is to measure their melting points or boiling points. Naphthalene melts at 80.5°C and camphor melts at 179.8°C, for example. Explain why this procedure is not so useful when distinguishing between ionic compounds such as NaCl and Al_2O_3.

4. Hydrogen chloride is a covalent compound that is a gas at room temperature. When cooled, it condenses to form a liquid. Neither HCl gas nor liquid HCl conducts electricity. But when HCl is dissolved in water, the result is hydrochloric acid, which conducts electricity very well. Explain why.

5. Which of the following would you expect to conduct an electric current?
(a) solid Na metal (b) liquid Na metal (c) solid NaCl
(d) liquid NaCl   (e) NaCl dissolved in water

6. Classify the following solids as molecular, network covalent, ionic, or metallic. Figures 9.1 and 5.10 may be helpful.
(a) $BaSO_4$ (b) $NaOH$ (c) Xe (d) I_2
(e) aluminum   (f) brass (mixture of Cu and Zn) (g) P_4 (h) P_4O_{10}

7. Which of the following solids are held together by an extended network of covalent bonds?
(a) sodium chloride   (b) graphite (c) gold
(d) calcium carbonate   (e) diamond (f) dry ice (solid CO_2)

8. Which force must be overcome to sublime dry ice, solid CO_2?
(a) metallic bonding (b) ionic bonding (c) covalent bonding
(d) dispersion forces

9. Which force must be overcome to melt solid argon?
(a) metallic bonding (b) ionic bonding (c) covalent bonding
(d) dispersion forces

10. Which of the following categories is most likely to contain a compound that is a poor conductor of electricity when solid but a very good conductor when molten?
(a) molecular solids (b) covalent solids (c) ionic solids
(d) metallic solids

Lattice Energies and the Strength of the Ionic Bond

11. Define the term *lattice energy*.

12. The lattice energy of NaCl refers to which of the following reactions?
(a) $2 Na(s) + Cl_2(s) \rightarrow 2 NaCl(s)$ (b) $NaCl(s) \rightarrow Na(g) + Cl(g)$
(c) $Na(g) + Cl(g) \rightarrow NaCl(g)$ (d) $NaCl(s) \rightarrow Na^+(g) + Cl^-(g)$
(e) $Na^+(g) + Cl^-(g) \rightarrow NaCl(s)$

13. Which of the following salts has the largest lattice energy?
(a) LiF (b) LiCl (c) LiBr (d) LiI

14. Which of the following salts has the largest lattice energy?
(a) NaCl (b) NaI (c) KI (d) MgO (e) MgS

426 CHAPTER 9 SOLIDS

15. Explain the following trends in lattice energies.

$$MgF_2 > MgCl_2 > MgBr_2 > MgI_2$$
$$BeF_2 > MgF_2 > CaF_2 > SrF_2 > BaF_2$$

16. Use the *CRC Handbook of Chemistry and Physics*[2] to determine the solubility in water of NaF, NaCl, NaBr, and NaI. Describe the relationship between the solubilities of the salts and their lattice energies.

17. Use lattice energies to explain why MgO is much less soluble in water than is CaO.

18. Using the enthalpy data in Appendix B.14 and knowing that the enthalpy change for the reaction $O(g) + 2\ e^- \rightarrow O^{2-}(g)$ is $+448$ kJ/mol$_{rxn}$, calculate the lattice enthalpies for MgO, CaO, and BaO. Explain the trend you observe.

Physical Properties That Result from the Structure of Metals

19. Explain why metals are solids at room temperature.

20. Explain why metals are malleable and ductile.

21. Explain why metals conduct heat and electricity.

The Structure of Metals and Other Monatomic Solids

22. Describe the difference in the way planes of atoms stack to form *hexagonal closest packed, cubic closest packed, body-centered cubic,* and *simple cubic* structures.

23. Explain why the structure of polonium is called *simple cubic;* why the structure of iron is called *body-centered cubic;* and why the structure of cobalt is called *hexagonal closest packed.*

24. Determine the coordination numbers of the metal atoms in each of the following structures.
    (a) cubic closest packed aluminum   (b) hexagonal closest packed magnesium
    (c) body-centered cubic chromium   (d) simple cubic polonium

25. In which of the following structures would a xenon atom form the largest number of induced dipole–induced dipole interactions?
    (a) simple cubic   (b) body-centered cubic
    (c) cubic closest packed (d) hexagonal closest packed

26. Sodium crystallizes in a structure in which the coordination number is eight. Which structure best describes the crystal?
    (a) simple cubic   (b) body-centered cubic
    (c) cubic closest packed (d) hexagonal closest packed

Unit Cells: Calculating Metallic and Ionic Radii of Atoms and Ions

27. Chromium metal ($d = 7.20$ g/cm^3) crystallizes in a body-centered cubic unit cell. Calculate the volume of the unit cell and the radius of a chromium atom.

28. The metallic radius of a vanadium atom is 0.1321 nm. What is the density of vanadium if the metal crystallizes in a body-centered cubic unit cell?

29. Calculate the atomic radius of an Ar atom, assuming that argon crystallizes at low temperature in a face-centered cubic unit cell with a density of 1.623 g/cm^3.

[2]CRC Press, Boca Raton, Florida.

30. Silver crystallizes in a face-centered cubic unit cell with an edge length of 0.40862 nm. Calculate the density of Ag metal in grams per cubic centimeter.

31. Potassium crystallizes in a cubic unit cell with an edge length of 0.5247 nm. The density of potassium is 0.856 g/cm^3. Determine whether the element crystallizes in a simple cubic, a body-centered cubic, or a face-centered cubic unit cell.

32. Determine whether calcium crystallizes in a simple cubic, a body-centered cubic, or a face-centered cubic unit cell, assuming that the cell edge length is 0.5582 nm and the density of the metal is 1.55 g/cm^3.

33. Determine whether molybdenum crystallizes in a simple cubic, a body-centered cubic, or a face-centered cubic unit cell, assuming that the cell edge length is 0.3147 nm and the density of the metal is 10.2 g/cm^3.

34. Which of the following metals crystallizes in a face-centered cubic unit cell with an edge length of 0.3608 nm if the density of the metal is 8.95 g/cm^3?
    (a) Na   (b) Ca   (c) Tl   (d) Cu   (e) Au

35. Barium crystallizes in a body-centered cubic structure in which the cell edge length is 0.5025 nm. Calculate the shortest distance between neighboring barium atoms in the crystal.

36. NaH crystallizes in a structure similar to that of NaCl. If the cell edge length in the crystal is 0.4880 nm, what is the average length of the Na—H bond?

37. TlI crystallizes in a structure similar to that of CsCl with a cell edge length of 0.4198 nm. Calculate the average Tl—I bond length in the crystal. If the ionic radius of an I$^-$ ion is 0.216 nm, what is the ionic radius of the Tl$^+$ ion?

38. Calculate the ionic radius of the Cs$^+$ ion, assuming that the cell edge length for CsCl is 0.4123 nm and the ionic radius of a Cl$^-$ ion is 0.181 nm.

39. CdO crystallizes in a cubic unit cell with a cell edge length of 0.4695 nm. Calculate the number of Cd^{2+} and O^{2-} ions per unit cell, assuming that the density of the crystal is 8.15 g/cm^3.

40. LiF crystallizes in a cubic unit cell with a cell edge length of 0.4017 nm. Calculate the number of Li$^+$ and F$^-$ ions per unit cell, assuming that the density of the salt is 2.640 g/cm^3.

41. Iron ($d = 7.86$ g/cm^3) crystallizes in a body-centered cubic unit cell at room temperature. Calculate the radius of an iron atom in the crystal. At temperatures above 910°C, iron prefers a face-centered cubic structure. If we assume that the change in the size of the iron atom is negligible when the metal is heated to 910°C, what is the density of iron in the face-centered cubic structure? Use the calculation to predict whether iron should expand or contract when it changes from the body to the face-centered structure.

Integrated Problems

42. From the enthalpy data in Appendix B.14 calculate the enthalpy change required to break the bond in the following: F$_2$(g), Cl$_2$(g), Br$_2$(g), and I$_2$(g). Compare the enthalpies to the enthalpy required to *melt* each of the halogens given in Table 9.1. What conclusions can you reach concerning the forces that hold the atoms together and those that hold the molecules to one another?

43. From the data in Appendix B calculate the lattice energy of BaCl$_2$. Compare the lattice energy of BaCl$_2$ to that of NaCl and account for any differences.

44. From the data in Appendix B calculate the lattice energy of BeCl$_2$ and compare it to that of BaCl$_2$. Explain any differences.

45. The enthalpies of fusion of the alkali metals are given below.

Metal	ΔH_{fus} (kJ/mol$_{rxn}$)
Li	2.9
Na	2.6
K	2.4
Rb	2.2

Identify the type of solid formed and explain the trend seen in the enthalpy required to melt the solids. Arrange the solids in order of increasing melting point.

46. For each of the properties (A through J) listed below choose the appropriate electronegativity characteristic [(a) through (f) below] for a binary compound.
   A. A good conductor of electricity
   B. A hard material that conducts electricity in the melted state
   C. An insulator
   D. A material that conducts electricity when dissolved in water
   E. A semiconductor
   F. A ceramic
   G. A molecular crystal
   H. A metallic compound
   I. An ionic compound
   J. A hard material that is an insulator
   (a) large ΔEN, low EN for both atoms
   (b) small ΔEN, high EN for both atoms
   (c) moderate ΔEN, moderate EN for both atoms
   (d) moderate ΔEN, high EN for both atoms
   (e) small ΔEN, low EN for both atoms
   (f) large ΔEN

47. Explain why binary materials may be separated into the ionic, covalent, semimetal, or metallic regions by a bond type triangle.

48. Classify the following binary compounds as primarily metallic, molecular, network covalent, or ionic. Figures 9.1 and 5.10 may be helpful.
   (a) B_2H_6 (b) B_4C (c) InAs   (d) HgI_2
   (e) Hg_2Na_3 (f) K_2S (g) Cd_3Mg (h) KBr
   (i) MgH_2 (j) GaS   (k) LiH   (l) Be_3P_2

CHAPTER
10
AN INTRODUCTION TO KINETICS AND EQUILIBRIUM

10.1   REACTIONS THAT DON'T GO TO COMPLETION

Suppose that you were asked to describe the steps involved in calculating the mass of the finely divided white solid produced when a 2.00-g strip of magnesium metal is burned. You might organize your work as follows.

- Assume that the magnesium reacts with oxygen in the atmosphere when it burns.
- Predict that the formula of the product is MgO.
- Use the formula to generate the following balanced equation.

$$2 \, Mg(s) + O_2(g) \longrightarrow 2 \, MgO(s)$$

- Use the mass of a mole of magnesium to convert grams of magnesium into moles of magnesium.

$$2.00 \, \text{g Mg} \times \frac{1 \, \text{mol Mg}}{24.31 \, \text{g Mg}} = 0.0823 \, \text{mol Mg}$$

- Use the balanced equation to convert moles of magnesium into moles of magnesium oxide.

$$0.0823 \text{ mol Mg} \times \frac{2 \text{ mol MgO}}{2 \text{ mol Mg}} = 0.0823 \text{ mol MgO}$$

- Use the mass of a mole of magnesium oxide to convert moles of MgO into grams of MgO.

$$0.0823 \text{ mol MgO} \times \frac{40.30 \text{ g MgO}}{1 \text{ mol MgO}} = 3.32 \text{ g MgO}$$

Before you read any further, ask yourself, "How confident am I in the answer?"

Before we can trust the answer we have to consider whether there are any hidden assumptions behind the calculation and then check the validity of the assumptions. Three assumptions have been made in the calculation.

- We assumed that the metal strip was pure magnesium.
- We assumed that the magnesium reacted only with the oxygen in the atmosphere to form MgO, ignoring the possibility that some of the magnesium might react with the nitrogen in the atmosphere to form Mg_3N_2.
- We assumed that the reaction didn't stop until all of the magnesium metal had been consumed.

It is relatively easy to correct for the fact that the starting material may not be pure magnesium by weight. We can also correct for the fact that as much as 5% of the product of the reaction is Mg_3N_2 instead of MgO. But it is the third assumption that is of particular importance in this chapter.

The assumption that all chemical reactions go to completion is fostered by calculations, such as predicting the amount of MgO produced by burning a known amount of magnesium. It is reinforced by demonstrations such as the reaction in which a copper penny dissolves in concentrated nitric acid, which seems to continue until the copper penny disappears.

However, chemical reactions don't always go to completion. The following equation provides an example of a chemical reaction that seems to stop prematurely.

$$2 \, NO_2(g) \rightleftharpoons N_2O_4(g)$$

At 25°C, when 1 mol of NO_2 is added to a 1.00-L flask, the reaction seems to stop when 95% of the NO_2 has been converted to N_2O_4. Once it has reached that point, the reaction doesn't go any further. As long as the reaction is left at 25°C, about 5% of the NO_2 that was present initially will remain in the flask. Reactions that seem to stop before the limiting reagent is consumed are said to reach **equilibrium.**

It is useful to recognize the difference between reactions that come to equilibrium and those that stop when they run out of the limiting reagent. The reaction between a copper penny and nitric acid is an example of a reaction that continues until it has essentially run out of the limiting reagent. We indicate this by writing the equation for the reaction with a single arrow pointing from the reactants to the products.

$$Cu(s) + 4 \, HNO_3(aq) \longrightarrow Cu^{2+}(aq) + 2 \, NO_3^-(aq) + 2 \, NO_2(g) + 2 \, H_2O(l)$$

We indicate that a reaction comes to equilibrium by writing a pair of arrows pointing in opposite directions between the two sides of the equation.

$$2 NO_2(g) \rightleftharpoons N_2O_4(g)$$

To work with reactions that come to equilibrium, we need a way to specify the amount of each reactant or product that is present in the system at any moment in time. By convention, this information is specified in terms of the concentration of each component of the system in units of moles per liter. This quantity is indicated by a symbol that consists of the formula for the reactant or product written in parentheses. For example,

(NO_2) = concentration of NO_2 in moles per liter at some moment in time

We then need a way to describe the system when it is at equilibrium. This is done by writing the symbols for each component of the system in square brackets.

$[NO_2]$ = concentration of NO_2 in moles per liter if, and only if, the reaction is at equilibrium

The fact that some reactions come to equilibrium raises a number of interesting questions.

- Why do the reactions seem to stop before all of the starting materials are converted to the products of the reaction?
- What is the difference between reactions that seem to go to completion and reactions that reach equilibrium?
- Is there any way to predict whether a reaction will go to completion or reach equilibrium?
- How does a change in the conditions of the reaction influence the amount of product formed?

Before we can understand how and why a chemical reaction comes to equilibrium, we have to build a model of the factors that influence the rate of a chemical reaction. We shall then apply the model to the simplest of all chemical reactions, those that occur in the gas phase. The next chapter includes interactions between the components of the reaction and the solvent in which it is run.

Checkpoint

If $N_2O_4(g)$ is placed into an evacuated container at 25°C, what chemical species will be in the container after a period of time?

10.2 GAS PHASE REACTIONS

The simplest chemical reactions are those that occur in the gas phase in a single step, such as the following reaction in which a compound known as *cis*-2-butene is converted into its isomer, *trans*-2-butene.

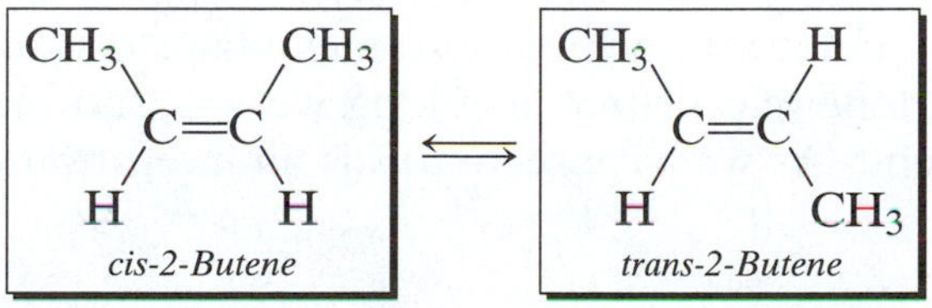

The reaction involves rotating one end of the C=C double bond relative to the other. Rotation around C=C bonds, however, doesn't occur at room temperature. The two isomers have different properties. *cis*-2-Butene melts at −139°C, whereas the trans form melts at −106°C. The cis isomer boils above 4°C, whereas the trans form boils at 1°C. We can obtain a pure sample of the starting material—*cis*-2-butene—and then watch as it is transformed to the product—*trans*-2-butene—when we heat the sample.

A sample that initially contained 1.000 mole of the cis isomer in a 10.0-L flask is at a temperature of 400°C. Table 10.1 shows the amount of both *cis*- and *trans*-2-butene present in the system with respect to time. Initially, there is no *trans*-2-butene in the system. With time, however, the amount of *cis*-2-butene gradually decreases as the compound is converted into the trans isomer.

TABLE 10.1   Experimental Data for the Isomerization of *cis*-2-Butene to *trans*-2-Butene at 400°C in a 10.0-L Flask

Time	Moles of cis-2-butene	Moles of trans-2-butene
0	1.000	0
5.00 days	0.919	0.081
10.00 days	0.848	0.152
15.00 days	0.791	0.209
20.00 days	0.741	0.259
40.00 days	0.560	0.440
60.00 days	0.528	0.472
120 days	0.454	0.546
1 year	0.441	0.559
2 years	0.441	0.559
3 years	0.441	0.559

The amount of the trans isomer present at any moment in time can be calculated from the amount of the cis isomer that remains in the system and vice versa. Because atoms are conserved, the total number of moles of the two isomers must always be the same as the number of moles that were present at the start of the reaction. In this case, we started with 1.000 mol of the cis isomer. At any moment in time, the number of moles of the cis isomer that remain in the system is $1.000 - x$, and the number of moles of the trans isomer is equal to x.

$$n_{cis\text{-}2\text{-butene}} = 1.000 - x$$
$$n_{trans\text{-}2\text{-butene}} = x$$

Figure 10.1 shows a plot of the number of moles of the cis and trans isomers of 2-butene versus time. Figure 10.1 clearly shows that there is no change in the number of moles once the reaction reaches the point at which the system contains 0.441 mol of *cis*-2-butene and 0.559 mol of *trans*-2-butene. No matter how long we wait, no more *cis*-2-butene is converted into *trans*-2-butene. As we have seen, this is an indication that the reaction comes to equilibrium.

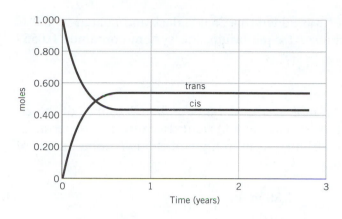

FIGURE 10.1 How the number of moles of the cis and trans isomers of 2-butene change with time at 400°C. Initially there is no *trans*-2-butene, but as time passes the cis isomer concentration decreases and the trans isomer concentration increases.

Chemical reactions at equilibrium are described in terms of the number of moles per liter of each component of the system, not the moles of each component. As we have seen, this quantity is represented by a symbol that consists of the chemical formula for that component written in brackets. For example,

[*cis*-2-butene] = concentration of *cis*-2-butene at equilibrium in moles per liter

[*trans*-2-butene] = concentration of *trans*-2-butene at equilibrium in moles per liter

The concentrations of *cis*- and *trans*-2-butene at equilibrium depend on the initial conditions of the experiment. But, at a given temperature, the ratio of the equilibrium concentrations of the two components of the reaction is always the same. It doesn't matter whether we start with a great deal of *cis*-2-butene or a relatively small amount, or whether we start with a pure sample of *cis*-2-butene or one that already contains some of the trans isomer. When the reaction reaches equilibrium at 400°C, the concentration of the trans isomer divided by that of the cis isomer concentration is always 1.27.

The equation that describes the relationship between the concentrations of the two components of the reaction at equilibrium is known as the **equilibrium constant expression,** where K_c is the **equilibrium constant** for the reaction.

$$K_c = \frac{[\textit{trans}\text{-2-butene}]}{[\textit{cis}\text{-2-butene}]}$$

The subscript c in the equilibrium constant indicates that the constant has been calculated from the concentrations of the reactants and products in units of moles per liter.

Checkpoint

If 1.0 mole of *trans*-2-butene is placed into an empty flask at 400°C, what will be the equilibrium ratio of *trans*-2-butene to *cis*-2-butene?

Exercise 10.1

Calculate the equilibrium constant for the conversion of *cis*-2-butene to *trans*-2-butene for the following sets of experimental data.

(a) A 5.00-mol sample of the cis isomer was added to a 10.0-L flask and heated to 400°C until the reaction came to equilibrium. At equilibrium, the system contained 2.80 mol of the trans isomer.

(b) A 0.100-mol sample of the cis isomer was added to a 25.0-L flask and heated to 400°C until the reaction came to equilibrium. At equilibrium, the system contained 0.0559 mol of the trans isomer.

Solution

(a) If 5.00 mol of cis-2-butene forms 2.80 mol of trans-2-butene, then 2.20 mol of the cis isomer must remain after the reaction reaches equilibrium. Because the experiment was done in a 10.0-L flask, the equilibrium concentrations of the two components of the reaction have the following values.

$$[trans\text{-}2\text{-}butene] = \frac{2.80 \text{ mol}}{10.0 \text{ L}} = 0.280 \ M$$

$$[cis\text{-}2\text{-}butene] = \frac{2.20 \text{ mol}}{10.0 \text{ L}} = 0.220 \ M$$

The equilibrium constant, K_c, for the reaction is therefore 1.27.

$$K_c = \frac{[trans\text{-}2\text{-}butene]}{[cis\text{-}2\text{-}butene]} = \frac{0.280 \ M}{0.220 \ M} = 1.27$$

(b) The system comes to equilibrium at the following concentrations of the cis and trans isomers.

$$[trans\text{-}2\text{-}butene] = \frac{0.0559 \text{ mol}}{25.0 \text{ L}} = 0.00224 \ M$$

$$[cis\text{-}2\text{-}butene] = \frac{0.0441 \text{ mol}}{25.0 \text{ L}} = 0.00176 \ M$$

Even though the equilibrium concentrations of the two isomers are very different from the values obtained in the previous experiment, the ratio of the concentrations is exactly the same.

$$K_c = \frac{[trans\text{-}2\text{-}butene]}{[cis\text{-}2\text{-}butene]} = \frac{0.00224 \ M}{0.00176 \ M} = 1.27$$

10.3 THE RATE OF A CHEMICAL REACTION

Experiments such as the one that gave us the data in Table 10.1 are classified as measurements of **chemical kinetics** (from a Greek stem meaning "to move"). The goal of the experiments is to describe the **rate of reaction,** that is, the rate at which the reactants are transformed into the products of the reaction.

The term *rate* is often used to describe the change in a quantity that occurs per unit of time. The rate of inflation, for example, is the change in the average cost of a collection of standard items per year. The rate at which an object travels through space is the distance traveled per unit of time, such as miles per hour or kilometers per second. In chemical kinetics, the distance traveled is the change in the concentration of one of the components of the reaction. The rate of a reaction is therefore the change in the concentration of one of the components, $\Delta(X)$, that occurs during a given time, Δt. Parentheses are used here because the concentration of X is changing—the reaction is not at equilibrium. In general, Δ represents *final* conditions minus initial conditions. Therefore, $\Delta(X)$ represents the final concentration X_{final} (or the concentration after the passage of some time t) minus the initial concentration $X_{initial}$. Because the concentration of reactant X decreases as the reaction proceeds $\Delta(X)$ is negative. Thus to have a positive rate a minus sign is used in the equation.

$$\text{Rate} = \frac{-\Delta(X)}{\Delta t}$$

Let's use the data in Table 10.1 to calculate the rate at which *cis*-2-butene is transformed into *trans*-2-butene during each of the following periods.

- During the first time interval, when the number of moles of *cis*-2-butene in the 10.0-L flask falls from 1.000 to 0.919.
- During the second interval, when the amount of *cis*-2-butene falls from 0.919 to 0.848 moles.
- During the third interval, when the amount of *cis*-2-butene falls from 0.848 to 0.791 moles.

Before we can calculate the rate of the reaction during each of the time intervals, we have to remember that the rate of reaction is defined in terms of changes in the number of moles per liter of one of the components of the reaction, not the number of moles of that reactant. Thus, we have to recognize that 1.000 mol of *cis*-2-butene in a 10.0-L flask corresponds to a concentration of 0.1000 M.

During the first time period, the rate of the reaction is 1.6×10^{-3} moles per liter per day.

$$\text{Rate} = \frac{-\Delta(X)}{\Delta t} = \frac{-(0.0919 \ M - 0.1000 \ M)}{(5.00 \ \text{days} - 0 \ \text{days})} = 1.6 \times 10^{-3} \ M/\text{day}$$

During the second time period, the rate of reaction is slightly smaller.

$$\text{Rate} = \frac{-\Delta(X)}{\Delta t} = \frac{-(0.0848 \ M - 0.0919 \ M)}{(10.00 \ \text{days} - 5.00 \ \text{days})} = 1.4 \times 10^{-3} \ M/\text{day}$$

During the third time period, the rate of reaction is even smaller.

$$\text{Rate} = \frac{-\Delta(X)}{\Delta t} = \frac{-(0.0791 \ M - 0.0848 \ M)}{(15.00 \ \text{days} - 10.00 \ \text{days})} = 1.1 \times 10^{-3} \ M/\text{day}$$

The calculations illustrate an important point: The rate of the reaction isn't constant; it changes with time. As for most reactions, the rate of the reaction gradually decreases as the starting materials are consumed, which means that the rate of reaction changes while it is being measured.

We can minimize the error this introduces into our measurements by measuring the rate of reaction over periods of time that are short compared with the time it takes for the reaction to occur. We might try, for example, to measure the infinitesimally small change in concentration, $d(X)$, that occurs over an infinitesimally short period of time, dt. This ratio is known as the *instantaneous rate of reaction*.

$$\text{Rate} = \frac{-d(X)}{dt}$$

The instantaneous rate of reaction at any moment in time can be calculated from a graph of the concentration of the reactant (or product) versus time. Figure 10.2 shows how the rate of reaction for the isomerization of *cis*-2-butene can be calculated from such a graph. The rate of reaction at any moment in time is equal to the slope of a tangent drawn to the curve at that moment.

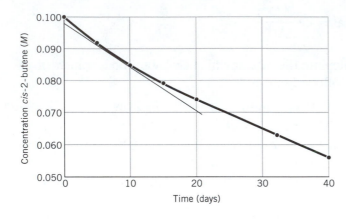

FIGURE 10.2 The rate of reaction at a given time for the isomerization of *cis*-2-butene is the slope of a tangent drawn to the concentration curve at that particular point in time.

An interesting result is obtained when the instantaneous rate of reaction is calculated at various points along the curve in Figure 10.2. The rate of reaction at every point on the curve is directly proportional to the concentration of *cis*-2-butene at that moment in time.

$$Rate = k(cis\text{-}2\text{-butene})$$

This equation, which is determined from experimental data, describes the rate of the reaction. It is called the **rate law** for the reaction. The proportionality constant k is known as the **rate constant.**

Checkpoint

The following data were obtained for the rate constant for the decomposition of a metabolite, a reactant that supplies building material or energy in a biochemical system.

Temperature (°C)	Rate Constant (s⁻¹)
15	2.5×10^{-2}
20	4.5×10^{-2}
25	8.1×10^{-2}
30	1.6×10^{-1}

What do the data suggest happens to the rate of the reaction as the temperature increases?

10.4 THE COLLISION THEORY OF GAS PHASE REACTIONS

One way to understand why some reactions come to equilibrium is to consider a simple gas phase reaction that occurs in a single step, such as the transfer of a chlorine atom from $ClNO_2$ to NO to form NO_2 and ClNO.

$$ClNO_2(g) + NO(g) \rightleftharpoons NO_2(g) + ClNO(g)$$

This reaction can be understood by writing the Lewis structures for all four components of the reaction. Because they contain an odd number of electrons, both NO and NO_2 can combine with a neutral chlorine atom to form a molecule in which all of the electrons are paired. The reaction therefore involves the transfer of a chlorine atom from one molecule to another, as shown in Figure 10.3.

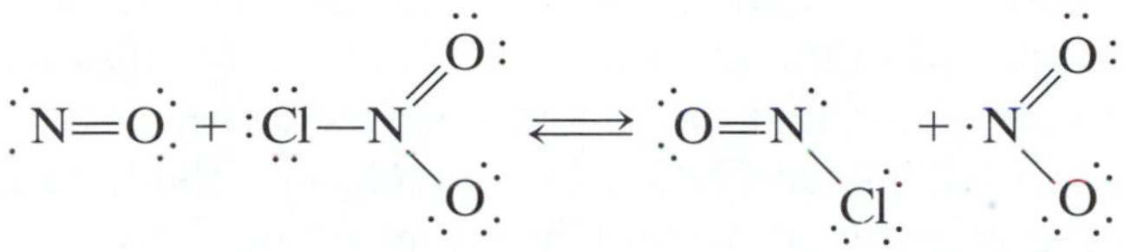

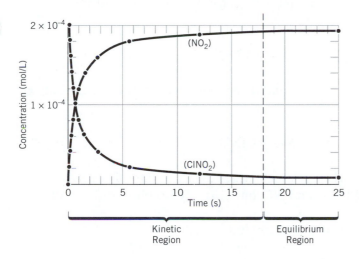

FIGURE 10.3 The reaction between $ClNO_2$ and NO to form NO_2 and ClNO is a simple, one-step reaction that involves the transfer of a chlorine atom.

Figure 10.4 combines a plot of the disappearance of the $ClNO_2$ consumed in the reaction with a plot of the appearance of NO_2 formed in the reaction. The data in Figure 10.4 are consistent with the following rate law for the reaction.

$$Rate = k(ClNO_2)(NO)$$

According to the rate law, the rate at which $ClNO_2$ and NO are converted into NO_2 and ClNO is proportional to the product of the concentrations of the two reactants. Initially, the rate of reaction is fast. As the reactants are converted to products, however, the $ClNO_2$ and NO concentrations become smaller, and the reaction slows down.

FIGURE 10.4 Plot of the change in the concentration of $ClNO_2$ superimposed on a plot of the change in concentration of NO_2 as $ClNO_2$ reacts with NO to produce NO_2 and ClNO. The graph can be divided into a kinetic and an equilibrium region.

We might expect the reaction to stop when it runs out of either $ClNO_2$ or NO. In practice, the reaction seems to stop before this happens. This is a very fast reaction—the concentration of $ClNO_2$ drops by a factor of 2 in less than a second. And yet, no matter how long we wait, some residual $ClNO_2$ and NO remain in the reaction flask.

Figure 10.4 divides the plot of the change in the concentrations of NO_2 and $ClNO_2$ into a **kinetic region** and an **equilibrium region.** The kinetic region marks the period during which the concentrations of the components of the reaction are constantly changing. The equilibrium region is the period after which the reaction seems to stop, when there is no further significant change in the concentrations of the components of the reaction.

The fact that the following reaction seems to stop before all of the reactants are consumed can be explained with the **collision theory** of chemical reactions.

$$ClNO_2(g) + NO(g) \rightleftharpoons NO_2(g) + ClNO(g)$$

The collision theory assumes that $ClNO_2$ and NO molecules must collide before a chlorine atom can be transferred from one molecule to the other. That assumption explains why the rate of the reaction is proportional to the concentration of both $ClNO_2$ and NO.

$$\text{Rate} = k(ClNO_2)(NO)$$

The number of collisions per second between $ClNO_2$ and NO molecules depends on their concentrations. As $ClNO_2$ and NO are consumed in the reaction, the number of collisions per second between the molecules becomes smaller, and the reaction slows down.

Suppose that we start with a mixture of $ClNO_2$ and NO, but no NO_2 or ClNO. The only reaction that can occur at first is the transfer of a chlorine atom from $ClNO_2$ to NO.

$$ClNO_2(g) + NO(g) \longrightarrow NO_2(g) + ClNO(g)$$

Eventually, NO_2 and ClNO build up in the reaction flask, and those molecules begin to collide as well. Collisions between the molecules can result in the transfer of a chlorine atom in the opposite direction.

$$ClNO_2(g) + NO(g) \longleftarrow NO_2(g) + ClNO(g)$$

The collision theory of chemical reactions assumes that the rate of a simple, one-step reaction is proportional to the product of the concentrations of the substances consumed in that reaction. The rate of the forward reaction is therefore proportional to the product of the concentrations of the two starting materials.

$$\text{Rate}_{forward} = k_f(ClNO_2)(NO)$$

The rate of the reverse reaction, on the other hand, is proportional to the concentrations of the products of the reaction.

$$\text{Rate}_{reverse} = k_r(NO_2)(ClNO)$$

Initially, the rate of the forward reaction is much larger than the rate of the reverse reaction, because the system contains $ClNO_2$ and NO but virtually no NO_2 and ClNO.

$$\text{Initially:}\qquad \text{Rate}_{forward} \gg \text{Rate}_{reverse}$$

As $ClNO_2$ and NO are consumed, the rate of the forward reaction slows down. At the same time, NO_2 and ClNO accumulate, and the reverse reaction speeds up.

The system eventually reaches a point at which the rates of the forward and reverse reactions are the same.

$$\text{Eventually:}\qquad \text{Rate}_{forward} = \text{Rate}_{reverse}$$

At this point, the concentrations of reactants and products do not change. $ClNO_2$ and NO are consumed in the forward reaction at the rate at which they are produced in the reverse reaction. The same thing happens to NO_2 and ClNO. When the rates of the forward and reverse reactions are the same, there is no longer any change in the concentrations of the reactants or products of the reaction. In other words, the reaction is at equilibrium.

We can now see that there are two ways to define equilibrium.

- A system in which there is no apparent change in the concentrations of the reactants and products of a reaction.
- A system in which the rates of the forward and reverse reactions are the same.

The first definition is based on the results of experiments that tell us that some reactions seem to stop prematurely—they reach a point at which no more reactants are converted to products before the limiting reagent is consumed. The other definition is based on a theoretical model of chemical reactions that explains why reactions reach equilibrium.

Checkpoint

If the rate of the reverse reaction is greater than the rate of the forward reaction at a given moment in time, does this mean that the reverse rate constant is greater than the forward rate constant?

10.5   EQUILIBRIUM CONSTANT EXPRESSIONS

Reactions don't stop when they come to equilibrium. But the rate of the forward and reverse reactions are the same, so there is no net change in the concentrations of the starting materials or products, and the reactions appear to stop on the macroscopic scale. Chemical equilibrium is an example of a *dynamic* balance between the forward and reverse reactions, not a static balance.

Let's look at the logical consequences of the assumption that the reaction between $ClNO_2$ and NO eventually reaches equilibrium for a fixed temperature.

$$ClNO_2(g) + NO(g) \rightleftharpoons NO_2(g) + ClNO(g)$$

The rates of the forward and reverse reactions are the same when the system is at equilibrium.

At equilibrium: $Rate_{forward} = Rate_{reverse}$

Substituting the rate laws for the forward and reverse reactions into the equality gives the following result.

At equilibrium: $k_f(ClNO_2)(NO) = k_r(NO_2)(ClNO)$

But this equation is valid only when the system is at equilibrium at a given temperature, so we should replace the $(ClNO_2)$, (NO), (NO_2), and $(ClNO)$ terms with symbols which indicate that the reaction is at equilibrium. By convention, we use square brackets for this purpose. The equation describing the balance between the forward and reverse reactions when the system is at equilibrium should therefore be written as follows.

At equilibrium: $k_f[ClNO_2][NO] = k_r[NO_2][ClNO]$

Rearranging the equation gives the following result.

$$\frac{k_f}{k_r} = \frac{[NO_2][ClNO]}{[ClNO_2][NO]}$$

Because k_f and k_r are constants, the ratio of k_f divided by k_r must also be a constant. This ratio is the **equilibrium constant** for the reaction, K_c. As we have seen, the ratio of the concentrations of the reactants and products is known as the **equilibrium constant expression.**

Equilibrium constant expression

$$K_c = \frac{k_f}{k_r} = \frac{[NO_2][ClNO]}{[ClNO_2][NO]}$$

Equilibrium constant

No matter what combination of concentrations of reactants and products we start with, the reaction reaches equilibrium when the ratio of the concentrations defined by the equilibrium constant expression is equal to the equilibrium constant for the reaction at the chosen temperature. We can start with a lot of $ClNO_2$ and very little NO, or a lot of NO and very little $ClNO_2$. It doesn't matter. When the reaction reaches equilibrium, the relationship between the concentrations of the reactants and products described by the equilibrium constant expression will always be the same. At 25°C, this reaction always reaches equilibrium when the ratio of the concentrations is 1.3×10^4. K_c is always reported without units. However, any calculations using K_c require the concentration of the products and reactants to be in units of molarity (moles/liter).

$$K_c = \frac{[NO_2][ClNO]}{[ClNO_2][NO]} = 1.3 \times 10^4$$

What happens if we approach equilibrium from the other direction? We start with a system that contains the products of the reaction—NO_2 and ClNO—and then let the reaction come to equilibrium. The rate laws for the forward and reverse reactions will still be the same.

$$\text{Rate}_{forward} = k_f(ClNO_2)(NO)$$
$$\text{Rate}_{reverse} = k_r(NO_2)(ClNO)$$

Now, however, the rate of the forward reaction initially will be much smaller than the rate of the reverse reaction.

$$\text{Initially:} \qquad \text{Rate}_{forward} \ll \text{Rate}_{reverse}$$

But, as time passes, the rate of the reverse reaction will slow down and the rate of the forward reaction will speed up until they become equal. At that point, the reaction will have reached equilibrium.

$$\text{At equilibrium:} \qquad k_f[ClNO_2][NO] = k_r[NO_2][ClNO]$$

Rearranging the equation gives us the same equilibrium constant expression.

$$K_c = \frac{k_f}{k_r} = \frac{[NO_2][ClNO]}{[ClNO_2][NO]}$$

We get the same equilibrium constant expression and the same equilibrium constant no matter whether we start with only reactants, only products, or a mixture of reactants and products.

Exercise 10.2

The rate constants for the forward and reverse reactions in the following equilibrium have been measured. At 25°C, k_f is 7.3×10^3 liters per mole·second and k_r is 0.55 liters per mole·second. Calculate the equilibrium constant for the reaction.

$$ClNO_2(g) + NO(g) \rightleftharpoons NO_2(g) + ClNO(g)$$

Solution

We start by recognizing that the rates of the forward and reverse reactions at equilibrium are the same.

At equilibrium: $Rate_{forward} = Rate_{reverse}$

We then substitute the rate laws for the reaction into the equality.

At equilibrium: $k_f[ClNO_2][NO] = k_r[NO_2][ClNO]$

We then rearrange the equation to get the equilibrium constant expression for the reaction.

$$K_c = \frac{k_f}{k_r} = \frac{[NO_2][ClNO]}{[ClNO_2][NO]}$$

The equilibrium constant for the reaction is therefore equal to the rate constant for the forward reaction divided by the rate constant for the reverse reaction.

$$K_c = \frac{k_f}{k_r} = \frac{7.3 \times 10^3 \text{ L/mol·s}}{0.55 \text{ L/mol·s}} = 1.3 \times 10^4$$

Checkpoint

If, for a given reaction, the forward rate constant is twice that of the reverse rate constant, what will be the value of the equilibrium constant?

Any reaction that reaches equilibrium, no matter how simple or complex, has an equilibrium constant expression that satisfies the following rules.

RULES FOR WRITING EQUILIBRIUM CONSTANT EXPRESSIONS

- Even though chemical reactions that reach equilibrium occur in both directions, the chemical species on the right side of the equation are assumed to be the "products" of the reaction and the chemical species on the left side of the equation are assumed to be the "reactants."
- The products of the reaction are always written above the line, in the numerator.
- The reactants are always written below the line, in the denominator.
- For systems in which all species are either gaseous or aqueous, the equilibrium constant expression contains a term for every reactant and every product of the reaction.

- The numerator of the equilibrium constant expression is the product of the concentrations of the "products" of the reaction raised to a power equal to the coefficient for the component in the balanced equation for the reaction.
- The denominator of the equilibrium constant expression is the product of the concentrations of the "reactants" raised to a power equal to the coefficient for the component in the balanced equation for the reaction.

Exercise 10.3

Write equilibrium constant expressions for the following reactions.
(a) $2 NO_2(g) \rightleftharpoons N_2O_4(g)$
(b) $2 SO_3(g) \rightleftharpoons 2 SO_2(g) + O_2(g)$
(c) $N_2(g) + 3 H_2(g) \rightleftharpoons 2 NH_3(g)$

Solution

In each case, the equilibrium constant expression is the product of the concentrations of the species on the right side of the equation divided by the product of the concentrations of those on the left side of the equation. All concentrations are raised to the power equal to the coefficient for the species in the balanced equation.

(a) $K_c = \dfrac{[N_2O_4]}{[NO_2]^2}$ (b) $K_c = \dfrac{[SO_2]^2[O_2]}{[SO_3]^2}$ (c) $K_c = \dfrac{[NH_3]^2}{[N_2][H_2]^3}$

What happens to the magnitude of the equilibrium constant for a reaction when we turn the equation around? Consider the following reaction, for example.

$$ClNO_2(g) + NO(g) \rightleftharpoons NO_2(g) + ClNO(g)$$

The equilibrium constant expression for the equation is written as follows.

$$K_c = \frac{[NO_2][ClNO]}{[ClNO_2][NO]} = 1.3 \times 10^4 \quad \text{(at 25°C)}$$

Because this is an equilibrium reaction, it can also be represented by an equation written in the opposite direction.

$$NO_2(g) + ClNO(g) \rightleftharpoons ClNO_2(g) + NO(g)$$

The equilibrium constant expression is now written as follows.

$$K_c' = \frac{[ClNO_2][NO]}{[NO_2][ClNO]}$$

Each of the equilibrium constant expressions is the inverse of the other. We can therefore calculate K_c' by dividing K_c into 1.

$$K_c' = \frac{1}{K_c} = \frac{1}{1.3 \times 10^4} = 7.7 \times 10^{-5}$$

We can also calculate equilibrium constants by combining two or more reactions for which the values of K_c are known. Assume, for example, that we know the equilibrium constants for the following gas phase reactions at 200°C.

$$N_2(g) + O_2(g) \rightleftharpoons 2\,NO(g) \qquad K_{c1} = 2.3 \times 10^{-19}$$
$$2\,NO(g) + O_2(g) \rightleftharpoons 2\,NO_2(g) \qquad K_{c2} = 3 \times 10^6$$

We can combine the reactions to obtain an overall equation for the reaction between N_2 and O_2 to form NO_2.

$$\begin{aligned}
N_2(g) + O_2(g) &\rightleftharpoons 2\,\cancel{NO(g)} \\
+ \quad 2\,\cancel{NO(g)} + O_2(g) &\rightleftharpoons 2\,NO_2(g) \\
\hline
N_2(g) + 2\,O_2(g) &\rightleftharpoons 2\,NO_2(g)
\end{aligned}$$

The equilibrium constant expression for the overall reaction is equal to the product of the equilibrium constant expressions for the two steps in the reaction.

$$\frac{[NO_2]^2}{[N_2][O_2]^2} = \frac{\cancel{[NO]^2}}{[N_2][O_2]} \times \frac{[NO_2]^2}{\cancel{[NO]^2}[O_2]}$$

The equilibrium constant for the overall reaction is therefore equal to the product of the equilibrium constants for the individual reactions.

$$K_c = K_{c1} \times K_{c2} = (2.3 \times 10^{-19})(3 \times 10^6) = 7 \times 10^{-13}$$

Exercise 10.4

The sugar glucose is metabolized in the body to produce a compound called glucose 6-phosphate. One possible way the reaction could proceed is by the direct reaction of inorganic phosphate, labeled P_i by biochemists, with glucose.

$$glucose(aq) + P_i(aq) \rightleftharpoons glucose\ 6\text{-phosphate}(aq)$$

Write the equilibrium constant for the reaction in terms of concentrations.

Solution

No matter whether a reaction occurs in the gas phase or, as in this reaction, in a cell in the body, the same principles apply. All the species in the reaction exist in a cellular solution, and hence all can be expressed as concentrations in moles per liter.

$$K_c = \frac{[glucose\ 6\text{-phosphate}]}{[glucose][P_i]}$$

10.6 REACTION QUOTIENTS: A WAY TO DECIDE WHETHER A REACTION IS AT EQUILIBRIUM

We now have a model that describes what happens when a reaction reaches equilibrium. At the molecular level, the rate of the forward reaction is equal to the rate of the reverse reaction. Since the reaction proceeds in both directions at the same rate, there is no apparent change in the concentrations of the reactants or the products on the macroscopic scale (i.e., the level of objects visible to the naked eye). The model can also be used to predict the direction in which a reaction, not at equilibrium, has to shift to reach equilibrium.

If the concentrations of the reactants are too large for the reaction to be at equilibrium, the rate of the forward reaction will be faster than the reverse reaction, and some of the reactants will be converted to products until equilibrium is achieved. Conversely, if the concentrations of the reactants are too small, the rate of the reverse reaction will exceed that of the forward reaction, and the reaction will convert some of the excess products back into reactants until the system reaches equilibrium.

We can determine the direction in which a reaction has to shift to reach equilibrium by calculating the **reaction quotient (Q_c)** for the reaction. The reaction quotient is written in the same way as the equilibrium constant, but in Q_c the concentrations refer to any moment in time. In the equilibrium constant expression only the equilibrium concentrations are used. To illustrate how we use the reaction quotient, consider the following gas phase reaction.

$$H_2(g) + I_2(g) \rightleftharpoons 2\ HI(g)$$

The equilibrium constant expression for the reaction is written as follows.

$$K_c = \frac{[HI]^2}{[H_2][I_2]} = 60 \qquad (\text{at } 350°C)$$

By analogy, we can write the expression for the reaction quotient as follows.

$$Q_c = \frac{(HI)^2}{(H_2)(I_2)}$$

Q_c can take on any value between zero and infinity. If the system contains a great deal of HI and very little H_2 and I_2, the reaction quotient is very large. If the system contains relatively little HI and a great deal of H_2 and/or I_2, the reaction quotient is very small.

At any moment in time, there are three possibilities.

1. **Q_c is smaller than K_c.** The system contains too much reactant and not enough product to be at equilibrium. The value of Q_c must increase in order for the reaction to reach equilibrium. Thus, the reaction has to convert some of the reactants into products to come to equilibrium.
2. **Q_c is equal to K_c.** If this is true, then the reaction is at equilibrium.
3. **Q_c is larger than K_c.** The system contains too much product and not enough reactant to be at equilibrium. The value of Q_c must become smaller before the reaction can come to equilibrium. Thus, the reaction must convert some of the products into reactants to reach equilibrium.

Exercise 10.5

Assume that the concentrations of H_2, I_2, and HI can be measured for the following reaction at any moment in time.

$$H_2(g) + I_2(g) \rightleftharpoons 2\ HI(g) \qquad K_c = 60 \qquad T = 350°C$$

For each of the following sets of concentrations, determine whether the reaction is at equilibrium. If it isn't, decide in which direction it must go to reach equilibrium.

(a) $(H_2) = (I_2) = (HI) = 0.010\ M$

(b) $(HI) = 0.30\ M$, $(H_2) = 0.010\ M$; $(I_2) = 0.15\ M$

(c) $(H_2) = (HI) = 0.10\ M$, $(I_2) = 0.0010\ M$

Solution

(a) The best way to decide whether the reaction is at equilibrium is to compare the reaction quotient with the equilibrium constant for the reaction.

$$Q_c = \frac{(HI)^2}{(H_2)(I_2)} = \frac{(0.010)^2}{(0.010)(0.010)} = 1.0 < K_c$$

The reaction quotient in this case is smaller than the equilibrium constant. The only way to get the system to equilibrium is to increase the magnitude of the reaction quotient. This can be done by converting some of the H_2 and I_2 into HI. The reaction therefore has to shift to the right to reach equilibrium.

(b) The reaction quotient for this set of concentrations is equal to the equilibrium constant for the reaction.

$$Q_c = \frac{(HI)^2}{(H_2)(I_2)} = \frac{(0.30)^2}{(0.010)(0.15)} = 60 = K_c$$

The reaction is therefore at equilibrium.

(c) The reaction quotient for this set of concentrations is larger than the equilibrium constant for the reaction.

$$Q_c = \frac{(HI)^2}{(H_2)(I_2)} = \frac{(0.10)^2}{(0.10)(0.0010)} = 1.0 \times 10^2 > K_c$$

To reach equilibrium, the concentrations of the reactants and products must be adjusted until the reaction quotient is equal to the equilibrium constant. This involves converting some of the HI back into H_2 and I_2. Thus, the reaction has to shift to the left to reach equilibrium.

Exercise 10.6

For the metabolism of glucose to glucose 6-phosphate given in Exercise 10.4, the equilibrium constant is 6×10^{-3} at 25°C. If the P_i (phosphate) concentration in the cell is approximately $1 \times 10^{-2}\ M$ and that of glucose 6-phosphate is about $1 \times 10^{-4}\ M$, what concentration of glucose is necessary to make the reaction proceed to the formation of glucose 6-phosphate?

Solution

In Exercise 10.4 we saw that

$$K_c = \frac{[\text{glucose 6-phosphate}]}{[\text{glucose}][P_i]}$$

We now know that K_c is 6×10^{-3}. If Q_c is smaller than K_c, glucose will react with P_i to give more product. If $Q_c = K_c$ the reaction will be at equilibrium. We first find the equilibrium concentration of glucose.

$$K_c = 6 \times 10^{-3} = \frac{1 \times 10^{-4}}{[\text{glucose}][1 \times 10^{-2}]}$$

$$[\text{glucose}] = 2 \; M$$

Thus, if the concentration of glucose is greater than 2 M, Q_c will be less than K_c and the reaction will proceed to the right. Is this a reasonable concentration of glucose to be maintained in the cell? It certainly is not. A sugar concentration of 2 M could never be tolerated by the body. Indeed, the blood sugar concentration is closer to 5×10^{-3} M. This means that the reaction we have just hypothesized for the metabolism of glucose by the body does not occur. How then is glucose metabolized? Another species, adenosine triphosphate (ATP), reacts with the glucose to provide an alternate way to convert glucose to glucose 6-phosphate. An application of equilibrium principles has allowed us to reject a possible direct reaction of glucose and seek another way to carry out the reaction.

10.7   CHANGES IN CONCENTRATION THAT OCCUR AS A REACTION COMES TO EQUILIBRIUM

The relative sizes of Q_c and K_c for a reaction tell us whether the reaction is at equilibrium at any moment in time. If it isn't, the relative sizes of Q_c and K_c tell us the direction in which the reaction must shift to reach equilibrium. Now we need a way to predict how far the reaction has to go to reach equilibrium.

Suppose that you are faced with the following problem.

Phosphorus pentachloride decomposes to phosphorus trichloride and chlorine when heated.

$$PCl_5(g) \rightleftharpoons PCl_3(g) + Cl_2(g)$$

The equilibrium constant for the reaction is 0.030 at 250°C. Assume that the initial concentration of PCl₅ is 0.100 mol/L and there is no PCl₃ or Cl₂ in the system when we start. Calculate the concentrations of PCl₅, PCl₃, and Cl₂ at equilibrium.

The first step toward solving the problem involves organizing the information so that it provides clues as to how to proceed. The problem contains four chunks of information: (1) a balanced equation, (2) an equilibrium constant for the reaction, (3) a description of the initial conditions, and (4) an indication of the goal of the calculation, namely, to figure out the equilibrium concentrations of the three components of the reaction.

The following format offers a useful way to summarize the information.

	$PCl_5(g) \rightleftharpoons$	$PCl_3(g) +$	$Cl_2(g)$	$K_c = 0.030$
Initial	0.100 M	0	0	
Equilibrium	?	?	?	

We start with the balanced equation and the equilibrium constant for the reaction and then add what we know about the initial and equilibrium concentrations of the various components of the reaction. Initially, the flask contains 0.100 mol/L of PCl_5 and no PCl_3 or Cl_2. Our goal is to calculate the equilibrium concentrations of the three substances.

Before we do anything else, we have to decide whether the reaction is at equilibrium. We can do this by comparing the reaction quotient for the initial conditions with the equilibrium constant for the reaction.

$$Q_c = \frac{(PCl_3)(Cl_2)}{(PCl_5)} = \frac{(0)(0)}{0.100} = 0$$

Although the equilibrium constant is small ($K_c = 3.0 \times 10^{-2}$), the reaction quotient is even smaller ($Q_c = 0$). The only way for the reaction to get to equilibrium is for some of the PCl_5 to decompose into PCl_3 and Cl_2.

Because the reaction isn't at equilibrium, one thing is certain: the concentrations of PCl_5, PCl_3, and Cl_2 will all change as the reaction comes to equilibrium. Because the reaction has to shift to the right to reach equilibrium, the PCl_5 concentration will become smaller, while the PCl_3 and Cl_2 concentrations will become larger.

At first glance, the problem appears to have three unknowns: the equilibrium concentrations of PCl_5, PCl_3, and Cl_2. Because it is difficult to solve a problem with three unknowns, we should look for relationships that can reduce the problem's complexity. One way of achieving that goal is to look at the relationship between the changes that occur in the concentrations of PCl_5, PCl_3, and Cl_2 as the reaction approaches equilibrium.

Exercise 10.7

Calculate the increase in the PCl_3 and Cl_2 concentrations that occur as the following reaction comes to equilibrium if the concentration of PCl_5 decreases by 0.042 mol/L.

$$PCl_5(g) \rightleftharpoons PCl_3(g) + Cl_2(g)$$

Solution

The decomposition of PCl_5 has a 1:1:1 stoichiometry, as shown in Figure 10.5.

$$PCl_5(g) \rightleftharpoons PCl_3(g) + Cl_2(g)$$

For every mole of PCl_5 that decomposes, we get 1 mol of PCl_3 and 1 mol of Cl_2. Thus, the change in the concentration of PCl_5 that occurs as the reaction comes to equilibrium is equal to the changes in the PCl_3 and Cl_2 concentrations. If 0.042 mol/L of PCl_5 is consumed as the reaction comes to equilibrium, 0.042 mol/L each of PCl_3 and Cl_2 must be formed at the same time.

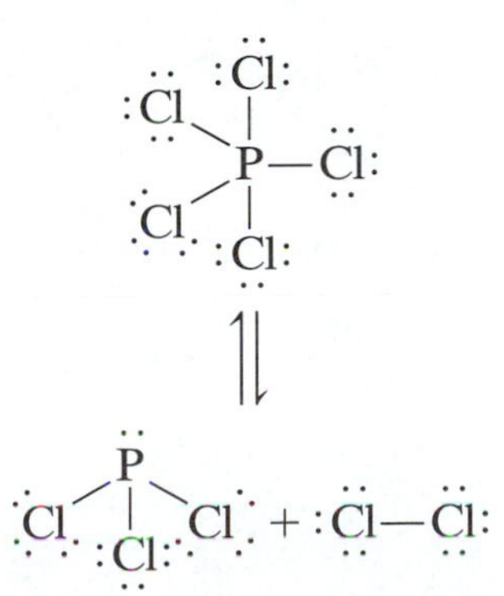

FIGURE 10.5 The decomposition of PCl_5 to form PCl_3 and Cl_2 is a reversible reaction with a 1:1:1 stoichiometry.

Checkpoint

Suppose 1.00 mole of $PCl_3(g)$ and 1.00 mole of $Cl_2(g)$ are placed into an empty 1.00-L container. If at equilibrium the concentration of PCl_3 has decreased by 0.96 mol/L, what are the concentrations of PCl_5 and Cl_2?

Exercise 10.7 raises an important point. There is a relationship between the *change in the concentrations* of the three components of the reaction as it comes to equilibrium because of the stoichiometry of the reaction.

It would be useful to have a symbol to represent the change that occurs in the concentration of one of the components of a reaction as it goes from the initial conditions to equilibrium. For chemical reactions we let $\Delta(X)$ represent the magnitude of the change that occurs in the molar concentration of X as the reaction comes to equilibrium. For example, $\Delta(PCl_5)$ is the magnitude of the change in the concentration of PCl_5 that occurs as the compound decomposes to form PCl_3 and Cl_2.

$$\underset{\substack{\text{PCl}_5 \text{ consumed} \\ \text{as the reaction} \\ \text{comes to} \\ \text{equilibrium}}}{\Delta(PCl_5)} = \underset{\substack{\text{Initial} \\ \text{concentration}}}{(PCl_5)_i} - \underset{\substack{\text{Concentration at} \\ \text{equilibrium}}}{[PCl_5]}$$

Rearranging the equation, we find that the concentration of PCl_5 at equilibrium is equal to the initial concentration of PCl_5 minus the amount of PCl_5 consumed as the reaction comes to equilibrium.

$$\underset{\substack{\text{Concentration at} \\ \text{equilibrium}}}{[PCl_5]} = \underset{\substack{\text{Initial} \\ \text{concentration}}}{(PCl_5)_i} - \underset{\substack{\text{PCl}_5 \text{ consumed} \\ \text{as reaction comes} \\ \text{to equilibrium}}}{\Delta(PCl_5)}$$

We can then define $\Delta(PCl_3)$ and $\Delta(Cl_2)$ as the changes that occur in the concentrations of PCl_3 and Cl_2 as the reaction comes to equilibrium. The concentrations of both substances at equilibrium will be larger than their initial concentrations.

$$[PCl_3] = (PCl_3)_i + \Delta(PCl_3)$$
$$[Cl_2] = (Cl_2)_i + \Delta(Cl_2)$$

Because of the 1:1:1 stoichiometry of the reaction, the magnitude of the change in the concentration of PCl_5 as the reaction comes to equilibrium is equal to the changes in the concentrations of PCl_3 and Cl_2, as we saw in Exercise 10.7.

$$\Delta(PCl_5) = \Delta(PCl_3) = \Delta(Cl_2)$$

We can therefore rewrite the equations that define the equilibrium concentrations of PCl_5, PCl_3, and Cl_2 in terms of a single unknown: ΔC.

$$[PCl_5] = (PCl_5)_i - \Delta C$$
$$[PCl_3] = (PCl_3)_i + \Delta C$$
$$[Cl_2] = (Cl_2)_i + \Delta C$$

Substituting what we know about the initial concentrations of PCl_5, PCl_3, and Cl_2 into the equations gives the following result.

$$[PCl_5] = 0.100 - \Delta C$$
$$[PCl_3] = [Cl_2] = 0 + \Delta C$$

We can now summarize what we know about the reaction as follows.

	$PCl_5(g)$	$\rightleftharpoons$	$PCl_3(g)$	$+$	$Cl_2(g)$
Initial	0.100 M		0		0
Equilibrium	$0.100 - \Delta C$		ΔC		ΔC

We now have only one unknown, ΔC, and we need only one equation to solve for one unknown. The obvious equation to turn to is the equilibrium constant expression for the reaction.

$$K_c = \frac{[PCl_3][Cl_2]}{[PCl_5]} = 0.030$$

Substituting what we know about the equilibrium concentrations of PCl_5, PCl_3, and Cl_2 into the equation gives the following result.

$$\frac{[\Delta C][\Delta C]}{[0.100 - \Delta C]} = 0.030$$

The equation can be expanded and then rearranged to give a quadratic equation:

$$\Delta C^2 + 0.030\Delta C - 0.0030 = 0$$

which can be solved with the quadratic formula.

$$\Delta C = \frac{-b \pm \sqrt{b^2 - 4ac}}{2a} = \frac{-(0.030) \pm \sqrt{(0.030)^2 - 4(1)(-0.0030)}}{2(1)}$$

$$\Delta C = 0.042 \quad \text{or} \quad -0.072$$

Although two answers come out of the calculation, only the positive root makes any physical sense because we can't have a negative concentration. Thus, the change in the concentrations of PCl_5, PCl_3, and Cl_2 as the reaction comes to equilibrium is 0.042 mol/L.

$$\Delta C = 0.042 \ M$$

By placing the value of ΔC back into the equations that define the equilibrium concentrations of PCl_5, PCl_3, and Cl_2, the following results are obtained.

$$[PCl_5] = 0.100 - 0.042 = 0.058 \ M$$
$$[PCl_3] = [Cl_2] = 0 + 0.042 = 0.042 \ M$$

In other words, slightly less than half of the PCl_5 present initially decomposes into PCl_3 and Cl_2 when the reaction comes to equilibrium.

To check whether the results of the calculation represent legitimate values for the equilibrium concentrations of the three components of this reaction, we can substitute the values into the equilibrium constant expression.

$$K_c = \frac{[PCl_3][Cl_2]}{[PCl_5]} = \frac{[0.042][0.042]}{[0.058]} = 0.030$$

The results must be legitimate because the equilibrium constant calculated from the concentrations is equal to the value of K_c given in the problem.

Exercise 10.8

Suppose 1.00 mol of butane were initially placed in a 1.00-L flask containing no isobutane at 25°C. What would be the equilibrium concentrations of butane and isobutane given that $K_c = 2.5$?

Solution

$$\text{butane}(g) \rightleftharpoons \text{isobutane}(g)$$

	butane(g)	isobutane(g)
Initial	1.00 M	0 M
Equilibrium	1.00 − ΔC	ΔC

$$K_c = \frac{[\text{isobutane}]}{[\text{butane}]} = \frac{[\Delta C]}{[1.00 - \Delta C]} = 2.5$$
$$\Delta C = 0.71 = [\text{isobutane}]$$
$$1.00 - \Delta C = 0.29 = [\text{butane}]$$

10.8   HIDDEN ASSUMPTIONS THAT MAKE EQUILIBRIUM CALCULATIONS EASIER

Suppose that you were asked to solve a slightly more difficult problem.

Sulfur trioxide decomposes to give sulfur dioxide and oxygen with an equilibrium constant of 1.6×10^{-10} at 300°C.

$$2\ SO_3(g) \rightleftharpoons 2\ SO_2(g) + O_2(g)$$

Calculate the equilibrium concentrations of the three components of the system if the initial concentration of SO_3 is 0.100 M.

Once again, the first step in the problem involves building a representation of the information in the problem.

	2 $SO_3(g)$ $\rightleftharpoons$	2 $SO_2(g)$ +	$O_2(g)$	$K_c = 1.6 \times 10^{-10}$
Initial	0.100 M	0	0	
Equilibrium	?	?	?	

We then compare the reaction quotient for the initial conditions with the equilibrium constant for the reaction.

$$Q_c = \frac{(SO_2)^2(O_2)}{(SO_3)^2} = \frac{(0)^2(0)}{0.100^2} = 0$$

Because the initial concentrations of SO_2 and O_2 are zero, the reaction has to shift to the right to reach equilibrium. As might be expected, some of the SO_3 has to decompose to SO_2 and O_2.

The stoichiometry of the reaction is more complex than the reaction in the previous section, but the changes in the concentrations of the three components of the reaction are still related. For every 2 mol of SO_3 that decomposes we get 2 mol of SO_2 and 1 mol of O_2, as shown in Figure 10.6. We can incorporate this relationship into the format we used earlier by using the balanced equation for the reaction as a guide.

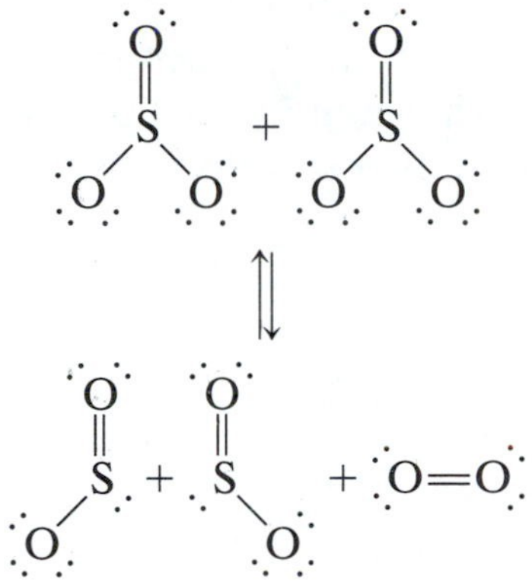

FIGURE 10.6 The stoichiometry of this reaction requires that the change in concentrations of both SO_3 and SO_2 must be twice as large as the change in the concentration of O_2 that occurs as the reaction comes to equilibrium.

The signs of the ΔC terms in the problem are determined by the fact that the reaction has to shift from left to right to reach equilibrium. The coefficients in the ΔC terms mirror the coefficients in the balanced equation for the reaction. Because twice as many moles of SO_2 are produced as moles of O_2, the change in the concentration of SO_2 as the reaction comes to equilibrium must be twice as large as the change in the concentration of O_2. Because 2 mol of SO_3 is consumed for every mole of O_2 produced, the change in the SO_3 concentration must be twice as large as the change in the concentration of O_2.

	$2\,SO_3(g)$	$\rightleftharpoons$	$2\,SO_2(g)$	$+\,O_2(g)$	$K_c = 1.6 \times 10^{-10}$
Initial	$0.100\ M$		0	0	
Change	$-2\Delta C$		$+2\Delta C$	$+\Delta C$	
Equilibrium	$0.100 - 2\Delta C$		$2\Delta C$	ΔC	

Substituting what we know about the problem into the equilibrium constant expression for the reaction gives the following equation.

$$K_c = \frac{[SO_2]^2[O_2]}{[SO_3]^2} = \frac{[2\Delta C]^2[\Delta C]}{[0.100 - 2\Delta C]^2} = 1.6 \times 10^{-10}$$

This equation is a bit more of a challenge to expand, but it can be rearranged to give the following cubic equation.

$$4\Delta C^3 - 6.4 \times 10^{-10}\Delta C^2 + 6.4 \times 10^{-11}\Delta C - 1.6 \times 10^{-12} = 0$$

Solving cubic equations is difficult, however. This problem is therefore an example of a family of problems that are difficult, if not impossible, to solve exactly. Such problems are solved with a general strategy that consists of making an assumption or approximation that turns them into simpler problems.

What assumption can be made to simplify the problem? Let's go back to the first thing we did after building a representation for the problem. We started our calculation by comparing the reaction quotient for the initial concentrations with the equilibrium constant for the reaction.

$$Q_c = \frac{(SO_2)^2(O_2)}{(SO_3)^2} = \frac{(0)^2(0)}{0.100^2} = 0$$

We then concluded that the reaction quotient ($Q_c = 0$) was smaller than the equilibrium constant ($K_c = 1.6 \times 10^{-10}$) and decided that some of the SO_3 would have to decompose in order for the reaction to come to equilibrium.

But what about the relative sizes of the reaction quotient and the equilibrium constant for the reaction? The initial values of Q_c and K_c are both relatively small, which means that the initial conditions are reasonably close to equilibrium, as shown in Figure 10.7. As a result, the reaction does not have far to go to reach equilibrium. It is therefore reasonable to assume that ΔC is relatively small in this problem.

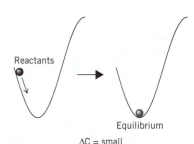

Reactants

Equilibrium

ΔC = small

FIGURE 10.7 When the initial conditions are close to equilibrium, the changes in the concentrations of the components of the reaction are often small enough compared with the initial concentrations to be ignored.

It is essential to understand the nature of the assumption being made. We are not assuming that ΔC is zero. If we did that, some of the unknowns would disappear from the equation! We are only assuming that ΔC is so small compared with the initial concentration of SO_3 that it doesn't make a significant difference when $2\Delta C$ is subtracted from that number. We can write the assumption as follows.

$$0.100 - 2\Delta C \approx 0.100$$

Let's now go back to the equation we are trying to solve.

$$\frac{[2\Delta C]^2[\Delta C]}{[0.100 - 2\Delta C]^2} = 1.6 \times 10^{-10}$$

By making the assumption that $2\Delta C$ is very much smaller than 0.100, we can replace the equation with the following approximate equation.

$$\frac{[2\Delta C]^2[\Delta C]}{0.100^2} \approx 1.6 \times 10^{-10}$$

Expanding this gives an equation that is much easier to solve for ΔC.

$$4\Delta C^3 \approx 1.6 \times 10^{-12}$$
$$\Delta C \approx 7.4 \times 10^{-5} \ M$$

Before we can go any further, we have to check our assumption that $2\Delta C$ is so small compared with 0.100 that it doesn't make a significant difference when it is subtracted from that number. Is the assumption valid? Is $2\Delta C$ small enough compared with 0.100 to be ignored?

$$0.100 - 2(0.000074) \approx 0.100$$

Using the rules of significant figures, when 0.00015 is subtracted from the initial concentration of 0.100, we are left with a result equal to the initial concentration. Therefore, the assumption is valid. In general, the change in concentration is sufficiently small to be negligible if the change divided by the initial concentration is less than 5%. In this example,

$$\frac{2 \ \Delta C}{0.100} \times 100 = \frac{2(7.4 \times 10^{-5})}{0.100} \times 100 = 0.15\% < 5\%$$

We can now use the approximate value of ΔC to calculate the equilibrium concentrations of SO_3, SO_2, and O_2.

$$[SO_3] = 0.100 - 2\Delta C \approx 0.100 \ M$$
$$[SO_2] = 2\Delta C \approx 1.5 \times 10^{-4} \ M$$
$$[O_2] = \Delta C \approx 7.4 \times 10^{-5} \ M$$

The equilibrium between SO_3 and mixtures of SO_2 and O_2 therefore strongly favors SO_3, not SO_2.

We can check the results of our calculation by substituting the results into the equilibrium constant expression for the reaction.

$$K_c = \frac{[SO_2]^2[O_2]}{[SO_3]^2} = \frac{[1.5 \times 10^{-4}]^2[7.4 \times 10^{-5}]}{[0.100]^2} = 1.7 \times 10^{-10}$$

The value of the equilibrium constant that comes out of the calculation agrees closely with the value given in the problem. Our assumption that $2\Delta C$ is negligibly small compared with the initial concentration of SO_3 is therefore valid, and we can feel confident in the answers it provides.

We can also use the equilibrium expression to solve for the concentration of products and reactants at equilibrium when we have an initial concentration of both a product and a reactant present. The same reaction of SO_3 used above can be used as an example.

Calculate the equilibrium concentrations of SO_3, SO_2, and O_2 if the initial concentrations are 0.100, 0, and 0.100 M, respectively.

	$2 \ SO_3(g)$	$\rightleftharpoons 2 \ SO_2(g) \ +$	$O_2(g)$	$K_c = 1.6 \times 10^{-10}$
Initial	0.100 M	0	0.100 M	
Change	$-2\Delta C$	$+2\Delta C$	$+\Delta C$	
Equilibrium	$0.100 - 2\Delta C$	$2\Delta C$	$0.100 + \Delta C$	

$$Q_c = \frac{(SO_2)^2(O_2)}{(SO_3)^2} = \frac{(0)^2(0.100)}{(0.100)^2} = 0 < K_c$$

The reaction proceeds to the right.

$$K_c = \frac{[SO_2]^2[O_2]}{[SO_3]^2} = \frac{[2\Delta C]^2[0.100 + \Delta C]}{[0.100 - 2\Delta C]^2} = 1.6 \times 10^{-10}$$

Because $\Delta C \ll 0.100\ M$, the above equation may be simplified as follows.

$$K_c = \frac{[2\Delta C]^2[0.100]}{(0.100)^2} = 1.6 \times 10^{-10}$$
$$4\Delta C^2 = 1.6 \times 10^{-11}$$
$$\Delta C = 2.0 \times 10^{-6}\ M$$

$$[SO_3] = 0.100 - 2\Delta C \approx 0.100\ M$$
$$[SO_2] = 2\Delta C \approx 4.0 \times 10^{-6}\ M$$
$$[O_2] = 0.100 + \Delta C \approx 0.100\ M$$

Note that the concentration of SO_2 ($4.0 \times 10^{-6}\ M$) produced in the calculation when O_2 was initially present is considerably less than the concentration of SO_2 ($1.5 \times 10^{-4}\ M$) produced in the previous calculations when there was no O_2 or SO_2 initially present.

Checkpoint

For which of the following equilibrium constants could it be safely assumed that ΔC is small?

$$A(g) \rightleftharpoons B(g) + C(g)$$

(a) $K = 1.0 \times 10^5$ (b) $K = 1.0 \times 10^{-5}$
(c) $K = 1.0 \times 10^{-1}$ (d) $K = 1.0 \times 10^{-10}$

Exercise 10.9

The equilibrium constant for the reaction

$$H_2(g) + Br_2(g) \rightleftharpoons 2\ HBr(g)$$

is 2.18×10^6 at 730°C.

(a) One mole of $H_2(g)$ and one mole of $Br_2(g)$ are placed into an empty container at 730°C and allowed to react. Make an estimate of how many moles of HBr would be produced. Do no calculations. Explain your estimate.

(b) If 3.20 moles of HBr is placed into an empty 12.0-L container at 730°C, what will be the concentrations of HBr, Br_2, and H_2 at equilibrium? State all assumptions. A calculation is required.

Solution

(a) The reaction will proceed to the right, and because the equilibrium constant is very large, 2.18×10^6, the reaction proceeds almost completely to the formation of products. Thus, it is not unreasonable to assume that very little reactant remains. If one mole each of H_2 and Br_2 completely react, 2 moles of HBr will be formed. At equilibrium we expect that almost all the product that can be formed will be formed.

(b) The original reaction will proceed to the left as shown below.

$$2\,HBr(g) \rightleftharpoons H_2(g) + Br_2(g)$$

For the reaction set up in this way,

$$K_c = \frac{1}{2.18 \times 10^6} = 4.59 \times 10^{-7} = \frac{[H_2][Br_2]}{[HBr]^2}$$

	$2\,HBr(g)$	$\rightleftharpoons$	$H_2(g)$	$+\ Br_2(g)$
Initial	$\dfrac{3.20\ mol}{12.0\ L} = 0.267\ M$		0	0
Change	$-2\Delta C$		ΔC	ΔC
Equilibrium	$0.267\ M - 2\Delta C$		ΔC	ΔC

K_c is small so that $2\Delta C$ is assumed to be negligible in comparison to 0.267.

$$4.59 \times 10^{-7} = \frac{[\Delta C][\Delta C]}{[0.267]^2}$$
$$\Delta C = 1.81 \times 10^{-4}\ M$$

Is the assumption valid?

$$\frac{2(1.81 \times 10^{-4}\ M)}{0.267\ M} \times 100 = 0.136\% < 5\%$$

The equilibrium concentrations are

$$[HBr] = 0.267 - 2\Delta C \approx 0.267\ M$$
$$[Br_2] = \Delta C = [H_2] = 1.81 \times 10^{-4}\ M$$

Note that in part (b) of the exercise you were asked to solve for the concentrations of all of the species. A calculation was necessary to determine all of the concentrations. In part (a) an estimation was sufficient. We could also have predicted that the equilibrium concentration of HBr in part (b) changed very little from its initial value because the equilibrium constant is so small. If we need to know the concentrations of H_2 and Br_2, however, a calculation is necessary.

10.9 THE EFFECT OF TEMPERATURE ON AN EQUILIBRIUM CONSTANT

The temperature at which the reaction was run has been reported each time an equilibrium constant has been given in this chapter. If the equilibrium constant is really constant, why do we have to worry about the temperature of the reaction?

The value of K_c for a reaction is constant at a given temperature, but it can change with temperature. Consider the reaction between nitrogen and oxygen to produce nitrogen monoxide

$$N_2(g) + O_2(g) \rightleftharpoons 2\,NO(g)$$

The equilibrium constant for the reaction increases with temperature as shown in Table 10.2. As the equilibrium constant becomes larger with increasing temperature, the equilibrium concentrations will change. Because the equilibrium constant is a ratio of product concentrations divided by reactant concentrations, an increase in the equilibrium constant dictates a shift in the reaction equilibrium toward the product, NO. Because equilibrium constants are temperature dependent, it is important to make sure that the correct equilibrium constant is used when calculating the extent of a chemical reaction.

TABLE 10.2   Temperature Dependence of the Equilibrium Constant for the Formation of NO

Temperature (°C)	K_c
25	4.3×10^{-31}
227	2.7×10^{-18}
727	7.5×10^{-9}
1727	4.0×10^{-4}
2727	1.5×10^{-2}

The equilibrium constant for the reaction at 25°C is so low that virtually no nitrogen monoxide is formed by the reaction between the atmospheric gases, N_2 and O_2. However, at high temperatures, such as those that result from a lightning flash or from combustion inside the cylinder of an automobile engine, the equilibrium constant is orders of magnitude larger. This results in the formation of NO in the atmosphere during electrical storms and from human activities such as driving a car. NO further reacts with O_2 in the atmosphere to produce NO_2, which can combine with water to produce nitric acid, HNO_3. The nitric acid then falls with rain to produce acid rain.

10.10   LE CHÂTELIER'S PRINCIPLE

In 1884 the French chemist and engineer Henry-Louis Le Châtelier proposed one of the central concepts of chemical equilibria. **Le Châtelier's principle** can be stated as follows:

> **A change in one of the variables that describes a system at equilibrium produces a shift in the position of the equilibrium that counteracts the effect of the change.**

Our attention so far has been devoted to describing what happens as a system comes to equilibrium. Le Châtelier's principle describes what happens to a system when something momentarily takes it away from equilibrium. This section focuses on three ways in which we can change the conditions of a chemical reaction at equilibrium: (1) changing the concentration of one of the components of the reaction, (2) changing the pressure on the system, and (3) changing the temperature at which the reaction is run.

Changes in Concentration

To illustrate what happens when we change the concentration of one of the reactants or products of a reaction at equilibrium, let's consider the following system at 25°C in a 1.0-L flask.

$$CH_3CH_2CH_2CH_3(g) \rightleftharpoons CH_3\overset{\overset{\displaystyle CH_3}{|}}{C}HCH_3(g) \qquad K_c = 2.5$$

Butane   Isobutane

Initial	0.50 *M*	0
Equilibrium	$0.50 - \Delta C$	ΔC

The equilibrium concentrations are found as before:

$$K_c = 2.5 = \frac{[\text{isobutane}]}{[\text{butane}]} = \frac{[\Delta C]}{[0.50 - \Delta C]}$$
$$[\text{isobutane}] = \Delta C = 0.36 \ M$$
$$[\text{butane}] = 0.50 - \Delta C = 0.14 \ M$$

Now suppose that to the equilibrium mixture 0.20 mol isobutane is added. At the instant of addition the concentration of each species becomes

$$\text{butane}(g) \rightleftharpoons \text{isobutane}(g)$$
$$0.14 \ M \qquad 0.36 + 0.20 = 0.56 \ M$$

The reaction quotient, Q_c, at the time of the addition is given by the following equation:

$$Q_c = \frac{(\text{isobutane})}{(\text{butane})} = \frac{0.56}{0.14} = 4.0$$

The system must readjust to establish the equilibrium ratio of 2.5, and so isobutane will decompose to form butane. When equilibrium is reestablished,

$$\text{butane}(g) \rightleftharpoons \text{isobutane}(g)$$
$$0.14 + \Delta C \qquad 0.56 - \Delta C$$

$$K_c = \frac{[\text{isobutane}]}{[\text{butane}]} = \frac{[0.56 - \Delta C]}{[0.14 + \Delta C]} = 2.5$$
$$\Delta C = 0.060$$

$$[\text{isobutane}] = 0.56 - 0.060 = 0.50 \ M$$
$$[\text{butane}] = 0.14 + 0.060 = 0.20 \ M$$

The calculation can be checked to determine if the ratio [isobutane]/[butane] correctly gives K_c.

$$K_c = \frac{0.50}{0.20} = 2.5$$

By comparing the new equilibrium concentrations with those obtained before adding isobutane we can see the effect of increasing the concentration of the isobutane on the equilibrium mixture.

Before	*After*
[isobutane] = 0.36 *M*	[isobutane] = 0.50 *M*
[butane] = 0.14 *M*	[butane] = 0.20 *M*

The addition of isobutane shifted the equilibrium in such a way as to produce more butane. Addition of more butane would have the opposite effect, causing the equilibrium to shift toward the products.

Changes in Pressure

Sometimes it is convenient to discuss gas phase chemical reactions in terms of the partial pressures of individual species rather than their concentrations. However, this makes no difference to the general conclusions about equilibrium that have been discussed. The reason is that partial pressures are related directly to concentrations through the ideal gas law equation.

The effect of changing the pressure on a gas phase reaction depends on the stoichiometry of the reaction. We can demonstrate this by looking at the result of increasing the total pressure on the following reaction at equilibrium.

$$N_2(g) + 3\,H_2(g) \rightleftharpoons 2\,NH_3(g)$$

Let's start with a system that initially contains 2.5 atm of N_2 and 7.5 atm of H_2 at 500°C, allow the reaction to come to equilibrium, and then increase the pressure by a factor of about 10. When this is done, we get the following results.

Before Compression	*After Compression*
$P_{NH_3} = 0.12$ atm	$P_{NH_3} = 8.4$ atm
$P_{N_2} = 2.4$ atm	$P_{N_2} = 21$ atm
$P_{H_2} = 7.3$ atm	$P_{H_2} = 62$ atm

Before the system was compressed, the partial pressure of NH_3 was only about 1% of the total pressure. After the system is compressed, the partial pressure of NH_3 is almost 10% of the total. The data provide another example of Le Châtelier's principle. A reaction at equilibrium was subjected to a stress—an increase in the total pressure on the system. The reaction then shifted in the direction that minimized the effect of the stress. The reaction shifted toward the products because this reduced the number of molecules in the gaseous mixture. This in turn decreased the total pressure exerted by the gases, as shown in Figure 10.8.

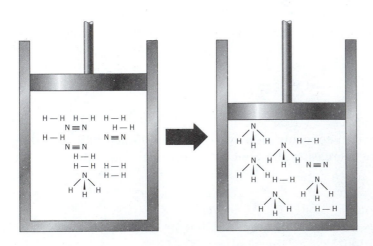

FIGURE 10.8 Because the total number of molecules in the system decreases when N_2 reacts with H_2 to form NH_3, shifting the equilibrium toward NH_3 decreases the total pressure of the gaseous mixture.

Whenever the pressure exerted on a system containing gaseous reactants or products is changed, the equilibrium will shift to minimize the pressure change. If pressure is increased the equilibrium will be shifted in the direction of fewer moles of gas. If the pressure is decreased the equilibrium will be shifted to produce more moles of gas.

Checkpoint

Which way will the equilibrium for the following reaction shift if more P_2 is added to the system at equilibrium? Which way will the equilibrium shift if the pressure is increased? The temperature remains constant.

$$2 P_2(g) \rightleftharpoons P_4(g)$$

Changes in Temperature

Changes in the concentrations of the reactants or products of a reaction shift the position of the equilibrium, but they do not change the equilibrium constant for the reaction. Similarly, a change in the pressure on a reaction shifts the position of the equilibrium without changing the magnitude of the equilibrium constant. Changes in the temperature of the system, however, affect the position of the equilibrium by changing the magnitude of the equilibrium constant for the reaction.

Chemical reactions usually give off heat to their surroundings or absorb heat from their surroundings. If we consider heat to be one of the reactants or products of a reaction, we can understand the effect of changes in temperature on the equilibrium. Increasing the temperature of a reaction that gives off heat is the same as adding more of one of the products of the reaction. It places a stress on the reaction, which must be alleviated by converting some of the products back to reactants.

The reaction in which NO_2 dimerizes to form N_2O_4 provides an example of the effect of changes in temperature on the equilibrium constant for a reaction. The reaction is exothermic.

$$2 NO_2(g) \rightleftharpoons N_2O_4(g) \qquad \Delta H° = -57.2 \text{ kJ/mol}_{rxn}$$

Thus, raising the temperature of the system is equivalent to adding excess product to the system. The equilibrium constant therefore decreases with increasing temperature. For an endothermic reaction, such as the formation of NO from N_2 and O_2 discussed in Section 10.9, an increase in temperature will cause an increase in the equilibrium constant.

Exercise 10.10

Predict the effect of the following changes on the decomposition of SO_3 to form SO_2 and O_2.

$$2 SO_3(g) \rightleftharpoons 2 SO_2(g) + O_2(g) \qquad \Delta H° = 197.84 \text{ kJ/mol}_{rxn}$$

(a) increasing the temperature of the reaction
(b) increasing the pressure on the reaction
(c) adding more O_2 when the reaction is at equilibrium
(d) removing O_2 from the system when the reaction is at equilibrium

Solution

(a) Because this is an endothermic reaction, which absorbs heat from its surroundings, an increase in the temperature of the reaction leads to an increase in the equilibrium constant and therefore a shift in the position of the equilibrium toward the products.

(b) There is a net increase in the number of molecules in the system as the reactants are converted to products, which leads to an increase in the pressure of the system. The system can minimize the effect of an increase in pressure by shifting the position of the equilibrium toward the reactants, thereby converting some of the SO_2 and O_2 to SO_3.

(c) Adding more O_2 to the system will shift the position of the equilibrium toward the reactants.

(d) Removing O_2 from the system has the opposite effect to (c); it shifts the equilibrium toward the products of the reaction.

Exercise 10.11

Exercise 10.6 showed that the reaction for the direct metabolism of glucose to glucose 6-phosphate

$$\text{glucose}(aq) + P_i(aq) \rightleftharpoons \text{glucose 6-phosphate}(aq)$$

was not possible in the body. However, the enthalpy change for the reaction is $+35$ kJ/mol$_{rxn}$. What would happen to the equilibrium constant for the reaction if the temperature were increased? How can the body increase its temperature?

Solution

For an endothermic reaction an increase in temperature will increase the equilibrium constant. Thus an increase in body temperature will increase the equilibrium constant for the reaction and the reaction could become feasible. The body increases its temperature through fevers.

10.11   EQUILIBRIUM REACTIONS THAT INVOLVE PURE SOLIDS AND LIQUIDS

If we add solid LiF to a liter of water about 2 g of the compound will dissolve. Continued addition of LiF will result in a buildup of solid LiF on the bottom of the flask. The chemical equation that describes this process is

$$\text{LiF}(s) \rightleftharpoons \text{Li}^+(aq) + \text{F}^-(aq)$$

How do we write the equilibrium constant expression for such a reaction? The ions Li^+ and F^- are dissolved in water, so we can give their equilibrium concentrations in moles per liter, $[\text{Li}^+]$ and $[\text{F}^-]$. LiF(s) is more difficult to deal with. We could calculate the concentration of LiF, but what would this mean? LiF is a solid, not dissolved in the water, and therefore the concentration of LiF is the number of moles of LiF in a liter of the solid LiF, not in a liter of water. The concentration of LiF in the solid can be calculated to be 102 M. This concentration of LiF is a constant; the number of moles of LiF in a given volume of LiF is always the same.

For any pure solid or liquid the concentration does not vary regardless of the amount of pure solid or liquid present. If the volume of LiF, for example, lying on the bottom of the flask were 0.0200 L, this would be equivalent to 2.04 mol of LiF. The concentration of that quantity of LiF is 2.04 mol/0.0200 L = 102 M.

Hence, we incorporate this constant into the equilibrium constant and write

$$K_c = \frac{[Li^+][F^-]}{[LiF]} = \frac{[Li^+][F^-]}{102}$$
$$102 K_c = K = [Li^+][F^-]$$

The *new* equilibrium constant symbol K ($K = K_c[\text{solid}]$ or $K = K_c[\text{liquid}]$) for any reaction involving a pure solid or liquid is often given a special subscript to further identify the type of reaction that it describes. For example, K for the dissolution of sparingly soluble solids such as LiF is subscripted as K_{sp}, where the "sp" refers to the solubility product.

K_{sp} can be used to predict whether a solid **precipitate** will form when solutions containing ions are mixed. If a solution of $LiNO_3$ (Li^+, NO_3^-) is added to a solution of NaF (Na^+, F^-), the initial concentrations of Li^+ and F^- in the mixture can be used to determine a reaction quotent, Q.

$$Q = (Li^+)(F^-)$$

If $Q > K_{sp}$ then a precipitate of solid LiF will form.

$$Li^+(aq) + F^-(aq) \rightleftharpoons LiF(s)$$

Exercise 10.12

If 2 g of LiF will dissolve in 1 liter of water, what is the equilibrium constant for the dissolution reaction?

$$LiF(s) = Li^+(aq) + F^-(aq)$$

Solution

First we write the equilibrium constant expression: $K_{sp} = [Li^+][F^-]$. Next we need to know the molar concentration of Li^+ and F^-. We know that 2 g of LiF dissolved. This corresponds to 0.08 moles of LiF.

$$\frac{2 \text{ g LiF}}{25.9 \text{ g/mol LiF}} = 0.08 \ M \text{ LiF}$$

For each mole of LiF that dissolves, one mole of Li^+ and one mole of F^- are formed. Thus if 0.08 mol of LiF dissolves, then 0.08 mol of Li^+ and 0.08 mol of F^- are formed. The concentrations of both Li^+ and F^- are therefore 0.08 mol/liter = 0.08 M. Therefore, the equilibrium constant can be determined.

$$K_{sp} = [0.08][0.08] = 6 \times 10^{-3}$$

Exercise 10.13

Solid magnesium hydroxide, $Mg(OH)_2$, is added to a liter of pure water to form a saturated solution. Use the K_{sp} for $Mg(OH)_2$ to estimate the molar concentration of each ion in solution and the solubility of the $Mg(OH)_2$ measured in units of g/100 mL. The K_{sp} of $Mg(OH)_2$ is 1.8×10^{-11}.

$$Mg(OH)_2(s) \rightleftharpoons Mg^{2+}(aq) + 2 OH^-(aq)$$

Solution

The equilibrium expression that describes the above equilibrium is

$$K_{sp} = [Mg^{2+}][OH^-]^2 = 1.8 \times 10^{-11}$$

For every mole of $Mg(OH)_2$ that dissolves there will be an equal number of moles of Mg^{2+} and twice as many moles of OH^- produced. If the solid is dissolved in pure water the initial concentrations of Mg^{2+} and OH^- are assumed to be essentially zero. Since the solid is dissolved in a liter of water the molar concentrations of Mg^{2+} and OH^- at equilibrium can be described as ΔC and $2\Delta C$, respectively. These variables may then be substituted into the equilibrium expression.

$$1.8 \times 10^{-11} = (\Delta C)(2\Delta C)^2$$
$$1.8 \times 10^{-11} = 4\Delta C^3$$
$$1.7 \times 10^{-4} = \Delta C$$

$$[Mg^{2+}] = \Delta C = 1.7 \times 10^{-4} \ M$$
$$[OH^-] = 2\Delta C = 2(1.7 \times 10^{-4}) = 3.4 \times 10^{-4} \ M$$

The molar concentration of the dissolved $Mg(OH)_2$ is ΔC, $1.7 \times 10^{-4} \ M$. We then convert this to solubility in units of g/100 mL.

Moles of $Mg(OH)_2$ in 100 mL = $(1.7 \times 10^{-4} \ M) \times (0.100 \ L) = 1.7 \times 10^{-5}$ moles $Mg(OH)_2$
Mass $Mg(OH)_2$/100 mL = $[1.7 \times 10^{-5} \ mol \ Mg(OH)_2] \times (58.32 \ g/mol) = 9.9 \times 10^{-4} \ g/100 \ mL$

The solubility of $Mg(OH)_2$ is 9.9×10^{-4} g/100 mL. Note that the number of moles of $Mg(OH)_2$ that dissolves is the same as the number of moles of Mg^{2+} formed but is one-half the number of moles of OH^- formed.

Checkpoint
If K_{sp} for the dissolution of LiF(s) is 6×10^{-3}, calculate the concentration of the ions formed.

Checkpoint
If the concentration of Mg^{2+} is 1.7×10^{-4} M and that of OH^- is 2.4×10^{-4} M in a solution, will a precipitate form?

10.12 LE CHÂTELIER'S PRINCIPLE AND THE HABER PROCESS

Ammonia has been produced commercially from N_2 and H_2 ever since 1913, when Badische Anilin und Soda Fabrik (BASF) built a plant that used the Haber process to make 30 metric tons of synthetic ammonia per day.

$$N_2(g) + 3 \ H_2(g) \rightleftharpoons 2 \ NH_3(g) \qquad \Delta H° = -92.2 \ kJ/mol_{rxn}$$

Until that time, the principal source of nitrogen for use in farming had been animal and vegetable waste. Today, almost 20 million tons of ammonia worth $2.5 billion is produced in the United States each year, about 80% of which is used for fertilizers. Ammonia is usually applied directly to the fields as a liquid at or near its boiling point of $-33.35°C$. By using this so-called anhydrous ammonia, farmers can apply a fertilizer that contains 82% nitrogen by weight.

The Haber process is an example of the use of Le Châtelier's principle to optimize the yield of an industrial chemical. An increase in the pressure at which the reaction is run favors the products of the reaction because there is a net reduction in the number of mole-

A photograph of the first high-pressure reactor for the synthesis of ammonia by the Haber process.

cules in the system as N_2 and H_2 combine to form NH_3. Because the reaction is exothermic, the equilibrium constant increases as the temperature of the reaction decreases.

Table 10.3 shows the mole percent of NH_3 at equilibrium when the reaction is run at different combinations of temperature and pressure. The mole percent of NH_3 under a particular set of conditions is equal to the number of moles of NH_3 at equilibrium divided by the total number of moles of all three components of the reaction times 100. As the data in Table 10.3 demonstrate, the best yields of ammonia are obtained at low temperatures and high pressures.

TABLE 10.3   Mole Percentage of NH_3 at Equilibrium

Temperature (°C)	Pressure (atm)			
	200	300	400	500
400	38.74	47.85	58.87	60.61
450	27.44	35.93	42.91	48.84
500	18.86	26.00	32.25	37.79
550	12.82	18.40	23.55	28.31
600	8.77	12.97	16.94	20.76

Unfortunately, low temperatures slow down the rate of the reaction, and the cost of building plants rapidly escalates as the pressure at which the reaction is run is increased. When commercial plants are designed, a temperature is chosen that allows the reaction to proceed at a reasonable rate without decreasing the equilibrium concentration of the product by too much. The pressure is also adjusted so that it favors the production of ammonia without excessively increasing the cost of building and operating the plant. The optimum conditions for running the reaction at present are a pressure between 140 and 340 atm and a temperature between 400°C and 600°C.

Despite all efforts to optimize reaction conditions, the percentages of hydrogen and nitrogen converted to ammonia are still relatively small. Another form of Le Châtelier's principle is therefore used to drive the reaction to completion. Periodically, the reaction mixture is cycled through a cooling chamber. The boiling point of ammonia (BP = −33°C) is much higher than that of either hydrogen (BP = −252.8°C) or nitrogen (BP = −195.8°C). Ammonia can be removed from the reaction mixture, forcing the equilibrium to the right. The remaining hydrogen and nitrogen gases are then recycled through the reaction chamber, where they react to produce more ammonia.

Exercise 10.14

An increase in temperature from 25°C to 47°C increases the equilibrium constant for the glucose metabolism reaction of Exercise 10.4 from 6×10^{-3} to 2×10^{-2}. If the P_i and glucose 6-phosphate concentrations remain the same as at 25°C, what would be the new equilibrium glucose concentration at 47°C? Does the increased temperature make this a likely process to occur in a cell?

Solution

The new K_c expression is

$$K_c = 2 \times 10^{-2} = \frac{[\text{glucose 6-phosphate}]}{[\text{glucose}][P_i]}$$

$$= \frac{1 \times 10^{-4}}{[\text{glucose}][1 \times 10^{-2}]}$$

$$[\text{glucose}] = 5 \times 10^{-1} \ M$$

As indicated in Exercise 10.6 the blood sugar concentration is about $5 \times 10^{-3} \ M$ in the body. Even though an increase in body temperature to 47°C will cause the reaction to proceed to the right, an unrealistically high glucose concentration is still needed to permit the direct conversion of glucose.

KEY TERMS

Approximation methods	Equilibrium region	Rate constant
Collision theory	Haber process	Rate law
Equilibrium	Kinetic region	Rate of reaction
Equilibrium constant, K_c	K_{sp}	Reaction quotient, Q_c
Equilibrium constant expression	Le Châtelier's principle	
	Precipitate	

PROBLEMS

Chemical Reactions Don't Always Go to Completion

1. Describe the difference between reactions that go to completion and reactions that come to equilibrium.

2. Define the terms *equilibrium, equilibrium constant, equilibrium constant expression,* and *reaction quotient.*

3. Describe the meaning of the symbol $[NO_2]$.

The Rates of Chemical Reactions

4. Describe how the rate of a chemical reaction is analogous to other rate processes, such as the rate at which a car travels or the rate of inflation.

5. Translate the following equation into an English sentence that carries the same meaning:

$$\text{Rate of reaction} = \frac{-\Delta(X)}{\Delta t}$$

6. Sketch a graph of what happens to the concentrations of N_2, H_2, and NH_3 versus time as the following reaction comes to equilibrium.

$$N_2(g) + 3\,H_2(g) \rightleftharpoons 2\,NH_3(g)$$

Assume that the initial concentrations of N_2 and H_2 are both 1.00 mol/L and that no NH_3 is present initially. Label the kinetic and the equilibrium regions of the graph.

A Collision Theory Model for Gas Phase Reactions

7. Describe how the collision theory model can be used to explain the fact that the rate at which $ClNO_2$ reacts with NO to form ClNO and NO_2 is proportional to the product of the concentrations of $ClNO_2$ and NO.

8. Use the fact that the rate of a chemical reaction is proportional to the product of the concentrations of the reagents consumed in that reaction to explain why reversible reactions inevitably come to equilibrium.

Writing Equilibrium Constant Expressions

9. Which of the following is the correct equilibrium constant expression for the reaction?

$$Cl_2(g) + 3\,F_2(g) \rightleftharpoons 2\,ClF_3(g)$$

(a) $K_c = \dfrac{2[ClF_3]}{[Cl_2] + 3[F_2]}$

(b) $K_c = \dfrac{[Cl_2] + 3[F_2]}{2[ClF_3]}$

(c) $K_c = \dfrac{[ClF_3]}{[Cl_2][F_2]}$

(d) $K_c = \dfrac{[ClF_3]^2}{[Cl_2][F_2]^3}$

(e) $K_c = \dfrac{[Cl_2][F_2]^3}{[ClF_3]^2}$

10. Which of the following is the correct equilibrium constant expression for the reaction?

$$2\,NO_2(g) \rightleftharpoons 2\,NO(g) + O_2(g)$$

(a) $K_c = \dfrac{[NO_2]}{[NO][O_2]}$

(b) $K_c = \dfrac{[NO][O_2]}{[NO_2]}$

(c) $K_c = \dfrac{[NO_2]^2}{[NO]^2[O_2]}$

(d) $K_c = \dfrac{[NO]^2[O_2]}{[NO_2]^2}$

(e) $K_c = \dfrac{[2\,NO]^2[O_2]}{[2\,NO_2]^2}$

11. Write equilibrium constant expressions for the following reactions.
    (a) $O_2(g) + 2\,F_2(g) \rightleftharpoons 2\,OF_2(g)$
    (b) $2\,SO_2(g) + O_2(g) \rightleftharpoons 2\,SO_3(g)$
    (c) $2\,SO_3(g) + 2\,Cl_2(g) \rightleftharpoons 2\,SO_2Cl_2(g) + O_2(g)$

12. Write equilibrium constant expressions for the following reactions.
    (a) $2\,NO(g) + 2\,H_2(g) \rightleftharpoons N_2(g) + 2\,H_2O(g)$
    (b) $2\,NOCl(g) \rightleftharpoons 2\,NO(g) + Cl_2(g)$
    (c) $2\,NO(g) + O_2(g) \rightleftharpoons 2\,NO_2(g)$

13. Write equilibrium constant expressions for the following reactions.
    (a) $2\,CO(g) + O_2(g) \rightleftharpoons 2\,CO_2(g)$
    (b) $CO_2(g) + H_2(g) \rightleftharpoons CO(g) + H_2O(g)$
    (c) $CO(g) + 2\,H_2(g) \rightleftharpoons CH_3OH(g)$

Calculating Equilibrium Constants

14. Calculate K_c for the following reaction at 400 K if 1.000 mol/L of NOCl decomposes at that temperature to give equilibrium concentrations of 0.0222 M NO, 0.0111 M Cl$_2$, and 0.978 M NOCl.

$$2\,NOCl(g) \rightleftharpoons 2\,NO(g) + Cl_2(g)$$

15. Taylor and Crist [*Journal of the American Chemical Society*, **63,** 1381 (1941)] studied the reaction between hydrogen and iodine to form hydrogen iodide.

$$H_2(g) + I_2(g) \rightleftharpoons 2\,HI(g)$$

They obtained the following data for the concentrations of H$_2$, I$_2$, and HI at equilibrium in units of moles per liter.

Trial	$[H_2]$	$[I_2]$	$[HI]$
I	0.0032583	0.0012949	0.015869
II	0.0046981	0.0007014	0.013997
III	0.0007106	0.0007106	0.005468

Calculate the value of K_c for each of the trials. Realizing that there will be deviation due to experimental error, is K_c constant for the reaction?

Combining Equilibrium Constant Expressions

16. Write equilibrium constant expressions for the following reactions.
    (a) $2\,NO_2(g) \rightleftharpoons 2\,NO(g) + O_2(g)$
    (b) $2\,NO(g) + O_2(g) \rightleftharpoons 2\,NO_2(g)$

Calculate the value of K_c at 500 K for reaction (a) assuming that the value of K_c for reaction (b) is 6.2×10^5.

17. Use the equilibrium constants for reactions (a) and (b) at 200°C to calculate the equilibrium constant for reaction (c) at that temperature.
  (a) $2 NO(g) \rightleftharpoons N_2(g) + O_2(g)$ $K_c = 4.3 \times 10^{18}$
  (b) $2 NO_2(g) \rightleftharpoons 2 NO(g) + O_2(g)$ $K_c = 3.4 \times 10^{-7}$
  (c) $2 NO_2(g) \rightleftharpoons N_2(g) + 2 O_2(g)$ $K_c = ?$

18. Use the equilibrium constants for reactions (a) and (b) at 1000 K to calculate the equilibrium constant for reaction (c), the water–gas shift reaction, at that temperature.
  (a) $CO(g) + \frac{1}{2} O_2(g) \rightleftharpoons CO_2(g)$ $K_c = 1.1 \times 10^{11}$
  (b) $H_2O(g) \rightleftharpoons H_2(g) + \frac{1}{2} O_2(g)$ $K_c = 7.1 \times 10^{-12}$
  (c) $CO(g) + H_2O(g) \rightleftharpoons CO_2(g) + H_2(g)$ $K_c = ?$

Reaction Quotient: A Way to Decide whether a Reaction Is at Equilibrium

19. Suppose that the reaction quotient (Q_c) for the following reaction at some moment in time is 1.0×10^{-8} and the equilibrium constant for the reaction (K_c) at the same temperature is 3×10^{-7}.

$$2 NO_2(g) \rightleftharpoons 2 NO(g) + O_2(g)$$

  Which of the following is a valid conclusion?
  (a) The reaction is at equilibrium.
  (b) The reaction must shift toward the products to reach equilibrium.
  (c) The reaction must shift toward the reactants to reach equilibrium.

20. Which of the following statements correctly describes a system for which Q_c is larger than K_c?
  (a) The reaction is at equilibrium.
  (b) The reaction must shift to the right to reach equilibrium.
  (c) The reaction must shift to the left to reach equilibrium.
  (d) The reaction can never reach equilibrium.

21. Under which set of conditions will the following reaction shift to the right to reach equilibrium?

$$2 SO_2(g) + O_2(g) \rightleftharpoons 2 SO_3(g)$$

  (a) $K_c < 1$ (b) $K_c > 1$ (c) $Q_c < K_c$ (d) $Q_c = K_c$ (e) $Q_c > K_c$

22. Carbon monoxide reacts with chlorine to form phosgene.

$$CO(g) + Cl_2(g) \rightleftharpoons COCl_2(g)$$

  The equilibrium constant, K_c, for the reaction is 1.5×10^4 at 300°C. Is the system at equilibrium at the following concentrations: $0.0040\ M\ COCl_2$, $0.00021\ M\ CO$, and $0.00040\ M\ Cl_2$? If not, in which direction does the reaction have to shift to reach equilibrium?

Changes in Concentration That Occur as a Reaction Comes to Equilibrium

23. Describe the relationship between the initial concentration of a reactant (X), the concentration of the reactant at equilibrium $[X]$, and the change in the concentration of X that occurs as the reaction comes to equilibrium, $\Delta(X)$.

24. Explain why the change in the N_2 concentration that occurs when the following reaction comes to equilibrium is related to the change in the H_2 concentration.

$$N_2(g) + 3 H_2(g) \rightleftharpoons 2 NH_3(g)$$

Derive an equation that describes the relationship between the changes in the concentrations of the two reagents.

25. When confronted with the task in the previous problem, beginning chemistry students often write the following equation.

$$\Delta(N_2) = 3\Delta(H_2)$$

Explain why the equation is wrong. Write the correct form of the relationship.

26. Calculate the changes in the CO and Cl_2 concentrations that occur if the concentration of $COCl_2$ decreases by 0.250 mol/L as the following reaction comes to equilibrium.

$$COCl_2(g) \rightleftharpoons CO(g) + Cl_2(g)$$

27. Calculate the changes in the N_2 and H_2 concentrations that occur if the concentration of NH_3 decreases by 0.234 mol/L as the following reaction comes to equilibrium.

$$2 NH_3(g) \rightleftharpoons N_2(g) + 3 H_2(g)$$

28. Which of the following equations describes the relationship between the magnitude of the changes in the NO_2 and O_2 concentrations as the following reaction comes to equilibrium?

$$2 NO(g) + O_2(g) \rightleftharpoons 2 NO_2(g)$$

(a) $\Delta(NO_2) = \Delta(O_2)$ (b) $\Delta(NO_2) = 2\Delta(O_2)$ (c) $\Delta(O_2) = 2\Delta(NO_2)$

29. Which of the following equations correctly describes the relationship between the changes in the Cl_2 and F_2 concentrations as the following reaction comes to equilibrium?

$$Cl_2(g) + 3 F_2(g) \rightleftharpoons 2 ClF_3(g)$$

(a) $\Delta(Cl_2) = \Delta(F_2)$ (b) $\Delta(Cl_2) = 2\Delta(F_2)$ (c) $\Delta(Cl_2) = 3\Delta(F_2)$
(d) $\Delta(F_2) = 2\Delta(Cl_2)$ (e) $\Delta(F_2) = 3\Delta(Cl_2)$

30. Which of the following describes the change that occurs in the concentration of H_2O when ammonia reacts with oxygen to form nitrogen oxide and water according to the following equation if the change in the NH_3 concentration is ΔC?

$$4 NH_3(g) + 5 O_2(g) \rightleftharpoons 4 NO(g) + 6 H_2O(g)$$

(a) ΔC (b) $1.5\Delta C$ (c) $2\Delta C$ (d) $4\Delta C$ (e) $6\Delta C$

31. Calculate the concentrations of H_2 and NH_3 at equilibrium if a reaction that initially contained 1.000 M concentrations of both N_2 and H_2 is found to have an N_2 concentration of 0.922 M at equilibrium.

	$N_2(g)$	+ 3 $H_2(g) \rightleftharpoons$ 2 $NH_3(g)$	
Initial	1.000 M	1.000 M	0 M
Equilibrium	0.922 M	?	?

32. Calculate the equilibrium constant for the reaction in the previous problem.

Hidden Assumptions That Make Equilibrium Calculations Easier

33. Some students have described the technique used in this chapter to simplify equilibrium problems as follows: "Assume that ΔC is zero." Explain why they are wrong. What is the correct way of describing the assumption?

34. Describe the advantage of setting up equilibrium problems so that ΔC is small compared with the initial concentrations.

35. Describe how to test whether ΔC is small enough compared with the initial concentrations to be legitimately ignored.

Gas Phase Equilibrium Problems

36. Calculate the concentrations of PCl_5, PCl_3, and Cl_2 that are present when the following gas phase reaction comes to equilibrium. Calculate the percent of the PCl_5 that decomposes when the reaction comes to equilibrium. Explain the difference between the results of this calculation and that obtained in Section 10.7.

$$PCl_5(g) \rightleftharpoons PCl_3 + Cl_2(g) \qquad K_c = 0.0013 \quad \text{(at 450 K)}$$
Initial $\quad$ 1.00 M $\qquad$ 0 $\qquad$ 0

37. Calculate the concentrations of PCl_5, PCl_3, and Cl_2 present when the following gas phase reaction comes to equilibrium. Calculate the percent decomposition in the reaction and explain any difference between the results of this calculation and the results obtained in Section 10.7.

$$PCl_5(g) \rightleftharpoons PCl_3(g) + Cl_2(g) \qquad K_c = 0.0013 \quad \text{(at 450 K)}$$
Initial $\quad$ 1.00 M $\qquad$ 0 $\qquad$ 0.20 M

38. Calculate the concentrations of NO, NO_2, and O_2 present when the following gas phase reaction reaches equilibrium.

$$2\,NO_2(g) \rightleftharpoons 2\,NO(g) + O_2(g) \qquad K_c = 3.4 \times 10^{-7} \quad \text{(at 200°C)}$$
Initial $\quad$ 0.100 M $\qquad$ 0 $\qquad$ 0

39. Calculate the concentrations of NO, NO_2, and O_2 present when the following gas phase reaction reaches equilibrium.

$$2\,NO_2(g) \rightleftharpoons 2\,NO(g) + O_2(g) \qquad K_c = 3.4 \times 10^{-7} \quad \text{(at 200°C)}$$
Initial $\quad$ 0.10 M $\qquad$ 0 $\qquad$ 0.050 M

40. Calculate the equilibrium concentrations of SO_3, SO_2, and O_2 present when 0.100 mol of SO_3 in a 250-mL flask at 300°C decomposes to form SO_2 and O_2.

$$2\,SO_3(g) \rightleftharpoons 2\,SO_2(g) + O_2(g) \qquad K_c = 1.6 \times 10^{-10} \quad \text{(at 300°C)}$$

41. Without calculations estimate the equilibrium concentration of SO_3 when a mixture of 0.100 mol of SO_2 and 0.050 mol of O_2 in a 250-mL flask at 300°C combine to form SO_3.

$$2\,SO_2(g) + O_2(g) \rightleftharpoons 2\,SO_3(g) \qquad K_c = 6.3 \times 10^9 \quad \text{(at 300°C)}$$

42. Calculate the equilibrium concentrations of N_2O_4 and NO_2 when 0.100 M N_2O_4 decomposes to form NO_2 at 25°C.

$$N_2O_4(g) \rightleftharpoons 2\,NO_2(g) \qquad K_c = 5.8 \times 10^{-5}$$

43. Without calculations estimate the equilibrium concentration of N_2O_4 present when 1.00 M NO_2 dimerizes to form N_2O_4 at 25°C.

$$N_2O_4(g) \rightleftharpoons 2\,NO_2(g) \qquad K_c = 5.8 \times 10^{-5}$$

44. Calculate the equilibrium concentrations of N_2, H_2, and NH_3 present when a mixture that was initially 0.10 M N_2, 0.10 M H_2, and 0.10 M NH_3 comes to equilibrium at 500°C.

$$N_2(g) + 3\,H_2(g) \rightleftharpoons 2\,NH_3(g) \qquad K_c = 0.040 \quad \text{(at 500°C)}$$

45. Calculate the equilibrium concentrations of CO, H_2O, CO_2, and H_2 present in the water–gas shift reaction at 800°C if the initial concentrations of CO and H_2O are 1.00 M.

$$CO(g) + H_2O(g) \rightleftharpoons CO_2(g) + H_2(g) \qquad K_c = 0.72 \quad \text{(at 800°C)}$$

46. What initial equal concentrations of CO and H_2O would be needed to reach an equilibrium concentration of 1.00 M CO_2 in the water–gas shift reaction described in the previous problem?

47. Calculate the equilibrium concentrations of N_2, O_2, and NO present when a mixture that was initially 0.100 M in N_2 and 0.090 M in O_2 comes to equilibrium at 600°C.

$$N_2(g) + O_2(g) \rightleftharpoons 2\,NO(g) \qquad K_c = 3.3 \times 10^{-10}$$

48. Sulfuryl chloride decomposes to sulfur dioxide and chlorine. Calculate the concentrations of the three components of the system at equilibrium if 6.75 g of SO_2Cl_2 in a 1.00-L flask decomposes at 25°C.

$$SO_2Cl_2(g) \rightleftharpoons SO_2(g) + Cl_2(g) \qquad K_c = 1.4 \times 10^{-5}$$

49. Without calculations estimate the concentrations of NO and NOCl at equilibrium if a mixture that was initially 0.50 M in NO and 0.10 M in Cl_2 combined to form nitrosyl chloride.

$$2\,NO(g) + Cl_2(g) \rightleftharpoons 2\,NOCl(g) \qquad K_c = 2.1 \times 10^3 \quad \text{(at 500 K)}$$

Le Châtelier's Principle

50. Le Châtelier's principle has been applied to many fields, ranging from economics to psychology to political science. Give an example of Le Châtelier's principle in a field outside the physical sciences.

51. Predict the effect of increasing the pressure at constant T on the following reactions at equilibrium.
    (a) $2\,SO_3(g) + 2\,Cl_2(g) \rightleftharpoons 2\,SO_2Cl_2(g) + O_2(g)$
    (b) $O_2(g) + 2\,F_2(g) \rightleftharpoons 2\,OF_2(g)$
    (c) $2\,NO(g) + O_2(g) \rightleftharpoons 2\,NO_2(g)$

52. Predict the effect of decreasing the pressure at constant T on the following reactions at equilibrium.
    (a) $N_2O_4(g) \rightleftharpoons 2\,NO_2(g)$
    (b) $N_2(g) + O_2(g) \rightleftharpoons 2\,NO(g)$
    (c) $NO(g) + NO_2(g) \rightleftharpoons N_2O_3(g)$

53. Predict the effect of increasing the concentration of the reagent indicated in boldface on each of the following reactions at equilibrium. T and P are constant.
    (a) $\mathbf{2\,NO_2(g)} \rightleftharpoons N_2O_4(g)$
    (b) $2\,SO_3(g) \rightleftharpoons 2\,SO_2(g) + \mathbf{O_2(g)}$
    (c) $\mathbf{PF_5(g)} \rightleftharpoons PF_3(g) + F_2(g)$

54. Predict the effect of decreasing the concentration of the boldface reagent on each of the following reactions at equilibrium. T and P are constant.
    (a) $N_2(g) + O_2(g) \rightleftharpoons \mathbf{2\,NO(g)}$
    (b) $\mathbf{3\,O_2(g)} \rightleftharpoons 2\,O_3(g)$
    (c) $Cl_2(g) + \mathbf{3\,F_2(g)} \rightleftharpoons 2\,ClF_3(g)$

55. Use Le Châtelier's principle to predict the effect of an increase in pressure on the solubility of a gas in water. T is constant.

56. List as many ways as possible of increasing the yield of ammonia in the Haber process.

$$N_2(g) + 3\,H_2(g) \rightleftharpoons 2\,NH_3(g)$$

57. Explain why an increase in pressure favors the formation of ammonia in the Haber process.

58. Predict how an increase in the volume of the container by a factor of 2 would affect the concentrations of ammonia and oxygen in the following reaction. T is constant.

$$4\,NH_3(g) + 5\,O_2(g) \rightleftharpoons 4\,NO(g) + 6\,H_2O(g)$$

Pure Liquids and Solids

59. Write an equation that describes the relationship between the concentrations of the Ag^+ and CrO_4^{2-} ions in a saturated solution of Ag_2CrO_4.

60. Write an equation that describes the relationship between the concentrations of the Bi^{3+} and S^{2-} ions in a saturated solution of Bi_2S_3.

61. Calculate the K_{sp} constant for the dissolution of strontium fluoride if the solubility of SrF_2 in water is 0.107 g/L.

62. Silver acetate, $Ag(CH_3CO_2)$, is marginally soluble in water. What is the K_{sp} for silver acetate if 1.190 g of $Ag(CH_3CO_2)$ dissolves in 99.40 mL of water?

63. People who have the misfortune of going through a series of X rays of the gastrointestinal tract are often given a suspension of solid barium sulfate in water to drink. $BaSO_4$ is used instead of other Ba^{2+} salts, which also reflect X rays, because it is relatively insoluble in water. (Thus the patient is exposed to the minimum amount of toxic Ba^{2+} ion.) What is the equilibrium constant for the dissolution of barium sulfate if 1.0 g of $BaSO_4$ dissolves in 400,000 g of water?

64. What is the solubility of silver sulfide in water in grams per 100 mL if the K_{sp} for Ag_2S is 6.3×10^{-50}?

65. What is the solubility in water for each of the following salts in grams per 100 mL?
    (a) Hg_2S ($K_{sp} = 1.0 \times 10^{-47}$)   (b) HgS ($K_{sp} = 4 \times 10^{-53}$)

Integrated Problems

66. Which of the following diagrams best represents the concentrations of the reactants and products for the following reaction at equilibrium? Explain what is wrong with each incorrect diagram. (⬤ represents isobutane, and ⬤ represents *n*-butane.)

$$\underset{\text{Isobutane}}{\overset{\text{CH}_3}{\underset{|}{\text{CH}_3\text{CHCH}_3(g)}}} \rightleftharpoons \underset{\text{n-Butane}}{\text{CH}_3\text{CH}_2\text{CH}_2\text{CH}_3(g)} \qquad K_c = 0.4$$

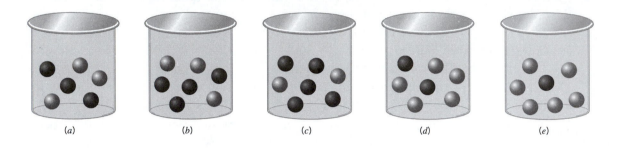

(a)   (b)   (c)   (d)   (e)

67. A sparingly soluble hypothetical ionic compound, MX_2, is placed into a beaker of distilled water. Which of the following diagrams *best* describes what happens in solution? Explain what is wrong with each incorrect diagram.

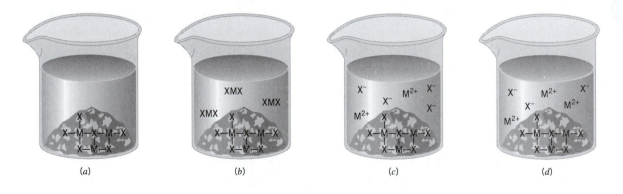

(a)   (b)   (c)   (d)

68. Describe the relationship between k_f and k_r for the following one-step reaction at equilibrium.

$$Z(g) + X(g) \underset{k_r}{\overset{k_f}{\rightleftharpoons}} Y(g) \qquad K_c = 1 \times 10^{-3}$$

Is $k_f = k_r$, $k_f < k_r$, or $k_f > k_r$? Explain your reasoning.

69. For the reaction $A \rightleftharpoons B$, match the graphs of concentration versus time to the appropriate set of rate constants.

$$\text{Rate}_{\text{forward}} = k_A(A) \qquad \text{Rate}_{\text{reverse}} = k_B(B)$$

(a) $k_A = k_B$ (b) $k_A = 1.0/s$, $k_B = 0.5/s$ (c) $k_A = 0.5/s$, $k_B = 1.0/s$

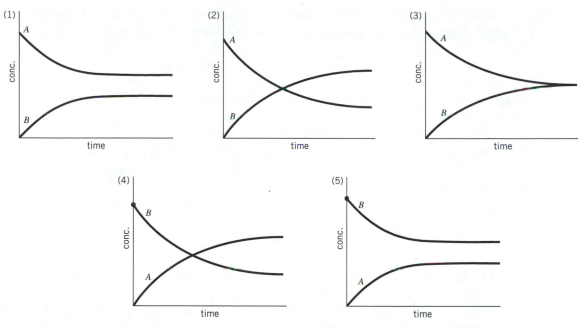

70. Write the equilibrium constant expression for the dissolving of strontium fluoride in water.

$$SrF_2(s) \rightleftharpoons Sr^{2+}(aq) + 2\,F^-(aq)$$

(a) When $SrF_2(s)$ is placed in water the compound dissolves to produce an equilibrium Sr^{2+} concentration of 5.8×10^{-4} moles per liter. What is K_{sp} for the reaction?

(b) If 50.00 mL of 0.100 M $Sr(NO_3)_2$ is mixed with 50.00 mL of 0.100 M NaF will a precipitate form? Explain your answer.

71. Several plots of concentration versus time for the reaction $A \rightleftharpoons B$, are given below. $K_c = 2$. Only one of the plots can be correct. Which one is it? Explain what is wrong with each of the incorrect plots.

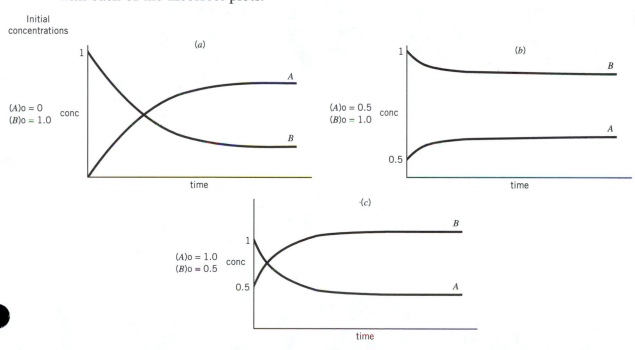

72. For the following reaction at 230°C the concentrations of the species at equilibrium were found to be as follows: $[NO] = 0.0542\ M$, $[O_2] = 0.127\ M$, $[NO_2] = 15.5\ M$.

$$2\ NO(g) + O_2(g) \rightleftharpoons 2\ NO_2(g)$$

   (a) What does it mean to say that a reaction has come to equilibrium? Must all reactions eventually come to equilibrium?
   (b) What is the equilibrium constant, K_c, for the reaction?
   (c) If, to these equilibrium concentrations, sufficient O_2 and NO_2 are added to increase $[O_2]$ and $[NO_2]$ to 1.127 and 16.5 M, respectively, while keeping $[NO]$ at 0.0542 M, in which direction will the reaction proceed?

73. The reaction below at a certain temperature has $K_c = 5.0 \times 10^{-9}$.

$$N_2F_4(g) \rightleftharpoons 2\ NF_2(g)$$

   (a) If 1.0 mole of N_2F_4 is placed in a 1.0-L flask with no NF_2 present, what will be the equilibrium concentrations of NF_2 and N_2F_4?
   (b) If 1.0 mole of NF_2 is placed in a 1.0-L flask with no N_2F_4 present, what will be approximately the equilibrium concentration of N_2F_4? No calculations are necessary.

74. The following equation represents a system at equilibrium.

$$Cl_2(g) \rightleftharpoons 2\ Cl(g)$$

   Which of the following could be a valid representation of this system? ($\bullet\!\bullet$ = Cl_2 and $\bullet$ = Cl.)

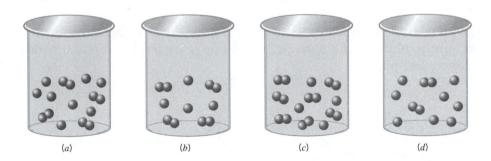

(a)   (b)   (c)   (d)   (e) all of the
                                                above

C H A P T E R
10
SPECIAL TOPICS

10A.1   A Rule of Thumb for Testing the Validity of Assumptions

10A.2   What Do We Do when the Approximation Fails?

10A.1   A RULE OF THUMB FOR TESTING THE VALIDITY OF ASSUMPTIONS

There was no doubt about the validity of the assumption that ΔC was small compared with the initial concentration of SO_3 in Section 10.8. The value of ΔC was so small that $2\Delta C$ was an order of magnitude smaller than the experimental error involved in measuring the initial concentration of SO_3.

In general, we can get some idea of whether ΔC might be small enough to be ignored by comparing the initial reaction quotient with the equilibrium constant for the reaction. If Q_c and K_c are both much smaller than 1, or both much larger than 1, the reaction doesn't have very far to go to reach equilibrium, and the assumption that ΔC is small enough to be ignored is probably legitimate.

This raises an interesting question: How do we decide whether it is valid to assume that ΔC is small enough to be ignored? The answer to the question depends on how much error we are willing to let into our calculation before we no longer trust the results. As a rule of thumb, chemists typically assume that ΔC is negligibly small so long as what is added to or subtracted from the initial concentrations of the reactants or products is less than 5% of the initial concentration. The best way to decide whether the assumption meets this rule of thumb in a particular calculation is to try it and see if it works.

Exercise 10A.1

Ammonia is made from nitrogen and hydrogen by the following reaction.

$$N_2(g) + 3\,H_2(g) \rightleftharpoons 2\,NH_3(g)$$

Assume that the initial concentration of N_2 is 0.100 mol/L and the initial concentration of H_2 is 0.100 mol/L. Calculate the equilibrium concentrations of the three components of the reaction at 500°C if the equilibrium constant for the reaction at that temperature is 0.040.

Solution

We start, as always, by building a representation for the problem based on the following general format.

	$N_2(g)$	$+$	$3\,H_2(g)$	$\rightleftharpoons$	$2\,NH_3(g)$	$K_c = 0.040$
Initial	0.100 M		0.100 M		0	
Equilibrium	?		?		?	

We then calculate the initial reaction quotient and compare it with the equilibrium constant for the reaction.

$$Q_c = \frac{(NH_3)^2}{(N_2)(H_2)^3} = \frac{0^2}{(0.100)(0.100)^3} = 0$$

The reaction quotient ($Q_c = 0$) is smaller than the equilibrium constant ($K_c = 0.040$), so the reaction has to shift to the right to reach equilibrium. This will result in a decrease in the concentrations of N_2 and H_2 and an increase in the NH_3 concentration.

The relationship between the changes in the concentrations of N_2, H_2, and NH_3 as the reaction comes to equilibrium is determined by the stoichiometry of the reaction, as shown in Figure 10.A1.

	$N_2(g)$	+	$3\ H_2(g)$	$\rightleftharpoons$	$2\ NH_3(g)$	$K_c = 0.040$
Initial	0.100 M		0.100 M		0	
Change	$-\Delta C$		$-3\Delta C$		$+2\Delta C$	
Equilibrium	$0.100 - \Delta C$		$0.100 - 3\Delta C$		$2\Delta C$	

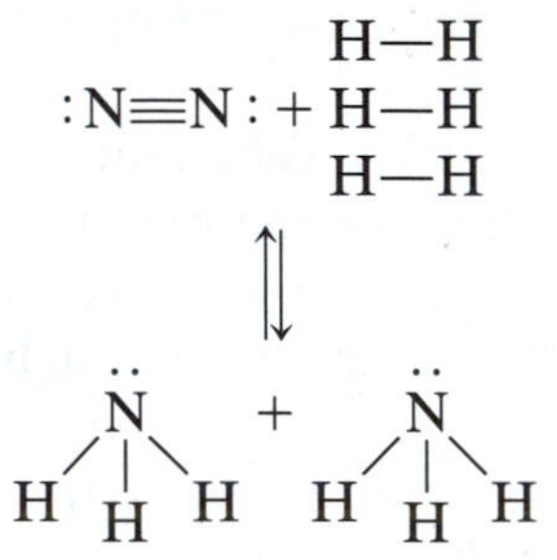

$$:N\equiv N: + \begin{array}{c} H-H \\ H-H \\ H-H \end{array}$$

$$\updownarrow$$

FIGURE 10A.1 The stoichiometry of this reaction determines the relationship between the magnitude of the changes in the concentrations of N_2, H_2, and NH_3 as this reaction comes to equilibrium.

Substituting this information into the equilibrium constant expression for the reaction gives the following equation.

$$K_c = \frac{[NH_3]^2}{[N_2][H_2]^3} = \frac{[2\Delta C]^2}{[0.100 - \Delta C][0.100 - 3\Delta C]^3} = 0.040$$

Since Q_c for the initial concentrations and K_c for the reaction are both smaller than 1, let's try the assumption that ΔC is small enough that subtracting it from 0.100—or even subtracting $3\Delta C$ from 0.100—doesn't make a significant change. This assumption gives the following approximate equation.

$$\frac{[2\Delta C]^2}{[0.100][0.100]^3} \approx 0.040$$

Solving the equation for ΔC gives the following result.

$$\Delta C \approx 1.0 \times 10^{-3}\ M$$

Now we have to check our assumptions. Is ΔC significantly smaller than 0.100? Yes, ΔC is about 1% of the initial concentration of N_2.

$$\frac{0.0010}{0.100} \times 100 = 1.0\%$$

Is $3\Delta C$ significantly smaller than 0.100? Once again, the answer is yes, $3\Delta C$ is only about 3% of the initial concentration of H_2.

$$\frac{3(0.0010)}{0.100} \times 100 = 3.0\%$$

We can therefore use the approximate value of ΔC to determine the equilibrium concentrations of N_2, H_2, and NH_3.

$$[NH_3] = 2\Delta C \approx 0.0020 \ M$$
$$[N_2] = 0.100 - \Delta C \approx 0.099 \ M$$
$$[H_2] = 0.100 - 3\Delta C \approx 0.097 \ M$$

Note that only 1% of the nitrogen is converted into ammonia under these conditions.

We can check the validity of our calculation by substituting the information back into the equilibrium constant expression:

$$K_c = \frac{[NH_3]^2}{[N_2][H_2]^3} \approx \frac{[0.0020]^2}{[0.099][0.097]^3} = 0.044$$

Once again, we have reason to accept the assumption that ΔC is small compared with the initial concentrations because the equilibrium constant calculated from the data agrees well with the value of K_c given in the problem.

10A.2   WHAT DO WE DO WHEN THE APPROXIMATION FAILS?

It is easy to envision a problem in which the assumption that ΔC is small compared with the initial concentrations can't possibly be valid. All we have to do is construct a problem in which there is a large difference between the values of Q_c for the initial concentrations and K_c for the reaction at equilibrium. Consider the following problem, for example.

Nitrogen oxide reacts with oxygen to form nitrogen dioxide.

$$2 \ NO(g) + O_2(g) \rightleftharpoons 2 \ NO_2(g)$$

The equilibrium constant for the reaction is 3.0×10^6 at 200°C. Assume initial concentrations of 0.100 M for NO and 0.050 M for O_2. Calculate the concentrations of the three components of the reaction at equilibrium.

We start, once again, by representing the information in the problem as follows.

	$2 \ NO(g)$ +	$O_2(g)$	$\rightleftharpoons$	$2 \ NO_2(g)$	$K_c = 3.0 \times 10^6$
Initial	0.100 M	0.050 M		0	
Equilibrium	?	?		?	

The first step is always the same: Compare the initial value of the reaction quotient with the equilibrium constant.

$$Q_c = \frac{(NO_2)^2}{(NO)^2(O_2)} = \frac{(0)^2}{(0.100)^2(0.050)} = 0$$

The relationship between the initial reaction quotient ($Q_c = 0$) and the equilibrium constant ($K_c = 3.0 \times 10^6$) tells us something we may already have suspected; the reaction must shift to the right to reach equilibrium.

Some might ask, "Why calculate the initial value of the reaction quotient for the reaction? Isn't it obvious that the reaction has to shift to the right to produce at least some NO_2?" Yes, it is. But calculating the value of Q_c for the reaction does more than tell us in which direction it has to shift to reach equilibrium. It also gives us an indication of how far the reaction has to go to reach equilibrium.

In this case, Q_c is so very much smaller than K_c for the reaction that we have to conclude that the initial conditions are very far from equilibrium. It would therefore be a mistake to assume that ΔC is small.

We can't assume that ΔC is negligibly small in this problem, but we can redefine the problem so that the assumption becomes valid. The key to achieving this goal is to remember the conditions under which we can assume that ΔC is small enough to be ignored. This assumption is only valid when Q_c is of the same order of magnitude as K_c (i.e., when Q_c and K_c are both much larger than 1 or much smaller than 1). We can solve problems for which Q_c isn't close to K_c by redefining the initial conditions so that Q_c becomes close to K_c (see Figure 10A.2). To show how this can be done, let's return to the problem given in this section.

Reactants

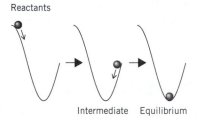

Intermediate   Equilibrium

Redefine problem
so that ΔC is small

FIGURE 10A.2 When the initial conditions are very far from equilibrium, it is often useful to redefine the problem. This involves driving the reaction as far as possible in the direction favored by the equilibrium constant. When the reaction returns to equilibrium from the intermediate conditions, changes in the concentrations of the components of the reaction are often small enough compared with the initial concentration to be ignored.

The equilibrium constant for the reaction between NO and O_2 to form NO_2 is much larger ($K_c = 3.0 \times 10^6$) than Q_c. This means that the equilibrium favors the products of the reaction. The best way to handle the problem is to drive the reaction as far as possible to the right, and then let it come back to equilibrium. Let's therefore define an intermediate set of conditions that correspond to what would happen if we push the reaction as far as possible to the right.

	$2\,NO(g)$ +	$O_2(g)$	$\rightleftharpoons$ $2\,NO_2(g)$	$K_c = 3.0 \times 10^6$
Initial	0.100 M	0.050 M	0	
Change	$-0.100\ M$	$-0.050\ M$	$+0.100\ M$	
Intermediate	0	0	0.100 M	

We can see where this gets us by calculating the reaction quotient for the intermediate conditions.

$$Q_c = \frac{(NO_2)^2}{(NO)^2(O_2)} = \frac{0.100^2}{(0)^2(0)} = \infty$$

The reaction quotient is now larger than the equilibrium constant, and the reaction has to shift back to the left to reach equilibrium. Some of the NO_2 must now decompose to form

NO and O_2. The relationship between the changes in the concentrations of the three components of this reaction is determined by the stoichiometry of the reaction, as shown in Figure 10.A3.

	$2\ NO(g)$	$+\ O_2(g)$	$\rightleftharpoons$	$2\ NO_2(g)$	$K_c = 3.0 \times 10^6$
Intermediate	0	0		0.100 M	
Change	$+2\Delta C$	$+\Delta C$		$-2\Delta C$	
Equilibrium	$2\Delta C$	ΔC		$0.100 - 2\Delta C$	

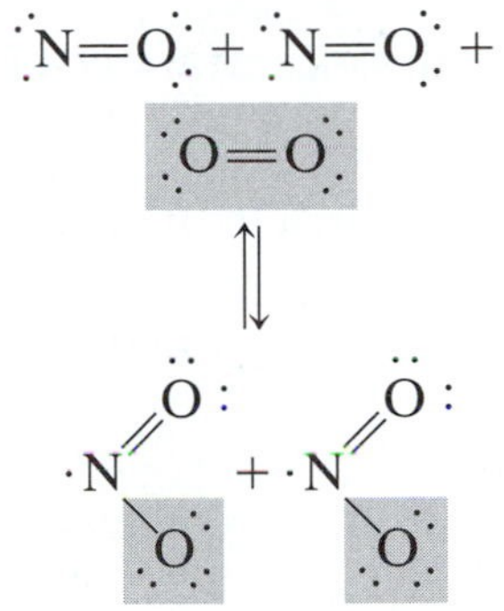

FIGURE 10A.3 Once again, the stoichiometry of the reaction determines the relationship between the magnitude of the changes in the concentrations of the three components of the reaction as it comes to equilibrium.

We now substitute what we know about the reaction into the equilibrium constant expression.

$$K_c = \frac{[NO_2]^2}{[NO]^2[O_2]} = \frac{[0.100 - 2\Delta C]^2}{[2\Delta C]^2[\Delta C]} = 3.0 \times 10^6$$

Because the reaction quotient for the intermediate conditions and the equilibrium constant are both relatively large, we can assume that the reaction doesn't have very far to go to reach equilibrium. In other words, we assume that $2\Delta C$ is small compared with the intermediate concentration of NO_2 and derive the following approximate equation.

$$\frac{0.100^2}{[2\Delta C]^2[\Delta C]} \approx 3.0 \times 10^6$$

We then solve the equation for an approximate value of ΔC.

$$\Delta C \approx 9.4 \times 10^{-4}\ M$$

We now check our assumption that $2\Delta C$ is small enough compared with the intermediate concentration of NO_2 to be ignored.

$$\frac{2(0.00094)}{0.100} \times 100 = 1.9\%$$

The value of $2\Delta C$ is less than 2% of the intermediate concentration of NO_2, which means that it can be legitimately ignored in the calculation.

Since the approximation is valid, we can use the new value of ΔC to calculate the equilibrium concentrations of NO_2, NO, and O_2.

$$[NO_2] = 0.100 - 2\Delta C \approx 0.098\ M$$
$$[NO] = 2\Delta C \approx 0.0019\ M$$
$$[O_2] = \Delta C \approx 0.00094\ M$$

The results of the calculation provide insight into the chemistry of the pollutants formed by an internal combustion engine. When a mixture of gasoline and air is burned, the N_2 and O_2 in air react to form NO, which can then react with oxygen to form NO_2.

$$N_2(g) + O_2(g) \rightleftharpoons 2\,NO(g)$$
$$2\,NO(g) + O_2(g) \rightleftharpoons 2\,NO_2(g)$$

Although the product of the reactions is often described as NO_x—to indicate that it is a mixture of NO and NO_2—our calculation suggests that the dominant product of the reaction would be NO_2, if the reaction comes to equilibrium.

In general, the assumption that ΔC is small compared with the initial concentrations of the reactants or products works best under the following conditions.

- When $K_c \ll 1$ and we approach equilibrium from left to right. (We start with excess reactants and form some products.)
- When $K_c \gg 1$ and we approach equilibrium from right to left. (We start with excess products and form some reactants.)

PROBLEMS

Hidden Assumptions That Make Equilibrium Calculations Easier

10A-1. Describe what happens if you make the assumption that ΔC is zero in the following equation.

$$\frac{[0.125 - \Delta C][2.40 - 2\Delta C]^2}{[0.200 + 2\Delta C]^2} = 1.3 \times 10^{-8}$$

Explain how to get around this problem.

A Rule of Thumb for Testing the Validity of Assumptions

10A-2. Explain why ΔC is relatively small when the reaction quotient (Q_c) is reasonably close to the equilibrium constant for the reaction (K_c).

10A-3. Explain why the assumption that ΔC is small compared with the initial concentrations of the reactants and products is doomed to fail when the reaction quotient (Q_c) is very different from the equilibrium constant for the reaction (K_c).

What Do We Do When the Approximation Fails?

10A-4. Describe the technique used to solve problems for which the reaction quotient is very different from the equilibrium constant.

10A-5. Before we can solve the following problem, we have to define a set of intermediate conditions under which the concentration of one of the reactants or products is zero.

	$2\,NO_2(g)$	$\rightleftharpoons$	$2\,NO(g) +$	$O_2(g)$	$K_c = 5.3 \times 10^{-6}$ at (250°C)
Initial	0.10 M		0.10 M	0.005 M	

Which of the following goals determines whether we push the reaction as far as possible to the right or as far as possible to the left?
(a) To make both Q_c and K_c large.
(b) To make both Q_c and K_c small.
(c) To bring Q_c as close as possible to K_c.
(d) To make the difference between Q_c and K_c as large as possible.

CHAPTER
11
ACIDS AND BASES

11.1 PROPERTIES OF ACIDS AND BASES

For more than 300 years, chemists have classified substances that behave like vinegar as **acids** and substances that have properties like wood ash as **bases** (or **alkalies**). The name "acid" comes from the Latin word *acidus,* which means "sour," and refers to the sharp odor and sour taste of many acids. Vinegar, for example, tastes sour because it is a dilute solution of acetic acid in water. Lemon juice tastes sour because it contains citric acid. Milk turns sour when it spoils because lactic acid is formed, and the unpleasant, sour odor of rotten meat or butter can be attributed to compounds such as butyric acid that form when fat spoils.

One of the characteristic properties of acids is the ability to dissolve most metals. Zinc metal, for example, rapidly dissolves in hydrochloric acid to form an aqueous solution of $ZnCl_2$ and hydrogen gas.

$$Zn(s) + 2\,HCl(aq) \longrightarrow Zn^{2+}(aq) + 2\,Cl^-(aq) + H_2(g)$$

Another characteristic property of acids is the ability to change the color of vegetable dyes, such as litmus. Litmus is a mixture of blue dyes that turns red in the presence of acid. Litmus has been used to test for acids for at least 300 years.

Bases also have characteristic properties. They taste bitter and often feel slippery. They change the color of litmus from red to blue, thereby reversing the change in color that occurs when litmus comes in contact with an acid. Bases become less alkaline when they are combined with acids, and acids lose their characteristic sour taste and ability to dissolve metals when they are mixed with alkalies.

11.2   THE ARRHENIUS DEFINITION OF ACIDS AND BASES

In 1887 Svante Arrhenius took a major step toward answering the important question, "What factors determine whether a compound is an acid or a base?" Arrhenius suggested that acids are neutral compounds that *ionize* when they dissolve in water to give H^+ ions and a corresponding negative ion. According to the model, hydrogen chloride is an acid because it ionizes when it dissolves in water to give hydrogen (H^+) and chloride (Cl^-) ions (Figure 11.1). This aqueous solution is known as hydrochloric acid and is often written as $HCl(aq)$.

$$HCl(g) \xrightarrow{H_2O} H^+(aq) + Cl^-(aq)$$

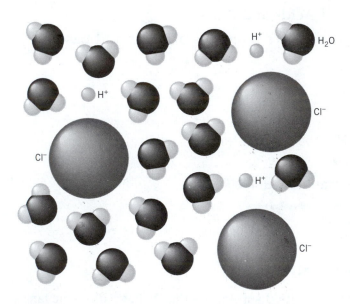

FIGURE 11.1 The Arrhenius model assumes that HCl dissociates into H^+ and Cl^- ions when it dissolves in water.

Arrhenius argued that bases are neutral compounds that dissociate in water to give OH^- and positive ions. NaOH is an Arrhenius base because it dissociates in water to give the hydroxide (OH^-) and sodium (Na^+) ions.

$$NaOH(s) \xrightarrow{H_2O} Na^+(aq) + OH^-(aq)$$

An **Arrhenius acid** is therefore any substance that ionizes when it dissolves in water to give H^+, or hydrogen ion. An **Arrhenius base** is any substance that gives OH^-, or hydroxide ion, when it dissolves in water. Arrhenius acids include compounds such as HCl, HCN, and H_2SO_4 that ionize in water to give the H^+ ion. Arrhenius bases include, but are not limited to, ionic compounds that contain the OH^- ion, such as NaOH, KOH, and $Ca(OH)_2$.

Checkpoint

Classify the following compounds as Arrhenius acids or bases: HNO_3, $Mg(OH)_2$, $HC_2H_3O_2$

$$HNO_3(aq) \xrightarrow{H_2O} H^+(aq) + NO_3^-(aq)$$

$$Mg(OH)_2(s) \xrightarrow{H_2O} Mg^{2+}(aq) + 2\ OH^-(aq)$$

$$HC_2H_3O_2(aq) \xrightarrow{H_2O} H^+(aq) + C_2H_3O_2^-(aq)$$

11.3   THE BRØNSTED–LOWRY DEFINITION OF ACIDS AND BASES

In 1923 Johannes Brønsted and Thomas Lowry independently proposed a more powerful set of definitions of acids and bases. The Brønsted, or Brønsted–Lowry, model is based on the assumption that acids donate H^+ ions to another ion or molecule, which acts as a base. According to the model, HCl doesn't dissociate in water to form H^+ and Cl^- ions. Instead, an H^+ ion is transferred from HCl to a water molecule to form H_3O^+, a **hydronium ion,** and Cl^- ion, as shown in Figure 11.2.

$$HCl(aq) + H_2O(l) \longrightarrow H_3O^+(aq) + Cl^-(aq)$$

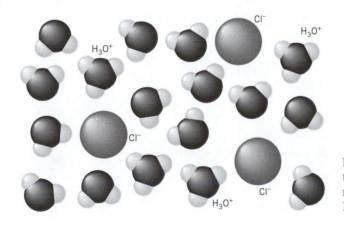

FIGURE 11.2 The Brønsted model assumes that HCl molecules donate an H^+ ion to water molecules to form H_3O^+ and Cl^- ions when HCl dissolves in water.

Because it is a proton, an H^+ ion is several orders of magnitude smaller than the smallest atom. As a result, the charge on an isolated H^+ ion is distributed over such a small amount of space that the H^+ ion is attracted toward any source of negative charge that exists in the solution. Thus, the instant that an H^+ ion is created in an aqueous solution, it bonds to the electronegative oxygen atom of a water molecule. The Brønsted model, in which H^+ ions are transferred from one ion or molecule to another, therefore seems more reasonable than the Arrhenius model, which assumes that H^+ ions exist in aqueous solution.

Even the Brønsted model is naive, however. Each H^+ ion that an acid donates to water is actually bound to four neighboring water molecules, as shown in Figure 11.3. A more realistic formula for the substance produced when an acid loses an H^+ ion is therefore $H(H_2O)_4^+$, or $H_9O_4^+$. For all practical purposes, however, this substance can be represented as the H_3O^+ ion.

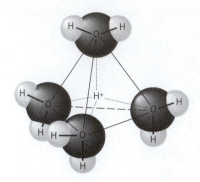

FIGURE 11.3 Structure of the $H(H_2O)_4^+$ ion formed when an acid reacts with water. For practical purposes, the ion can be thought of as an H_3O^+ ion.

The reaction between HCl and water provides the basis for understanding the definitions of a Brønsted acid and a Brønsted base. According to the model, when HCl dissociates in water,

$$HCl(aq) + H_2O(l) \longrightarrow H_3O^+(aq) + Cl^-(aq)$$

HCl acts as an H^+ ion donor and H_2O acts as an H^+ ion acceptor. A **Brønsted acid** is therefore any substance (such as HCl) that can donate an H^+ ion to a base. A **Brønsted base** is any substance (such as H_2O) that can accept an H^+ ion from an acid.

There are two ways of naming the H^+ ion. Some chemists call it a hydrogen ion; others call it a proton. As a result, Brønsted acids are known as either **hydrogen ion donors** or **proton donors.** Brønsted bases are **hydrogen ion acceptors** or **proton acceptors.**

From the perspective of the Brønsted model, reactions between acids and bases always involve the transfer of an H^+ ion from a proton donor to a proton acceptor. Acids can be neutral molecules.

$$\underset{Acid}{HCl(aq)} + \underset{Base}{NH_3(aq)} \longrightarrow Cl^-(aq) + NH_4^+(aq)$$

They can also be positive ions,

$$\underset{Acid}{NH_4^+(aq)} + \underset{Base}{OH^-(aq)} \rightleftharpoons NH_3(aq) + H_2O(l)$$

or negative ions.

$$\underset{Acid}{H_2PO_4^-(aq)} + \underset{Base}{H_2O(l)} \rightleftharpoons HPO_4^{2-}(aq) + H_3O^+(aq)$$

Brønsted bases can be identified from their Lewis structures. According to the Brønsted model, a base is any ion or molecule that can accept a proton. To understand the implications of this definition, look at how the prototypical base, the OH^- ion, accepts a proton.

$$H^+ + \;:\!\ddot{O}\!-\!H^- \longrightarrow H\!-\!\ddot{O}\!-\!H$$

The only way to accept an H^+ ion is to form a covalent bond to it. In order to form a covalent bond to an H^+ ion that has no valence electrons, the base must provide both of the electrons needed to form the bond. Thus, only compounds that have pairs of nonbonding valence electrons can act as H^+ ion acceptors, or Brønsted bases. The following compounds, for example, can all act as Brønsted bases because they all contain nonbonding pairs of electrons.

$$NH_3 \qquad H-\overset{..}{\underset{|}{N}}-H$$
$$\qquad\qquad\qquad H$$

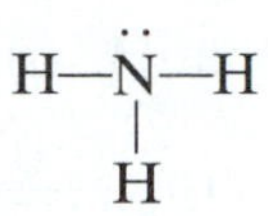

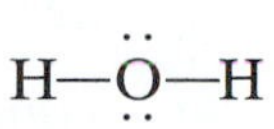

$$H_2O \qquad H-\overset{..}{\underset{..}{O}}-H$$

$$CO_3{}^{2-} \qquad \left[\, :\overset{..}{\underset{..}{O}}-\overset{\overset{\displaystyle :O:}{\|}}{C}-\overset{..}{\underset{..}{O}}: \,\right]^{2-}$$

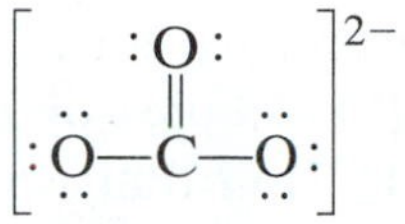

The Brønsted model includes any ion or molecule that contains one or more pairs of nonbonding valence electrons. There are many molecules and ions that satisfy the definition of a Brønsted base and relatively few, such as the following, that do not. Substances that do not behave as a Brønsted base have no nonbonding electron pairs.

$$CH_4 \qquad H-\overset{\displaystyle H}{\underset{\displaystyle H}{\underset{|}{\overset{|}{C}}}}-H$$

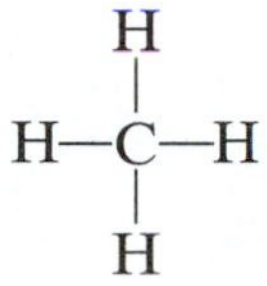

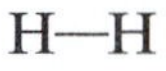

$$H_2 \qquad H-H$$

$$NH_4{}^+ \qquad \left[\, H-\overset{\displaystyle H}{\underset{\displaystyle H}{\underset{|}{\overset{|}{N}}}}-H \,\right]^+$$

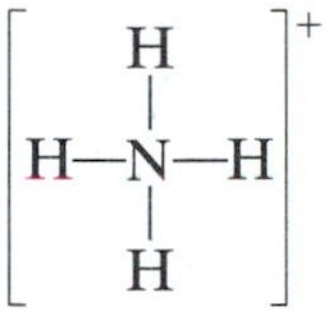

Exercise 11.1

Identify the reactant that behaves as a Brønsted acid and the reactant that behaves as a Brønsted base in each of the following reactions.
(a) $HF(aq) + OH^-(aq) \rightleftharpoons H_2O(l) + F^-(aq)$
(b) $HC_2H_3O_2(aq) + H_2O(l) \rightleftharpoons C_2H_3O_2{}^-(aq) + H_3O^+(aq)$
(c) $C_6H_5NH_2(aq) + HNO_3(aq) \rightleftharpoons C_6H_5NH_3{}^+(aq) + NO_3{}^-(aq)$

Solution

(a) acid: HF base: OH^-
(b) acid: $HC_2H_3O_2$ base: H_2O
(c) acid: HNO_3 base: $C_6H_5NH_2$

11.4 CONJUGATE ACID–BASE PAIRS

The Brønsted model of acids and bases creates a link between acids and bases. Every time a Brønsted acid acts as an H^+ ion donor, it forms a conjugate base. Imagine a generic acid, HA, where A is a symbol that represents any anion. When the acid donates an H^+ ion to water, one product of the reaction is the A^- ion, which is an H^+ ion acceptor or a Brønsted base.

$$\underset{Acid}{HA(aq)} + H_2O(l) \rightleftharpoons H_3O^+(aq) + \underset{Base}{A^-(aq)}$$

Conversely, every time a base gains an H^+ ion, the product is a Brønsted acid, HA.

$$A^-(aq) + H_2O(l) \rightleftharpoons HA(aq) + OH^-(aq)$$

$$\text{Base} \qquad\qquad\qquad\qquad \text{Acid}$$

Acids and bases in the Brønsted model therefore exist as **conjugate acid–base pairs** whose formulas are related by the gain or loss of a hydrogen ion.[1]

Our use of the symbols HA and A^- for a conjugate acid–base pair doesn't mean that all acids are electrically neutral molecules or that all bases are negative ions. It signifies only that the acid contains an H^+ ion that isn't present in the conjugate base. As noted earlier, Brønsted acids and bases can be electrically neutral molecules, positive ions, or negative ions. Various Brønsted acids and their conjugate bases are given in Table 11.1. Although the anions of strong acids are referred to as conjugate bases they are such very, very weak bases in water that they have essentially no base properties (e.g., Cl^-, Br^-, ClO_4^-). In like manner the cations of strong bases are referred to as conjugate acids although they have essentially no acid properties in water (e.g., Na^+, K^+, Ca^{2+}).

TABLE 11.1   Typical Brønsted Acids and Their Conjugate Bases

Acid	Base
H_3O^+	H_2O
H_2O	OH^-
HCl	Cl^-
H_2SO_4	HSO_4^-
HSO_4^-	SO_4^{2-}
NH_4^+	NH_3

It is important to recognize that some compounds can be both a Brønsted acid and a Brønsted base. H_2O and HSO_4^-, for example, can be found in both columns in Table 11.1. Water is the perfect example of this behavior because it simultaneously acts as an acid and a base when it reacts with itself to form the H_3O^+ and OH^- ions.

$$H_2O(l) + H_2O(l) \rightleftharpoons H_3O^+(aq) + OH^-(aq)$$

Checkpoint

Phosphoric acid, H_3PO_4, is a common additive in soft drinks. Write the chemical equation for the dissociation of phosphoric acid in water and predict the chemical formula of its conjugate base. Aniline, $C_6H_5NH_2$, is a base used in the manufacture of dyes. Write the chemical equation for the reaction of aniline in water and predict the chemical formula of its conjugate acid.

[1]The term *conjugate* comes from Latin stems meaning "joined together" and refers to things that are joined, particularly in pairs. It is therefore the perfect term to describe the relationship between Brønsted acids and bases.

The concept of conjugate acid–base pairs plays a vital role in explaining reactions between acids and bases. According to the Brønsted model, an acid always reacts with a base to form the conjugate base and conjugate acid. Consider the following reaction, for example.

$$HNO_3(aq) + NH_3(aq) \longrightarrow NH_4^+(aq) + NO_3^-(aq)$$

Acid   Base   Conjugate   Conjugate
                                    acid   base

In the course of the reaction, nitric acid donates an H^+ ion to form its conjugate base, the nitrate ion (NO_3^-). At the same time, ammonia acts as a base, accepting an H^+ ion to form its conjugate acid, the ammonium ion (NH_4^+).

The products of the reaction are often combined and written as an aqueous solution of an ionic compound, or salt.

$$HNO_3(aq) + NH_3(aq) \longrightarrow NH_4NO_3(aq)$$

Because the products of the reaction are neither as acidic as nitric acid nor as basic as ammonia, the reaction is often called a **neutralization reaction.** This doesn't imply that the products have no acid or base properties. It only suggests that the products are less acidic and less basic than the starting materials.

Water is often one of the products of a neutralization reaction. Consider the reaction between formic acid and sodium hydroxide, for example.

$$HCO_2H(aq) + NaOH(aq) \longrightarrow H_2O(l) + Na^+(aq) + HCO_2^-(aq)$$

Acid   Base   Conjugate   Conjugate
                                     acid   base

The salt produced in this reaction is sodium formate, $NaHCO_2$. The formate ion, HCO_2^-, is the conjugate base of formic acid, and water is the conjugate acid of the hydroxide ion in sodium hydroxide.

Exercise 11.2

Write the chemical equation for the reaction of chlorous acid, $HClO_2$, with potassium hydroxide. Identify the acid, base, conjugate acid, and conjugate base in the reaction.

Solution

$$HClO_2(aq) + KOH(aq) \longrightarrow H_2O(l) + K^+(aq) + ClO_2^-(aq)$$

acid: $HClO_2$ conjugate acid: H_2O
base: KOH   conjugate base: ClO_2^-

11.5   THE ROLE OF WATER IN THE BRØNSTED MODEL

The Lewis structure of water can help us understand why H_3O^+ and OH^- ions play such an important role in the chemistry of aqueous solutions. The Lewis structure suggests that the hydrogen and oxygen atoms in a water molecule are bound together by sharing a pair of electrons. But oxygen ($EN = 3.61$) is much more electronegative than hydrogen ($EN = 2.30$), so the electrons in the covalent bonds are not shared equally by the hydrogen and oxygen atoms. The electrons are drawn toward the oxygen atom in the center of the molecule and away from the hydrogen atoms on either end. As a result, the water molecule is **polar.** The oxygen atom carries a partial negative charge (-0.44), and each hydrogen atom carries a partial positive charge ($+0.22$) (see problem 79, Chapter 4).

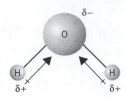

When some water molecules dissociate to form ions in water, positively charged H_3O^+ ions and negatively charged OH^- ions are formed.

$$H\!-\!\overset{..}{\underset{..}{O}}\!-\!H + H\!-\!\overset{..}{\underset{..}{O}}\!-\!H \longrightarrow \left[H\!-\!\overset{..}{\underset{\underset{H}{|}}{O}}\!-\!H\right]^+ + \left[:\!\overset{..}{\underset{..}{O}}\!-\!H\right]^-$$

The opposite reaction can also occur: H_3O^+ ions can combine with OH^- ions to form neutral water molecules.

$$H_3O^+(aq) + OH^-(aq) \longrightarrow 2\ H_2O(l)$$

The fact that water molecules dissociate to form H_3O^+ and OH^- ions, which can then recombine to form water molecules, is indicated by the following equation.

$$2\ H_2O(l) \rightleftharpoons H_3O^+(aq) + OH^-(aq)$$

The pair of arrows that separate the "reactants" and the "products" of this reaction indicate that the reaction occurs in both directions.

The Brønsted model explains water's role in acid–base reactions.

- Water dissociates to form ions by transferring an H^+ ion from one molecule acting as an acid to another molecule acting as a base.

$$\underset{Acid}{H_2O(l)} + \underset{Base}{H_2O(l)} \rightleftharpoons H_3O^+(aq) + OH^-(aq)$$

- Acids react with water by donating an H^+ ion to a neutral water molecule acting as a base to form the H_3O^+ ion.

$$\underset{Acid}{HF(aq)} + \underset{Base}{H_2O(l)} \rightleftharpoons H_3O^+(aq) + F^-(aq)$$

- Bases react with water by accepting an H^+ ion from a water molecule acting as an acid to form the OH^- ion.

$$\underset{Base}{NH_3(aq)} + \underset{Acid}{H_2O(l)} \rightleftharpoons NH_4^+(aq) + OH^-(aq)$$

- Water molecules can act as intermediates in acid–base reactions by gaining H^+ ions from the acid

$$HC_2H_3O_2(aq) + H_2O(l) \rightleftharpoons H_3O^+(aq) + C_2H_3O_2^-(aq)$$
$$\text{Acid} \qquad\qquad \text{Base}$$

and then losing the H^+ ions to the base.

$$NH_3(aq) + H_3O^+(aq) \rightleftharpoons NH_4^+(aq) + H_2O(l)$$
$$\text{Base} \qquad\qquad \text{Acid}$$

The reactions can be added to give an overall equation for an acid-base reaction.

$$HC_2H_3O_2(aq) + H_2O(l) \rightleftharpoons H_3O^+(aq) + C_2H_3O_2^-(aq)$$
$$NH_3(aq) + H_3O^+(aq) \rightleftharpoons NH_4^+(aq) + H_2O(l)$$
$$\overline{HC_2H_3O_2(aq) + H_2O(l) + NH_3(aq) + H_3O^+(aq) \rightleftharpoons H_3O^+(aq) + NH_4^+(aq)}$$
$$+ C_2H_3O_2^-(aq) + H_2O(l)$$

Note that $H_2O(l)$ and $H_3O^+(aq)$ appear on both sides of the sum of the chemical equations. These species are therefore omitted to give the following overall equation:

$$HC_2H_3O_2(aq) + NH_3(aq) \rightleftharpoons NH_4^+(aq) + C_2H_3O_2^-(aq)$$
$$\text{Acid} \qquad\quad \text{Base} \qquad\quad \text{Conjugate acid} \quad \text{Conjugate base}$$

11.6 TO WHAT EXTENT DOES WATER DISSOCIATE TO FORM IONS?

At 25°C, the density of water is 0.9971 g/cm^3, or 0.9971 g/mL. The concentration of pure water is therefore 55.35 M.

$$\frac{0.9971 \text{ g } H_2O}{1 \text{ mL}} \times \frac{1000 \text{ mL}}{1 \text{ L}} \times \frac{1 \text{ mol } H_2O}{18.015 \text{ g } H_2O} = 55.35 \text{ mol } H_2O/L$$

The concentration of the H_3O^+ and OH^- ions formed by the dissociation of pure neutral H_2O at that temperature is only 1.0×10^{-7} mol/L. The ratio of the concentration of the H_3O^+ (or OH^-) ions to the concentration of the neutral H_2O molecules is therefore 1.8×10^{-9}.

$$\frac{1.0 \times 10^{-7} \text{ } M \text{ } H_3O^+}{55.35 \text{ } M \text{ } H_2O} = 1.8 \times 10^{-9}$$

In other words, only about 2 parts per billion (ppb) of the water molecules dissociate into ions at room temperature.

It is difficult to imagine what 2 ppb means. One way to visualize the number is to assume that 2500 letters appear on a typical page of this book. If typographical errors occurred with a frequency of 2 ppb, the book would have to be 400,000 pages long to contain two errors. Figure 11.4 shows a model of 20 water molecules, one of which has dissociated to form H_3O^+ and OH^- ions. If this illustration were a high-resolution photograph of the structure of water, we would encounter a pair of H_3O^+ and OH^- ions on the average of only once for every 25 million such photographs.

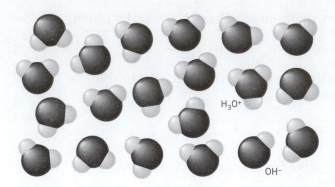

H_3O^+

OH^-

FIGURE 11.4 If this figure is thought of as a high-resolution snapshot of 20 H_2O molecules, one of which has dissociated to form H_3O^+ and OH^- ions, we would have to look at an average of 25 million such snapshots before we encountered another pair of the ions.

Water dissociates to form an equilibrium with H_3O^+ and OH^- ions.

$$2\,H_2O(l) \rightleftharpoons H_3O^+(aq) + OH^-(aq)$$

When this happens, the rate at which water molecules react to form the H_3O^+ and OH^- ions is equal to the rate at which the ions combine to form a pair of neutral water molecules. The extent to which water dissociates to form ions can be described in terms of the following equilibrium constant expression.

$$K_c = \frac{[H_3O^+][OH^-]}{[H_2O]^2}$$

As we have seen, the concentration of water at 25°C is 55.35 mol/L. The concentration of the H_3O^+ and OH^- ions, on the other hand, is only about 1.0×10^{-7} *M*. The concentration of H_2O molecules is therefore so much larger than the concentration of the H_3O^+ and OH^- ions that it remains effectively constant when water dissociates to form the ions. Chemists therefore rearrange the equilibrium constant expression for the dissociation of water to give the following equation.

$$K_c \times [H_2O]^2 = [H_3O^+][OH^-]$$

This effectively places two variables—the H_3O^+ and OH^- ion concentrations—on one side of the equation and two constants—K_c and $[H_2O]^2$—on the other side of the equation. We then replace the term on the left-hand side of the equation with a constant known as the **water dissociation equilibrium constant, K_w.**

$$K_w = [H_3O^+][OH^-]$$

Because the H_3O^+ and OH^- concentrations in pure water at 25°C are each 1.0×10^{-7} *M,* the value of K_w at that temperature is 1.0×10^{-14}.

$$[1.0 \times 10^{-7}][1.0 \times 10^{-7}] = 1.0 \times 10^{-14} \quad \text{(at 25°C)}$$

The equilibrium constant, K_w, is defined in terms of the dissociation of absolutely pure water. The value of this equilibrium constant, and therefore the resulting concentrations of H_3O^+ and OH^-, will vary slightly for actual laboratory solutions. Since these expressions only approximate actual solutions, equilibrium constants and concentrations in this text will be reported to only two significant figures, where it is understood that there is uncertainty associated with the second reported figure. However, the equilibrium constant expression is useful for approximations of real solutions of acids and bases dissolved in water. Regardless of the source of the H_3O^+ and OH^- ions in water, the product of the concentrations of the ions at equilibrium at 25°C can be approximated by 1.0×10^{-14}.

What happens to the H_3O^+ and OH^- ion concentrations that come from the dissociation of water when we add a strong acid to water? Suppose, for example, that we add enough acid to a beaker of water to raise the H_3O^+ concentration to 0.010 M. Note that there are now two sources of H_3O^+. One source is from the acid and the other is from the dissociation of water. According to Le Châtelier's principle, the additional H_3O^+ from the acid should drive the equilibrium between water and its ions to the left, reducing the number of H_3O^+ and OH^- ions which dissociate from the water.

$$2 H_2O(l) \rightleftharpoons H_3O^+(aq) + OH^-(aq)$$

Adding an acid to water therefore decreases the extent to which the water dissociates into ions.

Because the dissociation of water has shifted to the left, the concentration of H_3O^+ and OH^- ions from the dissociation will be less than their initial 1.0×10^{-7} M. The total concentration of H_3O^+ will be equal to the H_3O^+ from the acid (0.010 M) plus the H_3O^+ from the dissociation of water (1.0×10^{-7} M). However, the H_3O^+ concentration from the water is so small that it is negligible when compared to the amount from the acid. Therefore, when the system returns to equilibrium, the H_3O^+ ion concentration is still about 0.010 M. Furthermore, when the reaction returns to equilibrium, the product of the H_3O^+ and OH^- ion concentrations is once again approximated by K_w.

$$[H_3O^+][OH^-] = 1.0 \times 10^{-14}$$

If the concentration of the H_3O^+ ion is 0.010 M, the concentration of the OH^- ion when the system returns to equilibrium is only 1.0×10^{-12} M.

$$[OH^-] = \frac{K_w}{[H_3O^+]} = \frac{1.0 \times 10^{-14}}{0.010} = 1.0 \times 10^{-12} \ M$$

Some of the H_3O^+ ions in the solution come from the dissociation of water; others come from the acid that has been added to the water. All of the OH^- ions, on the other hand, come from the dissociation of water. In pure water, dissociation of the H_2O molecules gives us an OH^- ion concentration of 1.0×10^{-7} M. In the acid solution, the OH^- ion concentration is five orders of magnitude smaller. This means that adding enough acid to water to increase the concentration of the H_3O^+ ion to 0.010 M decreases the dissociation of water by a factor of about 100,000.

Adding an acid to water therefore has an effect on the concentration of both the H_3O^+ and OH^- ions. Some H_3O^+ is already present as a result of the dissociation of water. Adding an acid to water increases the concentration of the ion. However, adding an acid to water decreases the extent to which water dissociates and therefore leads to a significant decrease in the concentration of the OH^- ion.

As might be expected, the opposite effect is observed when a base is added to water. Because we are adding a base, the OH^- ion concentration increases. Once the system returns to equilibrium, the product of the H_3O^+ and OH^- ion concentrations is once again equal to K_w. The only way this can be achieved, of course, is by decreasing the concentration of the H_3O^+ ion.

Checkpoint

If an acid is added to water the equilibrium between water molecules and their ions $2 H_2O(l) \rightleftarrows H_3O^+(aq) + OH^-(aq)$ shifts to the left, decreasing the $[H_3O^+]$ from the dissociation of water. Does the K_w still equal 10^{-14}?

11.7   pH AS A MEASURE OF THE CONCENTRATION OF H_3O^+ ION

It can be difficult to work with the concentrations of H_3O^+ and OH^- ions in aqueous solutions because the range of concentrations is so large. Typical solutions in the laboratory have concentrations of H_3O^+ or OH^- ion as large as 0.1 M or as small as 1×10^{-14} M. The concentrations of the ions therefore cover a range of 14 orders of magnitude. The best way to bring into perspective the factor of 10^{14} that separates one end of the range from the other is to note that it is comparable to the difference between the value of three pennies and a national debt of \$3 trillion, or the difference between the radius of a gold atom and a distance of 8 miles.

In 1909, the Danish biochemist S. P. L. Sørenson proposed a way around the problem. Sørenson worked at a laboratory set up by the Carlsberg Brewery to apply scientific methods to the study of the fermentation reactions in brewing beer. Faced with the task of constructing graphs of the activity of the malt versus the H_3O^+ ion concentration, Sørenson suggested using logarithmic mathematics to condense the range of H_3O^+ and OH^- concentrations to a more convenient scale. By definition, the logarithm of a number is the power to which a base must be raised to obtain that number. The logarithm to the base 10 of 10^{-7}, for example, is -7.

$$\log(10^{-7}) = -7$$

Because the concentrations of the H_3O^+ and OH^- ions in aqueous solutions are usually smaller than 1 M, the logarithms of the concentrations are negative numbers. Because he considered positive numbers more convenient, Sørenson suggested that the sign of the logarithm should be changed after it had been calculated. He therefore introduced the *symbol "p" to indicate the negative of the logarithm of a number*. Thus, **pH** is the negative of the logarithm of the H_3O^+ ion concentration.

$$pH \approx -\log[H_3O^+]$$

However, the equation is valid only for very dilute solutions of acid in pure water. Most solutions used in the laboratory do not meet this criterion, so the above equation can only be used to approximate the pH of real solutions. Since pH values from this calculation are approximations, they normally are reported to no more than two digits after the decimal. In this text *calculated* pH values will be normally reported to only one digit past the decimal. However, experimentally measured values of pH using a pH meter can accurately determine the pH of a solution to two digits after the decimal.

Similarly, **pOH** is an approximation of the negative of the logarithm of the OH^- ion concentration.

$$pOH \approx -\log[OH^-]$$

Exercise 11.3

Approximate the pH of Pepsi Cola if the concentration of the H_3O^+ ion in the solution is 0.0035 *M*.

Solution

The pH of a solution is the negative of the logarithm of the H_3O^+ ion concentration.

$$
\begin{aligned}
pH &\approx -\log[H_3O^+] \\
&\approx -\log(3.5 \times 10^{-3}) \\
&\approx -(-2.5) = 2.5
\end{aligned}
$$

The advantage of the pH scale can be understood by constructing a graph that plots the concentration of H_3O^+ over three orders of magnitude (powers of 10). Figure 11.5*a* shows a scale with 10 equally spaced intervals representing 0 to 0.1 *M* H_3O^+. The smallest interval between 0 and 0.01 *M* is further divided into 10 additional units. The scale thus covers three orders of magnitude: 0 to 0.001, 0 to 0.01, and 0 to 0.1 *M* H_3O^+. The addition of a fourth order of magnitude to the scale would require a magnifying glass (0 to 0.0001 *M*) or an extension of the page (0 to 1 *M*). Several concentration values are shown in the $-\log[H_3O^+]$ plot in Figure 11.5*b*. In this plot the orders of magnitude are conveniently plotted as several equally spaced units. Because an order of magnitude on this scale is represented by only one unit, there is no problem in adding additional orders of magnitude.

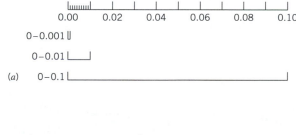

FIGURE 11.5 (*a*) Scale with 10 equally spaced intervals from 0 to 0.1 *M* H_3O^+. The interval between 0 and 0.01 is divided into smaller units. (*b*) Log scale with an H_3O^+ concentration ranging from 0.00001 to 0.1 *M*.

The utility of the pH scale is further illustrated in Figure 11.6*a*, which shows the possible combinations of H_3O^+ ion and OH^- concentrations in an aqueous solution. In this graph the concentration of OH^- is plotted on the vertical axis and the concentration of H_3O^+ is plotted on the horizontal axis. For the plot to be a reasonable size, it is restricted to one order of magnitude of concentration.

The concept of pH compresses the range of H_3O^+ concentrations into a scale that is much easier to handle. As the H_3O^+ concentration decreases from a molarity of roughly 1 to 10^{-14} (the typical range encountered in the laboratory), the pH of the solution increases from 0 to 14. The relationship between the H_3O^+ concentration, OH^- concentration, and the pH of a solution is shown in Table 11.2. The entire range of data in Table 11.2 is plotted in the graph of pH versus pOH in Figure 11.6*b*. Note that not only is the scale compressed in Figure 11.6*b*, but the shape of the graph has changed from a curve in

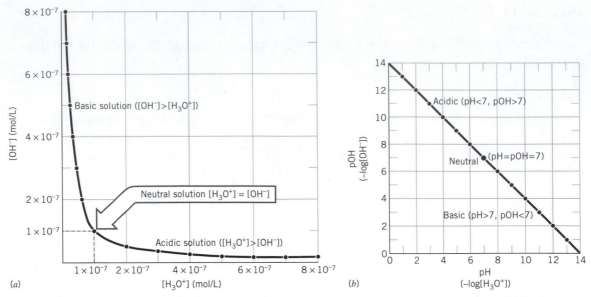

FIGURE 11.6 (*a*) A small fraction of data showing the relationship between the H_3O^+ and OH^- ion concentrations in pure water at equilibrium. Every point on the solid line represents a pair of H_3O^+ and OH^- concentrations when the solution is at equilibrium. (*b*) We can fit the entire range of H_3O^+ and OH^- concentrations from Table 11.2 on a single graph by plotting pH versus pOH. Any point on the solid line corresponds to a solution at equilibrium.

Figure 11.6*a* to a straight line in Figure 11.6*b*. This is the result of changing the scale of the plot to a log scale.

TABLE 11.2   Pairs of Equilibrium Concentrations of H_3O^+ and OH^- Ions with Their Respective pH Values in Water at 25°C

$[H_3O^+]$(mol/L)	$[OH^-]$(mol/L)	pH	
1	1×10^{-14}	0	
1×10^{-1}	1×10^{-13}	1	
1×10^{-2}	1×10^{-12}	2	
1×10^{-3}	1×10^{-11}	3	Acidic solution
1×10^{-4}	1×10^{-10}	4	
1×10^{-5}	1×10^{-9}	5	
1×10^{-6}	1×10^{-8}	6	
1×10^{-7}	1×10^{-7}	7	Neutral solution
1×10^{-8}	1×10^{-6}	8	
1×10^{-9}	1×10^{-5}	9	
1×10^{-10}	1×10^{-4}	10	
1×10^{-11}	1×10^{-3}	11	Basic solution
1×10^{-12}	1×10^{-2}	12	
1×10^{-13}	1×10^{-1}	13	
1×10^{-14}	1	14	

The concentration of the H_3O^+ ion in pure water at 25°C is 1.0×10^{-7} *M*. Thus, the pH of pure water is 7.

$$pH \approx -\log[H_3O^+] = -\log(1.0 \times 10^{-7} \ M) = 7.0$$

If the pH is given, the [H$_3$O$^+$] may be approximated by the following equation.

$$[H_3O^+] \approx 10^{-pH}$$

Solutions for which the concentrations of the H$_3$O$^+$ and OH$^-$ ions are equal are said to be *neutral*[2]. Solutions in which the concentration of the H$_3$O$^+$ ion is larger than 1×10^{-7} M at 25°C are described as *acidic*. Those in which the concentration of the H$_3$O$^+$ ion is smaller than 1×10^{-7} M are *basic*. Thus, at 25°C when the pH of a solution is less than 7, the solution is acidic. When the pH is more than 7, the solution is basic.

$$\left.\begin{array}{lll} \text{Acidic} & [H_3O^+] > 1 \times 10^{-7}\ M & pH < 7 \\ \text{Basic} & [H_3O^+] < 1 \times 10^{-7}\ M & pH > 7 \end{array}\right\} \text{ at } 25°C$$

Measurements of pH in the laboratory were historically done with **acid–base indicators,** which are weak acids or weak bases that change color when they gain or lose an H$^+$ ion. Acid–base indicators are still used in the laboratory for rough pH measurements. An example of an acid–base indicator is litmus, which turns pink in solutions whose pH is below 5 and turns blue when the pH is above 8. To a large extent, these indicators have been replaced by pH meters, which are more accurate. The actual measuring device in a pH meter is an electrode that consists of a resin-filled tube with a thin glass bulb at one end. When the electrode is immersed in a solution to be measured, it produces an electric potential that is directly proportional to the H$_3$O$^+$ ion concentration in the solution.

Adding an acid to water increases the H$_3$O$^+$ concentration and decreases the OH$^-$ concentration. Adding a base does the opposite. Regardless of what is added to water, however, the product of the concentrations of the ions at equilibrium is always 1.0×10^{-14} at 25°C.

$$[H_3O^+][OH^-] = 1.0 \times 10^{-14}$$

The relationship between the pH and pOH of an aqueous solution can be derived by taking the logarithm of both sides of the K_w expression.

$$\log([H_3O^+][OH^-]) = \log(1.0 \times 10^{-14})$$

The log of the product of two numbers is equal to the sum of their logs. Thus, the sum of the logs of the H$_3$O$^+$ and OH$^-$ ion concentrations is equal to the log of 10^{-14}.

$$\log[H_3O^+] + \log[OH^-] = -14.0$$

Both sides of the equation can now be multiplied by -1.

$$-\log[H_3O^+] - \log[OH^-] = 14.0$$

Substituting the definitions of pH and pOH into the equation gives the following result.

$$pH + pOH = 14.0$$

---

[2]The term *neutral* has been used previously in the text to describe particles with no net electric charge. Usually such particles have been referred to as *electrically neutral*. The term neutral in this chapter is used to describe a substance's acid or base properties. It is often necessary to examine the context of a sentence to determine use of the term *neutral*.

This equation can be used to convert from pH to pOH, and vice versa, for any aqueous solution at 25°C, regardless of how much acid or base has been added to the solution.

Checkpoint

Describe what happens to the pH of a solution as the concentration of the H_3O^+ ion in the solution increases.

Exercise 11.4

The pH values of samples of lemon juice and vinegar are quite similar: 2.2 and 2.5, respectively. Calculate the H_3O^+ and OH^- concentrations for both solutions and compare them to one another.

Solution

Lemon juice: $pH = 2.2 \approx -\log[H_3O^+]$ $\qquad$ $pOH = 11.8 \approx -\log[OH^-]$
$\qquad\qquad$ $[H_3O^+] \approx 10^{-2.2} = 6 \times 10^{-3}\ M$ $\qquad$ $[OH^-] \approx 10^{-11.8} = 2 \times 10^{-12}\ M$

Vinegar: $pH = 2.5 \approx -\log[H_3O^+]$ $\qquad$ $pOH = 11.5 \approx -\log[OH^-]$
$\qquad\qquad$ $[H_3O^+] \approx 10^{-2.5} = 3 \times 10^{-3}\ M$ $\qquad$ $[OH^-] \approx 10^{-11.5} = 3 \times 10^{-12}\ M$

Comparison: $\dfrac{\text{lemon juice}}{\text{vinegar}} = \dfrac{6 \times 10^{-3}\ M}{3 \times 10^{-3}\ M} = 2$

Note that the sample of lemon juice is twice as acidic as the vinegar sample, even though they have very similar pH values.

11.8   RELATIVE STRENGTHS OF ACIDS AND BASES

Many hardware stores sell muriatic acid—a 6 M solution of hydrochloric acid, $HCl(aq)$—to clean bricks and concrete and control the pH of swimming pools. Grocery stores sell vinegar, which is a 1 M solution of acetic acid, CH_3CO_2H. Although both substances are acids, you wouldn't use muriatic acid in salad dressing, and vinegar is ineffective in cleaning bricks or concrete.

The difference between the two acids can be explored with the apparatus shown in Figure 11.7. Two metal electrodes connected to a source of electricity are dipped into a beaker that contains a solution of one of these acids. If the solution can conduct an electric current, it completes the electric circuit and the lightbulb starts to glow. The intensity with which the bulb glows depends on the ability of the solution to conduct a current. This, in turn, depends on the concentration of the positive and negative ions in the solution that carry the electric current. As a result, the brightness of the lightbulb is directly related to the number of ions in solution.

When the electrodes are immersed in pure water, the lightbulb doesn't glow. This isn't surprising because the concentrations of H_3O^+ and OH^- ions in pure water are very small, only about $10^{-7}\ M$ at room temperature. When the electrodes are immersed in a 1 M solution of acetic acid, the bulb glows, but only dimly. When the electrodes are immersed in a 1 M solution of hydrochloric acid, the lightbulb glows very brightly. Although the concentration of the acid in both solutions is the same, 1 M, hydrochloric acid contains far more ions than the equivalent acetic acid solution.

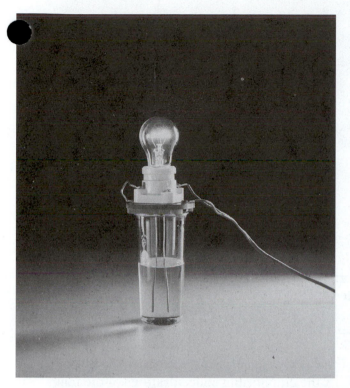 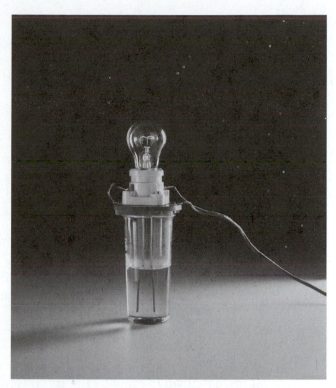

FIGURE 11.7 The conductivity apparatus shows that even though the solution of HCl on the left and the solution of acetic acid on the right have the same concentrations, the acid in the solution on the left with the brightly glowing bulb has dissociated to produce more ions.

The difference between these acids is a result of the different abilities of the two acids to donate a proton to water. The hydrochloric acid in muriatic acid is a **strong acid,** and the acetic acid in vinegar is a **weak acid.** Hydrochloric acid is strong because it is very good at transferring an H^+ ion to a water molecule. In a 6 M solution of hydrochloric acid, 99.996% of the HCl molecules react with water to form H_3O^+ and Cl^- ions.

$$HCl(aq) + H_2O(l) \longrightarrow H_3O^+(aq) + Cl^-(aq)$$

Only 0.004% of the HCl molecules—or 1 out of every 30,000—remain in solution as undissociated molecules. Hydrochloric acid is a strong acid indeed. Note that a strong acid is not the same thing as a concentrated acid. Concentration is measured as molarity. Even a weak acid can have a high concentration. A strong acid is one that dissociates essentially completely when dissolved in water. A very dilute solution of HCl will dissociate essentially 100% in water and is therefore a strong acid even though it has a low concentration.

Acetic acid (often abbreviated as HOAc) is a weak acid because it is not very good at transferring H^+ ions to water. In a 1 M solution, less than 0.4% of the CH_3CO_2H molecules react with water to form H_3O^+ and $CH_3CO_2^-$ ions.

$$CH_3CO_2H(aq) + H_2O(l) \rightleftharpoons H_3O^+(aq) + CH_3CO_2^-(aq)$$

More than 99.6% of the acetic acid molecules remain intact.

It would be useful to have a quantitative measure of the relative strengths of acids to replace the labels *strong* and *weak.* The extent to which an acid dissociates in water to produce the H_3O^+ ion is described in terms of an **acid dissociation equilibrium constant, K_a.**

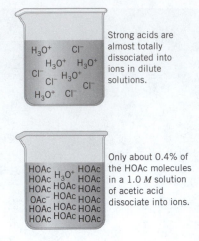

Strong acids are almost totally dissociated into ions in dilute solutions.

Only about 0.4% of the HOAc molecules in a 1.0 M solution of acetic acid dissociate into ions.

To understand the nature of this equilibrium constant, let's assume that the reaction between an acid and water can be represented by the following generic equation.

$$HA(aq) + H_2O(l) \rightleftharpoons H_3O^+(aq) + A^-(aq)$$

In other words, we will assume that some of the HA molecules react to form H_3O^+ and A^- ions, as shown in Figure 11.8. By convention, the equilibrium concentrations of the ions in units of moles per liter are represented by the symbols $[H_3O^+]$ and $[A^-]$. The concentration of the undissociated HA molecules that remain in solution is represented by the symbol $[HA]$.

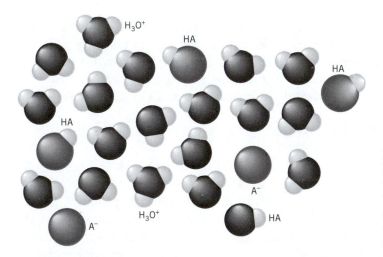

FIGURE 11.8 Some, but not all, of the HA molecules in a typical acid react with water to form H_3O^+ and A^- ions when the acid dissolves in water. In a strong acid, there are very few undissociated HA molecules. In a weak acid, most of the HA molecules remain.

The equilibrium constant expression for the reaction between HA and water would be written as follows.

$$K_c = \frac{[H_3O^+][A^-]}{[HA][H_2O]}$$

Like the equilibrium constant expression we first wrote for the dissociation of water, this is a legitimate equation. But most acid solutions are so dilute that the equilibrium concentration of H_2O is effectively the same as before the acid was added. Because the $[H_2O]$ does not change the equilibrium constant for the reaction may be written as follows.

$$\frac{[H_3O^+][A^-]}{[HA]} = [H_2O] \times K_c$$

The result is an equilibrium constant for the equation known as the *acid dissociation equilibrium constant, K_a.*

$$\frac{[H_3O^+][A^-]}{[HA]} = [H_2O] \, K_c = K_a$$

When a strong acid dissolves in water, it reacts extensively with water to form H_3O^+ and A^- ions. (Essentially 100% of the strong acid dissociates, leaving only a very small residual concentration of HA molecules in solution.) The product of the concentrations of the H_3O^+ and A^- ions is therefore much larger than the concentration of the HA molecules, so K_a for a strong acid is greater than 1. Hydrochloric acid, for example, has a K_a of roughly 1×10^6.

$$\frac{[H_3O^+][Cl^-]}{[HCl]} = 1 \times 10^6$$

Weak acids, on the other hand, react only slightly with water. The product of the concentrations of the H_3O^+ and A^- ions is therefore smaller than the concentration of the residual HA molecules. As a result, K_a for a weak acid is less than 1. Acetic acid, for example, has a K_a of only 1.8×10^{-5}.

$$\frac{[H_3O^+][CH_3CO_2^-]}{[CH_3CO_2H]} = 1.8 \times 10^{-5}$$

K_a can therefore be used to distinguish between strong acids and weak acids.

$$\begin{array}{ll} \text{Strong acids} & K_a > 1 \\ \text{Weak acids} & K_a \ll 1 \end{array}$$

The hydronium ion, H_3O^+, is defined as the strongest acid that is stable in water because the strong acids all dissociate essentially 100% to give H_3O^+. The equilibrium expression that describes the dissociation of H_3O^+ in water is defined in the same way as the K_a for strong and weak acids shown above.

$$H_3O^+(aq) + H_2O(l) \rightleftharpoons H_3O^+(aq) + H_2O(l)$$

$$K_c = \frac{[H_3O^+][H_2O]}{[H_3O^+][H_2O]} = 1$$

$$\frac{[H_3O^+][H_2O]}{[H_3O^+]} = [H_2O] \times K_c = K_a = 55$$

Because K_c is equal to 1 and $[H_2O]$ is 55, the K_a for the dissociation of H_3O^+ in water is equal to 55.

A list of common acids and their acid dissociation constants, K_a, is given in Table 11.3. A more complete list can be found in Appendix B.8. Also included in Appendix B.8 are

pK_a values ($pK_a = -\log K_a$). Commonly accepted K_a values are reported for the strong acids. Experimental determination of K_a values for strong acids is difficult, resulting in considerable variation in the values reported in the literature.

TABLE 11.3   Common Acids and Their Acid Dissociation Equilibrium Constants for the Loss of One Proton

	K_a
Strong Acids	
HI	3×10^9
HBr	1×10^9
HCl	1×10^6
H_2SO_4	1×10^3
$HClO_4$	1×10^8
H_3O^+	**55**
HNO_3	28
H_2CrO_4	9.6
Weak Acids	
H_3PO_4	7.1×10^{-3}
HF	7.2×10^{-4}
Citric acid	7.5×10^{-4}
CH_3CO_2H	1.8×10^{-5}
H_2S	1.0×10^{-7}
H_2CO_3	4.5×10^{-7}
H_3BO_3	7.3×10^{-10}
H_2O^a	1.8×10^{-16}

^{a}Note that this is the K_a of water as determined by the acid dissociation equilibrium expression, not the K_w described in Section 11.6.

Checkpoint

Write the equilibrium expressions that describe the K_a and the K_w of water. Show how the two expressions are different.

The ionization of bases in water can be described in the same manner as the dissociation of acids. A generic base, *B*, reacting with water can be described by the following chemical equation.

$$B(aq) + H_2O(l) \rightleftharpoons BH^+(aq) + OH^-(aq)$$

Strict adherence to the rules for writing equilibrium constant expressions gives the following equation.

$$K_c = \frac{[BH^+][OH^-]}{[B][H_2O]}$$

As we saw with the acids, the $[H_2O]$ will remain essentially constant, allowing the equilibrium to be described as follows.

$$K_c[H_2O] = \frac{[BH^+][OH^-]}{[B]} = K_b$$

The new equilibrium constant is the **base ionization equilibrium constant, K_b.** Just as K_a values describe the relative strengths of acids, K_b values describe the relative strengths of bases. The values of K_b for a limited number of bases are given in Appendix B, Table B.9. Also listed are the pK_b values ($pK_b = -\log K_b$).

Strong bases are defined as ions or molecules that completely ionize in water to produce OH^-. Examples of strong bases include the following.

Group IA metal hydroxides: LiOH, NaOH, KOH, RbOH, and CsOH

$$NaOH(s) \longrightarrow Na^+(aq) + OH^-(aq)$$

Soluble Group IIA metal hydroxides: $Ca(OH)_2$ (slightly soluble), $Sr(OH)_2$, and $Ba(OH)_2$

$$Ca(OH)_2(aq) \longrightarrow Ca^{2+}(aq) + 2\ OH^-(aq)$$

Soluble metal oxides: Li_2O, Na_2O, K_2O, and CaO

$$Li_2O(s) + H_2O(l) \longrightarrow 2\ Li^+(aq) + 2\ OH^-(aq)$$

Because the soluble Group IA and Group IIA metal hydroxides are strong bases and therefore dissociate essentially completely, there is very little tendency for the dissociation reaction to go in the reverse direction. This means that the cations formed by the dissociation of strong bases are such very weak conjugate acids that they have essentially no acid properties at all. The positively charged Group IA and IIA cations are referred to as neutral cations because they do not act as acids or bases.

Checkpoint
Which solution has the lowest pH, a solution with a low molarity of an acid with a small K_a or a solution with a low molarity of an acid with a high K_a?

11.9   RELATIVE STRENGTHS OF CONJUGATE ACID–BASE PAIRS

The relationship between the strength of an acid and its conjugate base can be understood by considering the implications of the fact that HCl is a strong acid. If it is a strong acid, HCl must be a good proton donor. HCl can only be a good proton donor, however, if the Cl^- ion is a poor proton acceptor. Thus, the Cl^- ion must be a weak base.

$$\underset{\textit{Strong acid}}{HCl(aq)}\ +\ H_2O(l) \longrightarrow H_3O^+(aq) + \underset{\textit{Weak base}}{Cl^-(aq)}$$

Because HCl is a strong acid it dissociates essentially completely, and there is little tendency for the reaction to go in the reverse direction. Thus, the chloride anion is such a very weak conjugate base that it has essentially no base properties. The anions that result from the dissociation of strong monoprotic acids (HI, HBr, HCl, HClO$_4$, and HNO$_3$) are referred to as neutral anions because they do not act as acids or bases.

Let's now consider the relationship between the strength of the ammonium ion (NH$_4^+$) and its conjugate base, ammonia (NH$_3$). The NH$_4^+$ ion is a weak acid because ammonia is a reasonably good base.

$$\underset{\textit{Weak acid}}{NH_4^+(aq)} + H_2O(l) \rightleftharpoons H_3O^+(aq) + \underset{\textit{Good base}}{NH_3(aq)}$$

The results of the two examples can be summarized in the form of the following general rules:

The stronger the acid, the weaker is the conjugate base.
The stronger the base, the weaker is the conjugate acid.

11.10   RELATIVE STRENGTHS OF PAIRS OF ACIDS AND BASES

We have already seen how the value of K_a can be used to decide whether an acid is a strong acid or a weak acid. At times it is also useful to compare the relative strengths of a pair of acids to decide which is stronger. Consider HCl and the H$_3$O$^+$ ion, for example.

$$HCl \qquad K_a \approx 1 \times 10^6$$
$$H_3O^+ \qquad K_a = 55$$

The K_a values suggest both are strong acids, but HCl is a stronger acid than the H$_3$O$^+$ ion.

We can now understand why such a high proportion of the HCl molecules in an aqueous solution reacts with water to form H$_3$O$^+$ and Cl$^-$ ions. The Brønsted model suggests that every acid–base reaction converts an acid into its conjugate base and a base into its conjugate acid.

$$\underset{\textit{Acid}}{HCl(aq)} + \underset{\textit{Base}}{H_2O(l)} \longrightarrow \underset{\textit{Acid}}{H_3O^+(aq)} + \underset{\textit{Base}}{Cl^-(aq)}$$

Note that there are two acids and two bases in the reaction. The stronger acid, however, is on the left side of the equation.

$$\underset{\textit{Stronger acid}}{HCl(aq)} + H_2O(l) \longrightarrow \underset{\textit{Weaker acid}}{H_3O^+(aq)} + Cl^-(aq)$$

What about the two bases: H$_2$O and the Cl$^-$ ion? The general rules given in the previous section suggest that the stronger of a pair of acids must form the weaker of a pair of conjugate bases. The fact that HCl is a stronger acid than the H$_3$O$^+$ ion implies that the Cl$^-$ ion is a weaker base than water.

$$\text{Acid strength} \qquad HCl > H_3O^+$$
$$\text{Base strength} \qquad Cl^- < H_2O$$

Thus, the equation for the reaction between HCl and water can be written as follows.

$$\underset{\substack{\textit{Stronger} \\ \textit{acid}}}{HCl(aq)} + \underset{\substack{\textit{Stronger} \\ \textit{base}}}{H_2O(l)} \longrightarrow \underset{\substack{\textit{Weaker} \\ \textit{acid}}}{H_3O^+(aq)} + \underset{\substack{\textit{Weaker} \\ \textit{base}}}{Cl^-(aq)}$$

It isn't surprising that 99.996% of the HCl molecules in a 6 M solution react with water to give H_3O^+ ions and Cl^- ions. The stronger of a pair of acids should react with the stronger of a pair of bases to form a weaker acid and a weaker base. *Strong acids with K_a values greater than the K_a of H_3O^+ (55) can be assumed to dissociate completely (100%) in water.*
Let's now look at the relative strengths of acetic acid and the H_3O^+ ion.

$$CH_3CO_2H \qquad K_a = 1.8 \times 10^{-5}$$
$$H_3O^+ \qquad K_a = 55$$

The K_a values suggest that acetic acid is a much weaker acid than the H_3O^+ ion, which explains why acetic acid is a weak acid in water. Once again, the reaction between the acid and water must convert the acid into its conjugate base and the base into its conjugate acid.

$$CH_3CO_2H(aq) + H_2O(l) \rightleftharpoons H_3O^+(aq) + CH_3CO_2^-(aq)$$
$$\text{Acid} \qquad \text{Base} \qquad \text{Acid} \qquad \text{Base}$$

But in this case, the stronger acid and the stronger base are on the right side of the equation.

$$CH_3CO_2H(aq) + H_2O(l) \rightleftharpoons H_3O^+(aq) + CH_3CO_2^-(aq)$$
$$\text{Weaker} \qquad \text{Weaker} \qquad \text{Stronger} \qquad \text{Stronger}$$
$$\text{acid} \qquad \text{base} \qquad \text{acid} \qquad \text{base}$$

As a result, only a few of the CH_3CO_2H molecules actually donate an H^+ ion to a water molecule to form the H_3O^+ and $CH_3CO_2^-$ ions. At equilibrium an acid–base reaction should lie predominantly on the side of the chemical equation with the weaker acid and weaker base.

The magnitude of K_a can also be used to explain why some compounds that qualify as Brønsted acids or bases don't act like acids or bases when they dissolve in water. As long as K_a for the acid is significantly larger than the value of K_a for water (1.8×10^{-16}), the acid will ionize to some extent. As the K_a value for the acid approaches the K_a for water, the compound becomes more like water in its acidity. Although it is still a Brønsted acid, it is so weak that we may be unable to detect the acidity in aqueous solution.

Measurements of the pH of dilute solutions are also good indicators of the relative strengths of acids and bases. Experimental values of the pH of 0.10 M solutions of a number of common acids and bases are given in Table 11.4. The equation pH $\approx -\log[H_3O^+]$ is a good model for the behavior of acids but is not exact. Table 11.4 shows that the actual pH of some 0.10 M acids vary from what would be calculated with the pH equation. For example, the pH equation would predict that the pH of a 0.10 M HCl solution would be 1.0, but the measured value is observed to be 1.1.

Checkpoint

HOBr is a weaker acid than HOCl. Which base, OBr^- or OCl^-, is strongest?

The Leveling Effect of Water

Since all strong acids dissociate essentially 100% in water to produce the same H_3O^+ ion, they all seem to have the same strength when dissolved in water, regardless of the value of K_a. This phenomenon is known as the **leveling effect** of water—the tendency of water to limit the strengths of strong acids and bases. More than 99% of the HCl molecules in hydrochloric acid react with water to form H_3O^+ and Cl^- ions, for example. Thus a 0.10 M

TABLE 11.4   pH of 0.10 M Solutions of Common Acids and Bases

Compound	pH
HCl (hydrochloric acid)	1.1
H_2SO_4 (sulfuric acid)	1.2
$NaHSO_4$ (sodium hydrogen sulfate)	1.4
H_2SO_3 (sulfurous acid)	1.5
H_3PO_4 (phosphoric acid)	1.5
HF (hydrofluoric acid)	2.1
CH_3CO_2H (acetic acid)	2.9
H_2CO_3 (carbonic acid)	3.8 (saturated solution)
H_2S (hydrogen sulfide)	4.1
NaH_2PO_4 (sodium dihydrogen phosphate)	4.4
NH_4Cl (ammonium chloride)	4.6
HCN (hydrocyanic acid)	5.1
Na_2SO_4 (sodium sulfate)	6.1
NaCl (sodium chloride)	6.4
H_2O (freshly boiled distilled water)	7.0
$NaCH_3CO_2$ (sodium acetate)	8.4
$NaHCO_3$ (sodium hyrogen carbonate)	8.4
Na_2HPO_4 (sodium hydrogen phosphate)	9.3
Na_2SO_3 (sodium sulfite)	9.8
NaCN (sodium cyanide)	11.0
NH_3 (aqueous ammonia)	11.1
Na_2CO_3 (sodium carbonate)	11.6
Na_3PO_4 (sodium phosphate)	12.0
NaOH (sodium hydroxide, lye)	13.0

solution of HCl produces approximately 0.10 M H_3O^+. In the same manner, a 0.10 M solution of the strong acid $HClO_4$ will also produce 0.10 M H_3O^+.

$$HCl(aq) + H_2O(l) \longrightarrow H_3O^+(aq) + Cl^-(aq)$$
$$HClO_4(aq) + H_2O(l) \longrightarrow H_3O^+(aq) + ClO_4^-(aq)$$

Thus, the strengths of strong acids are limited by the strength of the acid (H_3O^+) formed when water molecules accept an H^+ ion.

A similar phenomenon occurs in solutions of strong bases. Once the base reacts with water to form the OH^- ion, the solution cannot become any more basic. The strength of a strong base is limited by the strength of the base (OH^-) formed when water molecules lose an H^+ ion.

11.11   RELATIONSHIP OF STRUCTURE TO RELATIVE STRENGTHS OF ACIDS AND BASES

Several factors influence the probability of having a heart attack. Heart attacks occur more often among those who smoke than among those who don't, among the elderly more often than the young, among those who are overweight more than among those who aren't, and among those who never exercise more than among those who exercise regularly. It is possible to sort out the relative importance of these factors, however, by trying to keep as many as possible of the other factors constant.

The same approach can be used to sort out the factors that control the relative strengths of acids and bases. When an acid is dissolved in water, the interaction of the solvent with the acid and its conjugate base is very important in determining the degree of dissociation of the acid and thus the strength of the acid. These interactions are very complex and often difficult to quantify. The structure of the acid is also important in determining its strength. Although structure alone does not determine the strength of an acid, many trends can be rationalized from the Lewis structure of an acid. Three factors related to structure affect the acidity of the X—H bond in a binary acid: (1) the polarity of the bond, (2) the size of the atom X, and (3) the charge on the ion or molecule. Structure can also be used to help understand the relative acidity of the oxyacids, acids with an X—OH group.

The Polarity of the X—H Bond

When all the other factors are kept constant, acids become stronger as the X—H bond becomes *more polar*. Consider the following compounds, for example, which become more acidic as ΔEN, the difference between the electronegativities of the X and H atoms, increases (δ_H is the partial charge on the hydrogen atom). HF is the strongest of the four acids, and CH_4 is the weakest.

	K_a	ΔEN	δ_H	X—H Bond Length (nm)
HF	7.2×10^{-4}	1.9	0.29	0.101
H_2O	1.8×10^{-16}	1.3	0.22	0.103
NH_3	1×10^{-33}	0.8	0.14	0.107
CH_4	1×10^{-49}	0.2	0.05	0.114

In all four of the compounds the size of the central atom, X, to which the H is bonded is about the same. This can be seen from the similar bond lengths. As a general rule similar bond lengths will give similar bond strengths. However, it can also be seen in the above data that there are large differences in ΔEN among the four compounds. The molecule containing the central atom with the largest electronegativity, F, also has the H with the most positive partial charge, 0.29. As the central atom becomes more electronegative, electron density is pulled away from the hydrogen, resulting in a more positive partial charge on the hydrogen and a more polar H—X bond. In the case of HF, the difference between the electronegativities of the two atoms is so great that the partial charge on the hydrogen atom is significant even before the acid dissociates into ions.

When these compounds act as acids, an H—X bond is broken to form H^+ and X^- ions. From the data above it is apparent that the more polar this bond, the larger is the K_a, and hence the easier it is to form the ions. Thus, the more polar the bond, the stronger is the acid.

The data in Table 11.4 illustrate the magnitude of this effect. A 0.10 M HF solution is moderately acidic. Water is much less acidic, and the acidity of ammonia is so low that the chemistry of aqueous solutions of NH_3 is dominated by the ability of NH_3 to act as a base.

$$
\begin{array}{ll}
\text{HF} & \text{pH} = 2.1 \\
H_2O & \text{pH} = 7 \\
NH_3 & \text{pH} = 11.1
\end{array}
$$

The Size of the X Atom

On the basis of the discussion in the previous section, we might expect HF, HCl, HBr, and HI to become weaker acids as we go down the column of the periodic table, because the X—H bond becomes less polar. Experimentally, we find the opposite trend. The acids actually become stronger as we go down the column.

This can be explained by recognizing that the size of the X atom influences the acidity of the X—H bond. Acids become stronger as the X—H bond becomes weaker, and bonds generally become weaker as the atoms get larger (see Figure 11.9).

FIGURE 11.9 HI is a much stronger acid than HF because of the relative sizes of the fluorine and iodine atoms. The covalent radius of iodine is more than twice as large as fluorine, which means that the HI bond strength is less than that of HF. As a result, it is more difficult to separate the atoms in an HF molecule to form H^+ and F^- ions than it is to separate the atoms in an HI molecule to form H^+ and I^- ions.

The K_a data for HF, HCl, HBr, and HI reflect the fact that the enthalpy of atom combination (ΔH_{ac}°) of the compounds becomes less negative as the X atom becomes larger. A less negative enthalpy of atom combination indicates a lower bond strength and thus a stronger acid because the hydrogen is more easily lost. The data below show how bond strength (related to bond length and ΔH_{ac}°) is much more important than bond polarity for these acids.

	K_a	ΔH_{ac}°	δ_H	X—H Bond Length (nm)
HF	7.2×10^{-4}	-567.7 kJ/mol$_{rxn}$	0.29	0.101
HCl	1×10^6	-431.6 kJ/mol$_{rxn}$	0.11	0.136
HBr	1×10^9	-365.9 kJ/mol$_{rxn}$	0.08	0.151
HI	3×10^9	-298.0 kJ/mol$_{rxn}$	0.01	0.170

Checkpoint

Which acid is stronger, H_2O or H_2S? Why?

The Charge on the Acid or Base

The charge on a molecule or ion can influence its ability to act as an acid or a base. This is clearly shown when the pH of 0.1 M solutions of H_3PO_4 and the $H_2PO_4^-$, HPO_4^{2-}, and PO_4^{3-} ions are compared.

$$
\begin{array}{ll}
H_3PO_4 & pH = 1.5 \\
H_2PO_4^- & pH = 4.4 \\
HPO_4^{2-} & pH = 9.3 \\
PO_4^{3-} & pH = 12.0
\end{array}
$$

Compounds become less acidic and more basic as the negative charge increases.

$$
\begin{array}{ll}
\text{Acidity} & H_3PO_4 > H_2PO_4^- > HPO_4^{2-} \\
\text{Basicity} & H_2PO_4^- < HPO_4^{2-} < PO_4^{3-}
\end{array}
$$

The decrease in acidity with increasing negative charge occurs because it is much easier to remove a positive H^+ ion from a neutral H_3PO_4 molecule than it is to remove an

H^+ ion from a negatively charged $H_2PO_4^-$ ion. It is even harder to remove an H^+ ion from a negatively charged HPO_4^{2-} ion.

The increase in basicity with increasing negative charge can be similarly explained. There is a strong force of attraction between the negative charge on a PO_4^{3-} ion and the positive charge on an H^+ ion. As a result PO_4^{3-} is a reasonably good base. The force of attraction between the HPO_4^{2-} ion and an H^+ ion is smaller, because HPO_4^{2-} carries a smaller charge. The HPO_4^{2-} ion is therefore a weaker base than PO_4^{3-}. The charge on the $H_2PO_4^-$ ion is even smaller, so this ion is an even weaker base than HPO_4^{2-}.

Relative Strengths of Oxyacids

There is no difference in either the size of the atom bonded to hydrogen or the charge on the acid when we compare the O—H bonds in oxyacids of the same element, such as H_2SO_4 and H_2SO_3 or HNO_3 and HNO_2 (see Figure 11.10), yet there is a significant difference in the strengths of the acids.

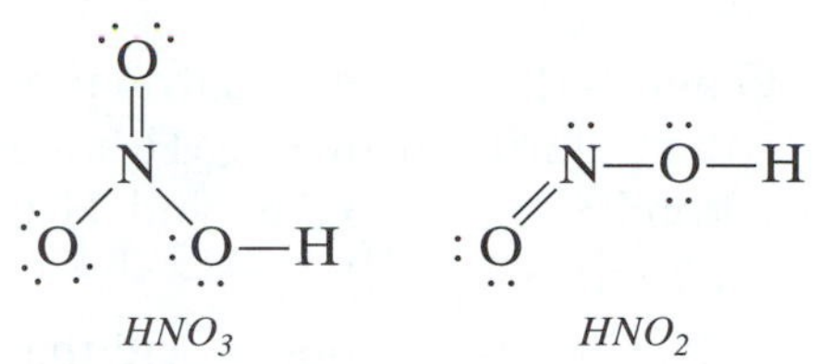

FIGURE 11.10 Lewis structures of HNO_3 and HNO_2. The O—H bond breaks to release H^+. The difference in acidity of the two compounds is related to the number of oxygens attached to the nitrogen.

Consider the following K_a data, for example.

H_2SO_4	$K_a \approx 1 \times 10^3$		HNO_3	$K_a \approx 28$
H_2SO_3	$K_a = 1.7 \times 10^{-2}$		HNO_2	$K_a = 5.1 \times 10^{-4}$

The acidity of the oxyacids increases significantly as the number of oxygen atoms attached to the central atom increases. H_2SO_4 is a much stronger acid than H_2SO_3, and HNO_3 is a much stronger acid than HNO_2.

This trend is easiest to see in the four oxyacids of chlorine. Because the O—H bond breaks to furnish H^+ in all four compounds, the size of the atom, an oxygen atom, bonded to the hydrogen is always the same. In addition, all four compounds have the same charge, zero. Therefore, it might seem difficult to explain such a large difference in the value of K_a for hypochlorous acid (HOCl) and perchloric acid ($HOClO_3$). Oxygen, however, is second only to fluorine in electronegativity, and it tends to draw electron density toward itself. As more oxygen atoms are added to the central atom, more electron density is drawn toward the oxygens. This draws electrons away from the O—H bond in the acids, as shown in Figure 11.11. The pulling of electron density away from the hydrogen results in the partial charge on the hydrogen, δ_H, becoming more positive, as can be seen above. Therefore, it is the polarity of the O—H bond that is important in determining the relative strengths of the oxyacids. The O—H bond in HOCl is less polar than in $HOClO_3$, so perchloric acid is the stronger acid.

Oxyacid	K_a	Number of Oxygen Atoms	δ_H
HOCl	2.9×10^{-8}	1	+0.24
HOClO	1.1×10^{-2}	2	+0.29
$HOClO_2$	5.0×10^2	3	+0.31
$HOClO_3$	1×10^8	4	+0.33

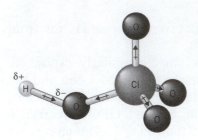

FIGURE 11.11 The large difference between the K_a values for HOCl and HOClO$_3$ is the result of the addition of oxygen atoms to the compounds. The electronegative oxygen atoms draw electron density toward themselves, resulting in electron density being pulled away from the hydrogen and toward the oxygen in the O—H bond. The net result is an increase in the polarity of the O—H bond, which leads to an increase in the acidity of the compound.

Exercise 11.5

For each of the following pairs, predict which species is the stronger acid and explain why.
(a) H_2O or NH_3 (b) NH_4^+ or NH_3
(c) NH_3 or PH_3 (d) H_2SeO_4 or H_2SeO_3

Solution

(a) Oxygen and nitrogen are about the same size, and H_2O and NH_3 are both neutral molecules. The difference between the compounds lies in the polarity of the X—H bond. Because oxygen is more electronegative, the O—H bond is more polar, and H_2O ($K_a = 1.8 \times 10^{-16}$) is a much stronger acid than NH_3 ($K_a = 1 \times 10^{-33}$).

(b) The difference between the species is the charge. The NH_4^+ ion is a stronger acid than NH_3 because it is easier to remove an H^+ ion from an NH_4^+ ion than from a neutral NH_3 molecule.

(c) PH_3 is a stronger acid than NH_3. Molecules become more acidic as the size of the atom holding the hydrogen atom increases. Longer bonds are weaker bonds and therefore more easily broken.

(d) The compounds are both oxyacids, which differ in the number of oxygen atoms attached to the selenium atom. More electron density is drawn away from the O—H bond in H_2SeO_4 because of the extra oxygen atom. Thus the O—H bond in H_2SeO_4 is more polar than in H_2SeO_3, and therefore H_2SeO_4 is a stronger acid.

Checkpoint
Which acid is stronger, HOCl or HOI? Why?

Four factors have been discussed that affect the relative strengths of acids. However, these represent general trends, not absolute rules. Because there are many interrelating factors that affect the strengths of acids, some acids are exceptions to one or more of the general trends.

The relative strengths of Brønsted bases can be predicted from the relative strengths of their conjugate acids combined with the general rule that the stronger of a pair of acids always has the weaker conjugate base.

Exercise 11.6

For each of the following pairs of compounds, predict which compound is the weaker base.
(a) OH^- or NH_2^- (b) NH_3 or NH_2^-
(c) NH_2^- or PH_2^- (d) NO_3^- or NO_2^-

Solution

(a) The OH^- ion is the conjugate base of water, and the NH_2^- ion is the conjugate base of ammonia. Because H_2O is a stronger acid than NH_3, the OH^- ion must be a weaker base than the NH_2^- ion.

(b) Because it carries a negative charge, the NH_2^- ion is a stronger base than NH_3. NH_3 is therefore the weaker base.

(c) PH_3 is a stronger acid than NH_3, which means the PH_2^- ion must be a weaker base than the NH_2^- ion.

(d) HNO_3 is a stronger acid than HNO_2, which means that the NO_3^- ion is the weaker base.

11.12 STRONG ACID pH CALCULATIONS

The simplest acid–base equilibria are those in which a strong acid (or base) is dissolved in water. Consider the calculation of the pH of a solution formed by adding a single drop of 2 M hydrochloric acid to 100 mL of water, for example. The key to the calculation is remembering that HCl is a strong acid ($K_a \approx 10^6$) and that acids as strong as HCl can be assumed to dissociate completely in water.

$$HCl(aq) + H_2O(l) \longrightarrow H_3O^+(aq) + Cl^-(aq)$$

The H_3O^+ ion concentration at equilibrium is therefore essentially equal to the initial concentration of the acid.

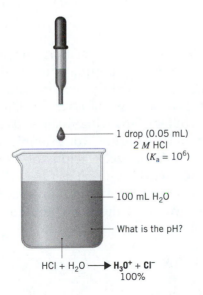

1 drop (0.05 mL)
2 M HCl
($K_a = 10^6$)

100 mL H_2O

What is the pH?

$HCl + H_2O \longrightarrow$ **$H_3O^+ + Cl^-$**
100%

A useful rule of thumb says that there are about 20 drops in each milliliter. One drop of 2 M HCl therefore has a volume of 0.05 mL, or 5×10^{-5} L. Now that we know the volume of the acid added to the beaker of water and the concentration of the acid, we can calculate the number of moles of HCl that were added to the water.

$$\frac{2 \text{ mol HCl}}{1 \text{ L}} \times 5 \times 10^{-5} \text{ L} = 1 \times 10^{-4} \text{ mol HCl}$$

Once we know the number of moles of HCl that have been added to the beaker and the volume of water in the beaker, we can calculate the concentration of solution prepared by adding one drop of 2 M HCl to 100 mL of water.

$$\frac{1 \times 10^{-4} \text{ mol HCl}}{0.100 \text{ L}} = 1 \times 10^{-3} \ M \text{ HCl}$$

According to the calculation, the initial concentration of HCl is $1 \times 10^{-3} \ M$.

If we assume that the acid dissociates completely, the H_3O^+ concentration at equilibrium is $1 \times 10^{-3} \ M$. The pH of the solution prepared by adding one drop of 2 M HCl to 100 mL of water can be approximated by the equation below to be 3.0.

$$pH \approx -\log[H_3O^+]$$
$$\approx -\log(1 \times 10^{-3}) = -(-3.0) = 3.0$$

Only in the cases of very dilute solutions is the theoretically calculated pH of an acid equal to the experimentally measured value. For the dilute solution of $1 \times 10^{-3} \ M$ HCl, the pH of 3.0 is a very good approximation. Using the above equation, a more concentrated solution of 0.10 M would theoretically have a pH of 1.0. However, as can be seen in Table 11.4, the measured pH of 0.10 M HCl is 1.1.

Checkpoint

When a strong acid is added to water, why can the pH usually be determined directly from the quantity of strong acid added?

11.13   WEAK ACID pH CALCULATIONS

Unlike the strong acids, which dissociate approximately 100% and leave essentially no molecular form of the acid in solution, the weak acids produce an equilibrium composed of significant amounts of both the dissociated and undissociated forms of the acid in solution. This equilibrium must be taken into account in calculating the amount of H_3O^+ produced in the partial dissociation of the weak acid. Equilibrium problems involving weak acids can be solved by applying the techniques developed in Chapter 10. Consider, for example, the dissociation of acetic acid, CH_3COOH, in water. The formula of acetic acid is often abbreviated as HOAc and that its conjugate base, the acetate ion ($C_2H_3O_2^-$) as OAc$^-$. The H_3O^+, OAc$^-$, and HOAc concentrations at equilibrium in an 0.10 M solution of acetic acid in water can be calculated by the following process. We start the calculation, as always, by building a representation of what we know about the reaction.

$$HOAc(aq) + H_2O(l) \rightleftharpoons H_3O^+(aq) + OAc^-(aq) \qquad K_a = 1.8 \times 10^{-5}$$

Initial	0.10 M		0	0
Equilibrium	?		?	?

We then compare the initial reaction quotient (Q_a) with the equilibrium constant (K_a) for the reaction. In this case, the initial reaction quotient is smaller than the equilibrium constant.

$$Q_a = \frac{(H_3O^+)(OAc^-)}{(HOAc)} = \frac{(0)(0)}{0.10} = 0 < K_a$$

We therefore conclude that the reaction must shift to the right to reach equilibrium.

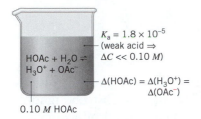

$K_a = 1.8 \times 10^{-5}$
(weak acid $\Rightarrow$
$\Delta C \ll 0.10\ M$)

$\Delta(HOAc) = \Delta(H_3O^+) =$
$\Delta(OAc^-)$

0.10 M HOAc

The balanced equation for the reaction states that we get one H_3O^+ ion and one OAc^- ion each time an HOAc molecule dissociates. This allows us to write equations for the equilibrium concentrations of the three components of the reaction. The change in the concentration of acetic acid as the reaction comes to equilibrium, ΔC, will be equal to the difference between the initial concentration of the acid, 0.10 M, and the concentration at equilibrium, [HOAc].

$$\Delta C = 0.10\ M - [\text{HOAc}]$$

Rearranging the equation gives the following result.

$$[\text{HOAc}] = 0.10\ M - \Delta C$$

Because there is very little H_3O^+ ion in pure water, we can assume that the initial concentration of this acid is effectively zero. Because we get one H_3O^+ ion and one OAc^- ion every time an acetic acid molecule dissociates, the concentrations of the ions at equilibrium are equal to the change in the concentration of acetic acid as the acid dissociates.

$$[H_3O^+] = \Delta C$$
$$[OAc^-] = \Delta C$$

We can therefore represent the problem that needs to be solved as follows.

$$\text{HOAc}(aq) + H_2O(l) \rightleftharpoons H_3O^+(aq) + OAc^-(aq) \qquad K_a = 1.8 \times 10^{-5}$$

	HOAc	H_3O^+	OAc^-
Initial	0.10 M	0	0
Equilibrium	0.10 − ΔC	ΔC	ΔC

Substituting what we know about the system at equilibrium into the K_a expression gives the following equation.

$$K_a = \frac{[H_3O^+][OAc^-]}{[\text{HOAc}]} = \frac{[\Delta C][\Delta C]}{[0.10 - \Delta C]} = 1.8 \times 10^{-5}$$

Although we could rearrange the equation and solve it with the quadratic formula, we can test the assumption that acetic acid is a relatively weak acid. In other words, we can assume that ΔC is very small compared with the initial concentration of acetic acid. If this assumption is true, then subtracting ΔC from the initial concentration of the acid (0.10 M) will not reduce that value.

$$\frac{[\Delta C][\Delta C]}{[0.10]} \approx 1.8 \times 10^{-5}$$
$$\frac{[\Delta C]^2}{[0.10]} \approx 1.8 \times 10^{-5}$$
$$\Delta C^2 \approx 1.8 \times 10^{-6}$$

We then solve the approximate equation for the value of ΔC.

$$\Delta C \approx 0.0013 \ M$$

We use the symbol $\approx$ to mean "approximately equal" until we are certain that our assumption to neglect the ΔC in the subtraction is valid. Is ΔC small enough to be ignored in this problem? Yes, because it is less than 5% of the initial concentration of acetic acid.

$$\frac{0.0013}{0.10} \times 100 = 1.3\%$$

The subtraction in the denominator of the equilibrium expression would be

$$[HOAc] = 0.10 \ M - 0.0013 \ M$$

Maintaining appropriate significant figures gives a result of 0.10 M. We can therefore use the value of ΔC to calculate the equilibrium concentrations of H_3O^+, OAc^-, and HOAc.

$$[HOAc] = 0.10 - \Delta C \approx 0.10 \ M$$
$$[H_3O^+] = [OAc^-] = \Delta C \approx 0.0013 \ M$$

Two assumptions were made in the above calculation.

- We assumed that the amount of acid that dissociates is small compared with the initial concentration of the acid.
- We assumed that enough acid dissociates to allow us to ignore the contribution to the H_3O^+ ion concentration from the dissociation of water.

In this case, both assumptions are valid. Only about 1.3% of the acetic acid molecules actually dissociate at equilibrium. But this gives us an H_3O^+ concentration that is four orders of magnitude larger than the concentration of the ion in pure water. Thus, we can legitimately ignore the dissociation of water.

The assumptions made in the calculation work for acids that are neither too strong nor too weak. Fortunately, many acids fall into that category. If the acid is too strong to assume that ΔC is small compared with the initial concentration, the problem can always be solved with the quadratic equation. If the acid is too weak to ignore the dissociation of water, variations in the pH of the water from one sample to another because of impurities dissolved in the water are probably as large as the effect on the pH of adding the very weak acid.

To illustrate the importance of paying attention to the assumptions that are made in pH calculations, consider the task of calculating the pH of a solution prepared by dissolving 1.0×10^{-8} mole of a strong acid, such as hydrochloric acid, in a liter of water. It is tempting to assume that the acid dissociates completely in water to give a solution with the following H_3O^+ ion concentration.

$$[H_3O^+] = 1.0 \times 10^{-8} \ M$$

If this concentration is used with the definition of pH we get the following result.

$$pH \approx -\log[H_3O^+] \approx -\log(1.0 \times 10^{-8}) = 8.0$$

The result of the calculation, however, is ludicrous. It is impossible to create a basic solution (pH > 7) by adding an acid to water!

The mistake, of course, was ignoring the concentration of the H_3O^+ ion from the dissociation of water. In pure water, we start with 1.0×10^{-7} M H_3O^+ ion. If we add 10^{-8} mol/L from the strong acid, the total concentration of H_3O^+ from both sources will be about 1.1×10^{-7} M. Thus, the pH of the solution is actually slightly less than 7.

$$\text{pH} \approx -\log[H_3O^+] \approx -\log(1.1 \times 10^{-7}) = 6.96$$

Two factors control the concentration of the H_3O^+ ion in a solution of a weak acid: (1) the acid dissociation equilibrium constant, K_a, of the acid and (2) the concentration of the solution. When we compare solutions of equivalent concentrations, the amount of the H_3O^+ ion at equilibrium increases as the value of K_a of the acid increases, as shown in Exercise 11.7.

Checkpoint

From the information given is it possible to determine which of the two following solutions has the highest pH? Explain your answer. Solution 1 has a low concentration of an acid with a large K_a. Solution 2 has a high concentration of an acid with a small K_a.

Exercise 11.7

Determine the approximate pH of 0.10 M solutions of the following acids.
(a) hypochlorous acid, HOCl, $K_a = 2.9 \times 10^{-8}$
(b) hypobromous acid, HOBr, $pK_a = 8.62$
(c) hypoiodous acid, HOI, $K_a = 2.3 \times 10^{-11}$

Solution

The first calculation can be set up as follows.

$$HOCl(aq) + H_2O(l) \rightleftharpoons H_3O^+(aq) + OCl^-(aq)$$

Initial	0.10 M	≈ 0	0
Equilibrium	$0.10 - \Delta C$	ΔC	ΔC

Substituting the information into the K_a expression gives the following equation.

$$K_a = \frac{[H_3O^+][OCl^-]}{[HOCl]} = \frac{[\Delta C][\Delta C]}{[0.10 - \Delta C]} = 2.9 \times 10^{-8}$$

Because the equilibrium constant for the reaction is relatively small, we can try the assumption that ΔC is small compared with the initial concentration of the acid.

$$\frac{[\Delta C][\Delta C]}{[0.10]} \approx 2.9 \times 10^{-8}$$

Solving the equation for ΔC gives the following result.

$$\Delta C \approx 5.4 \times 10^{-5} \ M$$

Both assumptions made in the calculation are legitimate. ΔC is small compared with the initial concentration of HOCl, but it is several orders of magnitude larger than the H_3O^+ ion concentration from the dissociation of water. We can therefore use the value of ΔC to approximate the pH of the solution.

$$pH \approx -\log[H_3O^+] \approx -\log(5.4 \times 10^{-5}) \approx 4.3$$

Repeating the calculation with the same assumptions for hypobromous and hypoiodous acid gives the following results. It is, however, necessary to first convert the pK_a of HOBr into K_a.

$$pK_a = -\log K_a = 8.62$$
$$K_a = 2.4 \times 10^{-9}$$

HOCl	$[H_3O^+] \approx 5.4 \times 10^{-5}\ M$	pH $\approx$ 4.3
HOBr	$[H_3O^+] \approx 1.5 \times 10^{-5}\ M$	pH $\approx$ 4.8
HOI	$[H_3O^+] \approx 1.5 \times 10^{-6}\ M$	pH $\approx$ 5.8

As expected, the H_3O^+ ion concentration at equilibrium—and therefore the pH of the solution—depends on the value of K_a for the acid. The H_3O^+ ion concentration decreases and the pH of the solution increases as the value of K_a becomes smaller. The next exercise shows how the H_3O^+ ion concentration at equilibrium depends on the initial concentration of the acid.

Exercise 11.8

Determine the approximate H_3O^+ ion concentration and pH of acetic acid solutions with the following concentrations: 1.0 M, 0.10 M, and 0.010 M.

Solution

The first calculation can be set up as follows.

$$HOAc(aq) + H_2O(l) \rightleftharpoons H_3O^+(aq) + OAc^-(aq) \qquad K_a = 1.8 \times 10^{-5}$$

Initial	1.0 M	≈ 0	0
Equilibrium	1.0 − ΔC	ΔC	ΔC

Substituting the information into the K_a expression gives the following equation.

$$K_a = \frac{[H_3O^+][OAc^-]}{[HOAc]} = \frac{[\Delta C][\Delta C]}{[1.0 - \Delta C]} = 1.8 \times 10^{-5}$$

We now assume that ΔC is small compared with the initial concentration of acetic acid.

$$\frac{[\Delta C][\Delta C]}{[1.0]} \approx 1.8 \times 10^{-5}$$

We then solve the equation for an approximate value of ΔC.

$$\Delta C \approx 0.0042$$

Once again, both of the characteristic assumptions of weak acid equilibrium calculations are legitimate in this case. ΔC is small compared with the initial concentration of acid, but large compared with the concentration of H_3O^+ ion from the dissociation of water. We can therefore use the value of ΔC to calculate the pH of the solution.

$$pH \approx -\log[H_3O^+] = -\log(4.2 \times 10^{-3}) = 2.4$$

Repeating the calculation for the different initial concentrations gives the following results.

1.0 M HOAc	$[H_3O^+] \approx 4.2 \times 10^{-3}$ M	pH $\approx$ 2.4
0.10 M HOAc	$[H_3O^+] \approx 1.3 \times 10^{-3}$ M	pH $\approx$ 2.9
0.010 M HOAc	$[H_3O^+] \approx 4.2 \times 10^{-4}$ M	pH $\approx$ 3.4

The concentration of the H_3O^+ ion in an aqueous solution of a weak acid gradually decreases and the pH of the solution increases as the solution becomes more dilute.

Checkpoint

Which of the above acetic acid solutions is most acidic?

11.14 BASE pH CALCULATIONS

With minor modifications, the techniques applied to equilibrium calculations for acids are valid for solutions of bases in water. Consider, for example, the approximation of the pH of an 0.10 M NH_3 solution. We can start by writing an equation for the reaction between ammonia and water.

$$NH_3(aq) + H_2O(l) \rightleftharpoons NH_4^+(aq) + OH^-(aq)$$

The ionization of NH_3 in water can be described by the following equilibrium expression.

$$K_b = \frac{[NH_4^+][OH^-]}{[NH_3]}$$

For NH_3, pK_b is 4.74.

$NH_3 + H_2O \rightleftharpoons$
$NH_4^+ + OH^-$

0.10 M NH_3 — $pK_b = 4.74$

$\Delta(NH_3) = \Delta(NH_4^+) = \Delta(OH^-)$

We can organize what we know about the equilibrium with the format we used for equilibria involving acids.

	$NH_3(aq)$	$+ H_2O(l) \rightleftharpoons$	$NH_4^+(aq) +$	$OH^-(aq)$	$pK_b = 4.74$
Initial	0.10 M		0	≈ 0	
Equilibrium	0.10 $- \Delta C$		ΔC	ΔC	

We convert pK_b into K_b and substitute the information into the equilibrium constant expression to give the following equation.

$$pK_b = -\log K_b = 4.74 \qquad K_b = 1.8 \times 10^{-5}$$

$$K_b = \frac{[NH_4^+][OH^-]}{[NH_3]} = \frac{[\Delta C][\Delta C]}{[0.10 - \Delta C]} = 1.8 \times 10^{-5}$$

K_b for ammonia is small enough to consider the assumption that ΔC is small compared with the initial concentration of the base.

$$\frac{[\Delta C][\Delta C]}{[0.10]} \approx 1.8 \times 10^{-5}$$

Solving the approximate equation gives the following result.

$$\Delta C \approx 1.3 \times 10^{-3}$$

That value of ΔC is small enough compared with the initial concentration of NH_3 to be ignored and yet large enough compared with the OH^- ion concentration in water to ignore the dissociation of water. ΔC can therefore be used to approximate the pOH of the solution.

$$pOH \approx -\log(1.3 \times 10^{-3}) = 2.9$$

This, in turn, can be used to calculate the pH of the solution.

$$pH = 14.0 - pOH = 11.1$$

Checkpoint

Which is more basic, a 1 M solution of a base that has a low K_b or a 1 M solution of a base that has a high K_b?

Relationship between K_a and K_b

Equilibrium problems involving bases can be solved if the value of K_b for the base is known. Values of K_b are listed in Table B.9 for only a limited number of compounds, however. The first step in many base equilibrium calculations therefore involves determining the value of K_b for the reaction from the value of K_a for the conjugate acid.

The relationship between K_a and K_b for a conjugate acid–base pair can be understood by using a specific example of the relationship. Consider a solution containing the acetate ion (OAc^-), for example. We already know the value of K_a for acetic acid.

$$HOAc(aq) + H_2O(l) \rightleftharpoons H_3O^+(aq) + OAc^-(aq) \qquad K_a = 1.8 \times 10^{-5}$$

What is the value of K_b for the conjugate base, OAc^-?

$$OAc^-(aq) + H_2O(l) \rightleftharpoons HOAc(aq) + OH^-(aq) \qquad K_b = ?$$

The K_a and K_b expressions for acetic acid and its conjugate base both contain the ratio of the equilibrium concentrations of the acid and its conjugate base. K_a is proportional to [OAc⁻] divided by [HOAc], and K_b is proportional to [HOAc] divided by [OAc⁻].

$$K_a = \frac{[H_3O^+][OAc^-]}{[HOAc]} \qquad K_b = \frac{[HOAc][OH^-]}{[OAc^-]}$$

Two changes need to be made to derive the K_b expression from the K_a expression: We need to remove the $[H_3O^+]$ term and introduce an $[OH^-]$ term. We can do this by multiplying the top and bottom of the K_a expression by the OH^- ion concentration.

$$K_a = \frac{[H_3O^+][OAc^-]}{[HOAc]} \times \frac{[OH^-]}{[OH^-]}$$

Rearranging the equation gives the following result.

$$K_a = \frac{[OAc^-]}{[HOAc][OH^-]} \times [H_3O^+][OH^-]$$

The two terms on the right side of the equation should look familiar. The first is the inverse of the K_b expression, and the second is the expression for K_w.

$$K_a = \frac{1}{K_b} \times K_w$$

Rearranging the equation gives the following result.

$$K_a \times K_b = K_w$$

According to this equation, the value of K_b for the reaction between the acetate ion and water can be calculated from the K_a for acetic acid.

$$K_b = \frac{K_w}{K_a} = \frac{1.0 \times 10^{-14}}{1.8 \times 10^{-5}} = 5.6 \times 10^{-10}$$

The relationship between K_a and K_b for a conjugate acid–base pair can be used to explain the following general rules given in Section 11.9.

The stronger the acid, the weaker is the conjugate base.
The stronger the base, the weaker is the conjugate acid.

The product of K_a for any Brønsted acid times K_b for its conjugate base is equal to K_w for water.

$$K_a \times K_b = K_w$$

But K_w is a relatively small number (1.0×10^{-14}). This means that when K_a is relatively large—a strong acid—K_b for the conjugate base must be relatively small—a weak base. Conversely, if K_b is large—a strong base—K_a for the conjugate acid must be small—a weak acid. As we discussed in Sections 11.8 and 11.9 the conjugate acid of strong bases (Group IA and IIA cations) and the conjugate bases of strong monoprotic acids (I^-, Br^-, Cl^-, ClO_4^-, NO_3^-) are so extremely weak they are referred to as neutral ions.

Acid–Base Properties of a Salt

A **salt** is an ionic compound with a cation other than H^+ and an anion other than OH^- or O^{2-}. Dilute solutions of soluble salts are assumed to completely ionize in water. The acid–base properties of a salt are therefore determined by the acid–base properties of the aqueous ions that result from the dissolution of the salt in water. To examine the acid–base properties of the salt potassium fluoride, KF, we begin with the dissociation of the salt in water (100% complete).

$$KF(aq) \longrightarrow K^+(aq) + F^-(aq)$$

Potassium ion, a Group IA cation, is neutral. However, F^- is the conjugate base of the weak acid HF. Because it is the conjugate base of a weak rather than strong acid, the fluoride ion will behave as a base. We therefore need a second chemical equation to show the acid–base properties of the salt.

$$F^-(aq) + H_2O(l) \rightleftharpoons HF(aq) + OH^-(aq)$$

Potassium fluoride is referred to as a basic salt because the fluoride ion behaves as a weak base. A solution of KF in water is therefore basic.

Ammonium nitrate, NH_4NO_3, is an example of an acidic salt. A dilute solution will ionize essentially completely in water.

$$NH_4NO_3(aq) \longrightarrow NH_4^+(aq) + NO_3^-(aq)$$

The nitrate anion, NO_3^-, is the conjugate base of the strong acid HNO_3 and is therefore neutral in its acid–base behavior. However, the ammonium cation, NH_4^+, is the conjugate acid of the weak base ammonia, NH_3, and is therefore expected to behave as an acid in water. Therefore, an NH_4NO_3 solution is acidic.

$$NH_4^+(aq) + H_2O(l) \rightleftharpoons H_3O^+(aq) + NH_3(aq)$$

Sodium chloride, NaCl, is an example of a neutral salt. Dilute solutions of sodium chloride ionize completely in water.

$$NaCl(aq) \longrightarrow Na^+(aq) + Cl^-(aq)$$

The resulting chloride ions are the conjugate base of a strong acid, and the sodium ion is from Group IA. Therefore, both of the resulting aqueous ions are neutral in terms of their acid–base properties and a NaCl solution is neutral.

Exercise 11.9

Calculate the HOAc, OAc^-, and OH^- concentrations at equilibrium in a 0.10 M NaOAc solution. (For HOAc, $K_a = 1.8 \times 10^{-5}$.)

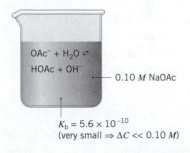

$OAc^- + H_2O \rightleftharpoons$
$HOAc + OH^-$

—— 0.10 M NaOAc

$K_b = 5.6 \times 10^{-10}$
(very small $\Rightarrow \Delta C \ll 0.10\ M$)

Solution

We start with a solution of an ionic salt—NaOAc—dissolved in water. We assume that the salt completely ionizes in water.

$$NaOAc(aq) \longrightarrow Na^+(aq) + OAc^-(aq)$$

Because Na^+ does not behave as an acid or base, only the acetate ion will significantly affect the pH:

	$OAc^-(aq) + H_2O(l) \rightleftharpoons HOAc(aq) + OH^-(aq)$			$K_b = ?$
Initial	0.10 M		0	≈ 0
Equilibrium	$0.10 - \Delta C$		ΔC	ΔC

The only other information we need to solve the problem is the value of K_b for the reaction between the acetate ion and water, which can be calculated from the value of K_a for acetic acid.

$$K_b = \frac{K_w}{K_a} = \frac{1.0 \times 10^{-14}}{1.8 \times 10^{-5}} = 5.6 \times 10^{-10}$$

Substituting what we know into the equilibrium constant expression gives the following equation.

$$K_b = \frac{[HOAc][OH^-]}{[OAc^-]} = \frac{[\Delta C][\Delta C]}{[0.10 - \Delta C]} = 5.6 \times 10^{-10}$$

Because the value of K_b for the base OAc^- is relatively small, we can feel reasonably confident that ΔC is small compared with the initial concentration of base.

$$\frac{[\Delta C][\Delta C]}{[0.10]} \approx 5.6 \times 10^{-10}$$

We can then solve the approximate equation the assumption generates.

$$\Delta C \approx 7.5 \times 10^{-6} \ M$$

ΔC is very much smaller than the initial concentration of the base, which means that one of the key assumptions of acid–base equilibrium calculations is legitimate. But $[OH^-]$ is still at least 75 times larger than the concentration of the OH^- ion in pure water, which means that we can also legitimately ignore the dissociation of water in the calculation. We can therefore use the value of ΔC to calculate the equilibrium concentrations of HOAc, OAc^-, and OH^-.

$$[OAc^-] = 0.10 - \Delta C \approx 0.10 \ M$$
$$[HOAc] = [OH^-] = \Delta C \approx 7.5 \times 10^{-6} \ M$$

Checkpoint

If K_a for HNO_2 is 5.1×10^{-4}, what is K_b for NO_2^-?

11.15   MIXTURES OF ACIDS AND BASES

So far, we have examined 0.10 M solutions of both acetic acid and sodium acetate.

$$0.10 \ M \ \text{HOAc} \quad [\text{H}_3\text{O}^+] \approx 1.3 \times 10^{-3} \ M \quad \text{pH} \approx 2.9$$
$$0.10 \ M \ \text{NaOAc} \quad [\text{OH}^-] \approx 7.5 \times 10^{-6} \ M \quad \text{pH} \approx 8.9$$

What would happen if we added enough sodium acetate to an acetic acid solution so that the solution is simultaneously 0.10 M in both HOAc and NaOAc?

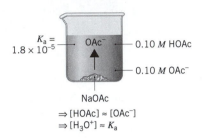

$$K_a = 1.8 \times 10^{-5}$$

OAc⁻

0.10 M HOAc

0.10 M OAc⁻

NaOAc

$\Rightarrow [\text{HOAc}] \approx [\text{OAc}^-]$
$\Rightarrow [\text{H}_3\text{O}^+] \approx K_a$

The first step toward answering the question is recognizing that there are two sources of the OAc⁻ ion in the solution. Acetic acid, of course, dissociates to give the H_3O^+ and OAc⁻ ions.

$$\text{HOAc}(aq) + \text{H}_2\text{O}(l) \rightleftharpoons \text{H}_3\text{O}^+(aq) + \text{OAc}^-(aq)$$

The salt, sodium acetate, however, also dissociates in water to give the OAc⁻ ion.

$$\text{NaOAc}(s) \xrightarrow{\text{H}_2\text{O}} \text{Na}^+(aq) + \text{OAc}^-(aq)$$

These reactions share a common ion: the OAc⁻ ion. Le Châtelier's principle predicts that adding sodium acetate to acetic acid should shift the equilibrium between HOAc and the H_3O^+ and OAc⁻ ions to the left. Thus, adding NaOAc reduces the extent to which HOAc dissociates. The mixture is therefore less acidic than 0.10 M HOAc, by itself.

In Exercise 11.9 we saw that OAc⁻ behaves as a base. The K_b for OAc⁻ is several orders of magnitude smaller than K_a for HOAc. Therefore, the reaction of OAc⁻ with water is negligible compared to the reaction of HOAc with water.

We can quantify the discussion by setting up the problem as follows.

	$\text{HOAc}(aq) + \text{H}_2\text{O}(l) \rightleftharpoons \text{H}_3\text{O}^+(aq) + \text{OAc}^-(aq)$			$K_a = 1.8 \times 10^{-5}$
Initial	0.10 M	≈ 0	0.10 M	
Equilibrium	?	?	?	

We don't have any basis for predicting whether the dissociation of water can be ignored in the calculation, so let's make that assumption and check its validity later.

If most of the H_3O^+ ion concentration at equilibrium comes from dissociation of the acetic acid, the reaction has to shift to the right to reach equilibrium.

	$\text{HOAc}(aq) + \text{H}_2\text{O}(l) \rightleftharpoons \text{H}_3\text{O}^+(aq) + \text{OAc}^-(aq)$			$K_a = 1.8 \times 10^{-5}$
Initial	0.10 M	≈ 0	0.10 M	
Equilibrium	$0.10 - \Delta C$	ΔC	$0.10 + \Delta C$	

Substituting the information into the K_a expression gives the following result.

$$K_a = \frac{[H_3O^+][OAc^-]}{[HOAc]} = \frac{[\Delta C][0.10 + \Delta C]}{[0.10 - \Delta C]} = 1.8 \times 10^{-5}$$

We assume that ΔC is small compared with the initial concentrations of HOAc and OAc^-.

$$\frac{[\Delta C][0.10]}{[0.10]} \approx 1.8 \times 10^{-5}$$

We then solve the approximate equation the assumption generates.

$$\Delta C \approx 1.8 \times 10^{-5} \, M$$

Are our two assumptions legitimate? Is ΔC small enough compared with 0.10 to be ignored? Is ΔC large enough so that the dissociation of water can be ignored? The answer to both questions is yes. We can therefore use that value of ΔC to calculate the H_3O^+ ion concentration at equilibrium and the pH of the solution.

$$\Delta C \approx 1.8 \times 10^{-5} \approx [H_3O^+]$$
$$pH \approx -\log(1.8 \times 10^{-5}) = 4.7$$

The solution is therefore acidic, but less acidic than HOAc by itself. Le Chatelier's principle predicts that the equilibrium for the dissociation of acetic acid in water would be shifted toward the reactants by the addition of acetate ion. This would result in a decrease in the amount of H_3O^+. In other words, the HOAc solution is less acidic due to the addition of a base, OAc^-.

11.16   BUFFERS AND BUFFER CAPACITY

Mixtures of a weak acid and its conjugate base are called **buffers.** The term *buffer* usually means "to lessen or absorb shock." The solutions are buffers because they lessen or absorb the drastic change in pH that occurs when small amounts of acids or bases are added to water.

In Section 11.12 we concluded that adding a single drop of 2 *M* HCl to 100 mL of pure water would change the pH by four units, from pH 7 to pH 3. If we add the same amount of HCl to a buffer solution that was 0.10 *M* in both acetic acid (HOAc) and a salt of the acetate ion (OAc^-), there would be no observable change in the pH of the solution. In fact, we can add 10 times as much 2 *M* HCl to the buffer solution without noticing a significant change in the pH of the solution. Thus, the pH of the buffer solution is truly "buffered" against the effect of small amounts of acid or base.

To understand how the buffer works, it is important to recognize that a buffer is made by mixing a weak acid and its conjugate base so that significant amounts of both a weak acid and weak base are present in solution.

$$\underset{\text{Weak acid}}{HOAc(aq)} + H_2O(l) \rightleftharpoons H_3O^+(aq) + \underset{\text{Conjugate base}}{OAc^-(aq)}$$

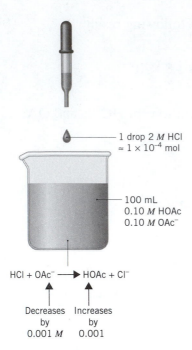

At any moment in time, the concentrations of acetic acid and the H_3O^+ and OAc^- ions in the solution are related by the following equation.

$$K_a = \frac{[H_3O^+][OAc^-]}{[HOAc]}$$

If we rearrange the equation, by solving for the H_3O^+ ion concentration at equilibrium, we get the following result:

$$[H_3O^+] = K_a \times \frac{[HOAc]}{[OAc^-]}$$

This equation suggests that the H_3O^+ ion concentration in the solution will stay more or less constant as long as there is no significant change in the ratio of the concentrations of the undissociated HOAc molecules and the OAc^- ions. The buffer solution was prepared by adding enough sodium acetate (NaOAc) to acetic acid (HOAc) to make the concentration of both the acid (HOAc) and its conjugate base (OAc^-) the same. (The solution is simultaneously 0.10 M HOAc and 0.10 M OAc^-.)

If a small amount of a strong acid, H_3O^+, is added to the solution, it will react with the OAc^- ion, which behaves as a Brønsted base by accepting a proton to form a bit more acetic acid.

$$OAc^-(aq) + H_3O^+(aq) \longrightarrow HOAc(aq) + H_2O(l)$$

The net result is a slight increase in the concentration of the undissociated HOAc molecules and a slight decrease in the concentration of the OAc^- ion. But, if only a small amount of acid is added to the solution, the ratio of the concentrations would be roughly the same. As a result, the H_3O^+ ion concentration in the solution remains essentially the same, which means there is no change in the pH of the solution. In other words, the addition of a strong acid, H_3O^+, to the buffer results in the strong acid being converted into an equivalent amount of weaker acid, HOAc.

If a small amount of a strong base, OH^-, is added to the solution, it will react with some of the undissociated HOAc molecules to form a bit more of the OAc^- ion.

$$HOAc(aq) + OH^-(aq) \longrightarrow OAc^-(aq) + H_2O(l)$$

But, once again, if we add only a small amount of the base, the ratio of the concentrations of HOAc and OAc^- in the solution will be more or less the same. This means that the pH of the solution is more or less the same. The addition of a strong base, OH^-, to the buffer results in the conversion of the strong base into an equivalent amount of weak base, OAc^-.

Because it contains both HOAc molecules and the OAc^- ion, the solution can resist a change in pH when reasonably small amounts of either a strong acid or a strong base are added to the solution. The concentrations of acetic acid and the acetate ion will change when the acid or base is added to the buffer solution, but the ratio of the concentrations of these two components of the equilibrium will not change by a large enough amount to produce a significant change in the pH of the solution.

As long as the concentrations of HOAc and the OAc^- ion in the buffer are larger than the amount of acid or base added to the solution, the pH remains constant. Table 11.5 shows experimental data that compare the effects of adding different amounts of hydrochloric acid to water and two buffer solutions of different concentrations. The first column gives the number of moles of HCl added per liter. The second column shows the effect of this much acid on the pH of water. The third column shows the effect on a buffer solution that contains 0.10 M HOAc and 0.10 M NaOAc. The fourth column shows the result of adding the acid to a more concentrated buffer that contains twice as much acetic acid and twice as much sodium acetate. Even when as much as 0.01 mol/L of acid is added to the first buffer, there is very little change in the pH of the buffer solution, as shown in Figure 11.12.

TABLE 11.5   Effect of Adding Hydrochloric Acid to Water and Two Buffer Solutions

Mol HCl/L Added to Each Solution	pH of Water	pH of Buffer 1 (0.10 M HOAc/ 0.10 M OAc⁻)	pH of Buffer 2 (0.20 M HOAc/ 0.20 M OAc⁻)
0	7	4.74	4.74
0.000001	6	4.74	4.74
0.00001	5	4.74	4.74
0.0001	4	4.74	4.74
0.001	3	4.74	4.74
0.01	2	4.66	4.70
0.1	1	2.72	4.27

By definition, the **buffer capacity** of a buffer is the amount of acid or base that can be added before the pH of the solution changes significantly. Note that the capacity of the second buffer solution in Table 11.5 is even greater than that of the first. Buffer 2 protects against quantities of acid or base as large as 0.1 mol/L.

Buffers can also be made from a weak base and its conjugate acid, such as ammonia and a salt of the ammonium ion.

$$\underset{\text{Conjugate acid}}{NH_4^+(aq)} + H_2O(l) \Longleftrightarrow H_3O^+(aq) + \underset{\text{Weak base}}{NH_3(aq)}$$

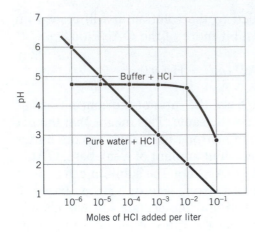

FIGURE 11.12 The pH of water is very sensitive to the addition of hydrochloric acid. Adding as little as 10^{-3} mol of HCl to a liter of water can drop the pH by 4 units. The pH of a buffer, however, remains constant until relatively large amounts of hydrochloric acid have been added.

There is an important difference, however, between this buffer and one made by mixing sodium acetate and acetic acid. Mixtures of HOAc and OAc$^-$ ion form an **acidic buffer,** with a pH below 7. Mixtures of NH_3 and the NH_4^+ ion form a **basic buffer,** with a pH above 7.

We can predict whether a buffer will be acidic or basic by comparing the values of K_a and K_b for the conjugate acid–base pair. K_a for acetic acid is significantly larger than K_b for the acetate ion. We therefore expect mixtures of the conjugate acid–base pair to be acidic.

$$HOAc \quad K_a = 1.8 \times 10^{-5}$$
$$OAc^- \quad K_b = 5.6 \times 10^{-10}$$

K_b for ammonia, on the other hand, is much larger than K_a for the ammonium ion. Mixtures of this acid–base pair are therefore basic.

$$NH_3 \quad K_b = 1.8 \times 10^{-5}$$
$$NH_4^+ \quad K_a = 5.6 \times 10^{-10}$$

In general, if K_a for the acid is larger than 1×10^{-7}, the buffer will be acidic. If K_b is larger than 1×10^{-7}, the buffer is basic.

Buffers are used extensively in the chemistry laboratory. Buffers can be purchased from chemical suppliers as solutions with known concentrations and pH values. They may also be purchased as packets of a mixture of a solid conjugate acid–base pair. Dissolving the packet in water yields a buffer with a pH equal to the value stated on the package. In addition, buffers may be prepared in the laboratory by mixing measured amounts of an appropriate conjugate acid–base pair and then adding strong acid or base to adjust the pH to the desired value. The pH of the buffer is normally monitored with a pH meter as the strong acid or strong base is added.

Checkpoint

The pH of a buffered solution is 7.4. What happens to the pH if an acid is added to the buffer? What happens to the pH if a base is added to the buffer? Does the resulting pH depend on how much acid or base is added?

Buffers in the Body

Buffers are very important in living organisms for maintaining the pH of biological fluids within the very narrow ranges necessary for the biochemical reactions of life processes.

Three primary buffer systems maintain the pH of blood in the human body within the limits 7.36 to 7.42. One of these buffers is composed of carbonic acid, H_2CO_3, and its conjugate base HCO_3^-, the hydrogen carbonate ion. Some of the CO_2 from respiring tissue that enters a red blood cell reacts with water and is converted to carbonic acid. The carbonic acid then can form an equilibrium with the hydrogen carbonate ion.

$$CO_2(g) + H_2O(l) \rightleftharpoons H_2CO_3(aq)$$
$$H_2CO_3(aq) + H_2O(l) \rightleftharpoons HCO_3^-(aq) + H_3O^+(aq) \quad \text{buffer reaction}$$

When the concentration of H_3O^+ in the blood is too high (too low a pH) a condition known as *acidosis* results. This can be caused by a disease such as diabetes mellitus, in which large amounts of acids are produced. As the concentration of H_3O^+ in blood increases the H_2CO_3/HCO_3^- equilibrium shifts to the left to maintain the proper pH.

A low concentration of H_3O^+ in the blood (too high a pH) is known as *alkalosis*. This condition can result during hyperventilation (rapid breathing) when excess amounts of CO_2 are lost from the body through the lungs. The loss of CO_2 shifts the CO_2/H_2CO_3 equilibrium to the left, resulting in a decrease in the concentration of carbonic acid in the blood. This disrupts the H_2CO_3/HCO_3^- buffer by shifting the equilibrium to the left and therefore reducing the amount of H_3O^+ in the blood. The resulting increase in pH is alkalosis.

Checkpoint
Lactic acid, $HC_3H_5O_3$, is produced in the body during strenuous exercise. Write a chemical reaction that can occur in blood to maintain the proper pH when lactic acid is produced.

11.17   BUFFER CAPACITY AND pH TITRATION CURVES

We can measure the concentration of a solution by a technique known as **titration.** The solution being studied is slowly added to a known quantity of a reagent with which it reacts until we observe something that tells us that exactly equivalent numbers of moles of the reagents are present. Titrations therefore depend on the existence of a class of compounds known as **indicators.** A common type of titration is an acid–base titration in which an acid of unknown concentration is titrated with a base that has an accurately known concentration. As a result of the titration it is possible to determine the concentration of the acid solution.

Phenolphthalein is an example of an indicator for acid–base reactions. When added to solutions that contain an acid, phenolphthalein is colorless. In the presence of a base, it has a light pink color. Phenolphthalein can therefore indicate when enough base has been added to consume the acid that was initially present in the solution because it turns from colorless to pink *just after* the point when equivalent numbers of moles of acid and base have been added to the solution.

The **end point** of an acid–base titration is the point at which the indicator turns color. The **equivalence** point is the point at which exactly enough base has been added to neutralize the acid. Chemists try to use indicators for which the end point is as close as possible to the point at which equivalent amounts of acid and base are present.

Phenolphthalein provides an example of how chemists compensate for the fact that the indicator doesn't always turn color exactly at the equivalence point of the titration. Acid–base titrations that use phenolphthalein as the indicator are stopped at the point at which a pink color appears and remains for 10 seconds as the solution is stirred. Because this is one drop before the indicator turns a permanent color, it is as close as we can get to the equivalence point of the titration.

Exercise 11.10

The reaction between sulfuric acid (H_2SO_4) and potassium hydroxide (KOH) can be described by the following equation.

$$H_2SO_4(aq) + 2\ KOH(aq) \longrightarrow 2\ K^+(aq) + SO_4^{2-}(aq) + 2\ H_2O(l)$$

A 25.00 mL sample of sulfuric acid solution requires 42.61 mL of a 0.1500 M KOH solution to titrate to the phenolphthalein end point. Calculate the concentration of the sulfuric acid solution.

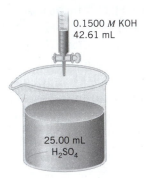

0.1500 M KOH
42.61 mL

25.00 mL
H_2SO_4

FIGURE 11.13 Titration of H_2SO_4 with KOH.

Solution

We only know the volume of the sulfuric acid solution, but we know both the volume and the concentration of the KOH solution, as shown in Figure 11.13. It therefore seems reasonable to start by calculating the number of moles of KOH in this solution.

$$\frac{0.1500 \text{ mol KOH}}{1\ \cancel{L}} \times 0.04261\ \cancel{L} = 6.392 \times 10^{-3} \text{ mol KOH}$$

We now know the number of moles of KOH consumed in the reaction, and we have a balanced chemical equation for the reaction that occurs when the two solutions are mixed. We can therefore calculate the number of moles of H_2SO_4 needed to consume that much KOH.

$$6.392 \times 10^{-3}\ \cancel{\text{mol KOH}} \times \frac{1 \text{ mol } H_2SO_4}{2\ \cancel{\text{mol KOH}}} = 3.196 \times 10^{-3} \text{ mol } H_2SO_4$$

We now know the number of moles of H_2SO_4 in the sulfuric acid solution and the volume of the solution. We can therefore calculate the number of moles of sulfuric acid per liter, or the molarity of the solution.

$$\frac{3.196 \times 10^{-3} \text{ mol } H_2SO_4}{0.02500 \text{ L}} = 0.1278\ M\ H_2SO_4$$

The sulfuric acid solution therefore has a concentration of 0.1278 mole per liter.

The solution conditions can be described during an acid–base titration by using a titration curve, a plot of solution pH versus amount of titrant added (see Figure 11.14). Buffers

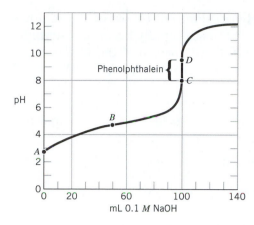

FIGURE 11.14 The pH of a 0.10 M solution of acetic acid as it is titrated with 0.10 M sodium hydroxide. The pH increases at first, as some of the HOAc is converted into OAc$^-$ ions. This creates a buffer, which resists changes in pH until enough NaOH has been added to completely react with the acetic acid. At this point the pH increases rapidly, then slowly levels off as the solution begins to behave more like a 0.10 M NaOH solution. Points A–D are explained in the text.

and buffering capacity have a lot to do with the shape of the titration curve in Figure 11.14, which shows the pH of a solution of acetic acid as it is titrated with a strong base, sodium hydroxide. Four points (A, B, C, and D) on the curve will be discussed in some detail.

Point A represents the pH at the start of the titration. The pH at this point is 2.9, the pH of a 0.10 M HOAc solution. As NaOH is added to the solution, some of the acid is converted to its conjugate base, the OAc$^-$ ion.

$$HOAc(aq) + OH^-(aq) \longrightarrow OAc^-(aq) + H_2O(l)$$

Point B is the point at which exactly half of the HOAc molecules have been converted to OAc$^-$ ions. At this point, the concentration of HOAc molecules is equal to the concentration of the OAc$^-$ ions. Because the ratio of the HOAc and OAc$^-$ concentrations is 1:1, the concentration of the H$_3$O$^+$ ion at point B is equal to the value of K_a for the acid.

$$\text{Point } B \qquad [H_3O^+] = K_a$$

This provides us with a way of measuring K_a for an acid. We can titrate a sample of the acid with a strong base, plot the titration curve, and then find the point along the curve at which exactly half of the acid has been consumed. The H$_3$O$^+$ ion concentration at that point will be equal to the value of K_a for the acid.

Point C is the *equivalence point* for the titration, the point at which enough base has been added to the solution to consume the acid present at the start of the titration. The goal of any titration is to find the equivalence point. What is actually observed is the *end point* of the titration—point D—the point at which the indicator changes color.

Every effort is made to bring the end point as close as possible to the equivalence point of a titration. Suppose that phenolphthalein is used as the indicator for this titration, for example. Phenolphthalein turns from colorless to pink as the pH of the solution changes from 8 to 10. If we wait until the phenolphthalein permanently turns pink, the end point will fall beyond the equivalence point for the titration. So we try to find the point at which adding one drop of base turns the entire solution to a pink color that fades in about 10 seconds while the solution is stirred. In theory, this is one drop before the end point of the titration and therefore closer to the equivalence point.

Note the shape of the pH titration curve in Figure 11.14. The pH rises rapidly at first because we are adding a strong base to a weak acid and the base neutralizes some of the acid. The curve then levels off, and the pH remains more or less constant as we add base because some of the HOAc present initially is converted into OAc$^-$ ions to form a buffer solution. The pH of the buffer solution stays relatively constant until most of the acid has been converted to its conjugate base. At that point the pH rises rapidly because essentially all of the HOAc in the system has been converted into OAc⁻ ions and the buffer is exhausted. The pH then gradually levels off as the solution begins to look like a 0.10 M NaOH solution.

The titration curve for a weak base, such as ammonia, titrated with a strong acid, such as hydrochloric acid, would be analogous to the curve in Figure 11.14. The principal difference is that the pH value is high at first and decreases as acid is added.

Checkpoint
Describe what happens to the pH of an acetic acid solution as NaOH is slowly added.

KEY TERMS

Acid dissociation
  equilibrium constant,
  K_a
Acidic buffer
Acids
Arrhenius acid/base
Base ionization
  equilibrium constant,
  K_b
Bases
Basic buffer

Brønsted acid/base
Buffer
Buffer capacity
Conjugate acid–base pair
End point
Equivalence point
Hydrogen ion
  acceptor/donor
Hydronium ion
Leveling effect

Neutralization reaction
pH
pOH
Proton acceptor/donor
Strong acid
Water dissociation
  equilibrium constant,
  K_w
Weak acid

PROBLEMS

Acids and Bases

1. Describe how acids, bases, and salts differ. Give examples of substances from daily life that fit each of the categories.

2. Acids (such as hydrochloric acid and sulfuric acid) react with bases (such as sodium hydroxide and ammonia) to form salts (such as sodium chloride, sodium sulfate, ammonium chloride, and ammonium sulfate). Write balanced equations for the reactions and describe how the salts differ from the acids and bases from which they are made.

3. Define the following terms; *acid, base, weak acid, strong acid, weak base,* and *strong base.*

4. Describe the difference between strong acids (such as hydrochloric acid) and weak acids (such as acetic acid) and the difference between strong bases (such as sodium hydroxide) and weak bases (such as ammonia).

5. Describe the Arrhenius definition of acids and bases, and give examples of acids and bases that fit this definition.

6. Which of the following compounds are Arrhenius acids?
   (a) HCl   (b) H_2SO_4 (c) H_2O (d) $FeCl_3$ (e) H_2S

7. Which of the following compounds are Arrhenius bases?
   (a) NaOH (b) MgO (c) Ca(OH)$_2$ (d) H$_2$O (e) NH$_3$

The Brønsted Definition of Acids and Bases

8. Write balanced equations to show what happens when hydrogen bromide dissolves in water to form an acidic solution and when ammonia dissolves in water to form a basic solution.

9. Give an example of an acid whose formula carries a positive charge, an acid that is electrically neutral, and an acid that carries a negative charge.

10. The following compounds are all oxyacids. In each case, the acidic hydrogen atoms are bound to oxygen atoms. Write the skeleton structures for the acids.
    (a) H$_3$PO$_4$ (b) HNO$_3$ (c) HClO$_4$ (d) H$_3$BO$_3$ (e) H$_2$CrO$_4$

11. Many insects leave small quantities of formic acid, HCO$_2$H, behind when they bite. Explain why the itching sensation can be relieved by treating the bite with an aqueous solution of ammonia (NH$_3$) or baking soda (NaHCO$_3$).

12. Describe how the Brønsted definition of an acid can be used to explain why compounds that contain hydrogen with an oxidation number of +1 are often acids.

13. Which of the following compounds can act as both a Brønsted acid and a Brønsted base?
    (a) NaHCO$_3$ (b) Na$_2$CO$_3$ (c) H$_2$CO$_3$ (d) CO$_2$ (e) H$_2$O

14. Which of the following compounds cannot be Brønsted bases?
    (a) H$_3$O$^+$ (b) MnO$_4^-$ (c) BH$_4^-$ (d) CN$^-$ (e) S^{2-}

15. Which of the following compounds cannot be Brønsted bases?
    (a) O$_2^{2-}$ (b) CH$_4$ (c) PH$_3$ (d) SF$_4$ (e) CH$_3^+$

16. Label the Brønsted acids and bases in the following reactions.
    (a) HSO$_4^-$(aq) + H$_2$O(l) $\leftrightarrows$ H$_3$O$^+$(aq) + SO$_4^{2-}$(aq)
    (b) CH$_3$CO$_2$H(aq) + OH$^-$(aq) $\leftrightarrows$ CH$_3$CO$_2^-$(aq) + H$_2$O(l)
    (c) CaF$_2$(s) + H$_2$SO$_4$(aq) $\leftrightarrows$ CaSO$_4$(aq) + 2 HF(aq)
    (d) HNO$_3$(aq) + NH$_3$(aq) $\leftrightarrows$ NH$_4$NO$_3$(aq)
    (e) LiCH$_3$(l) + NH$_3$(l) $\leftrightarrows$ CH$_4$(g) + LiNH$_2$(s)

Conjugate Acid–Base Pairs

17. Identify the conjugate acid–base pairs in the following reactions.
    (a) HCl(aq) + H$_2$O(l) $\leftrightarrows$ H$_3$O$^+$(aq) + Cl$^-$(aq)
    (b) HCO$_3^-$(aq) + H$_2$O(l) $\leftrightarrows$ H$_2$CO$_3$(aq) + OH$^-$(aq)
    (c) NH$_3$(aq) + H$_2$O(l) $\leftrightarrows$ NH$_4^+$(aq) + OH$^-$(aq)
    (d) CaCO$_3$(s) + 2 HCl(aq) $\leftrightarrows$ Ca^{2+}(aq) + 2 Cl$^-$(aq) + H$_2$CO$_3$(aq)

18. Which of the following is not a conjugate acid–base pair?
    (a) NH$_4^+$/NH$_3$ (b) H$_2$O/OH$^-$ (c) H$_3$O$^+$/OH$^-$ (d) CH$_4$/CH$_3^-$

19. Write Lewis structures for the following Brønsted acids and their conjugate bases.
    (a) formic acid, HCO$_2$H (b) methanol, CH$_3$OH

20. Write Lewis structures for the following Brønsted bases and their conjugate acids.
    (a) methylamine, CH$_3$NH$_2$ (b) acetate ion, CH$_3$CO$_2^-$

21. Identify the conjugate base of each of the following Brønsted acids.
    (a) H$_3$O$^+$ (b) H$_2$O (c) OH$^-$ (d) NH$_4^+$

22. Identify the conjugate base of each of the following Brønsted acids.
    (a) HPO$_4^{2-}$ (b) H$_2$PO$_4^-$ (c) HCO$_3^-$ (d) HS$^-$

23. Identify the conjugate acid of each of the following Brønsted bases.
    (a) O^{2-} (b) OH^- (c) H_2O (d) NH_2^-

24. Identify the conjugate acid of each of the following Brønsted bases.
    (a) Na_2HPO_4 (b) $NaHCO_3$ (c) Na_2SO_4 (d) $NaNO_2$

25. Which of the following ions is the conjugate base of a strong acid?
    (a) OH^- (b) HSO_4^- (c) NH_2^- (d) S^{2-} (e) H_3O^+

26. Use Lewis structures to predict the products of the following acid–base reactions.

$$NaOCH_3(aq) + NaHCO_3(aq) \longrightarrow$$
$$NH_4Cl(aq) + NaSH(aq) \longrightarrow$$

27. Write balanced chemical equations for the following acid–base reactions.
    (a) $Al_2O_3(s) + HCl(aq) \rightarrow$ (b) $CaO(s) + H_2SO_4(aq) \rightarrow$
    (c) $Na_2O(s) + H_2O(l) \rightarrow$ (d) $MgCO_3(s) + HCl(aq) \rightarrow$
    (e) $Na_2O_2(s) + H_3PO_4(aq) \rightarrow$

The Relative Strengths of Acids and Bases

28. In the Brønsted model of acid–base reactions, what does the statement that HCl is a stronger acid than H_2O mean in terms of the following reaction?

$$HCl(aq) + H_2O(l) \longrightarrow H_3O^+(aq) + Cl^-(aq)$$

29. In the Brønsted model of acid–base reactions, what does it mean to say that NH_3 is a weaker acid than H_2O?

30. What can you conclude from experiments that suggest that the following reaction proceeds to the right, as written?

$$HBr(aq) + H_2O(l) \longrightarrow H_3O^+(aq) + Br^-(aq)$$

31. What can you conclude from experiments that suggest that the following reaction proceeds to the right, as written?

$$HCl(aq) + NH_3(aq) \longrightarrow NH_4^+(aq) + Cl^-(aq)$$

32. What can you conclude from experiments that suggest that the following reaction does not proceed to the right, as written?

$$HCO_2^-(aq) + H_2O(l) \not\longrightarrow HCO_2H(aq) + OH^-(aq)$$

33. Describe the relationship between the acid-dissociation equilibrium constant for an acid, K_a, and the strength of the acid.

34. Use the table of acid dissociation equilibrium constants in Appendix B.8 to classify the following acids as either strong or weak.
    (a) acetic acid, CH_3CO_2H (b) boric acid, H_3BO_3
    (c) chromic acid, H_2CrO_4 (d) formic acid, HCO_2H
    (e) hydrobromic acid, HBr

35. Which of the following is the weakest Brønsted acid?
    (a) $H_2S_2O_3$, $K_a = 0.3$ (b) H_2CrO_4, $K_a = 9.6$ (c) H_3BO_3, $K_a = 7.3 \times 10^{-10}$
    (d) C_6H_5OH, $K_a = 1.0 \times 10^{-10}$

Factors That Control the Relative Strengths of Acids and Bases

36. Explain the following observations.
    (a) H_2SO_4 is a stronger acid than HSO_4^-.
    (b) HNO_3 is a stronger acid than HNO_2.
    (c) H_2S is a stronger acid than H_2O.
    (d) H_2S is a stronger acid than PH_3.

37. Arrange the following in order of increasing acid strength.
    (a) H_2O (b) HCl (c) CH_4 (d) NH_3

38. Arrange the following in order of increasing base strength.
    (a) NH_3 (b) PH_3 (c) H_2O (d) H_2S

39. Which of the following is the strongest Brønsted acid?
    (a) H_3O^+ (b) HF   (c) NH_3 (d) $NaHSO_4$ (e) NaOH

40. Which of the following is the weakest Brønsted acid?
    (a) H_2Se (b) H_2S (c) H_2O (d) SH^- (e) OH^-

41. Which of the following is the weakest Brønsted acid?
    (a) $HClO_4$ (b) H_3PO_4 (c) H_2CO_3 (d) H_2SO_4 (e) $Al(OH)_3$

42. Which of the following is the strongest Brønsted base?
    (a) NH_2^- (b) PH_2^- (c) CH_3^- (d) SiH_3^-

43. Which of the following is the strongest Brønsted base?
    (a) H_2O (b) SH^- (c) OH^- (d) S^{2-} (e) O^{2-}

44. Which of the following has the strongest conjugate base?
    (a) H_2O (b) H_2S (c) NH_3 (d) PH_3 (e) CH_4

pH as a Measure of the Relative Strengths of Acids and Bases

45. What happens to the concentration of the H_3O^+ ion when a strong acid is added to water? What happens to the concentration of the OH^- ion? What happens to the pH of the solution?

46. What happens to the OH^- ion concentration when a strong base is added to water? What happens to the concentration of the H_3O^+ ion? What happens to the pH of the solution?

47. Which of the following solutions is the most acidic?
    (a) 0.10 M acetic acid, pH = 2.9
    (b) 0.10 M hydrogen sulfide, pH = 4.1
    (c) 0.10 M sodium acetate, pH = 8.4
    (d) 0.10 M ammonia, pH = 11.1

48. Which of the following solutions is the most acidic?
    (a) 0.10 M H_3PO_4, pH = 1.4
    (b) 0.10 M $H_2PO_4^-$, pH = 4.4
    (c) 0.10 M HPO_4^{2-}, pH = 9.3
    (d) 0.10 M PO_4^{3-}, pH = 12.0

49. Which of the following compounds could dissolve in water to give a 0.10 M solution with a pH of about 5?
    (a) NH_3 (b) NaCl   (c) HCl   (d) KOH   (e) NH_4Cl

The Acid–Base Chemistry of Water

50. Explain why the concentrations of the H_3O^+ ion and the OH^- ion in pure water are the same.

51. The dissociation of water is an endothermic reaction.

$$2 \, H_2O(l) \rightleftharpoons H_3O^+(aq) + OH^-(aq) \qquad \Delta H° = 55.84 \text{ kJ/mol}_{rxn}$$

Use Le Châtelier's principle to predict what should happen to the fraction of water molecules that dissociate into ions as the temperature of water increases.

52. Use Le Châtelier's principle to explain why adding either an acid or a base to water suppresses the dissociation of water.

53. Explain why it is impossible for water to be at equilibrium when it contains large quantities of both the H_3O^+ and OH^- ions.

pH and pOH

54. Approximate the number of H_3O^+ and OH^- ions in 1.00 mL of pure water.

55. Approximate the pH and pOH of a 0.035 M HCl solution.

56. Approximate the pH and pOH of a solution that contains 0.568 g of HCl per 250 mL of solution.

57. Approximate the H_3O^+ and OH^- ion concentrations in a solution that has a pH of 3.72.

58. Explain why the pH of a solution decreases when the H_3O^+ ion concentration increases.

Acid Dissociation Equilibrium Constants

59. Which of the following factors influences the value of K_a for the dissociation of formic acid?

$$HCO_2H(aq) + H_2O(l) \rightleftharpoons HCO_2^-(aq) + H_3O^+(aq)$$

(a) temperature (b) pressure (c) pH
(d) the initial concentration of HCO_2H
(e) the initial concentration of the HCO_2^- ion

60. Which of the following solutions is the most acidic?
(a) 0.10 M CH_3CO_2H, $K_a = 1.8 \times 10^{-5}$
(b) 0.10 M HCO_2H, $K_a = 1.8 \times 10^{-4}$
(c) 0.10 M $ClCH_2CO_2H$, $K_a = 1.4 \times 10^{-3}$
(d) 0.10 M Cl_2CHCO_2H, $K_a = 5.1 \times 10^{-2}$

61. Which of the following compounds is the strongest base?
(a) $CH_3CO_2^-$ (for CH_3CO_2H, $K_a = 1.8 \times 10^{-5}$)
(b) HCO_2^- (for HCO_2H, $K_a = 1.8 \times 10^{-4}$)
(c) $ClCH_2CO_2^-$ (for $ClCH_2CO_2H$, $K_a = 1.4 \times 10^{-3}$)
(d) $Cl_2CHCO_2^-$ (for Cl_2CHCO_2H, $K_a = 5.1 \times 10^{-2}$)

62. List acetic acid, chlorous acid, hydrofluoric acid, and nitrous acid in order of increasing strength if 0.10 M solutions of the acids contain the following equilibrium concentrations.

Acid	[HA]	[A⁻]	[H₃O⁺]
HOAc	0.099 M	0.0013 M	0.0013 M
HOClO	0.072 M	0.028 M	0.028 M
HF	0.092 M	0.0081 M	0.0081 M
HNO$_2$	0.093 M	0.0069 M	0.0069 M

Strong Acids

63. How would you explain to a younger brother or sister taking chemistry for the first time why it is wrong to assume that the pH of a 10^{-8} M HCl solution is 8?

64. Explain why the H_3O^+ ion concentration in a strong acid solution depends on the concentration of the solution, but not on the value of K_a for the acid.

65. Approximate the pH of a 0.056 M solution of hydrochloric acid.

66. Nitric acid is often grouped with sulfuric acid and hydrochloric acid as one of the strong acids. Calculate the pH of 0.10 M nitric acid, assuming that it is a strong acid that dissociates completely. Calculate the pH of the solution using the value of K_a for the acid (for HNO_3, $K_a = 28$).

Hidden Assumptions in Weak Acid Calculations

67. Describe the two assumptions that are commonly made in weak acid equilibrium problems. Describe how you can test whether the assumptions are valid for a particular calculation.

68. Explain why the techniques used to calculate the equilibrium concentrations of the components of a weak acid solution can't be used for either strong acids or very weak acids.

Factors That Influence the H_3O^+ Ion Concentration in Weak Acid Solutions

69. Explain why the H_3O^+ ion concentration in a weak acid solution depends on both the value of K_a for the acid and the concentration of the acid.

70. Which of the following solutions has the largest H_3O^+ ion concentration?
(a) 0.10 M HOAc   (b) 0.010 M HOAc   (c) 0.0010 M HOAc

Weak Acids

71. Approximate the percentage of HOAc molecules that dissociate in 0.10 M, 0.010 M, and 0.0010 M solutions of acetic acid. What happens to the percent ionization as the solution becomes more dilute? (For HOAc, $K_a = 1.8 \times 10^{-5}$.)

72. Formic acid (HCO_2H) was first isolated by the destructive distillation of ants. In fact, the name comes from the Latin word for ants, *formi*. Approximate the HCO_2H, HCO_2^-, and H_3O^+ concentrations in a 0.100 M solution of formic acid in water. (For HCO_2H, $K_a = 1.8 \times 10^{-4}$.)

73. Hydrogen cyanide (HCN) is a gas that dissolves in water to form hydrocyanic acid. Approximate the H_3O^+, HCN, and CN^- concentrations in a 0.174 M solution of hydrocyanic acid. (For HCN, $K_a = 6 \times 10^{-10}$.)

74. The first disinfectant used by Joseph Lister was called *carbolic acid*. The substance is now known as *phenol*. Approximate the H_3O^+ ion concentration in a 0.0167 M solution of phenol (PhOH). (For PhOH, $K_a = 1.0 \times 10^{-10}$.)

75. Approximate the concentration of acetic acid that would give an H_3O^+ ion concentration of 2.0×10^{-3} M. (For HOAc, $K_a = 1.8 \times 10^{-5}$.)

76. Calculate the value of K_a for ascorbic acid (vitamin C) if 2.8% of the ascorbic acid molecules in a 0.100 M solution dissociate.

77. Calculate the value of K_a for nitrous acid (HNO_2) if a 0.100 M solution is 7.1% dissociated at equilibrium.

Bases

78. Which of the following equations correctly describes the relationship between K_b for the formate ion (HCO_2^-) and K_a for formic acid (HCO_2H)?
    (a) $K_a = K_w \times K_a$ (b) $K_b = K_a/K_w$ (c) $K_b = K_w/K_a$
    (d) $K_b = K_w + K_a$ (e) $K_b = K_w - K_a$

79. Use the relationship between K_a for an acid and K_b for its conjugate base to explain why strong acids have weak conjugate bases and weak acids have strong conjugate bases.

80. Approximate the HCO_2H, OH^-, and HCO_2^- ion concentrations in a solution that contains 0.020 mol of sodium formate ($NaHCO_2$) in 250 mL of solution. (For HCO_2H, $K_a = 1.8 \times 10^{-4}$.)

81. Approximate the OH^-, HOBr, and OBr^- ion concentrations in a solution that contains 0.050 mole of sodium hypobromite (NaOBr) in 500 mL of solution. (For HOBr, $K_a = 2.4 \times 10^{-9}$.)

82. Approximate the pH of a 0.756 M solution of NaOAc. (For HOAc, $K_a = 1.8 \times 10^{-5}$.)

83. A solution of NH_3 dissolved in water is known as both aqueous ammonia and ammonium hydroxide. Use the value of K_b for the following reaction to explain why aqueous ammonia is the better name.

$$NH_3(aq) + H_2O(l) \rightleftharpoons NH_4^+(aq) + OH^-(aq) \qquad K_b = 1.8 \times 10^{-5}$$

84. At 25°C, a 0.10 M aqueous solution of methylamine (CH_3NH_2) is 6.8% ionized.

$$CH_3NH_2(aq) + H_2O(l) \rightleftharpoons CH_3NH_3^+(aq) + OH^-(aq)$$

Calculate K_b for methylamine. Is methylamine a stronger base or a weaker base than ammonia?

85. What is the molarity of an aqueous ammonia solution that has an OH^- ion concentration of 1.0×10^{-3} M?

86. What is the pH of a 0.10 M solution of methylamine? See problem 84.

87. Calculate K_b for hydrazine (H_2NNH_2) if the pH of a 0.10 M aqueous solution of the rocket fuel is 10.54.

88. Approximate the pH of a 0.016 M aqueous solution of calcium acetate, $Ca(OAc)_2$. (For HOAc, $K_a = 1.8 \times 10^{-5}$.)

Buffers, Buffering Capacity, and pH Titration Curves

89. Explain how buffers resist changes in pH.

90. Explain why a mixture of HOAc and NaOAc is an acidic buffer, but a mixture of NH_3 and NH_4Cl is a basic buffer. (For HOAc, $K_a = 1.8 \times 10^{-5}$; for NH_4^+, $K_a = 5.6 \times 10^{-10}$.)

91. Which of the following mixtures would make the best buffer?
    (a) HCl and NaCl   (b) NaOAc and NH_3 (c) HOAc and NH_4Cl
    (d) NaOAc and NH_4Cl (e) NH_3 and NH_4Cl

92. Which of the following solutions is an acidic buffer?
    (a) 0.10 M HCl and 0.10 M NaOH
    (b) 0.10 M HCl and 0.10 M NaCl
    (c) 0.10 M HCO_2H and 0.10 M $NaHCO_2$
    (d) 0.10 M NH_3 and 0.10 M NH_4Cl

93. Which of the following solutions is a basic buffer?
    (a) 0.10 M HCl and 0.10 M NaOH
    (b) 0.10 M HCl and 0.10 M NaCl
    (c) 0.10 M HCO_2H and 0.10 M $NaHCO_2$
    (d) 0.10 M NH_3 and 0.10 M NH_4Cl

94. What is the best way to increase the capacity of a buffer made by dissolving $NaHCO_2$ in an aqueous solution of HCO_2H?
    (a) Increase the concentration of HCO_2H.
    (b) Increase the concentration of $NaHCO_2$.
    (c) Increase the concentrations of both HCO_2H and $NaHCO_2$.
    (d) Increase the ratio of the concentration of HCO_2H to the concentration of $NaHCO_2$.
    (e) Increase the ratio of the concentration of $NaHCO_2$ to the concentration of HCO_2H.

95. Determine the approximate pH of a buffer formed by a solution containing 0.10 M formic acid and 0.1 M sodium formate.

96. Determine the approximate pH of a buffer formed by a solution containing 0.10 M ammonia and 0.1 M ammonium chloride.

97. Determine the approximate pH of a buffer formed by adding 15 g of benzoic acid and 10 g of sodium benzoate to water to form 1.0 liter of solution.

98. Describe in detail the experiment you would use to measure the value of K_a for formic acid, HCO_2H.

99. Sketch a titration curve of 0.10 M NH_3 ($K_b = 1.8 \times 10^{-5}$) reacting with a 0.10 M HCl solution. Label the equivalence point, the end point, and the point at which the OH^- ion concentration is equal to K_b for the base.

Integrated Problems

100. Explain why pure (nonpolluted) rainwater has a pH of 5.6. Explain why boiling water to drive off the CO_2 raises the pH. Explain why the presence of SO_2, SO_3, NO, NO_2, and similar compounds in the atmosphere gives rise to acid rain.

101. On April 10, 1974, at Pitlochny, Scotland, the rain was found to have a pH of 2.4. Calculate the H_3O^+ ion concentration in the rain and compare it with the H_3O^+ concentration in 0.10 M acetic acid, which has a pH of 2.9.

102. Which of the following statements are true for a 1.0 M solution of the weak acid CH_3COOH in water? Explain what is wrong with any incorrect answers.
    (a) pH = 1.00
    (b) $[H_3O^+] \gg [CH_3COO^-]$
    (c) $[H_3O^+] = [CH_3COO^-]$
    (d) pH < 1.00

103. Which of the following statements are true for a 1.0 M solution of the strong acid HCl in water? Explain what is wrong with any incorrect answers.
    (a) $[Cl^-] > [H_3O^+]$
    (b) pH = 0.00
    (c) $[H_3O^+] = 1.0\ M$
    (d) $[HCl] = 1.0\ M$

104. Which of the following statements are true for a 1.0 M solution of the weak base NH_3 in water? Explain what is wrong with any incorrect answers.
    (a) $[OH^-] = [H_3O^+]$
    (b) $[NH_4^+] = [OH^-]$

(c) pH < 7.00

(d) [NH$_3$] = 1.0 M

105. The pH of a 0.10 M solution of formic acid is 2.37.

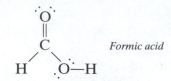

Formic acid

(a) Which of the two hydrogens in the formic acid structure is the "acidic" hydrogen? Explain.

(b) Write a chemical equation that describes what happens when formic acid is placed in water.

(c) Calculate the K_a of formic acid.

(d) Estimate the pH of a 0.10 M solution of the following acid in water. Explain your estimate.

106. Methylamine, CH$_3$NH$_2$, is a weak base with $K_b = 3.6 \times 10^{-4}$.

(a) Write a chemical equation that describes what happens when methylamine is placed into water.

(b) What is the pOH of a 0.10 M solution of methylamine?

(c) What is the pH of a 0.10 M solution of methylamine?

107. Rank the following solutions in order of increasing pH. Explain your reasoning.

(a) 0.10 M HOBrO$_2$

(b) 0.10 M HOBrO

(c) 0.10 M HBr

(d) 0.10 M NaBr

(e) 0.10 M NaBrO$_2$

(f) 0.10 M NaBrO$_3$

108. Describe a neutral (in terms of acid–base characteristics) solution. The K_w for the self-hydrolysis of H$_2$O at room temperature (25°C) is 1×10^{-14} and that at body temperature (37°C) is 2.5×10^{-14}.

(a) Describe the quantitative difference in the self-hydrolysis of pure water at 25°C and 37°C.

(b) Determine the approximate pH of pure water at 25°C and 37°C.

(c) Again describe a neutral solution. Is this the same description that you gave at the beginning of the problem?

109. Match each of the following to its corresponding particulate representation. Water molecules are not shown. Explain your reasoning.
    (a) a dilute solution of a strong acid
    (b) a concentrated solution of a weak acid
    (c) a good buffer

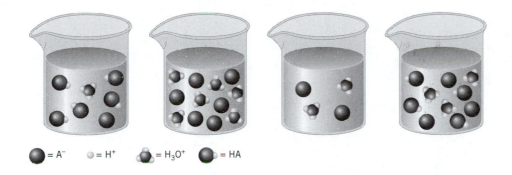

$\bullet = A^-$ $\circ = H^+$ $\unicode{x2B22} = H_3O^+$ $\unicode{x25CF}\!\!\!\circ = HA$

110. When NH_4Cl is dissolved in water the following species will be present: NH_4^+, Cl^-, H_2O, OH^-, H_3O^+, and NH_3. Using the symbols $\ll$, $<$, and $\approx$, arrange the species in order of increasing concentration in the solution.

111. The weak acid HB can be represented as $\circ\!\bullet$, where $\bullet$, represents B^- and $\circ$ represents H^+. Match the conjugate base of HB and the conjugate acid of HB with its appropriate particulate representation.
    (a) $\circ\!\bullet\!\circ$
    (b) $\bullet\!\circ\!\bullet$
    (c) $\bullet$
    (d) $\bullet\!\circ\!\bullet\!\circ$
    (e) $\circ$

C H A P T E R
11
SPECIAL TOPICS

11A.1   Diprotic Acids

11A.2   Diprotic Bases

11A.3   Triprotic Acids

11A.4   Compounds That Could Be Either Acids or Bases

11A.1   DIPROTIC ACIDS

The acid equilibrium problems discussed so far have focused on a family of compounds known as **monoprotic acids.** Each monoprotic acid has a single H^+ ion, or proton, it can donate when it acts as a Brønsted acid. Hydrochloric acid (HCl), acetic acid (CH_3CO_2H or HOAc), and nitric acid (HNO_3) are all monoprotic acids.

Several important acids can be classified as **polyprotic acids,** which can lose more than one H^+ ion when they act as Brønsted acids. **Diprotic acids,** such as sulfuric acid (H_2SO_4), carbonic acid (H_2CO_3), hydrogen sulfide (H_2S), chromic acid (H_2CrO_4), and oxalic acid ($H_2C_2O_4$), have two acidic hydrogen atoms. **Triprotic acids,** such as phosphoric acid (H_3PO_4) and citric acid ($C_6H_8O_7$), have three.

There is usually a large difference in the ease with which polyprotic acids lose the first and second (or second and third) protons. When sulfuric acid is classified as a strong acid, it is often assumed that it loses both of its protons when it reacts with water. That isn't a legitimate assumption. Sulfuric acid is a strong acid because K_a for the loss of the first proton is much larger than 1. We therefore assume that essentially all the H_2SO_4 molecules in an aqueous solution lose the first proton to form HSO_4^-, or hydrogen sulfate ion.

$$H_2SO_4(aq) + H_2O(l) \longrightarrow H_3O^+(aq) + HSO_4^-(aq \qquad K_{a1} = 1 \times 10^3$$

But K_a for the loss of the second proton is only 10^{-2}, and only 10% of the H_2SO_4 molecules in a 1 M solution lose a second proton.

$$HSO_4^-(aq) + H_2O(l) \rightleftharpoons H_3O^+(aq) + SO_4^{2-}(aq) \qquad K_{a2} = 1.2 \times 10^{-2}$$

H_2SO_4 only loses both H^+ ions when it reacts with a base stronger than water, such as ammonia.

Table 11A.1 gives values of K_a for some common polyprotic acids. The large difference between the values of K_a for the sequential loss of protons by a polyprotic acid is important because it means we can assume that the acids dissociate one step at a time, an assumption known as **stepwise dissociation.**

Let's look at the consequence of the assumption that polyprotic acids lose protons one step at a time by examining the chemistry of a saturated solution of H_2S in water. Hydrogen sulfide is the foul-smelling gas that gives rotten eggs their unpleasant odor. It is an excellent source of the S^{2-} ion, however, and is therefore commonly used in introductory

TABLE 11A.1   Acid Dissociation Equilibrium Constants for Common Polyprotic Acids

Acid	K_{a1}	K_{a2}	K_{a3}
Sulfuric acid (H_2SO_4)	1.0×10^3	1.2×10^{-2}	
Chromic acid (H_2CrO_4)	9.6	3.2×10^{-7}	
Oxalic acid ($H_2C_2O_4$)	5.4×10^{-2}	5.4×10^{-5}	
Sulfurous acid (H_2SO_3)	1.7×10^{-2}	6.4×10^{-8}	
Phosphoric acid (H_3PO_4)	7.1×10^{-3}	6.3×10^{-8}	4.2×10^{-13}
Citric acid ($C_6H_8O_7$)	7.5×10^{-4}	1.7×10^{-5}	4.0×10^{-7}
Carbonic acid (H_2CO_3)	4.5×10^{-7}	4.7×10^{-11}	
Hydrogen sulfide (H_2S)	1.0×10^{-7}	1.3×10^{-13}	

chemistry laboratories. H_2S is a weak acid that dissociates in steps. Some of the H_2S molecules lose a proton in the first step to form HS^-, or hydrogen sulfide ion.

$$\text{First step} \qquad H_2S(aq) + H_2O(l) \rightleftharpoons H_3O^+(aq) + HS^-(aq)$$

A small fraction of the HS^- ions formed in the reaction then go on to lose additional H^+ ions in a second step.

$$\text{Second step} \qquad HS^-(aq) + H_2O(l) \rightleftharpoons H_3O^+(aq) + S^{2-}(aq)$$

Since there are two steps in the reaction, we can write two equilibrium constant expressions.

$$K_{a1} = \frac{[H_3O^+][HS^-]}{[H_2S]} = 1.0 \times 10^{-7}$$

$$K_{a2} = \frac{[H_3O^+][S^{2-}]}{[HS^-]} = 1.3 \times 10^{-13}$$

Although each of the equations contains three terms, there are only four unknowns—$[H_3O^+]$, $[H_2S]$, $[HS^-]$, and $[S^{2-}]$—because the $[H_3O^+]$ and $[HS^-]$ terms appear in both equations. The $[H_3O^+]$ term represents the total H_3O^+ ion concentration from both steps and therefore must have the same value in both equations. Similarly, the $[HS^-]$ term, which represents the balance between the HS^- ions formed in the first step and the HS^- ions consumed in the second step, must have the same value for both equations.

It takes four equations to solve for four unknowns. We already have two equations: the K_{a1} and K_{a2} expressions. We are going to have to find either two more equations or a

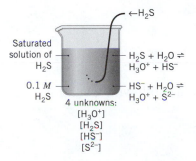

pair of assumptions that can generate two equations. We can base one assumption on the fact that the value of K_{a1} for H_2S is almost a million times larger than the value of K_{a2}.

$$K_{a1} \gg K_{a2}$$

This means that only a small fraction of the HS^- ions formed in the first step go on to dissociate in the second step. If this is true, most of the H_3O^+ ions in the solution come from the dissociation of H_2S, and most of the HS^- ions formed in this reaction remain in solution. As a result, we can assume that the H_3O^+ and HS^- ion concentrations are more or less equal.

$$\text{First assumption} \qquad [H_3O^+] \approx [HS^-]$$

We need one more equation, and therefore one more assumption. Note that H_2S is a weak acid ($K_{a1} = 1.0 \times 10^{-7}$, $K_{a2} = 1.3 \times 10^{-13}$). Thus, we can assume that most of the H_2S that dissolves in water will still be present when the solution reaches equilibrium. In other words, we can assume that the equilibrium concentration of H_2S is approximately equal to the initial concentration. Because a saturated solution of H_2S in water has an initial concentration of about 0.10 M, we can summarize the second assumption as follows.

$$\text{Second assumption} \qquad [H_2S] \approx 0.10 \ M$$

We now have four equations in four unknowns.

$$K_{a1} = \frac{[H_3O^+][HS^-]}{[H_2S]} = 1.0 \times 10^{-7}$$
$$K_{a2} = \frac{[H_3O^+][S^{2-}]}{[HS^-]} = 1.3 \times 10^{-13}$$
$$[H_3O^+] \approx [HS^-]$$
$$[H_2S] \approx 0.10 \ M$$

Since there is always a unique solution to four equations in four unknowns, we are now ready to calculate the H_3O^+, H_2S, HS^-, and S^{2-} concentrations at equilibrium in a saturated solution of H_2S in water.

Because K_{a1} is so much larger than K_{a2} for the acid, we can invoke the stepwise dissociation assumption. We assume that we can work with the equilibrium expression for the first step without worrying about the second step for the moment. We therefore start with the expression for K_{a1}.

$$K_{a1} = \frac{[H_3O^+][HS^-]}{[H_2S]} = 1.0 \times 10^{-7}$$

We then invoke one of our assumptions.

$$[H_2S] \approx 0.10 \ M$$

Substituting the approximation into the K_{a1} expression gives the following equation.

$$\frac{[H_3O^+][HS^-]}{[0.10]} \approx 1.0 \times 10^{-7}$$

We then invoke the other assumption.

$$[H_3O^+] \approx [HS^-] \approx \Delta C$$

Substituting that approximation into the K_{a1} expression gives the following result.

$$\frac{[\Delta C][\Delta C]}{[0.10]} \approx 1.0 \times 10^{-7}$$

We now solve the approximate equation for ΔC.

$$\Delta C \approx 1.0 \times 10^{-4}$$

If our assumptions are valid, we are three-fourths of the way to our goal. We know the H_2S, H_3O^+, and HS^- concentrations.

$$[H_2S] \approx 0.10 \ M$$
$$[H_3O^+] \approx [HS^-] \approx 1.0 \times 10^{-4} \ M$$

Having extracted the values of three unknowns from the first equilibrium expression, we turn to the second equilibrium expression.

$$K_{a2} = \frac{[H_3O^+][S^{2-}]}{[HS^-]} = 1.3 \times 10^{-13}$$

Substituting the known values of the H_3O^+ and HS^- ion concentrations into the expression gives the following equation.

$$\frac{[\cancel{1.0 \times 10^{-4}}][S^{2-}]}{[\cancel{1.0 \times 10^{-4}}]} = 1.3 \times 10^{-13}$$

Because the equilibrium concentrations of the H_3O^+ and HS^- ions are more or less the same, the S^{2-} ion concentration at equilibrium is approximately equal to the value of K_{a2} for this acid.

$$[S^{2-}] \approx 1.3 \times 10^{-13} \ M$$

It is now time to check our assumptions. Is the dissociation of H_2S small compared with the initial concentration? Yes. The HS^- and H_3O^+ ion concentrations obtained from the calculation are $1.0 \times 10^{-4} \ M$, which is 0.1% of the initial concentration of H_2S. The following assumption is therefore valid.

$$[H_2S] \approx 0.10 \ M$$

Is the difference between the S^{2-} and HS^- ion concentrations large enough to allow us to assume that essentially all of the H_3O^+ ions at equilibrium are formed in the first step and that essentially all of the HS^- ions formed in this step remain in solution? Yes. The S^{2-} ion concentration obtained from the calculation is 10^9 times smaller than the HS^- ion concentration. Thus, our other assumption is also valid.

$$[H_3O^+] \approx [HS^-]$$

We can therefore summarize the concentrations of the various components of the equilibrium as follows.

$$[H_2S] \approx 0.10 \ M$$
$$[H_3O^+] \approx [HS^-] \approx 1.0 \times 10^{-4} \ M$$
$$[S^{2-}] \approx 1.3 \times 10^{-13} \ M$$

11A.2 DIPROTIC BASES

The techniques we have used with diprotic acids can be extended to diprotic bases. The only challenge is calculating the values of K_b for the base. Suppose we are given the following problem:

Calculate the H_2CO_3, HCO_3^-, CO_3^{2-}, and OH^- concentrations at equilibrium in a solution that is initially 0.10 M in Na_2CO_3. (For H_2CO_3, $K_{a1} = 4.5 \times 10^{-7}$ and $K_{a2} = 4.7 \times 10^{-11}$.)

Sodium carbonate dissociates into its ions when it dissolves in water.

$$Na_2CO_3(aq) \xrightarrow{H_2O} 2 \ Na^+(aq) + CO_3^{2-}(aq)$$

The carbonate ion then acts as a base toward water, picking up a pair of protons (one at a time) to form the bicarbonate ion, HCO_3^-, and then eventually carbonic acid, H_2CO_3.

$$CO_3^{2-}(aq) + H_2O(l) \rightleftharpoons HCO_3^-(aq) + OH^-(aq) \qquad K_{b1} = ?$$
$$HCO_3^-(aq) + H_2O(l) \rightleftharpoons H_2CO_3(aq) + OH^-(aq) \qquad K_{b2} = ?$$

The first step in solving the problem involves determining the values of K_{b1} and K_{b2} for the carbonate ion. We start by comparing the K_b expressions for the carbonate ion with the K_a expressions for carbonic acid.

$$K_{b1} = \frac{[HCO_3^-][OH^-]}{[CO_3^{2-}]} \qquad K_{a2} = \frac{[H_3O^+][CO_3^{2-}]}{[HCO_3^-]}$$

$$K_{b2} = \frac{[H_2CO_3][OH^-]}{[HCO_3^-]} \qquad K_{a1} = \frac{[H_3O^+][HCO_3^-]}{[H_2CO_3]}$$

The expressions for K_{b1} and K_{a2} have something in common: they both depend on the concentrations of the HCO_3^- and CO_3^{2-} ions. The expressions for K_{b2} and K_{a1} also have something in common: they both depend on the HCO_3^- and H_2CO_3 concentrations. We can therefore calculate K_{b1} from K_{a2} and K_{b2} from K_{a1}.

We start by multiplying the top and bottom of the K_{a1} expression by the OH^- ion concentration to introduce the $[OH^-]$ term.

$$K_{a1} = \frac{[H_3O^+][HCO_3^-]}{[H_2CO_3]} \times \frac{[OH^-]}{[OH^-]}$$

We then group terms in the equation as follows.

$$K_{a1} = \frac{[HCO_3^-]}{[H_2CO_3][OH^-]} \times [H_3O^+][OH^-]$$

The first term in the equation is the inverse of the K_{b2} expression, and the second term is the K_w expression.

$$K_{a1} = \frac{1}{K_{b2}} \times K_w$$

Rearranging the equation gives the following result.

$$K_{a1}K_{b2} = K_w$$

Similarly, we can multiply the top and bottom of the K_{a2} expression by the OH^- ion concentration.

$$K_{a2} = \frac{[H_3O^+][CO_3^{2-}]}{[HCO_3^-]} \times \frac{[OH^-]}{[OH^-]}$$

Collecting terms gives the following equation.

$$K_{a2} = \frac{[CO_3^{2-}]}{[HCO_3^-][OH^-]} \times [H_3O^+][OH^-]$$

The first term in the equation is the inverse of K_{b1}, and the second term is K_w.

$$K_{a2} = \frac{1}{K_{b1}} \times K_w$$

The equation can therefore be rearranged as follows.

$$K_{a2}K_{b1} = K_w$$

We can now calculate the values of K_{b1} and K_{b2} for the carbonate ion.

$$K_{b1} = \frac{K_w}{K_{a2}} = \frac{1.0 \times 10^{-14}}{4.7 \times 10^{-11}} = 2.1 \times 10^{-4}$$

$$K_{b2} = \frac{K_w}{K_{a1}} = \frac{1.0 \times 10^{-14}}{4.5 \times 10^{-7}} = 2.2 \times 10^{-8}$$

We are finally ready to do the calculations. We start with the K_{b1} expression because the CO_3^{2-} ion is the strongest base in the solution and is therefore the best source of the OH^- ion.

$$K_{b1} = \frac{[HCO_3^-][OH^-]}{[CO_3^{2-}]}$$

The difference between K_{b1} and K_{b2} for the carbonate ion is large enough to suggest that most of the OH^- ions come from the first step and most of the HCO_3^- ions formed in the reaction remain in solution.

$$[OH^-] \approx [HCO_3^-] \approx \Delta C$$

The value of K_{b1} is small enough to assume that ΔC is small compared with the initial concentration of the carbonate ion. If this is true, the concentration of the CO_3^{2-} ion at equilibrium will be roughly equal to the initial concentration of Na_2CO_3, which was described in the statement of the problem as 0.10 M.

$$[CO_3^{2-}] \approx 0.10 \; M$$

Substituting this information into the K_{b1} expression gives the following result.

$$\frac{[\Delta C][\Delta C]}{[0.10]} \approx 2.1 \times 10^{-4}$$

The approximate equation can now be solved for ΔC.

$$\Delta C \approx 0.0046 \; M$$

We then use the value of ΔC to calculate the equilibrium concentrations of the OH^-, HCO_3^-, and CO_3^{2-} ions.

$$[CO_3^{2-}] \approx 0.10 \; M$$
$$[OH^-] \approx [HCO_3^-] \approx \Delta C \approx 0.0046 \; M$$

We now turn to the K_{b2} expression.

$$K_{b2} = \frac{[H_2CO_3][OH^-]}{[HCO_3^-]}$$

Substituting what we know about the OH^- and HCO_3^- ion concentrations into the equation gives the following result.

$$\frac{[H_2CO_3][0.0046]}{[0.0046]} \approx 2.2 \times 10^{-8}$$

According to this equation, the H_2CO_3 concentration at equilibrium is approximately equal to K_{b2} for the carbonate ion.

$$[H_2CO_3] \approx 2.2 \times 10^{-8} \; M$$

Summarizing the results of our calculations allows us to test the assumptions made generating the results.

$$[CO_3^{2-}] \approx 0.10 \ M$$
$$[OH^-] \approx [HCO_3^-] \approx 4.6 \times 10^{-3} \ M$$
$$[H_2CO_3] \approx 2.2 \times 10^{-8} \ M$$

All of our assumptions are valid. The extent of the reaction between the CO_3^{2-} ion and water to give the HCO_3^- ion is less than 5% of the initial concentration of Na_2CO_3. Furthermore, most of the OH^- ion comes from the first step, and most of the HCO_3^- ion formed in the first step remains in solution.

11A.3   TRIPROTIC ACIDS

Our techniques for working diprotic acid and diprotic base equilibrium problems can be applied to triprotic acids and bases as well. To illustrate this, let's calculate the H_3O^+, H_3PO_4, $H_2PO_4^-$, HPO_4^{2-}, and PO_4^{3-} concentrations at equilibrium in a 0.100 M H_3PO_4 solution, for which $K_{a1} = 7.1 \times 10^{-3}$, $K_{a2} = 6.3 \times 10^{-8}$, and $K_{a3} = 4.2 \times 10^{-13}$.

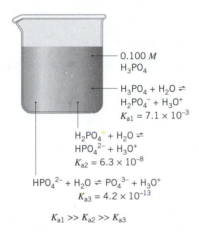

0.100 M H_3PO_4

$H_3PO_4 + H_2O \rightleftharpoons$ $H_2PO_4^- + H_3O^+$
$K_{a1} = 7.1 \times 10^{-3}$

$H_2PO_4^- + H_2O \rightleftharpoons$ $HPO_4^{2-} + H_3O^+$
$K_{a2} = 6.3 \times 10^{-8}$

$HPO_4^{2-} + H_2O \rightleftharpoons PO_4^{3-} + H_3O^+$
$K_{a3} = 4.2 \times 10^{-13}$

$K_{a1} \gg K_{a2} \gg K_{a3}$

Let's assume that the acid dissociates by steps and analyze the first step—the most extensive reaction.

$$H_3PO_4(aq) + H_2O(l) \xrightleftharpoons{K_{a1}} H_2PO_4^-(aq) + H_3O^+(aq)$$
$$K_{a1} = \frac{[H_3O^+][H_2PO_4^-]}{[H_3PO_4]} = 7.1 \times 10^{-3}$$

We now assume that the difference between K_{a1} and K_{a2} is large enough that most of the H_3O^+ ions come from the first step and most of the $H_2PO_4^-$ ions formed in the step remain in solution.

$$[H_3O^+] \approx [H_2PO_4^-] \approx \Delta C$$

Substituting the assumption into the K_{a1} expression gives the following equation.

$$\frac{[\Delta C][\Delta C]}{[0.100 - \Delta C]} = 7.1 \times 10^{-3}$$

The assumption that ΔC is small compared with the initial concentration of the acid fails in this problem. But we don't really need the assumption because we can use the quadratic formula to solve the equation.

$$\Delta C = 0.023 \ M$$

We can then use the value of ΔC to obtain the following information.

$$[H_3PO_4] \approx 0.100 - \Delta C \approx 0.077 \ M$$
$$[H_3O^+] \approx [H_2PO_4^-] \approx 0.023 \ M$$

We now turn to the second strongest acid in the solution.

$$K_{a2} = \frac{[H_3O^+][HPO_4^{2-}]}{[H_2PO_4^-]} = 6.3 \times 10^{-8}$$

Substituting what we know about the H_3O^+ and $H_2PO_4^-$ ion concentrations into the expression gives the following equation.

$$\frac{[\cancel{0.023}][HPO_4^{2-}]}{[\cancel{0.023}]} = 6.3 \times 10^{-8}$$

If our assumptions so far are correct, the HPO_4^{2-} ion concentration in the solution is equal to K_{a2}.

$$[HPO_4^{2-}] \approx 6.3 \times 10^{-8}$$

We have only one more equation, the equilibrium expression for the weakest acid in the solution.

$$K_{a3} = \frac{[H_3O^+][PO_4^{3-}]}{[HPO_4^{2-}]} = 4.2 \times 10^{-13}$$

Substituting what we know about the concentrations of the H_3O^+ and HPO_4^{2-} ions into the expression gives the following equation.

$$\frac{[0.023][PO_4^{3-}]}{[6.3 \times 10^{-8}]} \approx 4.2 \times 10^{-13}$$

The equation can be solved for the phosphate ion concentration at equilibrium.

$$[PO_4^{3-}] \approx 1.2 \times 10^{-18} \ M$$

Summarizing the results of the calculations helps us check the assumptions made along the way.

$$[H_3PO_4] \approx 0.077 \ M$$
$$[H_3O^+] \approx [H_2PO_4^-] \approx 0.023 \ M$$
$$[HPO_4^{2-}] \approx 6.3 \times 10^{-8} \ M$$
$$[PO_4^{3-}] \approx 1.2 \times 10^{-18} \ M$$

The only approximation used in working the problem was the assumption that the acid dissociates one step at a time. Is the difference between the concentrations of the $H_2PO_4^-$ and HPO_4^{2-} ions large enough to justify the assumption that essentially all of the H_3O^+ ions come from the first step? Yes. Is it large enough to justify the assumption that essentially all of the $H_2PO_4^-$ formed in the first step remains in solution? Yes.

You may never encounter an example of a polyprotic acid for which the difference between successive values of K_a are too small to allow us to assume stepwise dissociation. The assumption works even when we might expect it to fail.

Exercise 11A.1

Calculate the H_3O^+, H_3Cit, H_2Cit^-, $HCit^{2-}$, and Cit^{3-} concentrations in a 1.00 M solution of citric acid. (For H_3Cit, $K_{a1} = 7.5 \times 10^{-4}$, $K_{a2} = 1.7 \times 10^{-5}$, and $K_{a3} = 4.0 \times 10^{-7}$.)

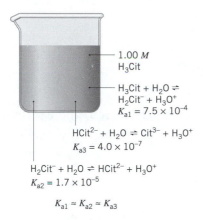

1.00 M
H_3Cit

$H_3Cit + H_2O \rightleftharpoons$
$H_2Cit^- + H_3O^+$
$K_{a1} = 7.5 \times 10^{-4}$

$HCit^{2-} + H_2O \rightleftharpoons Cit^{3-} + H_3O^+$
$K_{a3} = 4.0 \times 10^{-7}$

$H_2Cit^- + H_2O \rightleftharpoons HCit^{2-} + H_3O^+$
$K_{a2} = 1.7 \times 10^{-5}$

$K_{a1} \approx K_{a2} \approx K_{a3}$

Solution

At first glance, K_{a1}, K_{a2}, and K_{a3} for the acid might seem to be too close to each other to assume stepwise dissociation. The only way to tell for sure is to try the assumption and see whether it works. We start with the expression for the first step.

$$K_{a1} = \frac{[H_3O^+][H_2Cit^-]}{[H_3Cit]} = 7.5 \times 10^{-4}$$

Even though we might be suspicious of its validity, let's assume that most of the H_3O^+ ions at equilibrium come from the citric acid. Let's also assume that most of the H_2Cit^- ions formed in the first step remain in solution.

If the assumptions are valid, the concentrations of the H_3O^+ and H_2Cit^- ions at equilibrium are equal.

$$[H_3O^+] \approx [H_2Cit^-] \approx \Delta C$$

Substituting the assumption into the K_{a1} expression gives the following equation.

$$\frac{[\Delta C][\Delta C]}{[H_3Cit]} \approx 7.5 \times 10^{-4}$$

Because citric acid is a weak acid, we can assume that the equilibrium concentration of the undissociated H_3Cit molecules is roughly equal to the initial concentration of the acid.

$$[H_3Cit] \approx 1.00 \ M$$

Substituting this assumption into the K_{a1} expression gives the following result.

$$\frac{[\Delta C][\Delta C]}{[1.00]} \approx 7.5 \times 10^{-4}$$

We can now solve the approximate equation for ΔC.

$$\Delta C \approx 0.027 \ M$$

The value of ΔC can now be used to calculate the following concentrations.

$$[H_3Cit] \approx 1.00 \ M$$
$$[H_3O^+] \approx [H_2Cit^-] \approx 0.027 \ M$$

We can now turn to the K_{a2} expression.

$$K_{a2} = \frac{[H_3O^+][HCit^{2-}]}{[H_2Cit^-]} = 1.7 \times 10^{-5}$$

Substituting what we know about the H_3O^+ and H_2Cit^- concentrations into the expression gives the following equation.

$$\frac{[0.027][HCit^{2-}]}{[0.027]} \approx 1.7 \times 10^{-5}$$

If our assumptions are valid, the $HCit^{2-}$ ion concentration is equal to K_{a2} for the acid.

$$[HCit^{2-}] \approx 1.7 \times 10^{-5} \ M$$

We now turn to the K_{a3} expression.

$$K_{a3} = \frac{[H_3O^+][Cit^{3-}]}{[HCit^{2-}]} = 4.0 \times 10^{-7}$$

Substituting the approximate values for the H_3O^+ and $HCit^{2-}$ ion concentrations into the expression gives the following equation.

$$\frac{[0.027][Cit^{3-}]}{[1.7 \times 10^{-5}]} \approx 4.0 \times 10^{-7}$$

We now solve for the Cit^{3-} ion concentration.

$$[Cit^{3-}] \approx 2.5 \times 10^{-10} \ M$$

Summarizing the results of the calculation allows us to test the assumptions made in deriving them.

$$[H_3Cit] \approx 1.00 \ M$$
$$[H_3O^+] \approx [H_2Cit^-] \approx 0.027 \ M$$
$$[HCit^{2-}] \approx 1.7 \times 10^{-5} \ M$$
$$[Cit^{3-}] \approx 2.5 \times 10^{-10} \ M$$

Is it legitimate to assume that the amount of H_3Cit that dissociates in the first step is small compared with the initial concentration of the acid? Yes. Is it legitimate to assume that most of the H_3O^+ ion comes from the first step? Yes. Is it legitimate to assume that most of the H_2Cit^- ion formed in the first step remains in solution? Yes. Since all of the assumptions made in the calculation are valid, the results are also valid.

11A.4   COMPOUNDS THAT COULD BE EITHER ACIDS OR BASES

Sometimes the hardest part of a calculation is deciding whether the compound is an acid or a base. Consider sodium bicarbonate, for example, which dissolves in water to give the bicarbonate ion.

$$NaHCO_3(s) \xrightarrow{H_2O} Na^+(aq) + HCO_3^-(aq)$$

In theory, the bicarbonate ion can act as either a Brønsted acid or a Brønsted base toward water.

$$HCO_3^-(aq) + H_2O(l) \rightleftharpoons H_3O^+(aq) + CO_3^{2-}(aq)$$
$$HCO_3^-(aq) + H_2O(l) \rightleftharpoons H_2CO_3(aq) + OH^-(aq)$$

Which reaction predominates? Is the HCO_3^- ion more likely to act as an acid or as a base?

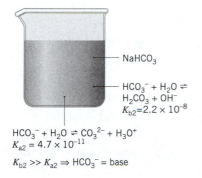

NaHCO₃

$HCO_3^- + H_2O \rightleftharpoons$
$H_2CO_3 + OH^-$
$K_{b2} = 2.2 \times 10^{-8}$

$HCO_3^- + H_2O \rightleftharpoons CO_3^{2-} + H_3O^+$
$K_{a2} = 4.7 \times 10^{-11}$

$K_{b2} \gg K_{a2} \Rightarrow HCO_3^- = base$

We can answer those questions by comparing the equilibrium constants for the reactions. The equilibrium in which the HCO_3^- ion acts as a Brønsted acid is described by K_{a2} for carbonic acid.

$$K_{a2} = \frac{[H_3O^+][CO_3^{2-}]}{[HCO_3^-]} = 4.7 \times 10^{-11}$$

The equilibrium in which the HCO_3^- acts as a Brønsted base is described by K_{b2} for the carbonate ion.

$$K_{b2} = \frac{[H_2CO_3][OH^-]}{[HCO_3^-]} = 2.2 \times 10^{-8}$$

Since K_{b2} is significantly larger than K_{a2}, the HCO_3^- ion is a stronger base than it is an acid.

Exercise 11A.2

Phosphoric acid, H_3PO_4, is obviously an acid. The phosphate ion, PO_4^{3-}, is a base. Predict whether solutions of the $H_2PO_4^-$ and HPO_4^{2-} ions are more likely to be acidic or basic.

Solution

Let's start by looking at the stepwise dissociation of phosphoric acid.

$$H_3PO_4(aq) + H_2O(l) \rightleftharpoons H_3O^+(aq) + H_2PO_4^-(aq) \qquad K_{a1} = 7.1 \times 10^{-3}$$
$$H_2PO_4^-(aq) + H_2O(l) \rightleftharpoons H_3O^+(aq) + HPO_4^{2-}(aq) \qquad K_{a2} = 6.3 \times 10^{-8}$$
$$HPO_4^{2-}(aq) + H_2O(l) \rightleftharpoons H_3O^+(aq) + PO_4^{3-}(aq) \qquad K_{a3} = 4.2 \times 10^{-13}$$

We can then look at the steps by which the PO_4^{3-} ion picks up protons from water to form phosphoric acid.

$$PO_4^{3-}(aq) + H_2O(l) \rightleftharpoons HPO_4^{2-}(aq) + OH^-(aq) \qquad K_{b1} = ?$$
$$HPO_4^{2-}(aq) + H_2O(l) \rightleftharpoons H_2PO_4^-(aq) + OH^-(aq) \qquad K_{b2} = ?$$
$$H_2PO_4^-(aq) + H_2O(l) \rightleftharpoons H_3PO_4(aq) + OH^-(aq) \qquad K_{b3} = ?$$

Applying the procedure used for sodium carbonate in Section 11A.2 gives the following relationships among the values of the six equilibrium constants.

$$K_{a1}K_{b3} = K_w$$
$$K_{a2}K_{b2} = K_w$$
$$K_{a3}K_{b1} = K_w$$

We can predict whether the $H_2PO_4^-$ ion should be an acid or a base by looking at the values of the K_a and K_b constants that characterize its reactions with water.

$$H_2PO_4^-(aq) + H_2O(l) \rightleftharpoons H_3O^+(aq) + HPO_4^{2-}(aq) \qquad K_{a2} = 6.3 \times 10^{-8}$$
$$H_2PO_4^-(aq) + H_2O(l) \rightleftharpoons H_3PO_4(aq) + OH^-(aq) \qquad K_{b3} = 1.4 \times 10^{-12}$$

Because K_{a2} is significantly larger than K_{b3}, solutions of the $H_2PO_4^-$ ion in water should be acidic.

We can use the same method to decide whether HPO_4^{2-} is an acid or a base when dissolved in water. We start by looking at the values of K_a and K_b that characterize its reactions with water.

$$HPO_4^{2-}(aq) + H_2O(l) \rightleftharpoons H_3O^+(aq) + PO_4^{3-}(aq) \qquad K_{a3} = 4.2 \times 10^{-13}$$
$$HPO_4^{2-}(aq) + H_2O(l) \rightleftharpoons H_2PO_4^-(aq) + OH^-(aq) \qquad K_{b2} = 1.6 \times 10^{-7}$$

Because K_{b2} is significantly larger than K_{a3}, solutions of the HPO_4^{2-} ion in water should be basic.

Measurements of the pH of 0.10 M solutions of the species yield the following results, which are consistent with the predictions of this exercise.

H_3PO_4	$H_2PO_4^-$	HPO_4^{2-}	PO_4^{3-}
pH = 1.5	pH = 4.4	pH = 9.3	pH = 12.0

PROBLEMS

Polyprotic Acids

11A.1. Calculate the H_3O^+, CO_3^{2-}, HCO_3^-, and H_2CO_3 concentrations at equilibrium in a solution that initially contained 0.100 mol of carbonic acid per liter. (For H_2CO_3, $K_{a1} = 4.5 \times 10^{-7}$ and $K_{a2} = 4.7 \times 10^{-11}$.)

11A.2. Calculate the equilibrium concentrations of the important components of a 0.250 M malonic acid ($HO_2CCH_2CO_2H$) solution. Use the symbol H_2M as an abbreviation for malonic acid and assume stepwise dissociation of the acid. (For H_2M, $K_{a1} = 1.4 \times 10^{-5}$ and $K_{a2} = 2.1 \times 10^{-8}$.)

11A.3. Check the validity of the assumption in the previous problem that malonic acid dissociates in a stepwise fashion by comparing the concentrations of the HM^- and M^{2-} ions obtained in the problem. Is this assumption valid?

11A.4. Glycine, the simplest of the amino acids found in proteins, is a diprotic acid with the formula $HO_2CCH_2NH_3^+$. If we symbolize glycine as H_2G^+, we can write the following equations for its stepwise dissociation.

$$H_2G^+(aq) + H_2O(l) \rightleftharpoons HG(aq) + H_3O^+(aq) \qquad K_{a1} = 4.5 \times 10^{-3}$$
$$HG(aq) + H_2O(l) \rightleftharpoons G^-(aq) + H_3O^+(aq) \qquad K_{a2} = 2.5 \times 10^{-10}$$

Calculate the concentrations of H_3O^+, G^-, HG, and H_2G^+ in a 2.00 M solution of glycine in water.

11A.5. Which of the following equations accurately describes a 2.0 M solution of glycine?
(a) $[H_3O^+] \approx [H_2G^+]$ (b) $[H_3O^+] < [H_2G^+]$ (c) $[H_3O^+] > [HG]$
(d) $[H_3O^+] \approx [HG]$ (e) $[H_3O^+] < [HG]$

11A.6. Which of the following equations results from the fact that glycine is a weak diprotic acid?
(a) $[G^-] \approx [H_2G^+]$ (b) $[G^-] \approx K_{a2}$ (c) $[G^-] \approx [HG]$
(d) $[G^-] > [HG]$ (e) $[G^-] \approx [H_3O^+]$

11A.7. Oxalic acid ($H_2C_2O_4$) has been implicated in diseases such as gout and kidney stones. Calculate the H_3O^+, $H_2C_2O_4$, $HC_2O_4^-$, and $C_2O_4^{2-}$ concentrations in a 1.25 M solution of oxalic acid.

$$H_2C_2O_4(aq) + H_2O(l) \rightleftharpoons H_3O^+(aq) + HC_2O_4^-(aq) \qquad K_{a1} = 5.4 \times 10^{-2}$$
$$HC_2O_4^-(aq) + H_2O(l) \rightleftharpoons H_3O^+(aq) + C_2O_4^{2-}(aq) \qquad K_{a2} = 5.4 \times 10^{-5}$$

Polyprotic Bases

11A.8. Which of the following sets of equations can be used to calculate K_{b1} and K_{b2} for sodium oxalate ($Na_2C_2O_4$) from K_{a1} and K_{a2} for oxalic acid ($H_2C_2O_4$)?
(a) $K_{b1} = K_w \times K_{a1}$ and $K_{b2} = K_w \times K_{a2}$
(b) $K_{b1} = K_w \times K_{a2}$ and $K_{b2} = K_w \times K_{a1}$
(c) $K_{b1} = K_w/K_{a1}$ and $K_{b2} = K_w/K_{a2}$
(d) $K_{b1} = K_w/K_{a2}$ and $K_{b2} = K_w/K_{a1}$
(e) $K_{b1} = K_{a1}/K_w$ and $K_{b2} = K_{a2}/K_w$

11A.9. Calculate the pH of a 0.028 M solution of sodium oxalate ($Na_2C_2O_4$) in water.

$$H_2C_2O_4(aq) + H_2O(l) \rightleftharpoons H_3O^+(aq) + HC_2O_4^-(aq) \qquad K_{a1} = 5.4 \times 10^{-2}$$
$$HC_2O_4^-(aq) + H_2O(l) \rightleftharpoons H_3O^+(aq) + C_2O_4^{2-}(aq) \qquad K_{a2} = 5.4 \times 10^{-5}$$

11A.10. Calculate the H_3O^+, OH^-, H_2CO_3, HCO_3^-, and CO_3^{2-} concentrations in a 0.150 M solution of sodium carbonate (Na_2CO_3) in water. (For H_2CO_3, $K_{a1} = 4.5 \times 10^{-7}$ and $K_{a2} = 4.7 \times 10^{-11}$.)

Compounds That Could Be Either Acids or Bases

11A.11. According to the data in Table 11.4, the pH of a 0.10 M $NaHCO_3$ solution is 8.4. Explain why the HCO_3^- ion forms solutions that are basic, not acidic.

11A.12. Which of the following statements concerning sodium hydrogen sulfide ($NaSH$) is correct? (For H_2S, $K_{a1} = 1.0 \times 10^{-7}$ and $K_{a2} = 1.3 \times 10^{-13}$.)
   (a) NaSH is an acid because K_{a1} for H_2S is much larger than K_{a2}.
   (b) NaSH is an acid because K_{a1} for H_2S is smaller than K_{b1} for Na_2S.
   (c) NaSH is a base because K_{b1} for Na_2S is much larger than K_{b2}.
   (d) NaSH is a base because K_{b2} for Na_2S is larger than K_{a2} for H_2S.

11A.13. Calculate the pH of 0.100 M solutions of H_3PO_4, NaH_2PO_4, Na_2HPO_4, and Na_3PO_4. (For H_3PO_4, $K_{a1} = 7.1 \times 10^{-3}$, $K_{a2} = 6.3 \times 10^{-8}$, and $K_{a3} = 4.2 \times 10^{-13}$.)

11A.14. Predict whether an aqueous solution of sodium hydrogen sulfate ($NaHSO_4$) should be acidic, basic, or neutral. (For H_2SO_4, $K_{a1} = 1 \times 10^3$ and $K_{a2} = 1.2 \times 10^{-2}$.)

11A.15. Predict whether an aqueous solution of sodium hydrogen sulfite ($NaHSO_3$) should be acidic, basic, or neutral. (For H_2SO_3, $K_{a1} = 1.7 \times 10^{-2}$ and $K_{a2} = 6.4 \times 10^{-8}$.)

12

OXIDATION–REDUCTION REACTIONS

12.1 COMMON OXIDATION–REDUCTION REACTIONS

Oxidation–reduction or **redox reactions** are the most common type of chemical reactions encountered on a daily basis. Most of the reactions used as sources of either heat or work are redox reactions. All combustion reactions are oxidation–reduction reactions. Combustion reactions include burning a log in the fireplace, burning gasoline in a car's engine, and burning natural gas or fuel oil in a home furnace. When one mole of natural gas (methane, CH_4) burns an oxidation–reduction reaction occurs that releases more than 800 kJ of energy.

$$CH_4(g) + 2\ O_2(g) \longrightarrow CO_2(g) + 2\ H_2O(g)$$

Corrosion is an oxidation–reduction reaction that involves the deterioration of metals resulting from reactions with chemical substances in the environment. The most common example of corrosion is the rusting of iron metal. Iron in the presence of both oxygen and water is oxidized to give a hydrated form of iron(II) oxide.

$$2\ Fe(s) + O_2(aq) + 2\ H_2O(l) \longrightarrow 2\ FeO{\cdot}H_2O(s)$$

The $FeO{\cdot}H_2O$ formed in the reaction is further oxidized by O_2 dissolved in water to give a hydrated form of iron(III) or ferric oxide.

$$4\ FeO{\cdot}H_2O(s) + O_2(aq) + 2\ H_2O(l) \longrightarrow 2\ Fe_2O_3{\cdot}3\ H_2O(s)$$

Because the reactions occur only in the presence of both water and oxygen, cars tend to rust where water collects. It is estimated that the United States spends $10 billion per year to protect metal from corrosion (painting) and to replace metal that has corroded.

The most important oxidation–reduction reactions occur in biological systems. Green plants convert carbon dioxide and water into carbohydrates and oxygen by a series of oxidation–reduction reactions called **photosynthesis.**

$$6 \text{ CO}_2(g) + 6 \text{ H}_2\text{O}(l) \longrightarrow \text{C}_6\text{H}_{12}\text{O}_6(aq) + 6 \text{ O}_2(g)$$

In animals the reverse process takes place through a sequence of redox reactions that produce energy on the cellular level.

$$\text{C}_6\text{H}_{12}\text{O}_6(aq) + 6 \text{ O}_2(g) \longrightarrow 6 \text{ CO}_2(g) + 6 \text{ H}_2\text{O}(l)$$

Additional common oxidation–reduction reactions include the production of electrical energy with batteries, the removal of stains from clothes with bleach, and photography.

Oxidation–reduction reactions of ionic compounds involve the transfer of electrons. The element or compound that gains electrons is said to undergo **reduction.** The element or compound that loses electrons undergoes **oxidation.** The reaction below takes place when magnesium is burned in air to produce the brilliant light associated with emergency flares. The magnesium metal loses electrons, which are accepted by the oxygen to form the ionic compound magnesium oxide.

$$2 \text{ Mg} + \text{O}_2 \longrightarrow 2 \text{ [Mg}^{2+}\text{][O}^{2-}\text{]}$$

Oxidation
Reduction

Redox reactions are often studied by considering the processes of oxidation and reduction separately. The oxidation or reduction reactions written by themselves are referred to as **half reactions.** In actual chemical reactions reduction and oxidation must always accompany one another. The electrons that are gained by the substance being reduced are lost by the substance being oxidized. The number of electrons gained and lost in a chemical reaction must be equal because no electrons can be created or destroyed during a reaction. In the above reaction the 4 electrons lost by the two magnesium atoms are gained by the oxygen molecule. The reduction and oxidation half-reactions for the above redox reactions are shown below.

Oxidation half-reaction $\quad 2 \text{ Mg} \rightleftharpoons 2 \text{ Mg}^{2+} + 4 \text{ } e^-$
Reduction half-reaction $\quad \text{O}_2 + 4 \text{ } e^- \rightleftharpoons 2 \text{ O}^{2-}$

The above definitions of oxidation and reduction do not accurately characterize redox reactions involving covalent molecules in which there is only a partial transfer of electron density. An example of such a reaction is given below.

$$\text{H}_2(g) + \text{Cl}_2(g) \rightleftharpoons 2 \text{ HCl}(g)$$

$$\text{H} \colon \text{H} + \colon \ddot{\text{Cl}} \colon \ddot{\text{Cl}} \colon \rightleftharpoons 2 \text{ }^{\delta+}\text{H} \colon \ddot{\text{Cl}} \colon^{\delta-}$$

In the reaction of hydrogen and chlorine to form hydrogen chloride, the reactants have pure covalent bonds in which a pair of electrons are equally shared. The product, HCl, has a polar covalent bond in which electron density is drawn away from the hydrogen and toward the more electronegative chlorine. Therefore, as the reaction progresses chlorine atoms gain electron density and are therefore reduced. However, each chlorine atom does *not* gain a full electron from the hydrogen to form H^+ and Cl^-. Chemists therefore developed the concept of **oxidation numbers** to describe oxidation–reduction reactions in which less than full electrons are actually gained or lost.

Checkpoint

Is it possible to have oxidation without reduction?

12.2   DETERMINING OXIDATION NUMBERS

When assigning oxidation numbers to atoms in a compound, the compound is treated as if it were ionic, and the shared electrons in each bond are assigned to the more electronegative element. Frequently chemists refer to an atom that has been assigned an oxidation number as having a particular **oxidation state.** Two methods for assigning oxidation numbers of atoms were introduced in Chapter 5. In Section 5.14 Lewis structures were used in which all bonding electrons between nonequivalent atoms were assigned to the more electronegative atom. Oxidation numbers were then assigned to each atom in the structure using the equation:

$$OX_a = V_a - Y_a$$

where OX_a is the oxidation number of atom a, V_a is the number of valence electrons of the atom, and Y_a is the number of electrons assigned to the atom in the Lewis structure after assigning bonding electrons to the more electronegative atom. This method of assigning oxidation numbers is particularly useful for organic compounds (compounds containing primarily C and H) in which carbon atoms have several different oxidation numbers. For example, 1-propanol ($CH_3CH_2CH_2OH$) has three carbon atoms, each assigned a different oxidation number.

$$
\begin{array}{ccccc}
 & H & H & H & \\
 & | & | & | & \\
H- & C- & C- & C- & O-H \\
 & | & | & | & \\
 & H & H & H &
\end{array}
$$

Assigning bonding electrons to the most electronegative element in each bond yields

$$
\begin{array}{ccccc}
H & H & H & & \\
\cdot\cdot & \cdot\cdot & \cdot\cdot & & \cdot\cdot \\
H & :C_a\cdot & \cdot C_b\cdot & \cdot C_c & :O: \quad H \\
 & H & H & H & \cdot\cdot
\end{array}
$$

The oxidation state of each of the carbons can now be determined:

$$OX_a = V_a - Y_a$$
$$OX_{Ca} = 4 - 7 = -3$$
$$OX_{Cb} = 4 - 6 = -2$$
$$OX_{Cc} = 4 - 5 = -1$$

Carbon has low (more negative) oxidation numbers when bonded to hydrogen. As the number of bonds between carbon and hydrogen decrease, as in the case of C_a (3 bonds to H) and C_b (2 bonds to H), the oxidation number increases ($C_a = -3$ and $C_b = -2$). In addition, an increase in the number of bonds to an electronegative element such as oxygen increases the oxidation number of carbon ($C_c = -1$).

A second method for assigning oxidation numbers was introduced in Section 5.13. This series of rules for assigning oxidation numbers allow chemists to treat covalent molecules as though complete electrons are transferred in a redox reaction even though we realize this is not the case. The free atoms are used as a reference for measuring the extent of oxidation or reduction and are therefore assigned an oxidation number of zero. If an atom loses electrons to become a positively charged cation, it is assigned an oxidation number equal to the charge on the cation (the number of electrons lost). Similarly, when electrons are gained by an atom to form an anion, the atom is assigned an oxidation number equal to the charge (the number of electrons gained).

When assigning oxidation numbers to inorganic species it is often easier to apply the rules than to draw the Lewis structure and do the calculation. For example, we can use the rules to find the oxidation number of nitrogen in the nitrate anion. In electrically neutral molecules or polyatomic ions the sum of the oxidation numbers of the individual atoms must be equal to the overall charge on the molecule or ion. In the polyatomic nitrate ion, NO_3^-, oxygen is assigned its typical oxidation number of -2. The oxidation state of nitrogen can be determined by adding the oxidation numbers of the individual atoms to equal the charge on the nitrate ion.

$$OX_N + 3(OX_O) = -1$$
$$OX_N + 3(-2) = -1$$
$$OX_N = +5$$

Checkpoint

What are the oxidation numbers for all the atoms in the following compounds?

$FeCl_3$, CH_4, O_2, MnO_4^-, $\underset{\underset{O}{\|}}{HCNH_2}$

12.3   RECOGNIZING OXIDATION–REDUCTION REACTIONS

The most powerful model of oxidation–reduction reactions is based on the following definitions:

Oxidation occurs when the oxidation number of an atom increases.
Reduction occurs when the oxidation number of an atom decreases.

Using this model of oxidation–reduction we can describe a reaction such as the following in which electrons are transferred in the formation of an ionic compound. The oxidation number for each atom is given directly beneath the atom.

$$2\,Mg + O_2 \longrightarrow 2\,[Mg^{2+}][O^{2-}]$$

Magnesium's oxidation number increases from zero to $+2$ indicating that magnesium is oxidized. Oxygen's oxidation number decreases from zero to -2 indicating that it is reduced.

This model for redox reactions is particularly useful for describing reactions such as the following in which there is a transfer of electron density but not a full electron. The oxidation numbers are given beneath each atom.

$$H_2(g) + Cl_2(g) \rightleftharpoons 2\,HCl(g)$$

In this reaction the oxidation number of hydrogen increases from zero to $+1$ indicating that hydrogen is oxidized. The chlorine is reduced because its oxidation number decreases from zero to -1. Oxidation–reduction reactions may be identified by recognizing chemical reactions that lead to a change in the oxidation number of one or more atoms.

Exercise 12.1

Indicate which of the following reactions are oxidation–reduction reactions. Indicate what species are being oxidized and what are being reduced.

(a) $Cu(s) +2\ Ag^+(aq) \rightleftharpoons Cu^{2+}(aq) + 2\ Ag(s)$
(b) $CH_3CO_2H(aq) + OH^-(aq) \rightleftharpoons CH_3CO_2^-(aq) + H_2O(l)$
(c) $SF_4(g) + F_2(g) \rightleftharpoons SF_6(g)$
(d) $CuSO_4(aq) + BaCl_2(aq) \rightleftharpoons BaSO_4(s) + CuCl_2(aq)$

Solution

(a) This is an oxidation–reduction reaction. The copper is oxidized from copper 0 to copper $+2$. The silver is reduced from silver $+1$ to silver 0.

(b) This is a Brønsted acid–base reaction. There is no change in the oxidation numbers of any of the atoms.

(c) This is an oxidation–reduction reaction. The SF_4 is oxidized due to the change in the oxidation number of sulfur. Sulfur changes from $+4$ in SF_4 to $+6$ in SF_6. The fluorine molecule, F_2, is reduced as the oxidation number decreases from zero in F_2 to -1 in SF_6.

(d) There is no change in oxidation number of any of the elements in this reaction.

Oxidation–reduction reactions of organic compounds can often be quickly recognized by a change in the number of bonds between carbon and hydrogen or oxygen (or another electronegative element). When carbon is oxidized in an organic molecule, the number of carbon–hydrogen bonds decrease and/or the number of carbon–oxygen bonds increase. For example, when a bottle of wine is opened and allowed to sit in contact with air, the wine will eventually turn to vinegar. The reaction involves the oxidation of the alcohol in wine, ethanol, to form acetic acid.

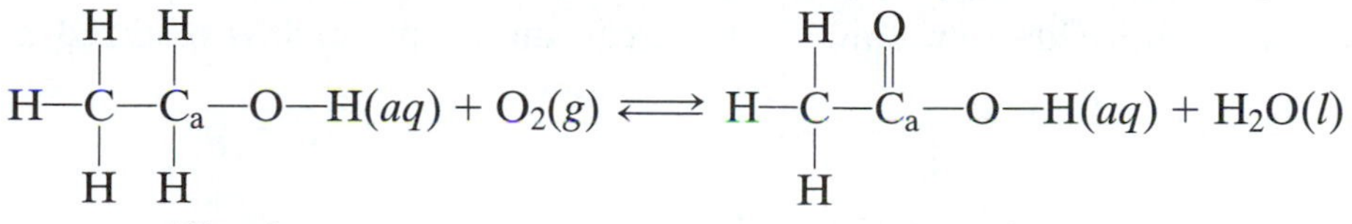

Ethanol   *Acetic acid*

In this reaction C_a has two bonds to hydrogen and one bond to oxygen in ethanol. In the product, acetic acid, C_a has no bonds to hydrogen and a single and double bond to oxygen. The carbon atom, C_a, has been oxidized. Even though only the one carbon atom within the molecule was oxidized we still refer to the entire molecule, ethanol, as having been oxidized. The molecular oxygen is reduced.

We can examine the reaction more quantitatively by assigning oxidation numbers to the atoms that are oxidized or reduced.

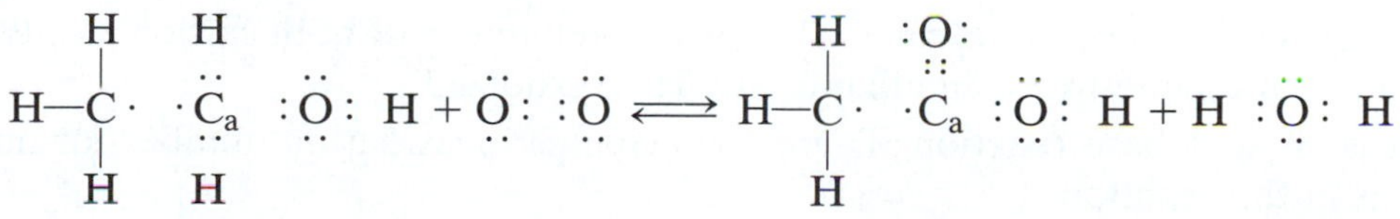

After assigning electrons in each bond of the Lewis structure to the more electronegative atom, the carbon atom, C_a, has five electrons in ethanol and only one electron in acetic acid.

$$OX_{C_a \text{ in ethanol}} = 4 - 5 = -1$$
$$OX_{C_a \text{ in acetic acid}} = 4 - 1 = +3$$

The oxidation number of the carbon has increased from -1 to $+3$. At the same time the oxidation number of oxygen has decreased from zero in elemental oxygen to -2 for the oxygen atom in acetic acid and the oxygen in water.

$$OX_{\text{elemental oxygen}} = 6 - 6 = 0$$
$$OX_{\text{oxygen in acetic acid}} = 6 - 8 = -2$$
$$OX_{\text{oxygen in water}} = 6 - 8 = -2$$

Many times in the chemical literature oxidation–reduction reactions of organic compounds are not written as balanced chemical equations as shown above but instead show only the initial organic reactant and the final organic product. In the laboratory acetic acid can be prepared from ethanol by oxidation with potassium dichromate, $K_2Cr_2O_7$. The reaction is often written as

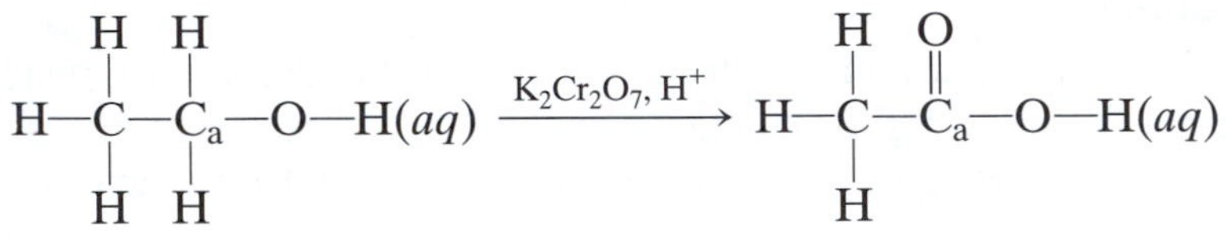

Only the reactant ethanol and the product acetic acid are shown. The substance that undergoes the reduction half-reaction (or oxidation half-reaction if the organic compound is reduced) plus any additional reaction conditions are written above the arrow. Note that the number of hydrogen and oxygen atoms on both sides of the equation are not balanced.

Exercise 12.2

Identify which of the following reactions (unbalanced) represent oxidation or reduction of an organic compound. Indicate whether the organic compound is oxidized or reduced. Describe the change in oxidation number of the carbon atom which is oxidized or reduced.

(a) $H_2C{=}CH_2(g) \xrightarrow{H_2} H_3C{-}CH_3(g)$
    Ethene   *Ethane*

(b) 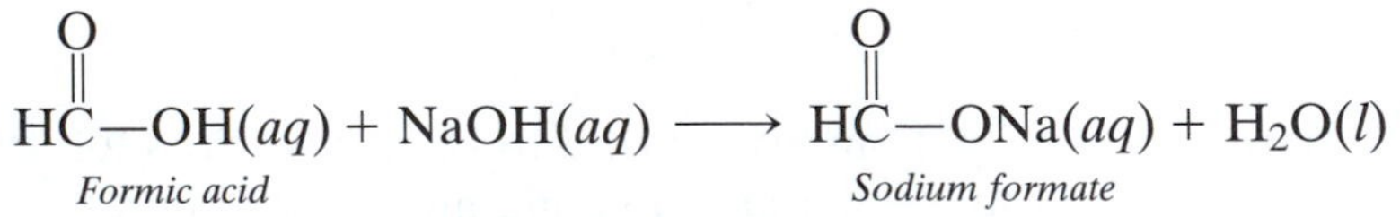
    Formic acid   *Sodium formate*

(c) 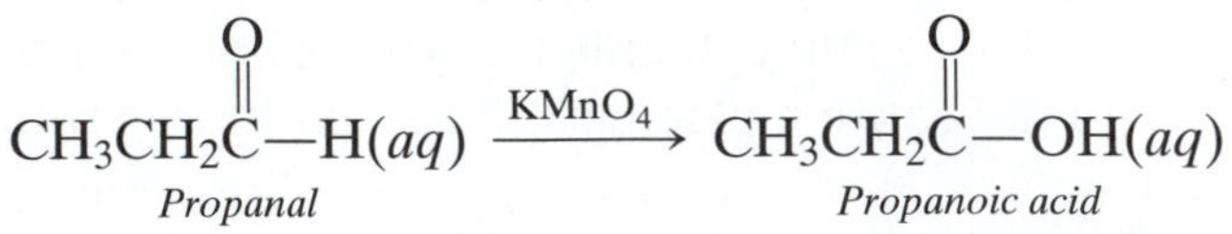
    Propanal   *Propanoic acid*

Solution

(a) The ethene is reduced to ethane. The oxidation number of both carbon atoms changes from -2 in ethene to -3 in ethane. The H_2 is oxidized.

(b) This is an acid–base reaction. There is no change in oxidation number for any of the atoms in the reaction.

(c) Propanal is oxidized to propanoic acid. The oxidation number of the carbon bonded to oxygen changes from +1 in propanal to +3 in propanoic acid. The KMnO₄ is reduced.

The method of listing redox reactions by showing only one reactant and its product is commonly used to describe biochemical processes. An example, shown in Figure 12.1, describes the citric acid (or Krebs) cycle. The citric acid cycle, a vital metabolic pathway, is made up of a series of reactions that serve as the principal source of metabolic energy for an organism. Nutrients from carbohydrates, proteins, and lipids are oxidized by the citric acid cycle to yield carbon dioxide and water. The carbon in carbon dioxide is in a +4 oxidation state, carbon's highest oxidation state. Carbon in this state has been completely oxidized.

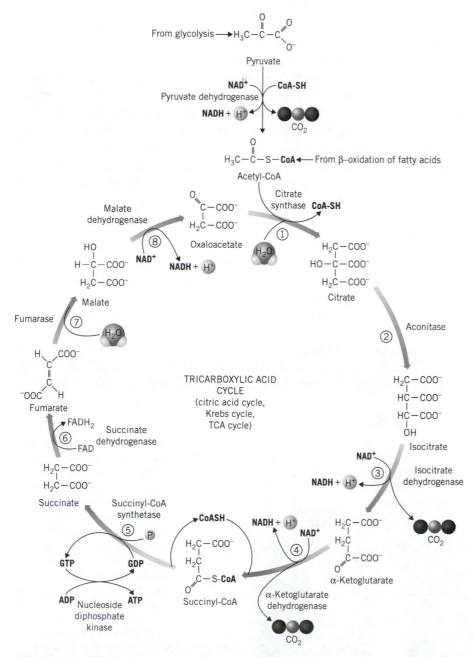

FIGURE 12.1 The citric acid cycle. Reprinted from R. H. Garrett and C. M. Grisham, *Biochemistry,* Harcourt Brace, New York, 1994, p. 602.

We will examine just one of the reactions from the cycle. Reaction 3 represents the first oxidation–reduction reaction in the cycle in which isocitrate is oxidized to α-ketoglutarate. The circled numbers are the oxidation numbers of the various atoms.

$$
\begin{array}{c}
\overset{(-2)}{H_2}\overset{(+3)}{C}-\overset{}{COO^-} \\
\overset{(-1)}{|}\;\;\overset{(+3)}{} \\
\overset{}{HC_b}-\overset{}{C_a}OO^- + NAD^+ \\
\overset{(0)}{|}\;\;\overset{(+3)}{} \\
\overset{}{HC_c}-\overset{}{COO^-} \\
| \\
OH \\
\textit{Isocitrate}
\end{array}
\rightleftharpoons
\begin{array}{c}
\overset{(-2)}{H_2}\overset{(+3)}{C}-\overset{}{COO^-} \\
\overset{(-2)}{|} \\
\overset{}{H_2C_b} \;\;\;\;\; + \overset{(+4)}{C_a}O_2 + NADH \\
\overset{(+2)}{|} \\
\overset{}{C_c}\overset{(+3)}{} \\
O\diagup\;\;\diagdown COO^- \\
\textit{α-Ketoglutarate}
\end{array}
$$

The oxidation number of each carbon is shown in the above chemical equation. Three carbon atoms (labeled C_a, C_b, and C_c) undergo changes in oxidation number during the reaction. Carbon atoms C_a and C_c are oxidized from +3 to +4 and 0 to +2, respectively. Carbon atom C_b is reduced from −1 to −2. There is a +3 increase in oxidation numbers and a −1 decrease for a net change within the molecule of +2, indicating that the isocitrate has been oxidized. Accompanying this oxidation is the reduction of NAD^+ (nicotinamide adenine dinucleotide) to produce its reduced form, NADH. NAD^+ undergoes a −2 change in oxidation number to balance the +2 change of isocitrate. The α-ketoglutarate and NADH produced in the reaction now participate in additional metabolic reactions.

12.4 VOLTAIC CELLS

A closer examination of oxidation–reduction reactions can be simplified by physically separating the process of oxidation from the process of reduction. This can be accomplished using a **voltaic cell** (also known as a **galvanic cell**). A voltaic cell is an electrochemical cell in which a redox reaction produces an electric current. In other words, the cell serves as a simple battery.

Figure 12.2 shows a voltaic cell formed by immersing a strip of zinc metal into a 1 M $Zn(NO_3)_2$ ion solution and immersing a piece of silver wire into a solution of 1 M $AgNO_3$ ion. The zinc metal and silver wire are connected with an electrical conductor. The circuit

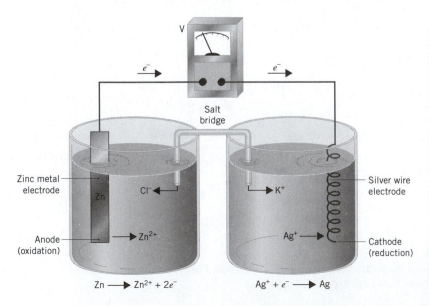

FIGURE 12.2 The reaction between zinc metal and silver ions produces an electric potential resulting in the flow of electrons from the zinc electrode to the silver electrode. K^+ ions move from the salt bridge into the solution to replace silver ions as the silver ions are converted to silver metal at the cathode. Cl^- ions move from the salt bridge into solution to balance the charge of the Zn^{2+} produced from the zinc metal at the anode. NO_3^- is not shown because it does not participate in the chemical reaction.

is completed with a **salt bridge,** a U-tube filled with a saturated solution of a soluble salt such as KCl. Oxidation (loss of electrons) takes place in the left beaker at the zinc electrode.

$$Zn(s) \rightleftharpoons Zn^{2+}(aq) + 2\ e^- \tag{12.1}$$

As the reaction proceeds the zinc metal electrode dissolves away as the zinc metal is converted into soluble zinc ions. Electrons are not stable in solution, and therefore the electrons produced cause a buildup of negative charge on the surface of the zinc metal, thus causing the electrode to be negative. A chemical equation in which electrons appear as a product or reactant are called **half-reactions.** Equation 12.1 represents an oxidation half-reaction. An oxidation half-reaction cannot occur without an accompanying reduction half-reaction.

Reduction takes place at the silver electrode, and the reduction half-reaction is

$$Ag^+(aq) + e^- \rightleftharpoons Ag(s) \tag{12.2}$$

Silver ions in solution migrate to the surface of the silver electrode and accept electrons. As a result the soluble silver ions are converted into silver metal that plates onto the surface of the silver electrode. The silver electrode supplies electrons for the reduction reaction and is positive relative to the zinc electrode. Because the two electrodes are connected by an electrical conductor, the excess electrons at the zinc electrode will move toward the positively charged silver electrode. The movement of electrons through a conductor produces an electric current. The voltaic cell is therefore acting as a battery.

In the voltaic cell in Figure 12.2 electrons flow from the zinc electrode to the silver. This electric current can be used to light a lightbulb or run an electric motor. If a voltmeter is connected between the two electrodes a **cell potential** for the voltaic cell can be measured. The cell potential is the potential of the cell to do work on its surrounding by driving an electric current through a wire. By definition, 1 joule of energy is produced when 1 coulomb of electrical charge is transported across a potential of 1 volt.

$$1\ V = \frac{1\ J}{1\ C}$$

The cell potential is measured in units of volts. The cell potential in Figure 12.2 is 1.5624 V. The chemical driving force for the reaction in the cell produces the electric potential. The study of these cells is therefore called electrochemistry.

One problem remains to be addressed. As the zinc metal is oxidized it forms zinc ions, which dissolve in solution, producing a positively charged solution. At the silver electrode silver ions are being reduced to silver metal, which causes the solution to become negatively charged. Electrically charged solutions are not stable. Therefore, negatively charged chloride ions will diffuse (move from an area of high concentration to an area of lower concentration) from the salt bridge into the zinc solution to balance the excess positive charge. Positively charged potassium ions will diffuse into the silver solution to yield an electrically neutral solution.

The electrode at which oxidation takes place in an electrochemical cell is called the **anode.** The electrode at which reduction occurs is called the **cathode.** In the voltaic cells shown in Figure 12.2, the zinc electrode is the anode and the silver electrode is the cathode.

Checkpoint

What is the sign of the anode in Figure 12.2? What is the sign of the cathode?

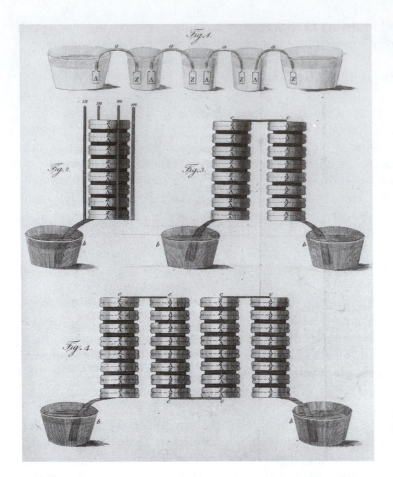

Illustrations of the voltaic pile prepared from strips of zinc and silver metal separated by moistened cardboard published by Volta in 1800.

Balancing Oxidation–Reduction Equations

The overall oxidation–reduction reaction in the voltaic cell shown in Figure 12.2 consists of the oxidation of zinc (reaction 12.1) and the reduction of silver (reaction 12.2). Using the law of conservation of mass it would be tempting to write the following complete oxidation–reduction reaction:

$$Ag^+(aq) + Zn(s) \rightleftharpoons Zn^{2+}(aq) + Ag(s) \tag{12.3}$$

There are an equal number of silver and zinc atoms on both sides of the equation, so atoms are conserved. However, there is not a balance of electric charge on the two sides of the equation. The implication of reaction 12.3 as written is that for every atom of Zn that is oxidized, two electrons will be released. At the same time the reduction of a silver ion requires only one electron. If the reaction proceeded as written in reaction 12.3, twice as many electrons would be produced at the zinc electrode as are used at the silver electrode. Because neither matter nor charge is created or destroyed in a chemical reaction, more electrons cannot be produced than are used. In a balanced chemical reaction not only must atoms be conserved but also electrons. Writing a balanced oxidation–reduction reaction requires that the number of electrons gained during reduction must equal the number of electrons lost during oxidation. Therefore, for every zinc atom that produces two electrons, *two* silver ions must be reduced.

$$
\begin{array}{r}
Zn(s) \rightleftharpoons Zn^{2+}(aq) + 2\,e^- \\
2 \times [Ag^+(aq) + e^- \rightleftharpoons Ag(s)] \\
\hline
Zn(s) + 2\,Ag^+(aq) \rightleftharpoons Zn^{2+}(aq) + 2\,Ag(s)
\end{array}
\tag{12.4}
$$

This process of writing a balanced oxidation–reduction reaction is known as the **half-reaction method.** The individual half-reactions are written and then multiplied by coefficients such that the electrons lost are equal to the electrons gained. The two half-reactions may then be added together to give the balanced oxidation–reduction reaction.

Checkpoint

Is the following oxidation–reduction reaction balanced? Explain your answer.

$$Zn^{2+}(aq) + Cr(s) \rightleftharpoons Cr^{3+}(aq) + Zn(s)$$

12.5   OXIDIZING AND REDUCING AGENTS

So far we have focused on what happens when a particular substance undergoes a change in oxidation state in an oxidation–reduction reaction. Let's now consider the role that each substance plays in the reaction.

When zinc metal reacts with Ag^+ ions, the zinc metal donates electrons to the Ag^+ ions and therefore reduces the ions. Zinc therefore acts as a **reducing agent** in the reaction. The same electrons that are lost by the zinc are gained by the silver.

$$\underset{\substack{Reducing \\ agent}}{Zn(s)} + 2\,Ag^+(aq) \rightleftharpoons Zn^{2+}(aq) + 2\,Ag(s)$$

The Ag^+ ions, on the other hand, gain electrons from the zinc metal and thereby oxidizes the zinc. The Ag^+ ion is therefore an **oxidizing agent.**

$$Zn(s) + 2\,\underset{\substack{Oxidizing \\ agent}}{Ag^+(aq)} \rightleftharpoons Zn^{2+}(aq) + 2\,Ag(s)$$

In general, oxidizing and reducing agents can be defined as follows.

Oxidizing agents **gain electrons (or electron density).**
Reducing agents **lose electrons (or electron density).**

Thus, an oxidizing agent undergoes a reduction half reaction and a reducing agent undergoes an oxidation half reaction.

Zinc metal is the reducing agent in the voltaic cell shown in Figure 12.2 because it reduces Ag^+ ions by providing electrons to the silver electrode. The Ag^+ ion is the oxidizing agent in the reaction because it oxidizes the zinc metal by removing electrons from the metal.

When a reducing agent loses electrons, it forms an oxidizing agent that could gain electrons if the reaction were reversed.

$$\underset{\substack{Reducing \\ agent}}{Zn} \rightleftharpoons \underset{\substack{Oxidizing \\ agent}}{Zn^{2+}} + 2\,e^-$$

Conversely, when an oxidizing agent gains electrons, it forms a reducing agent. This reducing agent could lose electrons if the reaction went in the opposite direction.

$$\underset{\substack{Oxidizing \\ agent}}{Ag^+} + e^- \rightleftharpoons \underset{\substack{Reducing \\ agent}}{Ag}$$

Oxidizing agents and reducing agents are therefore linked or coupled in much the same way that Brønsted acids and Brønsted bases are linked or coupled (see Section 11.4). Every oxidizing agent has a conjugate reducing agent, and vice versa.

The relationship between the strength of a reducing agent and its conjugate oxidizing agent can be understood by extending the argument used in Section 11.9 to link the strengths of Brønsted acids and bases. If zinc metal is a relatively good reducing agent, we must conclude that the Zn^{2+} ion is a relatively weak oxidizing agent.

$$Zn(s) \rightleftharpoons Zn^{2+}(aq) + 2\ e^-$$

Strong reducing agent — Weak oxidizing agent

If, on the other hand, the Ag^+ ion is a relatively good oxidizing agent, then silver metal must be a relatively weak reducing agent.

$$Ag^+(aq) + e^- \rightleftharpoons Ag(s)$$

Strong oxidizing agent — Weak reducing agent

As was seen in Figure 5.11 many atoms have more than one oxidation number. The ability of an element to behave as an oxidizing or reducing agent is related to its oxidation state.

When an element is in its highest oxidation state it can act only as an oxidizing agent.

When an element is in its lowest oxidation state it can act only as a reducing agent.

When an element is in an intermediate oxidation state it can act as an oxidizing or reducing agent.

Carbon can have oxidation numbers of $-4, -3, -2, -1, 0, +1, +2, +3$, and $+4$. In carbon dioxide, CO_2, carbon has an oxidation number of $+4$. The carbon in carbon dioxide cannot be further oxidized. Carbon dioxide can only be reduced. In other words, the carbon in carbon dioxide can only act as a oxidizing agent. Conversely in the molecule methane, CH_4, carbon has an oxidation number of -4 and can only undergo an oxidation reaction. Methane can only act as a reducing agent in a redox reaction. Elemental carbon in the form of graphite (zero oxidation number) can be either oxidized or reduced depending on the substance with which it is reacting.

Checkpoint
Identify the oxidizing and reducing agents in the following oxidation–reduction reaction.

$$Sn(s) + 4\ HNO_3(aq) \rightleftharpoons SnO_2(s) + 4\ NO_2(g) + 2\ H_2O(l)$$

12.6 RELATIVE STRENGTHS OF OXIDIZING AND REDUCING AGENTS

Several questions remain concerning the voltaic cell shown in Figure 12.2. Why is the zinc oxidized and the silver reduced? Why not the reverse process? Why does the reaction spontaneously occur? The following rule can be used to predict whether a redox reaction should occur and what substance should be oxidized and reduced if a reaction does occur.

Oxidation–reduction reactions should occur when they convert the stronger of a pair of oxidizing agents and the stronger of a pair of reducing agents into a weaker oxidizing agent and a weaker reducing agent.

Given that reaction 12.4 occurs in the voltaic cell, zinc must be a stronger reducing agent than silver metal and silver ion must be a stronger oxidizing agent than Zn^{2+} ion.

$$Zn(s) + 2 \, Ag^+(aq) \rightleftharpoons Zn^{2+} + 2 \, Ag(s)$$

| *Stronger reducing agent* | *Stronger oxidizing agent* | *Weaker oxidizing agent* | *Weaker reducing agent* |

On the basis of many such experiments, the common oxidation–reduction half-reactions have been organized into a table in which the strongest reducing agents are at one end and the strongest oxidizing agents are at the other, as shown in Table 12.1. By convention, all of the half-reactions are written as reduction half-reactions. The relationship between the half-reactions is quantified by listing the **reduction potential** for each half-reaction. Potential is defined as a measure of the driving force behind an electrochemical reaction and is measured in units of volts, V. Standard reduction potentials for additional species are listed in Appendix B.12.

The cell potential is dependent on the concentrations of any species present in solution, the partial pressures of any gases involved in the reaction, and the temperature at which the reaction is run. To provide a basis for comparing one half-reaction with another, the following set of **standard conditions** for electrochemical measurements has been defined:

1. For solutions the standard potential is based on a standard state of 1 M concentration.
2. All gases have a partial pressure of 1 atm.

Although measurements of standard potentials can be made at any temperature, they are often taken at 25°C. Potentials for reduction half-reactions measured under these conditions are referred to as **standard reduction potentials, $E°_{Red}$**. The $E°_{Red}$ can be interpreted as the tendency for the reduction half-reaction to occur. The more positive the potential the greater the tendency for the reaction to occur. The standard reduction potentials found in Table 12.1 and Appendix B.12 are experimentally determined and are measured relative to a standard hydrogen electrode. The standard hydrogen half reaction has been assigned a reduction potential of zero.

The $E°_{Red}$ values for the voltaic cell reaction can be found in Table 12.1.

$$Zn^{2+} + 2 \, e^- \rightleftharpoons Zn \qquad E° = -0.7628 \, V$$
$$Ag^+ + e^- \rightleftharpoons Ag \qquad E° = +0.7996 \, V$$

Because the silver reduction half-reaction has a much more positive potential ($+0.7996$ V) than the zinc half-reaction (-0.7628 V), the silver ion is much more likely to undergo the reduction half-reaction. If the silver ion is to be reduced and act as an oxidizing agent then something must be oxidized. The zinc half-reaction in Table 12.1 is written as a reduction half-reaction and therefore must be reversed to be an oxidation half-reaction. *When a reaction is reversed the magnitude of its potential remains the same but its sign changes.*

$$Zn \rightleftharpoons Zn^{2+} + 2 \, e^- \qquad E° = +0.7628 \, V$$

The oxidation half-reaction now has a positive potential, indicating that it is much more likely to occur than the reduction half-reaction of Zn^{2+} ion. Zinc metal will undergo the

TABLE 12.1   Standard Reduction Potentials, E°_{Red}

	Half-Reaction[a]	E°_{Red}	
	$K^+ + e^- \rightleftharpoons K$	−2.924	Best
	$Ca^{2+} + 2\,e^- \rightleftharpoons Ca$	−2.76	reducing
	$Na^+ + e^- \rightleftharpoons Na$	−2.7109	agents
	$Mg^{2+} + 2\,e^- \rightleftharpoons Mg$	−2.375	
	$Al^{3+} + 3\,e^- \rightleftharpoons Al$	−1.706	
	$Mn^{2+} + 2\,e^- \rightleftharpoons Mn$	−1.29	
	$Zn^{2+} + 2\,e^- \rightleftharpoons Zn$	−0.7628	
	$Cr^{3+} + 3\,e^- \rightleftharpoons Cr$	−0.74	
	$S + 2\,e^- \rightleftharpoons S^{2-}$	−0.508	
	$Cr^{3+} + e^- \rightleftharpoons Cr^{2+}$	−0.41	
	$Fe^{2+} + 2\,e^- \rightleftharpoons Fe$	−0.409	
	$Co^{2+} + 2\,e^- \rightleftharpoons Co$	−0.28	
	$Ni^{2+} + 2\,e^- \rightleftharpoons Ni$	−0.23	
	$Sn^{2+} + 2\,e^- \rightleftharpoons Sn$	−0.1364	
	$Pb^{2+} + 2\,e^- \rightleftharpoons Pb$	−0.1263	
	$Fe^{3+} + 3\,e^- \rightleftharpoons Fe$	−0.036	
	$2\,H^+ + 2\,e^- \rightleftharpoons H_2$	0.0000 . . .	
Oxidizing	$Sn^{4+} + 2\,e^- \rightleftharpoons Sn^{2+}$	0.15	↑
power	$Cu^{2+} + e^- \rightleftharpoons Cu^+$	0.158	Reducing
increases	$Cu^{2+} + 2\,e^- \rightleftharpoons Cu$	0.3402	power
↓	$O_2 + 2\,H_2O + 4\,e^- \rightleftharpoons 4\,OH^-$	0.401	increases
	$Cu^+ + e^- \rightleftharpoons Cu$	0.522	
	$MnO_4^- + 2\,H_2O + 3\,e^- \rightleftharpoons MnO_2 + 4\,OH^-$	0.588	
	$O_2 + 2\,H^+ + 2\,e^- \rightleftharpoons H_2O_2$	0.682	
	$Fe^{3+} + e^- \rightleftharpoons Fe^{2+}$	0.770	
	$Hg_2^{2+} + 2\,e^- \rightleftharpoons 2\,Hg$	0.7961	
	$Ag^+ + e^- \rightleftharpoons Ag$	0.7996	
	$Hg^{2+} + 2\,e^- \rightleftharpoons Hg$	0.851	
	$HNO_3 + 3\,H^+ + 3\,e^- \rightleftharpoons NO + 2\,H_2O$	0.96	
	$Br_2(aq) + 2\,e^- \rightleftharpoons 2\,Br^-$	1.087	
	$CrO_4^{2-} + 8\,H^+ + 3\,e^- \rightleftharpoons Cr^{3+} + 4\,H_2O$	1.195	
	$O_2 + 4\,H^+ + 4\,e^- \rightleftharpoons 2\,H_2O$	1.229	
	$Cr_2O_7^{2-} + 14\,H^+ + 6\,e^- \rightleftharpoons 2\,Cr^{3+} + 7\,H_2O$	1.33	
	$Cl_2(g) + 2\,e^- \rightleftharpoons 2\,Cl^-$	1.3583	
	$PbO_2 + 4\,H^+ + 2\,e^- \rightleftharpoons Pb^{2+} + 2\,H_2O$	1.467	
	$MnO_4^- + 8\,H^+ + 5\,e^- \rightleftharpoons Mn^{2+} + 4\,H_2O$	1.491	
	$Au^+ + e^- \rightleftharpoons Au$	1.68	
Best	$Co^{3+} + e^- \rightleftharpoons Co^{2+}$	1.842	
oxidizing	$O_3(g) + 2\,H^+ + 2\,e^- \rightleftharpoons O_2(g) + H_2O$	2.07	
agents	$F_2(g) + 2\,H^+ + 2\,e^- \rightleftharpoons 2\,HF(aq)$	3.03	

[a]In the tabulations of standard reduction potential the symbol H^+ is used instead of H_3O^+.

oxidation half-reaction. Therefore zinc metal is a stronger reducing agent than silver metal and Ag^+ is a better oxidizing agent than Zn^{2+}.

The same reasoning can be extended to Table 12.1. The strongest reducing agents (substances that tend to undergo an oxidation half-reaction) are at the top of the table. The strongest reducing agent listed in Table 12.1 is potassium metal. Note that the strongest reducing agent is not K^+ ion. The half-reaction is written as a reduction. Since a reducing agent must itself be oxidized, the half-reaction must be reversed.

$$ K \rightleftharpoons K^+ + e^- \qquad E° = +2.924 \text{ V} $$

Writing the half-reaction as an oxidation changes the sign of the potential listed in the table and gives a large positive potential. This indicates the strong tendency for this reaction to occur. Potassium is one of the most reactive metals—it bursts into flame when added to water, for example. Experimentally we find that neutral metal atoms act as reducing agents and are oxidized in all of their chemical reactions. Thus potassium readily gives up its electron to form K^+ ion.

The strongest oxidizing agent is found at the bottom of Table 12.1.

$$ F_2(g) + 2\ H^+ + 2\ e^- \rightleftharpoons 2\ HF(aq) \qquad E° = +3.03 \text{ V} $$

The large positive potential associated with this reduction half-reaction indicates its strong tendency to occur. Fluorine molecules therefore act as oxidizing agents by readily accepting electrons and undergoing this half-reaction. Fluorine is the most electronegative element in the periodic table. It should not be surprising to find that fluorine has a strong tendency to gain electrons and therefore be reduced.

Exercise 12.3

Arrange the following oxidizing and reducing agents in order of increasing strength.

Reducing agents Cl^-, Cu, H_2, HF, Pb, and Zn
Oxidizing agents Cr^{3+}, $Cr_2O_7^{2-}$, Cu^{2+}, H^+, O_2, O_3, and Na^+

Solution

According to Table 12.1, these reducing agents become stronger in the following order.

$$ HF < Cl^- < Cu < H_2 < Pb < Zn $$

The oxidizing agents become stronger in the following order.

$$ Na^+ < Cr^{3+} < H^+ < Cu^{2+} < O_2 < Cr_2O_7^{2-} < O_3 $$

Exercise 12.4

Use Table 12.1 to predict whether the following oxidation–reduction reactions should occur as written.

(a) $2\ Ag(s) + S(s) \rightarrow Ag_2S(s)$

(b) $2\ Ag(s) + Cu^{2+}(aq) \rightarrow 2\ Ag^+(aq) + Cu(s)$

(c) $MnO_4^-(aq) + 3\ Fe^{2+}(aq) + 2\ H_2O(l) \rightarrow MnO_2(s) + 3\ Fe^{3+}(aq) + 4\ OH^-(aq)$

(d) $MnO_4^-(aq) + 5\ Fe^{2+}(aq) + 8\ H^+(aq) \rightarrow Mn^{2+}(aq) + 5\ Fe^{3+}(aq) + 4\ H_2O(l)$

Solution

(a) No. The S^{2-} ion of Ag_2S is a better reducing agent than Ag, and the Ag^+ ion of Ag_2S is a better oxidizing agent than S.

(b) No. Cu is a better reducing agent than Ag, and the Ag^+ ion is a better oxidizing agent than Cu^{2+} ion.

(c) No. MnO_4^- in basic solution is not a strong enough oxidizing agent to oxidize Fe^{2+} to Fe^{3+}.

(d) Yes. MnO_4^- in acidic solution is a strong enough oxidizing agent to oxidize Fe^{2+} to Fe^{3+}.

12.7 STANDARD CELL POTENTIALS

The voltaic cell in Figure 12.2 has a potential of 1.5624 V. The concentrations of the Ag^+ ions and Zn^{2+} ions are both 1 M in the cell. Since these concentrations are standard conditions, this potential is referred to as a **standard cell potential, $E°$**. The larger the difference between the oxidizing and reducing strengths of the reactants and products, the larger is the cell potential. To obtain a relatively large cell potential, we have to react a strong reducing agent with a strong oxidizing agent. The overall cell potential for a reaction must be the sum of the potentials for the oxidation and reduction half-reactions.

$$E°_{reaction} = E°_{Ox} + E°_{Red}$$

The standard cell potential for the voltaic cell in Figure 12.2 can be calculated from standard half-cell potentials of the two half-reactions.

$$
\begin{array}{lll}
Zn(s) \rightleftharpoons Zn^{2+}(aq) + 2\,e^- & E°_{Ox} = +0.7628 \text{ V} \\
\underline{2\,Ag^+(aq) + 2\,e^- \rightleftharpoons 2\,Ag(s)} & \underline{E°_{Red} = +0.7996 \text{ V}} \\
Zn(s) + 2\,Ag^+(aq) \rightleftharpoons Zn^{2+}(aq) + 2\,Ag(s) & E°_{reaction} = +1.5624 \text{ V}
\end{array}
$$

Note that the units of half-cell potentials are volts, not volts per mole or volts per electron. When combining half-reactions it is only necessary to add the two half-cell potentials. We do not multiply the potentials by the integers used to balance the number of electrons transferred in the reaction. In the case of the silver half-reaction it was necessary to multiply the half-reaction by two in order for the electrons gained by the silver to equal the electrons lost by the zinc. However, the standard half-cell potential, $E°_{Red}$ was *not* multiplied by two. The value +0.7996 V was taken directly from Table 12.1.

The *magnitude* of the cell potential is a measure of the driving force behind a reaction. The larger the value of the cell potential, the further the reaction is from equilibrium. The *sign* of the cell potential tells us the direction in which the reaction must shift to reach equilibrium. The fact that $E°$ is positive for the zinc–silver cell tells us that when the system is present at standard conditions, it has to shift to the right to reach equilibrium. Reactions for which $E°$ is positive therefore have equilibrium constants that favor the formation of the products of the reaction. A reaction with a positive $E°$ is referred to as **spontaneous.**

Exercise 12.5

Use the overall cell potentials to predict which of the following reactions are spontaneous.

(a) $Cu(s) + 2\,Ag^+(aq) \longrightarrow Cu^{2+}(aq) + 2\,Ag(s)$ $E° = 0.46$ V

(b) $2\,Fe^{3+}(aq) + 2\,Cl^-(aq) \longrightarrow 2\,Fe^{2+}(aq) + Cl_2(g)$ $E° = -0.59$ V

(c) $2\,Fe^{3+}(aq) + 2\,I^-(aq) \longrightarrow 2\,Fe^{2+}(aq) + I_2(aq)$ $E° = 0.24$ V

(d) $2 H_2O_2(aq) \longrightarrow 2 H_2O(l) + O_2(aq)$ $E° = 1.09$ V

(e) $Cu(s) + 2 H^+(aq) \longrightarrow Cu^{2+}(aq) + H_2(g)$ $E° = -0.34$ V

Solution

Any reaction for which the overall cell potential is positive is spontaneous. Reactions (a), (c), and (d) are therefore spontaneous.

Exercise 12.6

Use the standard cell potential for the following reaction

$$Cu(s) + 2 H^+(aq) \rightleftharpoons Cu^{2+}(aq) + H_2(g) \qquad E° = -0.34 \text{ V}$$

to predict the standard cell potential for the opposite reaction.

$$Cu^{2+}(aq) + H_2(g) \rightleftharpoons Cu(s) + 2 H^+(aq) \qquad E° = ?$$

Solution

Turning the reaction around doesn't change the relative strengths of Cu^{2+} and H^+ ions as oxidizing agents, or copper metal and H_2 as reducing agents. The *magnitude* of the potential therefore must remain the same. But turning the equation around changes the *sign* of the cell potential and can therefore turn an unfavorable reaction into one that is spontaneous, or vice versa. The standard cell potential for the reduction of Cu^{2+} ions by H_2 gas is therefore $+0.34$ V.

$$Cu^{2+}(aq) + H_2(g) \rightleftharpoons Cu(s) + 2 H^+(aq) \qquad E° = -(-0.34 \text{ V}) = +0.34 \text{ V}$$

Checkpoint

What happens to the cell potential when the direction of the reaction is reversed?

Exercise 12.7

Use cell potential data to explain why copper metal does not dissolve in a 1 *M* solution of a typical strong acid, such as hydrochloric acid,

$$Cu(s) + 2 H^+(aq) \not\longrightarrow$$

but will dissolve in 1 *M* nitric acid.

$$3 Cu(s) + 2 HNO_3(aq) + 6 H^+(aq) \longrightarrow 3 Cu^{2+}(aq) + 2 NO(g) + 4 H_2O(l)$$

Solution

Copper does not dissolve in a typical strong acid because the overall cell potential for the oxidation of copper metal to Cu^{2+} ions coupled with the reduction of H^+ ions to H_2 is negative.

$$
\begin{array}{ll}
Cu \rightleftharpoons Cu^{2+} + 2e^- & E°_{Ox} = -(0.34 \text{ V}) \\
2 H^+ + 2e^- \rightleftharpoons H_2 & E°_{Red} = 0.000 \ldots \text{ V} \\
\hline
Cu(s) + 2 H^+(aq) \rightleftharpoons Cu^{2+}(aq) + H_2(g) & E° = E°_{Ox} + E°_{Red} = -0.34 \text{ V}
\end{array}
$$

Copper dissolves in nitric acid because the reaction at the cathode now involves the reduction of nitric acid to NO gas, and the potential for that half-reaction is strong enough to overcome the half-cell potential for oxidation of copper metal to Cu^{2+} ions.

$$\frac{\begin{array}{r} 3\,(Cu \rightleftharpoons Cu^{2+} + 2e^-) \\ 2(HNO_3 + 3\,H^+ + 3e^- \rightleftharpoons NO + 2\,H_2O) \end{array}}{3\,Cu(s) + 2\,HNO_3(aq) + 6\,H^+(aq) \rightleftharpoons 3\,Cu^{2+}(aq) + 2\,NO(g) + 4\,H_2O(l)}$$

$$\begin{array}{r} E^\circ_{Ox} = -(0.34\ \text{V}) \\ \underline{E^\circ_{Red} = 0.96\ \text{V}} \\ E^\circ = E^\circ_{Ox} + E^\circ_{Red} = 0.62\ \text{V} \end{array}$$

12.8 ELECTROCHEMICAL CELLS AT NONSTANDARD CONDITIONS

The potential of a voltaic cell is dependent on the concentrations of any species present in solution, the partial pressures of any gases involved in the reaction, and the temperature at which the reaction is run. All of the redox reactions discussed thus far in this chapter have been under *standard conditions* (1 *M* concentration of aqueous reactants and products and 1 atm partial pressure of gaseous reactants and products). Not all redox reactions take place under standard conditions. What may not be obvious is that for reactions which are initially at standard conditions the reaction conditions will change as the reaction progresses.

$$Zn(s) + 2\,Ag^+(aq) \rightleftharpoons Zn^{2+}(aq) + 2\,Ag(s)$$

As the reaction goes forward—in Figure 12.2 as zinc ions are formed and silver ions are consumed—the driving force behind the reaction must become weaker. Le Châtelier's principle tells us that if the reaction were at equilibrium, decreasing the concentration of Ag^+ and increasing the concentration of Zn^{2+} would shift the equilibrium back toward the reactants. Even when the reaction is not at equilibrium but is occurring in the forward direction, a decrease in the concentration of reactants and an increase in the concentration of the products decrease the tendency of the reaction to occur. Therefore, the cell potential, E_{cell} (no longer called the *standard cell potential, E°*) must become smaller.

This raises an interesting question: When does the cell potential become zero? Since the cell potential measures the driving force of the reaction, the potential becomes zero when the reaction reaches equilibrium. In the zinc–silver cell, when equilibrium is reached there is no net change in the amount of zinc metal or silver ions in the system. Therefore, no electrons flow from the anode to the cathode. If there is no longer a net flow of electrons, the cell cannot do electrical work. Its potential for doing work (or cell potential) must therefore be zero. A dead battery is at equilibrium and its potential is zero.

Checkpoint

Would you expect the following reaction run under nonstandard conditions to have a potential greater than, equal to, or less than its standard potential, E°? Explain your reasoning.

$$Ni(s) + Cu^{2+}(aq) \rightleftharpoons Ni^{2+}(aq) + Cu(s)$$

Initial concentrations of Cu^{2+} and Ni^{2+} are 1.5 and 0.010 *M*, respectively.

12.9   BATTERIES

Until recently, the market for disposable batteries was dominated by batteries based on the *Leclanché cell,* which was introduced in 1860. The cathode (electrode where reduction occurs) in the cell is a mixture of solid MnO_2 and carbon. The anode (electrode where oxidation occurs) is a sheet of zinc metal that is amalgamated with a trace of mercury. The electrolytic solution is a mixture of NH_4Cl and $ZnCl_2$. The cell reactions that occur during the discharge of a manganese dioxide–zinc primary battery can be written as follows.

$$\text{Cathode (+)} \quad 3\,MnO_2 + 2\,H_2O + 4\,e^- \rightleftharpoons Mn_3O_4 + 4\,OH^-$$
$$\text{Anode (−)} \quad Zn + 2\,OH^- \rightleftharpoons ZnO + H_2O + 2\,e^-$$
$$\text{Overall reaction} \quad 2\,Zn(s) + 3\,MnO_2(s) \rightleftharpoons Mn_3O_4(s) + 2\,ZnO(s)$$

The initial cell potential of a single dry cell is 1.54 V. The popularity of the Leclanché cell is based on the fact that it is relatively inexpensive, available in many voltages and sizes, and suitable for intermittent use.

In 1949, the first "alkaline" dry cell battery was produced. This cell also uses a mixture of MnO_2 and carbon as the cathode and amalgamated zinc as the anode. However, KOH is used as the electrolyte. Alkaline dry cells (see Figure 12.3) have several advantages. They can be used over a wider range of temperatures because the electrolyte is more stable. They also require very little electrolyte and can therefore be very compact. More importantly, the batteries maintain a constant voltage for a longer period of time and therefore last longer.

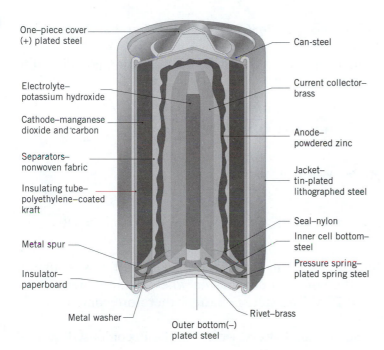

One–piece cover (+) plated steel

Electrolyte– potassium hydroxide

Cathode–manganese dioxide and carbon

Separators– nonwoven fabric

Insulating tube– polyethylene–coated kraft

Metal spur

Insulator– paperboard

Metal washer

Outer bottom(−) plated steel

Can-steel

Current collector– brass

Anode– powdered zinc

Jacket– tin-plated lithographed steel

Seal–nylon

Inner cell bottom– steel

Pressure spring– plated spring steel

Rivet–brass

FIGURE 12.3 Cutaway drawing of a typical alkaline battery.

Lead–Acid Battery

The lead storage battery similar to the type used in a car was first demonstrated to the French Academy of Sciences by Gaston Planté in 1860. It contained nine cells in parallel and was able to deliver extraordinarily large currents. The cells were constructed from lead plates separated by layers of flannel that were immersed in a 10% H_2SO_4 solution. As might be expected, the system soon became known as the *lead–acid battery.*

In spite of many studies of the charging and discharging of lead electrodes in acid solution in the last 140 years, there is still some doubt about the exact mechanism of the reactions in the lead–acid storage battery shown in Figure 12.4. The most common model suggests that the following reaction occurs at the PbO_2 plate.

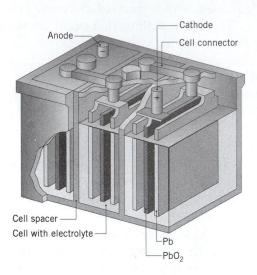

FIGURE 12.4 Cutaway drawing of a typical lead–acid battery.

$$PbO_2(s) + 4\ H^+(aq) + SO_4^{2-}(aq) + 2\ e^- \underset{charge}{\overset{discharge}{\rightleftharpoons}} 2\ H_2O(l) + PbSO_4(aq)$$

When the battery discharges, PbO_2 is reduced to $PbSO_4$. When it is charged, the reaction is reversed. The standard electrode potential for the half-reaction is 1.682 V.

The reaction at the lead metal plate can be described by the following equation, for which the standard potential is 0.3858 V.

$$Pb(s) + SO_4^{2-}(aq) \underset{charge}{\overset{discharge}{\rightleftharpoons}} PbSO_4(aq) + 2\ e^-$$

The total cell reaction can therefore be written as follows.

$$PbO_2(s) + Pb(s) + 4\ H^+(aq) + 2\ SO_4^{2-}(aq) \underset{charge}{\overset{discharge}{\rightleftharpoons}} 2\ PbSO_4(aq) + 2\ H_2O(l)$$

The magnitude of the standard potential for a single cell in lead–lead batteries is slightly larger than 2 V, although the sign of the potential depends on whether the battery is being charged or discharged. The typical 12-V lead storage battery therefore contains six 2-V cells.

During the charging of a lead–acid storage battery, water can be decomposed into its elements.

$$2\ H_2O(l) \rightleftharpoons 2\ H_2(g) + O_2(g)$$

Lead storage batteries therefore represent a potential threat of hydrogen explosions. Because such explosions can spray the 10% sulfuric acid electrolyte onto the individual who is working on the battery, safety goggles should be worn when working with these batteries.

Nicad Batteries

Another battery that has become increasingly popular is the nickel–cadmium or *nicad battery*. This type of battery is commonly used in watches and calculators. The history of the nicad battery traces back to the beginning of the twentieth century. After approximately 10,000 experiments, Thomas Edison found that he could make a rechargeable battery from a nickel–iron cell. The active materials in the cell were $Ni(OH)_2$ and iron metal that was partially oxidized to Fe^{2+}. In the mid 1920s, a German manufacturer found that the risk of hydrogen explosions in the batteries was significantly reduced when the iron metal was replaced with cadmium.

The chemistry of nickel–cadmium batteries is still not fully understood. A charge–discharge mechanism for the system might be written as follows.

$$2\ NiO(OH){\cdot}xH_2O + Cd + 2\ H_2O \underset{charge}{\overset{discharge}{\rightleftharpoons}} 2\ Ni(OH)_2{\cdot}yH_2O + Cd(OH)_2$$

When the batteries discharge, the nickel is reduced from +3 to +2 and the cadmium is oxidized from 0 to +2, with a net potential of 1.29 V. Both the oxidized and reduced forms of nickel contain water trapped in the crystals, but the amount of water of hydration in the two forms is different.

Fuel Cells

Ever since the start of the space program, enthusiasm has run high for batteries that are *fuel cells*. By definition, a fuel cell is an electrochemical cell in which a reaction with oxygen is used to generate electrical energy. Fuel cells therefore require the continuous feed of a fuel and oxygen which produces a low-voltage direct current as long as fuel is provided.

So far, it has not been practical to burn hydrocarbons, such as natural gas, in a fuel cell. The fuel cells used in the space program burn hydrogen gas to give water, with a cell potential of 1.10 V, as shown in Figure 12.5.

$$2\ H_2(g) + O_2(g) \rightleftharpoons 2\ H_2O(g) \qquad E° = 1.10\ V$$

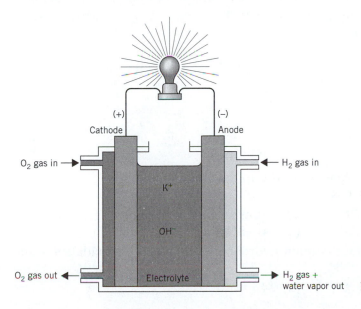

FIGURE 12.5 A hydrogen–oxygen fuel cell.

Fuel cells can also be constructed to burn either ammonia or hydrazine.

$$4 \, NH_3(g) + 3 \, O_2(g) \rightleftharpoons 2 \, N_2(g) + 6 \, H_2O(g) \qquad E° = 1.17 \, V$$

KEY TERMS

Anode	Oxidation number	Salt bridge
Battery	Oxidation state	Spontaneous
Cathode	Oxidizing agent	Standard cell potentials
Cell potential	Positive electrode	Standard conditions
Half-reaction	Redox reactions	Voltaic cell
Negative electrode	Reducing agent	
Oxidation	Reduction	

PROBLEMS

Oxidation–Reduction Reactions

1. Organic chemists often think about oxidation in terms of the loss of a pair of hydrogen atoms. Ethyl alcohol, CH_3CH_2OH, is oxidized to acetaldehyde, CH_3CHO, for example, by removing a pair of hydrogen atoms. Show that this reaction obeys the general rule that oxidation is any process in which the oxidation number of an atom increases.

2. Organic chemists often think about reduction in terms of the gain of a pair of hydrogen atoms. Ethylene, C_2H_4, is reduced when it reacts with H_2 to form ethane, C_2H_6, for example. Show that this reaction obeys the general rule that reduction is any process in which the oxidation number of an atom decreases.

3. Step 6 of the citric acid cycle in Figure 12.1 is the conversion of succinate to fumarate. Show that this reaction is an oxidation–reduction reaction.

4. Step 7 in the citric acid cycle in Figure 12.1 is the conversion of fumarate to malate. Is this an oxidation–reduction reaction? Why or why not?

5. One of the first steps in the corrosion of iron is

$$2 \, Fe(s) + O_2(aq) + 2 \, H_2O(l) \rightleftharpoons 2 \, Fe^{2+}(aq) + 4 \, OH^-(aq)$$

Is this a redox reaction?

Assigning Oxidation Numbers

6. Find the oxidation number of chromium in the following compounds.
   (a) CrO_4^{2-} (b) $Cr_2O_7^{2-}$ (c) CrO_2

7. Calculate the oxidation number of the aluminum atom in the following compounds.
   (a) $Al(OH)_3$ (b) $AlCl_3$ (c) $AlCl_4^-$ (d) $Al_2(SO_4)_3$

8. Which of the following compounds contain hydrogen with a negative oxidation number?
   (a) HCl (b) NH_4^+ (c) LiH (d) C_2H_6 (e) H_2O

9. Arrange the following compounds in order of increasing oxidation number of the carbon atom.
   (a) C   (b) CO   (c) CO_2 (d) H_2CO (e) CH_3OH (f) CH_4

10. Carbon can have any oxidation number between -4 and $+4$. Calculate the oxidation number of carbon in the following compounds.

    (a) 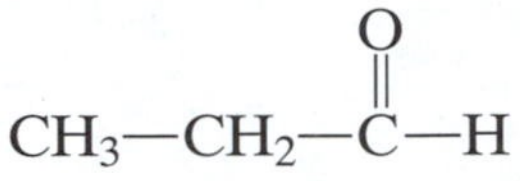 $CH_3-CH_2-\overset{\overset{\textstyle O}{\|}}{C}-H$

    (b) 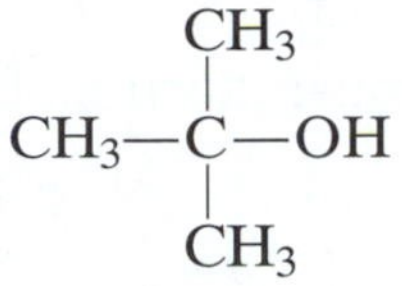 $CH_3-\overset{\overset{\textstyle CH_3}{|}}{\underset{\underset{\textstyle CH_3}{|}}{C}}-OH$

    (c)

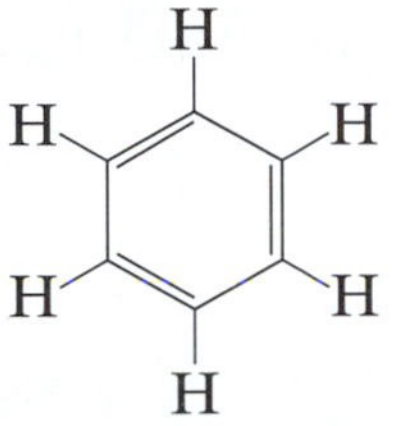

11. Sulfur can have any oxidation number between $+6$ and -2. Calculate the oxidation number of sulfur in the following compounds.
    (a) Na_2S (b) MgS
    (c) CS_2 (d) H_2SO_4
    (e) SF_6 (f) $BaSO_3$
    (g) $NaHSO_4$ (h) SCl_2

12. Calculate the oxidation number of each atom in the following compounds.

    (a) 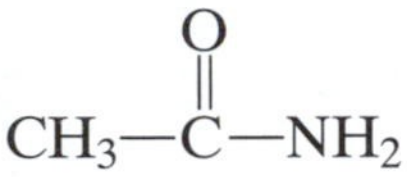 $CH_3-\overset{\overset{\textstyle O}{\|}}{C}-NH_2$

    (b) $CH_2{=}CH-CH_2-OH$

    (c) 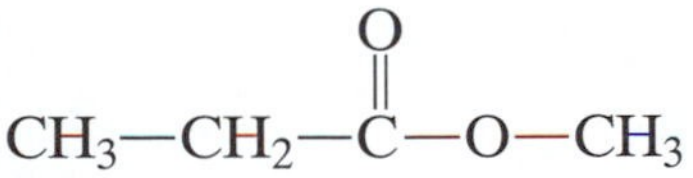 $CH_3-CH_2-\overset{\overset{\textstyle O}{\|}}{C}-O-CH_3$

13. What is the relationship between the oxidation number of chlorine in each of the following acids and the relative strength of the acids? $HClO_4$, $HClO_3$, $HClO_2$, $HClO$

Recognizing Oxidation–Reduction Reactions

14. Which statement(s) correctly describes the following reaction?

    $$3\ Sn^{2+}(aq) + Cr_2O_7{}^{2-}(aq) + 14\ H^+(aq) \rightleftharpoons 3\ Sn^{4+}(aq) + 2\ Cr^{3+}(aq) + 7\ H_2O(l)$$

    (a) Both the Sn^{2+} and H^+ ions are oxidizing agents.
    (b) The $Cr_2O_7{}^{2-}$ ion is the oxidizing agent.
    (c) The Sn^{2+} ion is reduced.
    (d) The Sn^{4+} ion must be a weak reducing agent.
    (e) None of the above are true.

15. Hydronium ion participates in reactions other than acid–base reactions. Which of the following involves oxidation or reduction of H_3O^+?
    (a) $Mg(s) + 2\ H_3O^+(aq) \rightleftharpoons Mg^{2+}(aq) + H_2(g) + 2\ H_2O(l)$
    (b) $H_3O^+(aq) + NH_3(aq) \rightleftharpoons NH_4{}^+(aq) + H_2O(l)$

16. Oxidation–reduction reactions are often used to prepare organic compounds. Determine if the organic reactants are oxidized or reduced. The reactions are *not* written as complete balanced reactions. They show only the major organic reactant and product.

    (a) $CH_3OH(aq) \xrightarrow{K_2Cr_2O_7,\ H^+} H\overset{\displaystyle O}{\overset{\|}{C}}H(aq)$

    (b) $CH_3CH_2CH_2OH(aq) \xrightarrow{K_2Cr_2O_7,\ H^+} CH_3CH_2\overset{\displaystyle O}{\overset{\|}{C}}OH(aq)$

17. Decide whether each of the following reactions involves oxidation–reduction. If it does, identify what is oxidized and what is reduced.

    (a) $4\ CH_3\overset{\displaystyle O}{\overset{\|}{C}}CH_3 + LiAlH_4 + 4\ H_2O \rightleftharpoons 4\ CH_3\overset{\displaystyle OH}{\overset{|}{C}}HCH_3 + LiOH + Al(OH)_3$

    (b) $CH_3CH_2OH \xrightarrow{H_2SO_4} CH_2{=}CH_2 + H_2O$

    (c) $CH_3\overset{\displaystyle O}{\overset{\|}{C}}{-}OH + CH_3NH_2 \rightleftharpoons CH_3\overset{\displaystyle O}{\overset{\|}{C}}{-}O^- + CH_3NH_3{}^+$

18. Use Lewis structures to explain what happens when CO_2 reacts with water to form carbonic acid, H_2CO_3. Is this an oxidation–reduction reaction?

19. Use Lewis structures to explain why reduction of the Hg^{2+} ion to the +1 oxidation state gives a diatomic $Hg_2{}^{2+}$ ion.

20. Use Lewis structures to show that the following is a reduction half-reaction.

$$NO_3{}^-(aq) + 4\ H^+(aq) + 3\ e^- \rightleftharpoons NO(g) + 2\ H_2O(l)$$

21. Use Lewis structures to explain what happens in the following reaction.

$$2\ SO_3{}^{2-}(aq) + O_2(g) \rightleftharpoons 2\ SO_4{}^{2-}(aq)$$

Common Oxidizing Agents and Reducing Agents

22. What makes Cl_2, O_2, $CrO_4{}^{2-}$, and $MnO_4{}^-$ good oxidizing agents?

23. What makes Na, Al, and Zn good reducing agents?

24. Which of the following can't be an oxidizing agent?
    (a) Cl^- (b) Br_2 (c) Fe^{3+} (d) Zn (e) CaH_2

25. Which of the following can't be a reducing agent?
    (a) H_2 (b) Cl_2 (c) Fe^{3+} (d) Al (e) LiH

26. Which of the following can be both an oxidizing agent and a reducing agent?
    (a) H_2 (b) I_2 (c) H_2O_2 (d) P_4 (e) S_8

27. Identify the oxidizing agents on both sides of the following unbalanced equation.

$$CrO_4{}^{2-}(aq) + PH_3(g) \underset{}{\overset{OH^-}{\rightleftharpoons}} Cr(OH)_4{}^-(aq) + P_4(aq)$$

The Relative Strengths of Oxidizing and Reducing Agents

28. Which of the following transition metals is the strongest reducing agent?
    (a) Cr (b) Mn (c) Fe (d) Co (e) Ni

29. Which of the following solutions is the strongest oxidizing agent?
    (a) MnO_4^- in acid (b) MnO_4^- in base
    (c) MnO_2 in acid (d) CrO_4^{2-} in acid
    (e) $Cr_2O_7^{2-}$ in acid

30. Which of the following oxidation–reduction reactions should occur as written?
    (a) $Co(s) + 2\ Cr^{3+}(aq) \rightarrow Co^{2+}(aq) + 2\ Cr^{2+}(aq)$
    (b) $H_2(g) + Zn^{2+}(aq) \rightarrow Zn(s) + 2\ H^+(aq)$
    (c) $Sn(s) + 2\ H^+(aq) \rightarrow Sn^{2+}(aq) + H_2(g)$
    (d) $3\ Na(l) + AlCl_3(l) \rightarrow 3\ NaCl(l) + Al(l)$

31. Which of the following oxidation–reduction reactions should occur as written? See Appendix B.12.
    (a) $PbO_2(s) + Mn^{2+}(aq) \rightarrow Pb^{2+}(aq) + MnO_2(s)$
    (b) $5\ CrO_4^{2-}(aq) + 3\ Mn^{2+}(aq) + 16\ H^+(aq) \rightarrow$
    $$5Cr^{3+}(aq) + 3\ MnO_4^-(aq) + 8\ H_2O(l)$$
    (c) $2\ H_2O_2(aq) \rightarrow 2\ H_2O(l) + O_2(g)$

32. Use Table 12.1 to predict the products of the following oxidation–reduction reactions.
    (a) $HNO_3(aq) + Mn(s) \rightarrow$
    (b) $Mg(s) + CrCl_3(aq) \rightarrow$

33. Which of the following pairs of ions can't coexist in aqueous solution?
    (a) Cr^{2+} and MnO_4^- (b) Fe^{3+} and $Cr_2O_7^{2-}$ (c) Ag^+ and Cu^{2+}
    (d) Fe^{2+} and Cr^{3+}

34. Which of the following pairs of ions can't coexist in aqueous solution? See Appendix B.12.
    (a) Na^+ and $S_2O_8^{2-}$ (b) Hg^{2+} and Cl^- (c) Cr^{2+} and I_3^-

35. Solutions of the Sn^{2+} ion are unstable because they are slowly oxidized by air to the Sn^{4+} ion. Explain why adding small pieces of tin metal to a solution of the Sn^{2+} can produce a solution with a reasonable shelf life.

36. On an exam, students were asked to use half-reactions to write a balanced equation for the following reaction.

$$NO_3^-(aq) + H^+(aq) + Cu(s) \rightleftharpoons Cu^{2+}(aq) + NO(g) + H_2O(l)$$

One student wrote the following half-reactions.

Oxidation $\quad Cu(s) \rightleftharpoons Cu^{2+}(aq) + 2\ e^-$
Reduction $\quad NO_3^-(aq) + 4\ H^+(aq) + 3\ e^- \rightleftharpoons NO(g) + 2\ H_2O(l)$

The student then wrote the following overall equation.

$$NO_3^-(aq) + 4\ H^+(aq) + Cu(s) \rightleftharpoons Cu^{2+}(aq) + NO(g) + 2\ H_2O(l)$$

Explain why the student received no credit for this answer.

Electrical Work from Spontaneous Oxidation–Reduction Reactions

37. Describe an experiment that would allow you to determine the relative strengths of copper and iron metal as reducing agents.

38. Explain why oxidation and reduction half-reactions have to be physically separated for an oxidation–reduction reaction to do work.

39. Describe the function of a salt bridge in an electrochemical cell. Explain what happens when the salt bridge is removed from the system and why.

Voltaic Cells from Spontaneous Oxidation–Reduction Reactions

40. Describe the relationships between pairs of the following terms: *cathode, anode, cation,* and *anion.*
41. Explain why cations move from the salt bridge into the solution containing the cathode of a voltaic cell.
42. Explain what happens when oxidation occurs at the anode and reduction occurs at the cathode in a voltaic cell.

Standard Cell Potentials for Voltaic Cells

43. Describe the difference between E and $E°$ for a cell.
44. Describe the conditions under which the cell potential must be determined in order for the measurement to be equal to $E°$.
45. Describe what the magnitude of $E°$ for an oxidation–reduction reaction tells us about the reaction.
46. Explain why cell potentials measure the relative strengths of a pair of oxidizing agents and the relative strengths of a pair of reducing agents but not their absolute strengths.

Predicting Spontaneous Oxidation–Reduction Reactions from the Sign of $E°$

47. Describe what the sign of $E°$ for an oxidation–reduction reaction tells us about the reaction.
48. Describe what happens to the sign and magnitude of $E°$ for an oxidation–reduction reaction when the direction in which the reaction is written is reversed.
49. Which of the following oxidation–reduction reactions should occur as written when run under standard conditions?
    (a) $Al(s) + Cr^{3+}(aq) \rightarrow Al^{3+}(aq) + Cr(s)$ $E° = 0.966$ V
    (b) $3\ Cr^{2+}(aq) \rightarrow Cr(s) + 2\ Cr^{3+}(aq)$ $E° = 0.33$ V
    (c) $Fe(s) + Cr^{3+}(aq) \rightarrow Cr(s) + Fe^{3+}(aq)$ $E° = -0.70$ V
    (d) $3\ H_2(g) + 2\ Cr^{3+}(aq) \rightarrow 2\ Cr(s) + 6\ H^+(aq)$ $E° = -0.74$
50. Which of the following oxidation–reduction reactions should occur as written when run under standard conditions?
    (a) $2\ Fe^{2+}(aq) + H_2O_2(aq) \rightarrow 2\ Fe^{3+}(aq) + 2\ OH^-(aq)$ $E° = 0.11$ V
    (b) $2\ Fe^{2+}(aq) + Cl_2(aq) \rightarrow 2\ Fe^{3+}(aq) + 2\ Cl^-(aq)$ $E° = 0.588$ V
    (c) $2\ Fe^{2+}(aq) + Br_2(aq) \rightarrow 2\ Fe^{3+}(aq) + 2\ Br^-(aq)$ $E° = 0.317$ V
    (d) $2\ Fe^{2+}(aq) + I_2(aq) \rightarrow 2\ Fe^{3+}(aq) + 2\ I^-(aq)$ $E° = -0.235$ V
51. Use the results of the previous problem to determine the relative strengths of H_2O_2, Cl_2, Br_2, I_2, and the Fe^{3+} ion as oxidizing agents.

Standard Reduction Half-cell Potentials

52. What do the following half-cell reduction potentials tell us about the relative strengths of zinc and copper metal as reducing agents? What do they tell us about the relative strengths of Zn^{2+} and Cu^{2+} ions as oxidizing agents?

$$Zn^{2+} + 2\ e^- \rightleftharpoons Zn \qquad E° = -0.7628 \text{ V}$$
$$Cu^{2+} + 2\ e^- \rightleftharpoons Cu \qquad E° = 0.3402 \text{ V}$$

Predicting Standard Cell Potentials

53. Predict which of the following reactions should occur spontaneously as written when run under standard conditions.
    (a) $Zn(s) + 2 H^+(aq) \rightarrow Zn^{2+}(aq) + H_2(g)$
    (b) $Cr(s) + 3 Fe^{3+}(aq) \rightarrow Cr^{3+}(aq) + 3 Fe^{2+}(aq)$
    (c) $Mn(s) + Mg^{2+}(aq) \rightarrow Mn^{2+}(aq) + Mg(s)$

54. Predict which of the following reactions should occur spontaneously as written when run under standard conditions. See Appendix B.12.
    (a) $HNO_2(aq) + HClO(aq) \rightarrow NO_3^-(aq) + Cl^-(aq) + 2 H^+(aq)$
    (b) $2 ClO_2^-(aq) \rightarrow ClO^-(aq) + ClO_3^-(aq)$
    (c) $3 Cu(s) + 2 HNO_3(aq) + 6 H^+(aq) \rightarrow 3 Cu^{2+}(aq) + 2 NO(g) + 4 H_2O(l)$

55. Describe what happens inside the voltaic cell that employs the following chemical reaction to produce a potential.

$$Zn(s) + Cu^{2+}(aq) \rightleftharpoons Zn^{2+}(aq) + Cu(s)$$

56. Calculate $E°$ for the following reaction and predict whether the reaction should occur spontaneously as written when run under standard conditions.

$$6 Fe^{2+}(aq) + Cr_2O_7^{2-}(aq) + 14 H^+(aq) \longrightarrow 6 Fe^{3+}(aq) + 2 Cr^{3+}(aq) + 7 H_2O(l)$$

57. Calculate $E°$ for the following reaction and predict whether the reaction should occur spontaneously as written when run under standard conditions. See Appendix B.12.

$$Al^{3+}(aq) + Fe(s) \longrightarrow Fe^{3+}(aq) + Al(s)$$

58. Some metals react with water at room temperature, and others do not. Calculate $E°$ for the following reaction and predict whether the reaction should occur spontaneously as written when run under standard conditions. See Appendix B.12.

$$Ca(s) + 2 H_2O(l) \longrightarrow Ca^{2+}(aq) + 2 OH^-(aq) + H_2(g)$$

59. Calculate $E°$ for the following reaction to predict whether the Cu^+ ion should spontaneously undergo disproportionation under standard conditions.

$$2 Cu^+(aq) \longrightarrow Cu(s) + Cu^{2+}(aq)$$

60. Use standard half-cell reduction potentials to predict whether an $Fe^{2+}(aq)$ solution should react with $H^+(aq)$ solution to produce $Fe^{3+}(aq)$ and $H_2(g)$ under standard conditions.

Using Standard Half-Cell Potentials to Understand Chemical Reactions

61. Use half-cell potentials to predict which of the following metals will be most reactive with strong acids (H^+).
    (a) Al (b) Mg (c) Zn

62. Exercise 12.7 described why Cu metal dissolves in nitric acid but not in hydrochloric acid. Use half-cell potentials to explain why Zn metal will dissolve in both acids.

63. Explain why copper, gold, mercury, platinum, and silver can be found in the metallic state in nature.

64. Which of the following is the strongest reducing agent?
    (a) Zn (b) Fe (c) H_2 (d) Cu (e) Ag

65. Which of the following is the strongest oxidizing agent? See Appendix B.12.
    (a) H_2O_2 in OH^- (b) H_2O_2 in H^+ (c) Na (d) O_2 in H^+ (e) Al

66. Which of the following is the strongest reducing agent? See Appendix B.12.
    (a) H^+ (b) H_2 (c) H^- (d) H_2O (e) O_2

67. Which is the better oxidizing agent? See Appendix B.12.
    (a) H^- or K (b) Sn or Fe^{2+} (c) Ag^+ or Au^+

Integrated Problems

68. Oxalic acid reacts with the chromate ion in acidic solution to give CO_2 and Cr^{3+} ions.

$$10\ H^+(aq) + 3\ H_2C_2O_4(aq) + 2\ CrO_4^{2-}(aq) \xrightarrow{H^+} 6\ CO_2(g) + 2\ Cr^{3+}(aq) + 8\ H_2O(l)$$

If 10.0 mL of oxalic acid consumes 40.0 mL of 0.0250 M CrO_4^{2-} solution, what is the molarity of the oxalic acid solution?

69. A 2.50 g sample of bronze was dissolved in sulfuric acid.

$$Cu(s) + 2\ H_2SO_4(aq) \rightleftharpoons CuSO_4(aq) + SO_2(g) + 2\ H_2O(l)$$

The $CuSO_4$ formed in the reaction was mixed with KI to form CuI.

$$2\ CuSO_4(aq) + 5\ I^-(aq) \rightleftharpoons 2\ CuI(s) + I_3^-(aq) + 2\ SO_4^{2-}(aq)$$

The I_3^- formed in the reaction was then titrated with $S_2O_3^{2-}$.

$$I_3^-(aq) + 2\ S_2O_3^{2-}(aq) \rightleftharpoons 3\ I^-(aq) + S_4O_6^{2-}(aq)$$

If 31.5 mL of 1.00 M $S_2O_3^{2-}$ ion was consumed in the titration, what was the percent by weight of copper in the original sample of bronze?

70. A 20.00 mL sample of a $K_2C_2O_4$ solution was acidified and then titrated with an acidic 0.256 M $KMnO_4$ solution. What is the molarity of the oxalate solution if it took 14.6 mL of the MnO_4^- solution to reach the end point of the titration?

$$5\ C_2H_2O_4(aq) + 2\ MnO_4^-(aq) + 6\ H^+(aq) \rightleftharpoons$$
$$10\ CO_2(g) + 2\ Mn^{2+}(aq) + 8\ H_2O(l)$$

71. Calculate the number of grams of ferrous chloride ($FeCl_2$) that can be oxidized by 3.2 mL of 3.0 M $KMnO_4$.

$$5\ Fe^{2+}(aq) + MnO_4^-(aq) + 8\ H^+(aq) \rightleftharpoons Mn^{2+}(aq) + 4\ H_2O(l) + 5\ Fe^{3+}(aq)$$

72. The stoichiometry of the reaction between hydrogen peroxide and the MnO_4^- ion depends on the pH of the solution in which the reaction is run.

$$2\ MnO_4^-(aq) + 5\ H_2O_2(aq) + 6\ H^+(aq) \rightleftharpoons 2\ Mn^{2+}(aq) + 5\ O_2(g) + 8\ H_2O(l)$$
$$2\ MnO_4^-(aq) + 3\ H_2O_2(aq) \rightleftharpoons$$
$$2\ MnO_2(s) + 3\ O_2(g) + 2\ OH^-(aq) + 2\ H_2O(l)$$

Assume that it took 32.45 mL of 0.145 M KMnO$_4$ to titrate 28.46 mL of 0.248 M H$_2$O$_2$. Was the reaction run in acidic or basic solution?

73. A 1.893 g sample of oxalic acid (H$_2$C$_2$O$_4$·2H$_2$O) was dissolved in about 20 mL of water, acidified, and then diluted to a total volume of 50.00 mL. It took 19.35 mL of an acidic MnO$_4^-$ ion solution to titrate a 25.00 mL aliquot of the oxalic acid solution to the MnO$_4^-$ end point. The MnO$_4^-$ solution was then used to standardize an H$_2$O$_2$ solution of unknown concentration. If it took 15.37 mL of the MnO$_4^-$ ion solution to titrate a 25.00 mL aliquot of the H$_2$O$_2$ solution, what was the concentration of the hydrogen peroxide solution?

$$5 \; H_2C_2O_4(aq) + 2 \; MnO_4^-(aq) + 6 \; H^+(aq) \rightleftharpoons 10 \; CO_2(g) + 2 \; Mn^{2+}(aq) + 8 \; H_2O(l)$$

74. Which of the following reactions are oxidation–reduction reactions? In each case identify what is oxidized and what is reduced.
    (a) the production of iron from its ore

$$Fe_2O_3(s) + 3 \; CO(g) \rightleftharpoons 2 \; Fe(l) + 3 \; CO_2(g)$$

    (b) the production of silicon by heating Ca$_3$(PO$_4$)$_2$ with carbon and sand

$$2 \; Ca_3(PO_4)_2(l) + 6 \; SiO_2(l) + 10 \; C(s) \rightleftharpoons P_4(g) + 6 \; CaSiO_3(l) + 10 \; CO(g)$$

    (c) the formation of barium sulfate

$$ZnSO_4(aq) + BaCl_2(aq) \rightleftharpoons ZnCl_2(aq) + BaSO_4(s)$$

75. An electrochemical cell is constructed from a Zn/Zn^{2+} electrode and a Cr/Cr^{3+} electrode as shown below. The voltage measured for the cell is 0.02 V. When the switch is closed the chromium electrode increases in mass.

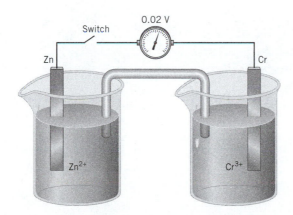

    (a) Which electrode is the anode and which is the cathode?
    (b) What is the direction of electron flow?
    (c) Which electrode is positive? Which electrode is negative?
    (d) What are the chemical reactions occurring at each electrode?
    (e) What is the overall chemical reaction of the cell?
    (f) If the Zn/Zn^{2+} couple is replaced by a standard hydrogen electrode (SHE) and the SHE is then connected to the Cr/Cr^{3+} couple, what will be the direction of the electron flow and the spontaneous overall chemical reaction?

76. An iron/tin galvanic cell is constructed. An iron strip is placed into a 1.0 M solution of $Fe(NO_3)_3$, and this electrode is connected by a salt bridge to an electrode consisting of a strip of tin in a 1.0 M $Sn(NO_3)_2$ solution. The tin strip is observed to dissolve.
    (a) Roughly sketch the galvanic cell.
    (b) Identify the anode and cathode.
    (c) Which electrode is positive and which is negative?
    (d) What is the overall spontaneous chemical reaction?
    (e) What is the standard potential of the cell?

77. The heroine of a movie has been captured by the villains and thrown into a storage room of an old mansion. There are steel bars on the windows, a silver door knob on the door, brass hinges on the skylight, and exposed copper plumbing in the room. In the corner of the room she finds a container of muriatic acid (6 M HCl), used for cleaning bricks. Suggest some options for her escape. Explain your reasoning.

C H A P T E R
12
SPECIAL TOPICS I

12AI.1   ELECTROLYTIC CELLS

The cells discussed so far have one thing in common: they all use a spontaneous chemical reaction to drive an electric current through an external circuit. These voltaic cells are important because they are the basis for the batteries that fuel modern society. But they are not the only kind of electrochemical cell. It is also possible to construct a cell that does work on a chemical system by driving an electric current through the system. These cells are called **electrolytic cells.** Electrolysis is used to drive an oxidation–reduction reaction in the direction in which it doesn't occur spontaneously.

12AI.2   THE ELECTROLYSIS OF MOLTEN NaCl

An idealized cell for the electrolysis of sodium chloride is shown in Figure 12AI.1. A source of direct current is connected to a pair of inert electrodes immersed in molten sodium chloride. The Na^+ ions flow toward the negative electrode and the Cl^- ions flow toward the positive electrode.

When Na^+ ions collide with the negative electrode, the battery carries a large enough potential to force the ions to pick up electrons to form sodium metal.

$$\text{Negative electrode (cathode)} \qquad Na^+ + e^- \rightleftharpoons Na$$

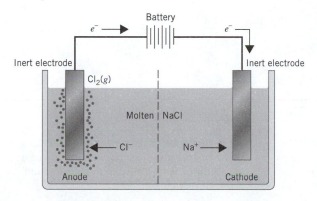

FIGURE 12AI.1 Electrolysis of molten sodium chloride involves using an electric current to reduce Na^+ ions to sodium metal at the cathode and to oxidize Cl^- ions to Cl_2 gas at the anode.

585

Cl^- ions that collide with the positive electrode are oxidized to Cl_2 gas, which bubbles off at this electrode.

$$\text{Positive electrode (anode)} \qquad 2\,Cl^- \rightleftharpoons Cl_2 + 2\,e^-$$

The net effect of passing an electric current through the molten salt in the electrolytic cell is to decompose sodium chloride into its elements, sodium metal and chlorine gas.

Electrolysis of NaCl
Cathode $(-)$	$Na^+ + e^- \rightleftharpoons Na$
Anode $(+)$	$2\,Cl^- \rightleftharpoons Cl_2 + 2\,e^-$

This example explains why the process is called **electrolysis.** The suffix *-lysis* comes from a Greek stem meaning "to loosen or split up." Electrolysis literally uses an electric current to split a compound into its elements.

$$2\,NaCl(l) \xrightarrow{\text{electrolysis}} 2\,Na(l) + Cl_2(g)$$

This example also illustrates the difference between voltaic cells and electrolytic cells. Voltaic cells use the energy given off in a spontaneous reaction to do electrical work. Electrolytic cells use electrical work as source of energy to drive the reaction in the opposite direction.

The dotted vertical line in the center of Figure 12AI.1 represents a diaphragm that keeps the Cl_2 gas produced at the anode from coming into contact with the sodium metal generated at the cathode. The function of the diaphragm can be understood by turning to a more realistic drawing of the commercial Downs cell used to electrolyze sodium chloride, shown in Figure 12AI.2.

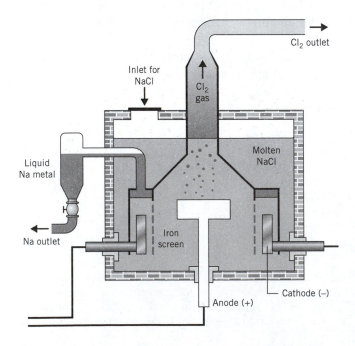

FIGURE 12AI.2 Cross section of the Downs cell used for the electrolysis of a molten mixture of calcium chloride and sodium chloride.

Chlorine gas that forms on the graphite anode inserted into the bottom of the Downs cell bubbles through the molten sodium chloride into a funnel at the top of the cell. Sodium metal that forms at the cathode floats up through the molten sodium chloride into a sodium-collecting ring, from which it is periodically drained. The diaphragm that separates the two electrodes is a screen of iron gauze, which prevents the explosive reaction between sodium metal and chlorine to form sodium chloride that would occur if the products of the electrolysis reaction came in contact.

The feedstock for the Downs cell is a 3:2 mixture by mass of $CaCl_2$ and $NaCl$. The mixture is used because it has a melting point of 580°C, whereas pure sodium chloride has to be heated to more than 800°C before it melts.

12AI.3   THE ELECTROLYSIS OF AQUEOUS NaCl

Figure 12AI.3 shows an idealized drawing of a cell in which an aqueous solution of sodium chloride is electrolyzed. Once again, the Na^+ ions migrate toward the negative electrode and the Cl^- ions migrate toward the positive electrode. But now there are two substances that can be reduced at the cathode: Na^+ ions and water molecules.

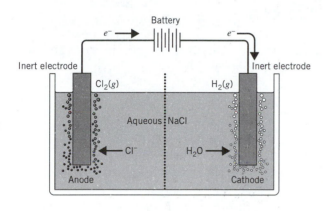

FIGURE 12AI.3 Electrolysis of aqueous sodium chloride results in reduction of water to form H_2 gas at the cathode and oxidation of Cl^- ions at the anode.

Cathode (−)
$$Na^+ + e^- \rightleftharpoons Na \qquad\qquad E°_{Red} = -2.71 \text{ V}$$
$$2\,H_2O + 2\,e^- \rightleftharpoons H_2 + 2\,OH^- \quad E°_{Red} = -0.83 \text{ V}$$

Because it is much easier to reduce water than Na^+ ions, the only product formed at the cathode is hydrogen gas.

$$\text{Cathode (−)} \qquad 2\,H_2O(l) + 2\,e^- \rightleftharpoons H_2(g) + 2\,OH^-(aq)$$

There are also two substances that can be oxidized at the anode: Cl^- ions and water molecules.

Anode (+)
$$2\,Cl^- \rightleftharpoons Cl_2 + 2\,e^- \qquad\qquad E°_{Ox} = -1.36 \text{ V}$$
$$2\,H_2O \rightleftharpoons O_2 + 4\,H^+ + 4\,e^- \quad E°_{Ox} = -1.23 \text{ V}$$

The standard potentials for the two half-reactions are so close to each other that we might expect to see a mixture of Cl_2 and O_2 gas collect at the anode. In practice, the only product of the reaction is Cl_2.

$$\text{Anode (+)} \qquad 2\,Cl^- \rightleftharpoons Cl_2 + 2\,e^-$$

At first glance, it would seem easier to oxidize water ($E^\circ_{Ox} = -1.23$ volts) than Cl^- ions ($E^\circ_{Ox} = -1.36$ volts). It is worth noting, however, that these are standard half-cell potentials, and the cell is never allowed to reach standard conditions. The solution is typically 25% NaCl by mass, which significantly decreases the potential required to oxidize the Cl^- ion. The pH of the cell is also kept very high, which decreases the oxidation potential for water. The deciding factor is a phenomenon known as **overvoltage,** which is the extra voltage that must be applied to a reaction to get it to occur at the rate at which it would occur in an ideal system. Under ideal conditions, a potential of 1.23 volts is large enough to oxidize water to O_2 gas. Under real conditions, however, it can take a much larger voltage to initiate the reaction. (The overvoltage for the oxidation of water can be as large as 1 volt.) By carefully choosing the electrode to maximize the overvoltage for the oxidation of water and then carefully controlling the potential at which the cell operates, we can ensure that only chlorine is produced in the reaction.

In summary, electrolysis of aqueous solutions of sodium chloride does not give the same products as electrolysis of molten sodium chloride. Electrolysis of molten NaCl decomposes the compound into its elements.

$$2\,NaCl(l) \xrightarrow{\text{electrolysis}} 2\,Na(l) + Cl_2(g)$$

Electrolysis of aqueous NaCl solutions gives a mixture of hydrogen and chlorine gas and an aqueous sodium hydroxide solution, as shown in Figure 12AI.4.

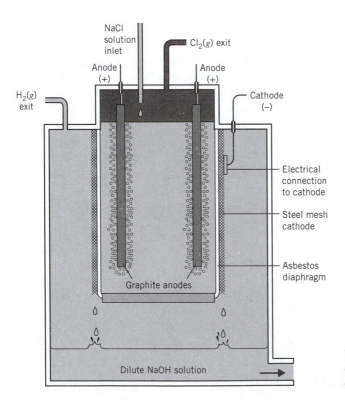

FIGURE 12AI.4 About half of the commercial cells for the electrolysis of aqueous sodium chloride use the diaphragm cell shown here.

$$2\,NaCl(aq) + 2\,H_2O(l) \xrightarrow{\text{electrolysis}} 2\,Na^+(aq) + 2\,OH^-(aq) + H_2(g) + Cl_2(g)$$

Because the demand for chlorine is much larger than the demand for sodium, electrolysis of aqueous sodium chloride is a more important process commercially. Electrolysis of an aqueous NaCl solution has two other advantages. It produces H_2 gas at the cathode, which can be collected and sold. It also produces NaOH, which can be drained from the bottom of the electrolytic cell and sold.

The dotted vertical line in Figure 12AI.3 represents a diaphragm that prevents the Cl_2 produced at the anode in the cell from coming into contact with the NaOH that accumulates at the cathode. When the diaphragm is removed from the cell, the products of the electrolysis of aqueous sodium chloride react to form sodium hypochlorite, which is the first step in the preparation of hypochlorite bleaches, such as Clorox.

$$Cl_2(g) + 2\ OH^-(aq) \rightleftharpoons Cl^-(aq) + OCl^-(aq) + H_2O(l)$$

12AI.4   ELECTROLYSIS OF WATER

A standard apparatus for the electrolysis of water is shown in Figure 12AI.5.

$$2\ H_2O(l) \xrightarrow{\text{electrolysis}} 2\ H_2(g) + O_2(g)$$

A pair of inert electrodes are sealed in opposite ends of a container designed to collect the H_2 and O_2 gas given off in the reaction. The electrodes are then connected to a battery or another source of electric current.

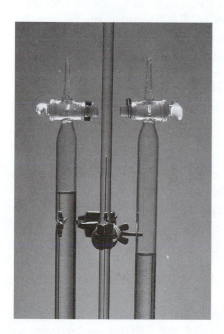

Electrolysis of an aqueous Na_2SO_4 solution results in the reduction of water to form H_2 gas at the cathode and the oxidation of water to form O_2 gas at the anode.

By itself, water is a very poor conductor of electricity. We therefore add an electrolyte to water to provide ions that can flow through the solution, thereby completing the electric circuit. The electrolyte must be soluble in water. It should also be relatively inexpensive. Most importantly, it must contain ions that are harder to oxidize or reduce than water.

$$2\ H_2O + 2\ e^- \rightleftharpoons H_2 + 2\ OH^- \qquad E^\circ_{Red} = -0.83\ V$$
$$2\ H_2O \rightleftharpoons O_2 + 4\ H^+ + 4\ e^- \qquad E^\circ_{Ox} = -1.23\ V$$

According to Appendix B.12, the following cations are harder to reduce than water: Li^+, Rb^+, K^+, Cs^+, Ba^{2+}, Sr^{2+}, Ca^{2+}, Na^+, and Mg^{2+}. Two of the cations are more likely candidates than the others because they form inexpensive, soluble salts: Na^+ and K^+.

Appendix B.12 suggests that the SO_4^{2-} ion might be the best anion to use because it is one of the most difficult anions to oxidize. The potential for oxidation of the ion to the peroxydisulfate ion is -2.05 volts.

$$2\ SO_4^{2-} \rightleftharpoons S_2O_8^{2-} + 2\ e^- \qquad E^{\circ}_{Ox} = -2.05\ V$$

When an aqueous solution of either Na_2SO_4 or K_2SO_4 is electrolyzed in the apparatus shown in Figure 12AI.5, H_2 gas collects at one electrode and O_2 gas collects at the other.

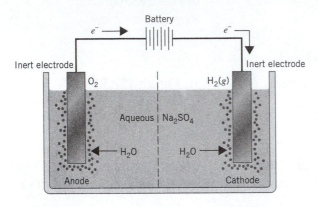

FIGURE 12AI.5 Electrolysis of an aqueous Na_2SO_4 solution results in the reduction of water to form H_2 gas at the cathode and oxidation of water to form O_2 gas at the anode.

What would happen if we added an indicator such as bromothymol blue to the apparatus? Bromothymol blue turns yellow in acidic solutions (pH $<$ 6) and turns blue in basic solutions (pH $>$ 7.6). According to the equations for the two half-reactions, the indicators should turn yellow at the anode and blue at the cathode.

$$\text{Cathode } (-) \quad 2\ H_2O + 2\ e^- \rightleftharpoons H_2 + 2\ OH^-$$
$$\text{Anode } (+) \quad 2\ H_2O \rightleftharpoons O_2 + 4\ H^+ + 4\ e^-$$

12AI.5 FARADAY'S LAW

Humphry Davy's experiments with electrochemistry led to his discovery of sodium, potassium, magnesium, calcium, barium, and strontium. It has been said, however, that his greatest discovery was Michael Faraday. Faraday was apprenticed to a bookbinder in London at the age of 13. His exposure to works brought into the shop for binding excited his interest in science, and he started attending lectures on science and doing experiments in chemistry and electricity on his own. He was eventually hired by Davy as a secretary and scientific assistant, and thereby began a lifelong commitment to research in both chemistry and physics.

Faraday's early research on electrolysis led him to propose a relationship between the amount of current passed through a solution and the mass of the substance decomposed or produced by the current. Faraday's law of electrolysis can be stated as follows.

The amount of a substance consumed or produced at one of the electrodes in an electrolytic cell is directly proportional to the amount of electricity that passes through the cell.

In order to use Faraday's law we need to recognize the relationship between current, time, and the amount of electric charge that flows through a circuit. By definition, one coulomb of charge is transferred when a 1 ampere (A) current flows for 1 second.

$$1\ C = 1\ A \cdot s$$

To illustrate how Faraday's law can be used, let's calculate the number of grams of sodium metal that will form at the cathode when a 10.0 A current is passed through molten sodium chloride for a period of 4.00 h.

We start by calculating the amount of electric charge that flows through the cell.

$$10.0 \text{ A} \times 4.00 \text{ h} \times \frac{60 \text{ min}}{1 \text{ h}} \times \frac{60 \text{ s}}{1 \text{ min}} \times \frac{1 \text{ C}}{1 \text{ A} \cdot \text{s}} = 1.44 \times 10^5 \text{ C}$$

Before we can use this information, we need a bridge between the macroscopic quantity and the phenomenon that occurs on the atomic scale. This bridge is represented by Faraday's constant, which describes the number of coulombs of charge carried by a mole of electrons.

$$\frac{6.022137 \times 10^{23} \text{ } e^-}{1 \text{ mol}} \times \frac{1.60217733 \times 10^{-19} \text{ C}}{1 \text{ } e^-} = 96485.31 \text{ C/mol}$$

Thus, the number of moles of electrons transferred when 1.44×10^5 C of electric charge flows through the cell can be calculated as follows.

$$1.44 \times 10^5 \text{ C} \times \frac{1 \text{ mol } e^-}{96,485 \text{ C}} = 1.49 \text{ mol } e^-$$

According to the balanced equation for the reaction that occurs at the cathode of the cell, we get one mole of sodium for every mole of electrons.

$$\text{Cathode } (-) \qquad \text{Na}^+ + e^- \rightleftharpoons \text{Na}$$

Thus, we get 1.49 mol, or 34.3 g, of sodium in 4.00 h.

$$1.49 \text{ mol Na} \times \frac{22.990 \text{ g Na}}{1 \text{ mol Na}} = 34.3 \text{ g Na}$$

The consequences of the calculation are interesting. We would have to run the electrolysis for more than 2 days to prepare a pound of sodium.

Exercise 12AI.1

Calculate the volume of H_2 gas at 25°C and 1.00 atm that will collect at the cathode when an aqueous solution of Na_2SO_4 is electrolyzed for 2.00 h with a 10.0 A current.

Solution

We start by calculating the amount of electrical charge that passes through the solution.

$$10.0 \text{ A} \times 2.00 \text{ h} \times \frac{60 \text{ min}}{1 \text{ h}} \times \frac{60 \text{ s}}{1 \text{ min}} \times \frac{1 \text{ C}}{1 \text{ A} \cdot \text{s}} = 7.20 \times 10^4 \text{ C}$$

We then calculate the number of moles of electrons that carry this charge.

$$7.20 \times 10^4 \text{ C} \times \frac{1 \text{ mol } e^-}{96,485 \text{ C}} = 0.746 \text{ mol } e^-$$

The balanced equation for the reaction that produces H_2 gas at the cathode indicates that we get a mole of H_2 gas for every 2 mol of electrons.

$$\text{Cathode } (-) \qquad 2\,H_2O + 2\,e^- \rightleftharpoons H_2 + 2\,OH^-$$

We therefore get one mole of H_2 gas at the cathode for every 2 mol of electrons that flow through the cell.

$$0.746 \text{ mol } e^- \times \frac{1 \text{ mol } H_2}{2 \text{ mol } e^-} = 0.373 \text{ mol } H_2$$

We now have the information we need to calculate the volume of the gas produced in this reaction.

$$V = \frac{nRT}{P} = \frac{(0.373 \text{ mol})(0.08206 \text{ L} \cdot \text{atm/mol} \cdot \text{K})(298 \text{ K})}{1.00 \text{ atm}} = 9.12 \text{ L}$$

We can extend the general pattern outlined in this section to answer questions that might seem impossible at first glance.

Exercise 12AI.2

Determine the oxidation number of the chromium in an unknown salt if electrolysis of a molten sample of the salt for 1.50 hours with a 10.0 A current deposits 9.71 grams of chromium metal at the cathode.

Solution

We start, as before, by calculating the number of moles of electrons that passed through the cell during electrolysis.

$$10.0 \text{ A} \times 1.50 \text{ h} \times \frac{60 \text{ min}}{1 \text{ h}} \times \frac{60 \text{ s}}{1 \text{ min}} \times \frac{1 \text{ C}}{1 \text{ A} \cdot \text{s}} = 5.40 \times 10^4 \text{ C}$$

$$5.40 \times 10^4 \text{ C} \times \frac{1 \text{ mol } e^-}{96,485 \text{ C}} = 0.560 \text{ mol } e^-$$

Since we don't know the balanced equation for the reaction at the cathode in this cell, it doesn't seem obvious how we are going to use this information. We therefore write down this result in a conspicuous location, then go back to the original statement of the question to see what else can be done.

The problem tells us the mass of chromium deposited at the cathode. We need to transform the mass into the number of moles of chromium metal generated.

$$9.71 \text{ g Cr} \times \frac{1 \text{ mol Cr}}{51.996 \text{ g Cr}} = 0.187 \text{ mol Cr}$$

We now know the number of moles of chromium metal produced and the number of moles of electrons it took to produce the metal. We might therefore look at the relationship between the moles of electrons consumed in the reaction and the moles of chromium produced.

$$\frac{0.560 \text{ mol } e^-}{0.187 \text{ mol Cr}} = \frac{3}{1}$$

Three moles of electrons are consumed for every mole of chromium metal produced. The only way to explain this is to assume that the net reaction at the cathode involves reduction of Cr^{3+} ions to chromium metal.

$$\text{Cathode } (-) \qquad Cr^{3+} + 3\,e^- \rightleftharpoons Cr$$

Thus, the oxidation number of chromium in the unknown salt must be +3.

12AI.6 GALVANIC CORROSION AND CATHODIC PROTECTION

The model for electrochemical reactions developed in this chapter can be used to explain the behavior observed when the corrosion of iron metal is studied.

- **Iron rusts very slowly in contact with dry air.** Iron metal rusts in contact with dry air because iron atoms on the surface of the metal slowly react with oxygen in the atmosphere to form a mixture of oxides.

$$2\,Fe(s) + O_2(g) \rightleftharpoons 2\,FeO(s)$$
$$4\,Fe(s) + 3\,O_2(g) \rightleftharpoons 2\,Fe_2O_3(s)$$
$$3\,Fe(s) + 2\,O_2(g) \rightleftharpoons Fe_3O_4(s)$$

The reaction is uneven, and there are a number of holes in the oxide layer that forms on the surface of the metal. Oxygen atoms migrate through the holes to the metal surface below the oxide layer. Corrosion therefore continues slowly, but surely. At sites where the metal was deformed when it was worked or shaped, the iron oxide often breaks off, exposing a fresh metal surface to further corrosion and pitting.

- **Iron corrodes on contact with strong acids.** Iron metal dissolves in strong acid, as would be expected from its position in Table 12.1. If the acid is boiled to remove any dissolved O_2 gas, the acid reacts with iron to form solutions of the Fe^{2+} ion. This can be understood by comparing the overall potential for the reaction between iron metal and acid to form Fe^{2+} ions with the potential for the reaction to form Fe^{3+} ions.

$$Fe(s) + 2\,H^+(aq) \rightleftharpoons Fe^{2+}(aq) + H_2(g) \qquad E^\circ = 0.409\ V$$
$$2\,Fe(s) + 6\,H^+(aq) \rightleftharpoons 2\,Fe^{3+}(aq) + 3\,H_2(g) \qquad E^\circ = 0.036\ V$$

If O_2 is present, the Fe^{2+} formed when the iron dissolves is oxidized to Fe^{3+}.

$$4\,Fe^{2+}(aq) + O_2(g) + 4\,H^+(aq) \rightleftharpoons 4\,Fe^{3+}(aq) + 2\,H_2O(l) \qquad E^\circ = 0.46\ V$$

- **Iron does not rust in contact with water that doesn't contain dissolved O_2 gas.** The standard potential for the reaction between iron metal and the H^+ ion is favorable.

$$Fe(s) + 2\,H^+(aq) \rightleftharpoons Fe^{2+}(aq) + H_2(g) \qquad E^\circ = 0.409\ V$$

However, the concentration of the H^+ ion in water is far from the standard state of 1 M. The cell potential for the reaction is lowered due to the low concentration of H^+. This is a large enough change to make the reaction slightly unfavorable.

- **Iron rusts more rapidly in contact with water that contains dissolved O_2 than it does in dry air.** When water and O_2 are both present, the oxidation and reduction reactions no longer have to occur at the same point on the metal surface. The iron metal now simultaneously acts as the anode, the cathode, and the circuit through which electrons flow, as shown in Figure 12AI.6. At one point on the metal surface, iron metal is oxidized to Fe^{2+} ions.

$$\text{Anode} \qquad \text{Fe} \rightleftharpoons \text{Fe}^{2+} + 2\,e^- \qquad E^{\circ}_{Ox} = 0.409 \text{ V}$$

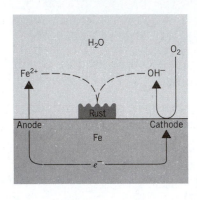

FIGURE 12AI.6 When iron metal is immersed in water that contains dissolved oxygen, oxidation of the iron and reduction of O_2 can occur at different points on the metal surface. The net effect of the reaction is the formation of Fe^{2+} and OH^- ions, which diffuse toward each other to form a complex that is further oxidized to form rust, $Fe_2O_3 \cdot 3H_2O$.

The electrons released in the reaction flow through the metal to another point, at which O_2 dissolved in the water is reduced.

$$\text{Cathode} \qquad \text{O}_2 + 2\,\text{H}_2\text{O} + 4\,e^- \rightleftharpoons 4\,\text{OH}^- \qquad E^{\circ}_{Red} = 0.401 \text{ V}$$

The net result is the formation of an aqueous solution of the Fe^{2+} and OH^- ions.

$$2\,\text{Fe}(s) + \text{O}_2(g) + 2\,\text{H}_2\text{O}(l) \rightleftharpoons 2\,\text{Fe}^{2+}(aq) + 4\,\text{OH}^- \qquad E^{\circ} = 0.810 \text{ V}$$

The Fe^{2+} and OH^- ions produced in the half-reactions diffuse toward each other and eventually combine to form a hydrated iron(II) oxide that undergoes further oxidation to form rust, $Fe_2O_3 \cdot 3H_2O$.

- **Iron rusts even more rapidly when it comes in contact with copper metal.** A voltaic cell is created at the point at which the two metals touch, as shown in Figure 12AI.7. The iron metal acts as the anode, giving up electrons to form Fe^{2+} ions.

$$\text{Anode} \qquad \text{Fe} \rightleftharpoons \text{Fe}^{2+} + 2\,e^- \qquad E^{\circ}_{Ox} = 0.409 \text{ V}$$

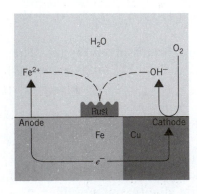

FIGURE 12AI.7 When iron comes in contact with copper metal, a galvanic cell is created in which the iron is oxidized to Fe^{2+} ions. The electrons given off are transferred to the copper metal, where they are used to reduce O_2 to OH^- ions.

The electrons flow to the copper metal, which becomes the cathode at which O_2 dissolved in water is reduced.

$$\text{Cathode} \qquad O_2 + 2\,H_2O + 4\,e^- \rightleftharpoons 4\,OH^- \qquad E^\circ_{Red} = 0.401 \text{ V}$$

Once again, the Fe^{2+} ions produced at the anode diffuse toward the OH^- ions given off at the cathode and eventually combine with the OH^- ions to form rust. This process is called *galvanic corrosion* because the overall reaction is similar to what would happen if the reaction were run in a galvanic cell.

Copper metal doesn't corrode when it touches iron because copper metal is a weaker reducing agent than iron. We can see this by imagining the voltaic cell that would be created if copper acted as the anode and the electrons flowed to iron, which would act as the cathode at which O_2 was reduced.

$$\text{Anode} \qquad Cu \rightleftharpoons Cu^{2+} + 2\,e^- \qquad\qquad E^\circ_{Ox} = -0.340 \text{ V}$$
$$\text{Cathode} \qquad O_2 + 2\,H_2O + 4\,e^- \rightleftharpoons 4\,OH^- \qquad E^\circ_{Red} = 0.401 \text{ V}$$

The overall potential for the reaction is much smaller than the potential for the reaction that occurs when iron acts as the anode.

$$Cu(s) + O_2(g) + 2\,H_2O(l) \rightleftharpoons Cu^{2+}(aq) + 4\,OH^-(aq) \qquad E^\circ = 0.061 \text{ V}$$
$$Fe(s) + O_2(g) + 2\,H_2O(l) \rightleftharpoons Fe^{2+}(aq) + 4\,OH^-(aq) \qquad E^\circ = 0.810 \text{ V}$$

The iron therefore corrodes before the copper.

An interesting question is raised when Figures 12AI.6 and 12AI.7 are compared. The overall reaction for the corrosion of iron is the same, regardless of whether copper metal is in contact with the iron metal. Thus, the overall cell potential for the reactions must be the same. Why, then, does iron corrode so much faster when it comes in contact with copper metal?

The answer seems to rest with the concept of overvoltage introduced in Section 12AI.3. The actual potential for the reduction of O_2 on a metal surface is usually larger than the value reported in tables of standard cell potentials. The overvoltage for the reduction of O_2 on copper metal, however, is significantly smaller than the overvoltage on iron. Since overvoltage is defined as the extra voltage that must be applied to a reaction to get it to occur at the same rate, oxidation of iron metal in contact with copper should occur much more rapidly than the corresponding reaction on an isolated iron metal surface.

- **Iron does not rust when it is in contact with zinc or magnesium metal.** Iron can be protected from corrosion if it is allowed to stay in contact with a metal that is a better reducing agent, as shown in Figure 12AI.8. This process is called **cathodic protection** because it involves making iron the cathode of a voltaic cell. The zinc or magnesium metal acts as the anode of a voltaic cell.

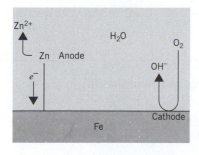

FIGURE 12AI.8 The corrosion of iron can be prevented by cathodic protection. The iron is either coated with zinc to form galvanized iron or connected to a piece of zinc or magnesium metal. Zinc and magnesium are both stronger reducing agents than iron, so they are oxidized instead of the iron. The iron therefore becomes the cathode at which O_2 is reduced, which protects the iron from corrosion.

$$\text{Anode} \qquad \text{Zn} \rightleftharpoons \text{Zn}^{2+} + 2\,e^- \qquad E°_{Ox} = 0.763 \text{ V}$$

The electrons given off in the reaction flow through the wire to the iron metal, which acts as the cathode at which O_2 is reduced.

$$\text{Cathode} \qquad O_2 + 2\,H_2O + 4\,e^- \rightleftharpoons 4\,OH^- \qquad E°_{Red} = 0.401 \text{ V}$$

The overall potential for the reaction is much larger than the potential for the reaction that would occur if iron were the anode.

$$2\,\text{Zn}(s) + O_2(g) + 2\,H_2O(l) \rightleftharpoons 2\,\text{Zn}^{2+}(aq) + 4\,OH^-(aq) \qquad E° = 1.164 \text{ V}$$

Thus, oxidation of the iron won't occur until the anode is either completely consumed or the electrical contact between the iron and zinc metal is broken. Cathodic protection is the theory behind "galvanized" iron, which is iron metal coated with a thin layer of zinc. Until virtually all of the zinc has corroded, the iron metal that provides the structure for the material remains intact.

PROBLEMS

Electrolytic Cells

12AI.1. Explain why electrolysis of aqueous NaCl produces H_2 gas, not sodium metal at the cathode, and Cl_2 gas, not O_2 gas, at the anode.

12AI.2. Calculate the standard cell potential necessary to electrolyze molten NaCl to form sodium metal and chlorine gas.

12AI.3. Calculate the standard cell potential necessary to electrolyze an aqueous solution of NaCl.

12AI.4. Which of the following statements best describes what happens when $MgCl_2$ is electrolyzed?
(a) Mg metal forms at the anode.
(b) Mg^{2+} ions are oxidized at the cathode.
(c) Cl_2 gas is formed at the anode by the oxidation of Cl^-.
(d) Cl^- ions flow toward the cathode.

12AI.5. Explain why an electrolyte, such as Na_2SO_4, has to be added to water before the water can be electrolyzed to H_2 and O_2.

12AI.6. Describe what would happen if $NiSO_4$ were used instead of Na_2SO_4 when water was electrolyzed.

12AI.7. Why does the solution around the cathode become basic when an aqueous solution of Na_2SO_4 is electrolyzed?

12AI.8. What are the most likely products of the electrolysis of an aqueous solution of KBr?
(a) $K^+(aq) + Br^-(aq)$ (b) $K(s) + Br_2(l)$
(c) $OH^-(aq) + H_2(g) + Br_2(l)$ (d) $K^+(aq) + OH^-(aq) + O_2(g) + Br_2(l)$

12AI.9. Describe what happens during the electrolysis of the following solutions.
(a) $FeCl_3(aq)$ (b) $NaI(aq)$ (c) $K_2SO_4(aq)$ (d) $H_2SO_4(aq)$

12AI.10. Why can't aluminum metal be prepared by the electrolysis of an aqueous solution of the Al^{3+} ion?

12AI.11. Which of the following reactions occurs at the anode during electrolysis of an aqueous Na_2SO_4 solution?
(a) $SO_4^{2-}(aq) \rightleftharpoons SO_2(g) + O_2(g) + 2\,e^-$
(b) $2\,H_2O(l) \rightleftharpoons O_2(g) + 4\,H^+(aq) + 4\,e^-$
(c) $2\,H_2O(l) + O_2(g) + 4\,e^- \rightleftharpoons 4\,OH^-(aq)$
(d) $SO_4^{2-}(aq) + 4\,H^+(aq) + 2\,e^- \rightleftharpoons SO_2(g) + 2\,H_2O(l)$

Faraday's Law

12AI.12. Describe an experiment that uses Faraday's law to determine the amount of electric current passing through a solution.

12AI.13. Calculate the mass of silver metal that can be prepared with 1.0000 C of electric charge.

12AI.14. The rate at which electric power is delivered is measured in units of kilowatts. One watt is the power delivered by a 1 A current at a potential of 4.06 V. Calculate the number of kilowatt-hours of electricity it takes to prepare 1.00 mole of sodium metal by electrolysis of molten NaCl.

12AI.15. If a 5.0 A current were used, how long would it take to prepare 1.0 mole of sodium metal? of magnesium metal? of aluminum metal?

12AI.16. Predict the products of the electrolysis of an aqueous solution of LiBr and calculate the number of grams of each product formed by electrolysis of the solution for 1.0 h with a 2.5 A current.

12AI.17. Calculate the number of coulombs necessary to produce a metric tonne (1000 kilograms) of Cl_2 gas.

12AI.18. Calculate the mass of copper metal deposited after a 2.0 M solution of $CuSO_4$ has been electrolyzed for 2.5 h with a 4.5 A current.

12AI.19. Calculate the ratio of the mass of O_2 to the mass of H_2 produced when an aqueous Na_2SO_4 solution is electrolyzed for 2.56 h with a 1.34 A current.

12AI.20. Under a particular set of conditions, electrolysis of an aqueous $AgNO_3$ solution generated 1.00 g of silver metal. How many grams of I_2 would be produced by the electrolysis of an aqueous solution of NaI under the same conditions?

12AI.21. Calculate the ratio of the mass of Br_2 that collects at the anode to the mass of aluminum metal plated out at the cathode when molten $AlBr_3$ is electrolyzed for 10.5 h with a 20.0 A current.

12AI.22. Assume that aqueous solutions of the following salts are electrolyzed for 20.0 min with a 10.0 A current. Which solution will deposit the most grams of metal at the cathode?
(a) $ZnCl_2$ (b) $ZnBr_2$ (c) WCl_6 (d) $ScBr_3$ (e) $HfCl_4$

12AI.23. Which of the following compounds will give the most grams of metal when a 10.0 A current is passed through molten samples of the salts for 2.0 h?
(a) KCl (b) $CaCl_2$ (c) $ScCl_3$

12AI.24. An electric current is passed through a series of three cells filled with aqueous solutions of $AgNO_3$, $Cu(NO_3)_2$, and $Fe(NO_3)_3$. How many grams of each metal will be deposited by a 1.58 A current flowing for 3.54 h?

12AI.25. What are the values of x and y in the formula Ce_xCl_y if electrolysis of an aqueous solution of the salt for 16.5 h with a 1.00 A current deposits 21.6 g of cerium metal?

12AI.26. What is the oxidation number of the manganese in an unknown salt if electrolysis of an aqueous solution of the salt for 30 min with a 10.0 A current generates 2.56 g of manganese metal?

12AI.27. Gold forms compounds in the $+1$ and $+3$ oxidation states. What is the oxidation number of gold in a compound that deposits 1.53 g of gold metal when electrolyzed for 15 min with a 2.50 A current?

Galvanic Corrosion and Cathodic Protection

12AI.28. Use cell potentials to explain why iron oxidizes to form Fe_2O_3.

12AI.29. Use cell potentials to explain why iron dissolves in strong acids.

12AI.30. Explain why the H^+ ion concentration in water is too small for iron to dissolve in water from which all oxygen has been removed.

12AI.31. Use cell potentials to explain why iron dissolves in water that contains oxygen.

12AI.32. Explain why it is a mistake to describe rust as iron(III) oxide, Fe_2O_3.

12AI.33. When iron rusts, the metal surface acts as the anode of a voltaic cell. Explain why the iron metal becomes the cathode of a voltaic cell when it is connected to a piece of magnesium or zinc.

12AI.34. Explain why galvanized iron, which has a thin coating of zinc metal, corrodes much more slowly than iron by itself.

C H A P T E R
12
SPECIAL TOPICS II

12AII.1   ELECTROCHEMICAL CELLS AT NONSTANDARD CONDITIONS: THE NERNST EQUATION

In 1836, a chemistry professor from London named John Frederic Daniell developed the first voltaic cell stable enough to be used as a battery. The following chemical equation describes the chemical reaction of this Daniell cell.

$$Zn(s) + Cu^{2+}(aq) \rightleftharpoons Zn^{2+}(aq) + Cu(s)$$

What happens when the above oxidation–reduction reaction is used to do work?

- The zinc electrode becomes lighter as zinc atoms are oxidized to Zn^{2+} ions, which go into solution.
- The copper electrode becomes heavier as Cu^{2+} ions in the solution are reduced to copper metal that plates out on the copper electrode.
- The concentration of Zn^{2+} ions at the anode increases, and the concentration of the Cu^{2+} ions at the cathode decreases.
- Negative ions flow from the salt bridge toward the anode to balance the charge on the Zn^{2+} ions produced at that electrode.
- Positive ions flow from the salt bridge toward the cathode to compensate for the Cu^{2+} ions consumed in the reaction.

An important property of the cell is missing from the list. Over a period of time, the cell runs down and eventually has to be replaced. Let's assume that our cell is initially a standard cell in which the concentrations of the Zn^{2+} and Cu^{2+} ions are both 1 M. As the reaction goes forward—as zinc metal is consumed and copper metal is produced—the driving force behind the reaction must become weaker. Therefore, the cell potential must become smaller.

In Section 12.7, we showed that the magnitude of the cell potential measures the driving force behind a reaction. The cell potential is therefore zero if and only if the reaction is at equilibrium. When the reaction is at equilibrium, there is no net change in the amount of zinc metal or copper ions in the system, so no electrons flow from the anode to the cathode. If there is no longer a net flow of electrons, the cell can no longer do electrical work. Its potential for doing work must therefore be zero.

In 1889 Hermann Walther Nernst showed that the potential for an electrochemical reaction is described by the following equation.

$$E = E° - \frac{RT}{nF} \ln Q_c$$

In the **Nernst equation,** E is the cell potential at some moment in time, $E°$ is the cell potential when the reaction is at standard conditions, R is the ideal gas constant in units of joules per mole, T is the temperature in Kelvin, n is the number of moles of electrons transferred in the balanced equation for the reaction, F is the charge on a mole of electrons, and Q_c is the reaction quotient at that moment in time. The symbol ln indicates a natural logarithm, which is the log to the base e, where e is an irrational number equal to 2.71828. . . .

Two terms in the Nernst equation are constants: R and F. The ideal gas constant expressed in joules is 8.314 J/mol·K, and the charge on a mole of electrons can be calculated from Avogadro's number and the charge on a single electron.

$$\frac{6.022137 \times 10^{23} \ e^-}{1 \ mol} \times \frac{1.60217733 \times 10^{-19} \ C}{1 \ e^-} = 96,485.31 \ C/mol$$

The temperature is usually 25°C. Substituting this information into the Nernst equation gives the following.

$$E = E° - \frac{0.02569}{n} \ln Q_c$$

Three of the remaining terms in this equation are characteristics of a particular reaction: n, $E°$, and Q_c. The standard potential for the Daniell cell is 1.10 V. Two moles of electrons are transferred from zinc metal to Cu^{2+} ions in the balanced equation for the reaction, so n is 2 for the cell. Because we never include the concentrations of solids in either reaction quotient or equilibrium constant expressions, Q_c for the reaction is equal to the concentration of the Zn^{2+} ion divided by the concentration of the Cu^{2+} ion.

$$Q_c = \frac{(Zn^{2+})}{(Cu^{2+})}$$

Substituting what we know about the Daniell cell into the Nernst equation gives the following result, which represents the cell potential for the Daniell cell at 25°C at any moment in time.

$$E = 1.10 \ V - \frac{0.02569}{2} \ln \frac{(Zn^{2+})}{(Cu^{2+})}$$

Figure 12AII.1 shows a plot of the potential for the Daniell cell as a function of the natural logarithm of the reaction quotient. When the reaction quotient is very small, the cell potential is positive and relatively large. This isn't surprising, because the reaction is far from equilibrium and the driving force behind the reaction should be relatively large. When the reaction quotient is very large, the cell potential is negative. This means that the reaction would have to shift back toward the reactants to reach equilibrium.

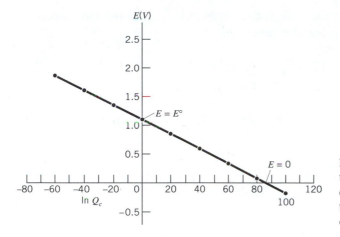

FIGURE 12AII.1 Plot of $\ln Q_c$ versus cell potential for the Daniell cell. The cell potential is equal to $E°$ when the cell is at standard conditions. The cell potential is equal to zero if and only if the reaction is at equilibrium.

The Nernst equation can be used to calculate the potential of a cell that operates at nonstandard conditions. For our purposes, however, the most important property of the Nernst equation is the fact that we can use it to measure the equilibrium constant for a reaction. To understand how this is done, we have to recognize what happens to the cell potential as an oxidation–reduction reaction comes to equilibrium. Because the reaction is at equilibrium, the reaction quotient is equal to the equilibrium constant ($Q_c = K_c$) and the overall cell potential for the reaction is zero ($E = 0$). At equilibrium, the Nernst equation therefore takes the following form.

$$\text{At equilibrium} \qquad 0 = E° - \frac{RT}{nF} \ln K_c$$

Rearranging the equation gives the following result.

$$\text{At equilibrium} \qquad nFE° = RT \ln K_c$$

According to this equation, we can calculate the equilibrium constant for any oxidation–reduction reaction from its standard cell potential.

$$K_c = e^{nFE°/RT}$$

Exercise 12AII.1

Calculate the potential at 25°C for the following reaction in which the concentration of Ag^+ is 4.8×10^{-3} M and the concentration of Cu^{2+} is 2.4×10^{-2} M.

$$Cu(s) + 2\,Ag^+(aq) \rightleftharpoons Cu^{2+}(aq) + 2\,Ag(s)$$

Solution

We start by looking up the standard potentials for the two half-reactions and calculating the standard cell potential.

Oxidation	$Cu \rightleftharpoons Cu^{2+} + 2\,e^-$	$E°_{Ox} = -(0.3402\ \text{V})$	
Reduction	$Ag^+ + e^- \rightleftharpoons Ag$	$E°_{Red} = 0.7996\ \text{V}$	

$$E° = E°_{Red} + E°_{Ox} = 0.4594\ \text{V}$$

We now set up the Nernst equation for the cell, noting that n is 2 because two electrons are transferred in the balanced equation for the reaction.

$$E = E^\circ - \frac{0.02569}{2} \ln \left[\frac{(Cu^{2+})}{(Ag^+)^2} \right]$$

Substituting the concentrations of the Cu^{2+} and Ag^+ ions into the equation gives the value for the cell potential.

$$E = 0.4594 \text{ V} - \frac{0.02569}{2} \ln \frac{(0.024)}{(0.0048)^2} = 0.37 \text{ V}$$

Exercise 12AII.2

Calculate the equilibrium constant at 25°C for the reaction between zinc metal and acid.

$$Zn(s) + 2 H^+(aq) \rightleftharpoons Zn^{2+}(aq) + H_2(g)$$

Solution

We start by calculating the overall standard potential for the reaction.

$$\begin{array}{ll} Zn \rightleftharpoons Zn^{2+} + 2 e^- & E^\circ_{Ox} = -(-0.7628 \text{ V}) \\ 2 H^+ + 2 e^- \rightleftharpoons H_2 & E^\circ_{Red} = 0.0000 \text{ V} \\ \hline Zn(s) + 2 H^+(aq) \rightleftharpoons Zn^{2+}(aq) + H_2(g) & E^\circ = 0.7628 \text{ V} \end{array}$$

We then substitute into the Nernst equation the implications of the fact that the reaction is at equilibrium: $Q_c = K_c$ and $E = 0$.

$$0 = E^\circ - \frac{RT}{nF} \ln K_c$$

We then rearrange the equation

$$nFE^\circ = RT \ln K_c$$

and solve for the natural logarithm of the equilibrium constant.

$$\ln K_c = \frac{nFE^\circ}{RT}$$

Substituting what we know about the reaction into the equation gives the following result.

$$\ln K_c = \frac{(2)(96{,}485 \text{ C})(0.7628 \text{ V})}{(8.314 \text{ J/mol·K})(298 \text{ K})} = 59.4$$

The equilibrium constant for the reaction can therefore be calculated by raising e to the power of 59.4.

$$K_c = e^{59.4} = 6 \times 10^{25}$$

This is a very large equilibrium constant, which means that equilibrium lies heavily on the side of the products. The equilibrium constant is so large that the equation for the reaction is written as if it proceeds to completion.

$$Zn(s) + 2\,H^+(aq) \longrightarrow Zn^{2+}(aq) + H_2(g)$$

12AII.2 BALANCING OXIDATION–REDUCTION EQUATIONS

A trial-and-error approach to balancing chemical equations was introduced in Chapter 2. This approach involves playing with the equation—adjusting the ratio of the reactants and products—until the following goals have been achieved.

Goals for Balancing Chemical Equations

- The same number of atoms of each element is found on both sides of the equation and therefore mass is conserved. (Mass is conserved because atoms are neither created nor destroyed in a chemical reaction.)
- The sum of the positive and negative charges is the same on both sides of the equation and therefore charge is conserved. (Charge is conserved because electrons are neither created nor destroyed in the reaction.)

At times, the equation is too complex to be solved by trial and error. Consider the following reaction, for example.

$$3\,Cu(s) + 8\,HNO_3(aq) \rightleftharpoons 3\,Cu^{2+}(aq) + 2\,NO(g) + 6\,NO_3^-(aq) + 4\,H_2O(l)$$

Other times, more than one balanced equation can be written. The following are just a few of the balanced equations that can be written for the reaction between the permanganate ion and hydrogen peroxide, for example.

$$2\,MnO_4^-(aq) + H_2O_2(aq) + 6\,H^+(aq) \rightleftharpoons 2\,Mn^{2+}(aq) + 3\,O_2(g) + 4\,H_2O(l)$$
$$2\,MnO_4^-(aq) + 3\,H_2O_2(aq) + 6\,H^+(aq) \rightleftharpoons 2\,Mn^{2+}(aq) + 4\,O_2(g) + 6\,H_2O(l)$$
$$2\,MnO_4^-(aq) + 5\,H_2O_2(aq) + 6\,H^+(aq) \rightleftharpoons 2\,Mn^{2+}(aq) + 5\,O_2(g) + 8\,H_2O(l)$$
$$2\,MnO_4^-(aq) + 7\,H_2O_2(aq) + 6\,H^+(aq) \rightleftharpoons 2\,Mn^{2+}(aq) + 6\,O_2(g) + 10\,H_2O(l)$$

Equations such as these have to be balanced by a more systematic approach than trial and error.

12AII.3 REDOX REACTIONS IN ACIDIC SOLUTIONS

A powerful technique for balancing oxidation–reduction equations involves dividing the reactions into separate oxidation and reduction half-reactions. We then balance the half-reactions, one at a time, and combine them so that electrons are neither created nor destroyed in the reaction.

The half-reaction technique is particularly useful for balancing equations for reactions such as the oxidation of sulfur dioxide by the dichromate ion in acidic solution.

$$SO_2(aq) + Cr_2O_7^{2-}(aq) \xrightarrow{H^+} SO_4^{2-}(aq) + Cr^{3+}(aq)$$

The reason why this equation is inherently more difficult to balance has nothing to do with the ratio of moles of SO_2 to moles of $Cr_2O_7^{2-}$. It results from the fact that the solvent takes an active role in both half-reactions. Let's therefore use this reaction to illustrate the steps in the half-reaction method for balancing equations.

STEP 1: Write a skeleton equation for the reaction.   We already know the skeleton equation for the reaction.

$$SO_2 + Cr_2O_7^{2-} \longrightarrow SO_4^{2-} + Cr^{3+}$$

STEP 2: Assign oxidation numbers to atoms on both sides of the equation.   When the rules for assigning oxidation numbers described in Section 5.13 are applied, we get the following results.

$$\underset{+4\ -2}{SO_2} + \underset{+6\ \ -2}{Cr_2O_7^{2-}} \longrightarrow \underset{+6\ \ -2}{SO_4^{2-}} + \underset{+3}{Cr^{3+}}$$

STEP 3: Determine which atoms are oxidized and which are reduced.   In the course of the reaction, there is an increase in the oxidation number of the sulfur atom and a decrease in the oxidation number of the chromium. Thus, sulfur is oxidized and chromium is reduced.

$$\underset{+4}{SO_2} + \underset{+6}{Cr_2O_7^{2-}} \longrightarrow \underset{+6}{SO_4^{2-}} + \underset{+3}{Cr^{3+}}$$

Oxidation

Reduction

STEP 4: Divide the reaction into oxidation and reduction half-reactions and balance the half-reactions.   The reaction can be divided into the following half-reactions.

$$\text{Oxidation} \quad \underset{+4}{SO_2} \longrightarrow \underset{+6}{SO_4^{2-}}$$

$$\text{Reduction} \quad \underset{+6}{Cr_2O_7^{2-}} \longrightarrow \underset{+3}{Cr^{3+}}$$

It doesn't matter which half-reaction we balance first, so let's start with reduction. Because the $Cr_2O_7^{2-}$ ion contains two chromium atoms that must be reduced from the $+6$ to the $+3$ oxidation state, six electrons are consumed in the half-reaction.

$$\text{Reduction} \quad Cr_2O_7^{2-} + 6\,e^- \longrightarrow 2\,Cr^{3+}$$

This raises an interesting question: What happens to the oxygen atoms when the chromium atoms are reduced? The seven oxygen atoms in the $Cr_2O_7^{2-}$ ion are formally in the -2 oxidation state. If these atoms were released into solution in the -2 oxidation state when the chromium was reduced, they would be present as O^{2-} ions. But the O^{2-} ion is a very strong base, and the reaction is being run in a strong acid. Any O^{2-} ions liberated when the dichromate ion is reduced would instantly react with the H^+ ions in the acid solution to form water. Because there are formally seven oxygen atoms in a -2 oxidation state in the dichromate ion, 14 H^+ ions are consumed and seven H_2O molecules are produced in this half-reaction.

$$\text{Reduction} \quad Cr_2O_7^{2-} + 14\,H^+ + 6\,e^- \rightleftharpoons 2\,Cr^{3+} + 7\,H_2O$$

We can now turn to the oxidation half-reaction, and start by noting that two electrons are given off when sulfur is oxidized from the +4 to the +6 oxidation state.

$$\text{Oxidation} \qquad SO_2 \longrightarrow SO_4^{2-} + 2\,e^-$$

The charge on both sides of the equation is balanced by remembering that the reaction is run in acid, which contains H^+ ions and H_2O molecules. We can therefore add H^+ ions or H_2O molecules to either side of the equation, as needed. The only way to balance the charge on both sides of the equation is to add four H^+ ions to the products of the reaction.

$$\text{Oxidation} \qquad SO_2 \longrightarrow SO_4^{2-} + 2\,e^- + 4\,H^+$$

We can then balance the number of hydrogen and oxygen atoms on both sides of the equation by adding a pair of H_2O molecules to the reactants.

$$\text{Oxidation} \qquad SO_2 + 2\,H_2O \rightleftharpoons SO_4^{2-} + 2\,e^- + 4\,H^+$$

STEP 5: Combine the two half-reactions so that electrons are neither created nor destroyed. Six electrons are consumed in the reduction half-reaction and two electrons are given off in the oxidation half-reaction. We can combine the half-reactions so that electrons are conserved by multiplying the reduction half-reaction by 3.

$$
\begin{array}{l}
Cr_2O_7^{2-} + 14\,H^+ + 6\,e^- \rightleftharpoons 2\,Cr^{3+} + 7\,H_2O \\
3(SO_2 + 2\,H_2O \rightleftharpoons SO_4^{2-} + 2\,e^- + 4\,H^+) \\
\hline
Cr_2O_7^{2-} + 3\,SO_2 + 14\,H^+ + 6\,H_2O \rightleftharpoons 2\,Cr^{3+} + 3\,SO_4^{2-} + 12\,H^+ + 7\,H_2O
\end{array}
$$

STEP 6: Balance the remainder of the equation by inspection, if necessary. Although the equation appears balanced, we are not quite finished with it. We can simplify the equation by subtracting 12 H^+ ions and 6 H_2O molecules from each side to generate the following balanced equation.

$$Cr_2O_7^{2-}(aq) + 3\,SO_2(aq) + 2\,H^+(aq) \rightleftharpoons 2\,Cr^{3+}(aq) + 3\,SO_4^{2-}(aq) + H_2O(l)$$

Checkpoint

On an exam, students were asked to use half-reactions to write a balanced equation for the reaction between SO_2 and the $Cr_2O_7^{2-}$ ion in the presence of a strong acid.

$$SO_2(g) + Cr_2O_7^{2-}(aq) \xrightarrow{H^+} SO_4^{2-}(aq) + Cr^{3+}(aq)$$

One student wrote the following half-reactions.

$$
\begin{array}{ll}
\text{Oxidation} & SO_2 + 4\,OH^- \rightleftharpoons SO_4^{2-} + 2\,H_2O + 2\,e^- \\
\text{Reduction} & Cr_2O_7^{2-} + 7\,H_2O + 6\,e^- \rightleftharpoons 2\,Cr^{3+} + 14\,OH^-
\end{array}
$$

The student then wrote the following overall equation.

$$3\,SO_2(g) + Cr_2O_7^{2-}(aq) + H_2O(l) \rightleftharpoons 3\,SO_4^{2-}(aq) + 2\,Cr^{3+}(aq) + 2\,OH^-(aq)$$

Explain why the instructor gave no credit for this answer, even though the overall equation is balanced.

Exercise 12AII.3

As we have seen, an endless number of balanced equations can be written for the reaction between the permanganate ion and hydrogen peroxide in acidic solution to form the Mn^{2+} ion and oxygen gas.

$$MnO_4^-(aq) + H_2O_2(aq) \xrightarrow{H^+} Mn^{2+}(aq) + O_2(g)$$

Use the half-reaction method to determine the correct stoichiometry for the reaction.

Solution

STEP 1: Write a skeleton equation for the reaction.

$$MnO_4^- + H_2O_2 \longrightarrow Mn^{2+} + O_2$$

STEP 2: Assign oxidation numbers to atoms on both sides of the equation.

$$MnO_4^- + H_2O_2 \longrightarrow Mn^{2+} + O_2$$
$$\;\;+7 \;\; -2 \quad\; +1 \; -1 \qquad\qquad +2 \qquad\quad 0$$

STEP 3: Determine which atoms are oxidized and which are reduced.

$$MnO_4^- + H_2O_2 \longrightarrow Mn^{2+} + O_2$$
$$+7 \qquad\quad -1 \qquad\qquad +2 \qquad\; 0$$

$$\text{Reduction}$$
$$\text{Oxidation}$$

STEP 4: Divide the reaction into oxidation and reduction half-reactions and balance the half-reactions. The reaction can be divided into the following half-reactions.

$$\text{Oxidation} \qquad H_2O_2 \longrightarrow O_2$$
$$\qquad\qquad\qquad\quad -1 \qquad\quad 0$$
$$\text{Reduction} \qquad MnO_4^- \longrightarrow Mn^{2+}$$
$$\qquad\qquad\qquad\quad +7 \qquad\qquad +2$$

Let's balance the reduction half-reaction first. It takes five electrons to reduce manganese from +7 to +2.

$$\text{Reduction} \qquad MnO_4^- + 5\,e^- \longrightarrow Mn^{2+}$$

Because the reaction is run in acid, we can add H^+ ions or H_2O molecules to either side of the equation, as needed. There are two hints that tell us which side of the equation gets H^+ ions and which side gets H_2O molecules. The only way to balance the charge on both sides of the equation is to add eight H^+ ions to the left side of the equation.

$$\text{Reduction} \qquad MnO_4^- + 8\,H^+ + 5\,e^- \longrightarrow Mn^{2+}$$

The only way to balance the number of oxygen atoms is to add four H_2O molecules to the right side of the equation.

$$\text{Reduction} \qquad MnO_4^- + 8\,H^+ + 5\,e^- \rightleftharpoons Mn^{2+} + 4\,H_2O$$

To balance the oxidation half-reaction, we have to remove two electrons from a pair of oxygen atoms in the -1 oxidation state to form a neutral O_2 molecule.

$$\text{Oxidation}\qquad H_2O_2 \longrightarrow O_2 + 2\,e^-$$

We can then add a pair of H^+ ions to the products to balance both charge and mass in the half-reaction.

$$\text{Oxidation}\qquad H_2O_2 \rightleftharpoons O_2 + 2\,H^+ + 2\,e^-$$

STEP 5: Combine the two half-reactions so that electrons are neither created nor destroyed. Two electrons are given off during oxidation and five electrons are consumed during reduction. We can combine the half-reactions so that electrons are conserved by using the lowest common multiple of 5 and 2.

$$2(MnO_4^- + 8\,H^+ + 5\,e^- \rightleftharpoons Mn^{2+} + 4\,H_2O)$$
$$\underline{5(H_2O_2 \rightleftharpoons O_2 + 2\,H^+ + 2\,e^-)}$$
$$2\,MnO_4^- + 5\,H_2O_2 + 16\,H^+ \rightleftharpoons 2\,Mn^{2+} + 5\,O_2 + 10\,H^+ + 8\,H_2O$$

STEP 6: Balance the remainder of the equation by inspection, if necessary. The simplest balanced equation for the reaction is obtained when 10 H^+ ions are subtracted from each side of the equation derived in the previous step.

$$2\,MnO_4^-(aq) + 5\,H_2O_2(aq) + 6\,H^+(aq) \rightleftharpoons 2\,Mn^{2+}(aq) + 5\,O_2(g) + 8\,H_2O(l)$$

12AII.4   REDOX REACTIONS IN BASIC SOLUTIONS

Half-reactions are also valuable for balancing equations in basic solutions. The key to success with such reactions is recognizing that basic solutions contain H_2O molecules and OH^- ions. We can therefore add water molecules or hydroxide ions to either side of the equation, as needed.

In the previous section, we obtained the following equation for the reaction between the permanganate ion and hydrogen peroxide in an acidic solution.

$$2\,MnO_4^-(aq) + 5\,H_2O_2(aq) + 6\,H^+(aq) \rightleftharpoons 2\,Mn^{2+}(aq) + 5\,O_2(g) + 8\,H_2O(l)$$

It might be interesting to see what happens when the reaction occurs in a basic solution.

Exercise 12AII.4

Write a balanced equation for the reaction between the permanganate ion and hydrogen peroxide in a basic solution to form manganese dioxide and oxygen.

$$MnO_4^-(aq) + H_2O_2(aq) \xrightarrow{OH^-} MnO_2(s) + O_2(g)$$

Solution

STEP 1: Write a skeleton equation for the reaction.

$$MnO_4^- + H_2O_2 \longrightarrow MnO_2 + O_2$$

STEP 2: Assign oxidation numbers to atoms on both sides of the equation.

$$\underset{+7\ -2}{MnO_4^-} + \underset{+1\ -1}{H_2O_2} \longrightarrow \underset{+4\ -2}{MnO_2} + \underset{0}{O_2}$$

STEP 3: Determine which atoms are oxidized and which are reduced.

$$\underset{+7}{MnO_4^-} + \underset{-1}{H_2O_2} \longrightarrow \underset{+4}{MnO_2} + \underset{0}{O_2}$$

Reduction

Oxidation

STEP 4: Divide the reaction into oxidation and reduction half-reactions and balance the half-reactions.   The reaction can be divided into the following half-reactions.

Reduction $\underset{+7}{MnO_4^-} \longrightarrow \underset{+4}{MnO_2}$

Oxidation $\underset{-1}{H_2O_2} \longrightarrow \underset{0}{O_2}$

Let's start by balancing the reduction half-reaction. It takes three electrons to reduce manganese from +7 to the +4 oxidation state.

Reduction $MnO_4^- + 3\ e^- \longrightarrow MnO_2$

We now try to balance either the number of atoms or the charge on both sides of the equation. Because the reaction is run in basic solution, we can add either OH^- ions or H_2O molecules to either side of the equation, as needed. The only way to balance the net charge of -4 on the left side of the equation is to add four OH^- ions to the products.

Reduction $MnO_4^- + 3\ e^- \longrightarrow MnO_2 + 4\ OH^-$

We can then balance the number of hydrogen and oxygen atoms by adding two H_2O molecules to the reactants.

Reduction $MnO_4^- + 3\ e^- + 2\ H_2O \rightleftharpoons MnO_2 + 4\ OH^-$

We now turn to the oxidation half-reaction. Two electrons are lost when hydrogen peroxide is oxidized to form O_2 molecules.

Oxidation $H_2O_2 \longrightarrow O_2 + 2\ e^-$

We can balance the charge on the half-reaction by adding a pair of OH^- ions to the reactants.

Oxidation $H_2O_2 + 2\ OH^- \longrightarrow O_2 + 2\ e^-$

The only way to balance the number of hydrogen and oxygen atoms is to add two H_2O molecules to the products.

Oxidation $H_2O_2 + 2\ OH^- \rightleftharpoons O_2 + 2\ H_2O + 2\ e^-$

STEP 5: Combine the two half-reactions so that electrons are neither created nor destroyed. Two electrons are given off during the oxidation half-reaction and three electrons are consumed in the reduction half-reaction. We can therefore combine the half-reactions by using the lowest common multiple of 2 and 3.

$$2(MnO_4^- + 3\,e^- + 2\,H_2O \rightleftharpoons MnO_2 + 4\,OH^-)$$
$$3(H_2O_2 + 2\,OH^- \rightleftharpoons O_2 + 2\,H_2O + 2\,e^-)$$
$$\overline{2\,MnO_4^- + 3\,H_2O_2 + 6\,OH^- + 4\,H_2O \rightleftharpoons 2\,MnO_2 + 3\,O_2 + 8\,OH^- + 6\,H_2O}$$

STEP 6: Balance the remainder of the equation by inspection, if necessary. The simplest balanced equation is obtained by subtracting four H_2O molecules and six OH^- ions from each side of the equation derived in the previous step.

$$2\,MnO_4^-(aq) + 3\,H_2O_2(aq) \rightleftharpoons 2\,MnO_2(s) + 3\,O_2(g) + 2\,OH^-(aq) + 2\,H_2O(l)$$

Note that the ratio of moles of MnO_4^- to moles of H_2O_2 consumed is different in acidic and basic solutions. This difference results from the fact that MnO_4^- is reduced all the way to Mn^{2+} in acid, but the reaction stops at MnO_2 in base.

12AII.5 MOLECULAR REDOX REACTIONS

Lewis structures can play a vital role in understanding oxidation–reduction reactions of complex molecules. Consider the following reaction, for example, which is used in the Breathalyzer to determine the amount of ethyl alcohol or ethanol on the breath of individuals who are suspected of driving while under the influence.

$$3\,CH_3CH_2OH(g) + 2\,Cr_2O_7^{2-}(aq) + 16\,H^+(aq) \rightleftharpoons$$
$$3\,CH_3CO_2H(aq) + 4\,Cr^{3+}(aq) + 11\,H_2O(l)$$

We could balance the oxidation half-reaction in terms of the molecular formulas of the starting material and the product of the half-reaction.

$$\text{Oxidation} \qquad C_2H_6O \longrightarrow C_2H_4O_2$$

It is easier to understand what happens in the reaction, however, if we assign oxidation numbers to each of the atoms in the Lewis structures shown in Figure 12AII.2. Oxidation numbers are assigned using the method developed in Sections 5.14 and 12.2.

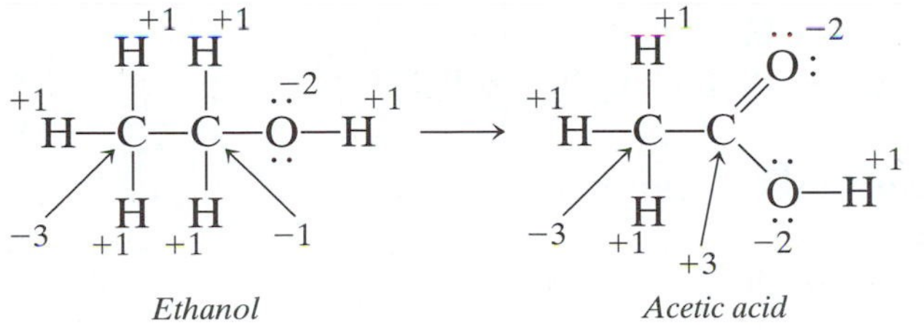

Ethanol   *Acetic acid*

FIGURE 12AII.2 Oxidation of ethanol to acetic acid.

The carbon atom in the CH_3— group in ethanol is assigned an oxidation state of -3, and the carbon atom in the —CH_2OH group has an oxidation state of -1. The carbon in the CH_3— group in the acetic acid formed in the reaction has the same oxidation state as it did in the starting material: -3. There is a change in the oxidation number of the other carbon atom, however, from −1 to +3. The oxidation half-reaction therefore formally corresponds to the loss of four electrons by one of the carbon atoms.

$$\text{Oxidation} \qquad CH_3CH_2OH \longrightarrow CH_3CO_2H + 4 \, e^-$$

Because the reaction is run in acidic solution, we can add H^+ and H_2O as needed to balance the equation.

$$\text{Oxidation} \qquad CH_3CH_2OH + H_2O \rightleftharpoons CH_3CO_2H + 4 \, e^- + 4 \, H^+$$

The other half of the reaction involves a six-electron reduction of the $Cr_2O_7^{2-}$ ion in acidic solution to form a pair of Cr^{3+} ions.

$$\text{Reduction} \qquad Cr_2O_7^{2-} + 6 \, e^- \longrightarrow 2 \, Cr^{3+}$$

Adding H^+ ions and H_2O molecules as needed gives the following balanced equation for the reduction half-reaction.

$$\text{Reduction} \qquad Cr_2O_7^{2-} + 14 \, H^+ + 6 \, e^- \rightleftharpoons 2 \, Cr^{3+} + 7 \, H_2O$$

We are now ready to combine the two half-reactions by recognizing that electrons are neither created nor destroyed in the reaction.

$$3(CH_3CH_2OH + H_2O \rightleftharpoons CH_3CO_2H + 4 \, e^- + 4 \, H^+)$$
$$2(Cr_2O_7^{2-} + 14 \, H^+ + 6 \, e^- \rightleftharpoons 2 \, Cr^{3+} + 7 \, H_2O)$$
$$\overline{3 \, CH_3CH_2OH + 2 \, Cr_2O_7^{2-} + 28 \, H^+ + 3 \, H_2O \rightleftharpoons}$$
$$3 \, CH_3CO_2H + 4 \, Cr^{3+} + 12 \, H^+ + 14 \, H_2O$$

Simplifying the equation by removing 3 H_2O molecules and 12 H^+ ions from both sides of the equation gives the balanced equation for the reaction.

$$3 \, CH_3CH_2OH(g) + 2 \, Cr_2O_7^{2-}(aq) + 16 \, H^+(aq) \rightleftharpoons$$
$$3 \, CH_3CO_2H(aq) + 4 \, Cr^{3+}(aq) + 11 \, H_2O(l)$$

Exercise 12AII.5

Methyllithium (CH_3Li) can be used to form bonds between carbon and either main-group metals or transition metals.

$$HgCl_2(s) + 2 \, CH_3Li(l) \rightleftharpoons Hg(CH_3)_2(l) + 2 \, LiCl(s)$$
$$WCl_6(s) + 6 \, CH_3Li(l) \rightleftharpoons W(CH_3)_6(l) + 6 \, LiCl(s)$$

It can also be used to form bonds between carbon and other nonmetals

$$PCl_3(l) + 3 \, CH_3Li(l) \rightleftharpoons P(CH_3)_3(l) + 3 \, LiCl(s)$$

or between carbon atoms.

$$CH_3Li(l) + H_2CO(g) \rightleftharpoons [CH_3CH_2OLi] \xrightarrow{H^+} CH_3CH_2OH(l)$$

Use Lewis structures to explain the stoichiometry of the following oxidation–reduction reaction, which is used to synthesize methyllithium.

$$CH_3Br(l) + 2\ Li(s) \rightleftharpoons CH_3Li(l) + LiBr(s)$$

Solution

The active metals are reducing agents in all of their chemical reactions. In this reaction, lithium metal reduces a carbon atom from an oxidation state of -2 to -4, as shown in Figure 12AII.3. The lithium is oxidized from a zero oxidation state in its elemental form to $+1$ in methyllithium. Therefore, the carbon atom has gained two electrons and each lithium atom has lost one electron. Thus, 2 mol of lithium is consumed for each mole of CH_3Br in the reaction. One of the Li^+ ions formed in the reaction combines with the negatively charged carbon atom to form methyllithium, and the other precipitates from solution with the Br^- ion as $LiBr$.

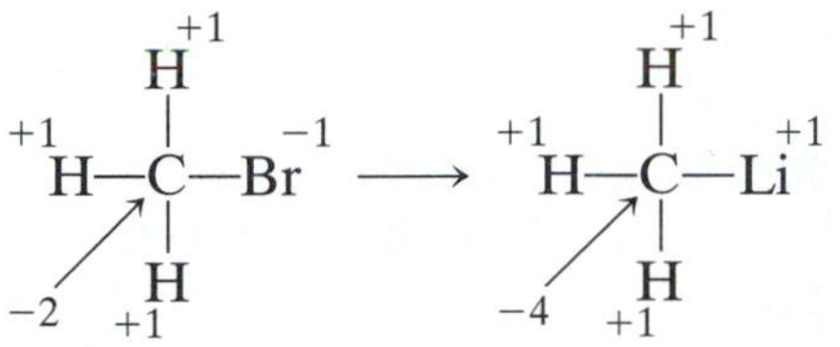

FIGURE 12AII.3 Reduction of CH_3Br with lithium metal.

$$CH_3Br(l) + 2\ Li(s) \rightleftharpoons CH_3Li(l) + LiBr(s)$$

PROBLEMS

The Nernst Equation

12AII.1. Explain why there is only one value of $E°$ for a cell at a given temperature but many different values of E.

12AII.2. Explain why the cell potential becomes smaller as the cell comes closer to equilibrium.

12AII.3. What does it mean when E for a cell is equal to zero? What does it mean when $E°$ is equal to zero?

12AII.4. Which of the following describes an oxidation–reduction reaction at equilibrium?
(a) $E = 0$ (b) $E° = 0$ (c) $E = E°$ (d) $Q_c = 0$ (e) $\ln K_c = 0$

12AII.5. Write the Nernst equation for the following half-cell and calculate the half-cell potential, assuming that the H^+ ion concentration is $10^{-7}\ M$ and that the partial pressure of the H_2 gas is 1 atm.

$$2\ H^+(aq) + 2\ e^- \rightleftharpoons H_2(g)$$

12AII.6. Describe how increasing the pH of the following half-reaction affects the half-cell reduction potential.

$$MnO_4^-(aq) + 8\ H^+(aq) + 5\ e^- \rightleftharpoons Mn^{2+}(aq)$$

Does the permanganate ion become a stronger oxidizing agent or a weaker oxidizing agent as the solution becomes more basic?

12AII.7. Write the Nernst equation for the following reaction and calculate the cell potential, assuming that the Al^{3+} ion concentration is 1.2 M and the Fe^{3+} ion concentration is 2.5 M.

$$Al^{3+}(aq) + Fe(s) \rightleftharpoons Al(s) + Fe^{3+}(aq)$$

12AII.8. Calculate $E°$ for the following reaction. See Appendix B.12.

$$2\ Fe^{2+}(aq) + H_2O_2(aq) \rightleftharpoons 2\ Fe^{3+}(aq) + 2\ OH^-(aq)$$

Write the Nernst equation for the reaction and calculate the cell potential for a system in a pH 10 buffer for which all other components are present at standard conditions.

12AII.9. Assume that we start with a Daniell cell at standard conditions.

$$Zn(s) + Cu^{2+}(aq) \rightleftharpoons Zn^{2+}(aq) + Cu(s)$$

Calculate the cell potential under the following set of conditions.
(a) Ninety-nine percent of the zinc metal and Cu^{2+} ions have been consumed.
(b) The reaction has reached 99.99% completion.
(c) The reaction has reached 99.9999% completion.
(d) The Cu^{2+} ion concentration is only $1 \times 10^{-8}\ M$.
(e) The reaction has reached equilibrium.

12AII.10. Calculate the standard cell potential for the voltaic cell built around the following oxidation–reduction reaction. See Appendix B.12.

$$2\ MnO_4^-(aq) + 5\ H_2O_2(aq) + 6\ H^+(aq) \rightleftharpoons$$
$$2\ Mn^{2+}(aq) + 8\ H_2O(l) + 5\ O_2(g)$$

Use the Nernst equation to predict the effect on the cell potential of an increase in the pH of the solution. Predict the effect of an increase in the H_2O_2 concentration.

12AII.11. What would be the effect of an increase in temperature on the potential of the Daniell cell? Would it increase, decrease, or remain the same?

12AII.12. Predict the standard cell potential for the following reaction at 25°C and 250°C.

$$Zn(s) + Fe^{2+}(aq) \rightleftharpoons Zn^{2+}(aq) + Fe(s)$$

12AII.13. The standard cell potential, $E°$, is zero for any cell in which the reaction at the cathode is the opposite of the reaction at the anode.

$$Cu(s) + Cu^{2+}(1.0\ M) \rightleftharpoons Cu^{2+}(1.0\ M) + Cu(s) \qquad E° = 0$$

A small voltage can be obtained, however, from a cell in which the Cu^{2+} ion concentrations are different in the two half-cells. Determine the anode and the cathode of the following cell and calculate the overall cell potential for this concentration cell.

$$Cu(s) + Cu^{2+}(2.50\ M) \rightleftharpoons Cu^{2+}(0.18\ M) + Cu(s)$$

12AII.14. Calculate the equilibrium constant at 25°C for the Daniell cell. Use the results of the calculation to explain why a single arrow rather than a double-headed arrow could be used in the following equation.

$$Zn(s) + Cu^{2+}(aq) \longrightarrow Zn^{2+}(aq) + Cu(s)$$

The Half-Reaction Method of Balancing Redox Equations

12AII.15. The chemistry of nitrogen is unusual because so many of its compounds undergo reactions that give nitrogen in its elemental form. Balance the following oxidation–reduction reaction in which ammonia is converted into molecular nitrogen in basic solution.
(a) $CuO(s) + NH_3(g) \rightarrow Cu(s) + N_2(g) + H_2O(l)$

12AII.16. Reactions in which a reagent undergoes both oxidation and reduction are known as *disproportionation* reactions. Write balanced equations for the following disproportionation reactions in acidic solution.
(a) $H_2O_2(aq) \rightarrow H_2O(l) + O_2(g)$
(b) $NO_2(g) + H_2O(l) \rightarrow HNO_3(aq) + NO(g)$

12AII.17. The term *conproportionation* has been proposed to describe reactions that are the opposite of disproportionation reactions. Write balanced equations for the following conproportionation reactions in acidic solution.
(a) $HIO_3(aq) + HI(aq) \rightarrow I_2(aq)$
(b) $SO_2(g) + H_2S(g) \rightarrow S_8(s) + H_2O(l)$

12AII.18. Acids that are also oxidizing agents play an important role in the preparation of small quantities of the halogens in the laboratory. Write balanced equations for the following oxidation–reduction equations in acidic solution.
(a) $HCl(aq) + HNO_3(aq) \rightarrow NO(g) + Cl_2(g)$
(b) $HBr(aq) + H_2SO_4(aq) \rightarrow SO_2(g) + Br_2(aq)$
(c) $HCl(aq) + MnO_4^-(aq) \rightarrow Cl_2(g) + Mn^{2+}(aq)$

12AII.19. The product of the reaction between copper metal and nitric acid depends on the concentration of the acid. In dilute nitric acid, NO is formed. Concentrated nitric acid produces NO_2. Write balanced equations for the following oxidation–reduction reactions between copper metal and nitric acid.
(a) $Cu(s) + HNO_3(aq) \rightarrow Cu^{2+}(aq) + NO(g)$
(b) $Cu(s) + HNO_3(aq) \rightarrow Cu^{2+}(aq) + NO_2(g)$

12AII.20. Use half-reactions to balance the following oxidation–reduction equation that seems to give only a single product in acid solution.

$$H_2S(g) + H_2O_2(aq) \longrightarrow S_8(s)$$

12AII.21. Some of the earliest matches consisted of white phosphorus and potassium chlorate glued to wooden sticks. Write a balanced equation for the following oxidation–reduction reaction, which occurred when the match was rubbed against a hard surface.

$$P_4(s) + KClO_3(s) \longrightarrow P_4O_{10}(s) + KCl(s)$$

Redox Reactions in Acidic Solutions

12AII.22. Write balanced equations for the following reactions, in which an acidic solution of the CrO_4^{2-} or $Cr_2O_7^{2-}$ ion is used as an oxidizing agent.
(a) $I^-(aq) + CrO_4^{2-}(aq) \rightarrow I_2(aq) + Cr^{3+}(aq)$
(b) $Fe^{2+}(aq) + Cr_2O_7^{2-}(aq) \rightarrow Fe^{3+}(aq) + Cr^{3+}(aq)$
(c) $H_2S(g) + CrO_4^{2-}(aq) \rightarrow S_8(s) + Cr^{3+}(aq)$

12AII.23. Write balanced equations for the following oxidation–reduction reactions, which involve an acidic solution of the MnO_4^- ion.
(a) $Fe^{2+}(aq) + MnO_4^-(aq) \rightarrow Fe^{3+}(aq) + Mn^{2+}(aq)$
(b) $S_2O_3^{2-}(aq) + MnO_4^-(aq) \rightarrow SO_4^{2-}(aq) + Mn^{2+}(aq)$
(c) $Mn^{2+}(aq) + PbO_2(s) \rightarrow MnO_4^-(aq) + Pb^{2+}(aq)$
(d) $SO_2(g) + MnO_4^-(aq) \rightarrow H_2SO_4(aq) + Mn^{2+}(aq)$

12AII.24. Write a balanced equation for the following reaction, which occurs in acidic solution.

$$Br_2(aq) + SO_2(g) \longrightarrow H_2SO_4(aq) + HBr(aq)$$

12AII.25. The two most common oxidation states of chromium are Cr(III) and Cr(VI). Balance the following oxidation–reduction equations that involve one of these oxidation states of chromium. Assume that the reactions occur in acidic solution.
(a) $Cr(s) + O_2(g) + H^+(aq) \rightarrow Cr^{3+}(aq)$
(b) $Fe^{2+}(aq) + Cr_2O_7^{2-}(aq) \rightarrow Fe^{3+}(aq) + Cr^{3+}(aq)$
(c) $BrO_3^-(aq) + Cr^{3+}(aq) \rightarrow Br_2(aq) + HCrO_4^-(aq)$

Redox Reactions in Basic Solutions

12AII.26. Balance the following oxidation–reduction equations that occur in basic solution.
(a) $NO(g) + MnO_4^-(aq) \rightarrow NO_3^-(aq) + MnO_2(s)$
(b) $NH_3(aq) + MnO_4^-(aq) \rightarrow N_2(g) + MnO_2(s)$

12AII.27. Most organic compounds react with MnO_4^- ion in the presence of acid to form CO_2 and water. MnO_4^- in the presence of base, however, is a weaker and therefore more gentle oxidizing agent. Balance the following redox reactions in which the MnO_4^- ion is used to oxidize an organic compound.
(a) $CH_3OH(aq) + MnO_4^-(aq) \xrightarrow{H^+} CO_2(g) + Mn^{2+}(aq)$
(b) $CH_3OH(aq) + MnO_4^-(aq) \xrightarrow{OH^-} HCO_2H(aq) + MnO_2(s)$

12AII.28. The Cr^{3+} ion is not soluble in basic solutions because it precipitates as $Cr(OH)_3$. Chromium metal will dissolve in excess base, however, because it can form a soluble $Cr(OH)_4^-$ complex ion. $Cr(OH)_3$ can also be dissolved by oxidizing the chromium to the +6 oxidation state. Write balanced equations for the following reactions that occur in basic solution.
(a) $Cr(s) + OH^-(aq) \rightarrow Cr(OH)_4^-(aq) + H_2(g)$
(b) $Cr(OH)_3(s) + H_2O_2(aq) \rightarrow CrO_4^{2-}(aq)$

12AII.29. Write balanced equations for the following reactions to see whether there is any difference between the stoichiometry of the reaction in which the sulfite ion is oxidized to the sulfate ion when the reaction is run in acidic versus basic solution.

$$SO_3^{2-}(aq) + O_2(g) \xrightarrow{H^+} SO_4^{2-}(aq)$$

$$SO_3^{2-}(aq) + O_2(g) \xrightarrow{OH^-} SO_4^{2-}(aq)$$

12AII.30. Use half-reactions to balance the following disproportionation reactions that occur in basic solution.

(a) $Cl_2(g) + OH^-(aq) \rightarrow Cl^-(aq) + OCl^-(aq)$
(b) $Cl_2(g) + OH^-(aq) \rightarrow Cl^-(aq) + ClO_3^-(aq)$
(c) $P_4(s) + OH^-(aq) \rightarrow PH_3(g) + H_2PO_2^-(aq)$
(d) $H_2O_2(aq) \rightarrow O_2(g) + H_2O(l)$

Industrial Redox Reactions

12AII.31. Write a balanced equation for the oxidation–reduction reaction used to extract gold from its ores.

$$Au(s) + CN^-(aq) + O_2(g) \longrightarrow Au(CN)_2^-(aq) + OH^-(aq)$$

12AII.32. Write a balanced equation for the Raschig process, in which ammonia reacts with the hypochlorite ion in a basic solution to form hydrazine.

$$NH_3(aq) + OCl^-(aq) \longrightarrow N_2H_4(aq) + Cl^-(aq)$$

13

CHEMICAL THERMODYNAMICS

13.1 SPONTANEOUS CHEMICAL AND PHYSICAL PROCESSES

No one is surprised when a cup of coffee gradually loses heat to its surroundings as it cools, or when the ice in a glass of lemonade melts as it absorbs heat from its surroundings. But we would be surprised if a cup of coffee suddenly grew hotter until it boiled, or if the water in a glass of lemonade suddenly froze on a hot summer's day.

A piece of zinc metal dissolves in a strong acid to give bubbles of hydrogen gas.

$$Zn(s) + 2 H^+(aq) \longrightarrow Zn^{2+}(aq) + H_2(g)$$

But if we saw a video in which H_2 bubbles formed on the surface of a solution and then sank through the solution until they disappeared, while a strip of zinc metal formed in the middle of the solution, we would conclude that the video was being run backward.

Many chemical and physical processes proceed in a direction in which they are said to be **spontaneous.** This raises a question: What factor or factors determine the direction in which a process is spontaneous? What drives a chemical reaction, for example, in one direction and not the other?

So many spontaneous processes are exothermic that it is not unreasonable to assume that one of the driving forces that determines whether a reaction is spontaneous is a tendency to give off heat. The following are all examples of spontaneous chemical reactions that are exothermic.

The oxidation of ethyl alcohol by the body:

$$CH_3CH_2OH(l) + 3\ O_2(g) \longrightarrow 2\ CO_2(g) + 3\ H_2O(l) \qquad \Delta H° = -1367\ kJ/mol_{rxn}$$

The rusting of iron:

$$4\ Fe(s) + 3\ O_2(g) \longrightarrow 2\ Fe_2O_3(s) \qquad \Delta H° = -1648.4\ kJ/mol_{rxn}$$

The reaction between adipic acid and hexamethylenediamine to produce nylon:

$$H_2NCH_2CH_2CH_2CH_2CH_2CH_2NH_2 + HO{-}\overset{\displaystyle O}{\overset{\|}{C}}{-}CH_2CH_2CH_2CH_2{-}\overset{\displaystyle O}{\overset{\|}{C}}{-}OH \longrightarrow$$

$${-}CH_2{-}\overset{\displaystyle O}{\overset{\|}{C}}{-}NH{-}CH_2CH_2CH_2CH_2CH_2CH_2{-}NH{-}\overset{\displaystyle O}{\overset{\|}{C}}{-}CH_2CH_2CH_2CH_2{-}\overset{\displaystyle O}{\overset{\|}{C}}{-}NH{-}CH_2{-}$$

There are also spontaneous processes, however, that absorb energy from their surroundings. At 100°C, water boils spontaneously even though the process is endothermic.

$$H_2O(l) \rightleftharpoons H_2O(g) \qquad \Delta H°_{373} = 40.88\ kJ/mol_{rxn}$$

Ammonium nitrate dissolves spontaneously in water, even though heat is absorbed when the reaction takes place.

$$NH_4NO_3(s) \xrightarrow{H_2O} NH_4^+\ (aq) + NO_3^-\ (aq) \qquad \Delta H° = 28.1\ kJ/mol_{rxn}$$

It is also important to recognize that the temperature of the system can play a role in determining the direction in which a process is spontaneous. At temperatures below 0°C, water spontaneously freezes.

$$T < 0°C \qquad H_2O(l) \longrightarrow H_2O(s)$$

At temperatures above 0°C, the process is spontaneous in the opposite direction.

$$T > 0°C \qquad H_2O(s) \longrightarrow H_2O(l)$$

Thus, the tendency of a spontaneous reaction or process to give off heat can't be the only driving force. There must be another factor that helps determine whether a reaction or process is spontaneous. This factor, known as **entropy,** is a measure of the disorder of the system.

Checkpoint

When a teaspoon of table salt is poured into a glass of water the salt readily dissolves. Is this a spontaneous process? Heat is absorbed during the process. Why does salt dissolve?

13.2 ENTROPY AS A MEASURE OF DISORDER

Perhaps the best way to understand entropy as a driving force in nature is to conduct a simple experiment with a new deck of cards. Open the deck, remove the jokers, and then turn the deck so that you can read the cards. The top card will be the ace of spades, followed by the two, three, and four of spades, and so on, as shown in Figure 13.1. Now divide the cards in half, shuffle the deck, and note that the deck becomes more disordered. The more often the deck is shuffled, the more disordered it becomes.

FIGURE 13.1 A deck of cards, fresh from the manufacturer, is a perfectly ordered system. When the cards are shuffled, they become more and more disordered. The driving force behind this process is a tendency for systems to move toward greater disorder.

In 1877 Ludwig Boltzmann provided a basis for understanding why a deck of cards becomes more disordered when shuffled when he introduced the concept of the entropy of a system as a measure of the amount of disorder in the system. A deck of cards fresh from the manufacturer is perfectly ordered, and the entropy of the system is zero. When the deck is shuffled, the entropy of the system increases as the deck becomes more disordered.

There are 8.066×10^{67} different ways of organizing a deck of cards. The probability of obtaining any particular sequence of cards when the deck is shuffled is therefore 1 in 8.066×10^{67}. In theory, it is possible to shuffle a deck of cards until the cards fall into perfect order. But it isn't very likely!

The relationship between the number of equivalent ways of describing a system and the amount of disorder in the system can be demonstrated with another analogy based on a deck of cards. There are 2,598,960 different hands that could be dealt in a game of five-card poker. More than half of the hands are essentially worthless. Winning hands are much rarer. Only 3,744 combinations correspond to a "full house," for example. Table 13.1 gives the number of equivalent combinations of cards for each category of poker hand, which is the probability for the category. As the hand becomes more disordered, the probability increases, and the hand becomes intrinsically less valuable.

TABLE 13.1   Number of Equivalent Combinations for Various Types of Poker Hands

Hand	Number of Combinations
Royal flush (AKQJ10 in one suit)	4
Straight flush (five cards in sequence in one suit)	36
Four of a kind	624
Full house (three of a kind plus a pair)	3,744
Flush (five cards in the same suit)	5,108
Straight (five cards in sequence)	10,200
Three of a kind	54,912
Two pairs	123,552
One pair	1,098,240
No pairs	1,302,540
Total	2,598,960

13.3   ENTROPY AND THE SECOND LAW OF THERMODYNAMICS

Thermodynamics is the study of energy and its transformations. The first law of thermodynamics (Chapter 7) provides us with ways of understanding heat changes in chemical reactions and the **second law of thermodynamics** allows us to determine the direction of processes. Experience has shown us that only those processes that produce a net increase in the disorder of the universe are allowed in nature. The role that disorder plays in determining the direction in which a natural process occurs is summarized by the second law, which states that in order for a process to occur the entropy of the universe must increase. The term *universe* means everything that might conceivably have its entropy altered as a result of the process.

To see how these laws affect the direction of a process consider the following two experiments. Imagine that an ice cube is dropped into one beaker of warm water at the same time a red-hot penny is dropped into another beaker of water. The first law of thermodynamics tells us that energy changes in the ice cube and penny will be balanced by energy changes in the water. However, the first law does not tell us which substance loses and which substance gains energy. Experience, on the other hand, tells us that the water in the beaker with the ice cube will become cooler and the water in the beaker with the penny will become hotter.

Although they seem to go in opposite directions, the directions in which the processes occur are determined by the same driving force. Both processes occur because they lead to a net increase in the disorder of the universe.

At the moment the ice cube is dropped into the water, the temperature of the ice cube is lower than the water that surrounds it. Thus, the average kinetic energy of the molecules in the ice cube is significantly smaller than the average kinetic energy of the molecules in the water. During the melting of the ice cube, the rigid structure of the ice breaks down. After the ice cube melts, the average kinetic energy of the H_2O molecules is the same throughout the system, which is another way of stating that the temperature becomes uniform. The exchange of energy has occurred in such a way that, after the ice cube melts, the universe is more disordered. The colder water molecules of ice have been warmed and released from the solid structure thus becoming more disordered. The warmer molecules of liquid water have cooled and become more ordered. Overall, though, there is more disorder due to the breakup of the solid than there is order from the cooling of the warm water.

A similar process is at work when a red-hot penny is dropped into a beaker of water. Initially, the average kinetic energy of the atoms in the metal is significantly larger than the average kinetic energy of the molecules in the water. Eventually, the penny and the water that surrounds it will reach the same temperature. The penny cools, the water warms, and the universe has gone to a more disordered state. The reverse process in which the penny becomes warmer and the water colder does not occur because that process would result in a more ordered universe. The atoms in the hot penny have acquired more order in cooling while the cold water molecules have become more random (disordered) due to warming.

Because neither of these processes could occur unless a net increase in the disorder of the universe occurred, we can conclude that these colder objects produce more disorder when brought to a common temperature than do the warmer objects. In other words, there is a total gain in disorder produced by the melting of the ice or the cooling of the penny.

Checkpoint

A hot brick is thrown into cold water. According to the second law of thermodynamics could the brick get hotter and the water get cooler? See Checkpoint in Section 7.4.

Like the first law of thermodynamics introduced in Section 7.1, the second law describes the results of our experiences while observing the world in which we live. Neither law can be derived mathematically, but no experience that contradicts either law has yet been found. At a time when all other areas of physics seemed open to question, Einstein wrote about thermodynamics as follows, "Thermodynamics is the only science about which I am firmly convinced that, within the framework of the applicability of its basic principles, it will never be overthrown." The thermodynamic term used to quantify discussions of changes in the amount of disorder in the universe is *entropy*, which has been given the symbol S. When a process produces an increase in the disorder of the universe, the entropy change, ΔS is *positive*.

It is important to recognize that the second law of thermodynamics describes what happens to the entropy of the *universe*, not the system in which we are interested. The change in the entropy of the universe that accompanies a chemical reaction is equal to the sum of the changes in the entropy of everything that might undergo an entropy change. Chemists divide these changes into what are called the *system* and the *surroundings*.

$$\Delta S_{univ} = \Delta S_{sys} + \Delta S_{surr}$$

Another way of stating the second law of thermodynamics is that for an event to occur spontaneously, ΔS_{univ} must be positive.

That doesn't mean, however, that the change in the entropy of either the system or its surroundings can't be negative. Consider what happens, once again, when a red-hot penny is dropped into a beaker of water. As the penny cools, the random thermal motion of the metal atoms in the penny decreases. As a result, the metal atoms become more ordered.

At the same time, however, the random thermal motion of the water molecules that surround the penny increases, which means that the immediate surroundings become more disordered. Although the change in the entropy of the system (penny) is negative, the change in the entropy of the surroundings (water) is positive enough that the overall change in the entropy of the universe is positive. Thus, the overall disorder of the universe increases.

The second law suggests that a process is likely to occur if ΔS_{univ} is positive. Any process, however, for which ΔS_{univ} is negative is highly improbable. Entropy, like enthalpy, is a state function (Section 7.5). Thus, the value of ΔS for a process depends only on the initial and final conditions of the system.

Checkpoint

When liquid water freezes more order is introduced. Does this violate the second law of thermodynamics?

Chemists generally choose the chemical reaction as the system in which they are interested. The term *surroundings* is then used to describe the environment that is altered as a consequence of the reaction. Unless otherwise specified, the symbol ΔS will always refer to the system. For chemists the system is usually a chemical reaction.

$\Delta S > 0$ **implies that the system becomes** *more disordered* **during the reaction.**

$\Delta S < 0$ **implies that the system becomes** *less disordered* **during the reaction.**

The following generalizations can help us decide when a chemical reaction leads to an increase in the disorder of the system.

- Solids have a much more regular structure than liquids. Liquids are therefore more disordered than solids.
- The particles in a gas are in a state of constant, random motion. Gases are therefore more disordered than the corresponding liquids.
- Any process that increases the number of gaseous particles in the system increases the amount of disorder.

Exercise 13.1

Which of the following processes will lead to an increase in the entropy of the system?
(a) The production of a fertilizer, NH_3:

$$N_2(g) + 3\,H_2(g) \rightleftharpoons 2NH_3(g)$$

(b) The evaporation of water:

$$H_2O(l) \rightleftharpoons H_2O(g)$$

(c) The reaction to produce lime (CaO) from limestone $CaCO_3$:

$$CaCO_3(s) \rightleftharpoons CaO(s) + CO_2(g)$$

(d) The formation of a single long chain polymeric molecule called nylon from adipic acid and hexamethylenediamine:

$$H_2NCH_2CH_2CH_2CH_2CH_2CH_2NH_2(aq) + HO\overset{\overset{O}{\|}}{C}-CH_2CH_2CH_2CH_2-\overset{\overset{O}{\|}}{C}-OH(l) \longrightarrow$$

$$-CH_2-\overset{\overset{O}{\|}}{C}-NH-CH_2CH_2CH_2CH_2CH_2CH_2-NH-\overset{\overset{O}{\|}}{C}-CH_2CH_2CH_2CH_2-\overset{\overset{O}{\|}}{C}-NH-CH_2-(s)$$

Solution

(a) Because the total number of gaseous molecules decreases in the reaction, the system becomes more ordered. The entropy of the system therefore decreases.

(b) Gases are more disordered than the corresponding liquids, so the entropy of the system increases.

(c) Reactions in which a compound decomposes into two products generally leads to an increase in entropy because the system becomes more disordered. The increase in entropy in this reaction is even larger because the starting material is a solid and one of the products is a gas.

(d) Because many molecules of adipic acid and hexamethylenediamine combine to form one long chain nylon molecule, the system becomes more ordered. The entropy of the system therefore decreases.

Section 13.1 suggested that the sign of ΔH for a chemical reaction affects the direction in which the reaction occurs.

> **Exothermic reactions ($\Delta H < 0$) are often, but not always, spontaneous.**

This section introduced the notion that ΔS for a reaction can also play a role in the direction of the reaction.

> **Reactions that lead to an increase in the disorder of the system ($\Delta S > 0$) are often, but not always, spontaneous.**

To decide in which direction a chemical reaction will occur (as predicted by ΔS_{univ}), it is therefore important to consider the effect of changes in both enthalpy and entropy of the system that occur during the reaction.

Exercise 13.2

Use the Lewis structures of NO_2 and N_2O_4 and the stoichiometry of the following reaction to decide whether ΔH and ΔS favor the reactants or the products of the reaction.

$$2\ NO_2(g) \rightleftharpoons N_2O_4(g)$$

Solution

NO_2 has an unpaired electron in its Lewis structure.

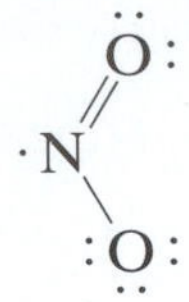

When a pair of NO_2 molecules collide in the proper orientation, a covalent bond can form between the nitrogen atoms to produce the dimer, N_2O_4.

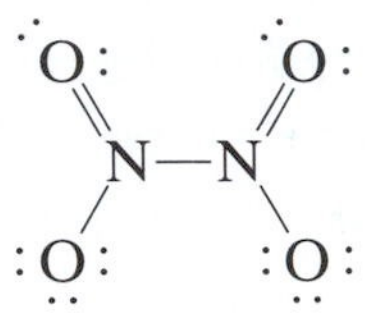

Because a new bond is formed, the reaction should be exothermic. ΔH for the reaction therefore favors the products.

The reaction leads to a decrease in the entropy of the system because the system becomes less disordered when two molecules come together to form a larger molecule. ΔS for the reaction therefore favors the reactants.

Checkpoint

Is it possible for a process to proceed if the entropy of the system decreases?

13.4   STANDARD-STATE ENTROPIES OF REACTION

When the change in the entropy of a system that accompanies a chemical reaction is measured under standard-state conditions, the result is the **standard-state entropy of reaction, $\Delta S°$.** By convention, the standard state for thermodynamic measurements is characterized by the following conditions.

> **Standard-state conditions:**
> **All solutions have concentrations of 1 *M*.**
> **All gases have partial pressures of 1 atm.**

Although standard-state entropies can be measured at any temperature, they are often measured at 25°C. Standard-state entropies of atom combination at 25°C for a variety of compounds are given in Appendix B.14.

13.5   THE THIRD LAW OF THERMODYNAMICS

The **third law of thermodynamics** defines absolute zero on the entropy scale.

> **The entropy of a perfect crystal is zero when the temperature of the crystal is equal to absolute zero (0 K).**

The crystal must be perfect, or else there will be some inherent disorder. It also must be at 0 K.

As the crystal warms to temperatures above 0 K, the particles in the crystal move more vigorously than at 0 K, generating more disorder. The entropy of the crystal gradually increases with temperature as the average kinetic energy of the particles increases. At the melting point, the entropy of the system increases abruptly as the compound is transformed into a liquid, which is not as well ordered as the solid. The entropy of the liquid gradually increases as the liquid becomes warmer because of the increase in the vibrational, rotational, and translational motion of the particles. At the boiling point, there is another abrupt increase in the entropy of the substance as it is transformed into a random, chaotic gas.

Matter is never at rest. Even at the coldest temperature, absolute zero, there is a residual motion of the atoms that make up all matter. On warming, the motion and hence the disorder of the atoms increases. Thermal entropies are a measure of the increase in disorder over that at absolute zero produced by warming the substance to a higher temperature.

Table 13.2 provides an example of the difference between the thermal entropy of a substance in the solid, liquid, and gaseous phases. Note that the units of entropy are joules per mole kelvin. A plot of the entropy of the system versus temperature is shown in Figure 13.2. The third law of thermodynamics describes the entropy, $S°$, of a single substance and how temperature affects the motion of the particles that make up that substance. For chemical reactions, chemists are interested in the difference in entropy, ΔS, between the products and the reactants at a fixed temperature.

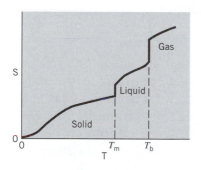

FIGURE 13.2 There is a gradual increase in the entropy of a solid when it is heated until the temperature of the system reaches the melting point of the solid. At this point, the entropy of the system increases abruptly as the solid melts to form a liquid. The entropy of the liquid gradually increases until the system reaches the boiling point of the liquid, at which point it increases abruptly once more. The entropy of the gas then increases gradually with temperature.

TABLE 13.2 Entropies of Solid, Liquid, and Gaseous Forms of Sulfur Trioxide

Compound	$S°$ (J/mol·K)
$SO_3(s)$	70.7
$SO_3(l)$	113.8
$SO_3(g)$	256.76

Checkpoint

Why is the entropy of a substance the lowest at absolute zero?

13.6   CALCULATING ENTROPY CHANGES FOR CHEMICAL REACTIONS

There is no way to measure directly the change in the entropy of the system that occurs during a chemical reaction. The value of ΔS for a reaction has to be obtained indirectly. One approach to determining ΔS for a reaction is based on the fact that entropy, like enthalpy, is a state function. ΔS is measured in units of $J/mol_{rxn}\cdot K$ for the reaction as written.

Thus, the change in the entropy of the system that occurs during a chemical reaction doesn't depend on the path used to transform the starting materials into the products of the reaction. Consider the following reaction, for example.

$$2\ NO_2(g) \rightleftharpoons N_2O_4(g)$$

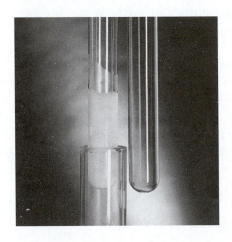

This photograph shows how the brown color of NO_2 gas in the tube on the right disappears as NO2 dimerizes to form N_2O_4 when an identical tube is immersed in liquid nitrogen on the left.

To gain some insight into the meaning of the entropy change that accompanies a reaction, let's approach this calculation the way we first approached calculations of ΔH for a reaction. We start by breaking all of the bonds in the starting materials to form isolated atoms in the gas phase. The entropy change produced by this process may be found from the entropy of atom combination (ac) values given in Appendix B, Table B.14. These entropy changes are designated by the symbol ΔS°_{ac}. The tabulated values, just as for the enthalpies of atom combination, refer to the process whereby 1 mol of the substance listed in the table is formed from its gaseous atoms. Thus if we want the entropy change for the breaking apart of a compound into its atoms, we must reverse the sign given to the entropies in the appendix.

$$2\ N(g)\ +\ 4\ O(g)$$

$$\Delta S^\circ = -(2 \times \Delta S^\circ_{ac})$$

$$2\ NO_2(g)$$

Because the overall reaction consumes 2 mol of NO_2 for every mole of N_2O_4 produced, the change in the entropy of the system that accompanies this step in the reaction must be multiplied by 2.

The dissociation of NO_2 into its gaseous atoms must lead to an increase in the disorder of the system. The ΔS°_{ac} for one mole of NO_2 is $-235\ J/mol_{rxn}\cdot K$. The value of ΔS° for this hypothetical bond breaking process is $+470\ J/mol_{rxn}\cdot K$.

Let's now consider what happens when the isolated atoms in the gas phase recombine to form N_2O_4.

$$2\,N(g) + 4\,O(g) \qquad \xrightarrow{\ \Delta S^\circ = \Delta S^\circ_{ac}\ } \qquad N_2O_4(g)$$

Six isolated atoms in the gas phase have to come together to form each N_2O_4 molecule. As a result, there is a significant decrease in the disorder in the system. The entropy of atom combination for N_2O_4 is large, -647 J/mol$_{rxn}$·K, and therefore ΔS° for this bond making process is both large and negative: $\Delta S^\circ = -647$ J/mol$_{rxn}$·K.

The overall entropy change for the dimerization of NO_2 to form N_2O_4 can be calculated by adding the values of ΔS° for the breaking and making of bonds in this hypothetical scheme for the reaction.

Bond breaking	470 J/mol$_{rxn}$·K
Bond making	$\underline{-647\ \text{J/mol}_{rxn}\text{·K}}$
	-177 J/mol$_{rxn}$·K

The overall entropy change in the reaction is negative, which should be expected because the reaction brings two isolated molecules together to form a more organized molecule.

The entropy of atom combination of a compound reveals much about the way the atoms are held together to form the compound. Consider, for example, the entropies of atom combination of gaseous cyclopentane and 1-pentene.

$$5\,C(g) + 10\,H(g) \longrightarrow \begin{array}{c} CH_2 \\ CH_2 \quad CH_2 \\ | \qquad | \\ CH_2 - CH_2 \end{array}(g)$$
Cyclopentane

$$5\,C(g) + 10\,H(g) \longrightarrow CH_2{=}CHCH_2CH_2CH_3(g)$$
1-Pentene

Both compounds have the same number of carbon and hydrogen atoms, but the way the atoms are arranged and the way they are bonded together differ as shown in Figure 13.3. The entropy of atom combination of cyclopentane is -1645 J/mol$_{rxn}$·K and that of 1-pentene is -1590 J/mol$_{rxn}$·K. What do the entropies tell us about the two compounds?

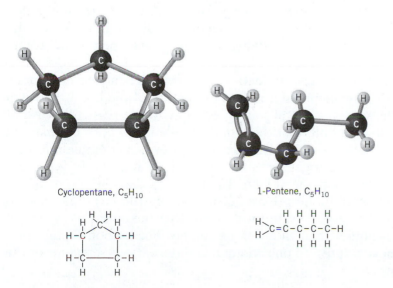

Cyclopentane, C_5H_{10} 1-Pentene, C_5H_{10}

FIGURE 13.3 Structures of cyclopentane and 1-pentene.

The entropy change accompanying a chemical reaction is essentially a measure of the removal or addition of constraints to atomic, molecular, or ionic motion. Thus entropies of atom combination are a measure of the freedom lost when species are formed from their gaseous atoms. There are several types of motion associated with molecules, as depicted in Figure 13.4. First, there is the movement of the molecule as an entity from one position in space to another, the type of motion described by the kinetic molecular theory of Chapter 6 and called translational motion by chemists. The greater the mass of a molecule, the greater its translational contribution to entropy. Molecules also tumble and twist as they move from place to place. This motion is called rotation and depends on the molecular weight and also on the shape of the molecule and the way the atoms are bonded together. Molecules that are symmetrical have less rotational freedom than molecules whose atoms are arranged in a less symmetric fashion.

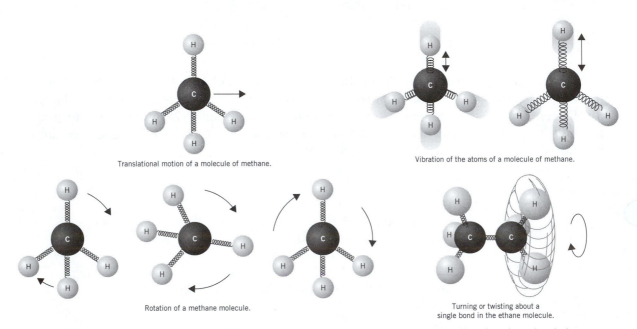

Translational motion of a molecule of methane.

Vibration of the atoms of a molecule of methane.

Rotation of a methane molecule.

Turning or twisting about a
single bond in the ethane molecule.

FIGURE 13.4 Molecules have several types of motion: translational, rotational, vibrational, and twisting.

There are also motions within molecules that the individual atoms possess even when bonded to other atoms. Vibrational motion occurs because the bond holding two atoms together behaves like a spring. When the atoms pull away from each other they are pulled back by the force of the bond, and a vibration motion begins just as when a mass on the end of a spring is pulled and then released. A turning motion occurs when atoms rotate around a single bond.

When atoms are formed into molecules, the entropy loss will depend on how many atoms the molecule contains, how symmetrical it is, and the nature of the bonds between the atoms. We shall examine how these general principles apply to the particular case of cyclopentane and 1-pentene. First we note that the atoms are identical, both in kind and number, for the two molecules. Thus any difference in entropy of atom combination must result from the way the atoms are held together in the molecule. The more negative entropy of atom combination for cyclopentane (Figure 13.3) tells us that the atoms of cyclopentane have lost more freedom of motion (have become more ordered) than those of 1-pentene during the formation process. This, in turn, means that the atoms of cyclopentane have a more constrained motion than those of 1-pentene. The carbon atoms composing 1-pentene are free to move relative to one another, and the various segments of the molecule twist around the single bonds between carbon atoms. Because cyclopentane is a cyclic structure, the twisting of segments of the molecule would require covalent bonds to break. Thus the carbon atoms in cyclopentane are restricted from undergoing that type of motion. The freedom of motion in 1-pentene accounts for the difference in entropies of atom combination of cyclopentane and 1-pentene.

The entropy change for the reaction

$$\text{Cyclopentane}(g) \rightleftharpoons \text{1-pentene}(g)$$

is calculated from:

Bond breaking	$\Delta S° =$	$1645 \text{ kJ/mol}_{rxn}\cdot\text{K}$
Bond making	$\Delta S° =$	$-1590 \text{ kJ/mol}_{rxn}\cdot\text{K}$
	$\Delta S° =$	$+55 \text{ kJ/mol}_{rxn}\cdot\text{K}$

The conversion of cyclopentane to 1-pentene is accompanied by an increase in entropy, which in turn means that there is greater freedom of motion (greater disorder) for the atoms in the molecule 1-pentene than for those of cyclopentane.

Checkpoint

What would be the sign of ΔS for the conversion of glucose from its cyclic into its open-chain form?

Glucose

Exercise 13.3

The entropy of atom combination of NO(g) is $-104 \text{ J/mol}_{rxn}\cdot\text{K}$ and those of $N_2(g)$ and $O_2(g)$ are -115 and $-117 \text{ J/mol}_{rxn}\cdot\text{K}$, respectively. Can you suggest why N_2 and O_2 lose more entropy than NO when formed from their atoms?

Solution

The formation processes

$N(g) + O(g) \longrightarrow NO(g)$	$\Delta S°_{ac} = -104 \text{ J/mol}_{rxn}\cdot\text{K}$
$N(g) + N(g) \longrightarrow N_2(g)$	$\Delta S°_{ac} = -115 \text{ J/mol}_{rxn}\cdot\text{K}$
$O(g) + O(g) \longrightarrow O_2(g)$	$\Delta S°_{ac} = -117 \text{ J/mol}_{rxn}\cdot\text{K}$

involve atoms that are about the same mass. Therefore the changes in entropy caused by molecular weight differences are about the same. Both N_2 and O_2 are more symmetric than NO. Thus formation of N_2 and O_2 results in the loss of more rotational freedom and a more negative $\Delta S°_{ac}$.

Checkpoint

The entropies of atom combination of butane and isobutane are -1469 and -1485 J/mol$_{rxn}$·K, respectively. How can you account for this difference?

$$CH_3CH_2CH_2CH_3 \qquad CH_3CHCH_3$$
$$\qquad\qquad\qquad\qquad\qquad |$$
$$\qquad\qquad\qquad\qquad\qquad CH_3$$

Butane   *Isobutane*

Entropies of atom combination give information on how atoms are bonded together. It is important to note, however, that for direct comparisons to be made, as was done with cyclopentane and 1-pentene, the number of reactant atoms must be the same. If more atoms are required to form the compound, more freedom of motion will be lost. The entropies of atom combination of $C_2H_6(g)$ and $C_2H_4(g)$ are -775 and -555 J/mol$_{rxn}$·K, respectively. C_2H_6 is formed from a total of eight atoms while C_2H_4 is formed from only six.

Entropy of atom combination data can provide a useful tool for studying a chemical reaction. There is always a positive entropy change when the bonds in the starting material are broken to produce isolated atoms in the gas phase. The gaseous atoms can then combine to reform the starting materials, or they can rearrange to form the products of the reaction. In either case, the system becomes less disordered. As a result, the entropy change is negative.

This raises an interesting question: Which process is favored by entropy, combination of the gaseous atoms to reform the starting materials, or rearrangement of the atoms to form the products of the reaction? Let's consider a concrete example, the two nitrogen atoms and four oxygen atoms in the example discussed at the beginning of Section 13.6. The atoms may combine to form either two molecules of NO_2 or one molecule of N_2O_4.

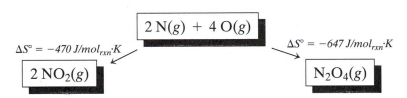

The second law of thermodynamics provides the basis on which we can analyze the system. The second law suggests that natural processes tend to maximum disorder. Entropy therefore favors the process in which the gaseous atoms combine to form the materials that involve the smallest loss of entropy. We can therefore conclude that for this reaction entropy favors the starting materials, not the products.

Exercise 13.4

Use entropies of atom combination to calculate the value of $\Delta S°$ at 298 K for the following reactions and explain the sign of $\Delta S°$ for each reaction.

(a) $CH_3OH(l) \rightleftharpoons CH_3OH(g)$

Compound	$\Delta S°_{ac}$ (J/mol$_{rxn}$·K)
$CH_3OH(l)$	-651.2
$CH_3OH(g)$	-538.19

(b) $N_2(g) + O_2(g) \rightleftharpoons 2\,NO(g)$

Compound	ΔS°_{ac} ($J/mol_{rxn} \cdot K$)
$N_2(g)$	-114.99
$O_2(g)$	-116.97
$NO(g)$	-103.59

Solution

(a) In order to make use of entropies of atom combination, we visualize the reaction as taking place by breaking all reactant bonds to form gaseous atoms and then combining those atoms to form products. Most reactions do not take place in this fashion, but because entropy is a state function, we can use this method to calculate changes in entropy. For liquid CH_3OH we imagine a process that involves breaking the intramolecular bonds and overcoming the intermolecular forces to produce isolated atoms in the gas phase. This leads to an increase in the entropy of the system by 651.2 $J/mol_{rxn} \cdot K$. If the atoms reassemble to form a mole of CH_3OH gas, the entropy of the system decreases by 538.19 $J/mol_{rxn} \cdot K$.

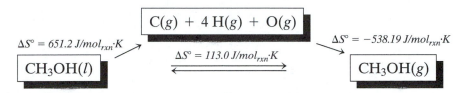

When we combine entropy of atom combination data for the steps in which bonds are broken and made we find that the overall entropy of reaction is positive.

Bond breaking	651.2 $J/mol_{rxn} \cdot K$
Bond making	-538.19 $J/mol_{rxn} \cdot K$
	113.0 $J/mol_{rxn} \cdot K$

The entropy of the system increases because we are transforming a relatively ordered liquid into a more disordered gas.

(b) The change in entropy of the system that accompanies the process by which a mole of N_2 and a mole of O_2 are transformed into isolated nitrogen and oxygen atoms in the gas phase is equal to the sum of the entropy changes required to break all of the bonds in the two starting materials.

Bond breaking:

$$
\begin{array}{ll}
N_2(g) \longrightarrow 2\,N(g) & \Delta S^{\circ} = 114.99\ J/mol_{rxn} \cdot K \\
O_2(g) \longrightarrow 2\,O(g) & \Delta S^{\circ} = 116.97\ J/mol_{rxn} \cdot K \\
\hline
N_2(g) + O_2(g) \longrightarrow 2\,N(g) + 2\,O(g) & \Delta S^{\circ} = 231.96\ J/mol_{rxn} \cdot K
\end{array}
$$

The magnitude of ΔS° for the step in which the gaseous atoms combine to form two moles of NO is equal to twice the entropy of atom combination for the product of the reaction. The sign of this step is negative because the system becomes more ordered.

Bond making:

$$2\,N(g) + 2\,O(g) \longrightarrow 2\,NO(g) \qquad \Delta S^{\circ}_{ac} = -207.18\ J/mol_{rxn} \cdot K$$

The overall change in entropy that accompanies the reaction can be calculated by combining the two steps in this hypothetical way of visualizing the reaction.

$$
\begin{array}{ll}
\text{Bond breaking} & 231.96 \text{ J/mol}_{rxn}\cdot\text{K} \\
\text{Bond making} & \underline{-207.18 \text{ J/mol}_{rxn}\cdot\text{K}} \\
& 24.78 \text{ J/mol}_{rxn}\cdot\text{K}
\end{array}
$$

$\Delta S°$ for the reaction is small but positive. As we saw in Exercise 13.3 the rotational freedom, resulting from symmetry, is greater for NO than for N_2 or O_2. Thus the small, positive entropy change for the reaction is in large part due to the gain in rotational freedom by the products, 2 NO(g).

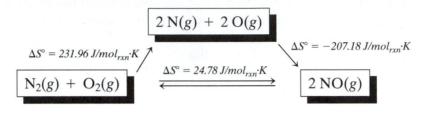

13.7   GIBBS FREE ENERGY

Section 13.3 noted that some reactions are spontaneous because they give off heat ($\Delta H < 0$) and others are spontaneous because they lead to an increase in the disorder of the system ($\Delta S > 0$). The following exercise shows how calculations of ΔH and ΔS can be used to determine the driving force behind a particular reaction.

Exercise 13.5

Use the enthalpy and entropy of atom combination data given below to calculate $\Delta H°$ and $\Delta S°$ for the following reaction and to decide in which direction each of these factors will drive the reaction.

$$N_2(g) + 3 H_2(g) \rightleftharpoons 2 NH_3(g)$$

Compound	$\Delta H°_{ac}$ (kJ/mol$_{rxn}$)	$\Delta S°_{ac}$ (J/mol$_{rxn}$·K)
$N_2(g)$	−945.41	−114.99
$H_2(g)$	−435.30	−98.74
$NH_3(g)$	−1171.76	−304.99

Solution

Both calculations can be based on the following diagram.

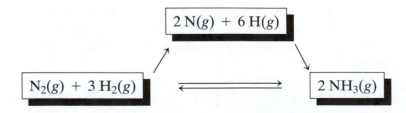

The enthalpy associated with breaking apart a mole of N_2 and 3 mol of H_2 molecules to form isolated nitrogen and hydrogen atoms in the gas phase is equal to the negative of the enthalpy of atom combination of N_2 ($\Delta H^{\circ}_{ac} = -945.41$ kJ/mol$_{rxn}$) plus three times the negative of the enthalpy of atom combination of H_2 ($\Delta H^{\circ}_{ac} = -435.30$ kJ/mol$_{rxn}$).

Bond breaking:

$$
\begin{array}{ll}
N_2(g) \longrightarrow 2\,N(g) & \Delta H^{\circ} = 945.41 \ \text{kJ/mol}_{rxn} \\
3\,H_2(g) \longrightarrow 6\,H(g) & \Delta H^{\circ} = 1305.9 \ \ \text{kJ/mol}_{rxn} \\
\hline
N_2(g) + 3\,H_2(g) \longrightarrow 2\,N(g) + 6\,H(g) & \Delta H^{\circ} = 2251.3 \ \ \text{kJ/mol}_{rxn}
\end{array}
$$

The magnitude of the enthalpy given off when the atoms come together to form 2 mol of NH_3 is equal to twice the enthalpy of atom combination of ammonia. The sign is now negative because this step in the reaction would be exothermic.

Bond making:

$$2\,N(g) + 6\,H(g) \longrightarrow 2\,NH_3 \qquad \Delta H^{\circ}_{ac} = -2343.52 \ \text{kJ/mol}_{rxn}$$

When the bond breaking and bond making steps are combined, we find that the overall reaction is exothermic ($\Delta H^{\circ} < 0$). The sum of the bond strengths in the products is greater than the sum of the bond strengths in the reactants and thus the products are favored by the enthalpy change.

$$
\begin{array}{r}
2251.3 \ \text{kJ/mol}_{rxn} \\
-2343.52 \ \text{kJ/mol}_{rxn} \\
\hline
-92.2 \ \text{kJ/mol}_{rxn}
\end{array}
$$

The entropy change associated with transforming a mole of N_2 and 3 mol of H_2 to nitrogen and hydrogen atoms can be found by a similar procedure.

Bond breaking:

$$
\begin{array}{ll}
N_2(g) \longrightarrow 2\,N(g) & \Delta S^{\circ} = 114.99 \ \text{J/mol}_{rxn}{\cdot}\text{K} \\
3\,H_2(g) \longrightarrow 6\,H(g) & \Delta S^{\circ} = 296.2 \ \text{J/mol}_{rxn}{\cdot}\text{K} \\
\hline
N_2(g) + 3\,H_2(g) \longrightarrow 2\,N(g) + 6\,H(g) & \Delta S^{\circ} = 411.2 \ \text{J/mol}_{rxn}{\cdot}\text{K}
\end{array}
$$

The entropy change associated with the formation of 2 mol of NH_3 can be calculated from the entropy of atom combination of ammonia.

Bond making:

$$2\,N(g) + 6\,H(g) \longrightarrow 2\,NH_3 \qquad \Delta S^{\circ}_{ac} = -609.98 \ \text{J/mol}_{rxn}{\cdot}\text{K}$$

The overall entropy of reaction is negative because the reaction transforms four moles of reactants into two moles of products. Thus the reactants are favored by the entropy change for the reaction.

$$
\begin{array}{r}
411.2 \ \text{J/mol}_{rxn}{\cdot}\text{K} \\
-609.98 \ \text{J/mol}_{rxn}{\cdot}\text{K} \\
\hline
-198.8 \ \text{J/mol}_{rxn}{\cdot}\text{K}
\end{array}
$$

This exercise raises an important question: What happens when one of the potential driving forces behind a chemical reaction is favorable and the other is not? We can answer this question by defining a new quantity known as the **Gibbs free energy (G)** of the system, which reflects the balance between the forces. The name of this quantity recognizes the contributions of J. Willard Gibbs, a professor of mathematical physics at Yale from 1871 until the early 1900s, who is considered by some to be the greatest scientist produced by the United States.

The Gibbs free energy of a system at any moment in time is defined as the enthalpy of the system minus the product of the temperature times the entropy of the system.

$$G = H - TS$$

The Gibbs free energy of the system is a state function because it is defined in terms of thermodynamic properties (enthalpy and entropy) that are state functions. At a given temperature the change in the Gibbs free energy of the system that occurs during a reaction is therefore equal to the change in the enthalpy of the system minus the product of the temperature times the change in entropy of the system.

$$\Delta G = \Delta H - T\Delta S$$

The beauty of the equation defining the free energy of a system is its ability to determine the relative importance of the enthalpy and entropy terms as driving forces behind a particular reaction. The change in the free energy of the system that occurs during a reaction measures the balance between the two driving forces that determine whether a reaction is spontaneous. As we have seen, the enthalpy and entropy have different sign conventions.

Favorable	*Unfavorable*
$\Delta H < 0$	$\Delta H > 0$
$\Delta S > 0$	$\Delta S < 0$

The entropy term is therefore subtracted from the enthalpy term when calculating ΔG for a reaction.

Because of the way the free energy of the system is defined, ΔG is negative for any reaction for which ΔH is negative and ΔS is positive. ΔG is therefore negative for any reaction that is favored by both the enthalpy and entropy terms. We can therefore conclude that any reaction for which ΔG is negative should be favorable, or spontaneous.

For favorable, or spontaneous, reactions, $\Delta G < 0$.

Conversely, ΔG is positive for any reaction for which ΔH is positive and ΔS is negative. Any reaction for which ΔG is positive is therefore unfavorable.

For unfavorable, or nonspontaneous, reactions, $\Delta G > 0$.

Reactions are classified as either **exothermic** ($\Delta H < 0$) or **endothermic** ($\Delta H > 0$) on the basis of whether they give off or absorb heat. Reactions can also be classified as **exergonic** ($\Delta G < 0$) or **endergonic** ($\Delta G > 0$) on the basis of whether the free energy of the system decreases or increases during the reaction.

Checkpoint

If ΔH for a process, such as dissolving table salt in water, is positive, what must be true for the process to occur?

When a reaction is favored by both enthalpy ($\Delta H < 0$) and entropy ($\Delta S > 0$), there is no need to calculate the value of ΔG to decide whether the reaction should proceed. The same can be said for reactions favored by neither enthalpy ($\Delta H > 0$) nor entropy ($\Delta S < 0$). Free energy calculations are important, however, for reactions favored by only one of these factors.

The change in the free energy of a system that occurs during a reaction can be measured under any set of conditions. If the data are collected under standard-state conditions, the result is the **standard-state free energy of reaction ($\Delta G°$).**

$$\Delta G° = \Delta H° - T\Delta S°$$

For gases the superscript circle represents 1 atm of pressure and for aqueous species a concentration of 1 M.

Exercise 13.6

Use enthalpy and entropy of atom combination data to calculate $\Delta H°$ and $\Delta S°$ for the following reaction at 25°C.

$$2\ NO_2(g) \rightleftharpoons N_2O_4(g)$$

Compound	$\Delta H°_{ac}$ (kJ/mol$_{rxn}$)	$\Delta S°_{ac}$ (J/mol$_{rxn}$·K)
NO_2	−937.86	−235.35
N_2O_4	−1932.93	−646.53

From the results of the calculation determine the value of $\Delta G°$ for the reaction and predict if the reaction will occur as written.

Solution

The following diagram can be used to calculate $\Delta H°$ and $\Delta S°$ for the reaction from enthalpy and entropy of atomization data.

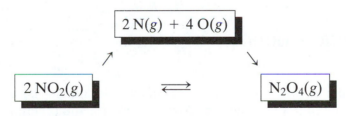

Enthalpy of reaction calculation:

Bond breaking	$2\ NO_2(g) \longrightarrow 2\ N(g) + 4\ O(g)$	$\Delta H° = 1875.7$ kJ/mol$_{rxn}$
Bond making	$2\ N(g) + 4\ O(g) \longrightarrow N_2O_4(g)$	$\Delta H°_{ac} = -1932.93$ kJ/mol$_{rxn}$
		$\Delta H° = -57.2$ kJ/mol$_{rxn}$

The overall reaction is exothermic, and the enthalpy change of reaction is therefore favorable.

Entropy of reaction calculation:

Bond breaking	$2\ NO_2(g) \longrightarrow 2\ N(g) + 4\ O(g)$	$\Delta S° = 470.70$ J/mol$_{rxn}$·K
Bond making	$2\ N(g) + 4\ O(g) \longrightarrow N_2O_4(g)$	$\Delta S°_{ac} = -646.53$ J/mol$_{rxn}$·K
		$\Delta S°_{ac} = -175.83$ J/mol$_{rxn}$·K

The reaction leads to a significant decrease in the disorder of the system, however, and is therefore not favored by the entropy of reaction.

To decide whether the reaction should proceed at 25°C we have to compare the $\Delta H°$ and $T\Delta S°$ terms to see which is larger. Before we can do this, we have to convert the temperature to kelvins.

$$T_K = 25°C + 273 = 298 \text{ K}$$

We also have to recognize that the units of $\Delta H°$ for the reaction are **kilojoules** and the units of $\Delta S°$ are **joules per kelvin.** At some point in the calculation, we have to convert these quantities to a consistent set of units. Perhaps the easiest way of doing this is to convert $\Delta S°$ to kilojoules. We then multiply the entropy term by the absolute temperature and subtract that quantity from the enthalpy term.

$$\begin{aligned}
\Delta G° &= \Delta H° - T\Delta S° \\
&= -57.2 \text{ kJ/mol}_{rxn} - (298 \text{ K})(-0.17583 \text{ kJ/mol}_{rxn}\cdot\text{K}) \\
&= -57.2 \text{ kJ/mol}_{rxn} + 52.4 \text{ kJ/mol}_{rxn} \\
&= -4.8 \text{ kJ/mol}_{rxn}
\end{aligned}$$

At 25°C, the standard-state free energy for the reaction is negative because the enthalpy term at that temperature is more negative than the entropy term, $T\Delta S°$. It is important to note that because standard-state conditions were used in the calculation, the conclusion that the reaction is spontaneous applies only when all gases are at 1 atm of pressure.

Exercise 13.7

In muscle cells undergoing vigorous exercise glucose is broken down into lactic acid. One possible way to produce the acid is the direct conversion reaction:

$$C_6H_{12}O_6(aq) \rightleftharpoons 2 \text{ CH}_3\text{CHOHCOOH}(aq)$$

Given the following thermodynamic data at 298.15 K, decide if the reaction is feasible under standard conditions and explain why or why not.

Compound	$\Delta H°_{ac}$ (kJ/mol$_{rxn}$)	$\Delta S°_{ac}$ (J/mol$_{rxn}\cdot$K)
$C_6H_{12}O_6(aq)$	−9670.0	−3013.9
$CH_3CHOHCOOH(aq)$	−4889.7	−1417

Solution

This is an example of how compiled thermochemical data can be used to determine if a reaction could occur. First we calculate $\Delta H°$ and $\Delta S°$ for the reaction using the thermodynamic data given above.

$$\Delta H° = (9670.0 \text{ kJ/mol}_{rxn}) - (2 \times 4889.7 \text{ kJ/mol}_{rxn}) = -109.4 \text{ kJ/mol}_{rxn}$$
$$\Delta S° = (3013.9 \text{ J/mol}_{rxn}\cdot\text{K}) - (2 \times 1417 \text{ J/mol}_{rxn}\cdot\text{K}) = 180 \text{ J/mol}_{rxn}\cdot\text{K}$$

$\Delta H°$ is negative, which means that the bonds in lactic acid are stronger than those in glucose. Thus lactic acid is favored by the enthalpy change. $\Delta S°$ is positive which means the molecular structure of lactic acid has more freedom of motion than that of glucose and hence lactic acid is favored by the entropy change. Both the enthalpy and entropy changes indicate that under standard conditions this reaction can proceed. We can also calculate $\Delta G°$ for the reaction.

$$\Delta G° = \Delta H° - T\Delta S° = -109.4 \text{ kJ/mol}_{rxn} - 298.15 \text{ K}(0.180 \text{ kJ/mol}_{rxn}\cdot\text{K}) = -163 \text{ kJ/mol}_{rxn}$$

13.8 THE EFFECT OF TEMPERATURE ON THE FREE ENERGY OF A REACTION

The balance between the contributions from the enthalpy and entropy terms to the free energy of a reaction depends on the temperature at which the reaction is run.

Exercise 13.8

Use the values of ΔH° and ΔS° calculated in Exercise 13.5 to predict whether the following reaction will occur at 25°C and 1 atm partial pressure for each gas.

$$N_2(g) + 3\ H_2(g) \rightleftharpoons 2\ NH_3(g)$$

Solution

According to Exercise 13.5, the reaction is favored by enthalpy but not by entropy.

$$\Delta H^\circ = -92.2 \text{ kJ/mol}_{rxn} \qquad \text{(favorable)}$$
$$\Delta S^\circ = -198.8 \text{ J/mol}_{rxn}\cdot\text{K} \qquad \text{(unfavorable)}$$

Before we can compare these terms to see which is larger, we have to incorporate into our calculation the temperature at which the reaction is run.

$$T_K = 25°C + 273 = 298 \text{ K}$$

We then multiply the entropy of reaction by the absolute temperature and subtract the $T\Delta S^\circ$ term from the ΔH° term.

$$
\begin{aligned}
\Delta G^\circ &= \Delta H^\circ - T\Delta S^\circ \\
&= (-92.2 \text{ kJ/mol}_{rxn}) - (298 \text{ K})(-0.1988 \text{ kJ/mol}_{rxn}\cdot\text{K}) \\
&= (-92.2 \text{ kJ/mol}_{rxn}) + 59.2 \text{ kJ/mol}_{rxn} \\
&= -33.0 \text{ kJ/mol}_{rxn}
\end{aligned}
$$

According to the calculation, the reaction should be spontaneous at 25°C, that is,

$$\Delta G^\circ = -33.0 \text{ kJ/mol}_{rxn}$$

Nitrogen and hydrogen gases both at 1 atm and 25°C should react to produce ammonia gas at 1 atm and 25°C.

What happens as we raise the temperature of the reaction? The equation used to define free energy suggests that the entropy term will become more important as the temperature increases, and may dominate the enthalpy term at high temperatures.

$$\Delta G^\circ = \Delta H^\circ - T\Delta S^\circ$$

Checkpoint

Is it possible to make a cement that will "set up" without releasing heat? Explain your reasoning.

Exercise 13.9

Predict whether the following reaction will occur at 500°C and 1 atm partial pressure of each gas.

$$N_2(g) + 3 H_2(g) \rightleftharpoons 2 NH_3(g)$$

Assume that the values of $\Delta H°$ and $\Delta S°$ used in Exercise 13.8 are still valid at that temperature.

Solution

We need to calculate the temperature on the Kelvin scale.

$$T_K = 500°C + 273 = 773 \text{ K}$$

We then multiply the entropy term by the absolute temperature and subtract the result from the value of $\Delta H°$ for the reaction.

$$\begin{aligned}
\Delta G°_{773} &= \Delta H°_{298} - T\Delta S°_{298} \\
&= (-92.2 \text{ kJ/mol}_{rxn}) - (773 \text{ K})(-0.1988 \text{ kJ/mol}_{rxn}\cdot\text{K}) \\
&= (-92.2 \text{ kJ/mol}_{rxn}) - (-154 \text{ kJ/mol}_{rxn}) \\
&= 62 \text{ kJ/mol}_{rxn}
\end{aligned}$$

Because the entropy term becomes more significant than the enthalpy term as the temperature increases, the reaction changes from one that proceeds at low temperatures to one that does not proceed at high temperatures. In other words, nitrogen and hydrogen at 1 atm each will not react to produce ammonia at 1 atm and 500°C.

13.9   BEWARE OF OVERSIMPLIFICATIONS

The calculation of $\Delta G°$ for the following reaction at 25°C in Exercise 13.8 is perfectly legitimate.

$$N_2(g) + 3 H_2(g) \rightleftharpoons 2 NH_3(g)$$

The enthalpy and entropy data used in the calculation were standard-state measurements made at 25°C, which can legitimately be used to predict the standard-state free energy of reaction at that temperature. The calculation of $\Delta G°_{773}$ for the reaction at 500°C in Exercise 13.9, however, was based on the assumption that the enthalpy and entropy data measured at 25°C are still valid at 500°C.

This is a useful assumption, but it isn't always a valid one. Changes in $\Delta H°$ and $\Delta S°$ are often small over moderate temperature ranges. When the temperature range over which the data are extrapolated is large, as it is in the case of Exercise 13.9, the results of the calculation should be taken with a grain of salt. They represent an estimate, not a prediction, of the magnitude of the effect of the change in temperature. In this case, the assumption that $\Delta H°$ and $\Delta S°$ are constant underestimates the change in the magnitude of $\Delta G°$ by about 20%. $\Delta G°_{773}$ for the reaction is 73 kJ/mol$_{rxn}$, not 62 kJ/mol$_{rxn}$.

13.10 STANDARD-STATE FREE ENERGIES OF REACTION

$\Delta G°$ for a reaction can also be calculated from tabulated standard-state free energy data. The data are tabulated in terms of **standard-state free energies of atom combination, $\Delta G°_{ac}$,** found in Appendix B, Table B.14. These values correspond to the formation of 1 mol of the substance listed in the table from its gaseous atoms. $\Delta G°_{ac}$ is a measure of the strength of the intermolecular forces and intramolecular bonds (ΔH) as well as a measure of the freedom lost on atom combination (ΔS). Consider the following reaction at 298 K and 1 atm partial pressure of each gas.

$$CH_4(g) + 2\,O_2(g) \longrightarrow CO_2(g) + 2\,H_2O(g)$$

Compound	$\Delta G°_{ac}$ (kJ/mol$_{rxn}$)
$CH_4(g)$	-1535.00
$O_2(g)$	-463.46
$CO_2(g)$	-1529.08
$H_2O(g)$	-866.80

The following diagram describes how free energy of atom combination data can be used to calculate the overall change in the free energy of the system that occurs in this reaction.

The bond breaking step in the reaction involves atomization of 2 mol of O_2 for each mole of CH_4 consumed in the reaction.

Bond breaking:

$$
\begin{aligned}
CH_4(g) &\longrightarrow C(g) + 4\,H(g) & \Delta G° &= 1535.00 \text{ kJ/mol}_{rxn} \\
2\,O_2(g) &\longrightarrow 4\,O(g) & \Delta G° &= \underline{926.92 \text{ kJ/mol}_{rxn}} \\
& & \Delta G° &= 2461.92 \text{ kJ/mol}_{rxn}
\end{aligned}
$$

The magnitude of the change in the free energy of the system as the isolated atoms in the gas phase combine to form the products of the reaction is equal to the sum of the free energy of atom combination of CO_2 and twice that of H_2O.

Bond making:

$$
\begin{aligned}
C(g) + 2\,O(g) &\longrightarrow CO_2(g) & \Delta G° &= -1529.08 \text{ kJ/mol}_{rxn} \\
4\,H(g) + 2\,O(g) &\longrightarrow 2\,H_2O(g) & \Delta G° &= \underline{-1733.6 \text{ kJ/mol}_{rxn}} \\
& & \Delta G° &= -3262.7 \text{ kJ/mol}_{rxn}
\end{aligned}
$$

When the results of the calculations are combined, we find that the overall free energy of reaction is negative. Free energies of atom combination include the enthalpy and entropy changes incurred when a compound is formed from its constituent atoms. Formation of the products of the reaction is therefore favored at 298 K and 1 atm.

$$
\begin{aligned}
&2461.92 \text{ kJ/mol}_{rxn} \\
&\underline{-3262.7 \text{ kJ/mol}_{rxn}} \\
&-800.8 \text{ kJ/mol}_{rxn}
\end{aligned}
$$

If all reactant molecules are atomized to gaseous atoms and if the gaseous atoms may recombine, what factors determine whether the recombination is to form the original reactants or the products? Those molecules having the largest total bond strengths are enthalpically preferred, whereas those molecules having the most freedom associated with their molecular structure are entropically preferred. There is a balance between the stability (enthalpy) and freedom (entropy) that determines the position of the equilibrium.

13.11 EQUILIBRIA EXPRESSED IN PARTIAL PRESSURES

Previously we used units of concentration to calculate equilibrium constants. This choice of units was indicated by the subscript c added to the symbols for the reaction quotients and equilibrium constants, to show that they were calculated from the concentrations of the components of the reaction.

We have also mentioned that equilibrium constants can be written in terms of pressures of gaseous species rather than concentrations. Gas phase equilibria are frequently described in terms of the partial pressures of the gases in the reaction. We can understand why this is possible by rearranging the ideal gas equation

$$PV = nRT$$

to give the following relationship between the pressure of a gas and its concentration in moles per liter

$$P = \frac{n}{V} \times RT$$

where n/V is the concentration in moles per liter. We can therefore characterize the following reaction

$$N_2(g) + 3\,H_2(g) \rightleftharpoons 2\,NH_3(g)$$

with an equilibrium constant defined in terms of units of concentration

$$K_c = \frac{[NH_3]^2}{[N_2][H_2]^3}$$

or an equilibrium constant defined in terms of partial pressures. K_p like K_c is always reported without units. However, any calculation involving K_p requires the partial pressures of the products and reactants to be in units of atmospheres or bars. This text will use atmospheres.

$$K_p = \frac{P_{NH_3}{}^2}{P_{N_2}P_{H_2}{}^3}$$

What is the relationship between K_p and K_c for a gas phase reaction? According to the rearranged version of the ideal gas equation, the pressure of a gas is equal to the concentration of the gas times the product of the ideal gas constant and the temperature in units of kelvins.

$$P = \left[\frac{n}{V}\right]RT$$

We can therefore calculate the value of K_p for a reaction by multiplying each of the terms in the K_c expression by RT.

$$K_p = \frac{P_{NH_3}{}^2}{P_{N_2}P_{H_2}{}^3} = \frac{([NH_3] \times RT)^2}{([N_2] \times RT)([H_2] \times RT)^3}$$

Collecting terms in this example gives the following result.

$$K_p = K_c \times (RT)^{-2}$$

In general, the value of K_p for a reaction can be calculated from K_c with the following equation.

$$K_p = K_c \times (RT)^{\Delta n}$$

In this equation, Δn is the difference between the number of moles of gaseous products and the number of moles of gaseous reactants in the balanced equation ($\Delta n = n_{prod} - n_{react}$).

Exercise 13.10

Calculate the value of K_p for the following reaction at 650°C if the value of K_c for the reaction at that temperature is 0.040.

$$N_2(g) + 3\,H_2(g) \rightleftharpoons 2\,NH_3(g)$$

Solution

The value of K_p for a reaction can be calculated from K_c with the following equation.

$$K_p = K_c \times (RT)^{\Delta n}$$

To use this equation, we need to know the value of Δn for the reaction. The balanced equation for the reaction generates 2 mol of gaseous products for every 4 mol of gaseous reactants consumed. The value of Δn for the reaction is therefore $2 - 4$, or -2.

$$K_p = K_c \times (RT)^{-2}$$

We now use the known value of the ideal gas constant and the temperature in kelvins to calculate the value of K_p for the reaction at the temperature given.

$$K_p = 0.040 \times [(0.08206\ \text{L·atm/mol·K})(923\ \text{K})]^{-2} = 7.0 \times 10^{-6}$$

The techniques for working problems using K_p expressions are the same as those described for K_c problems. The difference being that partial pressures are used instead of concentrations. The equilibrium constant calculated from tabulated standard free energies enthalpies, and entropies will give K_p for reactions which involve only gases, or which have only gases and pure liquids or solids as products and reactants. Calculations of an equilibrium constant from the tabulated data will give K_c for reactions that contain only aqueous species and pure liquids and solids. For reactions involving only gaseous species and pure liquids and solids, K_c may be converted into K_p by the relationship described above.

Exercise 13.11

For the reaction

$$H_2(g) + I_2(g) \rightleftharpoons 2\ HI(g)$$

K_p is 60 at 350°C. If initial pressures of H_2 and I_2 are 1.00 atm each, what will be the equilibrium pressures of HI, H_2, and I_2?

Solution

There is initially no HI present; therefore the reaction will proceed to the right to form HI. Representing the change in pressure which occurs as the reaction takes place with Δp gives the following relationships.

$$H_2(g) + I_2(g) \rightleftharpoons 2\ HI(g)$$

Initial	1.00 atm	1.00 atm	0 atm
Change	$-\Delta p$	$-\Delta p$	$2\Delta p$
Equilibrium	$1.00 - \Delta p$	$1.00 - \Delta p$	$2\Delta p$

$$K_p = \frac{P_{HI}{}^2}{P_{H_2}P_{I_2}} = \frac{(2\Delta p)^2}{(1.00 - \Delta p)(1.00 - \Delta p)} = 60$$

The equilibrium constant is large so that Δp will not be negligible compared to 1.00 atm.

$$\frac{4\Delta p^2}{(1.00 - \Delta p)^2} = 60$$

$$\frac{4\Delta p^2}{1.00 - 2\Delta p + \Delta p^2} = 60$$

$$60 - 120\Delta p + 56\Delta p^2 = 0$$

Using the quadratic equation solution gives

$$\Delta p = \frac{120 \pm \sqrt{120^2 - 4(56)(60)}}{2(56)} = \frac{120 \pm 31}{112}$$

If we choose the positive root, then Δp is 1.35, which means that more H_2 and I_2 have reacted than was initially present. Thus the negative root is used which gives a Δp of 0.79 atm. The equilibrium pressure of HI is $2\Delta p = 1.6$ atm, and those of H_2 and I_2 are $1.00 - \Delta p = 0.21$ atm.

13.12   INTERPRETING STANDARD-STATE FREE ENERGY OF REACTION DATA

We are now ready to ask the question: What does the value of $\Delta G°$ at 25°C tell us about the following reaction?

$$N_2(g) + 3\ H_2(g) \rightleftharpoons 2\ NH_3(g) \qquad \Delta G° = -33.0\ kJ/mol_{rxn}$$

By definition, the value of $\Delta G°$ for a reaction measures the difference between the free energies of the reactants and products *when all components of the reaction are present at standard-state conditions.*

$\Delta G°$ therefore describes this reaction only when all three components are present at 1 atm pressure. Note in the following equation that the reaction quotient, Q_p, is written using pressures, not concentrations as previously used in Section 10.6. This is because the tabulated values of the thermodynamic parameters for gases are given in units of pressure.

$$\text{Standard state} \qquad Q_p = \frac{P_{NH_3}{}^2}{P_{N_2}P_{H_2}{}^3} = \frac{(1)^2}{(1)(1)^3} = 1$$

The *sign* of $\Delta G°$ tells us the direction in which the reaction has to shift from standard-state conditions to come to equilibrium. The fact that $\Delta G°$ is negative for this reaction at 25°C means that the system under standard-state conditions at that temperature would have to shift to the right, converting some of the reactants into products, before it can reach equilibrium. The *magnitude* of $\Delta G°$ for a reaction tells us how far the standard state is from equilibrium. The larger the value of $\Delta G°$, the further the reaction has to go to get from the standard-state conditions to equilibrium.

Assume, for example, that we start with the following reaction under standard-state conditions, as shown in Figure 13.5.

$$N_2(g) + 3\,H_2(g) \rightleftharpoons 2\,NH_3(g)$$

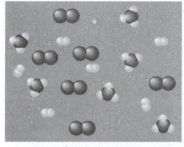

Standard state
$P_{H_2} = P_{N_2} = P_{NH_3} = 1$ atm

FIGURE 13.5 At standard-state conditions, the partial pressures of N_2, H_2, and NH_3 in this system are 1 atm. Thus Q_p for the system is equal to 1.

The value of ΔG at that moment in time will be equal to the standard-state free energy for the reaction, $\Delta G°$

When $Q_p = 1$, $\Delta G = \Delta G°$.

As the reaction gradually shifts to the right, converting N_2 and H_2 into NH_3, the value of ΔG for the reaction becomes more positive. If we could find some way to harness the tendency of the reaction to come to equilibrium, we could get the reaction to do work. The free energy of a reaction at any moment in time is therefore said to be a measure of the energy available to do work.

13.13   THE RELATIONSHIP BETWEEN FREE ENERGY AND EQUILIBRIUM CONSTANTS

When a reaction leaves the standard state because of a change in the ratio of the partial pressures of the products to the reactants, we have to describe the system in terms of nonstandard-state free energies of reaction. The difference between $\Delta G°$ and ΔG for a reaction is important. There is only one value of $\Delta G°$ for a reaction at a given temperature, but there are an infinite number of possible values of ΔG.

Figure 13.6 shows the relationship between ΔG and $\ln Q_p$ for the following reaction.

$$N_2(g) + 3\,H_2(g) \rightleftharpoons 2\,NH_3(g)$$

Data on the left side of Figure 13.6 correspond to relatively small values of Q_p. They therefore describe systems in which there is far more reactant than product. The sign of ΔG for these systems is negative and the magnitude of ΔG is large. The system is therefore relatively far from equilibrium, and the reaction must shift to the right to reach equilibrium.

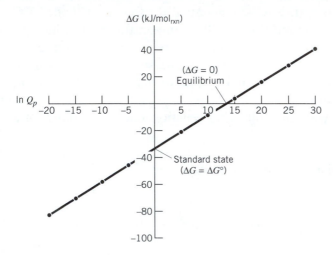

FIGURE 13.6 Plot of $\ln Q_p$ versus ΔG for the reaction in which N_2 and H_2 combine to form NH_3. $\Delta G = \Delta G°$ when the reaction quotient is 1 ($\ln Q_p = 0$). $\Delta G = 0$ when the reaction is at equilibrium.

Data on the far right-hand side of Figure 13.6 describe systems in which there is more product than reactant. The sign of ΔG is now positive and the magnitude of ΔG is moderately large. The sign of ΔG tells us that the reaction would have to shift to the left to reach equilibrium. The magnitude of ΔG tells us that we don't have quite so far to go to reach equilibrium.

The points at which the straight line in Figure 13.6 cross the horizontal and vertical axes of the diagram are particularly important. The straight line crosses the vertical axis when the reaction quotient for the system is equal to 1. That point therefore describes the standard-state conditions, and the value of ΔG at that point is equal to the standard-state free energy of reaction, $\Delta G°$.

When $Q_p = 1$, $\Delta G = \Delta G°$.

The point at which the straight line crosses the horizontal axis describes a system for which ΔG is equal to zero. Because there is no driving force behind the reaction, the system must be at equilibrium. The equilibrium constant when written in terms of pressures is given the symbol K_p.

When $Q_p = K_p$, $\Delta G = 0$.

Checkpoint

If $Q_p > K_p$ which way does the reaction proceed?

In general the relationship between the free energy of reaction at any moment in time (ΔG) and the standard-state free energy of reaction ($\Delta G°$) is described by the following equation.

$$\Delta G = \Delta G° + RT \ln Q$$

In this equation, R the ideal gas constant measured in joules is 8.314 J/mol$_{rxn}$·K, T is the temperature in kelvins, ln represents a logarithm to the base e, and Q is the reaction quotient measured in pressure or concentration units.

As we have seen, the driving force behind a chemical reaction is zero ($\Delta G = 0$) when the reaction is at equilibrium ($Q = K$).

$$0 = \Delta G° + RT \ln K$$

We can solve the equation for the relationship between $\Delta G°$ and K.

$$\Delta G° = -RT \ln K$$

This equation allows us to calculate the equilibrium constant for any reaction from the standard-state free energy of reaction, or vice versa.

To understand the relationship between $\Delta G°$ and K we must recognize that the magnitude of $\Delta G°$ tells us how far the standard state is from equilibrium. The smaller the absolute value of $\Delta G°$, the closer the standard state is to equilibrium. The larger the absolute value of $\Delta G°$, the further the reaction has to go to reach equilibrium. The relationship between $\Delta G°$ and the equilibrium constant for a chemical reaction is illustrated by the data in Table 13.3.

TABLE 13.3   Values of $\Delta G°$ and K for Common Reactions at 25°C

Reaction	$\Delta G°$ (kJ/mol$_{rxn}$)	K
$2\ SO_3(g) \rightleftharpoons 2\ SO_2(g) + O_2(g)$	141.7	1.5×10^{-25}
$H_2O(l) \rightleftharpoons H^+(aq) + OH^-(aq)$	79.9	1.0×10^{-14}
$AgCl(s) \overset{H_2O}{\rightleftharpoons} Ag^+(aq) + Cl^-(aq)$	55.7	1.7×10^{-10}
$HOAc(aq) \overset{H_2O}{\rightleftharpoons} H^+(aq) + OAc^-(aq)$	27.1	1.8×10^{-5}
$N_2(g) + 3\ H_2(g) \rightleftharpoons 2\ NH_3(g)$	−33.0	6×10^5
$HCl(g) \overset{H_2O}{\longrightarrow} H^+(aq) + Cl^-(aq)$	−35.9	2×10^6
$Cu^{2+}(aq) + 4\ NH_3(aq) \longrightarrow Cu(NH_3)_4{}^{2+}(aq)$	−70.6	2.3×10^{12}
$Zn(s) + Cu^{2+}(aq) \longrightarrow Zn^{2+}(aq) + Cu(s)$	−212.5	1.8×10^{37}

Exercise 13.12

Use the value of $\Delta G°$ for the following reaction obtained in Exercise 13.8 to calculate the equilibrium constant for the reaction at 25°C.

$$N_2(g) + 3\ H_2(g) \rightleftharpoons 2\ NH_3(g)$$

Solution

Exercise 13.8 gave the following value for $\Delta G°$ for this reaction at 25°C.

$$\Delta G° = -33.0 \text{ kJ/mol}_{rxn}$$

We now turn to the relationship between $\Delta G°$ for a reaction and the equilibrium constant for the reaction.

$$\Delta G° = -RT \ln K$$

Solving for the natural log of the equilibrium constant produces the following equation.

$$\ln K = -\frac{\Delta G°}{RT}$$

Substituting the known values of $\Delta G°$, R, and T into the equation gives the result:

$$\ln K = -\frac{(-33.0 \times 10^3 \text{ J/mol}_{rxn})}{(8.314 \text{ J/mol}_{rxn}\cdot K)(298 \text{ K})} = 13.3$$

The equilibrium constant for the reaction at 25°C is therefore 6×10^5.

$$K = e^{13.3} = 6 \times 10^5$$

It is easy to make mistakes when handling the sign of the relationship between $\Delta G°$ and K. It is therefore a good idea to check the final answer to see whether it makes sense. $\Delta G°$ for the reaction in this exercise is negative, and the equilibrium should lie on the side of the products. The equilibrium constant should therefore be much larger than 1, which it is.

Exercise 13.13

Use the value of $\Delta G°$ for the following reaction to calculate the acid dissociation equilibrium constant (K_a) at 25°C for formic acid.

$$HCO_2H(aq) \rightleftharpoons H^+(aq) + HCO_2^-(aq) \qquad \Delta G° = 21.3 \text{ kJ/mol}_{rxn}$$

Solution

We start with the relationship between $\Delta G°$ and the equilibrium constant for the reaction

$$\Delta G° = -RT \ln K$$

and solve for the natural logarithm of the equilibrium constant.

$$\ln K = -\frac{\Delta G°}{RT}$$

Substituting the known values of $\Delta G°$, R, and T into the equation gives the following result.

$$\ln K = -\frac{21.3 \times 10^3 \text{J/mol}_{rxn}}{(8.314 \text{ J/mol}_{rxn}\cdot K)(298 \text{ K})} = -8.60$$

We can now calculate the value of the equilibrium constant.

$$K = e^{-8.60} = 1.8 \times 10^{-4}$$

In this case $\Delta G°$ is positive and the equilibrium will favor the reactants; in other words, the equilibrium constant will be less than 1.

Exercise 13.14

The amino acid glutamine can be formed by the reaction of NH_3 with glutamate.

For this reaction the equilibrium constant has been determined to be 3.5×10^{-3} at 25°C.

(a) Calculate $\Delta G°$ for the reaction.

(b) What does this value for $\Delta G°$ signify?

(c) NH_3 is present in the body at a concentration of about 1×10^{-2} M. If a concentration of glutamine of 1×10^{-2} M was desired from the reaction, what would have to be the concentration of glutamate?

Solution

(a) $\Delta G° = -RT \ln K$
$$= -8.314 \text{ J/mol}_{rxn} \cdot \text{K}(298 \text{ K}) \ln 3.5 \times 10^{-3}$$
$$= 14 \times 10^3 \text{ J/mol}_{rxn}$$

(b) The large positive $\Delta G°$ means that the reaction does not proceed very far to the right. This does not mean, however, that the reaction cannot ever proceed spontaneously to make glutamine. It is ΔG that determines the spontaneous direction of the reaction, not $\Delta G°$.

(c) $\Delta G = \Delta G° + RT \ln Q = \Delta G° + RT \ln \dfrac{\text{(glutamine)}}{\text{(glutamate)}(NH_3)}$

$$\Delta G = (14 \times 10^3 \text{ J/mol}_{rxn}) + 8.314 \text{ J/mol}_{rxn} \cdot \text{K}(298 \text{ K}) \ln \dfrac{(1 \times 10^{-2})}{\text{(glutamate)}(1 \times 10^{-2})}$$

For the reaction to proceed, ΔG must be negative. In order to find what glutamate concentration will make ΔG negative, we first set $\Delta G = 0$. Then

$$\Delta G = 0 = (14 \times 10^3 \text{ J/mol}_{rxn}) + 8.314 \text{ J/mol}_{rxn} \cdot \text{K}(298 \text{ K}) \ln \dfrac{(1 \times 10^{-2})}{\text{(glutamate)}(1 \times 10^{-2})}$$

and (glutamate) $= 3 \times 10^2$ M at equilibrium. That is, in order for the reaction to proceed, the glutamate concentration would have to exceed 300 M! Thus this reaction could not be used for the biosynthesis of glutamine. Compare the results of this calculation to that of Exercise 10.6. Can you use the procedure of Exercise 10.6 to arrive at the same conclusion for the glutamate reaction?

13.14   THE TEMPERATURE DEPENDENCE OF EQUILIBRIUM CONSTANTS

When equilibrium constants were introduced in Chapter 10 we noted that they are not strictly constant because they change with temperature. We are now ready to understand why.

The standard-state free energy of reaction is a measure of how far the standard state is from equilibrium.

$$\Delta G° = -RT \ln K$$

But the magnitude of $\Delta G°$ depends on the temperature of the reaction.

$$\Delta G° = \Delta H° - T\Delta S°$$

As a result, the equilibrium constant must depend on the temperature of the reaction.

A good example of this phenomenon is the reaction in which NO_2 dimerizes to form N_2O_4.

$$2\ NO_2(g) \rightleftharpoons N_2O_4(g)$$

In Exercise 13.2 we noted that this reaction is favored by enthalpy because it forms a new bond, which makes the system more stable. We also noted that the reaction is not favored by entropy because it leads to a decrease in the disorder of the system.

NO_2 is a brown gas and N_2O_4 is colorless. We can therefore monitor the extent to which NO_2 dimerizes to form N_2O_4 by examining the intensity of the brown color in a sealed tube of the gas. What should happen to the equilibrium between NO_2 and N_2O_4 as the temperature is lowered?

For the sake of argument, let's assume that there is no significant change in the values of either $\Delta H°$ or $\Delta S°$ as the system is cooled. The contribution to the free energy of the reaction from the enthalpy term is therefore constant, but the contribution from the entropy term becomes less important in relation to the enthalpy term as the temperature is lowered.

$$\Delta G° = \Delta H° - T\Delta S°$$

Figure 13.7 shows what happens to the intensity of the brown color when a sealed tube containing NO_2 gas is immersed in liquid nitrogen. There is a drastic decrease in the amount of NO_2 in the tube as it is cooled to $-196°C$.

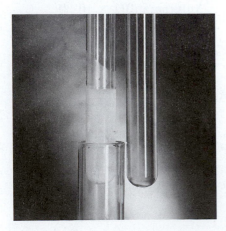

FIGURE 13.7 The equilibrium between NO_2 and N_2O_4 is influenced by two factors. $\Delta H°$ for the reaction favors N_2O_4, but $\Delta S°$ favors NO_2. At low temperatures, the $\Delta H°$ term is more important than the $\Delta S°$ term, and the characteristic brown color of NO_2 gas disappears as a tube containing NO_2 is moved into a liquid nitrogen bath ($-196°C$).

The dimerization of NO_2 to form N_2O_4 is favored by enthalpy ($\Delta H° = -57.21$ kJ/mol$_{rxn}$), but it isn't favored by entropy ($\Delta S° = -175.83$ J/mol$_{rxn}$·K). The reaction shifts to the right as we lower the temperature. As a result, the equilibrium constant for the reaction increases as we decrease the temperature of the system, as shown by the data in Table 13.4. Thus, to maximize the amount of product formed in the reaction, the reaction should be carried out at as low a temperature as possible.

TABLE 13.4   Temperature Dependence of the Equilibrium Constant for the Dimerization of NO_2

Temperature (°C)	K_c	K_p
100	2.1	0.069
25	170	6.95
0	1.4×10^3	62.5
-78	4.0×10^8	2.5×10^7

Exercise 13.15

Proteins serve two main purposes in the body. They are the structural elements of cells, and they facilitate the chemical reactions that must take place in living systems. Proteins consist of long sequences of amino acids. Even a small protein such as insulin has a molecular weight of 6×10^3 g/mol. Hemoglobin has a molecular weight of 6.5×10^4 g/mol. In the aqueous environment of the body the long-chain molecules fold as shown in Figure 13.8. Many proteins lose their folded structure when heated or exposed to a change in pH. When this happens the protein cannot carry out its normal biochemical function and the protein is said to have denatured.

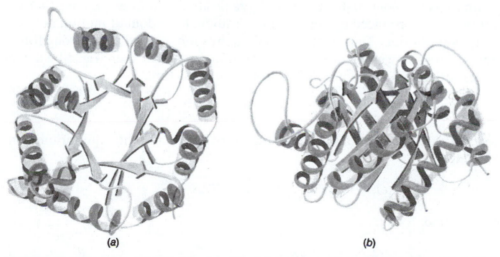

(a)   (b)

FIGURE 13.8 Top view (*a*) and side view (*b*) of the enzyme triose phosphate isomerase illustrate the folding and twisting of the protein chain within the molecule. Reprinted with permission from J. R. Holum, *Fundamentals of General, Organic, and Biological Chemistry,* John Wiley & Sons, Inc., New York, 1994, p. 598.

The equilibrium constant for the denaturation of trypsin is 7.20 at 50°C.

$$\text{Trypsin} \rightleftharpoons \text{denatured trypsin}$$

(a) Calculate $\Delta G°$ for the process.

(b) $\Delta H°$ for the denaturation of trypsin is $+278$ kJ/mol$_{rxn}$. What is $\Delta S°$ at 50°C?

(c) What does the sign and magnitude of $\Delta S°$ imply concerning the difference between denatured trypsin and trypsin?

(d) Under standard conditions will trypsin denature at 50°C?

(e) What is the maximum temperature at which trypsin is stable?

(f) Using $\Delta H°$ and $\Delta S°$ describe what happens when a protein such as trypsin denatures.

Solution

(a) $\Delta G° = -RT \ln K_c = -8.314$ J/mol$_{rxn}$·K$(323$ K$) \ln 7.20$
$$= -5.30 \text{ kJ/mol}_{rxn}$$

(b) $\Delta G° = \Delta H° - T\Delta S°$
$$-5.30 \text{ kJ/mol}_{rxn} = 278 \text{ kJ/mol}_{rxn} - (323 \text{ K})\Delta S°$$
$$\Delta S° = 877 \text{ J/mol}_{rxn}\text{·K}$$

(c) $\Delta S°$ is very large and positive. This means that much freedom of motion has been gained by the denaturing of trypsin.

(d) $\Delta G°$ is negative for the denaturation of trypsin at 50°C, so heating trypsin to that temperature will cause it to unfold.

(e) If $\Delta G°$ for the denaturation process is positive trypsin will not denature. To find this temperature we set $\Delta G° = 0 = \Delta H° - T\Delta S°$ and solve for T assuming $\Delta H°$ and $\Delta S°$ do not change much with temperature.

$$0 = 278 \text{ kJ/mol}_{rxn} - T(0.877 \text{ kJ/mol}_{rxn}\text{·K})$$
$$T = 317 \text{ K} = 44°C$$

At temperatures below 44°C trypsin will remain in its folded state.

(f) The positive sign of $\Delta H°$ shows that the protein is more tightly bonded in its folded structure than in its unfolded state. The very large positive sign of $\Delta S°$ means that the unfolding process has produced rotational and vibrational freedom of motion not possible in the folded position. Thus denaturation requires the disruption of strong forces of attraction to produce weaker ones. For denaturation to occur there must be a gain in entropy that offsets the disruption of the forces of attraction.

We can now develop the relationship between temperature and the equilibrium constant. We have seen that

$$\Delta G° = -RT \ln K = \Delta H° - T\Delta S°$$

Solving for $\ln K$ we find

$$\ln K = \frac{-\Delta H°}{RT} + \frac{\Delta S°}{R}$$

Because both $\Delta H°$ and $\Delta S°$ are essentially constant for moderate changes in temperature, this equation describes the change in K resulting from a change in temperature. Thus $\ln K$

is seen to depend on the magnitudes and signs of $\Delta H°$ and $\Delta S°$. However, the significance of the term $-\Delta H°/RT$ is dependent on temperature. For example, as temperature increases this term becomes smaller. If $\Delta H° < 0$, $-\Delta H°/RT$ is positive and contributes to making ln K large no matter what the sign of $\Delta S°$. But as temperature increases this term becomes a smaller contributor and ln K decreases. If $\Delta H° > 0$, $-\Delta H°/RT$ is negative and contributes to making ln K small. At higher temperatures this negative influence becomes less significant and ln K is larger than at low temperatures.

It is tempting to predict the effect of temperature on the equilibrium constant by considering only the expression

$$\Delta G° = \Delta H° - T\Delta S°$$

Suppose that in a given situation $\Delta H°$ is positive and $\Delta S°$ is negative. Then $\Delta G°$ would become more positive with increasing temperature, and it might be expected that the equilibrium constant would decrease. However, according to Le Châtelier's principle, a positive $\Delta H°$ means that the equilibrium constant should increase with increasing temperature. This dilemma is resolved by noting that it is not $\Delta G°$ but $\Delta G°/T$ which determines the change of the equilibrium constant with temperature:

$$\Delta G° = -RT \ln K \qquad \text{therefore} \qquad \ln K = \frac{-\Delta G°}{RT}$$

Note that this equation can **not** be used to describe the change in K as temperature is changed because $\Delta G°$ is **not** constant with respect to temperature (see Table 13.5).

The reaction of methane to produce ethane

$$2\ CH_4(g) \rightleftharpoons C_2H_6(g) + H_2(g)$$

has $\Delta H° = 64.94$ kJ/mol$_{rxn}$ and $\Delta S° = -12.24$ J/mol$_{rxn}$·K. If $\Delta H°$ and $\Delta S°$ are assumed not to vary appreciably with temperature, $\Delta G°$ will become more positive with increasing temperature but K will nevertheless increase.

$$\Delta G° = 64.94\ \text{kJ/mol}_{rxn} - T(-0.01224\ \text{J/mol}_{rxn}\text{·K})$$

The data in Table 13.5 illustrate the temperature dependence of the equilibrium constant for the above reaction, whose free energy change becomes more positive with increasing temperature.

TABLE 13.5   Temperature Dependence of the Equilibrium Constant for the Reaction of Methane to Produce Ethane

Temperature (K)	$\Delta G°$ (kJ/mol$_{rxn}$)	$\Delta G°/T$ (J/mol$_{rxn}$·K)	ln K	K
298	68.58	230	−27.7	9×10^{-13}
400	69.84	175	−21.0	8×10^{-10}
600	72.28	120	−14.4	6×10^{-7}

Checkpoint

If $\Delta H° > 0$ and $\Delta S° > 0$ for a reaction, which way will the equilibrium shift if the temperature is decreased? If $\Delta H° > 0$ and $\Delta S° < 0$ for a reaction which way will the equilibrium shift if the temperature is decreased?

KEY TERMS

Disorder

Endergonic ($\Delta G > 0$)

Endothermic

Enthalpy

Entropy (S)

Entropy of atom
combination

Exergonic ($\Delta G < 0$)

Exothermic

First law of
thermodynamics

Gibbs free energy (G)

Reaction quotient (Q)

Second law of
thermodynamics

Spontaneous

Standard-state conditions

Standard-state entropy
change ($\Delta S°$)

Standard-state free energy
of atom combination
($\Delta G°_{ac}$)

Standard-state free energy
of reaction ($\Delta G°$)

Third law of
thermodynamics

PROBLEMS

Spontaneous Chemical Reactions

1. Define *enthalpy* and *entropy.*
2. What determines if a reaction will be spontaneous?

Entropy as a Measure of Disorder

3. Describe the relationship between the entropy of a system and the number of equivalent ways in which the state of the system can be described.
4. What happens to the ability of a poker hand to win a game of poker as the entropy of the hand becomes larger?
5. Which poker hand has the highest entropy: one pair or two pairs?
6. Give examples of natural processes that lead to an increase in the entropy of the system.

Entropy and the Second Law of Thermodynamics

7. Which of the following reactions leads to an increase in the entropy of the system?
   (a) $H_2(g) + Cl_2(g) \rightleftarrows 2\, HCl(g)$ (b) $2\, NO_2(g) \rightleftarrows N_2O_4(g)$
   (c) $CaO(s) + CO_2(g) \rightleftarrows CaCO_3(s)$ (d) $2\, H_2(g) + O_2(g) \rightleftarrows 2\, H_2O(g)$
   (e) $2\, NH_3(g) \rightleftarrows N_2(g) + 3\, H_2(g)$
8. In which of the following processes is $\Delta S°$ negative?
   (a) $2\, H_2O_2(aq) \rightleftarrows 2\, H_2O(l) + O_2(g)$ (b) $CO_2(s) \rightleftarrows CO_2(g)$
   (c) $H_2O(l) \rightleftarrows H_2O(g)$ (d) $4\, Al(s) + 3\, O_2(g) \rightleftarrows 2\, Al_2O_3(s)$
   (e) $NaCl(s) \overset{H_2O}{\rightleftarrows} Na^+(aq) + Cl^-(aq)$
9. In which of the following processes is $\Delta S°$ positive?
   (a) $2\, NO(g) + Cl_2(g) \rightleftarrows 2\, NOCl(g)$ (b) $NaCl(s) \rightleftarrows NaCl(l)$
   (c) $O_2(g) \rightleftarrows 2\, O_3(g)$ (d) $C_2H_4(g) + H_2(g) \rightleftarrows C_2H_6(g)$
10. Which of the following processes should have the most positive value of $\Delta S°$?
    (a) $N_2(g) + O_2(g) \rightleftarrows 2\, NO(g)$ (b) $3\, C_2H_2(g) \rightleftarrows C_6H_6(l)$
    (c) $H_2O(l) \rightleftarrows H_2O(g)$ (d) $4\, Al(s) + 3\, O_2(g) \rightleftarrows 2\, Al_2O_3(s)$
11. Which of the following processes should have the most positive value of $\Delta S°$?

    (a) $H_2O(l) \rightleftarrows H_2O(s)$ (b) $NaNO_3(s) \overset{H_2O}{\rightleftarrows} Na^+(aq) + NO_3^-(aq)$
    (c) $2\, HCl(g) \rightleftarrows H_2(g) + Cl_2(g)$ (d) $2\, H_2(g) + O_2(g) \rightleftarrows 2\, H_2O(g)$

Standard-State Entropies of Reaction

12. One of the key steps toward transforming coal into a liquid fuel involves reducing carbon monoxide with H_2 gas to form methanol. Calculate $\Delta S°$ at 298 K for the following reaction to determine whether a favorable change in the entropy of the system might be a driving force behind the reaction.

$$CO(g) + 2\ H_2(g) \rightleftharpoons CH_3OH(l)$$

13. Tetraphosphorus decaoxide is often used as a dehydrating agent because of its tendency to pick up water. Calculate $\Delta S°$ at 25°C for the following reaction to determine whether entropy might be the driving force behind this reaction.

$$P_4O_{10}(s) + 6\ H_2O(l) \rightleftharpoons 4\ H_3PO_4(aq)$$

14. Calculate $\Delta S°$ at 25°C for the following reaction. Comment on both the sign and the magnitude of $\Delta S°$. The entropy of atom combination of $NH_4NO_2(s)$ is -949.5 J/mol $_{rxn} \cdot$K.

$$NH_4NO_2(s) \rightleftharpoons N_2(g) + 2\ H_2O(g)$$

15. Calculate $\Delta S°$ at 298 K for the following reaction. Is there more disorder in the products or reactants?

$$3\ O_2(g) \rightleftharpoons 2\ O_3(g)$$

16. Compare the entropies of atom combination for the various forms of elemental phosphorus. Explain why $\Delta S°_{ac}$ is more negative for $P(s)$ than $P_2(g)$. Why is $\Delta S°_{ac}$ for $P_4(g)$ more negative than for $P_2(g)$?

Compound	$\Delta S°_{ac}$ (J/mol$_{rxn} \cdot$K)
$P(s)$	-122.10
$P(g)$	0
$P_2(g)$	-108.257
$P_4(g)$	-372.79

Gibbs Free Energy

17. Explain the difference between ΔG and $\Delta G°$ for a chemical reaction.
18. What does it mean when ΔG for a reaction is zero? Can $\Delta G°$ for a reaction be zero?
19. Which of the following combinations of $\Delta H°$ and $\Delta S°$ always indicates a spontaneous reaction?
    (a) $\Delta H° > 0, \Delta S° < 0$ (b) $\Delta H° < 0, \Delta S° > 0$ (c) $\Delta H° > 0, \Delta S° > 0$
    (d) $\Delta H° < 0, \Delta S° < 0$ (e) $\Delta H° = 0, \Delta S° = 0$
20. Predict the signs of $\Delta H°$ and $\Delta S°$ for the following reactions without referring to a table of thermodynamic data, and explain your predictions.

	Hints
(a) $2\ H_2(g) + O_2(g) \rightleftharpoons 2\ H_2O(g)$	The reaction is explosive under suitable conditions.
(b) $2\ Na(s) + Cl_2(g) \rightleftharpoons 2\ NaCl(s)$	Is table salt stable?
(c) $N_2(g) + 3\ H_2(g) \rightleftharpoons 2\ NH_3(g)$	The reaction occurs under standard conditions.

(d) $2 Cu(NO_3)_2(s) \rightleftharpoons 2 CuO(s) + 4 NO_2(g) + O_2(g)$ The reaction occurs spontaneously to produce stable CuO.

21. Predict the signs of $\Delta H°$ and $\Delta S°$ for the following reaction without referring to a table of thermodynamic data, and explain your predictions.

$$NH_3(g) \underset{\longleftarrow}{\overset{H_2O}{\rightleftharpoons}} NH_3(aq)$$

Explain why the odor of NH_3 gas that collects above an NH_3 solution becomes more intense as the temperature increases.

22. Limestone decomposes to form lime and carbon dioxide when it is heated. Calculate $\Delta H°$ and $\Delta S°$ at 25°C for the decompzosition of limestone to identify the driving force behind the reaction.

$$CaCO_3(s) \rightleftharpoons CaO(s) + CO_2(g)$$

23. Calculate $\Delta H°$ and $\Delta S°$ at 298 K for the thermite reaction to identify the driving force behind the reaction.

$$Fe_2O_3(s) + 2 Al(s) \rightleftharpoons Al_2O_3(s) + 2 Fe(s)$$

24. The first step in extracting iron ore from pyrite, FeS_2, involves roasting the ore in the presence of oxygen to form iron(III) oxide and sulfur dioxide.

$$4 FeS_2(s) + 11 O_2(g) \rightleftharpoons 2 Fe_2O_3(s) + 8 SO_2(g)$$

Calculate $\Delta H°$ and $\Delta S°$ at 25°C for the reaction and determine the driving force behind the reaction.

25. Calculate $\Delta H°$ and $\Delta S°$ at 298 K for the following reaction. What is the major driving force behind the reaction?

$$2 KMnO_4(s) + 5 H_2O_2(aq) + 6 H^+(aq) \longrightarrow$$
$$2 K^+(aq) + 2 Mn^{2+}(aq) + 5 O_2(g) + 8 H_2O(l)$$

26. It is possible to make hydrogen chloride by reacting phosphorus pentachloride with water and boiling the HCl out of the solution. Calculate $\Delta H°$ and $\Delta S°$ for this reaction at 25°C. What is the driving force behind the reaction: the enthalpy of reaction, the entropy of reaction, or Le Châtelier's principle?

$$PCl_5(g) + 4 H_2O(l) \rightleftharpoons H_3PO_4(aq) + 5 HCl(g)$$

27. Which of the following processes is spontaneous when all species are in their standard states at 298 K?
   (a) $CH_3OH(l) \rightleftharpoons CH_3OH(g)$
   (b) $CH_3OH(l) \rightleftharpoons HCHO(g) + H_2(g)$
   (c) $2 CH_3OH(l) \rightleftharpoons 2 CH_4(g) + O_2(g)$
   (d) $CH_3OH(l) \leftrightarrows CO(g) + 2 H_2(g)$

28. Calculate $\Delta H°$ and $\Delta S°$ for the following reaction at 298 K. Explain why it is a mistake to use water to put out a fire that contains white-hot iron metal.

$$3 Fe(s) + 4 H_2O(l) \longrightarrow Fe_3O_4(s) + 4 H_2(g)$$

29. Use thermodynamic data to predict if sodium or silver will react with water at 298 K.

$$2\ Na(s) + 2\ H_2O(l) \longrightarrow 2\ Na^+(aq) + 2\ OH^-(aq) + H_2(g)$$
$$2\ Ag(s) + 2\ H_2O(l) \longrightarrow 2\ Ag^+(aq) + 2\ OH^-(aq) + H_2(g)$$

30. Use thermodynamic data at 25°C to explain why zinc reacts with 1 M acid, but not with water.

$$Zn(s) + 2\ H^+(aq) \longrightarrow Zn^{2+}(aq) + H_2(g)$$
$$Zn(s) + 2\ H_2O(l) \nrightarrow Zn^{2+}(aq) + 2\ OH^-(aq) + H_2(g)$$

31. Calculate $\Delta H°$ and $\Delta S°$ at 25°C for the following reaction. Why is ammonium nitrate a potential explosive?

$$2\ NH_4NO_3(s) \longrightarrow 2\ N_2(g) + O_2(g) + 4\ H_2O(g)$$

32. Silane, SiH_4, decomposes to form elemental silicon and hydrogen. Calculate $\Delta H°$ and $\Delta S°$ for this reaction at 25°C and predict the effect on the reaction of an increase in the temperature at which it is run.

$$SiH_4(g) \rightleftharpoons Si(s) + 2\ H_2(g)$$

33. For which of the following reactions would you expect $\Delta H°$ and $\Delta G°$ to be about the same?
   (a) $4\ Fe(s) + 3\ O_2(g) \rightarrow 2\ Fe_2O_3(s)$
   (b) $2\ Na(s) + 2\ H_2O(l) \rightarrow 2\ Na^+(aq) + 2\ OH^-(aq) + H_2(g)$
   (c) $Fe_2O_3(s) + 2\ Al(s) \rightarrow Al_2O_3(s) + 2\ Fe(s)$
   (d) $N_2O_4(g) \rightleftharpoons 2\ NO_2(g)$
   (e) $CaC_2(s) + 2\ H_2O(l) \rightleftharpoons Ca^{2+}(aq) + 2\ OH^-(aq) + C_2H_2(g)$

The Effect of Temperature on the Free Energy of Chemical Reactions

34. For each of the following reactions how will the $\Delta G°$ change as the temperature increases?
   (a) $2\ CO(g) + O_2(g) \rightleftharpoons 2\ CO_2(g)$
       $\Delta H° = -565.97$ kJ/mol$_{rxn}$, $\Delta S° = -173.00$ J/mol$_{rxn}$·K
   (b) $2\ H_2O(g) \rightleftharpoons 2\ H_2(g) + O_2(g)$
       $\Delta H° = 483.64$ kJ/mol$_{rxn}$, $\Delta S° = 90.01$ J/mol$_{rxn}$·K
   (c) $2\ N_2O(g) \rightleftharpoons 2\ N_2(g) + O_2(g)$
       $\Delta H° = -164.1$ kJ/mol$_{rxn}$, $\Delta S° = 148.66$ J/mol$_{rxn}$·K
   (d) $PbCl_2(s) \overset{H_2O}{\rightleftharpoons} Pb^{2+}(aq) + 2\ Cl^-(aq)$
       $\Delta H° = 23.39$ kJ/mol$_{rxn}$, $\Delta S° = -12.5$ J/mol$_{rxn}$·K

35. Explain why the $\Delta G°$ for the following reaction becomes more positive as the temperature increases.

$$N_2(g) + 3\ H_2(g) \rightleftharpoons 2\ NH_3(g)$$

36. Which of the following diagrams best describes the relationship between $\Delta G°$ and temperature for the following reaction?

$$2 H_2(g) + O_2(g) \rightleftharpoons 2 H_2O(g)$$

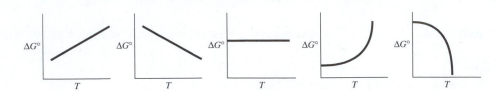

37. Calculate $\Delta G°$ for the following reaction from $\Delta H°$ and $\Delta S°$ data for the reaction at 298 K. Compare the result of this calculation with the value obtained from free energy of atom combination data.

$$CS_2(l) + 3 O_2(g) \rightleftharpoons CO_2(g) + 2 SO_2(g)$$

38. What is the sign of $\Delta G°$ for the following process at $-10°C$, $0°C$, and $10°C$ at 1 atm pressure?

$$H_2O(s) \longrightarrow H_2O(l)$$

39. What is the sign of $\Delta G°$ for the following process at $90°C$, $100°C$, and $110°C$ at 1 atm pressure of $H_2O(g)$?

$$H_2O(l) \longrightarrow H_2O(g)$$

40. For the process $H_2O(l) \rightleftharpoons H_2O(g)$ what are the signs of ΔH and ΔS?

Equilibria Expressed in Partial Pressures

41. Explain why pressure can be used instead of concentration to describe equilibrium constant expressions for gas phase reactions.

42. Which equation correctly describes the relationship between K_p and K_c for the following reaction?

$$Cl_2(g) + 3 F_2(g) \rightleftharpoons 2 ClF_3(g)$$

(a) $K_p = K_c$ (b) $K_p = K_c \times (RT)^{-1}$ (c) $K_p = K_c \times (RT)^{-2}$
(d) $K_p = K_c \times (RT)$ (e) $K_p = K_c \times (RT)^2$

43. Which equation correctly describes the relationship between K_p and K_c for the following reaction?

$$2 NO_2(g) \rightleftharpoons 2 NO(g) + O_2(g)$$

(a) $K_p = K_c$ (b) $K_p = 1/K_c$ (c) $K_p = K_c(RT)$ (d) $K_p = K_c(RT)^{-1}$

44. Calculate K_p for the decomposition of NOCl at 500 K if 27.3% of a 1.00 atm sample of NOCl decomposes to NO and Cl_2 at equilibrium.

$$2 NOCl(g) \rightleftharpoons 2 NO(g) + Cl_2(g)$$

45. Calculate the partial pressures of phosgene, carbon monoxide, and chlorine at equilibrium when a system that was initially 0.124 atm in $COCl_2$ decomposes at 300°C according to the following equation.

$$COCl_2(g) \rightleftharpoons CO(g) + Cl_2(g) \qquad K_p = 3.2 \times 10^{-3}$$

46. Without calculations, approximate the partial pressure of SO_3 that would be present at equilibrium when a mixture that was initially 0.490 atm in SO_2 and 0.245 atm in O_2 comes to equilibrium at 700 K.

$$2 SO_2(g) + O_2(g) \rightleftharpoons 2 SO_3(g) \qquad K_p = 6.7 \times 10^4$$

47. Calculate the partial pressures of SO_3, SO_2, and O_2 that would be present at equilibrium when a mixture that was initially 0.490 atm in SO_3 reacts at 700 K.

$$2 SO_3(g) \rightleftharpoons 2 SO_2(g) + O_2(g) \qquad K_p = 1.5 \times 10^{-5}$$

48. Approximate the partial pressures of N_2 and H_2 present at equilibrium when a mixture that was initially 0.50 atm in N_2, 0.60 atm in H_2, and 0.20 atm in NH_3 comes to equilibrium at a temperature at which K_p for the following reaction is 1.0×10^{-5}.

$$N_2(g) + 3 H_2(g) \rightleftharpoons 2 NH_3(g)$$

49. A mixture of NH_3, O_2, NO_2, and H_2O is initially 0.50 atm in each of the four gases. In which direction will the reaction proceed?

$$4 NO_2(g) + 6 H_2O(g) \rightleftharpoons 4 NH_3(g) + 7 O_2(g) \qquad K_p = 1.8 \times 10^{-28}$$

50. The partial pressures of N_2, H_2, and NH_3 present at equilibrium at 1000 K are 2.0, 6.0, and 0.018 atm, respectively. Calculate $\Delta G°$ for the following reaction.

$$2 NH_3(g) \rightleftharpoons N_2(g) + 3 H_2(g)$$

51. Calculate the concentrations of N_2, O_2, and NO present when a mixture that was initially 0.40 M N_2 and 0.60 M O_2 reacts to form NO at 700°C. (Hint: Look carefully at the symbol for the equilibrium constant for the reaction.)

$$N_2(g) + O_2(g) \rightleftharpoons 2 NO(g) \qquad K_p = 4.3 \times 10^{-9} \text{ (at 700°C)}$$

52. Industrial chemicals can be made from coal by a process that starts with the reaction between red-hot coal and steam to form a mixture of CO and H_2. This mixture, which is known as synthesis gas, can be converted to methanol (CH_3OH) in the presence of a ruthenium/cobalt catalyst. The methanol produced in the reaction can then be converted into a host of other products, ranging from acetic acid to gasoline. The partial pressure of methanol when a mixture of 1.00 atm CO and 2.00 atm H_2 comes to equilibrium at 65°C is 0.98 atm. Calculate the standard free energy of the reaction at 65°C.

$$CO(g) + 2 H_2(g) \rightleftharpoons CH_3OH(g)$$

The Relationship between Free Energy and Equilibrium Constants

53. Which of the following correctly describes a reaction at equilibrium?
    (a) $\Delta G = 0$ (b) $\Delta G° = 0$ (c) $\Delta G = \Delta G°$ (d) $Q = 0$ (e) $\ln K = 0$

54. Calculate the equilibrium constant at 25°C for the following reaction from the standard-state free energies of atom combination of the reactants and products.

$$CO(g) + 2\,H_2(g) \rightleftharpoons CH_3OH(g)$$

55. Calculate K_p at 1000 K for the following reaction. Describe the assumptions you had to make to do the calculation.

$$2\,CH_4(g) \rightleftharpoons C_2H_6(g) + H_2(g)$$

56. Calculate the equilibrium constant at 25°C for the following reaction.

$$2\,HI(g) + Cl_2(g) \rightleftharpoons 2\,HCl(g) + I_2(s)$$

57. Calculate $\Delta H°$, $\Delta S°$, and $\Delta G°$ for the following reactions at 298 K. Use the data to calculate the values of K_a for hydrochloric and acetic acid. Compare the results of the calculations with the K_a data in Table 11.3 in Chapter 11.

$$HCl(g) \rightleftharpoons H^+(aq) + Cl^-(aq)$$
$$CH_3CO_2H(aq) \rightleftharpoons H + (aq) + CH_3CO_2^-(aq)$$

58. Calculate $\Delta H°$, $\Delta S°$, and $\Delta G°$ for the following reaction at 298 K. Use the data to calculate the value of K_b for ammonia. Compare the result of the calculation with the value of K_b in Appendix B.9.

$$NH_3(aq) + H_2O(l) \rightleftharpoons NH_4^+(aq) + OH^-(aq)$$

The Temperature Dependence of Equilibrium Constants

59. Synthesis gas can be made by reacting red-hot coal with steam. Estimate the temperature at which the equilibrium constant for the reaction is equal to 1.

$$C(s) + H_2O(g) \rightleftharpoons CO(g) + H_2(g)$$

60. Calculate $\Delta H°$, $\Delta S°$, $\Delta G°$ and the equilibrium constant at 25°C for the following reaction. Predict the effect of an increase in the temperature of the system on the equilibrium constant for the reaction.

$$CO_2(g) + H_2(g) \rightleftharpoons CO(g) + H_2O(g)$$

61. What happens to the equilibrium constant for the following reaction as the temperature increases?

$$PCl_5(g) \rightleftharpoons PCl_3(g) + Cl_2(g)$$

62. At approximately what temperature does the equilibrium constant for the following reaction become larger than 1?

$$PCl_5(g) \rightleftharpoons PCl_3(g) + Cl_2(g)$$

63. For which of the following reactions would the equilibrium constant increase with increasing temperature?
   (a) $2 H_2(g) + O_2(g) \rightleftharpoons 2 H_2O(g)$
   (b) $2 HCl(g) \rightleftharpoons H_2(g) + Cl_2(g)$
   (c) $2 NH_3(g) \rightleftharpoons N_2(g) + 3 H_2(g)$

64. Calculate the equilibrium constant for the following reaction at 25°C, 200°C, 400°C, and 600°C. Explain any trend you observe. Assume $\Delta H°$ and $\Delta S°$ do not change with temperature.

$$2 SO_3(g) \rightleftharpoons 2 SO_2(g) + O_2(g)$$

65. Calculate $\Delta H°$ and $\Delta S°$ for the following reaction. Predict what will happen to the equilibrium constant of the reaction as the temperature increases.

$$Cu(s) + 2 H^+ \rightleftharpoons Cu^{2+}(aq) + H_2(g)$$

Integrated Problems

66. $\Delta S°$ for the following reaction is favorable even though $\Delta H°$ is not.

$$CH_3OH(l) \rightleftharpoons CH_3OH(g)$$

Assume that methanol boils at the temperature at which $\Delta G°$ for this reaction is equal to zero. Use the values of $\Delta H°$ and $\Delta S°$ at 25°C for the reaction to estimate the boiling point of methanol.

67. Use the equation that describes the relationship between ΔG and $\Delta G°$ for a reaction to construct a graph of ΔG versus $\ln Q_c$ at 25°C for the following reactions over a range of values of Q_c between 10^{-10} and 10^{10}.

$$HOAc(aq) \overset{H_2O}{\rightleftharpoons} H^+(aq) + OAc^-(aq) \qquad \Delta G° = 27.1 \text{ kJ}$$

$$HCl(g) \overset{H_2O}{\longrightarrow} H^+(aq) + Cl^-(aq) \qquad \Delta G° = -35.9$$

Explain the significance of the difference between the points at which the straight lines for the data cross the horizontal and vertical axes.

68. From the following $\Delta H°$ and $\Delta S°$ values predict whether each of the reactions would lead to $K > 1$ or $K < 1$ at 25°C. If a reaction has $K < 1$ at what temperature might K become greater than 1? Explain your reasoning.
   (a) $\Delta H° = 10 \text{ kJ/mol}_{rxn}$; $\Delta S° = 30 \text{ J/mol}_{rxn} \cdot K$
   (b) $\Delta H° = 2 \text{ kJ/mol}_{rxn}$; $\Delta S° = 100 \text{ J/mol}_{rxn} \cdot K$
   (c) $\Delta H° = -125 \text{ kJ/mol}_{rxn}$; $\Delta S° = 80 \text{ J/mol}_{rxn} \cdot K$
   (d) $\Delta H° = -10 \text{ kJ/mol}_{rxn}$; $\Delta S° = -30 \text{ J/mol}_{rxn} \cdot K$

CHAPTER
14
KINETICS

14.1   THE FORCES THAT CONTROL A CHEMICAL REACTION

The first half of this book divided chemical reactions into two categories: those that occurred and those that did not. We noted that sodium reacts with water, for example,

$$2\,Na(s) + H_2O(l) \longrightarrow 2\,Na^+(aq) + 2\,OH^-(aq) + H_2(g)$$

whereas magnesium metal does not.

$$Mg(s) + H_2O(l) \longrightarrow\!\!\!\times$$

The second half of this book introduced a more sophisticated view of chemical reactions. By introducing the idea that chemical reactions can come to equilibrium, we can construct a continuum of chemical reactions. Toward one end of the continuum are those reactions that only proceed to a limited extent. Reactions such as the following lie very heavily on the side of the reactants.

$$\text{HOAc}(aq) + \text{H}_2\text{O}(l) \longleftarrow \text{H}_3\text{O}^+(aq) + \text{OAc}^-(aq) \qquad K_a = 1.8 \times 10^{-5}$$

Toward the other end are reactions that proceed almost to completion.

$$\text{Zn}(s) + \text{Cu}^{2+}(aq) \longrightarrow \text{Zn}^{2+}(aq) + \text{Cu}(s) \qquad K = 1.8 \times 10^{37}$$

The previous chapter showed how thermodynamics can explain why some reactions occur and not others. In that chapter, we saw that thermodynamics is a powerful tool for predicting what should (or should not) happen in a chemical reaction. It is ideally suited for answering questions that begin, "What if . . . ?" Thermodynamics predicts, for example, that hydrogen should react with oxygen to form water.

$$2\,\text{H}_2(g) + \text{O}_2(g) \longrightarrow 2\,\text{H}_2\text{O}(g)$$

It also predicts that iron(III) oxide should react with aluminum metal to form aluminum oxide and iron metal.

$$\text{Fe}_2\text{O}_3(s) + 2\,\text{Al}(s) \longrightarrow \text{Al}_2\text{O}_3(s) + 2\,\text{Fe}(s)$$

Both of these reactions can, in fact, occur. We can experience the first by touching a balloon filled with H_2 gas with a candle tied to the end of a meter stick. We can demonstrate the other by igniting a mixture of Fe_2O_3 and powdered aluminum.

This doesn't mean that H_2 and O_2 burst spontaneously into flame the instant the gases are mixed. In the absence of a spark, mixtures of H_2 and O_2 can be stored for years with no detectable change. Nor does it mean that scrupulous attention must be paid to prevent Fe_2O_3 from coming in contact with aluminum metal, lest some violent reaction take place. In the absence of a source of heat, mixtures of Fe_2O_3 and powdered aluminum are so stable they are sold in 50-pound lots under the trade name Thermite.

It is a mistake to lose sight of the fact that thermodynamics can never tell us what will happen. It can only tell us what should or might happen. Reactions that behave as thermodynamics predicts they should are said to be under **thermodynamic control.** It's not surprising from a thermodynamic point of view, for example, that potassium metal bursts into flame when it comes in contact with water.

$$2\,\text{K}(s) + 2\,\text{H}_2\text{O}(l) \longrightarrow 2\,\text{K}^+(aq) + 2\,\text{OH}^-(aq) + \text{H}_2(g) \qquad \Delta G^\circ = -406.77 \text{ kJ/mol}_\text{rxn}$$

Other reactions are said to be under **kinetic control.** Thermodynamics predicts that these reactions should occur, but the rate of the reactions is negligibly slow. Examples of reactions under kinetic control include the reactions in which elemental sulfur and its compounds burn.

$$\text{S}_8(s) + 8\,\text{O}_2(g) \longrightarrow 8\,\text{SO}_2(g)$$
$$\text{CS}_2(g) + 3\,\text{O}_2(g) \longrightarrow \text{CO}_2(g) + 2\,\text{SO}_2(g)$$
$$2\,\text{H}_2\text{S}(g) + 3\,\text{O}_2(g) \longrightarrow 2\,\text{H}_2\text{O}(g) + 2\,\text{SO}_2(g)$$

Thermodynamics predicts that the reactions should form SO_3, not SO_2, because SO_3 is more stable than SO_2. Under normal conditions, however, the reactions invariably stop at SO_2, because the reaction between SO_2 and O_2 to form SO_3 is extremely slow.

To fully understand chemical reactions, we need to combine the predictions of thermodynamics with studies of the factors that influence the rates of chemical reactions. These factors fall under the general heading of *chemical kinetics* (from a Greek stem meaning "to move"), which is the subject of this chapter.

Checkpoint

Thermodynamics predicts that under normal circumstances carbon in the form of graphite is more stable than carbon in the form of diamond. Why do you not have to worry that a diamond ring will turn to graphite?

14.2 CHEMICAL KINETICS

Chemical kinetics can be defined as the study of the following questions.

- What is the rate at which the reactants are converted into the products of a reaction?
- What factors influence the rate of a reaction?
- What is the sequence of steps, or the mechanism, by which the reactants are converted into products?

As we saw in Chapter 10, the term *rate* is used to describe the change in a quantity that occurs per unit of time. The rate at which cars are sold, for example, is equal to the number of cars that change hands divided by the period of time during which this number is counted. The rate at which an object travels through space is the distance traveled per unit of time, such as miles per hour or kilometers per second. In chemical kinetics, the distance traveled is the change in the concentration of one of the components of the reaction. The rate of a reaction is therefore the change in the concentration of one of the reactants or products, $\Delta(X)$, that occurs during a given period of time, Δt.

$$\text{Rate} = \frac{(X)_{\text{final}} - (X)_{\text{initial}}}{t_{\text{final}} - t_{\text{initial}}} = \frac{\Delta(X)}{\Delta t}$$

One of the first kinetic experiments was done in 1850 by Ludwig Wilhemy, a lecturer in physics at the University of Heidelberg. Wilhemy studied the rate at which sucrose, or cane sugar, reacts with acid to form a mixture of fructose and glucose known as invert sugar.

$$\text{Sucrose}(aq) + H_3O^+(aq) \longrightarrow \underbrace{\text{glucose}(aq) + \text{fructose}(aq)}_{\textit{invert sugar}}$$

Wilhemy found that the rate at which sucrose is consumed in the reaction is directly proportional to its concentration when the pH and temperature of the solution are held constant. The rate of the reaction at any moment in time can therefore be described by the following equation, in which k is a proportionality constant and (sucrose) is the concentration of sucrose in units of moles per liter at that moment. Parentheses around the sucrose represent the molar concentration at any point during the reaction. Brackets around sucrose, [sucrose], indicate the concentration after equilibrium has been established.

$$Rate = k(sucrose)$$

Because the sucrose concentration decreases as sucrose is consumed in the reaction, the rate of the reaction gradually decreases.

14.3   IS THE RATE OF REACTION CONSTANT?

Acid–base titrations often use phenolphthalein as the end point indicator, as shown in Figure 14.1. You might not have noticed, however, what happens when a solution that contains phenolphthalein in the presence of excess base is allowed to stand for a few minutes. Although the solution initially has a pink color, it gradually turns colorless as the phenolphthalein reacts with the OH^- ion in a strongly basic solution.

FIGURE 14.1 Acid–base titration using phenolphthalein as the indicator.

Table 14.1 shows what happens to the concentration of phenolphthalein in a solution at 25°C that was initially 0.00500 M in phenolphthalein and 0.61 M in OH^- ion. As you can see when the data are plotted in Figure 14.2, the phenolphthalein concentration decreases by a factor of 10 over a period of about 230 seconds. Because the concentration of OH^- is so large relative to the phenolphthalein the rate of reaction is dependent only on the concentration of phenolphthalein.

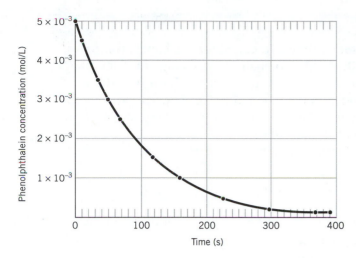

FIGURE 14.2 Plot of the concentration of phenolphthalein versus time for the reaction between the acid–base indicator and excess OH$^-$ ion.

To find the rate of reaction over any period of time, we divide the change in the concentration of phenolphthalein (PHTH) by the time interval over which the change occurred. Because phenolphthalein is consumed in the reaction, the change in the PHTH concentration, Δ(PHTH), is a negative number. (The concentration of phenolphthalein is smaller at the end of the time interval than the concentration at the beginning of the time interval.) By convention, rate data are reported as positive numbers. When a reactant is used to express the rate, we therefore add a negative sign to the equation that defines the rate of the reaction.

$$\text{Rate} = -\frac{\Delta(\text{PHTH})}{\Delta t}$$

TABLE 14.1   Experimental Data for the Reaction between Phenolphthalein and Excess Base

Concentration of Phenolphthalein (M)	Time (s)
0.00500	0.0
0.00450	10.5
0.00400	22.3
0.00350	35.7
0.00300	51.1
0.00250	69.3
0.00200	91.6
0.00150	120.4
0.00100	160.9
0.000500	230.3
0.000250	299.6
0.000150	350.7
0.000100	391.2

Checkpoint

Use Table 14.1 to determine how much time is required for the concentration of phenolphthalein to decrease from 0.00300 to 0.00250 M (a change of 0.00050 M). Determine the time required for the concentration to decrease from 0.00100 to 0.000500 M (a change of 0.00050 M). As the reaction proceeds what happens to the time required for the concentration to change by 0.00050 M? As the reaction proceeds what happens to the rate of the reaction?

Exercise 14.1

Use the data in Table 14.1 to calculate the rate at which phenolphthalein reacts with the OH^- ion during each of the following periods.

(a) During the first time interval, when the phenolphthalein concentration falls from 0.00500 M to 0.00450 M.

(b) During the second interval, when the concentration falls from 0.00450 M to 0.00400 M.

(c) During the third interval, when the concentration falls from 0.00400 M to 0.00350 M.

Solution

The rate of reaction is equal to the change in the phenolphthalein concentration divided by the length of time over which the change occurs.

(a) During the first time period, the rate of the reaction is 4.8×10^{-5} moles per liter per second.

$$\text{Rate} = -\frac{\Delta(X)}{\Delta t} = -\frac{(0.00450 \ M) - (0.00500 \ M)}{(10.5 - 0 \ s)} = 4.8 \times 10^{-5} \ M/s$$

(b) During the second time period, the rate of reaction is slightly lower.

$$\text{Rate} = -\frac{\Delta(X)}{\Delta t} = -\frac{(0.00400 \ M) - (0.00450 \ M)}{(22.3 - 10.5 \ s)} = 4.2 \times 10^{-5} \ M/s$$

(c) During the third time period, the rate of reaction is even lower.

$$\text{Rate} = -\frac{\Delta(X)}{\Delta t} = -\frac{(0.00350 \ M) - (0.00400 \ M)}{(35.7 - 22.3 \ s)} = 3.7 \times 10^{-5} \ M/s$$

14.4   INSTANTANEOUS RATES OF REACTION

Exercise 14.1 illustrates an important point: The rate of the reaction between phenolphthalein and the OH^- ion isn't constant; it changes with time. Like most reactions, the rate of the reaction gradually decreases as the reactants are consumed.

The rate of reaction not only changes from one time period to another, but it is also constantly changing. This means that the rate of reaction changes while it is being measured. To minimize the error this introduces into our measurements, it seems advisable to measure the rate of reaction over periods of time that are short compared with the time it takes for the reaction to occur. We might try, for example, to measure the infinitesimally small change in concentration, $d(X)$, that occurs over an infinitesimally short period of time, dt. If the time interval of the measurement is very short,

$$-\frac{\Delta(X)}{\Delta t} = -\frac{d(X)}{dt}$$

The ratio of these quantities is known as the **instantaneous rate of reaction.**

$$\text{Rate} = -\frac{d(X)}{dt}$$

The instantaneous rate of reaction can be calculated from a graph of the concentration of one of the components of the reaction versus time. Figure 14.3 shows how the instantaneous rate can be calculated for the reaction between phenolphthalein and excess base. The rate of reaction at any moment in time is equal to the slope of a tangent drawn to the curve at that moment.

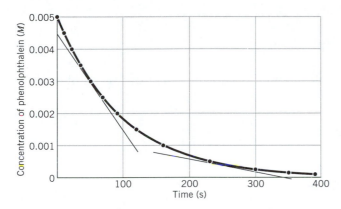

FIGURE 14.3 The instantaneous rate of reaction at any moment in time can be calculated from the slope of a tangent to the graph of concentration versus time. At a point 50 s after the beginning of the reaction, for example, the instantaneous rate of reaction is 2.7×10^{-5} mole per liter per second. At a point 250 s after the beginning of the reaction, the instantaneous rate of reaction is 4.1×10^{-6} mole per liter per second.

The instantaneous rate of reaction can be measured at any time between the moment at which the reactants are mixed and the time when the reaction reaches equilibrium. The *initial instantaneous rate of reaction* is the rate measured at the moment the reactants are mixed.

Checkpoint

What is the problem with using a large value for dt when determining the rate of a reaction using the following equation?

$$\text{Rate} = -\frac{d(X)}{dt}$$

14.5   RATE LAWS AND RATE CONSTANTS

An interesting result is obtained when the instantaneous rate of reaction is calculated at various points along the curve in Figure 14.3. The rate of reaction at every point on the curve is directly proportional to the concentration of phenolphthalein at that moment in time.

$$\text{Rate} = k(\text{phenolphthalein})$$

Because this equation is an experimental law that describes the rate of the reaction, it is called the **rate law** for the reaction. The proportionality constant, k, is known as the **rate constant.**

Checkpoint

Use the rate law for the reaction between phenolphthalein and excess base to explain why the rate of the reaction gradually slows down.

Exercise 14.2

Calculate the rate constant for the reaction between phenolphthalein and the OH^- ion if the instantaneous rate of reaction is 2.5×10^{-5} mol/L·s when the concentration of phenolphthalein is 0.00250 M.

Solution

We start with the rate law for the reaction.

$$\text{Rate} = k(\text{phenolphthalein})$$

We then substitute the known rate of reaction and the known concentration of phenolphthalein into the equation.

$$2.5 \times 10^{-5} \frac{\text{mol/L}}{\text{s}} = k(0.00250 \text{ mol/L})$$

Solving for the rate constant gives the following result.

$$k = 0.010 \text{ s}^{-1}$$

Exercise 14.3

Use the rate constant for the reaction between phenolphthalein and the OH^- ion to calculate the initial instantaneous rate of reaction for the data in Table 14.1.

Solution

Substituting the rate constant for the reaction from the preceding exercise and the initial concentration of phenolphthalein into the rate law for the reaction gives the following result.

$$\text{Rate} = k(\text{phenolphthalein})$$
$$= (0.010 \text{ s}^{-1})(0.00500 \text{ mol/L}) = 5.0 \times 10^{-5} \frac{\text{mol/L}}{\text{s}}$$

Because the rate of reaction is the change in the concentration of phenolphthalein divided by the time over which the change occurs, it is reported in units of moles per liter per second. Because the number of moles of phenolphthalein per liter is the molarity of the solution, the rate can also be reported in terms of the change in molarity per second: M/s.

Checkpoint
Describe the difference between the *rate* of a chemical reaction and the *rate constant*.
How do the rate and rate constant vary with respect to concentration and time?

14.6 A PHYSICAL ANALOG OF KINETIC SYSTEMS

A physical model of the kinetics of a chemical reaction can be constructed with a 50-mL buret and a 1-ft length of capillary tubing, as shown in Figure 14.4. The buret is filled with water, and the volume of water is monitored versus time as water gradually flows through the capillary tubing. A typical set of data obtained with the apparatus is given in Table 14.2.

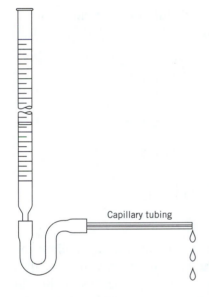

Capillary tubing

FIGURE 14.4 The rate at which water in the buret flows through the capillary tube is dependent on the diameter of the bore of the capillary tube.

TABLE 14.2 Volume of Water in Buret versus Time

Volume of Water in Buret (mL)	Time (s)
50	0
40	19
30	42
20	72
10	116
0	203

The volume of water in the buret versus time is plotted in Figure 14.5. This graph is similar to the plots of the concentration of phenolphthalein versus time in Figures 14.2 and 14.3. The average rate at which the first 10 mL of water drains from the buret is 0.53 mL per second.

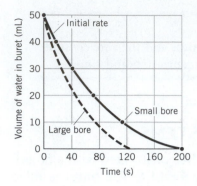

FIGURE 14.5 Plot of the volume of water in the buret of the apparatus shown in Figure 14.4 versus time. The solid line represents the rate of water flowing through a small bore tube, and the dashed line represents the rate of water flowing through a larger bore tube.

$$\text{First 10 mL} \qquad \text{Rate} = -\frac{(40 - 50)\ \text{mL}}{(19 - 0)\ \text{s}} = 0.53\ \text{mL/s}$$

But the average rate steadily decreases for the second, third, fourth, and fifth 10-mL portions.

$$\text{Second 10 mL} \qquad \text{Rate} = -\frac{(30 - 40)\ \text{mL}}{(42 - 19)\ \text{s}} = 0.43\ \text{mL/s}$$

$$\text{Third 10 mL} \qquad \text{Rate} = -\frac{(20 - 30)\ \text{mL}}{(72 - 42)\ \text{s}} = 0.33\ \text{mL/s}$$

$$\text{Fourth 10 mL} \qquad \text{Rate} = -\frac{(10 - 20)\ \text{mL}}{(116 - 72)\ \text{s}} = 0.23\ \text{mL/s}$$

$$\text{Fifth 10 mL} \qquad \text{Rate} = -\frac{(0 - 10)\ \text{mL}}{(203 - 116)\ \text{s}} = 0.11\ \text{mL/s}$$

If the rate at which water escaped from the buret were constant, we could calculate the rate at any moment in time and get the same answer. But the rate at which water escapes from the buret depends on the pressure pushing the water through the tube. As the height of the column of water in the buret decreases, less pressure is exerted on the water in the capillary tube, and the water flows through the tube at a slower rate.

Checkpoint

Explain why the rate at which the buret drains is constant when a source of compressed air is connected to the top of the buret.

Instead of calculating the rate averaged over a 20-s or 80-s period, we would do better to plot the data, draw a smooth curve through the points, and then calculate the instantaneous rate by determining the slope of a tangent to the curve at that moment in time.

Suppose the capillary tubing in Figure 14.4 is replaced by similar tubing of the same length but a larger bore. Now a smaller hydrostatic pressure is needed to force the water through the capillary. Consequently, if the initial volumes of water in the buret are the same, the rate of flow will be increased through the larger bore capillary. The larger bore capillary corresponds to a chemical reaction that has a fast rate, and the smaller bore is analogous to a chemical reaction that has a slower rate. Because the rates for both processes change, the only way to compare the rates is to compare them at the same instant in time (see Figure 14.5). A convenient way of doing this is to compare the initial rates.

14.7 THE RATE LAW VERSUS THE STOICHIOMETRY OF A REACTION

In the 1930s, Sir Christopher Ingold and co-workers at the University of London studied the kinetics of substitution reactions such as the following.

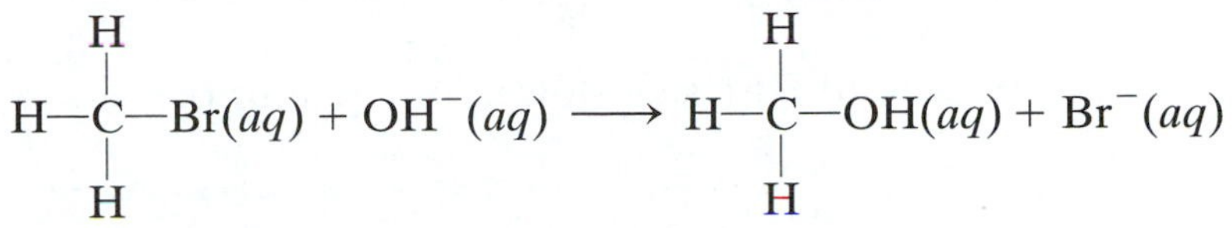

$$\underset{\substack{|\\H}}{\overset{\substack{H\\|}}{H-C-Br}}(aq) + OH^-(aq) \longrightarrow \underset{\substack{|\\H}}{\overset{\substack{H\\|}}{H-C-OH}}(aq) + Br^-(aq)$$

They found that the rate of the reaction is proportional to the concentrations of both reactants.

$$\text{Rate} = k(\text{CH}_3\text{Br})(\text{OH}^-)$$

When they ran a similar reaction on a slightly different starting material, they got similar products.

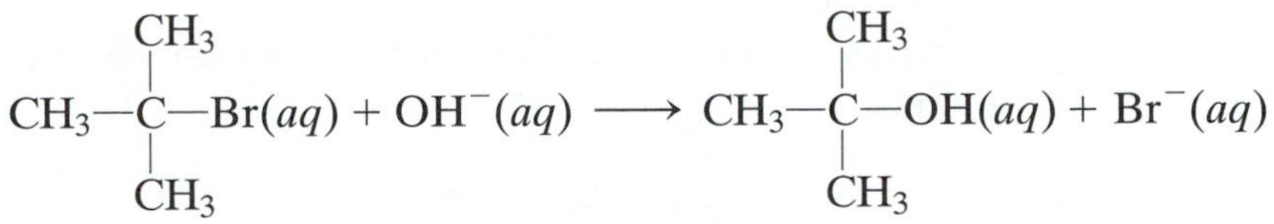

$$\underset{\substack{|\\CH_3}}{\overset{\substack{CH_3\\|}}{CH_3-C-Br}}(aq) + OH^-(aq) \longrightarrow \underset{\substack{|\\CH_3}}{\overset{\substack{CH_3\\|}}{CH_3-C-OH}}(aq) + Br^-(aq)$$

But now the rate of reaction was proportional to the concentration of only one of the reactants.

$$\text{Rate} = k((\text{CH}_3)_3\text{CBr})$$

These results illustrate an important point: *The rate law for a reaction can't be predicted from the stoichiometry of the reaction; it must be determined experimentally.* Sometimes, the rate law is consistent with what we expect from the stoichiometry of the reaction. The balanced equation for the following reaction, for example, involves two molecules of HI, and the rate law for the reaction depends on $(\text{HI})^2$.

$$2\,\text{HI}(g) \longrightarrow \text{H}_2(g) + \text{I}_2(g) \qquad \text{Rate} = k(\text{HI})^2 = -\frac{\Delta(\text{HI})}{\Delta t}$$

Often, however, the rate law doesn't follow the stoichiometry. The stoichiometry for the decomposition of N_2O_5, for example, might lead us to expect a rate law that is proportional to $(\text{N}_2\text{O}_5)^2$. The observed rate law, however, is proportional to (N_2O_5).

$$2\,\text{N}_2\text{O}_5(g) \longrightarrow 4\,\text{NO}_2(g) + \text{O}_2(g) \qquad \text{Rate} = k(\text{N}_2\text{O}_5) = -\frac{\Delta(\text{N}_2\text{O}_5)}{\Delta t}$$

Note that 2 moles of N_2O_5 disappear for every mole of oxygen formed and every 4 moles of NO_2 that are formed. Therefore the relationships between the rates of disappearance of N_2O_5 and the rates of formation of O_2 and NO_2 are

$$-\frac{\Delta(\text{N}_2\text{O}_5)}{\Delta t} = 2\,\frac{\Delta(\text{O}_2)}{\Delta t} \qquad \text{and} \qquad -\frac{\Delta(\text{N}_2\text{O}_5)}{\Delta t} = \frac{1}{2}\,\frac{\Delta(\text{NO}_2)}{\Delta t}$$

Checkpoint

What is the relationship between the rate of disappearance of HI and the rate of formation of H_2 in the reaction below?

$$2\ HI(g) \longrightarrow H_2(g) + I_2(g)$$

Is the rate of disappearance of HI faster, slower, or equal to the rate of formation of H_2?

14.8   ORDER AND MOLECULARITY

Some reactions occur in a single step. The reaction in which a chlorine atom is transferred from $ClNO_2$ to NO to form NO_2 and ClNO is a good example of a one-step reaction.

$$ClNO_2(g) + NO(g) \longrightarrow NO_2(g) + ClNO(g)$$

Other reactions occur by a series of individual steps. N_2O_5, for example, decomposes to NO_2 and O_2 by a three-step mechanism.

Step 1 $N_2O_5 \rightleftharpoons NO_2 + NO_3$
Step 2 $NO_2 + NO_3 \rightarrow NO_2 + NO + O_2$
Step 3 $NO + NO_3 \rightarrow 2\ NO_2$

The steps in a reaction are classified in terms of their **molecularity,** which describes the number of molecules consumed in that step. When a single molecule is consumed, the step is called **unimolecular.** When two molecules are consumed, it is **bimolecular.**

Exercise 14.4

Determine the molecularity of each step in the reaction by which N_2O_5 decomposes to NO_2 and O_2.

Solution

All we have to do is count the number of molecules consumed in each step in the reaction to decide that the first step is unimolecular and the other two steps are bimolecular.

Step 1 $N_2O_5 \rightleftharpoons NO_2 + NO_3$ (unimolecular)
Step 2 $NO_2 + NO_3 \rightarrow NO_2 + NO + O_2$ (bimolecular)
Step 3 $NO + NO_3 \rightarrow 2\ NO_2$ (bimolecular)

Reactions can also be classified in terms of their **order.** The decomposition of N_2O_5 is a **first-order reaction** because the rate of reaction depends on the concentration of N_2O_5 raised to the first power.

$$Rate = k(N_2O_5)$$

The decomposition of HI is a **second-order reaction** because the rate of reaction depends on the concentration of HI raised to the second power.

$$\text{Rate} = k(\text{HI})^2$$

When the rate of a reaction depends on more than one reagent, we classify the reaction in terms of the order of each reagent. The overall order is the sum of the orders of the individual reactants.

In general for a reaction

$$A + B \longrightarrow C + D$$

the rate law is written in terms of the concentrations of the reactants, each raised to an exponent. The exponent gives the order of the reaction with respect to a particular reactant. Thus, for the general reaction above

$$\text{Rate} = k(A)^m(B)^n$$

where k is the rate constant for the reaction, m is the order with respect to A, and n is the order with respect to B. The overall order of the reaction is given by $m + n$.

Exercise 14.5

Classify the order of the reaction between NO and O_2 to form NO_2.

$$2\,NO(g) + O_2(g) \longrightarrow 2\,NO_2(g)$$

The experimentally determined rate law for the reaction is

$$\text{Rate} = k(\text{NO})^2(\text{O}_2)$$

Solution

The reaction is first order in O_2, second order in NO, and third order overall.

When the rate of a reaction doesn't depend on the concentration of one or more of the substances consumed in that reaction, it is said to be **zero order.**[1] The rate law for the following reaction, for example, is first order in $(CH_3)_3CBr$ and zero order in the OH^- ion.

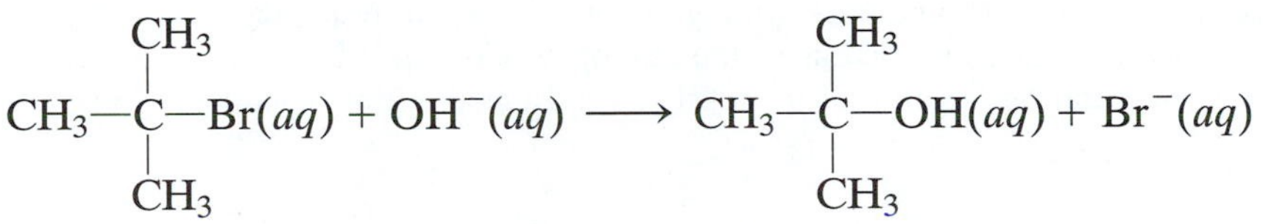

$$
\begin{array}{ccc}
\quad\ \text{CH}_3 & & \quad\ \text{CH}_3 \\
\ \ \ | & & \ \ \ | \\
\text{CH}_3\!-\!\text{C}\!-\!\text{Br}(aq) + \text{OH}^-(aq) & \longrightarrow & \text{CH}_3\!-\!\text{C}\!-\!\text{OH}(aq) + \text{Br}^-(aq) \\
\ \ \ | & & \ \ \ | \\
\quad\ \text{CH}_3 & & \quad\ \text{CH}_3
\end{array}
$$

$$\text{Rate} = k((\text{CH}_3)_3\text{CBr})^1(\text{OH}^-)^0 = k((\text{CH}_3)_3\text{CBr})$$

---

[1]In a zero-order reaction, the rate of the reaction literally depends on the concentration of the reagent to the zeroth power. Since any quantity raised to the zeroth power is equal to 1, the concentration of the reagent has no effect on the rate of the reaction.

It is important to recognize the difference between the molecularity and the order of a reaction. The molecularity of a reaction, or a step within a reaction, describes what happens on the molecular level. The order of a reaction describes what happens on the macroscopic scale. We determine the order of a reaction experimentally by watching the products of a reaction appear or the starting materials disappear. The molecularity of the reaction is something we deduce to explain the experimental results.

14.9   A COLLISION THEORY OF CHEMICAL REACTIONS

The **collision theory** of chemical reactions introduced in Section 10.4 can be used to explain the observed rate laws for both one-step and multistep reactions. The theory assumes that the rate of any step in a reaction depends on the frequency of collisions between the particles involved in that step.

Figure 14.6 provides a basis for understanding the implications of the collision theory for simple, one-step reactions, such as the following.

$$ClNO_2(g) + NO(g) \longrightarrow NO_2(g) + ClNO(g)$$

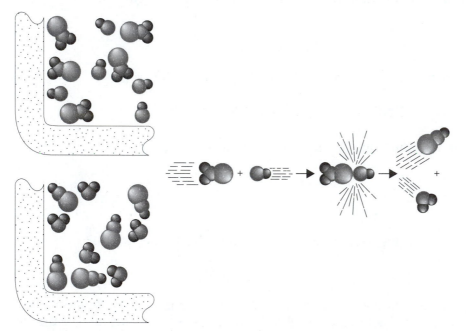

FIGURE 14.6 "Snapshot" of a small portion of a container in which $ClNO_2$ reacts with NO to form NO_2 and ClNO. The collision theory of reactions assumes that molecules must collide in order to react. Anything that increases the frequency of the collisions increases the rate of reaction, so the rate of reaction must be proportional to the concentrations of both of the reactants consumed in this reaction.

The kinetic molecular theory assumes that the number of collisions per second in a gas depends on the number of particles per liter. Thus, the rate at which NO_2 and ClNO are formed in the reaction should be directly proportional to the concentrations of both $ClNO_2$ and NO.

$$Rate = k(ClNO_2)(NO)$$

The collision theory suggests that the rate of any step in a reaction is proportional to the concentrations of the reagents consumed in that step. The rate law for a one-step reaction should therefore agree with the stoichiometry of the reaction. The following reaction, for example, occurs in a single step.

$$CH_3Br(aq) + OH^-(aq) \longrightarrow CH_3OH(aq) + Br^-(aq)$$

When the reactant molecules collide in the proper orientation, a pair of nonbonding electrons on the OH^- ion can be donated to the carbon atom at the center of the CH_3Br molecule, as shown in Figure 14.7. When this happens, a carbon–oxygen bond forms at the same time that the carbon–bromine bond is broken. The net result of the reaction is the substitution of an OH^- ion for a Br^- ion. Because the reaction occurs in a single step, which involves collisions between the two reactants, the rate of the reaction is proportional to the concentration of both reactants.

$$\text{Rate} = k(CH_3Br)(OH^-)$$

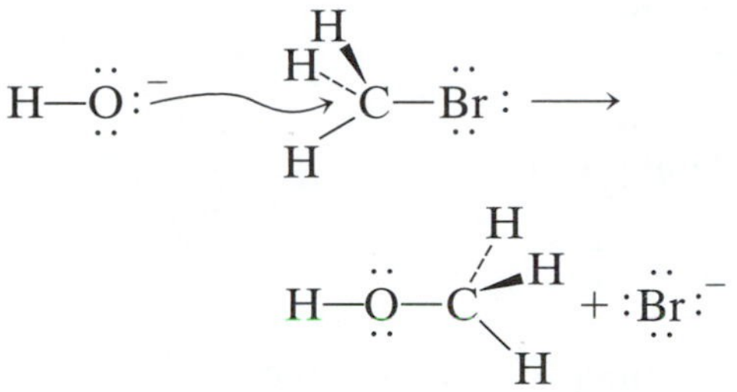

FIGURE 14.7 The reaction between CH_3Br and the OH^- ion occurs in a single step, which involves attack by the OH^- ion on the carbon atom. In the course of the reaction, a carbon–oxygen bond forms at the same time that the carbon–bromine bond is broken.

Not all reactions occur in a single step. The following reaction occurs in three steps, as shown in Figure 14.8.

$$(CH_3)_3CBr(aq) + OH^-(aq) \longrightarrow (CH_3)_3COH(aq) + Br^-(aq)$$

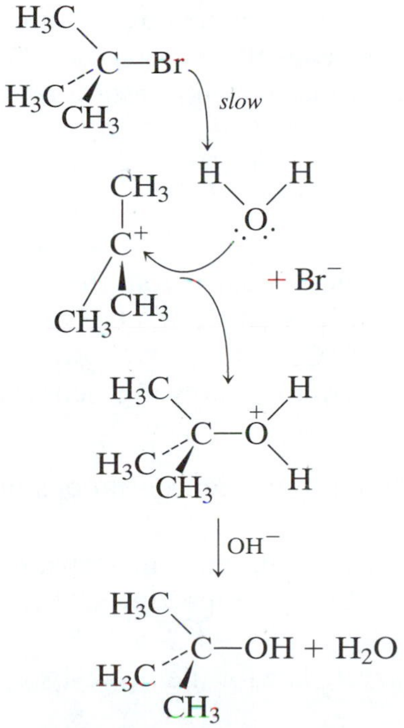

FIGURE 14.8 The reaction between $(CH_3)_3CBr$ and the OH^- ion follows a very different mechanism from that of CH_3Br. In the first step, the carbon–bromine bond breaks to form a positively charged $(CH_3)_3C^+$ ion and a Br^- ion. The $(CH_3)_3C^+$ ion then combines with water to give an intermediate that loses an H^+ ion to another OH^- molecule to form $(CH_3)_3COH$ and water.

In the first step, the $(CH_3)_3CBr$ molecule dissociates into a pair of ions.

$$CH_3-\underset{\underset{CH_3}{|}}{\overset{\overset{CH_3}{|}}{C}}-Br \longrightarrow CH_3-\underset{\underset{CH_3}{|}}{\overset{\overset{CH_3}{|}}{C}}^+ + Br^- \qquad \text{First step}$$

The positively charged $(CH_3)_3C^+$ ion then reacts with water in a second step.

$$CH_3-\underset{\underset{CH_3}{|}}{\overset{\overset{CH_3}{|}}{C}}^+ + H_2O \longrightarrow CH_3-\underset{\underset{CH_3}{|}}{\overset{\overset{CH_3}{|}}{C}}-OH_2^+ \qquad \text{Second step}$$

The product of the reaction then loses a proton to either the OH^- ion or water in the final step.

$$CH_3-\underset{\underset{CH_3}{|}}{\overset{\overset{CH_3}{|}}{C}}-OH_2^+ + OH^- \longrightarrow CH_3-\underset{\underset{CH_3}{|}}{\overset{\overset{CH_3}{|}}{C}}-OH + H_2O \qquad \text{Third step}$$

The second and third steps in the reaction are very much faster than the first.

$$
\begin{array}{ll}
(CH_3)_3CBr \longrightarrow (CH_3)_3C^+ + Br^- & \text{Slow step} \\
(CH_3)_3C^+ + H_2O \longrightarrow (CH_3)_3COH_2^+ & \text{Fast step} \\
(CH_3)_3COH_2^+ + OH^- \longrightarrow (CH_3)_3COH + H_2O & \text{Fast step}
\end{array}
$$

The overall rate of this reaction approximates the rate of the first step. The first step in this reaction is thus called the **rate-limiting step** because it literally limits the rate at which the products of the reaction can be formed. Because only one reagent is involved in the rate-limiting step, the overall rate of reaction is proportional to the concentration of only that reagent.

$$\text{Rate} = k((CH_3)_3CBr)$$

The rate law for this reaction therefore differs from what we would predict from the stoichiometry of the reaction. Although the reaction consumes both $(CH_3)_3CBr$ and OH^-, the rate of the reaction is proportional only to the concentration of $(CH_3)_3CBr$.

The rate laws for chemical reactions can be explained by the following general rules.

- The rate of any step in a reaction is directly proportional to the concentrations of the reagents consumed in that step.
- The overall rate law for a reaction is determined by the sequence of steps, or the **mechanism,** by which the reactants are converted into the products of the reaction.
- The overall rate law for a reaction is dominated by the rate law for the slowest step in the reaction.

14.10 THE MECHANISMS OF CHEMICAL REACTIONS

What happens when the first step in a multistep reaction is not the rate-limiting step? Consider the reaction between NO and O_2 to form NO_2, for example.

$$2\,NO(g) + O_2(g) \longrightarrow 2\,NO_2(g)$$

The reaction involves a two-step mechanism. The first step is a relatively fast reaction in which a pair of NO molecules combine to form a dimer, N_2O_2. The product of this step then undergoes a much slower reaction in which it combines with O_2 to form a pair of NO_2 molecules.

$$\text{Step 1} \qquad 2\,NO \rightleftharpoons N_2O_2 \quad \text{(fast)}$$
$$\text{Step 2} \qquad N_2O_2 + O_2 \longrightarrow 2\,NO_2 \quad \text{(slow)}$$

The net effect of these reactions is the transformation of two NO molecules and one O_2 molecule into a pair of NO_2 molecules.

$$
\begin{array}{l}
2\,NO \rightleftharpoons N_2O_2 \\
\underline{N_2O_2 + O_2 \longrightarrow 2\,NO_2 \quad \text{rate-limiting step}} \\
2\,NO + O_2 \longrightarrow 2\,NO_2
\end{array}
$$

In this reaction, the second step is the rate-limiting step. No matter how fast the first step takes place, the overall reaction cannot proceed any faster than the second step in the reaction. As we have seen, the rate of any step in a reaction is directly proportional to the concentrations of the reactants consumed in that step. The rate law for the second step in this reaction is therefore proportional to the concentrations of both N_2O_2 and O_2.

$$\text{Step 2} \qquad \text{Rate}_{2nd} = k(N_2O_2)(O_2)$$

Because the first step in the reaction is much faster, the overall rate of reaction is more or less equal to the rate of the rate-limiting second step.

$$\text{Rate}_{overall} \approx k(N_2O_2)(O_2)$$

This rate law is not very useful, because it is difficult to measure the concentrations of the highly reactive species called intermediates, such as N_2O_2, that are simultaneously formed and consumed in the reaction. It would be better to have an equation that related the overall rate of reaction to the concentrations of the original reactants.

Let's take advantage of the fact that the first step in the reaction is an equilibrium step.

$$\text{Step 1} \qquad 2\,NO \rightleftharpoons N_2O_2$$

The collision theory predicts that the rate of the forward reaction in this step will depend on the concentration of NO raised to the second power.

$$\text{Step 1} \qquad \text{Rate}_{forward} = k_f(NO)^2$$

The rate of the reverse reaction, however, will depend only on the concentration of N_2O_2.

$$\text{Step 1} \qquad \text{Rate}_{reverse} = k_r(N_2O_2)$$

Because the first step in the reaction is very much faster than the second, the first step should come to equilibrium. When that happens, the rate of the forward and reverse reactions for the first step are the same.

$$k_f(NO)^2 = k_r(N_2O_2)$$

Let's rearrange the equation to solve for one of the terms that appears in the rate law for the second step in the reaction

$$(N_2O_2) = \frac{k_f}{k_r}(NO)^2$$

Substituting this equation for (N_2O_2) in the rate law for the second step gives the following result.

$$\text{Rate}_{2nd} = \frac{k(k_f)}{k_r}(NO)^2(O_2) = k\,K_c(NO)^2(O_2)$$

Note that k_f/k_r is equal to the equilibrium constant for the reaction as shown in Section 10.5. Since k, k_f, and k_r are all constants, they can be replaced by a single constant, k', to give the experimental rate law for this reaction described in Exercise 14.5.

$$\text{Rate}_{overall} \approx \text{Rate}_{2nd} = k'(NO)^2(O_2)$$

14.11   ZERO-ORDER REACTIONS

The collision theory can explain why the rate law for a reaction can be zero order in one of the substances consumed in that reaction. All we do is assume that the reaction occurs in more than one step and that the rate-limiting step for the reaction doesn't involve that substance. The rate law for the following reaction, for example, is zero order in the OH^- ion because the OH^- ion isn't involved in the rate-limiting step.

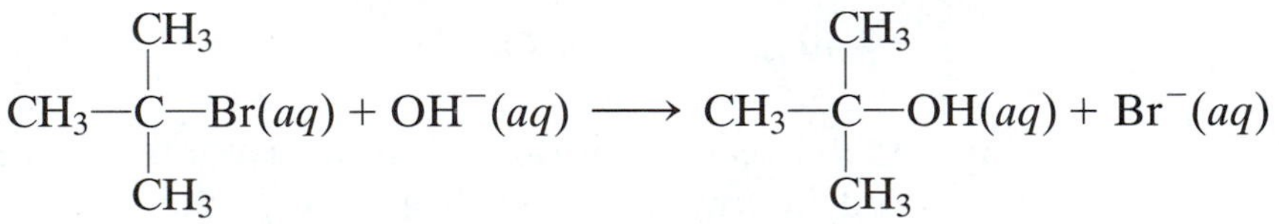

$$CH_3\!-\!\underset{\underset{\textstyle CH_3}{|}}{\overset{\overset{\textstyle CH_3}{|}}{C}}\!-\!Br(aq) + OH^-(aq) \longrightarrow CH_3\!-\!\underset{\underset{\textstyle CH_3}{|}}{\overset{\overset{\textstyle CH_3}{|}}{C}}\!-\!OH(aq) + Br^-(aq)$$

One question remains unanswered: How do we explain the fact that some reactions are zero order in all of the starting materials? Consider the following reaction, for example, in which nitrogen oxide decomposes to its elements on the surface of a piece of platinum metal.

$$2\,NO(g) \xrightarrow{\ Pt\ } N_2(g) + O_2(g)$$

In the presence of a large excess of NO, the rate of the reaction is zero order in NO.

$$\text{Rate} = k(NO)^0 = k$$

The decomposition of NO (or NO_2) into its elements is an important reaction because of the significant environmental problem that NO and NO_2 represent when the gases accumulate in the atmosphere. Most of the catalytic processes that have been studied to remove NO and NO_2 from the atmosphere are accompanied by the release of other pollutants, such as CO_2. Thus, the search for a solid surface on which NO can be induced to decompose into its elements is of vital importance.[2]

A simple analogy can help us understand why the decomposition of NO to its elements on a solid surface is zero order in NO, as long as a large excess of NO is present. Assume that you are waiting in line to eat at a popular restaurant. Until there is an empty table, no one else can be seated. The rate at which people are seated doesn't depend on the number of people standing in line. It only depends on the rate at which tables are vacated. The same thing is true for the decomposition of NO on the platinum metal surface. For NO to react, it must find a site on the metal at which it can bind to the metal surface. No matter how much NO is present, there are only a limited number of sites available at which reaction can occur, and only those NO molecules that occupy one of these sites can undergo a reaction. The reaction is therefore zero order until virtually all of the NO has been consumed.

Any chemical reaction that occurs by a mechanism by which the reactant has to bind to a catalyst that has a limited number of active sites will exhibit similar behavior. Consider the following reaction, for example.

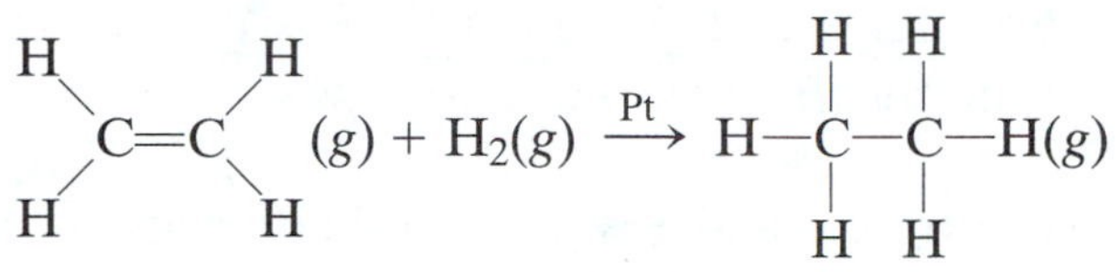

$$H_2C{=}CH_2(g) + H_2(g) \xrightarrow{\text{Pt}} H_3C{-}CH_3(g)$$

Both starting materials must bind to an active site on the platinum metal before the reaction can occur. When both gases are present in excess, the rate of the reaction is zero order in *both* of the starting materials.

$$\text{Rate} = k(C_2H_4)^0(H_2)^0 = k$$

Checkpoint

How does the dependence of rate on concentration differ between a zero-order and first-order reaction? Describe what happens to the rate of a reaction with respect to time for both a zero- and a first-order reaction.

14.12 DETERMINING THE ORDER OF A REACTION FROM RATES OF REACTION

The rate law for a reaction can be determined by studying what happens to the instantaneous rate of reaction when we start with differential initial concentrations of the reactants. To show how this is done, let's determine the rate law for the reaction in which hydrogen iodide decomposes to give a mixture of hydrogen and iodine in the gas phase.

$$2\,HI(g) \longrightarrow H_2(g) + I_2(g)$$

[2]For a review of the role that *catalysts* such as platinum metal can play in improving the environment, see the paper by John N. Armor in *Chemistry of Materials*, **6**, 730–738 (1994).

Data on initial rates of reaction for three experiments run at different initial concentrations of HI are given in Table 14.3.

TABLE 14.3   Rate of Reaction Data for the Decomposition of HI

Trial	Initial Concentration of HI (M)	Initial Instantaneous Rate of Reaction (M/s)
Trial 1	1.0×10^{-2}	4.0×10^{-6}
Trial 2	2.0×10^{-2}	1.6×10^{-5}
Trial 3	3.0×10^{-2}	3.6×10^{-5}

The only difference in the three experiments is the initial concentration of HI. Let's compare the experiments two at a time. The difference between trial 1 and trial 2 is a twofold increase in the initial HI concentration, which leads to a fourfold increase in the initial rate of reaction.

$$\frac{\text{Rate for trial 2}}{\text{Rate for trial 1}} = \frac{1.6 \times 10^{-5} \ M/s}{4.0 \times 10^{-6} \ M/s} = 4.0$$

The only difference between trials 1 and 3 is a threefold increase in the initial concentration of HI. The initial rate of reaction, however, increases by a factor of 9.

$$\frac{\text{Rate for trial 3}}{\text{Rate for trial 1}} = \frac{3.6 \times 10^{-5} \ M/s}{4.0 \times 10^{-6} \ M/s} = 9.0$$

The data suggest that the rate of the reaction is proportional to the *square of the HI concentration*. The reaction is therefore second order in HI, as noted in Section 14.7.

$$\text{Rate} = k(\text{HI})^2$$

Exercise 14.6

Given the following experimental data determine the rate law for the reaction.

$$\text{NH}_4^+(aq) + \text{NO}_2^-(aq) \longrightarrow \text{N}_2(g) + 2\ \text{H}_2\text{O}(l)$$

	Initial Concentration of NH_4^+ (M)	Initial Concentration of NO_2^- (M)	Initial Instantaneous Rate of Reaction (M/s)
Trial 1	5.00×10^{-2}	2.00×10^{-2}	2.70×10^{-7}
Trial 2	5.00×10^{-2}	4.00×10^{-2}	5.40×10^{-7}
Trial 3	1.00×10^{-1}	2.00×10^{-2}	5.40×10^{-7}

Solution

As shown in Section 14.8, the rate law for trial 1 can be written as

$$\text{Rate}_1 = k(\text{NH}_4^+)_1{}^m(\text{NO}_2^-)_1{}^n$$

and the rate law for trial two is

$$\text{Rate}_2 = k(\text{NH}_4^+)_2{}^m(\text{NO}_2^-)_2{}^n$$

The order with respect to NH_4^+ and NO_2^- can be found from the experimental data if we know how the initial rates of reaction change when the concentrations of only one of the reactants is varied. In trials 1 and 2 the (NH_4^+) remains constant while the (NO_2^-) is doubled. Thus, a ratio of trial 1 to trial 2 can be used to determine how the changing (NO_2^-) affects the initial rate. This allows the determination of the order, n, with respect to (NO_2^-).

$$\frac{\text{Rate}_1}{\text{Rate}_2} = \frac{k(\text{NH}_4^+)_1{}^m(\text{NO}_2^-)_1{}^n}{k(\text{NH}_4^+)_2{}^m(\text{NO}_2^-)_2{}^n}$$

$$\frac{2.70 \times 10^{-7}}{5.40 \times 10^{-7}} = \frac{k(5.00 \times 10^{-2})^m(2.00 \times 10^{-2})^n}{k(5.00 \times 10^{-2})^m(4.00 \times 10^{-2})^n}$$

Note that the rate constants and the (NH_4^+) terms are the same in the numerator and denominator and therefore cancel out of the equation.

$$\frac{2.70 \times 10^{-7}}{5.40 \times 10^{-7}} = \frac{(2.00 \times 10^{-2})^n}{(4.00 \times 10^{-2})^n} = \left(\frac{2.00 \times 10^{-2}}{4.00 \times 10^{-2}}\right)^n = \left(\frac{1}{2}\right)^n$$

$$\frac{1}{2} = \left(\frac{1}{2}\right)^n$$

$$n = 1$$

Similarly, we can use the ratio of trials 1 and 3 to find the order with respect to NH_4^+.

$$\frac{\text{Rate}_1}{\text{Rate}_3} = \frac{k(5.00 \times 10^{-2})^m(2.00 \times 10^{-2})^n}{k(1.00 \times 10^{-1})^m(2.00 \times 10^{-2})^n}$$

$$\frac{2.70 \times 10^{-7}}{5.40 \times 10^{-7}} = \left(\frac{5.00 \times 10^{-2}}{1.00 \times 10^{-1}}\right)^m$$

$$m = 1$$

The rate law is then, $\quad \text{Rate} = k(\text{NH}_4^+)^1(\text{NO}_2^-)^1$, and the overall order is $1 + 1 = 2$.

Checkpoint
If neither m nor n is known, why can't trials 2 and 3 be used to determine either order? Can trials 1 and 2 be used to determine the order with respect to NH_4^+?

Checkpoint
Determine the value of the rate constant for the reaction in Exercise 14.6.

14.13   THE INTEGRATED FORM OF FIRST-ORDER AND SECOND-ORDER RATE LAWS

Section 14.10 showed that the rate law for a reaction is a useful way of studying the mechanism of a chemical reaction. The form of the rate law discussed thus far is used to describe the rate and concentration at one instant in time. However, if we want to calculate the change in the concentration of a reactant or a product during a given time change, a new form of the rate law must be used, the **integrated form.** Both the rate law and its integrated form can be used to determine the reaction order and rate constant.

Let's start with the rate law for a reaction that is first order in the disappearance of a single reactant, X.

$$-\frac{d(X)}{dt} = k(X)$$

When the equation is rearranged and both sides are integrated, we get the following result.

Integrated Form of the First-Order Rate Law

$$\ln\left[\frac{(X)}{(X)_0}\right] = -kt$$

In this equation, (X) is the concentration of X at any moment in time, $(X)_0$ is the initial concentration of the reagent, k is the rate constant for the reaction, and t is the time since the reaction started.

To illustrate the power of the integrated form of the rate law for a reaction, we will use the equation to calculate how long it would take for the ^{14}C in a piece of charcoal to decay to half of its original concentration. All radioactive decay processes are first order. Therefore ^{14}C decays by first-order kinetics and it has a rate constant of $1.21 \times 10^{-4} \ y^{-1}$.

$$\text{Rate} = k(^{14}C)$$

The integrated form of the rate law would be written as follows.

$$\ln\left[\frac{(^{14}C)}{(^{14}C)_0}\right] = -kt$$

We are interested in the moment when the concentration of ^{14}C in the charcoal is half of its initial value.

$$(^{14}C) = \frac{1}{2}(^{14}C)_0$$

Substituting this relationship into the integrated form of the rate law gives the following equation.

$$\ln\left[\frac{\frac{1}{2}(^{14}C)_0}{(^{14}C)_0}\right] = -kt$$

We now simplify the equation

$$\ln\left(\frac{1}{2}\right) = -kt$$

and then solve for t.

$$t = -\frac{\ln(1/2)}{k} = \frac{0.693}{1.21 \times 10^{-4} \text{ y}^{-1}} = 5.73 \times 10^{3} \text{ y}$$

It therefore takes 5.73×10^{3} years for half of the ^{14}C in the sample to decay. This is called the **half-life** ($t_{1/2}$) of ^{14}C. In general, the half-life for a first-order kinetic process can be calculated from the rate constant as follows.

$$t_{1/2} = -\frac{\ln(1/2)}{k} = \frac{0.693}{k}$$

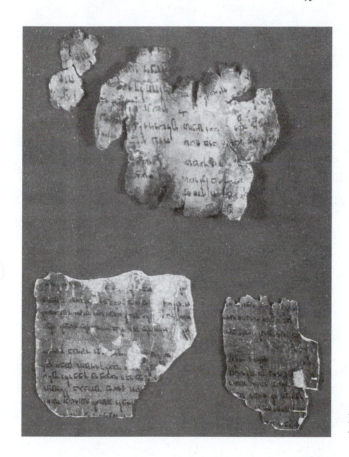

Fragments of the Dead Sea Scrolls, which were authenticated by ^{14}C dating.

Checkpoint

Which radioactive isotope would have the longer half-life, ^{15}O or ^{19}O?

$$^{15}\text{O} \qquad k = 5.63 \times 10^{-3} \text{ s}^{-1}$$
$$^{19}\text{O} \qquad k = 2.38 \times 10^{-2} \text{ s}^{-1}$$

Let's now turn to the rate law for a reaction that is second order in a single reactant, X.

$$\frac{-d(X)}{dt} = k(X)^{2}$$

The integrated form of the rate law for this reaction is written as follows.

Integrated Form of the Second-Order Rate Law

$$\frac{1}{(X)} - \frac{1}{(X)_{0}} = kt$$

Once again, (X) is the concentration of X at any moment in time, $(X)_0$ is the initial concentration of X, k is the rate constant for the reaction, and t is the time since the reaction started.

The half-life of a second-order reaction can be calculated from the integrated form of the second-order rate law.

$$\frac{1}{(X)} - \frac{1}{(X)_0} = kt$$

We start by asking, "How long would it take for the concentration of X to decay from its initial value, $(X)_0$, to a value half as large?"

$$\frac{1}{\frac{1}{2}(X)_0} - \frac{1}{(X)_0} = kt_{1/2}$$

The first step in simplifying the equation involves multiplying the top and bottom halves of the first term by 2.

$$\frac{2}{(X)_0} - \frac{1}{(X)_0} = kt_{1/2}$$

Subtracting the second term on the left side of the equation from the first gives the following result.

$$\frac{1}{(X)_0} = kt_{1/2}$$

We can now solve the equation for the half-life of the reaction.

$$t_{1/2} = \frac{1}{k(X)_0}$$

There is an important difference between the equations for calculating the half-lives of first-order and second-order reactions. The half-life of a first-order reaction is a constant, which is proportional to the rate constant for the reaction.

$$\textbf{First-Order Reaction} \qquad t_{1/2} = \frac{0.693}{k}$$

The half-life for a second-order reaction is inversely proportional to both the rate constant for the reaction and the initial concentration of the reactant that is consumed in the reaction.

$$\textbf{Second-Order Reaction} \qquad t_{1/2} = \frac{1}{k(X)_0}$$

Discussions of reaction half-lives are usually confined to first-order processes because this is the only reaction order for which the half-life is independent of reactant concentration. The equations that describe first- and second-order reactions are summarized in Table 14.4.

TABLE 14.4   Equations that Describe First- and Second-Order Reactions

Order	Rate Law	Integrated Form of the Rate Law	Half-Life
First	rate = $k(X)$	$\ln \dfrac{(X)}{(X)_0} = -kt$	$t_{1/2} = 0.693/k$
Second	rate = $k(X)^2$	$\dfrac{1}{(X)} - \dfrac{1}{(X)_0} = kt$	$t_{1/2} = \dfrac{1}{k(X)_0}$

14.14 DETERMINING THE ORDER OF A REACTION WITH THE INTEGRATED FORM OF RATE LAWS

The integrated form of the rate laws for first- and second-order reactions provides another way of determining the order of a reaction. We can start by assuming, for the sake of argument, that the reaction is first order in reactant X.

$$\text{Rate} = k(X)$$

We then test the assumption by checking concentration versus time data for the reaction to see whether they fit the integrated form of the first-order rate law.

$$\ln\left[\frac{(X)}{(X)_0}\right] = -kt$$

To see how this is done, let's rearrange the integrated form of the first-order rate law as follows:

$$\ln(X) - \ln(X)_0 = -kt$$

We then solve the equation for the natural logarithm of the concentration of X at any moment in time.

$$\ln(X) = \ln(X)_0 - kt$$

This equation contains two variables, $\ln(X)$ and t, and two constants, $\ln(X)_0$ and k. It can therefore be set up in terms of the equation for a straight line.

$$y = mx + b$$
$$\ln(X) = -kt + \ln(X)_0$$

If the reaction is first order in X, a plot of the natural logarithm of the concentration of X versus time will be a straight line with a slope equal to $-k$, as shown in Figure 14.9.

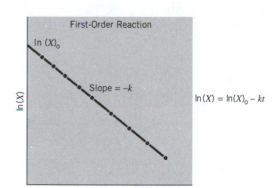

$$\ln(X) = \ln(X)_0 - kt$$

FIGURE 14.9 For an overall first order rate law, a plot of the natural log of the concentration of X versus the amount of time since the reaction started will be a straight line with a slope equal to the negative of the rate constant.

If the plot of $\ln(X)$ versus time isn't a straight line, the reaction can't be first order in X. We therefore assume, for the sake of argument, that it is second order in X.

$$\text{Rate} = k(X)^2$$

We then test the assumption by checking whether the experimental data fit the integrated form of the second-order rate law.

$$\frac{1}{(X)} - \frac{1}{(X)_0} = kt$$

This equation, like the one for first-order reactions, contains two variables, (X) and t, and two constants, $(X)_0$ and k. Thus, it also can be set up in terms of the equation for a straight line.

$$y = mx + b$$

$$\frac{1}{(X)} = kt + \frac{1}{(X)_0}$$

If the reaction is second order in X, a plot of the reciprocal of the concentration of X versus time will be a straight line with a slope equal to k, as shown in Figure 14.10. If the plot of $1/(X)$ versus time is not linear, the reaction cannot be second order.

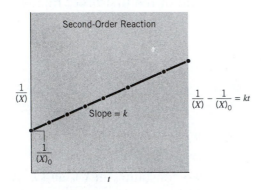

FIGURE 14.10 For an overall second-order reaction involving only X, a plot of the inverse of the concentration of X versus the amount of time since the reaction started will be a straight line with a slope equal to the rate constant, k.

Exercise 14.7

Use the data in Table 14.1 to determine whether the reaction between phenolphthalein and the OH^- ion is first-order or second-order with respect to the phenolphthalein. OH^- is in excess.

Solution

The first step in solving the problem involves calculating the natural log of the phenolphthalein concentration, $\ln(PHTH)$, and the reciprocal of the concentration, $1/(PHTH)$, for each point at which a measurement was taken.

(PHTH) (mol/L)	ln(PHTH)	1/(PHTH)	Time (s)
0.00500	−5.298	200	0.0
0.00450	−5.404	222	10.5
0.00400	−5.521	250	22.3
0.00350	−5.655	286	35.7
0.00300	−5.809	333	51.1
0.00250	−5.991	400	69.3
0.00200	−6.215	500	91.6
0.00150	−6.502	667	120.4
0.00100	−6.908	1.00×10^3	160.9
0.000500	−7.601	2.00×10^3	230.3
0.000250	−8.294	4.00×10^3	299.6
0.000150	−8.805	6.67×10^3	350.7
0.000100	−9.210	1.00×10^4	391.2

We then construct graphs of ln(PHTH) versus t (Figure 14.11) and 1/(PHTH) versus t (Figure 14.12). Only one of the graphs, Figure 14.11, gives a straight line. We therefore conclude that the data fit a first-order kinetic equation, as noted in Section 14.5.

$$\text{Rate} = k(\text{phenolphthalein})$$

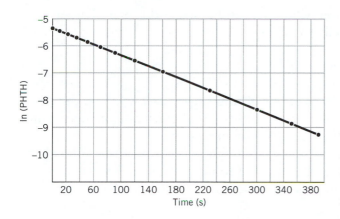

FIGURE 14.11 A plot of the natural log of the concentration of phenolphthalein versus time for the reaction between phenolphthalein and OH^- ion is a straight line, which shows that the reaction is first order in phenolphthalein.

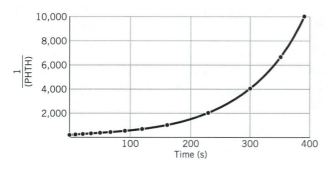

FIGURE 14.12 A plot of the reciprocal of the concentration of phenolphthalein versus time for the reaction between phenolphthalein and OH^- ion is *not* a straight line, which shows that the reaction is not second order in phenolphthalein.

14.15 REACTIONS THAT ARE FIRST ORDER IN TWO REACTANTS

What about reactions that are first order in two reactants, X and Y, and therefore second order overall?

$$\text{Rate} = k(X)(Y)$$

A plot of 1/(X) versus time won't give a straight line because the reaction is not second order in X. Unfortunately, neither will a plot of ln(X) versus time because the reaction isn't strictly first order in X. It is simultaneously first order in both X and Y.

One way around the problem is to turn the reaction into a **pseudo-first-order reaction** by making the concentration of one of the reactants so large that it is effectively constant. The rate law for the reaction is still first order in both reactants. But the initial concentration of one reactant is so much larger than the other that the rate of reaction seems to be sensitive only to changes in the concentration of the reagent present in limited quantities. The phenolphthalein/OH^- reaction described in Section 14.3 is pseudo-first order with respect to phenolphthalein because the OH^- is in excess.

Assume, for the moment, that the reaction is studied under conditions for which there is a large excess of Y. If this is true, the concentration of Y will remain essentially constant during the reaction. As a result, the rate of the reaction won't depend on the concentration of the excess reagent. Instead, it will appear to be first order in the other reactant, X. A plot of $\ln(X)$ versus time will therefore give a straight line.

$$\text{Rate} = k'(X)$$

If there is a large excess of X, the reaction will appear to be first order in Y. Under these conditions, a plot of $\ln(Y)$ versus time will be linear.

$$\text{Rate} = k''(Y)$$

The value of the rate constant obtained from either of these equations won't be the actual rate constant for the reaction. It will be the product of the rate constant for the reaction times the concentration of the reagent that is present in excess.

In our discussion of acid–base equilibria, we pointed out that the concentration of water is so much larger than any other component of the solutions that we can incorporate it into the equilibrium constant expression for the reaction of an acid or base ionizing in water.

$$K_a = \frac{[H_3O^+][A^-]}{[HA]} \qquad K_b = \frac{[BH^+][OH^-]}{[B]}$$

We now understand why this is done. Because the concentration of water is so large, the reaction between an acid or a base and water is a pseudo-first-order reaction that depends only on the concentration of the acid or base.

14.16 THE ACTIVATION ENERGY OF CHEMICAL REACTIONS

Only a small fraction of the collisions between reactant molecules convert the reactants into the products of the reaction. This can be understood by turning, once again, to the reaction between $ClNO_2$ and NO.

$$ClNO_2(g) + NO(g) \longrightarrow NO_2(g) + ClNO(g)$$

In the course of the reaction, a chlorine atom is transferred from one nitrogen atom to another. For the reaction to occur, the nitrogen atom in NO must collide with the chlorine atom in $ClNO_2$.

The reaction won't occur if the oxygen end of the NO molecule collides with the chlorine atom on $ClNO_2$.

No reaction

Nor will it occur if one of the oxygen atoms on $ClNO_2$ collides with the nitrogen atom on NO.

No reaction

Another factor that influences whether a reaction will occur is the energy the molecules have when they collide. Not all of the molecules have the same kinetic energy, as shown in Figure 14.13. This is important because the kinetic energy molecules carry when they collide is the principal source of the energy that must be invested in a reaction to get it started.

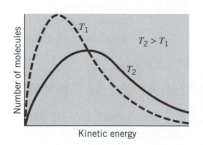

FIGURE 14.13 The kinetic molecular theory states that the average kinetic energy of a gas is proportional to the temperature of the gas, and nothing else. At any given temperature, however, some of the gas particles are moving faster than others.

The overall standard free energy for the reaction between $ClNO_2$ and NO is favorable.

$$ClNO_2(g) + NO(g) \longrightarrow NO_2(g) + ClNO(g) \qquad \Delta G° = -23.6 \text{ kJ/mol}_{rxn}$$

But, before the reactants can be converted into products, the collision energy of the system must overcome the **activation energy (E_a)** for the forward reaction, as shown in Figure 14.14. The vertical axis in Figure 14.14 represents the energy as a chlorine atom is transferred from $ClNO_2$ to NO. The horizontal axis represents the sequence of infinitesimally small changes that must occur to convert the reactants into the products of the reaction and is called the *reaction coordinate*.

The energy of activation for the reverse reaction, $E_{a \text{ reverse}}$, is the energy difference between the products and the top of the energy of activation curve as shown in Figure 14.14. The products must overcome the energy barrier in order to revert back to the reactants. Note that the enthalpy change, ΔH, for the reaction is also shown in this figure.

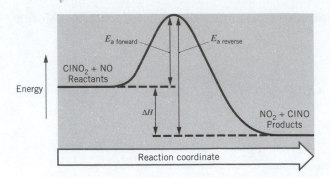

FIGURE 14.14 The forward activation energy for a reaction, E_a, is the change in the potential energy of the reactant molecules that must be overcome before the reaction can occur.

To understand why reactions have an activation energy, consider what has to happen in order for $ClNO_2$ to react with NO. First, and foremost, the two molecules have to collide with sufficient energy. Not only do they have to be brought together, they have to collide in exactly the right orientation relative to one another to ensure that a reaction can occur. Some energy also must be invested to begin breaking the $Cl—NO_2$ bond so that the $Cl—NO$ bond can form.

NO and $ClNO_2$ molecules that collide in the correct orientation, with enough kinetic energy to climb the activation energy barrier, can react to form NO_2 and $ClNO$. As the temperature of the system increases, the number of molecules that have enough energy to react when they collide also increases. The rate of reaction generally increases with temperature. As a rough rule, the rate of a reaction doubles for every 10°C increase in the temperature of the system.

Checkpoint

For an exothermic reaction $E_{a\ forward} < E_{a\ reverse}$ as shown in Figure 14.14. Is this also true for an endothermic reaction? Draw a reaction coordinate diagram to explain your reasoning.

14.17   CATALYSTS AND THE RATES OF CHEMICAL REACTIONS

Aqueous solutions of hydrogen peroxide are stable until we add a small quantity of the I^- ion, a piece of platinum metal, a few drops of blood, or a freshly cut slice of turnip, at which point the hydrogen peroxide rapidly decomposes.

$$2\ H_2O_2(aq) \longrightarrow 2\ H_2O(l) + O_2(g)$$

The reaction therefore provides the basis for understanding the effect of a catalyst on the rate of a chemical reaction. Four criteria must be satisfied in order for a substance to be classified as a **catalyst.**

- Catalysts increase the rate of reaction.
- Catalysts aren't consumed by the reaction.
- A small quantity of catalyst should be able to affect the rate of reaction for a large amount of reactant.
- Catalysts don't change the equilibrium constant, ΔH, or ΔS for the reaction.

The first criterion provides the basis for defining a catalyst as something that increases the rate of a reaction. The second reflects the fact that anything consumed in the reaction is a reactant, not a catalyst. The third criterion is a consequence of the second; because catalysts aren't consumed in the reaction, they can catalyze the reaction over and over again. The fourth criterion results from the fact that catalysts speed up the rates of the forward and reverse reactions equally, so the equilibrium constant for the reaction remains the same.

Catalysts increase the rates of reactions by providing a new mechanism that has a smaller activation energy, E_a, as shown in Figure 14.15. A larger proportion of the collisions that occur between reactants now have enough energy to overcome the activation energy for the reaction. As a result, the rate of reaction increases.

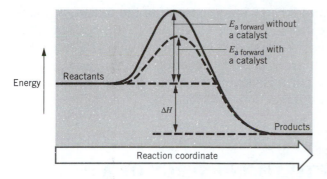

FIGURE 14.15 A catalyst increases the rate of a reaction by providing an alternative mechanism that has a smaller activation energy.

The effect of several catalysts on the activation energy for the decomposition of hydrogen peroxide and the relative rate of the reaction is summarized in Table 14.5. Adding a source of I^- ion to the solution decreases the activation energy by 25%, which increases the rate of the reaction by a factor of 2000. A piece of platinum metal decreases the activation energy even further, thereby increasing the rate of reaction by a factor of 40,000. The *catalase* enzyme in blood or turnips decreases the activation energy by almost a factor of 10, which leads to a 600 billion-fold increase in the rate of reaction.

TABLE 14.5   Effect of Catalysts on the Activation Energy for the Decomposition of Hydrogen Peroxide

Catalyst	$E_a(kJ/mol_{rxn})$	Relative Rate of Reaction
None	75.3	1
I^-	56.5	2.0×10^3
Pt	49.0	4.1×10^4
Catalase	8	6.3×10^{11}

The bombardier beetle uses the enzyme-catalyzed decomposition of hydrogen peroxide as a defensive mechanism. When attacked, it mixes the contents of a sac that is about 25% H_2O_2 with a suspension of a crystalline peroxidase enzyme in a turretlike mixing tube. The reaction is so exothermic that the mixture is vaporized and the beetle can direct a hot spray at its attacker. The enzyme causes the reaction to occur so rapidly that the beetle is instantly ready to defend itself when attacked.

To illustrate how a catalyst can decrease the activation energy for a reaction by providing another pathway for the reaction, let's look at the mechanism for the decomposition of hydrogen peroxide catalyzed by the I^- ion. In the presence of the I^- ion, the decomposition of H_2O_2 doesn't have to occur in a single step. It can occur in two steps, both of which are easier and therefore faster. In the first step, the I^- ion is oxidized by H_2O_2 to form the hypoiodite ion, OI^-.

$$H_2O_2(aq) + I^-(aq) \longrightarrow H_2O(l) + OI^-(aq)$$

In the second step, the OI^- ion is reduced to I^- by H_2O_2.

$$OI^-(aq) + H_2O_2(aq) \longrightarrow H_2O(l) + O_2(g) + I^-(aq)$$

There is no net change in the concentration of the I^- ion as a result of the reactions, therefore the I^- ion satisfies the criteria for a catalyst. Because H_2O_2 and I^- are both involved in the first step in the reaction, and because the first step in the reaction is the rate-limiting step, the overall rate of reaction is first order in both reagents. The OI^- is a reaction intermediate that is in principle isolatable, but it is used up to make the products and as a consequence it does not appear as one of the reaction products.

14.18   DETERMINING THE ACTIVATION ENERGY OF A REACTION

The rate of a reaction depends on the temperature at which it is run. As the temperature increases, the molecules move faster and therefore collide more frequently. The molecules also have more kinetic energy. Thus, the proportion of collisions that can overcome the activation energy for the reaction increases with temperature.

The relationship between temperature and the rate of a reaction can be explained by assuming that the rate constant depends on the temperature at which the reaction is run. In 1889, Svante Arrhenius showed that the relationship between temperature and the rate constant for a reaction obeyed the following equation.

$$k = Z\,e^{-E_a/RT}$$

In this equation, k is the rate constant for the reaction, Z is a proportionality constant that varies from one reaction to another, e is the base of natural logarithms, E_a is the activation energy for the reaction, R is the ideal gas constant in joules per mole kelvin, and T is the temperature in kelvins. For a given Z and T, the rate constant, k, is determined by E_a.

The **Arrhenius equation** can be used to determine the activation energy for a reaction. We start by taking the natural logarithm of both sides of the equation.

$$\ln k = \ln Z - \frac{E_a}{RT}$$

We then rearrange the equation to fit the equation for a straight line.

$$y = mx + b$$

$$\ln k = -\frac{E_a}{R}\left(\frac{1}{T}\right) + \ln Z$$

According to this equation, a plot of $\ln k$ versus $1/T$ should give a straight line with a slope of $-E_a/R$, as shown in Figure 14.16.

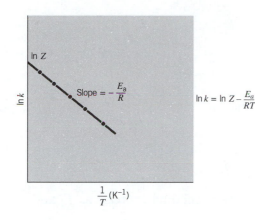

FIGURE 14.16 A plot of the natural log of the rate constant for a reaction at different temperatures versus the inverse of the temperature in kelvins is a straight line with a slope equal to $-E_a/R$.

By paying careful attention to the mathematics of logarithms, it is possible to derive another form of the Arrhenius equation that can be used to predict the effect of a change in temperature on the rate constant for a reaction.

$$\ln\left(\frac{k_1}{k_2}\right) = \frac{E_a}{R}\left(\frac{1}{T_2} - \frac{1}{T_1}\right)$$

Exercise 14.8

Use the following data to determine the forward activation energy for the decomposition of HI.

Temperature (K)	Rate Constant (M/s)
573	2.91×10^{-6}
673	8.38×10^{-4}
773	7.65×10^{-2}

Solution

We can determine the activation energy for a reaction from a plot of the natural log of the rate constants versus the reciprocal of the absolute temperature. We therefore start by calculating $1/T$ and the natural logarithm of the rate constants.

ln k	1/T (K^{-1})
-12.75	0.00175
-7.08	0.00149
-2.57	0.00129

When we construct a graph of the data, we get a straight line with a slope of -2.2×10^4 K. According to the Arrhenius equation, the slope of the line is equal to $-E_a/R$.

$$-2.2 \times 10^4 \text{ K} = -\frac{E_a}{8.314 \text{ J/mol·K}}$$

When the equation is solved, we get the following value for the activation energy for the reaction.

$$E_a = 1.8 \times 10^2 \text{ kJ/mol}$$

Exercise 14.9

Calculate the rate of decomposition of HI at 600°C.

Solution

We start with the following form of the Arrhenius equation.

$$\ln\left(\frac{k_1}{k_2}\right) = \frac{E_a}{R}\left(\frac{1}{T_2} - \frac{1}{T_1}\right)$$

We then pick any one of the three data points used in the preceding exercise as T_1 and allow the value of T_2 to be 873 K.

$$T_1 = 573 \text{ K} \qquad k_1 = 2.91 \times 10^{-6} \ M/s$$
$$T_2 = 873 \text{ K} \qquad k_2 = ?$$

Substituting what we know about the system into the equation given above gives the following result.

$$\ln\frac{2.91 \times 10^{-6}}{k_2} = \frac{1.8 \times 10^5 \text{ J/mol}}{8.314 \text{ J/mol·K}}\left(\frac{1}{873 \text{ K}} - \frac{1}{573 \text{ K}}\right)$$

We can simplify the right-hand side of the equation as follows.

$$\ln\frac{2.91 \times 10^{-6}}{k_2} = -13$$

We then take the antilog of both sides of the equation.

$$\frac{2.91 \times 10^{-6}}{k_2} = 2 \times 10^{-6}$$

Solving for k_2 gives the rate constant for the reaction at 600°C.

$$k_2 = 1 \ M/s$$

Increasing the temperature of the reaction from 573 K to 873 K therefore increases the rate constant for the reaction by a factor of almost a million.

14.19   THE KINETICS OF ENZYME-CATALYZED REACTIONS

Enzymes are proteins that catalyze reactions in living systems. In 1913, Lenor Michaelis and his student M. L. Menten studied the rate at which an enzyme isolated from yeast catalyzed the hydrolysis of sucrose into fructose and glucose.

$$\text{Sucrose}(aq) + \text{H}_2\text{O}(l) \xrightarrow{Enzyme} \text{fructose}(aq) + \text{glucose}(aq)$$

At first glance, this reaction seems similar to the reaction between sucrose and acid, which had been studied by Ludwig Wilhemy 60 years earlier.

$$\text{Sucrose}(aq) + H_3O^+(aq) \longrightarrow \text{fructose}(aq) + \text{glucose}(aq)$$

Wilhemy found that the reaction between sucrose and acid was first order in sucrose at constant pH.

$$\text{Rate} = k(\text{sucrose})$$

Michaelis and Menten, however, found that the initial rate of the enzyme-catalyzed reaction was first order in sucrose only at low concentrations of sucrose. At high concentrations, the initial rate of reaction did not increase when more sucrose was added to the system.

As we have seen, when the rate of reaction doesn't depend on the concentration of one of the reactants, the reaction is said to be *zero order* in that reactant. The enzyme-catalyzed hydrolysis of sucrose apparently changes from a first-order reaction to a zero-order reaction as the amount of sucrose in the system increases, as shown in Figure 14.17. There is a maximum initial rate of reaction, Rate_{max}, for the enzyme-catalyzed reaction that can't be exceeded no matter how much sucrose we add to the solution. Once the reaction reaches Rate_{max}, the only way to increase the rate of reaction is to add more enzyme to the solution.

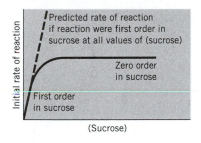

FIGURE 14.17 If the reaction in which an enzyme hydrolyzes sucrose to fructose and glucose were first order in the concentration of sucrose, the initial rate of reaction should be directly proportional to the initial concentration of sucrose. Michaelis and Menten, however, showed that the reaction is first order in sucrose at low concentrations of sucrose but zero order in sucrose at high concentrations.

Michaelis and Menten explained this behavior by assuming that the reaction proceeds through the following mechanism. In the first step, the enzyme (E) combines with sucrose (S) to form a complex, ES.

$$E + S \xrightarrow{k_1} ES$$

The complex can dissociate back to the initial reactants to form the enzyme and sucrose.

$$ES \xrightarrow{k_2} E + S$$

The enzyme–sucrose complex may also decompose to produce the products (P) of the reaction, thus regenerating the enzyme.

$$ES \xrightarrow{k_3} E + P$$

The slow step is the decomposition of ES. The rate of the overall reaction will be determined by the concentration of the complex:

$$\text{Rate}_{overall} \approx \text{Rate}_{3rd} = k_3(ES)$$

This means that the maximum rate of reaction will be seen when essentially all of the enzyme is tied up as the ES complex.

There is a limit to the rate at which the enzyme can consume sucrose. When there is a large excess of sucrose in the solution, every time the enzyme combines with a molecule of sucrose causing the sucrose to decompose, the enzyme will immediately pick up another sucrose molecule. No matter how much sucrose is added to the solution, the reaction can't occur any faster. The reaction no longer depends on the sucrose concentration and is therefore zero order in sucrose.

$$\text{Rate} = k(\text{sucrose})^0 = k$$

Enzyme-catalyzed reactions are therefore similar to the metal-catalyzed reactions described in Section 14.11. When there is enough reactant in the system so that essentially all of the catalytic sites are occupied at any moment in time, the reaction will exhibit zero-order kinetics. No matter how much reactant we add to the system, the reaction can't go any faster.

Checkpoint

Chapter 8 described the change in boiling point of a liquid with a change in external pressure exerted on the liquid. As a result the boiling point of water is lower at the top of a mountain than it is at sea level. Use the Arrhenius equation to describe why it takes longer to cook a hard-boiled egg at the top of a mountain than it does at sea level. Explain your reasoning.

KEY TERMS

Activation energy (E_a)

Arrhenius equation

Bimolecular

Catalyst

Chemical kinetics

Collision theory

Enzyme

First-order reaction

Half-life

Initial rate of reaction

Instantaneous rate of reaction

Integrated form of the rate laws

Kinetic control

Mechanism

Molecularity

Order

Pseudo-first-order reaction

Rate constant

Rate law

Rate-limiting step

Second-order reaction

Thermodynamic control

Unimolecular

Zero-order reaction

PROBLEMS

Chemical Kinetics Defined

1. Define the terms *thermodynamic control* and *kinetic control* and give an example of each.

2. Describe the difference between the terms *rate of reaction, rate law,* and *rate constant.*

3. Describe the difference between the rate of reaction measured over a finite period of time and the instantaneous rate of reaction. Explain the advantage of measuring the instantaneous rate of reaction.

4. Explain why the rate of each step in a reaction is proportional to the concentrations of the reagents consumed in that step.

5. Which of the following graphs best describes the relationship between the rate of a reaction and the temperature of the reaction?

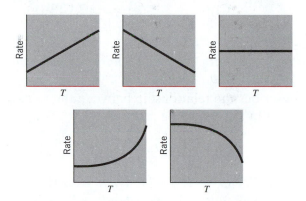

6. Describe what happens to the rate at which a reactant is consumed during the course of a reaction. (Does it increase, decrease, or remain the same?)

7. Explain why mixtures of H_2 and O_2 gas do not react when stored at room temperature for several years, whereas the reaction is complete within a few days at 300°C, within a few hours at 500°C, and almost instantaneously at 700°C.

Rate Laws and Rate Constants

8. What are the units of the rate constant for the following reaction?

$$N_2O_4(g) \rightleftharpoons 2\,NO_2(g)$$
$$\text{Rate} = k(N_2O_4)$$

9. What are the units of the rate constant for the following reaction?

$$2\,NO_2(g) \rightleftharpoons N_2O_4(g)$$
$$\text{Rate} = k(NO_2)^2$$

10. What are the units of the rate constant for the following reaction?

$$2\,Br^-(aq) + H_2O_2(aq) + 2\,H^+(aq) \rightleftharpoons Br_2(aq) + 2\,H_2O(aq)$$
$$\text{Rate} = k(Br^-)(H_2O_2)(H^+)$$

11. A reaction is said to be diffusion controlled when it occurs as fast as the reactants diffuse through the solution. The following is a good example of a diffusion-controlled reaction.

$$H_3O^+(aq) + OH^-(aq) \rightleftharpoons 2\,H_2O(aq)$$

Assume that the rate constant for the reaction is $1.4 \times 10^{11}\ M^{-1}\ s^{-1}$ at 25°C and that the reaction obeys the following rate law. Calculate the rate of reaction in a neutral solution (pH = 7.00).

$$\text{Rate} = k(H_3O^+)(OH^-)$$

Different Ways of Expressing the Rate of Reaction

12. Which equation describes the relationship between the rates at which Cl_2 and F_2 are consumed in the following reaction?

$$Cl_2(g) + 3\ F_2(g) \rightleftharpoons 2\ ClF_3(g)$$

(a) $-d(Cl_2)/dt = -d(F_2)/dt$ (b) $-d(Cl_2)/dt = 2[-d(F_2)/dt]$
(c) $2[-d(Cl_2)/dt] = -d(F_2)/dt$ (d) $-d(Cl_2)/dt = 3[-d(F_2)/dt]$
(e) $3[-d(Cl_2)/dt] = -d(F_2)/dt$

13. Which equation describes the relationship between the rates at which Cl_2 is consumed and ClF_3 is produced in the following reaction?

$$Cl_2(g) + 3\ F_2(g) \rightleftharpoons 2\ ClF_3(g)$$

(a) $-d(Cl_2)/dt = d(ClF_3)/dt$ (b) $-d(Cl_2)/dt = 2[d(ClF_3)/dt]$
(c) $2[-d(Cl_2)/dt] = d(ClF_3)/dt$ (d) $-d(Cl_2)/dt = 3[d(ClF_3)/dt]$
(e) $3[-d(Cl_2)/dt] = d(ClF_3)/dt$

14. Calculate the rate at which N_2O_4 is formed in the following reaction at the moment in time when NO_2 is being consumed at a rate of 0.0592 M/s.

$$2\ NO_2(g) \rightleftharpoons N_2O_4(g)$$

15. Calculate the rate for the formation of I_2 in the following reaction if the rate for the disappearance of HI is 0.039 $M\ s^{-1}$.

$$2\ HI(g) \rightleftharpoons H_2(g) + I_2(g)$$
$$\text{Rate} = k(HI)^2$$

16. Ammonia burns in the gas phase to form nitrogen oxide and water.

$$4\ NH_3(g) + 5\ O_2(g) \longrightarrow 4\ NO(g) + 6\ H_2O(g)$$

Derive the relationship between the rates at which NH_3 and O_2 are consumed in the reaction. Derive the relationship between the rates at which O_2 is consumed and H_2O is produced.

17. Calculate the rate of formation of NO and the rate of disappearance of NH_3 for the following reaction at the moment in time when the rate of formation of water is 0.040 M/s.

$$4\ NH_3(g) + 5\ O_2(g) \longrightarrow 4\ NO(g) + 6\ H_2O(g)$$

18. The instantaneous rate of disappearance of the MnO_4^- ion in the following reaction is 4.56×10^{-3} M/s at some moment in time.

$$10\ I^-(aq) + 2\ MnO_4^-(aq) + 16\ H^+(aq) \longrightarrow 2\ Mn^{2+}(aq) + 5\ I_2(aq) + 8\ H_2O(aq)$$

What is the rate of appearance of I_2 at the same moment?

The Rate Law versus the Stoichiometry of a Reaction

19. Describe the difference between the *stoichiometry* and the *mechanism* of a reaction.

20. Describe the conditions under which the rate law for a reaction is most likely to reflect the stoichiometry of the reaction.

21. Describe one or more factors that can make the rate law for a reaction differ from what the stoichiometry of the reaction leads us to expect.

Order and Molecularity

22. Describe the difference between unimolecular and bimolecular reactions. Give an example of each.

23. Which graph best describes the rate of the following reaction if the reaction is first order in N_2O_4?

$$N_2O_4(g) \rightleftharpoons 2 NO_2(g)$$

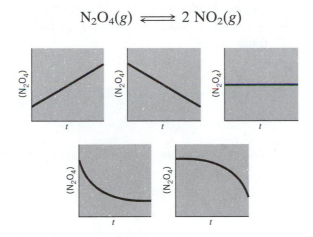

The Mechanisms of Chemical Reactions

24. Each of the following reactions was found experimentally to be second order overall. Which of them is most likely to be an elementary reaction that occurs in a single step?
    (a) $2 NO_2(g) + Cl_2(g) \rightarrow 2 NO_2Cl(g)$
    (b) $N_2O_3(g) \rightarrow NO(g) + NO_2(g)$
    (c) $3 O_2(g) \rightarrow 2 O_3(g)$
    (d) $2 NO(g) \rightarrow N_2(g) + O_2(g)$

25. NO reacts with H_2 according to the following equation.

$$2 NO(g) + 2 H_2(g) \rightleftharpoons N_2(g) + 2 H_2O(g)$$

    The mechanism for the reaction involves two steps.

$$2 NO + H_2 \longrightarrow N_2 + H_2O_2 \qquad \text{(slow step)}$$
$$H_2O_2 + H_2 \longrightarrow 2 H_2O \qquad \text{(fast step)}$$

    What is the experimental rate law for the reaction?

26. The following reaction is first order in both NO_2 and F_2.

$$2 NO_2(g) + F_2(g) \rightleftharpoons 2 NO_2F(g)$$
$$\text{Rate} = k(NO_2)(F_2)$$

    The rate law is consistent with which of the following mechanisms?
    (a) $2 NO_2 + F_2 \rightleftharpoons 2 NO_2F$

(b) $NO_2 + F_2 \rightleftharpoons NO_2F + F$ (fast step)
$\ NO_2 + F \rightarrow NO_2F$ (slow step)
(c) $NO_2 + F_2 \rightarrow NO_2F + F$ (slow step)
$\ NO_2 + F \rightarrow NO_2F$ (fast step)
(d) $F_2 \rightarrow 2F$ (slow step)
$\ 2 NO_2 + 2 F \rightarrow 2 NO_2F$ (fast step)

27. The disproportionation of NO to N_2O and NO_2 is third order in NO.

$$3 NO(g) \rightleftharpoons N_2O(g) + NO_2(g)$$
$$Rate = k(NO)^3$$

The rate law is consistent with which of the following mechanisms?
(a) $NO + NO + NO \rightarrow N_2O + NO_2$ (one-step reaction)
(b) $2 NO \rightarrow N_2O_2$ (slow step)
$\ N_2O_2 + NO \rightarrow N_2O + NO_2$ (fast step)
(c) $2 NO \rightleftharpoons N_2O_2$ (fast step)
$\ N_2O_2 + NO \rightarrow N_2O + NO_2$ (slow step)

28. Is a rate law for the reaction

$$I^-(aq) + OCl^-(aq) \longrightarrow Cl^-(aq) + OI^-(aq)$$

that is first order in I^- and second order in OCl^- consistent with the following mechanism?

$$OCl^- + H_2O \rightleftharpoons HOCl + OH^-   \text{(fast step)}$$
$$I^- + HOCl \longrightarrow HOI + Cl^-   \text{(slow step)}$$
$$HOI + OH^- \longrightarrow OI^- + H_2O   \text{(fast step)}$$

29. N_2O_5 decomposes to form NO_2 and O_2.

$$2 N_2O_5(g) \longrightarrow 4 NO_2(g) + O_2(g)$$

The rate law for the reaction is first order in N_2O_5.

$$Rate = k(N_2O_5)$$

Show how the rate law is consistent with the following three-step mechanism for the reaction.

Step 1 $N_2O_5 \rightleftharpoons NO_2 + NO_3$ (fast step)
Step 2 $NO_2 + NO_3 \rightarrow NO_2 + NO + O_2$ (slow step)
Step 3 $NO + NO_3 \rightarrow 2 NO_2$ (fast step)

30. The following is the experimental rate law for the reaction between hydrogen and bromine to form hydrogen bromide, HBr.

$$Rate = k(H_2)(Br_2)^{1/2}$$

Explain how the following mechanism is consistent with the rate law.

$$Br_2 \rightleftharpoons 2\ Br \quad (fast\ step)$$
$$Br + H_2 \longrightarrow HBr + H \quad (slow\ step)$$
$$H + Br_2 \longrightarrow HBr + Br \quad (fast\ step)$$
$$H + HBr \longrightarrow H_2 + Br_2 \quad (fast\ step)$$

The Relationship between the Rate Constants and the Equilibrium Constant for a Reaction

31. Describe the relationship between the forward and reverse rate constants and the equilibrium constant for a one-step reaction.

32. The following is a one-step reaction.

$$CH_3Cl(aq) + I^-(aq) \rightleftharpoons CH_3I(aq) + Cl^-(aq)$$

What is the equilibrium constant for the reaction if the rate constant for the forward reaction is $5.2 \times 10^{-7}\ M^{-1}\ s^{-1}$ and the rate constant for the reverse reaction is $1.5 \times 10^{-11}\ M^{-1}s^{-1}$?

Determining the Order of a Reaction from Initial Rates of Reaction

33. Use the following data to determine the rate law for the reaction between nitrogen oxide and chlorine to form nitrosyl chloride.

$$2\ NO(g) + Cl_2(g) \longrightarrow 2\ NOCl(g)$$

Initial (NO) (M)	Initial (Cl₂) (M)	Initial Rate of Reaction (M/s)
0.10	0.10	0.117
0.20	0.10	0.468
0.30	0.10	1.054
0.30	0.20	2.107
0.30	0.30	3.161

34. Use the results of the preceding problem to determine the rate constant for the reaction. Predict the initial instantaneous rate of reaction when the initial NO and Cl_2 concentrations are both 0.50 M.

35. Use the following data to determine the rate law for the reaction between nitrogen oxide and oxygen to form nitrogen dioxide. Hint: How are pressures related to concentrations?

$$2\ NO(g) + O_2(g) \rightleftharpoons 2\ NO_2(g)$$

Initial Pressure NO (mmHg)	Initial Pressure O₂ (mmHg)	Initial Rate of Reaction (mmHg/s)
100	100	0.355
150	100	0.800
250	100	2.22
150	130	1.04
150	180	1.44

36. Use the results of the preceding problem to determine the rate constant for the reaction. Predict the initial rate of reaction when the initial pressures for both NO and O_2 are 250 mmHg.

37. Use the following data to determine the rate law for the reaction between methyl iodide and the OH^- ion in aqueous solution to form methanol and the iodide ion.

$$CH_3I(aq) + OH^-(aq) \rightleftharpoons CH_3OH(aq) + I^-(aq)$$

Initial (CH_3I) (M)	Initial (OH^-) (M)	Initial Rate of Reaction (M/s)
1.35	0.10	8.78×10^{-6}
1.00	0.10	6.50×10^{-6}
0.85	0.10	5.53×10^{-6}
0.85	0.15	8.29×10^{-6}
0.85	0.25	1.38×10^{-5}

38. Use the results of the preceding question to determine the rate constant for the reaction. Predict the initial instantaneous rate of reaction when the initial CH_3I concentration is 0.10 M and the initial OH^- concentration is 0.050 M.

The Integrated Forms of the Rate Laws

39. Describe the kinds of problems that are best solved by using the rate law for a reaction, such as the following.

$$Rate = k(N_2O_5)$$

Describe the kinds of problems that are best solved with the integrated form of the rate law.

$$\ln\left[\frac{(N_2O_5)}{(N_2O_5)_0}\right] = -kt$$

40. Water was heated in a test tube to 75.0°C and then allowed to cool to room temperature. If the process followed first-order kinetics with a rate constant of 8.0×10^{-4} s^{-1}, what was the temperature of the water after 400 s?

41. In an acceleration test for a BMW 325es, the following speed-versus-time data were collected after the car shifted into fourth gear.

Speed (mph)	80.4	83.9	87.5	91.2	95.1	99.3
Time (s)	16	18	20	22	24	26

Are the data consistent with first-order kinetics? Predict the time at which the speed should reach 100 mph.

42. The reaction in which NO_2 forms a dimer is second order in NO_2.

$$2\,NO_2(g) \rightleftharpoons N_2O_4(g)$$
$$Rate = k(NO_2)^2$$

Calculate the rate constant for the reaction if it takes 0.0050 second for the initial concentration of NO_2 to decrease from 0.50 M to 0.25 M.

43. The decomposition of hydrogen peroxide is first order in H_2O_2.

$$2 H_2O_2(aq) \longrightarrow 2 H_2O(aq) + O_2(g)$$
$$\text{Rate} = k(H_2O_2)$$

How long will it take for half of the H_2O_2 in a 10-gallon sample to be consumed if the rate constant for the reaction is $5.6 \times 10^{-2} \text{ s}^{-1}$?

The Integrated Forms of the Rate Laws and Half-life Calculations

44. Calculate the rate constant for the following acid–base reaction if the half-life for the reaction is 0.0282 s at 25°C and the reaction is first order in the NH_4^+ ion.

$$NH_4^+(aq) + H_2O(l) \rightleftharpoons NH_3(aq) + H_3O^+(aq)$$

45. The following reaction is second order in NO_2.

$$2 NO_2(g) \rightleftharpoons N_2O_4(g)$$

What effect would doubling the initial concentration of NO_2 have on the half-life for the reaction?

46. The following reaction is first order in both reactants and therefore second order overall.

$$CH_3I(aq) + OH^-(aq) \rightleftharpoons CH_3OH(aq) + I^-(aq)$$

When the reaction is run in a buffer solution, however, it is pseudo-first order in CH_3I.

$$\text{Rate} = k(CH_3I)$$

What is the half-life of the reaction in a pH 10.00 buffer if the rate constant for this pseudo-first-order reaction is $6.5 \times 10^{-9} \text{ s}^{-1}$?

47. The age of a rock can be estimated by measuring the amount of ^{40}Ar trapped inside. The calculation is based on the fact that ^{40}K decays to ^{40}Ar by a first-order process. It also assumes that none of the ^{40}Ar produced by the reaction has escaped from the rock since the rock was formed.

$$^{40}K + e^- \longrightarrow {}^{40}Ar \qquad k = 5.81 \times 10^{-11} \text{ year}^{-1}$$

Calculate the half-life of the radioactive decay.

48. Another way of determining the age of a rock involves measuring the extent to which the ^{87}Rb in the rock has decayed to ^{87}Sr.

$$^{87}Rb \longrightarrow {}^{87}Sr + e^- \qquad k = 1.42 \times 10^{-11} \text{ year}^{-1}$$

What fraction of the ^{87}Rb would still remain in a rock after 1.19×10^{10} years?

49. ^{14}C measurements on the linen wrappings from the Book of Isaiah in the Dead Sea Scrolls suggest that the scrolls contain about 79.5% of the ^{14}C expected in living tissue. How old are the scrolls, if the half-life for the decay of ^{14}C is 5730 years?

50. The Lascaux cave near Montignac in France contains a series of cave paintings. Radiocarbon dating of charcoal taken from the site suggests an age of 15,520 years. What fraction of the ^{14}C present in living tissue is still present in the sample? (For ^{14}C, $t_{1/2}$ = 5730 years.)

51. A skull fragment found in 1936 at Baldwin Hills, California, was dated by ^{14}C analysis. Approximately 100 g of bone was cleaned and treated with 1 M HCl(aq) to destroy the mineral content of the bone. The bone protein was then collected, dried, and pyrolyzed. The CO_2 produced was collected and purified, and the ratio of ^{14}C to ^{12}C was measured. If the sample contained roughly 5.7% of the ^{14}C present in living tissue, how old was the skeleton? (For ^{14}C, $t_{1/2}$ = 5730 years.)

52. Charcoal samples from Stonehenge in England emit 62.3% of the disintegrations per gram of carbon per minute expected for living tissue. What is the age of the samples? (For ^{14}C, $t_{1/2}$ = 5730 years.)

53. A lump of beeswax was excavated in England near a collection of Bronze Age objects roughly 2500 to 3000 years of age. Radiocarbon analysis of the beeswax suggests an activity equal to roughly 90.3% of the activity observed for living tissue. Determine whether the beeswax was part of the hoard of Bronze Age objects. (For ^{14}C, $t_{1/2}$ = 5730 years.)

54. The activity of the ^{14}C in living tissue is 15.3 disintegrations per minute per gram of carbon. The limit for reliable determination of ^{14}C ages is 0.10 disintegration per minute per gram of carbon. Calculate the maximum age of a sample that can be dated accurately by radiocarbon dating. Assume the half-life of ^{14}C is 5730 years.

Determining the Order of a Reaction with the Integrated Forms of Rate Laws

55. Use the following data to determine the rate law for the decomposition of N_2O.

$$2 N_2O(g) \longrightarrow 2 N_2(g) + O_2(g)$$

(N_2O) (M)	0.100	0.086	0.079	0.075	0.066	0.059	0.049
Time (s)	0	80	120	160	240	320	480

56. Use the results of the preceding problem to calculate the rate constant for the reaction. Predict the concentration of N_2O after 900 s.

57. Use the following data to determine the rate law for the hydrolysis of the BH_4^- ion.

$$BH_4^-(aq) + 4 H_2O(l) \longrightarrow B(OH)_4^-(aq) + 4 H_2(g)$$

(BH_4^-) (M)	0.100	0.088	0.077	0.068	0.060	0.052	0.046
Time (h)	0	24	48	72	96	120	144

58. Use the results of the preceding question to calculate the half-life for the reaction.

59. Triphenylphosphine, PPh_3, reacts with nickel tetracarbonyl, $Ni(CO)_4$, to displace a molecule of carbon monoxide.

$$Ni(CO)_4 + PPh_3 \rightleftharpoons Ph_3PNi(CO)_3 + CO$$

The following data were obtained when the reaction was run at 25°C in the presence of a large excess of triphenylphosphine.

($Ni(CO)_4$) (M)	10.0	7.6	5.8	4.4	3.3	2.5
Time (s)	0	40	80	120	160	200

Use the data to determine whether the reaction is first order or second order in $Ni(CO)_4$.

60. The rate of the reaction in the preceding problem does not depend on the concentration of PPh_3. Combine this fact with the results of the preceding problem to determine whether the rate law for the reaction is consistent with the following mechanism.

$$Ni(CO)_4 \longrightarrow Ni(CO)_3 + CO \quad \text{(slow step)}$$
$$Ni(CO)_3 + PPh_3 \rightleftharpoons Ph_3PNi(CO)_3 \quad \text{(fast step)}$$

61. $Cr(NH_3)_5Cl^{2+}$ reacts with the OH^- ion in aqueous solution to displace Cl^- from the complex ion.

$$Cr(NH_3)_5Cl^{2+}(aq) + OH^-(aq) \rightleftharpoons Cr(NH_3)_5(OH)^{2+}(aq) + Cl^-(aq)$$

The following data were obtained when the reaction was run at 25°C in a buffer solution at constant pH.

$(Cr(NH_3)_5Cl^{2+})$ (M)	1.00	0.81	0.66	0.53	0.43	0.35
Time (min)	0	3	6	9	12	15

Use the data to determine whether the reaction is first order or second order in $Cr(NH_3)_5Cl^{2+}$.

62. The rate of the reaction in the preceding problem is proportional to the pH of the buffer solution in which the reaction is run. Each time the buffer is changed so that the OH^- ion concentration is doubled, the rate of reaction increases by a factor of 2. Combine this observation with the results of the preceding problem to determine the rate law for the reaction.

63. Show that the rate law derived in the preceding problem is consistent with the following mechanism

$$Cr(NH_3)_5Cl^{2+} + OH^- \longrightarrow Cr(NH_3)_4(NH_2)(Cl)^+ + H_2O \quad \text{(slow step)}$$
$$Cr(NH_3)_4(NH_2)(Cl)^+ \rightleftharpoons Cr(NH_3)_4(NH_2)^{2+} + Cl^- \quad \text{(fast step)}$$
$$Cr(NH_3)_4(NH_2)^{2+} + H_2O \rightleftharpoons Cr(NH_3)_5(OH)^{2+} \quad \text{(fast step)}$$

64. Dimethyl ether, CH_3OCH_3, decomposes at high temperatures as shown in the following equation.

$$CH_3OCH_3(g) \longrightarrow CH_4(g) + H_2(g) + CO(g)$$

The following data were obtained when the partial pressure of CH_3OCH_3 was studied as the compound decomposed at 500°C. Use the data to determine the order of the reaction. Hint: How is pressure related to concentration?

$P_{CH_3OCH_3}$ (mmHg)	312	278	251	227	157
Time (s)	0	390	777	1195	3155

Reactions That Are First Order in Two Reactants

65. The following reaction is first order in both CH_3I and OH^-.

$$CH_3I(aq) + OH^-(aq) \rightleftharpoons CH_3OH(aq) + I^-(aq)$$

Describe how to turn the reaction into one that is pseudo-first order in CH_3I.

Catalysts and the Rate of Chemical Reactions

66. Describe the four properties of a catalyst. Give an example of a catalyzed reaction and show how the catalyst meets the criteria.

A Collision Theory of Chemical Reactions

67. Describe the factors that determine whether a collision between two molecules will lead to a reaction.
68. Describe the relationship between the rate of a reaction and the activation energy for the reaction.

Determining the Activation Energy of a Reaction

69. Assume that the activation energy was measured for both the forward ($E_a = 120$ kJ/mol$_{rxn}$) and reverse ($E_a = 185$ kJ/mol$_{rxn}$) directions of a reversible reaction. What would be the activation energy for the reverse reaction in the presence of a catalyst that decreased the activation energy for the forward reaction to 90 kJ/mol$_{rxn}$?
70. The rate constant for the decomposition of N_2O_5 increases from 1.52×10^{-5} s^{-1} at 25°C to 3.83×10^{-3} s^{-1} at 45°C. Calculate the activation energy for the reaction.
71. Calculate the activation energy for the following reaction, if the rate constant for the reaction increases from 87.1 M^{-1} s^{-1} at 500 K to 1.53×10^3 M^{-1} s^{-1} at 650 K.

$$2\ NO_2(g) \rightleftharpoons 2\ NO(g) + O_2(g)$$

72. Calculate the activation energy for the decomposition of NO_2 from the temperature dependence of the rate constant for the reaction.

$$2\ NO_2(g) \longrightarrow N_2(g) + 2\ O_2(g)$$

Temperature (K)	319	329	352	381	389
k (M^{-1} s^{-1})	0.522	0.755	1.70	4.02	5.03

73. Calculate the rate constant at 780 K for the following reaction, if the rate constant for the reaction is 3.5×10^{-7} M^{-1} s^{-1} at 550 K and the activation energy is 183 kJ/mol$_{rxn}$.

$$2\ HI(g) \rightleftharpoons H_2(g) + I_2(g)$$

74. Calculate the rate constant at 75°C for the following reaction, if the rate constant for the reaction is 6.5×10^{-5} M^{-1} s^{-1} at 25°C and the activation energy is 92.9 kJ/mol$_{rxn}$.

$$CH_3I(aq) + OH^-(aq) \rightleftharpoons CH_3OH(aq) + I^-(aq)$$

The Kinetics of Enzyme-Catalyzed Reactions

75. Explain why the rate of the enzyme-catalyzed hydrolysis of sucrose is first order in sucrose at low concentrations of the substance.
76. Explain why the rate of enzyme-catalyzed reactions becomes zero order at very high concentrations of the substrate.

Integrated Problems

77. A 50-mL Mohr buret is connected to a 1-ft length of capillary tubing. The buret is then filled with water, and the volume of water versus time is monitored as the water gradually flows through the capillary tubing. Use the following data to determine whether the rate at which water flows through the capillary tubing fits first-order or second-order kinetics. Determine the rate constant for the process. (See buret in problem 79.)

V (mL)	50	40	30	20	10	0
t (s)	0	19	42	72	116	203

78. Predict the effect of doubling the length of the capillary tubing on the rate constant for the process described in the previous problem. Compare your prediction with the following experimental results. (See buret in problem 79.)

V (mL)	50	40	30	20	10	0
t (s)	0	43	97	171	282	523

79. A piece of glass tubing threaded through a one-hole rubber stopper was inserted in the mouth of the Mohr buret. A source of compressed air was then connected to the glass tubing. When a constant pressure was applied to the top of the buret, the following data were obtained.

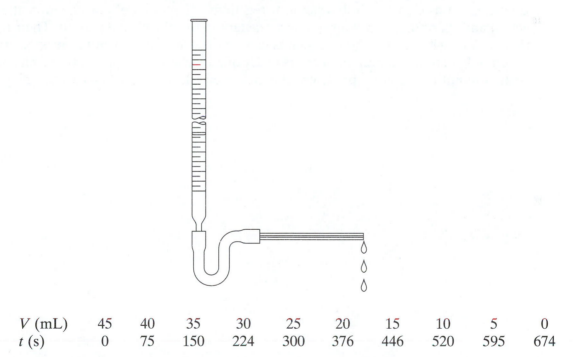

V (mL)	45	40	35	30	25	20	15	10	5	0
t (s)	0	75	150	224	300	376	446	520	595	674

Determine whether the data fit zero-order, first-order, or second-order kinetics.

80. For the reaction

$$2\ N_2O_5(g) \rightleftharpoons 4\ NO_2(g) + O_2(g)$$

the activation energy for the forward reaction is 200 kJ/mol$_{rxn}$. Using enthalpy of atom combination data from Appendix B, Table B.14, answer the following.
(a) What is the activation energy for the reverse reaction?
(b) If a catalyst is added and the activation energy for the forward reaction is reduced to 150 kJ/mol$_{rxn}$, what will be the activation energy for the reverse reaction?

81. From the reaction coordinate diagrams below, and assuming constant temperature and Z for all the diagrams, select the diagram for the conversion of reactants to products that has the following:
    (a) the smallest rate constant for an endothermic reaction
    (b) the largest rate constant for an exothermic reaction
    (c) the largest rate constant for a reverse reaction
    (d) the most rapid establishment of equilibrium
    Of diagrams II and IV, which is most likely to have an equilibrium constant greater than 1? Can you decide which of the two reactions proceeds the most rapidly?

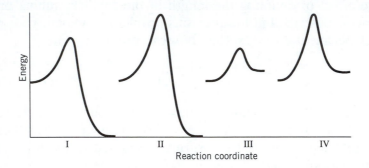

82. Enzymes act as catalysts in biochemical reactions. They are not consumed in the reaction and therefore do not appear as a reactant in the chemical equation. Their function is to provide a site where reactants can be brought together in the proper orientation to lead to a reaction. Predict the order of a biochemical reaction to which a very small amount of the appropriate enzyme has been added. Explain your reasoning.

C H A P T E R
14
SPECIAL TOPICS

14A.1   Deriving the Integrated Rate Laws

14A.1   DERIVING THE INTEGRATED RATE LAWS

To derive the integrated form of the first-order rate law, we start with the equation that describes the rate law for the reaction.

$$-\frac{d(X)}{dt} = k(X)$$

We then rearrange the equation as follows.

$$\frac{1}{(X)}d(X) = -k\,dt$$

Our goal is to integrate both sides of the equation. Mathematically, this is equivalent to finding the area under the curve that would be produced if the function were graphed. This process is indicated with integral signs, as follows.

$$\int \frac{1}{(X)}\,d(X) = \int -k\,dt$$

We are interested in the area under the curve between the time when the reaction starts ($t = 0$) and some later time (t).

$$\int_{X_0}^{X}\frac{1}{(X)}\,d(X) = \int_0^t -k\,dt$$

The integral of $(X)^{-1}\,d(X)$ is equal to the natural logarithm of (X). The integrated form of the first-order rate law can therefore be written as follows.

Integrated Form of the First-Order Rate Law

$$\ln\left[\frac{(X)}{(X)_0}\right] = -kt$$

When using this equation, remember that (X) is the concentration of the reactant at any moment in time, $(X)_0$ is the initial concentration of the reactant, k is the rate constant for the reaction, and t is the time since the reaction started.

The derivation of the integrated form of the second-order rate law also starts with the equation that defines the rate law of the reaction.

$$-\frac{d(X)}{dt} = k(X)^2$$

We start by rearranging the equation as follows.

$$-\frac{1}{(X)^2} \, d(X) = k \, dt$$

We then integrate both sides of the equation.

$$\int_{X_0}^{X} -\frac{1}{(X)^2} \, d(X) = \int_{0}^{t} k \, dt$$

The integral of $(X)^{-2} \, d(X)$ is $-(X)^{-1}$. The integrated form of the second-order rate law is therefore written as follows.

Integrated Form of the Second-Order Rate Law

$$\frac{1}{(X)} - \frac{1}{(X)_0} = kt$$

Once again, the (X) term is the concentration of X at any moment in time, $(X)_0$ is the initial concentration of X, k is the rate constant for the reaction, and t is the time since the reaction started.

CHAPTER
15
CHEMICAL ANALYSIS

How do chemists know what is in the food you eat or the water you drink? How do chemists determine the structure of molecules? How can chemistry be used to solve crimes? This chapter introduces the types of questions that chemists can answer, the strategies that chemists use to solve problems, and the tools that they use to do it. The focus of this chapter is on modern instrumental techniques. Case studies of real-world problems are used as examples of the wide variety of problems that chemists can solve. There are many types of instruments and methods of analysis that chemists use in the laboratory. We begin by looking at several ways in which similar instruments and methods can be categorized.

15.1   METHODS OF ANALYSIS

One way in which methods of chemical analysis can be categorized is by the type of information that they provide. **Quantitative** methods of analysis tell how much of the **analyte** (the specific chemical substance to be analyzed for) is present in a sample. **Qualitative** methods of analysis tell what chemical substances are present in a sample. **Structural analysis** determines the chemical structure of the analyte.

Methods of chemical analysis are also categorized as **wet** or **instrumental** methods of analysis. Wet chemical methods are based on chemical reactions. These techniques are divided into **volumetric, gravimetric,** and **qualitative** analysis methods. Volumetric methods are quantitative methods that measure the volume of a solution, such as in a titration. Gravimetric methods of analysis are quantitative methods based on a measurement of mass; they involve the precipitation of the analyte followed by drying and weighing of the precipitate. Qualitative analysis is based on specific observable reactions (color change, precipitant formation, evolution of gas) that occur when a chemical reagent is added to a sample that contains the analyte. Wet methods of analysis were the primary laboratory techniques available to chemists until the mid-twentieth century. Most traditional introductory chemistry laboratory experiments employ these techniques.

Instrumental methods of analysis involve the use of instruments for a measurement and are often based on physical rather than chemical properties of the analyte. Instrumental methods have largely replaced wet methods of analysis in the modern laboratory because they are faster, less labor intensive, and often more sensitive than wet chemistry methods. Wet chemistry methods of analysis, however, are still used to prepare samples for analysis.

15.2   SEPARATION OF MIXTURES

Most of the substances that we encounter on a daily basis are not pure substances but are mixtures consisting of several or many chemical compounds, as noted in Chapter 1. It is often necessary for chemists to separate a mixture into its pure components before the components can be identified or measured.

The most widely used method for separating components of a mixture is **chromatography.** Chromatography includes a wide variety of separation techniques. Some of these require no instrumentation, whereas others involve sophisticated computer-controlled instruments. In common to all types of chromatography is the use of a **stationary phase** and a **mobile phase.** The stationary phase is a solid or liquid coated on a solid that remains stationary, as its name implies. The mobile phase is a gas or liquid that moves either over or through the stationary phase. A sample mixture to be separated is carried through the stationary phase by the flowing mobile phase. Components of the mixture will interact with both the stationary phase and the mobile phase. Components that interact strongly with the mobile phase move rapidly with the flow of the mobile phase. Components which interact strongly with the stationary phase are retarded in their flow. This provides the mechanism for the separation of the mixture into its components. The principal difference between one type of chromatography and another is the makeup of the stationary and mobile phases.

15.3   THE OLIVE OIL CAPER[1]

Most cooks consider olive oil to be the finest vegetable oil in terms of its taste and aroma. Nutritionally it is considered superior to other vegetable oils because it is believed to decrease cholesterol levels in the blood, therefore reducing the risk of heart disease. As a result, olive oil is generally more expensive than other vegetable cooking oils. To increase profits, a few companies have been suspected of mixing less expensive corn, peanut, or soybean oils with their olive oil. In the late 1980s Richard Flor at the U.S. Customs Services laboratory in Washington, D.C., received a sample taken from a recently imported shipment of olive oil. The sample looked like olive oil but had an off-taste. However, taste is subjective and not admissible as evidence in court. With the assistance of his colleague Le Tiet Hecking, Flor developed a technique to determine whether the olive oil sample had been adulterated. Any difference in taste between oils must be based on differences in the chemical composition of the oil.

The Problem: How to Determine whether the Olive Oil Is Pure?

Oils are a complex mixture of triglycerides, compounds formed from glycerol and three fatty acids. Glycerol is an organic molecule consisting of a three-carbon chain with a hy-

---

[1]Robin Meadows, Making the grade, *Chem Matters*, **7,** 10–11 (December 1989). Excerpted with permission. Copyright © 1989 American Chemical Society.

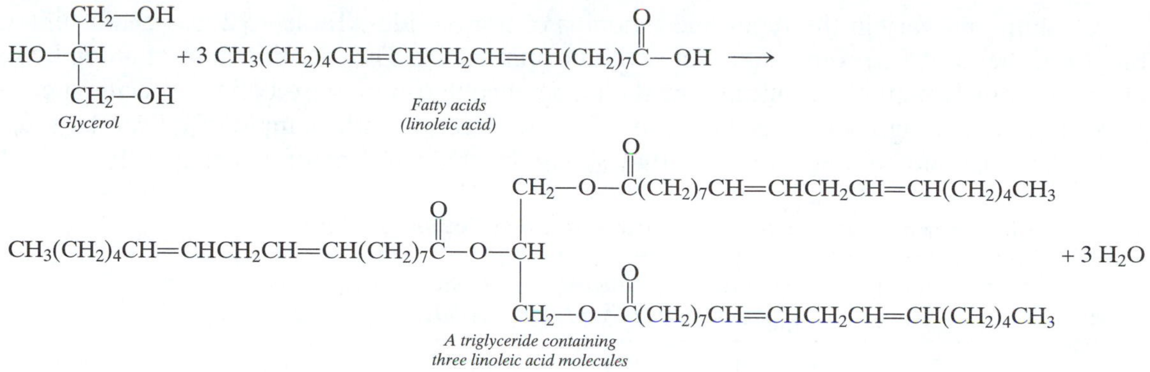

$$\begin{array}{l} CH_2-OH \\ HO-CH \\ CH_2-OH \end{array} \quad + 3\ CH_3(CH_2)_4CH{=}CHCH_2CH{=}CH(CH_2)_7\overset{\displaystyle O}{\overset{\|}{C}}-OH \longrightarrow$$

Glycerol   Fatty acids
                (linoleic acid)

$$CH_3(CH_2)_4CH{=}CHCH_2CH{=}CH(CH_2)_7\overset{O}{\overset{\|}{C}}-O-CH \quad\begin{array}{l} CH_2-O-\overset{O}{\overset{\|}{C}}(CH_2)_7CH{=}CHCH_2CH{=}CH(CH_2)_4CH_3 \\[2ex] \\[2ex] CH_2-O-\overset{O}{\overset{\|}{C}}(CH_2)_7CH{=}CHCH_2CH{=}CH(CH_2)_4CH_3 \end{array}\quad + 3\ H_2O$$

A triglyceride containing
three linoleic acid molecules

droxy functional group (OH) attached to each carbon. Fatty acids consist of long carbon chains with a carboxylic acid functional group (COOH), as shown above. Fatty acids differ from one another based on the number of carbons and the number and location of double bonds within the carbon chain. Table 15.1 lists common fatty acids found in triglycerides.

TABLE 15.1   Common Fatty Acids

Acid	Number of Carbons	Number of C–C Double Bonds	Formula
Myristic	14	0	$CH_3(CH_2)_{12}COOH$
Palmitic	16	0	$CH_3(CH_2)_{14}COOH$
Stearic	18	0	$CH_3(CH_2)_{16}COOH$
Oleic	18	1	$CH_3(CH_2)_7CH{=}CH(CH_2)_7COOH$
Linoleic	18	2	$CH_3(CH_2)_4CH{=}CHCH_2CH{=}CH(CH_2)_7COOH$

Triglycerides differ from one another based on the types of fatty acids in the molecule and the location of the fatty acids on the glycerol. They can be represented by a shorthand notation indicating the three fatty acids used to form them. The first triglyceride shown below was formed from two linoleic acids (L) and one oleic acid (O) and is designated LLO. The second triglyceride below was formed from the fatty acids linoleic (L), oleic (O), and palmitic (P) and is designated LOP.

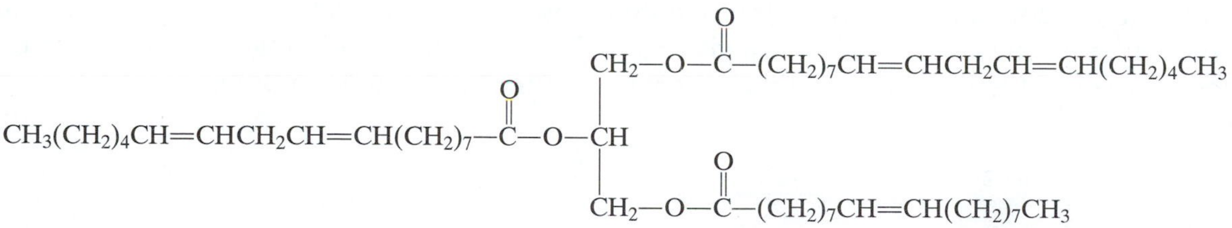

$$CH_3(CH_2)_4CH{=}CHCH_2CH{=}CH(CH_2)_7-\overset{O}{\overset{\|}{C}}-O-CH\quad\begin{array}{l} CH_2-O-\overset{O}{\overset{\|}{C}}-(CH_2)_7CH{=}CHCH_2CH{=}CH(CH_2)_4CH_3 \\[2ex] \\[2ex] CH_2-O-\overset{O}{\overset{\|}{C}}-(CH_2)_7CH{=}CH(CH_2)_7CH_3 \end{array}$$

LLO

$$CH_3(CH_2)_7CH{=}CH(CH_2)_7-\overset{O}{\overset{\|}{C}}-O-CH\quad\begin{array}{l} CH_2-O-\overset{O}{\overset{\|}{C}}-(CH_2)_7CH{=}CHCH_2CH{=}CH(CH_2)_4CH_3 \\[2ex] \\[2ex] CH_2-O-\overset{O}{\overset{\|}{C}}-(CH_2)_{14}CH_3 \end{array}$$

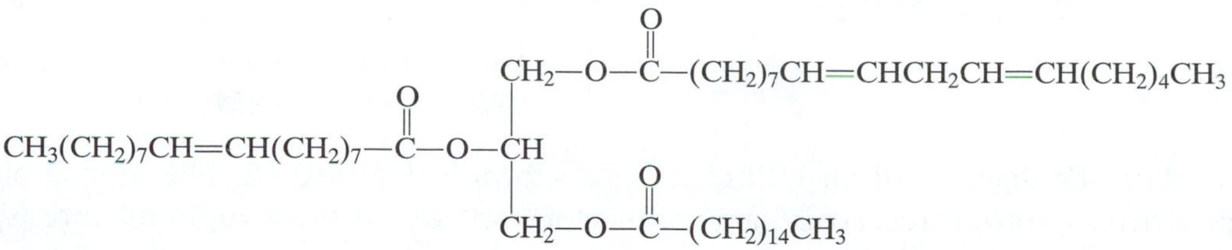

LOP

Cooking oils vary in the types and amounts of triglycerides. Table 15.2 lists characteristic ranges of the fatty acid composition of several common oils. The identification of an oil by its fatty acid composition is complicated by the range of composition based on the specific species of plant and growing conditions. However, if the triglycerides in a sample of oil can be separated from one another and identified they should be characteristic of a particular type of oil.

TABLE 15.2   Percent Fatty Acid Composition of Some Common Oils[a]

Oil	Myristic Acid	Palmitic Acid	Stearic Acid	Oleic Acid	Linoleic Acid
Olive	0–1	5–15	1–4	67–84	8–12
Peanut	—	7–12	2–6	30–60	20–38
Corn	1–2	1–11	3–4	25–35	50–60
Cottonseed	0–2	6–10	2–4	20–30	50–58
Soybean	1–2	6–10	2–4	20–30	50–58

[a]Adapted from John R. Holum, *Fundamentals of General, Organic, and Biological Chemistry,* 5th ed. John Wiley & Sons, New York, 1978, p. 570.

The Method of Analysis: High-Performance Liquid Chromatography

High-performance liquid chromatography (HPLC) can be used for the separation of a wide array of complex mixtures. The separation takes place in a chromatography **column** packed with an unreactive, finely ground solid coated with a liquid stationary phase. The liquid mobile phase is forced through the column under high pressure. As the mobile phase carries a mixture through the column, components in the mixture interact with the stationary phase. The interaction can be based on solubility, electrostatic attraction (attraction between positive and negative electric charges), adsorption (attractive forces at the interface between two immiscible phases), or other mechanisms. The interactions typically involve attractions resulting from intermolecular forces. Different components in the mixture will interact with the stationary phase to varying degrees, resulting in some components spending more time associated with the stationary phase than other components. The interaction retards the progress of the components through the column. Components of the mixture with relatively weak interactions with the stationary phase are moved through the column quite rapidly by the mobile phase. Components that interact strongly with the stationary phase remain on the column much longer. This results in the separation of a mixture into its components, with the separated components exiting the column at different times as shown in Figure 15.1. The time that it takes a particular component to move through the column is known as the **retention time** and is a characteristic property of that particular chemical substance for a given set of experimental conditions. Retention times can therefore be used for a qualitative analysis of the components of a mixture.

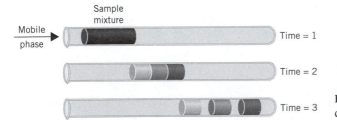

FIGURE 15.1 Separation of a mixture on a chromatography column.

A schematic diagram of an HPLC setup is shown in Figure 15.2. The mobile phase is pumped from a solvent reservoir past an injector port, where the sample mixture is intro-

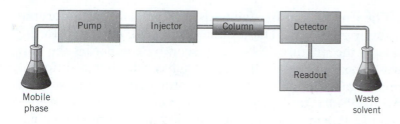

FIGURE 15.2 Diagram of a high-performance liquid chromatograph.

duced. High pressures are required to push the mobile phase and sample mixture through the column. As the mobile phase exits the column, it passes through a detector capable of measuring the presence of the components of the mixture. The detector produces an electrical response which is observed on a readout device. The mobile phase solvent and mixture components are collected for recycling or disposal. The retention time of a component must be compared to the retention time of pure known solutions for a qualitative identification of the component. The response of the detector (area under a peak in a chromatogram) is proportional to the amount of the component present. Samples of known composition and concentration, called standards, must be run for each component for a quantitative analysis.

The Solution: Analysis of Olive Oil

Flor and Hecking analyzed samples from a shipment of the suspect olive oil using HPLC. Chromatograms of a sample of good olive oil and olive oil from the suspect shipment are shown in Figure 15-3. The horizontal axis is a plot of time, and the vertical axis represents the response of the detector (in this case absorbance of electromagnetic radiation). Each separated peak represents a component of the mixture. Peaks corresponding to triglycerides

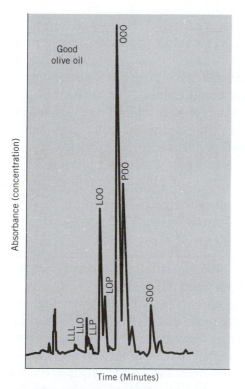
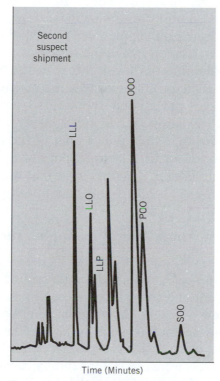

FIGURE 15.3 Chromatogram of olive oil sample on the left and sample suspected of adulteration on the right. Each peak is labeled to identify the triglyceride. Reprinted with permission from Robin Meadows *Chem Matters,* **7,** No. 4, 11, Copyright © 1989 American Chemical Society.

in the mixture have been identified using known standards and are labeled according to the fatty acids used to form the triglycerides. The area under a peak is related to the concentration of that triglyceride in the mixture. As can be seen by comparing the two chromatograms, the two mixtures appear to contain the same triglycerides but in different concentrations. From the chromatograms, Flor and Hecking estimated that the suspect olive oil had been adulterated with 25% to 30% corn oil, which has higher concentrations of the triglycerides LLL, LLO, and LLP than olive oil. They verified this by preparing mixtures of corn oil and olive oil in the lab. They found that a mixture of 28% corn oil and 72% olive oil produced a chromatogram essentially identical to the chromatogram of the suspect olive oil. These were results that could be presented as evidence in court.

Checkpoint
Using Figure 15.3 describe the differences in the relative amounts of LLL and OOO in the good olive oil and in the suspect sample.

15.4   THE GREAT APPLE SCARE OF '89

On February 26, 1989, the CBS investigative news program "60 Minutes" reported accusations from a Natural Resources Defense Council (NRDC) report entitled "Intolerable Risk: Pesticides in Our Children's Food." The focus of the report was on Alar (daminozide), a growth regulator used since the 1960s primarily on apples to keep the fruit on the tree longer, resulting in apples that were better shaped, redder, and firmer and had a longer shelf life. The NRDC, an activist group, charged that Alar posed an intolerable cancer risk particularly to children who ate apples or apple products. The NRDC further attacked the Environmental Protection Agency (EPA) for not having banned the use of Alar. Public reaction to the NRDC report and the media coverage resulted in a 31% decrease in apple juice sales, a 25% decrease in applesauce sales, and the removal of apples and apple products from school lunch cafeterias in the spring of 1989.

Alar was first identified as a risk factor by the EPA in 1985. Apple processors began testing for Alar in their products using a method that could detect Alar at the 1–2 ppm level (1 ppm, or 1 part per million, represents 1 part analyte in a million parts of sample and is equivalent to 1 mg of analyte in a 1 kg sample). The relative size of 1 ppm can be placed in perspective by noting that 1 ppm of the distance between New York City and Los Angeles is about 15 feet. The 1–2 ppm level was well below the 30 ppm maximum level allowed by the federal government at that time. After the 1985 EPA report the use of Alar by apple growers began a substantial decline.

The Problem: How Much Alar Is in Apple Products?

On May 14, 1989, "60 Minutes" expanded its report on Alar, describing analytical tests that had detected Alar in apple products down to the 0.02 ppm level. The next day Dr. C. Robert Binkley, vice president of technical services at Knouse Foods in Biglerville, Pennsylvania, met with his staff to view a tape of the program. Binkley announced that Knouse Foods would upgrade their testing facilities to match the level of detection described by "60 Minutes." Work was immediately begun to design two new laboratory facilities and an analytical services laboratory building. Orders were placed for new analytical instrumentation, and Maurine Rickard, head of analytical services, was sent to observe a laboratory in California where the new procedure was being used. The new procedure not only allowed the determination of Alar at lower detection limits but also minimized interferences from other chemical substances. By July 1989 the analytical services building was complete, new gas chromatograph/mass spectrometers were installed, and Rickard was testing all apple products at the new levels of detection. Knouse Foods was able to state that their products contained no detectable Alar.

Alar, daminozide, is an organic acid and is not known to be carcinogenic. However, hydrolysis of Alar produces unsymmetrical dimethylhydrazine (UDMH) (see Figure 15.4), which has been shown to cause cancer in animals in some laboratory tests. Alar is known to penetrate apple skins and therefore cannot be washed off. Heat treatment during apple processing could cause as much as 1% of any Alar present to decompose into unsymmetrical dimethylhydrazine. In addition, it has been estimated that 1% to 5% of any Alar present might be converted to the carcinogenic by-product under physiological conditions in the body.

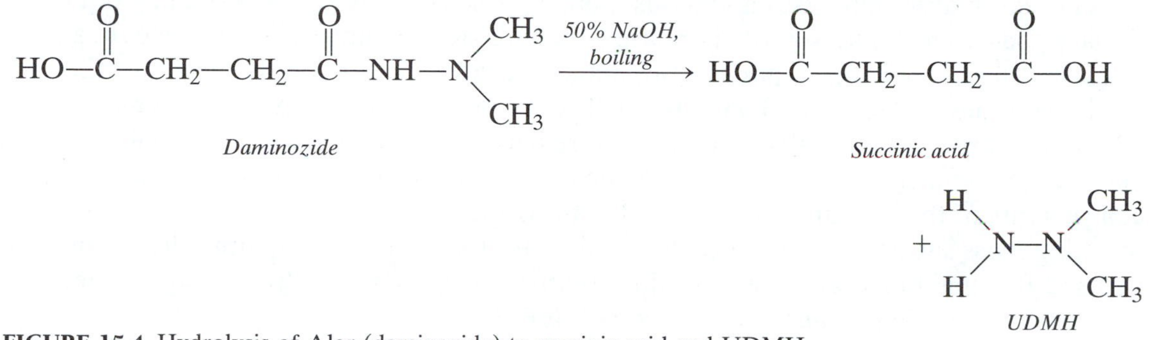

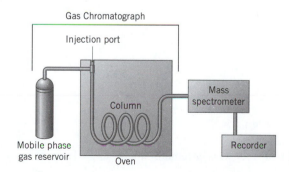

FIGURE 15.4 Hydrolysis of Alar (daminozide) to succinic acid and UDMH.

The Method of Analysis: Gas Chromatography/Mass Spectrometry

Tandem techniques are the combination of two or more types of analytical methods. The capabilities of the resulting instrument is often greater than the simple sum of the abilities of the two individual techniques. One of the most powerful and most widely used tandem techniques is **gas chromatography/mass spectrometry (GC/MS).** A gas chromatograph (GC), like the HPLC previously described, is a chromatographic instrument used to separate a mixture into its components (Figure 15.5). A sample mixture is injected into the GC and separated into its component compounds in a column. As each compound exits the GC column it moves into the mass spectrometer, which serves as a detector. Not only does the mass spectrometer produce a signal response to each component leaving the column, but it can also produce a spectrum that will allow for the qualitative identification of the component.

FIGURE 15.5 Diagram of a gas chromatograph.

The major limitation of GC is that the sample to be analyzed must be volatile and thermally stable so that it can be carried along by the gaseous mobile phase. Because Alar is not thermally stable, samples suspected of containing Alar must be derivatized (converted to another compound) before analysis. Alar is hydrolyzed to UDMH by the reaction shown in Figure 15.4. The UDMH is then reacted with salicylaldehyde to produce the stable product shown in Figure 15.6.

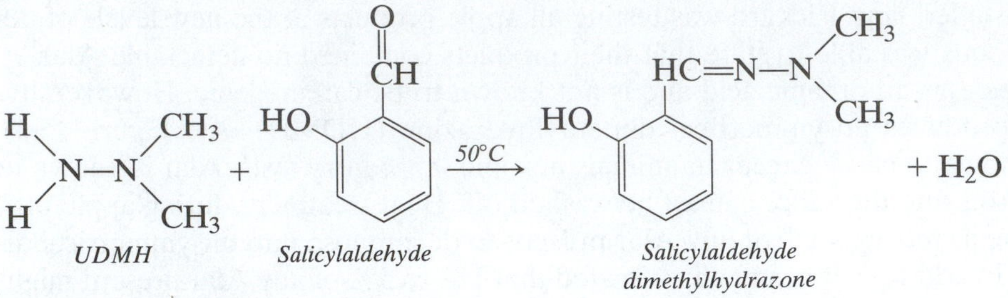

FIGURE 15.6 Derivatization of UDMH.

Gas chromatography uses a gaseous mobile phase (such as He or N_2) and a liquid stationary phase. The liquid stationary phase either is coated on unreactive particles in a metal or glass coiled tube called a **packed column** or is coated directly on the inside surface of a small-bore tube called a **capillary column.** The column is placed in an oven to maintain an adequate temperature for all components to remain in the vapor state. As the mobile phase exits the column it passes through a detector capable of measuring the presence of the components of the mixture. In this case the mass spectrometer serves as the detector. The resulting mass spectrum and the retention time of the components exiting the column can be used for qualitative analysis, and the response of the detector to each component can be used for quantitative analysis just as with HPLC.

Figure 15.7 shows a chromatogram of an apple juice sample. As with most biological samples the sample mixture contains many components, and thus the resulting chromatogram is quite complex with many peaks. When there are many peaks with similar retention times an exact qualitative identification is difficult. Coupling the GC with a mass spectrometer simplifies the analysis.

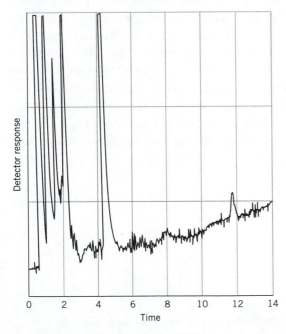

FIGURE 15.7 Gas chromatogram of apple juice.

The use of a mass spectrometer to determine the atomic masses of elements was described in Section 2.2. When coupled with a GC the mass spectrometer can be used to identify the chemical component responsible for each peak in the gas chromatogram. Bombardment of a component molecule by high energy electrons produces a **parent ion** by dislodging an electron from the molecule. The parent ion is often unstable and will tend to fragment into smaller ions. The fragmentation pattern is unique and characteristic of a given compound and therefore useful for qualitative analysis. The weakest bonds in a compound will tend to rupture most easily to form the fragments. Since a compound has a constant structure, the same bonds

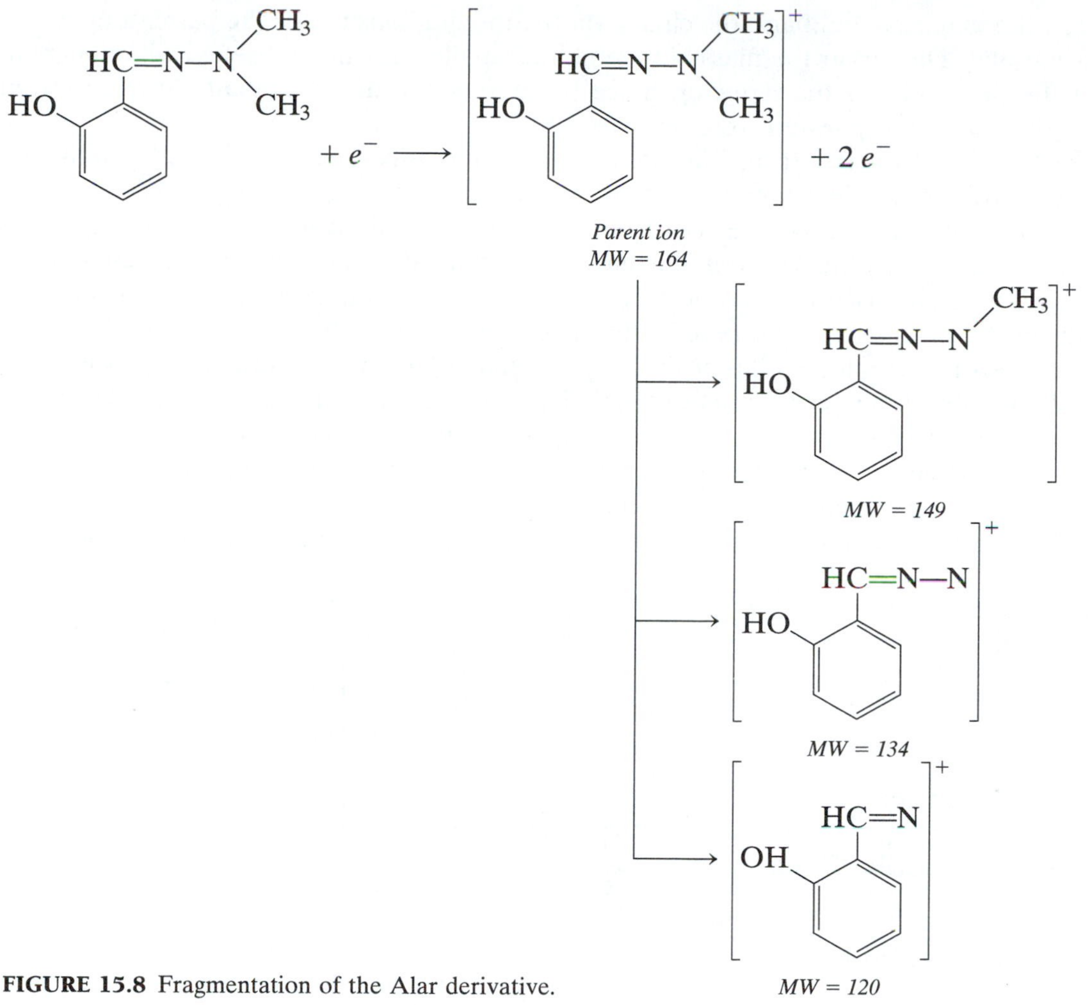

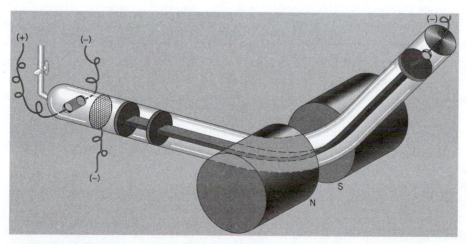

FIGURE 15.8 Fragmentation of the Alar derivative.

will always rupture, yielding the same fragmentation pattern. The formation of the parent ion and selected fragment ions of the Alar derivative are shown in Figure 15.8. The resulting mixture of parent ion and fragment ions can be separated according to their *mass to charge* ratio. Because most of the ions have a +1 charge the ions are in effect separated by mass.

The parent and fragment ions are accelerated through an evacuated tube. The method for separating the ions varies among different types of mass spectrometers. The mass spectrometer shown in Figure 15.9 uses a magnetic field to separate the ions. The interaction

FIGURE 15.9 Diagram of a mass spectrometer.

between the magnetic field and the charge on the moving ions bends the path along which the ions travel. The larger the mass of the ion, the smaller the angle through which its path is bent before it enters the detector. Since the parent ion and fragments have different masses they will be separated from one another.

The resulting mass spectrum is a plot of relative intensity (relative number of ions of a given type striking the detector) on the vertical axis versus mass to charge ratio on the horizontal axis. Figure 15.10 shows the mass spectrum for the Alar derivative. Several useful pieces of information can be obtained from the mass spectrum to assist in identifying the compound. Mass spectra can be used to determine the mass of a parent ion from the peak or group of peaks with the highest assigned mass in the spectrum (to the right on the x axis). In Figure 15.10 the mass represented by this peak is approximately 164 g/mol, which corresponds to the molecular weight of the Alar derivative. The selected fragmentation ions shown in Figure 15.8 have molecular weights that correspond to the masses assigned to peaks in the mass spectrum of Figure 15.10. Since the fragmentation pattern does not vary for constant experimental conditions, the mass spectrum of a suspected GC component can be compared to the mass spectrum of a standard, a known compound, in order to verify the identification of the component.

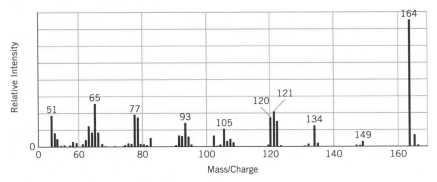

FIGURE 15.10 Mass spectrum of the Alar derivative.

Checkpoint

Suggest a structure for the fragment ion that would produce the peak shown at a mass value of 93 in Figure 15.10.

The Solution: Risk Assessment, the Government, the Media, and the Public

In response to the "60 Minutes" reports, the International Apple Institute took out full page ads in newspapers across the country describing the health benefits of apples. The NRDC launched a media campaign with actress Meryl Streep as their spokesperson calling for tighter pesticide regulations. The Environmental Protection Agency (EPA), Food and Drug Administration (FDA), and U.S. Department of Agriculture (USDA) issued a joint statement that apples were safe and posed no imminent health hazard to children. The EPA also charged that the study on which the NRDC based its report was misleading and based on poor data.

The Great Apple Scare of '89 became less concerned with Alar in apples than with the interrelationship between science, public opinion, and politics in the regulation of chemicals in foods and the environment. In June 1989 Uniroyal Chemical Company, the producer of Alar, announced that it would voluntarily halt the sale of Alar in the United States for food crop use. In addition it would recall and issue refunds on all Alar meant for food use. A significant factor in the controversy was the public's skepticism that industry and government agencies could be trusted to regulate the level of suspected carcinogens in the environment to protect the public. This distrust was fueled by the inexact process of risk assessment.

Risk assessment and the setting of acceptable limits of potentially hazardous chemicals in the environment is based on many assumptions. The NRDC risk estimate was 25 times greater than the EPA estimate. Each group accused the other of using inaccurate, out-of-date, or misinterpreted data. Two important questions arise: In their assessment of the risk of public exposure to Alar, why did the EPA and other agencies discount the study on which the NRDC report was based? Second, why is it so difficult to measure the risk stemming from exposure to pesticides and pesticide residues?

The key to the first question is the difference between advances that have been made in measurements of exposure to potentially hazardous substances (detection) and measurements of their biological effect (risk). Advances in technology have enabled us to detect substances at levels as low as a few picograms (10^{-12} g). Unfortunately, the bioassays used to determine whether the substances are hazardous are insensitive and subject to large experimental errors. As a result, studies of the effect of low dosages consistent with normal exposure to the substances can't be done. High doses are therefore used, and the data are then extrapolated to low dosage situations.

As a rule, toxicologists try to limit the dose in bioassays to the maximum tolerated dose (MTD) that causes no more than a 10% weight loss and few early deaths within the test animal population. In the study used by the NRDC the top dose of UDMH was 29 mg/kg of body weight per day. Because the dose used in the study was so high that many animals died early, the EPA concluded that it was not valid to estimate the risk of exposure to UDMH in humans. They noted that the average citizen's exposure to UDMH is only 0.000047 mg/kg and that other studies showed no significant increase in tumors at exposure levels up to 3 mg/kg.

Government regulation of additives and contaminants in food is further complicated by the Delaney clause enacted in 1958 which states that no additive found to induce cancer in humans or animals shall be deemed as safe. Prior to 1985 apple producers used less sensitive methods to assure that their products did not contain Alar above the accepted tolerance level. In 1989 more sophisticated techniques replaced those methods to allow the detection of smaller concentrations of Alar. As chemical analysis techniques become more sensitive and as limits of detection drop below the part per billion (ppb) level, the establishment of regulations for acceptable limits of suspected carcinogens will become more important.

A major theme for research in the twenty-first century will be to improve our ability to assess the risks of exposure to potentially hazardous substances. In their book *In Search of Safety: Chemicals and Cancer Risk* (Harvard University Press, 1988), John Graham, Laura Green, and Marc Roberts wrote, "as scientists discover more of the causal mechanisms underlying the health effects of chemicals . . . they will be able to make quantitative risk assessments with the reliability that the legal/political system pretends is possible today."

Exercise 15.1

Select the chromatographic method (GC or HPLC) that would be best for separating each of the following substances from a mixture.
(a) pentane, C_5H_{12}
(b) vitamin E

Vitamin E (α-tocopherol)

Solution

(a) Pentane is a relatively low molecular weight, nonpolar molecule that is easily volatilized and thus easily separated from a mixture by gas chromatography.

(b) Vitamin E has a high molecular weight and consequently would not be easily volatilized. However, it would be expected to be soluble in a nonpolar solvent and could therefore be separated from a mixture using high-performance liquid chromatography.

Exercise 15.2

A mixture of octane (C_8H_{18}), acetic acid ($CH_3\overset{\overset{\displaystyle O}{\|}}{C}OH$), and methyl chloride ($CH_3Cl$) is separated by high-performance liquid chromatography using a nonpolar stationary phase and a polar aqueous buffer as the mobile phase. Which component will come off the column first and which will come off last?

Solution

Since the mobile phase is a polar aqueous solution and the stationary phase is nonpolar, the primary intermolecular forces are hydrogen bonding in the mobile phase and dispersion forces in the stationary phase. Acetic acid will move through the column most rapidly with the mobile phase because it will be the most soluble in water as a result of its polarity and hydrogen bonding. Octane will be retained on the column longest since it is the only nonpolar component and will therefore interact with the stationary phase the most.

15.5   FIGHTING CRIME WITH CHEMISTRY

In 1987 results of forensic DNA tests of semen taken from a rape victim were used as part of the evidence presented in court to convict Tommi Lee Andrews of rape, sexual battery, aggravated battery, and armed burglary. This represented one of the first cases in the United States in which DNA tests were used to help obtain a conviction. In 1993 Kirk Bloodsworth was released from the Maryland House of Corrections after his 1985 conviction of rape and murder was overturned based on DNA evidence. Maryland state attorneys stated that Kirkwood would never even have been charged had DNA testing been available in 1985. DNA testing has become an important tool for forensic chemists in solving crimes.

The Problem: How Can DNA Be Used to Link a Suspect to a Crime Scene?

Deoxyribonucleic acid (DNA) is the most fundamental molecule of all living organisms. DNA molecules are found in the cells of all living organisms and are responsible for the storage of genetic information. This genetic information is what causes an organism's offspring to have characteristics common to its parents.

DNA is a polymer made up of repeating units known as nucleotides that consist of a phosphate, a sugar, and a nitrogen base. The sugar found in DNA is deoxyribose (Figure 15.11). There are four nitrogen bases found in DNA: adenine (A), guanine (G), cytosine (C), and thymine (T) (Figure 15.12). The phosphate and sugar form the backbone of the DNA polymer strand, with a nitrogen base branching from each sugar (Figure 15.13). It is the sequence of nucleotides along the DNA strand that determines the genetic code of the DNA. A human DNA strand contains several billion nucleotides. The segment of a DNA molecule that contains a sequence of nucleotides which codes for a specific trait is called a **gene.**

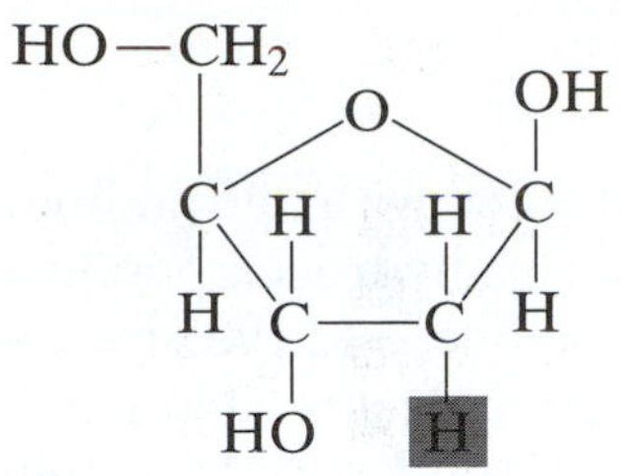

FIGURE 15.11 β-D-Deoxy-ribofuranose, the sugar on which DNA is built.

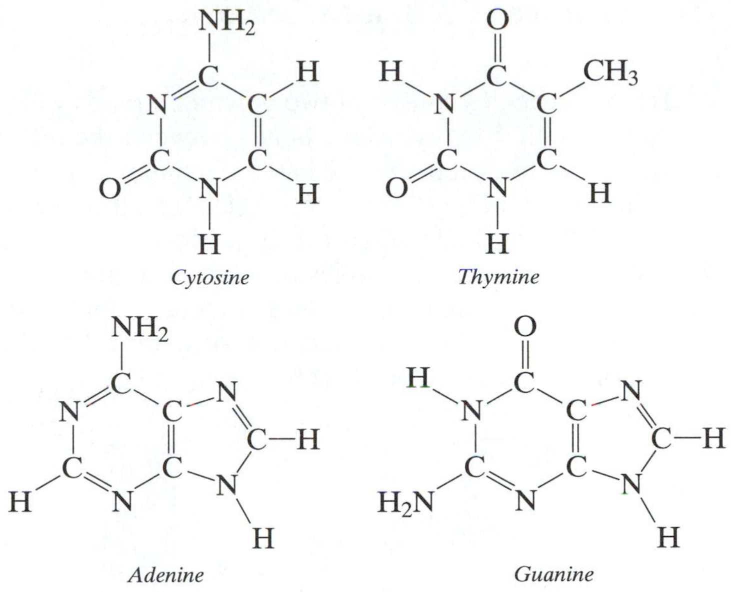

Cytosine   *Thymine*

Adenine   *Guanine*

FIGURE 15.12 The nitrogen bases found in DNA.

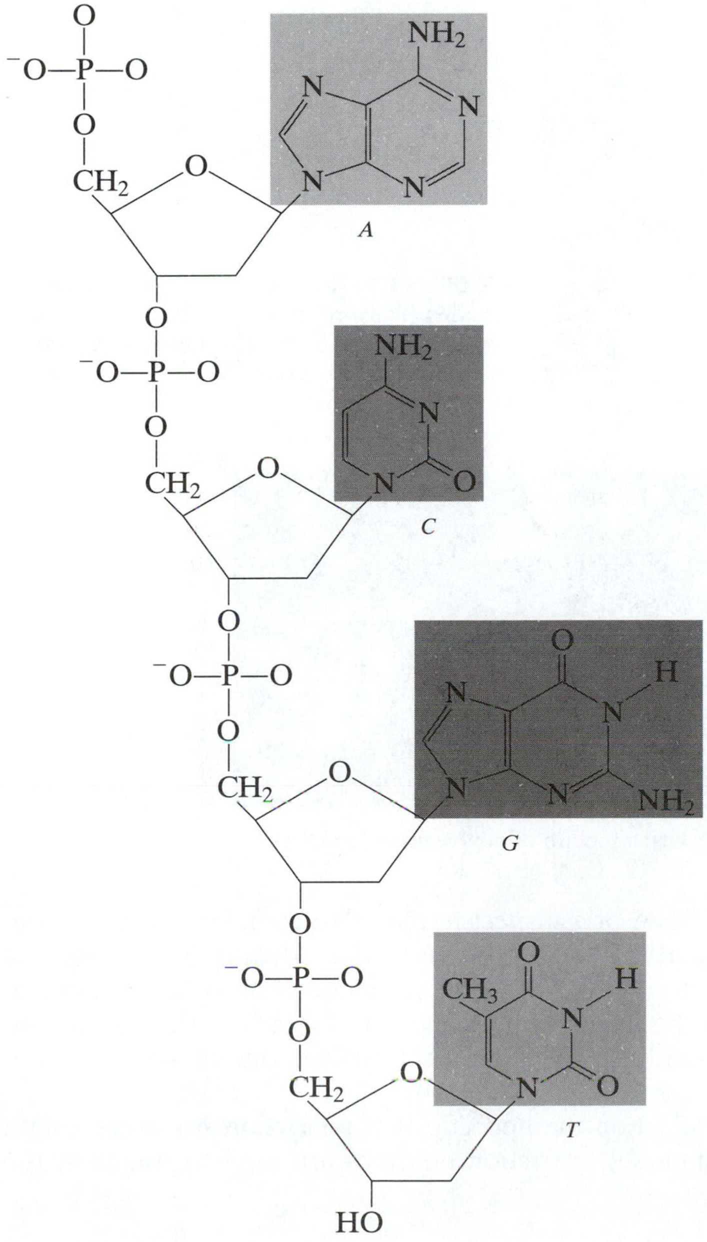

FIGURE 15.13 This tetranucleotide provides an example of a short segment of a single strand of DNA.

DNA molecules consist of two polymer strands which are intertwined with one another and held together by hydrogen bonds between the nitrogen bases. This hydrogen bonding always occurs between specific bases. Thymine always links with adenine, and cytosine always links to guanine (Figure 15.14). The resulting structure of the double-stranded DNA is referred to as a helix (Figure 15.15). Most of the sequences of nucleotides found within the members of a single species of plant or animal are identical. But there are always areas for each individual where the sequence differs, which is what makes each member of a species unique. With the exception of identical siblings, every human has a unique sequence of nucleotides and therefore unique DNA.

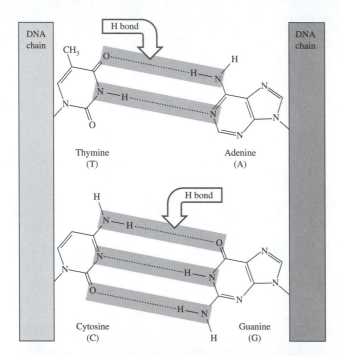

FIGURE 15.14 Two strands of DNA are held together by hydrogen bonds between specific pairs of nitrogen bases. This figure shows the hydrogen bonds between A and T and between G and C.

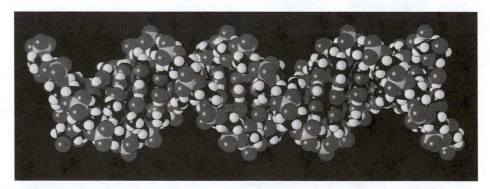

FIGURE 15.15 A computer model of the helical structure of double-stranded DNA.

Forensic scientists compare the DNA of a suspect to the DNA found in semen, blood, hair roots, or other body cells found at the scene of a crime. An individual's DNA can be used much like a fingerprint. Since trying to match the sequence of all of the billions of nucleotides in DNA samples would be an overwhelming task, forensic scientists analyze only those portions of the DNA strand that vary significantly from one individual to another.

The most powerful DNA identification method is called **restriction fragment length polymorphism (RFLP).** In this technique, restriction enzymes are used to fragment the

DNA at the location of a specific sequence of nucleotides. Since the sequence would occur at different places on the DNA of different individuals, the resulting fragments differ in size from one individual to another. The fragments (Figure 15.16a) are then separated according to molecular weight and electric charge by **electrophoresis.** A short strand of synthetic DNA called a probe, with a known sequence of nitrogen bases and containing radioactive isotopes, is added to the separated fragments. The probe will bind to DNA fragments that have the complementary series of nitrogen bases (A to T and G to C) as shown in Figure 15.16b. The purpose of the probe is to provide a means of detecting the complementary sections of DNA. The radioactive isotopes in the probe undergo radioactive decay, which leaves a pattern on photographic film that corresponds to the pattern of the separated DNA fragments that are attached to probes. The steps of RFLP analysis are shown in Figure 15.17.

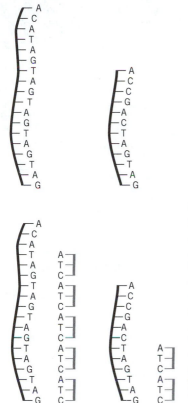

FIGURE 15.16 (a) The two DNA fragments shown can be separated from one another by electrophoresis. If they have the same electric charge the smaller fragment on the right will move faster through an electrophoresis gel than the larger fragment on the left, allowing them to be separated from one another. (b) Fragments of separated DNA which contain a specific sequence of bases (such as TAG shown in the diagram) can be detected using a radioactive probe (shown in a lighter shade). The probe molecules contain a specific sequence of bases (in this example, ATC) which allows them to bind to complementary base sequences (TAG) on the DNA fragments. This illustration has been simplified by showing repeating sequences of only three bases. Actual DNA fragments have sequences of 200 to 300 bases; probe molecules contain 20 to 100 bases. Reprinted with permission from Richard Saferstein *Chem Matters,* **9,** No. 3. Copyright © 1991 American Chemical Society.

The pattern of an electrophoresis gel used in a Minnesota murder investigation is shown in Figure 15.18. The pattern on the left labeled D corresponds to DNA from the suspect's blood. The patterns labeled "jeans" and "shirt" correspond to DNA from bloodstains found on the suspect's jeans and shirt. Pattern V shows results from the DNA of the victim's blood. The remaining two patterns are standards of known composition. The gel shows a good match between the victim's blood and the bloodstains on the suspect's clothes. The suspect later confessed to the murder.

Checkpoint

Describe the sequence of bases on a probe molecule designed to bind to a strand of DNA containing the following sequence of bases: TTCG.

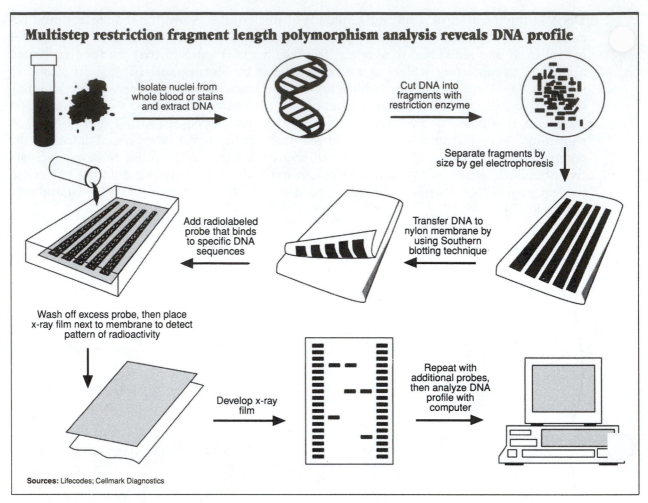

Multistep restriction fragment length polymorphism analysis reveals DNA profile

Isolate nuclei from whole blood or stains and extract DNA

Cut DNA into fragments with restriction enzyme

Separate fragments by size by gel electrophoresis

Add radiolabeled probe that binds to specific DNA sequences

Transfer DNA to nylon membrane by using Southern blotting technique

Wash off excess probe, then place x-ray film next to membrane to detect pattern of radioactivity

Develop x-ray film

Repeat with additional probes, then analyze DNA profile with computer

Sources: Lifecodes; Cellmark Diagnostics

FIGURE 15.17 Steps in restriction fragment length polymorphism analysis. Reprint with permission from P. Zurer, DNA profiling fast becoming an accepted tool for identification?, *C & E News,* **72,** 11 (1994). Copyright 1994 American Chemical Society.

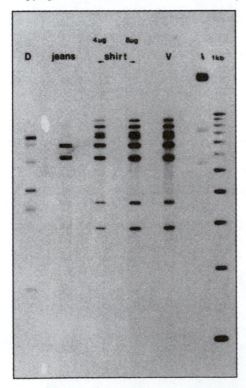

FIGURE 15.18 DNA fingerprints from a Minnesota murder case. This single gel was used to compare the DNA of blood taken from the suspect's jeans and shirt with blood samples taken from the suspect, D, and the victim, V. The sample from the jeans and the two shirt samples match the sample from the victim. From R. Saferstein, DNA Fingerprinting, *Chem Matters,* **9,** 12 (1991).

The Method of Analysis: Electrophoresis

Electrophoresis is a widely used technique in biochemistry and molecular biology for the separation of components in biological samples. The components to be separated must have an electric charge in aqueous solution. Amino acids, proteins, and DNA fragments are examples of components that can be separated by the technique. The sample is placed on a solid substrate such as paper, cellulose acetate, or polyacrylamide gel. An electric field is then applied across the solid substrate by means of a positive and negative electrode located at opposite ends of the substrate. Components with positive charges are attracted by and move toward the negative electrode. In the same manner negatively charged components move toward the positive electrode. The rate of migration of the components along or through the substrate is dependent on the magnitude of the charge and the size of the sample components. Thus, the components in the mixture can be separated. By using standards (samples of known composition) simultaneously with the mixture, the components in the mixture can be identified by comparing the distances of migration of the components and standards.

The Solution: Identification beyond a Reasonable Doubt

If all of the over 3 billion nucleotides in DNA were compared in a DNA test there could be no doubt of the match between two samples and therefore the identification of the individual from which a sample came. However, only a small portion of the DNA strand is actually used in the test. Even though the sections used are those that show the greatest variation among individuals, there is some probability of an incorrect match. The probability of a mismatch is the subject of considerable controversy, particularly in court cases. Some experts say a mismatch could occur only once in 10 million, whereas others place the probability at once in several thousand. Currently a great deal of research is being done to establish reliable databases for the calculation of good probability data and for the standardization of DNA testing techniques in forensic laboratories.

15.6   INTERACTION OF ELECTROMAGNETIC RADIATION WITH MATTER: SPECTROSCOPY

Spectroscopy is a method of analysis based on the interaction of electromagnetic radiation with matter. Two common interactions between electromagnetic radiation and matter are **reflection** and **refraction.** When visible light strikes a mirror it is reflected. When visible light passes through a prism it is refracted (bent) and broken up into its component wavelengths (colors). The different interactions of electromagnetic radiation with matter allow chemists to measure many properties of matter. This chapter is primarily concerned with a third type of interaction, namely, **absorption** of electromagnetic radiation. Chapter 3 described how electromagnetic radiation is absorbed by an atom in order for an electron to move from one energy level to a higher energy level. The energy of the radiation absorbed corresponds to the change in energy of the electron. Section 3.3 noted that the energy of the photons (E) of the electromagnetic radiation that produce this transition is related to the wavelength (λ) and frequency (v) by the following equation:

$$E = hv = hc/\lambda$$

Checkpoint

Which of the following electron configurations could correspond to an atom of zinc that has absorbed electromagnetic radiation?

$$\text{Zn} \quad [\text{Ar}] \, 4s^2 \, 3d^{10} \qquad \text{Zn} \quad [\text{Ar}] \, 4s^2 \, 3d^9 \, 4p^1 \qquad \text{Zn} \quad [\text{Ar}] \, 4s^2 \, 3d^9 \, 5s^1$$

Electromagnetic radiation passing through a sample can be described on a macroscopic scale by Figure 15.19, in which the intensity of electromagnetic radiation exiting the sample (I_T) is less than the initial intensity of electromagnetic radiation (I_0) striking it. Only the electromagnetic radiation with an energy corresponding to the difference in energy between two energy levels in the sample molecules is absorbed (I_A).

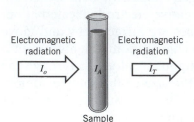

Electromagnetic radiation I_o I_A Electromagnetic radiation I_T

Sample

FIGURE 15.19 Absorbance of light by a sample.

The relationship between these three variables is given by

$$I_0 = I_T + I_A$$

The transmittance, T, and percent transmittance, $\%T$, are defined as follows:

$$T = I_T/I_0 \qquad \%T = (I_T/I_0) \times 100$$

The absorbed radiation is known as absorbance, A, and is defined as

$$A = \log(I_0/I_T)$$

Since (I_0/I_T) is equal to $1/T$, absorbance and transmittance are related by

$$A = \log(1/T) \qquad \text{and} \qquad A = \log(100/\%T)$$

The second equation is often algebraically rearranged to give a convenient method for converting between absorbance and percent transmittance.

$$A = \log 100 - \log \%T$$
$$A = 2 - \log \%T$$

Figure 15.20 shows a simplified diagram of an absorption spectrometer. The source emits electromagnetic radiation, which is dispersed into its component wavelengths by a dispersing element. A selected wavelength of electromagnetic radiation is aimed at the sample. The amount of radiation passing through the sample is measured by the detector. An electrical response from the detector that is proportional to the amount of light striking it is recorded on an output device.

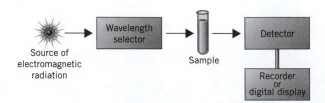

Source of electromagnetic radiation

Wavelength selector

Sample

Detector

Recorder or digital display

FIGURE 15.20 Diagram of a spectrometer.

Spectrophotometric methods of analysis are typically classified according to the region of the electromagnetic spectrum employed (see Figure 3.3). Like other methods of analysis, spectrophotometric methods can be utilized for quantitative, qualitative, and structural analyses. These techniques may also be classified as **elemental** or **molecular** analyses. In elemental analysis the sample is analyzed without regard for the molecule or compound in which the element is found. This is contrasted by molecular analysis which involves the analysis of a sample for the molecules present. It is important when choosing a spectroscopic technique to select the technique that gives the type of information that is needed.

15.7   THE FOX RIVER MYSTERY[2]

Fish kills, in which large numbers of fish die suddenly, occur occasionally in many bodies of water. They are often due to natural phenomena such as the stress of spawning or a lack of dissolved oxygen. However, in 1986 the number and size of fish kills on a section of the Fox River in Oshkosh, Wisconsin, became so serious that the Wisconsin Department of Natural Resources (DNR) began to monitor the dissolved oxygen in the water. In the spring of 1988 the largest kill occurred, over 30,000 fish. Measurements of the level of dissolved oxygen were found to be normal. Joe Ball, a water resource specialist with the DNR began to search for the cause of the kills.

The Problem: What Was Causing the Fish Kills?

Typical causes of large fish kills are pesticides, herbicides, and toxic materials discharged by industries. Ball began a systematic investigation looking at each facility located along the section of the river where the kills occurred. The water near a golf course was checked for herbicides and pesticides, discharge water from a sewage treatment plant was checked for chlorine levels, and water near local industries was checked for toxic organic compounds and heavy metals. But each analysis gave a negative result. Ball could not find the source of the pollution causing the fish kills. A testing facility for an outboard motor company was investigated next. The facility tested several powerful prototype motors on boats secured in position. No fuel leaks were found. The scientific literature was examined to determine the typical compounds found in the exhaust from motors. These included unburned volatile organics, carbon dioxide, water, and carbon monoxide. The level of volatile organics in the water at the testing facility was found to be within acceptable limits. The carbon dioxide and water from the exhaust do not harm fish. Carbon monoxide, although a lethal gas, is not very soluble in water and disperses rapidly into the air. However, Ball decided to investigate the carbon monoxide more thoroughly.

Carbon monoxide can bind to the hemoglobin in blood. Hemoglobin carries oxygen from the lungs to tissue in the body and then carries carbon dioxide back to the lungs to be exhaled. Carbon monoxide binds so tightly to hemoglobin that the hemoglobin is no longer capable of binding to oxygen and therefore results in suffocation. Figure 15.21 shows the heme group of a hemoglobin molecule. The heme group on the left is bound to oxygen, and the heme group on the right is bound to carbon monoxide.

Since carbon monoxide is difficult to detect in water, Ball decided to examine the blood of some of the dead fish and compare that to the blood of healthy fish. Tom Ecker, a chemist at the Wisconsin State Laboratory of Hygiene, was asked to do the analysis. He analyzed the blood of the fish using visible spectroscopy.

[2]Harvey Black, Fox River fish kill, *Chem Matters,* **8,** 6–9 (October 1990). Excerpted with permission. Copyright © 1990 American Chemical Society.

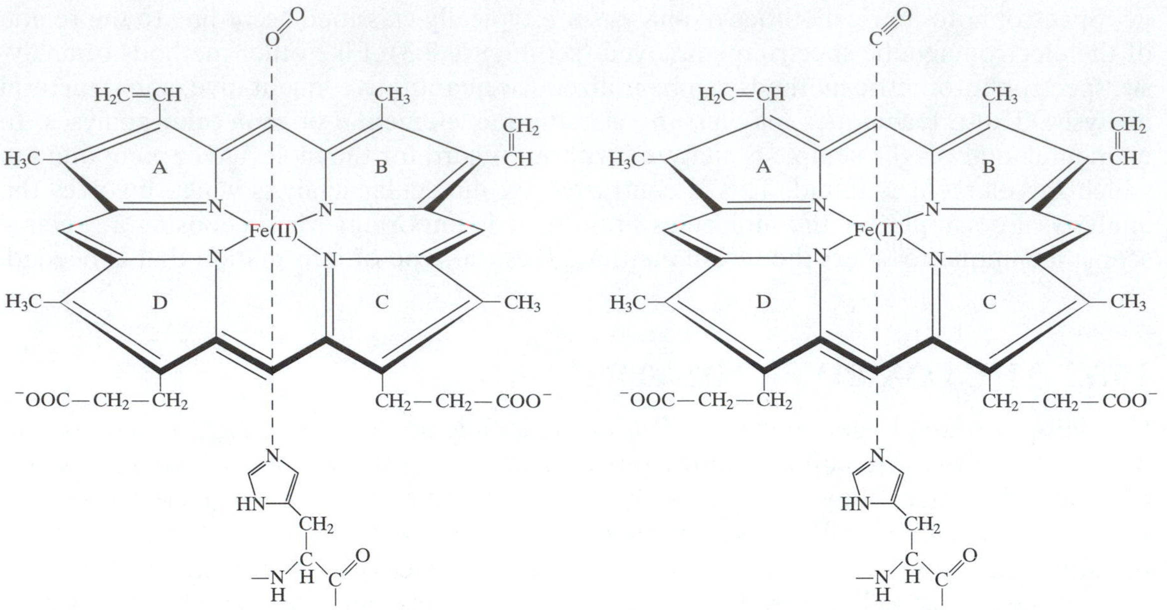

FIGURE 15.21 Heme group of a hemoglobin molecule. The iron atom in the heme group on the left is bound to an oxygen molecule. The iron atom in the right-hand heme is bound to carbon monoxide.

The Method of Analysis: Ultraviolet/Visible Spectroscopy

Ultraviolet/visible (UV/Vis) spectroscopy is an absorption method used for the analysis of molecules and ions. An absorption spectrum is obtained by measuring the absorbance of a sample at different wavelengths. Figure 15.22 shows absorbance spectra for blood samples from both dead and healthy fish. At wavelengths between 500 and 580 nm a large portion of the incident radiation is absorbed, and at wavelengths longer than 620 nm very little is absorbed. The peaks in the spectrum represent wavelengths of electromagnetic radiation that correspond in energy to possible electron transitions in the sample molecule and are therefore absorbed. Other wavelengths are transmitted by the sample. A normal hemoglobin molecule has a unique set of possible electron transitions, resulting in absorbance of specific wavelengths. The binding of carbon monoxide to hemoglobin causes small changes in the structure of hemoglobin and therefore results in slight changes in its electron transitions and its spectrum. As can be seen in the spectra in Figure 15.22, the peaks for hemoglobin to which carbon monoxide is bound are shifted slightly from the peaks of normal hemoglobin. The spectra obtained from the blood of the dead fish showed that some of the hemoglobin in the blood samples was bound to carbon monoxide.

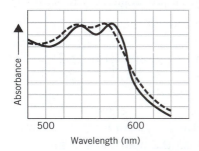

FIGURE 15.22 UV/visible spectra of hemoglobin in two blood samples. The solid line is the spectrum for hemoglobin bound to oxygen. The dashed line is the spectrum for hemoglobin bound to carbon monoxide. Reprinted with permission from Harvey Black *Chem Matters*, **8**, No. 3, 9, Copyright © 1990 American Chemical Society.

UV/Vis spectroscopy is used primarily for quantitative analysis to determine the concentration of an analyte in a sample. The relationship between concentration and absorbance is described by Beer's law,

$$A = \epsilon bc$$

where A is absorbance, ϵ is molar absorptivity, b is path length, and c is concentration. The equation shows the three variables that can influence absorbance.

The path length refers to the distance that electromagnetic radiation must pass through the sample. If the path length is increased the radiation will encounter more sample molecules as it passes through and therefore a greater quantity of radiation will be absorbed. Path length is usually measured in units of centimeters.

Absorbance measurements can be used for the determination of the concentration of a sample because of the direct relationship between absorbance and concentration described by Beer's law. As the concentration of a sample increases, radiation passing through the sample will encounter more molecules of the absorbing substance, causing more radiation to be absorbed. Concentration is measured in units of molarity. Absorbance is therefore directly proportional to both path length (b) and concentration (c). Beer's law is written as an equality using a proportionality constant, molar absorptivity (ϵ). Molar absorptivity is dependent on the substance being measured and the wavelength of electromagnetic energy being measured. The dependence of hemoglobin's molar absorptivity on wavelength can be seen in the spectrum of hemoglobin shown in Figure 15.22 in which the absorbance is seen to change with wavelength for a sample with a constant hemoglobin concentration and a constant pathlength. Molar absorptivity is measured in units of molarity^{-1} cm^{-1} (M^{-1} cm^{-1}).

Absorbance data are typically used for quantitative measurements by constructing a standard curve. The absorbances of several standard solutions containing known analyte concentrations are measured at a fixed wavelength. Figure 15.23 shows a plot of absorbance versus the concentration **(standard Beer's law plot)** for the solutions of hemoglobin. The straight-line plot indicates that Beer's law is obeyed. A straight line has a constant slope equal to $\Delta y/\Delta x$. For a Beer's law plot the slope is equal to $\Delta A/\Delta c$. Because the measurements are done at a fixed wavelength and a constant path length, ϵb is a constant. Therefore, slope $= \Delta A/\Delta c = \epsilon b$. The absorbance of an unknown sample can now be measured and the concentration of the unknown determined by using the standard plot.

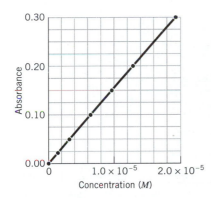

FIGURE 15.23 Beer's law standard curve of absorbance versus concentration of a series of hemoglobin standard solutions measured at 576 nm.

Checkpoint

The UV/Vis spectrum of hemoglobin bound to oxygen shown as the solid line in Figure 15.22 was obtained using a single sample of hemoglobin with a constant concentration and was measured in a sample container with a constant path length. The change in absorbance shown in Figure 15.22 must therefore be due to changes in hemoglobin's molar absorptivity from one wavelength to another. At which of the following wavelengths does hemoglobin bound to oxygen have the largest molar absorptivity?

500 nm   540 nm   590 nm

The Solution: The Culprit Is Identified

The results of Ecker's analysis showed that 60% to 70% of the hemoglobin molecules in the dead fish were bound to carbon monoxide. Additional tests verified the results. As a final test Ball placed a cage of fish in the river near the outboard motor testing site. A second cage of fish was placed at a control site. After just 8 hours the fish at the test site were dead or dying whereas the fish at the control site were healthy. The blood of the dead fish was analyzed and found to have 50% to 80% of its hemoglobin bound to carbon monoxide. This confirmed carbon monoxide from the test facility as the cause of the fish kills. The testing facility installed a piping system to exhaust the motors out through a stack rather than into the water. In addition the company paid the state of Wisconsin $40,000 for the fish kill and an additional $20,000 to fund further studies of the river.

Exercise 15.3

The absorbance of a hemoglobin solution is measured in a UV/Vis spectrometer using a cuvette with a 1.00 cm path length. The absorbance at 576 nm is found to be 0.175. Use the standard curve in Figure 15.23 to determine both the concentration of the hemoglobin solution and the molar absorptivity of hemoglobin.

Solution

The absorbance of 0.175 is located on the y axis of the spectrum in Figure 15.23. A horizontal line is drawn from this absorbance until it intersects with the Beer's law plot (see Figure 15.24). At this point a vertical line is dropped to the x axis to determine the concentration of hemoglobin that corresponds to the absorbance of 0.175. The concentration is found to be 1.13×10^{-5} M. The slope of the Beer's law plot ($\Delta y/\Delta x$) is equal to $\Delta A/\Delta c = \epsilon b$. Since the path length, b, is 1.00 cm in this measurement, the molar absorptivity is equal to the slope of the line. The molar absorptivity is therefore determined to be 15.5×10^3 M^{-1} cm^{-1}.

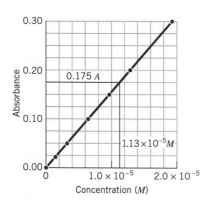

FIGURE 15.24 Determining an unknown concentration from a Beer's law standard curve.

15.8   AN OFF-COLOR FATTY ALCOHOL[3]

The Problem: What Is the Cause of the Off-color in the Fatty Alcohol?

A company that sells fatty alcohols to the rubber industry received a complaint from one of its customers that the fatty alcohol, stearyl alcohol ($C_{18}H_{37}OH$), had an off-color. The customer returned a sample of the off-color alcohol. The sample was then used to identify

[3]"Professional Analytical Chemists in Industry: A Short Course for Undergraduate Students in Problem Solving," Procter & Gamble.

the contaminants in the fatty alcohol that were causing the off-color. The contaminants were first separated from the fatty alcohol using chromatography. They were then examined using infrared spectroscopy.

The Method of Analysis: Infrared Spectroscopy

Infrared (IR) spectroscopy is a molecular absorbance technique used primarily for qualitative analysis and structural determination. In Chapter 6 you learned that all matter is in constant motion on the atomic level. The covalent bonds in molecules can be thought of as springs that hold the atoms together rather than the rigid sticks often used in molecular model kits. Two atoms connected by a covalent bond can vibrate just like two balls connected by a spring. The energy associated with the vibration will depend on the mass of the two atoms and the strength of the bond that connects them, as shown in Figure 15.25. Two atoms bonded together will therefore have their own characteristic vibrational frequency. The energy associated with this vibration falls in the infrared portion of the electromagnetic spectrum. Infrared radiation with a frequency that matches the frequency of vibration in a molecule will be absorbed by that molecule. Absorption of infrared radiation can be reported in units of wavelength, which is related to frequency by the relationship $\lambda = c/\nu$. Where c is the speed of electromagnetic radiation in a vacuum. Table 15.3 lists the expected wavelength range for absorption by a number of common covalently bonded atoms.

TABLE 15.3   Characteristic Infrared Wavelengths and Wavenumbers of Absorption

Bond Type	Range of Wavelengths (μm)	Range of Wavenumbers (cm^{-1})
—C—H	3.38–3.51	2960–2850
=C—H	3.23–3.33	3100–3000
≡C—H	3.03	3300
C=C	5.95–6.17	1680–1620
C≡C	4.49–4.76	2230–2100
O—H	2.74–4.00 (broad)	3650–2500 (broad)
N—H	2.94–3.13	3400–3200
C=O	5.56–6.13	1800–1630
C—O	7.69–10.00	1300–1000
C—N	7.35–9.80	1360–1020

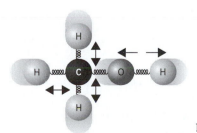

FIGURE 15.25 Vibration of atoms in a methanol molecule.

IR spectroscopy finds its greatest utility in structure determination of organic molecules. Organic chemicals are often classified by **functional groups,** an atom or group of atoms that give a compound its characteristic properties. Organic alcohols have a hydroxyl,

OH, as the functional group, whereas ketones have a carbonyl group, C=O, as shown in Figure 15.26. The use of infrared radiation as a probe of the structure of a molecule can be demonstrated with the infrared spectrum of alcohols. Since all organic alcohols contain the hydroxyl functional group, they all will have an O—H bond and, according to Table 15.3, will therefore have an absorption peak in the range between 2.7 and 4.0 μm.

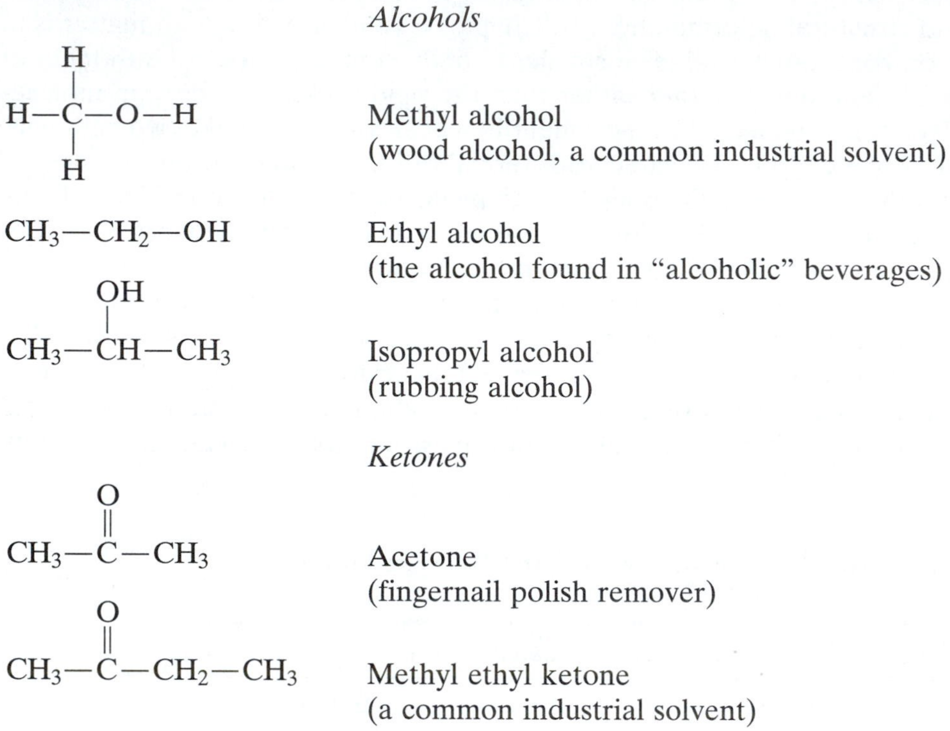

Alcohols

$$H-\underset{\underset{H}{|}}{\overset{\overset{H}{|}}{C}}-O-H$$

Methyl alcohol
(wood alcohol, a common industrial solvent)

$$CH_3-CH_2-OH$$

Ethyl alcohol
(the alcohol found in "alcoholic" beverages)

$$CH_3-\underset{\underset{}{|}}{\overset{\overset{OH}{|}}{CH}}-CH_3$$

Isopropyl alcohol
(rubbing alcohol)

Ketones

$$CH_3-\overset{\overset{O}{\|}}{C}-CH_3$$

Acetone
(fingernail polish remover)

$$CH_3-\overset{\overset{O}{\|}}{C}-CH_2-CH_3$$

Methyl ethyl ketone
(a common industrial solvent)

FIGURE 15.26 Common alcohols and ketones.

UV/Vis spectra are shown plotted as absorbance versus wavelength as shown in Figure 15.22. Infrared spectra, however, are plotted as percentage transmittance on the y axis rather than absorbance. Since absorbance and percentage transmittance have an inverse relationship (as more radiation is absorbed, less radiation is transmitted through the sample), infrared spectra have a baseline along the top edge of the spectra with absorption peaks pointing down. The x axis may be plotted as wavelength measured in micrometers (microns, 10^{-6} m) or as **wavenumbers,** $\bar{\nu}$, measured in cm^{-1}. Wavenumbers are the reciprocal of the wavelength measured in cm and are directly proportional to the frequency of infrared radiation as shown in the following equation.

$$\bar{\nu} = 1/\lambda = \nu/c$$

Checkpoint

NO and NO$_2$ are examined by infrared spectroscopy. Is the N–O bond strength greater in NO or NO$_2$? Which molecule would you expect to absorb infrared radiation at the greater frequency? Wavelength? Wavenumber?

Wavenumber ranges for absorption by several common covalently bonded atoms accompany the wavelength ranges given in Table 15.3. Often infrared spectra are plotted showing both wavelength and wavenumbers on the *x* axis. In the spectra shown in Figure 15.27 wavelength is plotted along the top of the spectra and wavenumbers along the bottom. Figure 15.27 shows the infrared spectra for two alcohols, ethyl alcohol and isopropyl alcohol. Note the absorption band for the hydroxyl group between 2.7 and 3.0 μm (3700–3330 cm^{-1}). Since both molecules have the same functional group, they both absorb infrared radiation of approximately the same frequency or wavelength. By using tables of characteristic wavelengths of absorbance, it is possible to identify the functional groups found in an organic molecule.

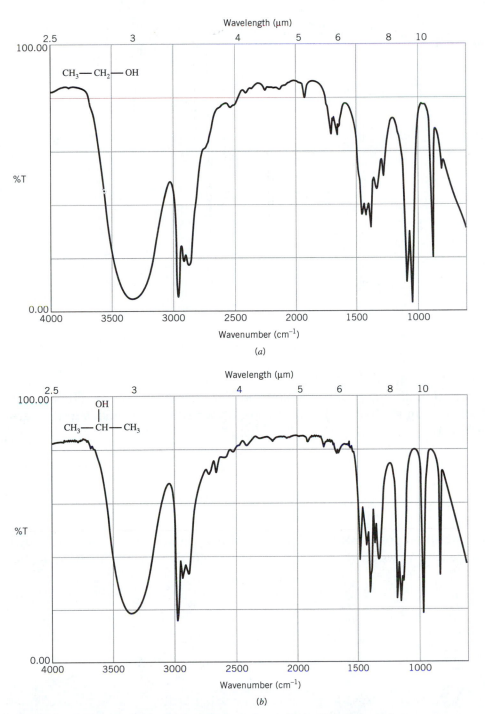

FIGURE 15.27 IR spectra of (*a*)ethyl alcohol and (*b*) isopropyl alcohol.

Exercise 15.4

The spectrum below is an infrared spectrum of acetone. Use Table 15.3 to identify the peak caused by the carbonyl group, C=O.

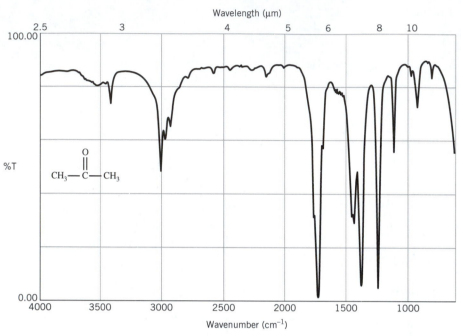

EXERCISE 15.4

Solution

Table 15.3 lists the wavelength range for a carbonyl absorbance as 5.56–6.13 μm (1800–1630 cm^{-1}). The peak at 5.85 μm (1710 cm^{-1}) in the acetone spectrum corresponds to the carbonyl group absorbance.

Exercise 15.5

A carboxylic acid functional group contains both a carbonyl group and a hydroxyl group. Which of the spectra below would you expect to be due to the carboxylic acid, acetic acid:

$$CH_3-\overset{\displaystyle O}{\overset{\|}{C}}-OH$$

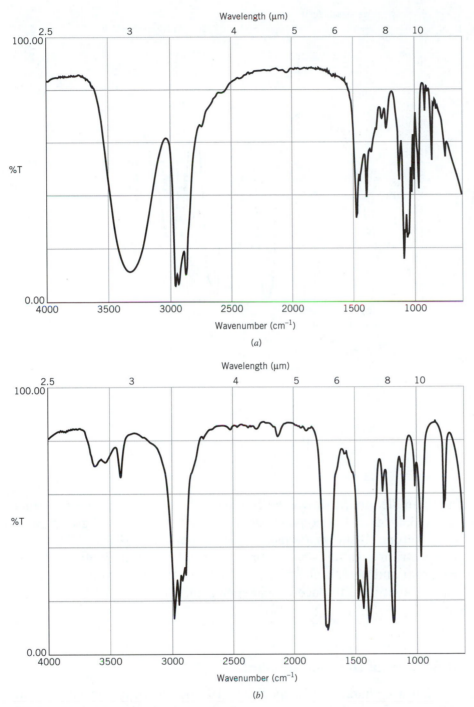

(a)

(b)

EXERCISE 15.5 (continued on next page)

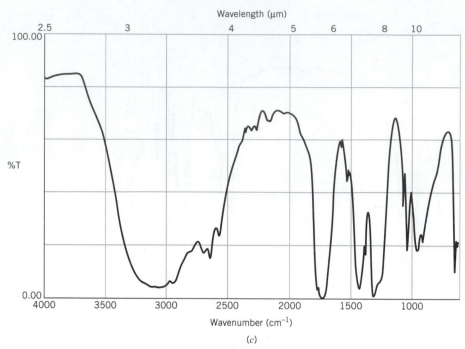

(c)

EXERCISE 15.5 (continued)

Solution

Spectrum *a* is of butanol, an alcohol. It shows the characteristic broad absorption of a hydroxyl group at 3.00 μm (3330 cm^{-1}) but no peak corresponding to a carbonyl group. Spectrum *b* is of methyl ethyl ketone. It shows a typical carbonyl peak at 5.82 μm (1720 cm^{-1}) but no hydroxyl peak. Spectrum *c* contains both a broad peak at 3.25 μm (3080 cm^{-1}) corresponding to an OH group and a peak at 5.85 μm (1710 cm^{-1}) corresponding to a carbonyl group. Spectrum *c* is therefore the acetic acid spectrum.

The Solution: Identification of the Contaminant

Because each compound has a characteristic spectrum, IR spectroscopy can be used for qualitative analysis. This technique allows the identification of the *molecules* that are present in a sample. Therefore, spectra of the unknown compounds can be matched to previously obtained spectra of known compounds. This is most easily accomplished in cases where the unknown is suspected to be one of a limited number of possibilities or when a database of known spectra can be searched by computer. Figure 15.28 shows the spectrum obtained for one of the contaminants in the fatty alcohol sample. This spectrum was compared to standard spectra and identified as long oil soya alkyd, a common component in paint. The customer with the off-color fatty alcohol was contacted and confirmed that painters had been working in the plant after the fatty alcohol was received and before the off-color was noted. The off-color was due to volatile organic compounds that had evaporated from the paint and been absorbed by the fatty alcohol.

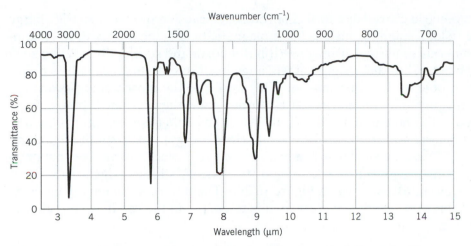

FIGURE 15.28 IR spectrum of contaminant from the fatty alcohol sample. The spectrum is consistent with a common component in paint. From "Professional Analytical Chemists in Industry: A Short Course for Undergraduate Students in Problem Solving," Proctor & Gamble.

15.9 THE SEARCH FOR NEW COMPOUNDS

Each year thousands of previously unknown naturally occurring chemical compounds are discovered. These new compounds often have interesting properties with many potential applications. An area of recent research has been the search for new compounds that are potential new medicinal drugs. Researchers have focused much of their search on sources that have been used for medicinal purposes by traditional native cultures. These diverse and sometimes exotic sources of new compounds include algae, plants, marine sponges, and frog or toad skin secretions. The native medicines often contain compounds which are bioactive and thus provide medicinal benefits. By isolating and identifying the bioactive compounds researchers hope to develop new drugs.

The Problem: What Are the Structures of the Bioactive Compounds in *Cryptolepis sanguinolenta?*

A team of scientists from Ghana, the University of Pittsburgh, and Burroughs Wellcome Company (a pharmaceutical company) investigated *Cryptolepis sanguinolenta,* a shrub found in West Africa.[4] The shrub has long been used by Ghanaian traditional healers for the treatment of fevers. In addition, an extract from the roots has been used at the Center for Scientific Research into Plant Medicine in Ghana for the treatment of malaria and urinary and upper respiratory tract infections. Investigation of the plant involved the isolation and purification of the bioactive components of the plant, followed by identification and structure determination. A major tool for structure determination of chemical compounds is nuclear magnetic resonance spectrometry.

The Method of Analysis: Nuclear Magnetic Resonance Spectroscopy

Nuclear magnetic resonance (NMR) spectroscopy is one of the most powerful tools available for the structure determination of organic molecules. NMR spectroscopy measures changes in the energy in the nuclei of atoms. The two nuclei that are most important for the determination of the structure of organic molecules are ^1H and ^{13}C.

[4]A. N. Tackie et al., *Journal of Natural Products,* **56,** 653–670 (1993).

Nuclei have an electric charge due to their protons. Because a moving electric charge creates a magnetic field, a spinning nucleus has a magnetic field associated with it. A spinning nucleus is analogous to a tiny bar magnet (it has a north and south pole). When such a magnet is placed close to another magnet, it will align itself so that its south pole is pointing toward the north pole of the other magnet. This would be the most stable (lowest energy) state of the magnet. The magnet could be turned until the two north poles of the magnets are aligned with one another. This would be an unstable (high energy) state, and the magnet when released will return to its original (low energy) north–south alignment, as shown in Figure 15.29a. In a similar way a spinning nucleus placed in an external magnetic field will either align with the magnetic field (low energy state) or against the magnetic field (high energy state), as shown in Figure 15.29b. The difference in energy between the two energy states is quite small and corresponds to energy in the radio wave region of the electromagnetic spectrum.

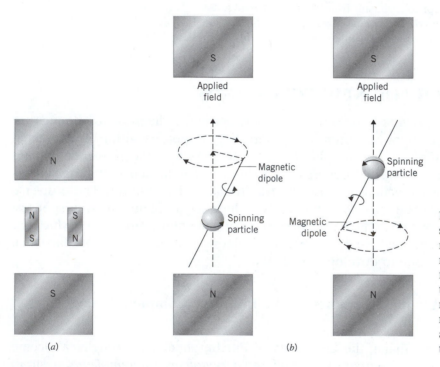

FIGURE 15.29 (a) Two small bar magnets in an magnetic field. One bar magnet is aligned with the field and the other against the field. (b) Two spinning nuclei in an external magnetic field. One nucleus is aligned with the field and the other against it.

In NMR spectroscopy the sample is placed in a strong external magnetic field and irradiated with radio waves. The waves have sufficient energy to cause the excitation of the nuclei to higher energy levels. The resulting NMR spectra are plotted as absorbance versus ppm, where ppm represents the parts per million shift in frequency from a reference frequency.

It may appear that NMR spectra would give relatively little useful information (i.e., one peak for the ^1H nuclei present and another peak for ^{13}C nuclei). However, each type of nucleus is capable of giving different signals, depending on the magnetic environment in which the nucleus is found within the molecule. The magnitude of the external magnetic field that the nucleus experiences depends not only on the external magnetic field of the spectrometer but also on magnetic fields generated within the structure of the molecule. Since moving electric charges generate magnetic fields, both moving electrons and spinning nuclei will generate local magnetic fields within the molecule. Thus, the resulting NMR spectrum gives information about the environments in which various nuclei are found. Figure 15.30 shows the ^1H (proton) NMR and ^{13}C NMR spectra of ethanol. Each group of peaks in the ^1H spectrum corresponds to hydrogen nuclei (protons) in a slightly different magnetic environment. Both ^1H and ^{13}C NMR spectra are often obtained for a molecule in order to provide as much data as possible for a structure determination.

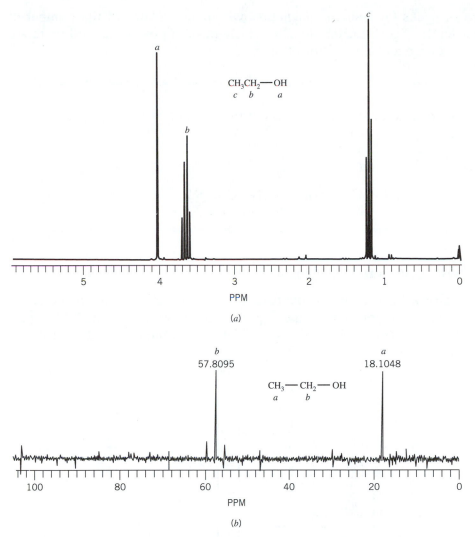

FIGURE 15.30 (a) ^{1}H NMR spectrum of ethanol and (b) ^{13}C NMR spectrum of ethanol.

NMR spectra give three principal pieces of information:

1. The location of a set of peaks along the frequency axis (*x* axis) is known as the chemical shift and gives information about the electron environment of the nuclei. In Figure 15.30a there are three groups of peaks labeled a, b, and c. These three groups of peaks correspond to the hydrogens labeled a, b, and c below. In the ^{13}C NMR spectrum shown in Figure 15.30b there are two peaks corresponding to the two carbon atoms.

$$H_c-\underset{\underset{H_c}{|}}{\overset{\overset{H_c}{|}}{C}}-\underset{\underset{H_b}{|}}{\overset{\overset{H_b}{|}}{C}}-O-H_a$$

2. The area under a group of peaks corresponds to the relative number of nuclei in each environment. Therefore, the area of the set of peaks labeled b in Figure 15.30a is twice as large as the area of peak a, because there is only one hydrogen in the position labeled a on the structure but there are two hydrogens in position b on the structure.

3. The pattern of peaks for a single magnetic environment relates to the number of neighboring nuclei. In Figure 15.30*a* there is a single peak at position a, and there are four peaks at position b and three at position c.

Checkpoint

Predict the number of sets (or groups) of peaks in the proton NMR spectra of acetone and acetic acid.

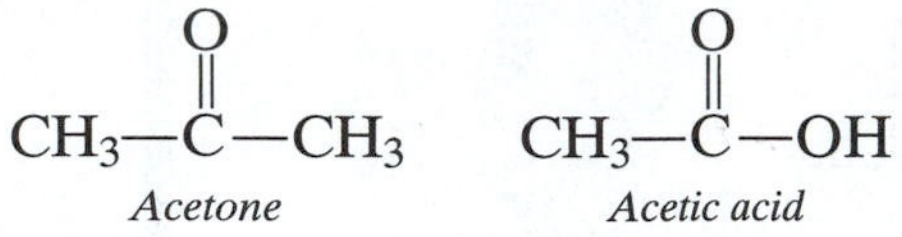

$$CH_3-\overset{\overset{\displaystyle O}{\|}}{C}-CH_3 \qquad CH_3-\overset{\overset{\displaystyle O}{\|}}{C}-OH$$

Acetone   *Acetic acid*

The Solution: A New Compound Is Identified

The investigation of *Cryptolepis sanguinolenta* resulted in the isolation of a new bioactive compound which was named cryptospirolepine. Figure 15.31 shows one of the NMR spectra used in the structure determination, and Figure 15.32 shows the deduced structure of cryptospirolepine.

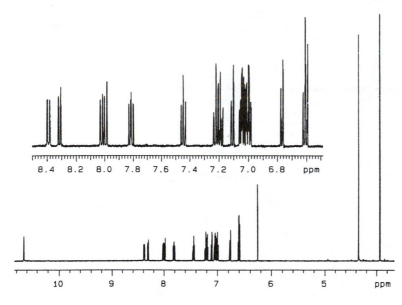

FIGURE 15.31 ^{1}H NMR spectrum of cryptospirolepine. The portion of the spectrum from 6.6 to 8.4 ppm has been expanded and is shown in the insert. From A. N. Tackie et al., *Journal of Natural Products,* **56,** 657 (1993).

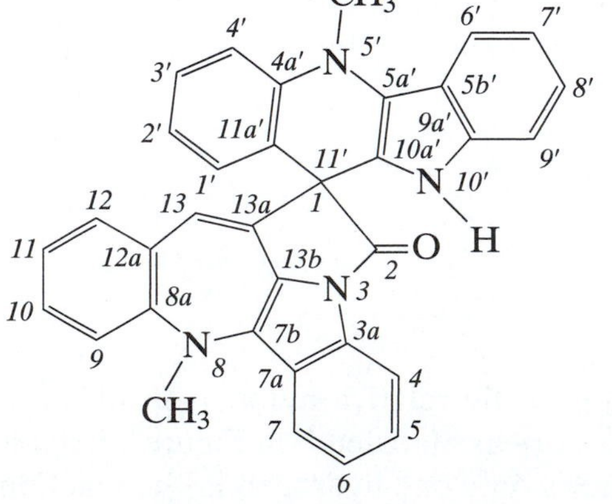

FIGURE 15.32 Structure of cryptospirolepine. Adapted from A. N. Tackie et al., *Journal of Natural Products,* **56,** 654 (1993).

15.10   THE SEARCH FOR THE NORTHWEST PASSAGE—THE FRANKLIN EXPEDITION[5]

On May 19, 1845, Sir John Franklin set sail from England in command of two ships, the H.M.S. *Terror* and the H.M.S. *Erebus*. The ships had a crew of 134 officers and men and were well provisioned with 61,987 kg of flour, 16,749 liters of liquor, 8000 tins (1- to 8-pound capacity) of meats, soups, and vegetables, and 4200 kg of lemon juice to prevent scurvy. The expedition's 3-year mission was to find the Northwest Passage, a navigatable sea route above North America to the Pacific Ocean. The ships never returned. Over the next 10 years rescue expeditions were sent in search of the Franklin expedition, but no survivors were ever found. The rescue expeditions brought back horror stories of starvation and cannibalism from the Inuit Indians of King William Island above the Arctic Circle. Later expeditions discovered skeletal remains and abandoned campsites. One expedition discovered skeletal remains by one of the ship's lifeboats mounted on a wooden sled for dragging the boat across the island. The lifeboat contained among other supplies silk handkerchiefs, scented soap, sponges, slippers, and combs. These were strange provisions for desperate men pulling a heavy boat.

The Problem: Why Did the Franklin Expedition Perish?

Although the fate of the Franklin expedition was not in doubt many questions remained. Why had well-trained and well-equipped explorers been unable to survive where the Inuit Indians had lived for centuries? Why had the crew taken such useless provisions in the lifeboat that they had tried to drag across the island? In 1981 Dr. Owen Beattie, a forensic anthropologist at the University of Alberta, set out to determine what had happened to the Franklin expedition. On King William Island he found bones that he suspected were skeletal remains of the Franklin expedition. He also collected bones of Inuit Indians and caribou from the same geographical area and time period as the Franklin expedition. These samples were subjected to a range of different tests, including tests for heavy metals using an atomic absorption spectrometer.

The Method of Analysis: Atomic Absorption Spectrometry

Atomic absorption (AA) spectrometry is used primarily for quantitative elemental analysis of metals. Unlike previously described spectroscopic techniques, AA spectrometry does not differentiate between a metal as a free atom, an aqueous metal ion, or a metal ion or atom in a compound. It instead measures how much of the element is present regardless of the form or forms in which it is found in the sample.

In AA spectrometry a flame or furnace is used to break compounds apart into their individual atoms. The flame or furnace evaporates the solvent, breaks all bonds to give free atoms, and momentarily suspends the atoms in the path of the electromagnetic radiation. The beam of electromagnetic radiation passes through the flame and is absorbed by the atoms as shown in Figure 15.33. Only those wavelengths of radiation that correspond in energy to changes in energy levels within the atom are absorbed.

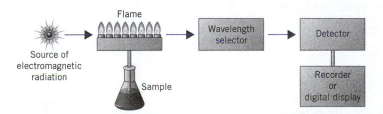

FIGURE 15.33 Diagram of atomic absorption spectrometer.

[5]Owen Beattie and John Geiger, *Frozen in Time, Unlocking the Secrets of the Franklin Expedition,* E. P. Dutton, New York, 1987.

The absorbance of radiation is related to the concentration by Beer's law, $A = \epsilon bc$. The absorbances of standards are measured and compared to the absorbance of the same element in the sample in order to find the concentration of an element in the sample.

The Solution: Surprising Results

Results of the analysis for the concentration of lead were quite unexpected. Table 15.4 shows these results. The lead levels in the bone samples from the Franklin expedition were much higher than those of contemporary Inuit Indians, caribou, or modern humans. Lead poisoning is known to cause anorexia, weakness, fatigue, irritability, stupor, paranoia, anemia, and abdominal pain. Lead poisoning could therefore have led to declining health in the crew and could have been a significant factor in the ability of the officers to make appropriate decisions as in the case of the provisions for the lifeboat. However, lead levels in bone only indicate exposure to lead at some time during the person's life. To establish whether the lead poisoning had occurred during the expedition, it would be necessary to have tissue samples to analyze.

TABLE 15.4   Lead Level in Bone Samples[a]

Sample	Mean Pb Concentration (ppm)
Franklin expedition	138.1
Inuit Indians	5.1
Caribou	2.0
Modern human	29.8

[a]W. A. Kowal, P. M. Krahn, and O. B. Beattie, Lead levels in human tissues from the Franklin forensic project, *Int. J. Environ. Anal. Chem.*, **35,** 119–126 (1989).

In the summers of 1984 and 1986 Beattie led expeditions to Beechey Island, where the Franklin expedition had spent the winter of 1845. It was known that three crewmen, John Hartnell, William Braine, and John Torrington, had died during that first winter and been buried on Beechey Island. Because the location was north of the Arctic Circle the graves would be beneath the permafrost and the bodies might still be preserved. Beattie received permission to exhume the bodies and take tissue samples. Results of the analysis of hair samples for lead content are given in Table 15.5. These results indicate that all three crewmen had been exposed to very high levels of lead during the time of the expedition. Lead concentrations of that level would certainly have had detrimental health effects.

TABLE 15.5   Lead Analysis of Hair Samples[a]

Sample	Mean Lead Concentration (ppm)
Torrington	565
Hartnell	326
Braine	225
Modern human	4

[a]W. A. Kowal, P. M. Krahn, and O. B. Beattie, Lead levels in human tissues from the Franklin forensic project, *Int. J. Environ. Anal. Chem.*, **35,** 119–126 (1989).

A final question remained. What was the source of the lead poisoning? Near the camp-site of the Franklin expedition on Beechey Island was a garbage dump where among other items the crew members had discarded the empty tin cans used for storing meat and vegetables. At the time of the Franklin expedition in 1845 the canning of foods in tins was relatively new. Beattie found that the tins had been sealed using lead solder. It seems highly probable that the solder was the source of the lead that caused lead poisoning of the crew and contributed to the loss of the Franklin expedition.

15.11   DEAD CATS[6]

Cats were dying in the Glen Oaks neighborhood. At first mostly stray cats were found dead, but then family pets began dying. Tissue samples of the dead animals were sent to Michael McClure, a chemist at the state veterinary diagnostic laboratory. There could be several causes for the deaths: bacteria, a virus, or a poison. Identifying the specific cause within any one of the three categories could be quite difficult. Virology and bacteria tests both gave negative results. This focused the search on some type of poison. The tissue samples were analyzed for heavy metals, such as lead, using atomic absorption spectroscopy. Chromatographic techniques were used to test for alkaloids (organic poisons found in many plants) and for common insecticides. However, all of the tests gave no clue to the cause of death.

The Problem: What Was Killing the Cats?

A local veterinarian called McClure and asked him to come to Glen Oaks to see another dead cat that had been found. In a vacant lot was the body of another dead cat and nearby was a dead crow. From the condition of the cat's body it was obvious that the crow had been eating the cat. This suggested that whatever had killed the cat had been passed to the crow. The poison had evidently acted quickly, since the crow was found beside the cat's body. Very few poisons can be passed on in this manner and act this quickly. The field of potential poisons was narrowing. McClure suspected "1080," a rodenticide known as sodium monofluoroacetate.

$$\text{F}-\overset{\overset{\displaystyle \text{H}}{|}}{\underset{\underset{\displaystyle \text{H}}{|}}{\text{C}}}-\overset{\overset{\displaystyle \text{O}}{\|}}{\text{C}}\diagdown_{\text{O}^-\text{Na}^+}$$

Method of Analysis: Potentiometry

Potentiometry is based on the relationship between the electric potential developed by an electrode in solution and the concentration of the analyte in solution. Chapter 12 introduced the use of electrodes to measure the potential of an oxidation–reduction reaction and described the potential as a measure of the driving force of a redox reaction. Section 12.8 described the dependence of electric potential on the concentration of the products and reactants. The relationship between potential and concentration allows the use of electrodes for the quantitative analysis of analytes capable of participating in a redox reaction. The most common potentiometric measurement is the measurement of pH using a pH me-

[6]M. W. McClure, The crow's warning, *Chem Matters*, **8,** 7–9 (April 1990). Excerpted with permission. Copyright © 1990 American Chemical Society.

ter. In this measurement the potential of a glass pH electrode is related to the negative log of the concentration of H_3O^+ in solution.

Potentiometric measurements are not limited to pH but can also be used for the determination of the concentrations of a variety of ions. An **ion-selective electrode (ISE)** produces a potential that is proportional to the concentration of a given ion. Ion-selective electrodes are available for the measurement of a number of different ions such as Na^+, K^+, NH_4^+, Ca^{2+}, S^{2-}, F^-, and NO_3^-. A potentiometric measurement requires the use of two electrodes (although they are both often combined into a single physical electrode probe) attached to a pH/pIon meter. The ion-selective electrode develops a potential that depends on the concentration of the analyte. The fluoride ion-selective electrode contains a crystal of LaF_3. A potential develops at the surface of the crystal that is proportional to the negative log of the concentration of F^- in contact with the crystal. The second electrode is a reference electrode, which maintains a constant potential that acts as a reference for the ion-selective electrode. Using standards of known F^- concentrations the pH/pIon meter is calibrated to give a direct reading of pF:

$$pF = -\log[F^-]$$

The Solution: The Fluoride Level Is High

A tissue sample was treated so that any 1080 present would be broken down to yield fluoride ion as one of the products. A fluoride ion-selective electrode was then used to measure the F^- concentration in the sample. Tissue samples normally have only a trace amount of fluoride ion, less than 5 ppm. The sample from the dead cat showed a concentration of 35 ppm fluoride ion, suggesting that indeed 1080 was the cause of death.

KEY TERMS

Absorbance

Absorption

Atomic absorption spectroscopy (AA)

Beer's law

Chromatography column

Chromatography

Deoxyribonucleic acid (DNA)

Electrophoresis

Fats

Fatty acids

Frequency

Functional group

Gas chromatography/mass spectrometry (GC/MS)

Gas chromatography (GC)

Gravimetric

High-performance liquid chromatography (HPLC)

Infrared spectrometry

Ion-selective electrode

Mass spectrometry (MS)

Mobile phase

Molar absorptivity

Nuclear magnetic resonance spectroscopy (NMR)

Parent ion

Pathlength

Percent transmittance

Potentiometry

Qualitative methods

Quantitative methods

Retention time

Risk assessment

Spectroscopy

Standard Beer's law plot

Standards

Stationary phase

Transmittance

Ultraviolet/visible spectroscopy (UV/Vis)

Volumetric

Wavenumber

Wavelength

PROBLEMS

1. Distinguish between *quantitative*, *qualitative*, and *structural* methods of analysis.
2. Distinguish between *volumetric* and *gravimetric* methods of analysis.

Chromatography

3. Describe how chromatography is used to separate a mixture. What functions do the stationary and mobile phases have in the separation?
4. Compare and contrast the techniques of gas chromatography and high-performance liquid chromatography. (What is the same and what is different about the two techniques?)
5. How are the intermolecular forces, described in Chapter 8, important in the separation of components on a chromatography column?
6. Hexane, $CH_3(CH_2)_4CH_3$, is found to have a long retention time when separated from a mixture. A column with a C_{18} (18-carbon chain) stationary phase and a methanol–water mobile phase is used for the separation. Explain why hexane has a long retention time.
7. You wish to chromatographically separate a mixture of the compounds shown below. Which of the two compounds will be most difficult to separate from one another? Why?

(a) $CH_3CH_2CH_2\overset{\displaystyle O}{\overset{\displaystyle \|}{C}}$—OH

(b) $CH_3\overset{\displaystyle CH_3}{\overset{\displaystyle |}{C}}HCH_2CH_2CH_2CH_2CH_2CH_2$—OH

(c) $CH_3CH_2CH_2\overset{\displaystyle CH_3}{\overset{\displaystyle |}{C}}HCH_2CH_2CH_2CH_3$

(d) CH_3CH_2—O—$CH_2CH_2CH_2CH_3$

(e) $CH_3\overset{\displaystyle CH_3}{\overset{\displaystyle |}{C}}HCH_2CH_2CH_2\overset{\displaystyle OH}{\overset{\displaystyle |}{C}}HCH_2CH_3$

8. The mixture below is separated using HPLC with a octadecylsiloxane (nonpolar) stationary phase and a methanol (CH_3OH)/water mobile phase. In what order will the components come off the column?

$$CH_3CH_2CH_2CH_2CH_2CH_2CH_3$$
Heptane

$$CH_3CH_2CH_2CH_2CH_2CH_2CH_2—OH$$
Heptanol

$$CH_3CH_2CH_2—O—CH_2CH_2CH_2CH_3$$
Butyl propyl ether

What could happen to the order in which the compounds come off a polar column if pentane ($CH_3CH_2CH_2CH_2CH_3$) is used as the mobile phase?

Mass Spectrometry

9. What is a mass spectrum? What information can be obtained from a mass spectrum?
10. What are the principal functions of the gas chromatograph and the mass spectrometer in a GC/MS system?
11. Which of the following compounds is responsible for the mass spectrum in Figure 15.34? Explain your reasoning.
    (a) octanol, $CH_3CH_2CH_2CH_2CH_2CH_2CH_2CH_2$—OH
    (b) heptane, $CH_3CH_2CH_2CH_2CH_2CH_2CH_3$
    (c) acetic acid, CH_3COOH

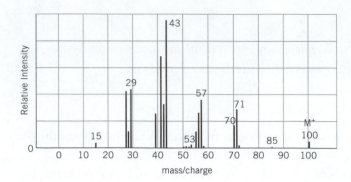

FIGURE 15.34 Mass spectrum to accompany problem 15.11.

Spectroscopy

12. Compare and contrast UV/Vis spectroscopy and atomic absorption spectroscopy. (What is the same and what is different about the two techniques?)

13. Give the Beer's law equation and define all of the terms.

14. Which of the following spectrophotometric techniques are used primarily for quantitative analysis and which are used for qualitative (or structure determination) analysis?
    (a) infrared   (b) NMR
    (c) UV/Vis   (d) atomic absorption

15. UV/Vis, infrared, and NMR spectroscopy all involve passing electromagnetic radiation through a sample and measuring the resulting absorbance or transmittance. What is different about the three techniques that allows them to provide different information about a molecule?

16. Explain why samples to be analyzed by IR and NMR spectroscopic techniques are normally separated into pure components before the analysis rather than being analyzed as a mixture.

17. The absorbance of a solution of copper nitrate is measured using UV/Vis spectrophotometry. What will happen to the absorbance for each of the conditions given below?
    (a) A cuvette with a longer pathlength is used.
    (b) The concentration of the copper nitrate is decreased.
    (c) The wavelength of the spectrometer is changed to a value where the molar absorptivity is less.

18. Match the following compounds with their IR spectra shown in Figure 15.35. Explain your reasoning.
    (a) octane, $CH_3CH_2CH_2CH_2CH_2CH_2CH_2CH_3$
    (b) 3-pentanol, $CH_3CH_2\overset{\displaystyle OH}{\underset{\displaystyle |}{C}}HCH_2CH_2$
    (c) isobutyric acid, $CH_3\underset{\displaystyle \underset{|}{CH_3}}{C}H\overset{\displaystyle \overset{O}{\|}}{C}-OH$
    (d) butyraldehyde, $CH_3CH_2CH_2\overset{\displaystyle \overset{O}{\|}}{C}H$

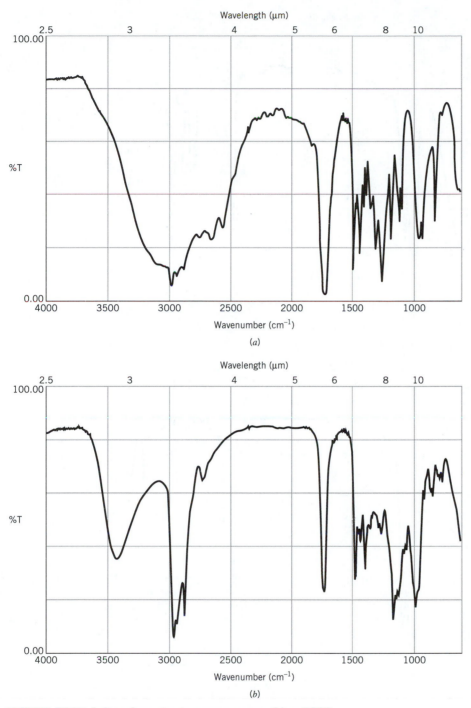

FIGURE 15.35 Infrared spectra to accompany problem 15.18.

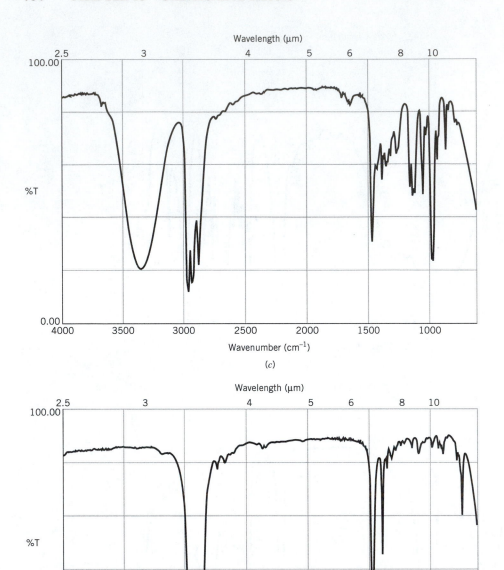

FIGURE 15.35 (continued)

19. Which of the following compounds is responsible for the NMR spectrum shown in Figure 15.36? Explain your reasoning.

(a) butyraldehyde, $CH_3CH_2CH_2\overset{\displaystyle O}{\overset{\|}{C}}H$

(b) octane, $CH_3CH_2CH_2CH_2CH_2CH_2CH_2CH_3$

(c) formic acid, $HC\overset{\overset{O}{\|}}{-}OH$

(d) methanol, CH_3OH

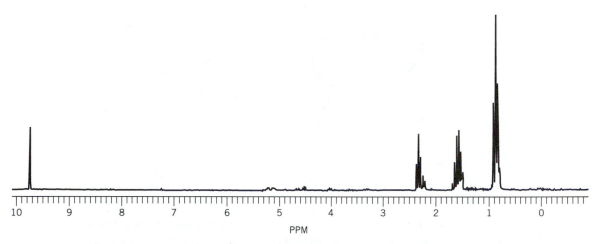

FIGURE 15.36 ^1H NMR spectrum to accompany problem 15.19.

20. Suggest appropriate structures for each of the following:
    (a) A compound with the molecular formula $C_4H_8O_2$ has three sets of peaks in its ^1H NMR spectrum with relative intensities of $2:3:3$ and two strong absorbance peaks in the IR at 5.65 μm (1770 cm^{-1}) and 8.05 μm (1240 cm^{-1}).
    (b) A compound with the formula $C_3H_6O_2$ has three sets of peaks in its ^1H NMR spectrum with relative intensities of $3:2:1$ and two strong IR absorption peaks: one a broad peak at 3.4 μm (2940 cm^{-1}) and the other a sharper peak at 5.8 μm (1720 cm^{-1}).

21. Titrations are a quantitative volumetric method of analysis in which the volume of titrant necessary to completely react with a sample, the equivalence point, is measured. Absorbance measurements are sometimes used to monitor the progress of a titration in order to determine the equivalence point. A plot of absorbance versus volume of titrant added to the sample will have an inflection point that corresponds to the equivalence point. A generalized chemical equation for the titration can be written as:

$$\text{Titrant} + \text{sample} \longrightarrow \text{products}$$

Which of the following plots of absorbance versus volume of titrant describe the titration when the spectrometer is set to a wavelength at which the sample absorbs but the titrant and products do not? Explain your reasoning. Which of the plots describe a titration in which the spectrometer is set to a wavelength at which the products absorb but not the sample or titrant? Explain your reasoning.

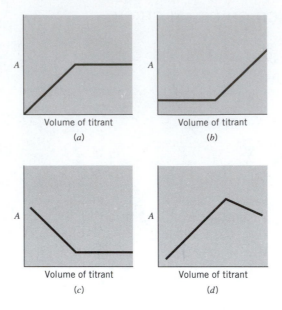

22. When making quantitative measurements of hemoglobin solutions using UV/Vis spectrophotometry, a wavelength of 576 nm will often be used. Explain why this wavelength is used as opposed to some other wavelength, such as 620 nm (see Figure 15.22).

23. Calculate the absorbance of a 5.2×10^{-4} M solution of the drug tolbutamine measured at 262 nm in a 1.00 cm cell. The molar absorptivity of tolbutamine at 262 nm is 703 M^{-1} cm^{-1}.

24. The concentration of iron in a solution can be determined using UV/Vis spectrophotometry by reacting the iron with 1,10-phenanthroline to produce a colored complex. The following data were obtained for a series of standard solutions of iron/1,10-phenanthroline complex measured in a 1.00-cm cell.

Iron Concentration (M)	Absorbance
0.50×10^{-4}	0.109
1.0×10^{-4}	0.218
2.0×10^{-4}	0.436
3.0×10^{-4}	0.656
4.0×10^{-4}	0.872

(a) Prepare a Beer's law plot using the above data.
(b) Calculate the concentration of an iron/1,10-phenanthroline complex solution that has an absorbance of 0.317.
(c) Calculate the molar absorptivity of the iron/1,10-phenanthroline complex.

25. Spectroscopic measurements can be used for kinetic studies by monitoring the absorbance of a reactant with respect to time. A reaction is found to have the following rate law with respect to a reactant, X.

$$\text{Rate} = k[X]$$

Which of the following plots of absorbance of X versus time corresponds to the above rate law? Explain your reasoning. Which of the plots corresponds to a rate law of $\text{Rate} = k[X]^0$? Explain.

(a)	(b)	(c)	(d)	(e)

Potentiometry

26. Enzyme electrodes are used for the quantitative analysis of samples of biological fluids. The electrodes are constructed by placing a membrane with immobilized enzyme directly onto an ion-selective electrode. An enzyme electrode that is sensitive to the concentration of urea is constructed by immobilizing the enzyme urease in a membrane placed onto an ammonium ion-selective electrode. Urease catalyzes the following reaction:

$$NH_2CONH_2(aq) + 2\ H_2O(l) + H^+(aq) \xrightarrow{urease} 2\ NH_4^+(aq) + HCO_3^-(aq)$$
$$\underset{Urea}{} \qquad\qquad\qquad\qquad\qquad\qquad \underset{Ammonium}{}$$

   Explain how this electrode can respond to the presence of urea. On the basis of the discussion of enzymes in Chapter 14, why would the electrode not give a response for urea without the presence of the enzyme?

Integrated Problems

27. Discuss why legislation requiring zero concentration of pesticides in food would be unrealistic.

28. Indicate which of the following are elemental methods of analysis and which are molecular methods.
    (a) mass spectrometry   (b) gas chromatography
    (c) UV/Vis spectroscopy   (d) atomic absorption spectrometry
    (e) NMR spectrometry

29. Suggest the steps that a chemist might take to find the cause of an unusual aftertaste in a shipment of ginger ale.

30. Suggest an appropriate method of analysis for each of the following problems and describe why the method is appropriate to solve the problem.
    (a) Separate the components in a sample of perfume.
    (b) Determine the concentration of calcium in a sample of drinking water.
    (c) Determine whether mercury is leaching into the soil at a landfill.
    (d) Identify an organic compound isolated from a sample of milk.
    (e) Determine the structure of an enzyme isolated from beef heart.
    (f) Analyze the fumes from the exhaust of a car.

31. Which of the techniques described in this chapter can be used to distinguish between the following isomers (compounds with the same molecular formula but different structures)? Explain your reasoning.

$$CH_3CH_2CH_2CH_2CH_2{-}OH \qquad CH_3CHCH_2CH_2CH_3$$
$$\qquad\qquad\qquad\qquad\qquad\qquad\quad |$$
$$\qquad\qquad\qquad\qquad\qquad\qquad\ OH$$

32. A lab assistant, by mistake, has washed all of the labels off a set of bottles that still contain chemicals. You know what was in the bottles but you don't know which bottle is which. Describe how NMR, IR, or a combination of the two techniques could be used to distinguish between the chemicals listed in each of the sets below. Explain your reasoning.

    (a) methanol, CH_3OH; ethanol, CH_3CH_2OH; and propanol, $CH_3CH_2CH_2OH$

    (b) propanol, $CH_3CH_2CH_2OH$; propanoic acid, CH_3CH_2COOH; and ethyl methyl ether, CH_3CH_2—O—CH_3

33. A mixture of the following compounds was separated by HPLC with a stationary non-polar phase and a CH_3OH/water mobile phase.

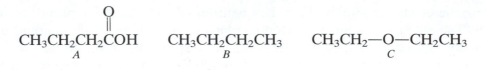

$$CH_3CH_2CH_2COH \qquad CH_3CH_2CH_2CH_3 \qquad CH_3CH_2—O—CH_2CH_3$$
$$A \qquad\qquad\qquad B \qquad\qquad\qquad\qquad C$$

    (a) Which of the following chromatograms would you expect to obtain? Explain your reasoning.

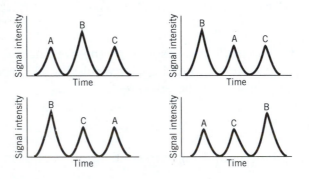

    (b) Use the correct chromatogram above to estimate the *relative* amounts of A, B, and C in the original mixture. Explain your reasoning.

    (c) Could IR spectroscopy be used to specifically identify A, B, and C? If so, what characteristics of the compounds would allow for this identification? Explain.

A P P E N D I X
A

A.1   SYSTEMS OF UNITS

All measurements contain a number that indicates the magnitude of the quantity being measured and a set of units that provide a basis for comparing the quantity with a standard reference. There are several systems of units, each containing units for properties such as length, volume, weight, and time.

English Units of Measurement

In the English system of units in use in the United States, the individual units are defined in an arbitrary way. There are 12 inches in a foot, 3 feet in a yard, and 1760 yards in a mile. There are 2 cups in a pint and 2 pints in a quart but 4 quarts in a gallon. There are 16 ounces in a pound but 32 ounces in a quart. The relationships between some of the common units in the English system are given in Table A.1.

TABLE A.1   The English System of Units

Length: inch (in.), foot (ft), yard (yd), mile (mi)

12 in. = 1 ft	5280 ft = 1 mi
3 ft = 1 yd	1760 yd = 1 mi

Volume: fluid ounce (oz), cup (c), pint (pt), quart (qt), gallon (gal)

2 c = 1 pt	32 oz = 1 qt
2 pt = 1 qt	4 qt = 1 gal

Weight: ounce (oz), pound (lb), ton

16 oz = 1 lb	2000 lb = 1 ton

Time: second (s), minute (min), hour (h), day (d), year (y)

60 s = 1 min	24 h = 1 d
60 min = 1 h	$365\frac{1}{4}$ d = 1 y

The Metric System

More than 300 years ago, the Royal Society of London discussed replacing the irregular English system of units with one based on decimals. It was not until the French Revolution, however, that a decimal-based system of units was adopted. This **metric system** is based on the fundamental units of measurement for length, volume, and mass shown in Table A.2.

TABLE A.2   The Fundamental Units of the Metric System

Length: meter (m)

1 m = 1.094 yd	1 yd = 0.9144 m

Volume: liter (L)

1 L = 1.057 qt	1 qt = 0.9464 L

Mass: gram (g)

1 g = 0.002205 lb	1 lb = 453.6 g

The principal advantage of the metric system is the ease with which the base units can be converted into a unit that is more appropriate for the quantity being measured. This is done by adding a prefix to the name of the base unit. The prefix *kilo-* (k), for example, implies multiplication by a factor of 1000. Thus, a kilometer is equal to 1000 meters.

$$1 \text{ km} = 1000 \text{ m}$$

The prefix *milli-* (m), on the other hand, means division by a factor of 1000. A milliliter (mL) is equal to 0.001 liters.

$$1 \text{ mL} = 0.001 \text{ L}$$

The common metric prefixes are given in Table A.3.

TABLE A.3 Metric System Prefixes

Prefix	Symbol	Meaning
femto-	f	$\times\ 1/1{,}000{,}000{,}000{,}000{,}000\ (10^{-15})$
pico-	p	$\times\ 1/1{,}000{,}000{,}000{,}000\ (10^{-12})$
nano-	n	$\times\ 1/1{,}000{,}000{,}000\ (10^{-9})$
micro-	μ	$\times\ 1/1{,}000{,}000\ (10^{-6})$
milli-	m	$\times\ 1/1{,}000\ (10^{-3})$
centi-	c	$\times\ 1/100\ (10^{-2})$
deci-	d	$\times\ 1/10\ (10^{-1})$
kilo-	k	$\times\ 1{,}000\ (10^{3})$
mega-	M	$\times\ 1{,}000{,}000\ (10^{6})$
giga-	G	$\times\ 1{,}000{,}000{,}000\ (10^{9})$
tera-	T	$\times\ 1{,}000{,}000{,}000{,}000\ (10^{12})$

A second advantage of the metric system is the link between the base units of length and volume. By definition, a liter is equal to the volume of a cube exactly 10 cm tall, 10 cm long, and 10 cm wide. Because the volume of that cube is 1000 cubic centimeters and a liter contains 1000 milliliters, 1 milliliter is equivalent to 1 cubic centimeter.

$$1\ \text{mL} = 1\ \text{cm}^3$$

The third advantage of the metric system is the link between the base units of volume and weight. The gram was originally defined as the mass of 1 mL of water at 4°C. (It is important to specify the temperature because water expands or contracts as the temperature changes.)

SI Units of Measurement

A series of international conferences on weights and measures has been held periodically since 1875 to refine the metric system. At the 11th conference, in 1960, a new system of units known as the **International System of Units** (abbreviated **SI** in all languages) was proposed as a replacement for the metric system. The seven base units for the SI system are given in Table A.4.

TABLE A.4 SI Base Units

Physical Quantity	Name of Unit	Symbol
Length	meter	m
Mass	kilogram	kg
Time	second	s
Temperature	kelvin	K
Electric current	ampere	A
Amount of substance	mole	mol
Luminous intensity	candela	cd

Derived SI Units

The units of every measurement in the SI system, no matter how simple or complex, should be derived from one or more of the seven base units. The preferred unit for volume is the cubic meter, for example, because volume has units of length cubed and the SI unit for length is the meter. The preferred unit for speed is meters per second because speed is the distance traveled divided by the time it takes to cover this distance.

SI unit of volume: m^3 SI unit of speed: m/s

Some of the common derived SI units are given in Table A.5.

TABLE A.5   Common Derived SI Units in Chemistry

Physical Quantity	Name of Unit	Symbol	
Density		kg/m^3	
Electric charge	coulomb	C	$(1 \text{ C} = 1 \text{ A·s})$
Electric potential	volt	V	$(1 \text{ V} = 1 \text{ J/C})$
Energy	joule	J	$(1 \text{ J} = 1 \text{ kg·m}^2/\text{s}^2)$
Force	newton	N	$(1 \text{ N} = 1 \text{ kg·m/s}^2)$
Frequency	hertz	Hz	$(1 \text{ Hz} = 1 \text{ s}^{-1})$
Pressure	pascal	Pa	$(1 \text{ Pa} = 1 \text{ N/m}^2)$
Velocity (speed)	meters per second	m/s	
Volume	cubic meter	m^3	

Non-SI Units

Strict adherence to SI units would require changing directions such as "add 250 mL of water to a 1-L beaker" to "add 0.00025 cubic meters of water to an 0.001-m^3 container." Because of this, a number of units that are not strictly acceptable under the SI convention are still in use. Some of the non-SI units are given in Table A.6.

TABLE A.6   Non-SI Units in Common Use

Physical Quantity	Name of Unit	Symbol	
Volume	liter	L	$(1 \text{ L} = 1 \times 10^{-3} \text{ m}^3)$
Length	angstrom	Å	$(1 \text{ Å} = 0.1 \text{ nm})$
Pressure	atmosphere	atm	$(1 \text{ atm} = 101.325 \text{ kPa})$
	torr	mmHg	$(1 \text{ mmHg} = 133.32 \text{ Pa})$
Energy	electron volt	eV	$(1 \text{ eV} = 1.602 \times 10^{-19} \text{ J})$
Temperature	degree Celsius	°C	$(T_C = T_K - 273.15)$
Concentration	molarity	M	$(1 \, M = 1 \text{ mol/L})$

A.2   UNCERTAINTY IN MEASUREMENT

There is a fundamental difference between stating that there are 12 inches in a foot and stating that the circumference of the earth at the equator is 24,903.01 miles. The first relationship is based on a *definition*. By convention, there are exactly 12 inches in 1 foot. The second relationship is based on a *measurement*. It reports the circumference of the Earth to within the limits of experimental error in an actual measurement.

Many unit factors are based on definitions. There are exactly 5280 feet in a mile and 2.54 centimeters in an inch, for example. Unit factors based on definitions are known with complete certainty. (There is no error or uncertainty associated with the numbers.) Measurements, however, are always accompanied by a finite amount of error or uncertainty, which reflects limitations in the techniques used to make them.

The first measurement of the circumference of the earth, in the third century B.C., for example, gave a value of 250,000 stadia, or 29,000 miles. As the quality of the instruments used to make the measurement improved, the amount of error gradually decreased. But it never disappeared. Regardless of how carefully measurements are made, they always contain an element of uncertainty.

Systematic and Random Errors

There are two sources of error in a measurement: (1) limitations in the sensitivity of the instruments used and (2) imperfections in the techniques used to make the measurement. These errors can be divided into two classes: systematic and random.

The idea of **systematic error** can be understood in terms of the bull's-eye analogy shown in Figure A.1. Imagine what would happen if you aimed at a target with a rifle whose sights were not properly adjusted. Instead of hitting the bull's-eye, you would systematically hit the target at another point. Your results would be influenced by a systematic error caused by an imperfection in the equipment being used.

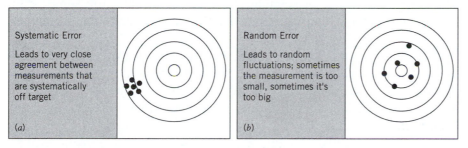

FIGURE A.1 (a) Systematic errors give results that are systematically too small or too large. (b) Random errors give results that fluctuate between being too small and too large.

Systematic error can also result from mistakes the individual makes while taking the measurement. In the bull's-eye analogy, a systematic error of this kind might occur if you flinched and pulled the rifle toward you each time it was fired.

To understand **random error,** imagine what would happen if you closed your eyes for an instant just before you fired the rifle. The bullets would hit the target more or less randomly. Some would hit too high, others would hit too low. Some would hit too far to the right, others too far to the left. Instead of an error that systematically gives a result too far in one direction, you now have a random error with random fluctuations.

Random errors most often result from limitations in the equipment or techniques used to make a measurement. Suppose, for example, that you wanted to collect 25 mL of a solution. You could use a beaker, a graduated cylinder, or a buret. Volume measurements made with a 50-mL beaker are accurate to within ±5 mL. In other words, you would be as likely to obtain 20 mL of solution (5 mL too little) as 30 mL (5 mL too much). You could decrease the amount of error by using a graduated cylinder that is capable of measurements to within ±1 mL. The error could be decreased even further by using a buret, which is capable of delivering a volume to within 1 drop, or ±0.05 mL.

Exercise A.1

Which of the following would lead to systematic errors, and which would produce random errors?

(a) Using a 1-qt milk carton to measure 1-L samples of milk.

(b) Using a balance that is sensitive to ±0.1 g to obtain 250-mg samples of vitamin C.

Solution

Procedure (a) would result in a systematic error. The volume would always be too small because a quart is slightly smaller than a liter. Procedure (b) would produce a random error because the equipment used to make the measurement is not sensitive enough.

Accuracy and Precision

To most people, *accuracy* and *precision* are synonyms, words that have the same or nearly the same meaning. In the physical sciences, there is an important difference between the terms. Measurements are **accurate** when they agree with the true value of the quantity being measured. They are **precise** when individual measurements of the same quantity agree.

The difference between accuracy and precision is similar to the difference between two terms from statistics: *validity* and *reliability*. Accuracy means the same thing as validity. A measurement is accurate, or valid, only if we get the correct answer. Precision is the same as reliability. A measurement is precise, or reliable, if we get essentially the same result each time we make the measurement. Measurements are therefore precise when they are reproducible.

To understand how systematic and random errors affect accuracy and precision, let's return to the bull's-eye analogy (see Figure A.2). Systematic errors influence the accuracy, but not the precision, of a measurement. It is possible to get measurements that are consistently the same, but systematically wrong. Random errors influence both the accuracy and the precision of the measurement.

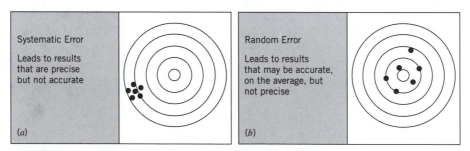

FIGURE A.2 (*a*) Systematic errors affect the accuracy of a measurement. The measurement may still be precise, but the results are systematically different from the correct answer. (*b*) Random errors can affect both the accuracy and precision of a measurement.

Systematic errors can be reduced by increasing the care and patience of the individual making the measurement or by improving the accuracy of the equipment used. Random errors can be reduced by averaging the results of many measurements of the same quantity because the precision of a series of measurements increases with the square root of the number of measurements.

A.3 SIGNIFICANT FIGURES

It is important to be honest when reporting a measurement, so that it doesn't appear more accurate than the equipment used to make the measurement allows. We can achieve this goal by controlling the number of digits, or **significant figures,** used to report the measurement.

Imagine what would happen if you determined the mass of an old copper penny on a postage scale and then on an analytical balance. The postage scale might give a mass of about 3 g, which means that the penny is closer to 3 g than either 2 g or 4 g. The analytical balance is inherently more sensitive; it can measure the mass of an object to the nearest ± 0.001 g, in which case you might find that the penny has a mass of 2.531 g.

Postage scale	3 ± 1 g
Analytical balance	2.531 ± 0.001 g

The postage scale gave a measurement that had only one reliable digit. That measurement is therefore said to be good to only one significant figure. The analytical balance gave four significant figures (2.531). The number of significant figures in a measurement is the number of digits that are known with some degree of confidence (2, 5, and 3) plus the last digit (1), which is generally an estimate or approximation. As we improve the sensitivity of the equipment used to make a measurement, the number of significant figures increases.

At first glance, it might seem that we can determine the number of significant figures by simply counting the digits in the measurement. Unfortunately zeros represent a problem. Zeros in a number can be classified in three ways: leading zeros, trailing zeros, and zeros between two significant figures.

- Leading zeros are never significant. Leading zeros are used only to set the position of the decimal point. Consider the numbers 0.0045, 0.045, 0.45, 4.5, and 45. In each case the relative error is the same. The number is known to an accuracy of ± 1 part in 45. Therefore, there are two significant figures in all five of these numbers.
- Zeros between two significant figures are always significant. In the number 3105 the zero is between 1 and 5 both of which are significant. Therefore, the zero is also significant, giving a total of four significant figures. The numbers 40.05, 0.0102, and 1706.2 have four, three, and five significant figures, respectively.
- Trailing zeros that are not needed to hold the decimal point are significant. For example 4.00 has three significant figures because the trailing zeros are not necessary to show the position of the decimal. Their function is to show that the number is known to 4.00 ± 0.01. Trailing zeros where there is no decimal point present a problem. It is often unclear how many zeros are significant. For example, does the measurement of 400 mL represent 400 ± 1 mL, 400 ± 10 mL, or 400 ± 100 mL? The only way to clearly show the number of significant figures in a measurement like this is to express the number in scientific notation. The measurement 400 mL can be written as 4.00×10^2 (three significant figures), 4.0×10^2 (two significant figures), or 4×10^2 (one significant figure) depending on the degree to which the measurement is known.

Addition and Subtraction with Significant Figures

What is the mass of a solution prepared by adding 0.507 g of salt to 150.0 g of water? If we attacked the problem without considering significant figures, we would simply add the two measurements.

150.0 g H_2O	(without using significant figures)
0.507 g salt	
150.507 g solution	

But this answer doesn't make sense. We only know the mass of the water to the nearest tenth of a gram, so we can only know the total mass of the solution to within ±0.1 g. Taking significant figures into account, we find that adding 0.507 g of salt to 150.0 g of water gives a solution with a mass of 150.5 g.

$$
\begin{array}{l}
150.0 \text{ g H}_2\text{O} \qquad\qquad \text{(using significant figures)} \\
\underline{0.507 \text{ g salt}} \\
150.5 \text{ g solution}
\end{array}
$$

Many of the calculations in this book are done by combining measurements with different degrees of accuracy and precision. The guiding principle in carrying out the calculations can be stated for addition and subtraction as follows:

When measurements are added or subtracted, the number of significant figures to the right of the decimal in the answer is determined by the measurement with the fewest digits to the right of the decimal.

Multiplication and Division with Significant Figures

For multiplication and division we count the total number of significant figures in each measurement, not just the number of decimal places.

When measurements are multiplied or divided, the answer can contain no more total significant figures than the measurement with the fewest total number of significant figures.

To illustrate this rule, let's calculate the cost of the copper in one of the old pennies that is pure copper. Let's assume that the penny has a mass of 2.531 g, that it is essentially pure copper, and that the price of copper is 67 cents per pound. We can start by converting from grams to pounds.

$$
2.531 \text{ g} \times \frac{1.000 \text{ lb}}{453.6 \text{ g}} = 0.005580 \text{ lb}
$$

We then use the price of a pound of copper to calculate the cost of the copper metal.

$$
0.005580 \text{ lb} \times \frac{67¢}{1.0 \text{ lb}} = 0.37¢
$$

There are four significant figures in both the mass of the penny (2.531) and the number of grams in a pound (453.6). But there are only two significant figures in the price of copper, so the final answer can only have two significant figures.

The Difference between Measurements and Definitions

Assume that you wanted to calculate the length in inches of a piece of wood 1.245 feet long. The calculation is easy to set up.

$$
1.245 \text{ ft} \times \frac{12 \text{ in.}}{1 \text{ ft}} = 14.94 \text{ in.}
$$

It is not so easy to decide how many significant figures the final answer should have. The original measurement (1.245 ft) has four significant figures, but there seem to be only two significant figures in the number of inches in a foot. Thus, it might seem that the answer should contain only two significant figures.

We can clear up this confusion by remembering that some unit factors are based on definitions. For example, 1 foot is defined as exactly 12 inches. Unit factors based on definitions have an infinite number of significant figures. The answer to this problem therefore contains four significant figures.

Rounding Off

When the answer to a calculation contains too many significant figures, it must be rounded off. Assume that the answer to a calculation is 1.247 and that the least accurate measurement has only three significant figures. The simplest way to round off this number would be to ignore the final digit and report the first three: 1.24. This approach has the disadvantage of introducing a systematic error into our calculations. Each time we round off, we would underestimate the value of the final answer.

There are 10 digits that can occur in the last decimal place in a calculation. One way of rounding off involves underestimating the answer for five of the digits (0, 1, 2, 3, and 4) and overestimating the answer for the other five (5, 6, 7, 8, and 9). This approach to rounding off is summarized as follows.

- If the digit is smaller than 5, drop the digit and leave the remaining number unchanged. Thus, 1.684 becomes 1.68.
- If the digit is 5 or larger, drop the digit and add 1 to the preceding digit. Thus, 1.247 becomes 1.25.

A.4 SCIENTIFIC NOTATION

Chemists routinely work with numbers that are extremely small. (The mass of an electron, for example, is 0.000,000,000,000,000,000,000,000,000,911 g.) They also work with numbers that are extremely large. (There are 10,300,000,000,000,000,000,000 carbon atoms in a 1-carat diamond.) There isn't a calculator made that will accept either of these numbers as they are written here. Before we can use these numbers, it is necessary to convert them to **scientific notation,** that is, to convert them to a number between 1 and 10 multiplied by 10 raised to some exponent.

Before we discuss how to translate numbers into scientific notation, it may be useful to review some of the basics of exponential mathematics.

- A number raised to the zero power is equal to 1.

$$10^0 = 1$$

- A number raised to the first power is equal to itself.

$$10^1 = 10$$

- A number raised to the nth power is equal to the product of that number times itself $n - 1$ times.

$$10^5 = 10 \times 10 \times 10 \times 10 \times 10 = 100,000$$

- Dividing by a number raised to some exponent is the same as multiplying by that number raised to an exponent of the opposite sign.

$$\frac{Z}{10^2} = Z \times 10^{-2} \qquad \frac{Z}{10^{-3}} = Z \times 10^3$$

The following rule can be used to convert numbers into scientific notation:

The exponent in scientific notation is equal to the number of times the decimal point must be moved to produce a number between 1 and 10.

According to the 1990 census, the population of Chicago was $6,070,000 \pm 1000$. Note that the population is known only to four significant figures because the error in the census is ± 1000. To convert the number to scientific notation we move the decimal point to the left six times and use the correct number of significant figures.

$$6070000 \pm 1000 = 6.070 \pm 0.001 \times 10^6$$

Atoms are very small, and therefore there are an enormous number of atoms in a small quantity of a given substance. In 1.0 gram of carbon we can calculate that there are 50,140,000,000,000,000,000,000 atoms. How can this number be reported to the correct number of significant figures in scientific notation? First we note that 1.0 gram contains only two significant figures. Thus the number of carbon atoms can be given to only two digits. The decimal point is moved to the left 22 times, and we write

$$50,140,000,000,000,000,000,000 = 5.0 \times 10^{22}$$

To convert numbers smaller than 1 into scientific notation, we have to move the decimal point to the right. The decimal point in 0.000985, for example, must be moved to the right four times. There are only three significant figures in this number so the number written in scientific notation can contain only three digits.

$$0.000985 = 9.85 \times 10^{-4}$$

Converting 0.000,000,000,000,000,000,000,000,000,911 g per electron into scientific notation involves moving the decimal point to the right 28 times.

$$0.0000000000000000000000000911 = 9.11 \times 10^{-28}$$

The mass of an electron was therefore listed in Table 1.3 as 9.11×10^{-28} grams.

The primary reason for converting numbers into scientific notation is to make calculations with unusually large or small numbers less cumbersome. But there is another important advantage to scientific notation. Because zeros are no longer used to set the decimal point, all of the digits in a number in scientific notation are significant, as shown by the following examples:

$$1.03 \times 10^{22} \qquad \text{three significant figures}$$
$$9.852 \times 10^{-5} \qquad \text{four significant figures}$$
$$2.0 \times 10^{-23} \qquad \text{two significant figures}$$

Exercise A.2

Convert the following numbers into scientific notation.
(a) 0.004694   (b) 1.98   (c) $4,679,000 \pm 100$

Solution

(a) 4.694×10^{-3} (b) 1.98×10^{0} (c) 4.6790×10^{6}

A.5   THE GRAPHICAL TREATMENT OF DATA

If the basis of science is a natural curiosity about the world that surrounds us, an important step in doing science is trying to find patterns in the observations and measurements that result from this curiosity.

Anyone who has played with a prism, or seen a rainbow, has watched what happens when white light is split into a spectrum of different colors. The difference between the blue and red light in the spectrum is the result of differences in the frequencies and wavelengths of the light, as shown in Table A.7.

TABLE A.7   Characteristic Wavelengths (λ) and Frequencies (ν) of Light of Different Colors

Color	Wavelength (m)	Frequency (s^{-1})
Violet	4.100×10^{-7}	7.312×10^{14}
Blue	4.700×10^{-7}	6.379×10^{14}
Green	5.200×10^{-7}	5.765×10^{14}
Yellow	5.800×10^{-7}	5.169×10^{14}
Orange	6.000×10^{-7}	4.997×10^{14}
Red	6.500×10^{-7}	4.612×10^{14}

There is an obvious pattern in the data: as the wavelength of the light becomes larger, the frequency becomes smaller. But recognizing this pattern is not enough. It would be even more useful to construct a mathematical equation that fits the data. This would allow us to calculate the frequency of light of any wavelength, such as blue-green light with a wavelength of 5.00×10^{-7} m. Or to calculate the wavelength of light of a known frequency, such as blue-violet light with a frequency of 7.00×10^{14} cycles per second.

The first step toward constructing an equation that fits the data involves plotting the data in different ways until we get a straight line. We might decide, for example, to plot the wavelengths on the vertical axis and the frequencies on the horizontal axis, as shown in Figure A.3. When we construct a graph, we should keep the following points in mind.

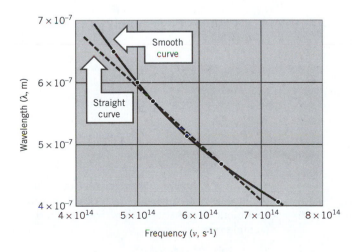

FIGURE A.3 Plot of wavelength (λ) in meters versus frequency (ν) in cycles per second for light of different colors. Note that the curve is almost, but not quite, a straight line.

- The scales of the graph should be chosen so that the data fill as much of the available space as possible.
- It isn't necessary to include the origin (0,0) on the graph. In fact, it may be more efficient to leave off the origin, so that the data fill the available space.
- Once the scales have been chosen and labeled, the data are plotted one point at a time.
- A straight line or a smooth curve is then drawn through as many points as possible. Because of experimental error, the line or curve may not pass through every data point.

If this process gives a straight line, we can conclude that the quantity plotted on the vertical axis (λ) is **directly proportional** to the quantity on the horizontal axis (v). We can then fit the graph to the equation for a straight line: $y = mx + b$.

$$\lambda = mv + b$$

The graph in Figure A.3 is not quite a straight line. Because the data are known to four significant figures, the deviation from a straight line is not the result of experimental error. We must therefore conclude that the wavelength and frequency of light are not directly proportional. We might therefore test whether the two sets of measurements are **inversely proportional.** In other words, we might try to fit the data to the following equation.

$$\lambda = m \left(\frac{1}{v} \right) + b$$

We can test this relationship by plotting the wavelengths in Table A.7 versus the inverse of the frequencies, as shown in Figure A.4.

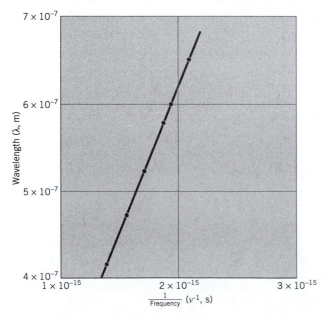

FIGURE A.4 A plot of wavelength (λ) versus the inverse of frequency ($1/v$) for light of different colors gives a straight line, within experimental error.

The graph in Figure A.4 gives a beautiful straight-line relationship. We can now calculate the slope of the line (m) and the y intercept (b), as shown in Figure A.5. When doing the calculation, it is important to choose points that are *not* original data points. When this is done, the slope of the line in Figure A.4 is found to be equal to the speed of light: 2.998×10^8 meters per second.

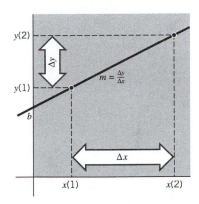

FIGURE A.5 To calculate the slope (m) of a straight line, divide the distance between two points on the vertical axis by the corresponding distance between two points along the horizontal axis. The y-intercept (b) can be determined by reading the value on the vertical axis that corresponds to a value of zero on the horizontal axis.

When the graph includes they origin, the y-intercept can be determined by reading the value on the vertical axis that corresponds to a value of zero on the horizontal axis. This isn't possible for the data in Figure A.4, however, because the origin was left off the graph. Another approach to determining the intercept starts by selecting a point on the straight line. We then combine values of y and x read from the plot for this point with the slope of the line to calculate the value of the intercept. When this approach is applied to the data in Figure A.4, we find that the intercept is equal to zero. The data in Figure A.4 therefore fit the following equation.

$$\lambda = (2.998 \times 10^8 \text{ m/s}) \left(\frac{1}{\nu} \right)$$

Rearranging the equation, we find that the product of the frequency times the wavelength of light is equal to the speed of light.

$$\nu\lambda = 2.998 \times 10^8 \text{ m/s}$$

Exercise A.3

The following data were obtained from a study of the relationship between the volume (V) and temperature (T) of a gas at constant pressure.

Volume (mL)	273.0	277.4	282.7	287.8	293.1	298.1
Temperature (°C)	0.0	5.0	10.0	15.0	20.0	25.0

Determine the temperature at which the volume of the gas should become equal to zero.

Solution

A plot of the data gives a straight line, as shown in Figure A.6. The equation for the straight line can be written as follows:

$$T = mV + b$$

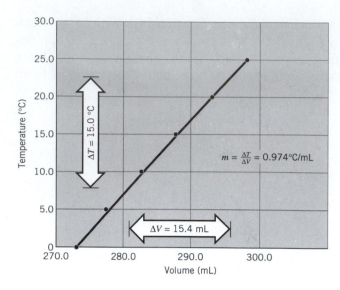

FIGURE A.6 Plot of the temperature versus volume data from Exercise A.3.

By choosing any two points on the curve, such as $T = 7.5°C$ and $T = 22.5°C$, and reading the volumes at those temperatures from the straight line that passes through the points, we can calculate the slope of the line.

$$m = \frac{\Delta T}{\Delta V} = \frac{22.5°C - 7.5°C}{295.5 \text{ mL} - 280.1 \text{ mL}} = 0.974°C/mL$$

The value of b for the equation can be calculated from the data for any point on the straight line, such as the point at which the temperature is 22.5°C and the volume is 295.5 mL.

$$b = T - mV = 22.5°C - \frac{0.974°C}{1 \text{ mL}} (295.5 \text{ mL}) = -265°C$$

According to these data, the volume of the gas should become zero when the gas is cooled until the temperature reaches about −265°C.

A P P E N D I X
B
TABLES

TABLE B.1   Values of Selected Fundamental Constants

Speed of light in a vacuum (c)	$c = 2.99792458 \times 10^8$ m/s
Charge on an electron (q_e)	$q_e = 1.60217733 \times 10^{-19}$ C
Rest mass of an electron (m_e)	$m_e = 9.109389 \times 10^{-28}$ g
	$m_e = 5.485799 \times 10^{-4}$ amu
Rest mass of a proton (m_p)	$m_p = 1.672623 \times 10^{-24}$ g
	$m_p = 1.00727647$ amu
Rest mass of a neutron (m_n)	$m_n = 1.674928 \times 10^{-24}$ g
	$m_n = 1.0086649$ amu
Faraday's constant (F)	$F = 96,485$ C/mol
Planck's constant (h)	$h = 6.626075 \times 10^{-34}$ J·s
Ideal gas constant (R)	$R = 0.0820568$ L·atm/mol·K
	$R = 8.31451$ J/mol·K
Atomic mass unit (amu)	1 amu $= 1.6605402 \times 10^{-24}$ g
Boltzmann's constant (k)	$k = 1.380658 \times 10^{-23}$ J/K
Avogadro's constant (N)	$N = 6.0221367 \times 10^{23}$ mol^{-1}
Rydberg constant (R_H)	$R_H = 1.0973715 \times 10^7$ m^{-1}
	$= 1.0973715 \times 10^{-2}$ nm^{-1}

TABLE B.2   Selected Conversion Factors

Energy	$1 \text{ J} = 0.2390 \text{ cal} = 10^7 \text{ erg}$
	$1 \text{ cal} = 4.184 \text{ J}$ (by definition)
	$1 \text{ eV/atom} = 1.6021793 \times 10^{-19} \text{ J/atom} = 96.485 \text{ kJ/mol}$
Temperature	$\text{K} = {}^{\circ}\text{C} + 273.15$
	${}^{\circ}\text{C} = 5/9({}^{\circ}\text{F} - 32)$
	${}^{\circ}\text{F} = 9/5({}^{\circ}\text{C}) + 32$
Pressure	$1 \text{ atm} = 760 \text{ mmHg}$ (by definition) $= 760 \text{ torr}$ (by definition)
	$1 \text{ atm} = 101.325 \text{ kPa} = 14.7 \text{ lb/in}^2$
Mass	$1 \text{ kg} = 2.2046 \text{ lb}$
	$1 \text{ lb} = 453.59 \text{ g} = 0.45359 \text{ kg}$
	$1 \text{ oz} = 0.06250 \text{ lb} = 28.350 \text{ g}$
	$1 \text{ ton} = 2000 \text{ lb} = 907.185 \text{ kg}$
	$1 \text{ tonne (metric)} = 1000 \text{ kg} = 2204.62 \text{ lb}$
Volume	$1 \text{ mL} = 0.001 \text{ L} = 1 \text{ cm}^3$ (by definition)
	$1 \text{ oz (fluid)} = 0.031250 \text{ qt} = 0.029573 \text{ L}$
	$1 \text{ qt} = 0.9463529 \text{ L}$
	$1 \text{ L} = 1.05672 \text{ qt}$
Length	$1 \text{ m} = 39.370 \text{ in.}$
	$1 \text{ mi} = 1.60934 \text{ km}$
	$1 \text{ in.} = 2.54 \text{ cm}$ (by definition)

TABLE B.3   The Vapor Pressure of Water

Temperature (°C)	Pressure (mmHg)	Temperature (°C)	Pressure (mmHg)	Temperature (°C)	Pressure (mmHg)	Temperature (°C)	Pressure (mmHg)
0	4.6	13	11.2	26	25.2	39	52.4
1	4.9	14	12.0	27	26.7	40	55.3
2	5.3	15	12.8	28	28.3	41	58.3
3	5.7	16	13.6	29	30.0	42	61.5
4	6.1	17	14.5	30	31.8	43	64.8
5	6.5	18	15.5	31	33.7	44	68.3
6	7.0	19	16.5	32	35.7	45	71.9
7	7.5	20	17.5	33	37.7	46	75.7
8	8.0	21	18.7	34	39.9	47	79.6
9	8.6	22	19.8	35	42.2	48	83.7
10	9.2	23	21.1	36	44.6	49	88.0
11	9.8	24	22.4	37	47.1	50	92.5
12	10.5	25	23.8	38	49.7		

TABLE B.4   Radii of Atoms and Ions

Element	Ionic Radius (nm)	Ionic Charge	Covalent Radius (nm)	Metallic Radius (nm)
Aluminum	0.050	(+3)	0.125	0.1431
Antimony	0.245	(−3)	0.141	
	0.09	(+3)		
	0.062	(+5)		
Arsenic	0.222	(−3)	0.121	0.1248
	0.058	(+3)		
	0.047	(+5)		
Astatine	0.227	(−1)		
	0.051	(+7)		
Barium	0.135	(+2)	0.198	0.2173
Beryllium	0.031	(+2)	0.089	0.1113
Bismuth	0.213	(−3)	0.152	0.1547
	0.096	(+3)		
	0.074	(+5)		
Boron	0.020	(+3)	0.088	0.083
Bromine	0.196	(−1)	0.1142	
	0.039	(+7)		
Cadmium	0.097	(+2)	0.141	0.1489
Calcium	0.099	(+2)	0.174	0.1973
Carbon	0.260	(−4)	0.077	
	0.015	(+4)		
Cesium	0.169	(+1)	0.235	0.2654
Chlorine	0.181	(−1)	0.099	
	0.026	(+7)		
Chromium	0.064	(+3)	0.117	0.1249
	0.052	(+6)		
Cobalt	0.074	(+2)	0.116	0.1253
	0.063	(+3)		
Copper	0.096	(+1)	0.117	0.1278
	0.072	(+2)		
Fluorine	0.136	(−1)	0.064	0.0717
	0.007	(+7)		
Francium	0.176	(+1)		0.27
Gallium	0.062	(+3)	0.125	0.1221
Germanium	0.272	(−4)	0.122	0.1225
	0.053	(+4)		
Gold	0.137	(+1)	0.134	0.1442
	0.091	(+3)		
Hydrogen	0.208	(−1)	0.0371	
	10^{-6}	(+1)		
Indium	0.081	(+3)	0.150	0.1626

(*continued*)

TABLE B.4   Radii of Atoms and Ions (continued)

Element	Ionic Radius (nm)	Ionic Charge	Covalent Radius (nm)	Metallic Radius (nm)
Iodine	0.216	(−1)	0.1333	
	0.050	(+7)		
Iron	0.076	(+2)	0.1165	0.1241
	0.064	(+3)		
Lead	0.215	(−4)	0.154	0.1750
	0.120	(+2)		
	0.084	(+4)		
Lithium	0.068	(+1)	0.123	0.152
Magnesium	0.065	(+2)	0.136	0.160
Manganese	0.080	(+2)	0.117	0.124
	0.046	(+7)		
Mercury	0.127	(+1)	0.144	0.160
	0.110	(+2)		
Molybdenum	0.062	(+6)	0.129	0.1362
Nickel	0.072	(+2)	0.115	0.1246
Nitrogen	0.171	(−3)	0.070	
	0.013	(+3)		
	0.011	(+5)		
Oxygen	0.140	(−2)	0.066	
	0.176	(−1)		
Phosphorus	0.212	(−3)	0.110	0.108
	0.042	(+3)		
	0.034	(+5)		
Polonium	0.230	(−2)	0.153	0.167
	0.056	(+6)		
Potassium	0.133	(+1)	0.2025	0.2272
Radium	0.140	(+2)		0.220
Rubidium	0.148	(+1)	0.216	0.2475
Scandium	0.081	(+3)	0.144	0.1606
Selenium	0.198	(−2)	0.117	
	0.069	(+4)		
	0.042	(+6)		
Silicon	0.271	(−4)	0.117	
	0.041	(+4)		
Silver	0.126	(+1)	0.134	0.1444
Sodium	0.095	(+1)	0.157	0.1537
Strontium	0.113	(+2)	0.192	0.2151
Sulfur	0.184	(−2)	0.104	
	0.037	(+4)		
	0.029	(+6)		

Element	Ionic Radius (nm)	Ionic Charge	Covalent Radius (nm)	Metallic Radius (nm)
Tellurium	0.221	(−2)	0.137	0.1432
	0.081	(+4)		
	0.056	(+6)		
Thallium	0.095	(+3)	0.155	0.1704
Tin	0.294	(−4)	0.140	0.1405
	0.102	(+2)		
	0.071	(+4)		
Titanium	0.090	(+2)	0.132	0.1448
	0.068	(+4)		
Tungsten	0.065	(+6)	0.130	0.1370
Uranium	0.083	(+6)		0.1385
Vanadium	0.059	(+5)	0.122	0.1321
Xenon			0.209	
Zinc	0.074	(+2)	0.125	0.1332
Zirconium	0.079	(+4)	0.145	0.167

TABLE B.5   Ionization Energies (kJ/mol)

Atomic Number	Symbol	I	II	III	IV	V	VI	VII
1	H	1,312.0						
2	He	2,372.3	5,250.3					
3	Li	520.2	7,297.9	11,814.6				
4	Be	899.4	1,757.1	14,848.3	21,005.9			
5	B	800.6	2,427.0	3,659.6	25,025.0	32,825.7		
6	C	1,086.4	2,352.6	4,620.4	6,222.5	37,829.4	47,275.6	
7	N	1,402.3	2,856.0	4,578.0	7,474.9	9,444.7	50,370.4	64,358.0
8	O	1,313.9	3,388.2	5,300.3	7,469.1	10,989.2	13,326.1	71,332
9	F	1,681.0	3,374.1	6,050.3	8,407.5	11,022.4	15,163.6	17,867.2
10	Ne	2,080.6	3,952.2	6,122	9,370	12,177	15,238	19,998
11	Na	495.8	4,562.4	6,912	9,543	13,352	16,610	20,114
12	Mg	737.7	1,450.6	7,732.6	10,540	13,629	17,994	21,703
13	Al	577.6	1,816.6	2,744.7	11,577	14,831	18,377	23,294
14	Si	786.4	1,577.0	3,231.5	4,355.4	16,091	19,784	23,785
15	P	1,011.7	1,903.2	2,912	4,956	6,273.7	21,268	25,397
16	S	999.58	2,251	3,361	4,564	7,012	8,495.4	27,105
17	Cl	1,251.1	2,297	3,822	5,158	6,540	9,362	11,017.9
18	Ar	1,520.5	2,665.8	3,931	5,771	7,238	8,780.8	11,994.9
19	K	418.8	3,051.3	4,411	5,877	7,975	9,648.5	11,343
20	Ca	589.8	1,145.4	4,911.8	6,474	8,144	10,496	12,320
21	Sc	631	1,235	2,389	7,099	8,844	10,720	13,310
22	Ti	658	1,310	2,652.5	4,174.5	9,573	11,516	13,590
23	V	650	1,413	2,828.0	4,506.5	6,294	12,362	14,489
24	Cr	652.8	1,592	2,987	4,740	6,690	8,738	15,540
25	Mn	717.4	1,509.0	3,248.3	4,940	6,990	9,220	11,508
26	Fe	759.3	1,561	2,957.3	5,290	7,240	9,600	12,100
27	Co	758	1,646	3,232	4,950	7,670	9,840	12,400
28	Ni	736.7	1,752.9	3,393	5,300	7,280	10,400	12,800
29	Cu	745.4	1,957.9	3,553	5,330	7,710	9,940	13,400
30	Zn	906.4	1,733.2	3,832.6	5,730	7,970	10,400	12,900
31	Ga	578.8	1,979	2,963	6,200			
32	Ge	762.1	1,537.4	3,302	4,410	9,020		
33	As	947	1,797.8	2,735.4	4,837	6,043	12,300	
34	Se	940.9	2,045	2,973.7	4,143.4	6,590	7,883	14,990
35	Br	1,139.9	2,100	3,500	4,560	5,760	8,550	9,938
36	Kr	1,350.7	2,350.3	3,565	5,070	6,240	7,570	10,710
37	Rb	403.0	2,632	3,900	5,070	6,850	8,140	9,570
38	Sr	549.5	1,064.5	4,120	5,500	6,910	8,760	10,200
39	Y	616	1,181	1,980	5,960	7,430	8,970	11,200
40	Zr	660	1,267	2,218	3,313	7,870		
41	Nb	664	1,382	2,416	3,960	4,877	9,899	12,100
42	Mo	684.9	1,558	2,621	4,480	5,910	6,600	12,230

Atomic Number	Symbol	I	II	III	IV	V	VI	VII
43	Tc	702	1,472	2,850				
44	Ru	711	1,617	2,747				
45	Rh	720	1,744	2,997				
46	Pd	805	1,874	3,177				
47	Ag	731.0	2,073	3,361				
48	Cd	867.7	1,631.4	3,616				
49	In	558.3	1,820.6	2,704	5,200			
50	Sn	708.6	1,411.8	2,943.0	3,930.2	6,974		
51	Sb	833.7	1,595	2,440	4,260	5,400	10,400	
52	Te	869.2	1,790	2,698	3,609	5,668	6,820	13,200
53	I	1,008.4	1,845.8	3,200				
54	Xe	1,170.4	2,046	3,100				
55	Cs	375.7	2,440					
56	Ba	502.9	965.23					
57	La	538.1	1,067	1,850.3				
58	Ce	527.8	1,047	1,949	3,547			
59	Pr	523	1,018	2,086	3,761	5,543		
60	Nd	530	1,035	2,130	3,900			
61	Pm	535	1,052	2,150	3,970			
62	Sm	543	1,068	2,260	3,990			
63	Eu	547	1,084	2,400	4,110			
64	Gd	592	1,167	1,990	4,250			
65	Tb	564	1,112	2,110	3,840			
66	Dy	572	1,126	2,200	4,000			
67	Ho	581	1,139	2,203	4,100			
68	Er	589	1,151	2,194	4,120			
69	Tm	596	1,163	2,285	4,120			
70	Yb	603	1,175	2,415	4,216			
71	Lu	524	1,340	2,022	4,360			
72	Hf	642	1,440	2,250	3,210			
73	Ta	761						
74	W	770						
75	Re	760	1,260	2,510	3,640			
76	Os	840						
77	Ir	880						
78	Pt	870	1,791.0					
79	Au	890.1	1,980					
80	Hg	1,007.0	1,809.7	3,300				
81	Tl	589.3	1,971.0	2,878				
82	Pb	715.5	1,450.4	3,081.4	4,083	6,640		
83	Bi	703.3	1,610	2,466	4,370	5,400	8,520	
84	Po	812						

(continued)

TABLE B.5   Ionization Energies (kJ/mol) (continued)

Atomic Number	Symbol	I	II	III	IV	V	VI	VII
85	At							
86	Rn	1,037.0						
87	Fr							
88	Ra	509.3	979.0					
89	Ac	498.8	1,170					
90	Th	587	1,110	1,930	2,780			
91	Pa	568						
92	U	584						
93	Np	597						
94	Pu	585						
95	Am	578.2						
96	Cm	581						
97	Bk	601						
98	Cf	608						
99	Es	619						
100	Fm	627						
101	Md	635						
102	No	642						

TABLE B.6   Electron Affinities

Atomic Number	Symbol	Electron Affinity (kJ/mol)	Atomic Number	Symbol	Electron Affinity (kJ/mol)
1	H	72.8	37	Rb	46.89
2	He	*	38	Sr	*
3	Li	59.8	39	Y	0
4	Be	*	40	Zr	50
5	B	27	41	Nb	96
6	C	122.3	42	Mo	96
7	N	−7	43	Tc	70
8	O	141.1	44	Ru	110
9	F	328.0	45	Rh	120
10	Ne	*	46	Pd	60
11	Na	52.7	47	Ag	125.7
12	Mg	*	48	Cd	*
13	Al	45	49	In	29
14	Si	133.6	50	Sn	121
15	P	71.7	51	Sb	101
16	S	200.42	52	Te	190.15
17	Cl	348.8	53	I	295.3
18	Ar	*	54	Xe	*
19	K	48.36	55	Cs	45.49
20	Ca	*	56	Ba	*
21	Sc	*	57–71	La–Lu	50
22	Ti	20	72	Hf	*
23	V	50	73	Ta	60
24	Cr	64	74	W	60
25	Mn	*	75	Re	14
26	Fe	24	76	Os	110
27	Co	70	77	Ir	150
28	Ni	111	78	Pt	205.3
29	Cu	118.3	79	Au	222.74
30	Zn	0	80	Hg	*
31	Ga	29	81	Tl	30
32	Ge	120	82	Pb	110
33	As	77	83	Bi	110
34	Se	194.96	84	Po	180
35	Br	324.6	85	At	270
36	Kr	*	86	Rn	*
			87	Fr	44.0

*These elements have negative electron affinities. Electron affinity is the negative of the enthalpy change for the reaction $X(g) + e^- \rightarrow X^-(g)$.

TABLE B.7   Electronegativities[a]

Atomic Number	Element	Electro-negativity	Atomic Number	Element	Electro-negativity
1	H	2.300	41	Nb	1.25
2	He	4.157	42	Mo	1.39
3	Li	0.912	43	Tc	1.52
4	Be	1.576	44	Ru	1.66
5	B	2.051	45	Rh	1.79
6	C	2.544	46	Pd	1.91
7	N	3.066	47	Ag	1.98
8	O	3.610	48	Cd	1.52
9	F	4.193	49	In	1.656
10	Ne	4.787	50	Sn	1.824
11	Na	0.869	51	Sb	1.984
12	Mg	1.293	52	Te	2.158
13	Al	1.613	53	I	2.359
14	Si	1.916	54	Xe	2.582
15	P	2.253	55	Cs	0.66
16	S	2.589	56	Ba	0.88
17	Cl	2.869	57–71		
18	Ar	3.242	72		
19	K	0.734	73		
20	Ca	1.034	74		
21	Sc	1.15	75		
22	Ti	1.25	76		
23	V	1.37	77		
24	Cr	1.45	78		
25	Mn	1.55	79		
26	Fe	1.67	80	Hg	1.76
27	Co	1.76	81		
28	Ni	1.86	82		
29	Cu	1.83	83		
30	Zn	1.59	84		
31	Ga	1.756	85		
32	Ge	1.994	86		
33	As	2.211	87		
34	Se	2.424	88		
35	Br	2.685	89		
36	Kr	2.966	90		
37	Rb	0.706	91		
38	Sr	0.963	92		
39	Y	1.00	93–103		
40	Zr	1.12			

[a]Allen electronegativities taken from L. C. Allen, *Int. J. Quant. Chem.,* **48,** 253–277 (1993); L. C. Allen, *J. Am. Chem. Soc.,* **111,** 9003–9014 (1989).

TABLE B.8   Acid Dissociation Equilibrium Constants

Compound	Dissociation Reaction	K_a	pK_a
Acetic acid	$CH_3CO_2H + H_2O \rightleftharpoons CH_3CO_2^- + H_3O^+$	1.75×10^{-5}	4.757
Ammonium ion	$NH_4^+ + H_2O \rightleftharpoons NH_3 + H_3O^+$	5.6×10^{-10}	9.25
Arsenic acid	$H_3AsO_4 + H_2O \rightleftharpoons H_2AsO_4^- + H_3O^+$	6.0×10^{-3}	2.22
	$H_2AsO_4^- + H_2O \rightleftharpoons HAsO_4^{2-} + H_3O^+$	1.0×10^{-7}	7.00
	$HAsO_4^{2-} + H_2O \rightleftharpoons AsO_4^{3-} + H_3O^+$	3.0×10^{-12}	11.52
Arsenous acid	$H_3AsO_3 + H_2O \rightleftharpoons H_2AsO_3^- + H_3O^+$	6.0×10^{-10}	9.22
	$H_2AsO_3^- + H_2O \rightleftharpoons HAsO_3^{2-} + H_3O^+$	3.0×10^{-14}	13.52
Benzoic acid	$C_6H_5CO_2H + H_2O \rightleftharpoons C_6H_5CO_2^- + H_3O^+$	6.3×10^{-5}	4.20
Boric acid	$H_3BO_3 + H_2O \rightleftharpoons H_2BO_3^- + H_3O^+$	7.3×10^{-10}	9.14
Carbonic acid	$H_2CO_3 + H_2O \rightleftharpoons HCO_3^- + H_3O^+$	4.5×10^{-7}	6.35
	$HCO_3^- + H_2O \rightleftharpoons CO_3^{2-} + H_3O^+$	4.7×10^{-11}	10.33
Chloric acid	$HClO_3 + H_2O \rightleftharpoons ClO_3^- + H_3O^+$	5.0×10^2	-2.70
Chloroacetic acid	$ClCH_2CO_2H + H_2O \rightleftharpoons ClCH_2CO_2^- + H_3O^+$	1.4×10^{-3}	2.85
Chlorous acid	$HClO_2 + H_2O \rightleftharpoons ClO_2^- + H_3O^+$	1.1×10^{-2}	1.96
Chromic acid	$H_2CrO_4 + H_2O \rightleftharpoons HCrO_4^- + H_3O^+$	9.6	-0.98
	$HCrO_4^- + H_2O \rightleftharpoons CrO_4^{2-} + H_3O^+$	3.2×10^{-7}	6.50
Citric acid	$H_3Cit + H_2O \rightleftharpoons H_2Cit^- + H_3O^+$	7.5×10^{-4}	3.13
	$H_2Cit^- + H_2O \rightleftharpoons HCit^{2-} + H_3O^+$	1.7×10^{-5}	4.77
	$HCit^{2-} + H_2O \rightleftharpoons Cit^{3-} + H_3O^+$	4.0×10^{-7}	6.40
Dichloroacetic acid	$Cl_2CHCO_2H + H_2O \rightleftharpoons Cl_2CHCO_2^- + H_3O^+$	5.1×10^{-2}	1.29
Formic acid	$HCO_2H + H_2O \rightleftharpoons HCO_2^- + H_3O^+$	1.8×10^{-4}	3.75
Glycine	$H_3N^+CH_2CO_2H + H_2O \rightleftharpoons$ $H_3N^+CH_2CO_2^- + H_3O^+$	4.5×10^{-3}	2.35
	$H_3N^+CH_2CO_2^- + H_2O \rightleftharpoons$ $H_2NCH_2CO_2^- + H_3O^+$	2.5×10^{-10}	9.60
Hydrazoic acid	$HN_3 + H_2O \rightleftharpoons N_3^- + H_3O^+$	1.9×10^{-5}	4.72
Hydrobromic acid	$HBr + H_2O \rightleftharpoons Br^- + H_3O^+$	1×10^9	-9
Hydrochloric acid	$HCl + H_2O \rightleftharpoons Cl^- + H_3O^+$	1×10^6	-6
Hydrocyanic acid	$HCN + H_2O \rightleftharpoons CN^- + H_3O^+$	6×10^{-10}	9.22
Hydrofluoric acid	$HF + H_2O \rightleftharpoons F^- + H_3O^+$	7.2×10^{-4}	3.14
Hydroiodic acid	$HI + H_2O \rightleftharpoons I^- + H_3O^+$	3×10^9	-9.5
Hydrogen peroxide	$H_2O_2 + H_2O \rightleftharpoons HO_2^- + H_3O^+$	2.2×10^{-12}	11.66
Hydrogen selenide	$H_2Se + H_2O \rightleftharpoons Hse^- + H_3O^+$	1.0×10^{-4}	4.00
Hydrogen sulfide	$H_2S + H_2O \rightleftharpoons HS^- + H_3O^+$	1.0×10^{-7}	7.00
	$HS^- + H_2O \rightleftharpoons S^{2-} + H_3O^+$	1.3×10^{-13}	12.89
Hypobromous acid	$HOBr + H_2O \rightleftharpoons OBr^- + H_3O^+$	2.4×10^{-9}	8.62
Hypochlorous acid	$HOCl + H_2O \rightleftharpoons OCl^- + H_3O^+$	2.9×10^{-8}	7.54
Hypoiodous acid	$HOI + H_2O \rightleftharpoons OI^- + H_3O^+$	2.3×10^{-11}	10.64
Iodic acid	$HIO_3 + H_2O \rightleftharpoons IO_3^- + H_3O^+$	0.16	0.80
Nitric acid	$HNO_3 + H_2O \rightleftharpoons NO_3^- + H_3O^+$	28	-1.45
Nitrous acid	$HNO_2 + H_2O \rightleftharpoons NO_2^- + H_3O^+$	5.1×10^{-4}	3.29
Oxalic acid	$H_2C_2O_4 + H_2O \rightleftharpoons HC_2O_4^- + H_3O^+$	5.4×10^{-2}	1.27
	$HC_2O_4^- + H_2O \rightleftharpoons C_2O_4^{2-} + H_3O^+$	5.4×10^{-5}	4.27

(continued)

TABLE B.8   Acid Dissociation Equilibrium Constants (continued)

Compound	Dissociation Reaction	K_a	pK_a
Perchloric acid	$HOClO_3 + H_2O \rightleftharpoons ClO_4^- + H_3O^+$	1×10^8	-8
Periodic acid	$H_5IO_6 + H_2O \rightleftharpoons H_4IO_6^- + H_3O^+$	2.3×10^{-2}	1.64
Phenol	$C_6H_5OH + H_2O \rightleftharpoons C_6H_5O^- + H_3O^+$	1.0×10^{-10}	10.00
Phosphoric acid	$H_3PO_4 + H_2O \rightleftharpoons H_2PO_4^- + H_3O^+$	7.1×10^{-3}	2.15
	$H_2PO_4^- + H_2O \rightleftharpoons HPO_4^{2-} + H_3O^+$	6.3×10^{-8}	7.20
	$HPO_4^{2-} + H_2O \rightleftharpoons PO_4^{3-} + H_3O^+$	4.2×10^{-13}	12.38
Phosphorous acid	$H_3PO_3 + H_2O \rightleftharpoons H_2PO_3^- + H_3O^+$	1.00×10^{-2}	2.00
	$H_2PO_3^- + H_2O \rightleftharpoons HPO_3^{2-} + H_3O^+$	2.6×10^{-7}	6.59
Sulfamic acid	$H_2NSO_3H + H_2O \rightleftharpoons H_2NSO_3^- + H_3O^+$	1.03×10^{-1}	0.987
Sulfuric acid	$H_2SO_4 + H_2O \rightleftharpoons HSO_4^- + H_3O^+$	1×10^3	-3
	$HSO_4^- + H_2O \rightleftharpoons SO_4^{2-} + H_3O^+$	1.2×10^{-2}	1.92
Sulfurous acid	$H_2SO_3 + H_2O \rightleftharpoons HSO_3^- + H_3O^+$	1.7×10^{-2}	1.77
	$HSO_3^- + H_2O \rightleftharpoons SO_3^{2-} + H_3O^+$	6.4×10^{-8}	7.19
Thiocyanic acid	$HSCN + H_2O \rightleftharpoons SCN^- + H_3O^+$	71	-1.85
Trichloroacetic acid	$Cl_3CCO_2H + H_2O \rightleftharpoons Cl_3CCO_2^- + H_3O^+$	0.22	0.66
Water	$H_2O + H_2O \rightleftharpoons OH^- + H_3O^+$	1.8×10^{-16}	15.75

TABLE B.9   Base Ionization Equilibrium Constants

Compound	Ionization Reaction	K_b	pK_b
Ammonia	$NH_3 + H_2O \rightleftharpoons NH_4^+ + OH^-$	1.8×10^{-5}	4.74
Aniline	$C_6H_5NH_2 + H_2O \rightleftharpoons C_6H_5NH_3^+ + OH^-$	4.0×10^{-10}	9.40
Butylamine	$CH_3(CH_2)_3NH_2 + H_2O \rightleftharpoons CH_3(CH_2)_3NH_3^+ + OH^-$	4.0×10^{-4}	3.40
Dimethylamine	$(CH_3)_2NH + H_2O \rightleftharpoons (CH_3)_2NH_2^+ + OH^-$	5.9×10^{-4}	3.23
Ethanolamine	$HOCH_2CH_2NH_2 + H_2O \rightleftharpoons HOCH_2CH_2NH_3^+ + OH^-$	3.3×10^{-5}	4.50
Ethylamine	$CH_3CH_2NH_2 + H_2O \rightleftharpoons CH_3CH_2NH_3^+ + OH^-$	4.4×10^{-4}	3.37
Hydrazine	$H_2NNH_2 + H_2O \rightleftharpoons H_2NNH_3^+ + OH^-$	1.2×10^{-6}	5.89
Hydroxylamine	$HONH_2 + H_2O \rightleftharpoons HONH_3^+ + OH^-$	1.1×10^{-8}	7.97
Methylamine	$CH_3NH_2 + H_2O \rightleftharpoons CH_3NH_3^+ + OH^-$	4.8×10^{-4}	3.32
Pyridine	$C_5H_5NH + H_2O \rightleftharpoons C_5H_5NH^+ + OH^-$	1.7×10^{-9}	8.77
Trimethylamine	$(CH_3)_3N + H_2O \rightleftharpoons (CH_3)_3NH^+ + OH^-$	6.3×10^{-5}	4.20
Urea	$H_2NCONH_2 + H_2O \rightleftharpoons H_2NCONH_3^+ + OH^-$	1.5×10^{-14}	13.82

TABLE B.10   Solubility Product Equilibrium Constants

Substance	K_{sp}	Substance	K_{sp}	Substance	K_{sp}
AgBr	5.0×10^{-13}	α-CoS	4.0×10^{-21}	MnCO$_3$	1.8×10^{-11}
AgCN	1.2×10^{-16}	β-CoS	2.0×10^{-25}	Mn(OH)$_2$	2×10^{-13}
Ag$_2$CO$_3$	8.1×10^{-12}	Cr(OH)$_3$	6.3×10^{-31}	MnS	3×10^{-13}
AgOH	2.0×10^{-8}	CuBr	5.3×10^{-9}	NiCO$_3$	6.6×10^{-9}
AgC$_2$H$_3$O$_2$	4.4×10^{-3}	CuCl	1.2×10^{-6}	NiC$_2$O$_4$	4×10^{-10}
Ag$_2$C$_2$O$_4$	3.4×10^{-11}	CuCN	3.2×10^{-20}	α-NiS	3.2×10^{-19}
AgCl	1.8×10^{-10}	CuCrO$_4$	3.6×10^{-6}	β-NiS	1.0×10^{-24}
Ag$_2$CrO$_4$	1.1×10^{-12}	CuCO$_3$	1.4×10^{-10}	γ-NiS	2.0×10^{-26}
AgI	8.3×10^{-17}	Cu(OH)$_2$	2.2×10^{-20}	PbBr$_2$	4.0×10^{-5}
Ag$_2$S	6.3×10^{-50}	CuI	1.1×10^{-13}	PbCO$_3$	7.4×10^{-14}
AgSCN	1.0×10^{-12}	Cu$_2$S	2.5×10^{-48}	PbC$_2$O$_4$	4.8×10^{-10}
Ag$_2$SO$_4$	1.4×10^{-5}	CuS	6.3×10^{-36}	PbCl$_2$	1.6×10^{-5}
Al(OH)$_3$	1.3×10^{-33}	CuSCN	4.8×10^{-15}	PbCrO$_4$	2.8×10^{-13}
AuCl	2.0×10^{-13}	FeCO$_3$	3.2×10^{-11}	PbF$_2$	2.7×10^{-8}
AuCl$_3$	3.2×10^{-23}	Fe$_2$C$_2$O$_4$	3.2×10^{-7}	Pb(OH)$_2$	1.2×10^{-15}
AuI	1.6×10^{-23}	Fe(OH)$_2$	8.0×10^{-16}	PbI$_2$	7.1×10^{-9}
AuI$_3$	5.5×10^{-46}	Fe(OH)$_3$	4×10^{-38}	PbS	8.0×10^{-28}
BaCO$_3$	5.1×10^{-9}	FeS	6.3×10^{-18}	PbSO$_4$	1.6×10^{-8}
BaC$_2$O$_4$	2.3×10^{-8}	Hg$_2$Br$_2$	5.6×10^{-23}	SnS	1.0×10^{-25}
BaCrO$_4$	1.2×10^{-10}	Hg$_2$(CN)$_2$	5×10^{-40}	Sn(OH)$_2$	1.4×10^{-28}
BaF$_2$	1.0×10^{-6}	Hg$_2$CO$_3$	8.9×10^{-17}	Sn(OH)$_4$	1×10^{-56}
Ba(OH)$_2$	5×10^{-3}	Hg$_2$(OAc)$_2$	3×10^{-11}	SrCO$_3$	1.1×10^{-10}
BaSO$_4$	1.1×10^{-10}	Hg$_2$C$_2$O$_4$	2.0×10^{-13}	SrC$_2$O$_4$	1.6×10^{-7}
Bi$_2$S$_3$	1×10^{-97}	HgC$_2$O$_4$	1×10^{-7}	SrCrO$_4$	2.2×10^{-5}
CaCO$_3$	2.8×10^{-9}	Hg$_2$Cl$_2$	1.3×10^{-18}	SrF$_2$	2.5×10^{-9}
CaC$_2$O$_4$	4×10^{-9}	Hg$_2$CrO$_4$	2.0×10^{-9}	SrSO$_4$	3.2×10^{-7}
CaCrO$_4$	7.1×10^{-4}	Hg$_2$I$_2$	4.5×10^{-29}	TlBr	3.4×10^{-6}
CaF$_2$	4.0×10^{-11}	Hg$_2$S	1.0×10^{-47}	TlCl	1.7×10^{-4}
Ca(OH)$_2$	5.5×10^{-6}	HgS	4×10^{-53}	TlI	6.5×10^{-8}
CdCO$_3$	5.2×10^{-12}	K$_2$NaCo(NO$_2$)$_6$	2.2×10^{-11}	Zn(CN)$_2$	2.6×10^{-13}
Cd(CN)$_2$	1.0×10^{-8}	MgCO$_3$	3.5×10^{-8}	ZnCO$_3$	1.4×10^{-11}
Cd(OH)$_2$	2.5×10^{-14}	MgC$_2$O$_4$	1×10^{-8}	ZnC$_2$O$_4$	2.7×10^{-8}
CdS	8×10^{-27}	MgF$_2$	6.5×10^{-9}	Zn(OH)$_2$	1.2×10^{-17}
CoCO$_3$	1.4×10^{-13}	Mg(OH)$_2$	1.8×10^{-11}	α-ZnS	1.6×10^{-24}
Co(OH)$_3$	1.6×10^{-44}	MgNH$_4$PO$_4$	2.5×10^{-13}	β-ZnS	2.5×10^{-22}

TABLE B.11   Complex Formation Equilibrium Constants

Equilibrium	K_f	Equilibrium	K_f
$Ag^+ + 2\ Br^- \rightleftharpoons AgBr_2^-$	2.1×10^7	$Fe^{2+} + 6\ CN^- \rightleftharpoons Fe(CN)_6^{4-}$	1×10^{35}
$Ag^+ + 2\ Cl^- \rightleftharpoons AgCl_2^-$	1.1×10^5	$Fe^{3+} + 6\ CN^- \rightleftharpoons Fe(CN)_6^{3-}$	1×10^{42}
$Ag^+ + 2\ CN^- \rightleftharpoons Ag(CN)_2^-$	1.3×10^{21}	$Fe^{3+} + SCN^- \rightleftharpoons Fe(SCN)^{2+}$	8.9×10^2
$Ag^+ + 2\ I^- \rightleftharpoons AgI_2^-$	5.5×10^{11}	$Fe^{3+} + 2\ SCN^- \rightleftharpoons Fe(SCN)_2^+$	2.3×10^3
$Ag^+ + 2\ NH_3 \rightleftharpoons Ag(NH_3)_2^+$	1.1×10^7	$Hg^{2+} + 4\ Br^- \rightleftharpoons HgBr_4^{2-}$	1×10^{21}
$Ag^+ + 2\ SCN^- \rightleftharpoons Ag(SCN)_2^-$	3.7×10^7	$Hg^{2+} + 4\ Cl^- \rightleftharpoons HgCl_4^{2-}$	1.2×10^{15}
$Ag^+ + 2\ S_2O_3^{2-} \rightleftharpoons Ag(S_2O_3)_2^{3-}$	2.9×10^{13}	$Hg^{2+} + 4\ CN^- \rightleftharpoons Hg(CN)_4^{2-}$	3×10^{41}
$Al^{3+} + 6\ F^- \rightleftharpoons AlF_6^{3-}$	6.9×10^{19}	$Hg^{2-} + 4\ I^- \rightleftharpoons HgI_4^{2-}$	6.8×10^{29}
$Al^{3+} + 4\ OH^- \rightleftharpoons Al(OH)_4^-$	1.1×10^{33}	$I_2 + I^- \rightleftharpoons I_3^-$	7.8×10^2
$Cd^{2+} + 4\ Cl^- \rightleftharpoons CdCl_4^{2-}$	6.3×10^2	$Ni^{2+} + 4\ CN^- \rightleftharpoons Ni(CN)_4^{2-}$	2×10^{31}
$Cd^{2+} + 4\ CN^- \rightleftharpoons Cd(CN)_4^{2-}$	6.0×10^{18}	$Ni^{2+} + 6\ NH_3 \rightleftharpoons Ni(NH_3)_6^{2+}$	5.5×10^8
$Cd^{2+} + 4\ I^- \rightleftharpoons CdI_4^{2-}$	2.6×10^5	$Pb^{2+} + 4\ Cl^- \rightleftharpoons PbCl_4^{2-}$	4×10^1
$Cd^{2+} + 4\ OH^- \rightleftharpoons Cd(OH)_4^{2-}$	4.2×10^8	$Pb^{2+} + 4\ I^- \rightleftharpoons PbI_4^{2-}$	3.0×10^4
$Cd^{2+} + 4\ NH_3 \rightleftharpoons Cd(NH_3)_6^{2+}$	1.3×10^7	$Sb^{3+} + 4\ Cl^- \rightleftharpoons SbCl_4^-$	5.2×10^4
$Co_2^+ + 6\ NH_3 \rightleftharpoons Co(NH_3)_6^{2+}$	1.3×10^5	$Sb^{3+} + 4\ OH^- \rightleftharpoons Sb(OH)^{4-}$	2×10^{38}
$Co_3^+ + 6\ NH_3 \rightleftharpoons Co(NH_3)_6^{3+}$	2×10^{35}	$Sn^{2+} + 4\ Cl^- \rightleftharpoons SnCl_4^{2-}$	3.0×10^1
$Co_2^+ + 4\ SCN^- \rightleftharpoons Co(SCN)_4^{2-}$	1×10^3	$Zn^{2+} + 4\ CN^- \rightleftharpoons Zn(CN)_4^{2-}$	5×10^{16}
$Cr^{3+} + 4\ OH^- \rightleftharpoons Cr(OH)_4^-$	8×10^{29}	$Zn^{2+} + 4\ OH^- \rightleftharpoons Zn(OH)_4^{2-}$	4.6×10^{17}
$Cu^{2+} + 4\ OH^- \rightleftharpoons Cu(OH)_4^{2-}$	3×10^{18}	$Zn^{2+} + 4\ NH_3 \rightleftharpoons Zn(NH_3)_4^{2+}$	2.9×10^9
$Cu^{2+} + 4\ NH_3 \rightleftharpoons Cu(NH_3)_4^{2+}$	2.1×10^{13}		

TABLE B.12   Standard Reduction Potentials

Half-reaction	$E°$ (V)
$3 N_2 + 2 H^+ + 2 e^- \rightleftharpoons 2 HN_3$	-3.1
$Li^+ + e^- \rightleftharpoons Li$	-3.045
$Rb^+ + e^- \rightleftharpoons Rb$	-2.925
$K^+ + e^- \rightleftharpoons K$	-2.924
$Cs^+ + e^- \rightleftharpoons Cs$	-2.923
$Ba^{2+} + 2 e^- \rightleftharpoons Ba$	-2.90
$Sr^{2+} + 2 e^- \rightleftharpoons Sr$	-2.89
$Ca^{2+} + 2 e^- \rightleftharpoons Ca$	-2.76
$Na^+ + e^- \rightleftharpoons Na$	-2.7109
$Mg(OH)_2 + 2 e^- \rightleftharpoons Mg + 2 OH^-$	-2.69
$Mg^{2+} + 2 e^- \rightleftharpoons Mg$	-2.375
$H_2 + 2 e^- \rightleftharpoons 2 H^-$	-2.23
$Al^{3+} + 3 e^- \rightleftharpoons Al$ (0.1 M NaOH)	-1.706
$Be^{2+} + 2 e^- \rightleftharpoons Be$	-1.70
$Ti^{2+} + 2 e^- \rightleftharpoons Ti$	-1.63
$Zn(CN)_4^{2-} + 2 e^- \rightleftharpoons Zn + 4 CN^-$	-1.26
$Zn(NH_3)_4^{2+} + 2 e^- \rightleftharpoons Zn + 4 NH_3$	-1.04
$Mn^{2+} + 2 e^- \rightleftharpoons Mn$	-1.029
$SO_4^{2-} + H_2O + 2 e^- \rightleftharpoons SO_3^{2-} + 2 OH^-$	-0.92
$Cr^{2+} + 2 e^- \rightleftharpoons Cr$	-0.91
$TiO_2 + 4 H^+ + 4 e^- \rightleftharpoons Ti + 2 H_2O$	-0.87
$2 H_2O + 2 e^- \rightleftharpoons H_2 + 2 OH^-$	-0.8277
$Zn^{2+} + 2 e^- \rightleftharpoons Zn$	-0.7628
$Cr^{3+} + 3 e^- \rightleftharpoons Cr$	-0.74
$2 SO_3^{2-} + 3 H_2O + 4 e^- \rightleftharpoons S_2O_3^{2-} + 6 OH^-$	-0.58
$PbO + H_2O + 2 e^- \rightleftharpoons Pb + 2 OH^-$	-0.576
$Ga^{3+} + 3 e^- \rightleftharpoons Ga$	-0.560
$S + 2 e^- \rightleftharpoons S^{2-}$	-0.508
$2 CO_2 + 2 H^+ + 2 e^- \rightleftharpoons H_2C_2O_4$	-0.49
$Ni(NH_3)_6^{2+} + 2 e^- \rightleftharpoons Ni + 6 NH_3$	-0.48
$Co(NH_3)_6^{2+} + 2 e^- \rightleftharpoons Co + 6 NH_3$	-0.422
$Cr^{3+} + e^- \rightleftharpoons Cr^{2+}$	-0.41
$Fe^{2+} + 2 e^- \rightleftharpoons Fe$	-0.409
$Cd^{2+} + 2 e^- \rightleftharpoons Cd$	-0.4026
$PbSO_4 + 2 e^- \rightleftharpoons Pb + SO_4^{2-}$	-0.356
$In^{3+} + 3 e^- \rightleftharpoons In$	-0.338
$Tl^+ + e^- \rightleftharpoons Tl$	-0.3363
$Ag(CN)_2^- + e^- \rightleftharpoons Ag + 2 CN^-$	-0.31
$Co^{2+} + 2 e^- \rightleftharpoons Co$	-0.28
$H_3PO_4 + 2 H^+ + 2 e^- \rightleftharpoons H_3PO_3 + H_2O$	-0.276
$Ni^{2+} + 2 e^- \rightleftharpoons Ni$	-0.23
$2 SO_4^{2-} + 4 H^+ + 2 e^- \rightleftharpoons S_2O_6^{2-} + 2 H_2O$	-0.224
$CO_2 + 2 H^+ + 2 e^- \rightleftharpoons HCO_2H$	-0.20

(continued)

TABLE B.12　Standard Reduction Potentials (continued)

Half-reaction	$E°$ (V)
$O_2 + 2\,H_2O + 2\,e^- \rightleftharpoons H_2O_2 + 2\,OH^-$	−0.146
$Sn^{2+} + 2\,e^- \rightleftharpoons Sn$	−0.1364
$Pb^{2+} + 2\,e^- \rightleftharpoons Pb$	−0.1263
$CrO_4{}^{2-} + 4\,H_2O + 3\,e^- \rightleftharpoons Cr(OH)_3 + 5\,OH^-$	−0.12
$WO_3 + 6\,H^+ + 6\,e^- \rightleftharpoons W + 3\,H_2O$	−0.09
$Ru^{3+} + e^- \rightleftharpoons Ru^{2+}$	−0.08
$O_2 + H_2O + 2\,e^- \rightleftharpoons HO_2{}^- + OH^-$	−0.076
$Fe^{3+} + 3\,e^- \rightleftharpoons Fe$	−0.036
$2\,H^+ + 2\,e^- \rightleftharpoons H_2$	0.0000000 . . .
$NO_3{}^- + H_2O + 2\,e^- \rightleftharpoons NO_2{}^- + 2\,OH^-$	0.01
$AgBr + e^- \rightleftharpoons Ag + Br^-$	0.0713
$S_4O_6{}^{2-} + 2\,e^- \rightleftharpoons 2\,S_2O_3{}^{2-}$	0.0895
$Sn^{4+} + 2\,e^- \rightleftharpoons Sn^{2+}$	0.15
$Cu^{2+} + e^- \rightleftharpoons Cu^+$	0.158
$ClO_4{}^- + H_2O + 2\,e^- \rightleftharpoons ClO_3{}^- + 2\,OH^-$	0.17
$SO_4{}^{2-} + 4\,H^+ + 2\,e^- \rightleftharpoons H_2SO_3 + H_2O$	0.20
$AgCl + e^- \rightleftharpoons Ag + Cl^-$	0.2223
$IO_3{}^- + 3\,H_2O + 6\,e^- \rightleftharpoons I^- + 6\,OH^-$	0.26
$Hg_2Cl_2 + 2\,e^- \rightleftharpoons 2\,Hg + 2\,Cl^-$	0.2682
$Cu^{2+} + 2\,e^- \rightleftharpoons Cu$	0.3402
$ClO_3{}^- + H_2O + 2\,e^- \rightleftharpoons ClO_2{}^- + 2\,OH^-$	0.35
$Ag(NH_3)_2{}^+ + e^- \rightleftharpoons Ag + 2\,NH_3$	0.373
$O_2 + 2\,H_2O + 4\,e^- \rightleftharpoons 4\,OH^-$	0.401
$H_2SO_3 + 4\,H^+ + 4\,e^- \rightleftharpoons S + 3\,H_2O$	0.45
$HgCl_4{}^{2-} + 2\,e^- \rightleftharpoons Hg + 4\,Cl^-$	0.48
$Cu^+ + e^- \rightleftharpoons Cu$	0.522
$I_3{}^- + 2\,e^- \rightleftharpoons 3\,I^-$	0.5338
$I_2 + 2\,e^- \rightleftharpoons 2\,I^-$	0.535
$MnO_4{}^- + 2\,H_2O + 3\,e^- \rightleftharpoons MnO_2 + 4\,OH^-$	0.588
$ClO_2{}^- + H_2O + 2\,e^- \rightleftharpoons ClO^- + 2\,OH^-$	0.59
$O_2 + 2\,H^+ + 2\,e^- \rightleftharpoons H_2O_2$	0.682
$Fe^{3+} + e^- \rightleftharpoons Fe^{2+}$	0.770
$Hg_2{}^{2+} + 2\,e^- \rightleftharpoons 2\,Hg$	0.7961
$Ag^+ + e^- \rightleftharpoons Ag$	0.7996
$Hg^{2+} + 2\,e^- \rightleftharpoons Hg$	0.851
$H_2O_2 + 2\,e^- \rightleftharpoons 2\,OH^-$	0.88
$ClO^- + H_2O + 2\,e^- \rightleftharpoons Cl^- + 2\,OH^-$	0.89
$2\,Hg^{2+} + 2\,e^- \rightleftharpoons Hg_2{}^{2+}$	0.905
$NO_3{}^- + 3\,H^+ + 2\,e^- \rightleftharpoons HNO_2 + H_2O$	0.94
$ClO_2 + e^- \rightleftharpoons ClO_2{}^-$	0.95
$NO_3{}^- + 4\,H^+ + 3\,e^- \rightleftharpoons NO + 2\,H_2O$	0.96
$Pd^{2+} + 2\,e^- \rightleftharpoons Pd$	0.987
$HNO_2 + H^+ + e^- \rightleftharpoons NO + H_2O$	0.99
$IO_3{}^- + 6\,H^+ + 6\,e^- \rightleftharpoons I^- + 3\,H_2O$	1.085

Half-reaction	$E°$ (V)
$Br_2(aq) + 2\,e^- \rightleftharpoons 2\,Br^-$	1.087
$Cr^{6+} + 3\,e^- \rightleftharpoons Cr^{3+}$	1.10
$ClO_3^- + 2\,H^+ + 2\,e^- \rightleftharpoons ClO_2^- + H_2O$	1.15
$ClO_4^- + 2\,H^+ + 2\,e^- \rightleftharpoons ClO_3^- + H_2O$	1.19
$2\,IO_3^- + 12\,H^+ + 10\,e^- \rightleftharpoons I_2 + 6\,H_2O$	1.19
$HCrO_4^- + 7\,H^+ + 3\,e^- \rightleftharpoons Cr^{3+} + 4\,H_2O$	1.195
$Pt^{2+} + 2\,e^- \rightleftharpoons Pt$	1.2
$MnO_2 + 4\,H^+ + 2\,e^- \rightleftharpoons Mn^{2+} + 2\,H_2O$	1.208
$O_2 + 4\,H^+ + 4\,e^- \rightleftharpoons 2\,H_2O$	1.229
$O_3 + H_2O + 2\,e^- \rightleftharpoons O_2 + 2\,OH^-$	1.24
$Tl^{3+} + 2\,e^- \rightleftharpoons Tl^+$	1.247
$ClO_2 + H^+ + e^- \rightleftharpoons HClO_2$	1.27
$2\,HNO_2 + 4\,H^+ + 4\,e^- \rightleftharpoons N_2O + 3\,H_2O$	1.27
$Au^{3+} + 2\,e^- \rightleftharpoons Au^+$	1.29
$Cr_2O_7^{2-} + 14\,H^+ + 6\,e^- \rightleftharpoons 2\,Cr^{3+} + 7\,H_2O$	1.33
$ClO_4^- + 8\,H^+ + 7\,e^- \rightleftharpoons \frac{1}{2}\,Cl_2 + 4\,H_2O$	1.34
$Cl_2 + 2\,e^- \rightleftharpoons 2\,Cl^-$	1.3583
$ClO_4^- + 8\,H^+ + 8\,e^- \rightleftharpoons Cl^- + 4\,H_2O$	1.37
$Au^{3+} + 3\,e^- \rightleftharpoons Au$	1.42
$ClO_3^- + 6\,H^+ + 6\,e^- \rightleftharpoons Cl^- + 3\,H_2O$	1.45
$PbO_2 + 4\,H^+ + 2\,e^- \rightleftharpoons Pb^{2+} + 2\,H_2O$	1.467
$2\,ClO_3^- + 12\,H^+ + 10\,e^- \rightleftharpoons Cl_2 + 6\,H_2O$	1.47
$HClO + H^+ + 2\,e^- \rightleftharpoons Cl^- + H_2O$	1.49
$MnO_4^- + 8\,H^+ + 5\,e^- \rightleftharpoons Mn^{2+} + 4\,H_2O$	1.491
$HClO_2 + 3\,H^+ + 4\,e^- \rightleftharpoons Cl^- + 2\,H_2O$	1.56
$2\,NO + 2\,H^+ + 2\,e^- \rightleftharpoons N_2O + H_2O$	1.59
$2\,HClO_2 + 6\,H^+ + 6\,e^- \rightleftharpoons Cl_2 + 4\,H_2O$	1.63
$2\,HClO + 2\,H^+ + 2\,e^- \rightleftharpoons Cl_2 + 2\,H_2O$	1.63
$HClO_2 + 2\,H^+ + 2\,e^- \rightleftharpoons HClO + H_2O$	1.64
$MnO_4^- + 4\,H^+ + 3\,e^- \rightleftharpoons MnO_2 + 2\,H_2O$	1.679
$Au^+ + e^- \rightleftharpoons Au$	1.68
$PbO_2 + SO_4^{2-} + 4\,H^+ + 2\,e^- \rightleftharpoons PbSO_4 + 2\,H_2O$	1.685
$N_2O + 2\,H^+ + 2\,e^- \rightleftharpoons N_2 + H_2O$	1.77
$H_2O_2 + 2\,H^+ + 2\,e^- \rightleftharpoons 2\,H_2O$	1.776
$Co^{3+} + e^- \rightleftharpoons Co^{2+}$	1.842
$S_2O_8^{2-} + 2\,e^- \rightleftharpoons 2\,SO_4^{2-}$	2.05
$O_3(g) + 2\,H^+ + 2\,e^- \rightleftharpoons O_2(g) + H_2O$	2.07
$F_2(g) + 2\,H^+ + 2\,e^- \rightleftharpoons 2\,HF(aq)$	3.03

TABLE B.13   Bond Dissociation Enthalpies

| | | | | | | *Single-Bond Dissociation Enthalpies (kJ/mol)* | | | | | | | |
	As	B	Br	C	Cl	F	H	I	N	O	P	S	Si
As	180		255	200	310	485	300	180		330			
B		300	370		445	645		270		525			
Br			195	270	220	240	370	180	250		270	215	330
C				350	330	490	415	210	305	360	265	270	305
Cl					240	250	431	210	190	205	330	270	400
F						160	569		280	215	500	325	600
H							435	300	390	464	325	370	320
I								150		200	180		230
N									160	165			330
O										140	370	423	464
P											210		
S												260	
Si													225

**Double- and Triple-Bond Dissociation
Enthalpies (kJ/mol)**

C=C	611	C=S	477
C≡C	837	N=N	418
C=O	745	N≡N	946
C≡O	1075	N=O	594
C=N	615	O=O	498
C≡N	891	S=O	523

TABLE B.14   Standard-State Enthalpies, Free Energies, and Entropies of Atom Combination, 298.15 K

Substance	ΔH°_{ac} (kJ/mol$_{rxn}$)	ΔG°_{ac} (kJ/mol$_{rxn}$)	ΔS°_{ac} (J/mol$_{rxn}\cdot$K)
Aluminum			
Al(s)	−326.4	−285.7	−136.21
Al(g)	0	0	0
Al^{3+}(aq)	−857	−771	−486.2
Al$_2$O$_3$(s)	−3076.0	−2848.9	−761.33
AlCl$_3$(s)	−1395.6	−1231.5	−549.46
AlF$_3$(s)	−2067.5	−1896.4	−574.36
Al$_2$(SO$_4$)$_3$(s)	−7920.1	−7166.9	−2525.9
Barium			
Ba(s)	−180	−146	−107.4
Ba(g)	0	0	0
Ba^{2+}(aq)	−718	−707	−160.6
BaO(s)	−983	−903	−260.88
Ba(OH)$_2\cdot$8 H$_2$O(s)	−9931.6	−8915	−3419
BaCl$_2$(s)	−1282	−1168	−376.96
BaCl$_2$(aq)	−1295	−1181	−378.04
BaSO$_4$(s)	−2929	−2673	−850.1
Ba(NO$_3$)$_2$(s)	−3612	−3217	−1229.4
Ba(NO$_3$)$_2$(aq)	−3573	−3231	−1140.7
Beryllium			
Be(s)	−324.3	−286.6	−126.77
Be(g)	0	0	0
Be^{2+}(aq)	−707.1	−666.3	−266.0
BeO(s)	−1183.1	−1026.6	−283.18
BeCl$_2$(s)	−1058.1	−943.6	−383.99
Bismuth			
Bi(s)	−207.1	−168.2	−130.31
Bi(g)	0	0	0
Bi$_2$O$_3$(s)	−1735.6	−1525.3	−705.8
BiCl$_3$(s)	−951.2	−800.2	−505.6
BiCl$_3$(g)	−837.8	−741.2	−323.79
Bi$_2$S$_3$(s)	−1393.7	−1191.8	−677.2
Boron			
B(s)	−562.7	−518.8	−147.59
B(g)	0	0	0
B$_2$O$_3$(s)	−3145.7	−2926.4	−736.10
B$_2$H$_6$(g)	−2395.7	−2170.4	−763.07
B$_5$H$_9$(l)	−4729.7	−4251.4	−1615.45
B$_5$H$_9$(g)	−4699.2	−4248.2	−1523.75
B$_{10}$H$_{14}$(s)	−8719.3	−7841.2	−2963.92

(continued)

TABLE B.14   Standard-State Enthalpies, Free Energies, and Entropies of Atom Combination, 298.15 K (continued)

Substance	ΔH°_{ac} (kJ/mol$_{rxn}$)	ΔG°_{ac} (kJ/mol$_{rxn}$)	ΔS°_{ac} (J/mol$_{rxn}$·K)
$H_3BO_3(s)$	−3057.5	−2792.7	−891.92
$BF_3(g)$	−1936.7	−1824.9	−375.59
$BCl_3(l)$	−1354.9	−1223.2	−442.7
$B_3N_3H_6(l)$	−4953.1	−4535.4	−1408.9
$B_3N_3H_6(g)$	−4923.9	−4532.8	−1319.84
Bromine			
$Br_2(l)$	−223.768	−164.792	−197.813
$Br_2(g)$	−192.86	−161.68	−104.58
$Br(g)$	0	0	0
$HBr(g)$	−365.93	−339.09	−91.040
$HBr(aq)$	−451.08	−389.60	−207.3
$BrF(g)$	−284.72	−253.49	−104.81
$BrF_3(g)$	−604.45	−497.56	−358.75
$BrF_5(g)$	−935.73	−742.6	−648.60
Calcium			
$Ca(s)$	−178.2	−144.3	−113.46
$Ca(g)$	0	0	0
$Ca^{2+}(aq)$	−721.0	−697.9	−208.0
$CaO(s)$	−1062.5	−980.1	−276.19
$Ca(OH)_2(s)$	−2097.9	−1912.7	−623.03
$CaCl_2(s)$	−1217.4	−1103.8	−380.7
$CaSO_4(s)$	−2887.8	−2631.3	−860.3
$CaSO_4 \cdot 2\,H_2O(s)$	−4256.7	−4383.2	−1553.8
$Ca(NO_3)_2(s)$	−3557.0	−3189.0	−1234.5
$CaCO_3(s)$	−2849.3	−2639.5	−703.2
$Ca_3(PO_4)_2(s)$	−7278.0	−6727.9	−1843.5
Carbon			
$C(graphite)$	−716.682	−671.257	−152.36
$C(diamond)$	−714.787	−668.357	−155.719
$C(g)$	0	0	0
$CO(g)$	−1076.377	−1040.156	−121.477
$CO_2(g)$	−1608.531	−1529.078	−266.47
$COCl_2(g)$	−1428.0	−1318.9	−366.02
$CH_4(g)$	−1662.09	−1534.997	−430.684
$HCHO(g)$	−1509.72	−1412.01	−329.81
$H_2CO_3(aq)$	−2599.14	−2396.02	−683.3
$HCO_3^-(aq)$	−2373.83	−2156.47	−664.8
$CO_3^{2-}(aq)$	−2141.33	−1894.26	−698.16
$CH_3OH(l)$	−2075.11	−1882.25	−651.2
$CH_3OH(g)$	−2037.11	−1877.94	−538.19
$CCl_4(l)$	−1338.84	−1159.19	−602.49
$CCl_4(g)$	−1306.3	−1154.57	−509.04
$CHCl_3(l)$	−1433.84	−1265.20	−567.3

Substance	ΔH°_{ac} (kJ/mol_{rxn})	ΔG°_{ac} (kJ/mol_{rxn})	ΔS°_{ac} $(J/mol_{rxn} \cdot K)$
$CHCl_3(g)$	−1402.51	−1261.88	−472.69
$CH_2Cl_2(l)$	−1516.80	−1356.37	−540.1
$CH_2Cl_2(g)$	−1487.81	−1354.98	−447.69
$CH_3Cl(g)$	−1572.15	−1444.08	−432.9
$CS_2(l)$	−1184.59	−1082.49	−342.40
$CS_2(g)$	−1156.93	−1080.64	−255.90
$HCN(g)$	−1271.9	−1205.43	−224.33
$CH_3NO_2(l)$	−2453.77	−2214.51	−805.89
$C_2H_2(g)$	−1641.93	−1539.81	−344.68
$C_2H_4(g)$	−2251.70	−2087.35	−555.48
$C_2H_6(g)$	−2823.94	−2594.82	−774.87
$CH_3CHO(l)$	−2745.43	−2515.35	−775.9
$CH_3CO_2H(l)$	−3286.8	−3008.86	−937.4
$CH_3CO_2H(g)$	−3234.55	−2992.96	−814.7
$CH_3CO_2H(aq)$	−3288.06	−3015.42	−918.5
$CH_3CO_2^-(aq)$	−3070.66	−2785.03	−895.8
$CH_3CH_2OH(l)$	−3266.12	−2968.51	−1004.8
$CH_3CH_2OH(g)$	−3223.53	−2962.22	−882.82
$CH_3CH_2OH(aq)$	−3276.7	−2975.37	−1017.0
$C_6H_6(l)$	−5556.96	−5122.52	−1464.1
$C_6H_6(g)$	−5523.07	−5117.36	−1367.7

Chlorine

$Cl_2(g)$	−243.358	−211.360	−107.330
$Cl(g)$	0	0	0
$Cl^-(aq)$	−288.838	−236.908	−108.7
$ClO_2(g)$	−517.5	−448.6	−230.47
$Cl_2O(g)$	−412.2	−345.2	−225.24
$Cl_2O_7(l)$	−1750	—	—
$HCl(g)$	−431.64	−404.226	−93.003
$HCl(aq)$	−506.49	−440.155	−223.4
$ClF(g)$	−255.15	−223.53	−106.06

Chromium

$Cr(s)$	−396.6	−351.8	−150.73
$Cr(g)$	0	0	0
$CrO_3(s)$	−1733.6	—	—
$CrO_4^{2-}(aq)$	−2274.4	−2006.47	−768.51
$Cr_2O_3(s)$	−2680.4	−2456.89	−751.0
$Cr_2O_7^{2-}(aq)$	−4027.7	−3626.8	−1214.5
$(NH_4)_2Cr_2O_7$	−7030.7	—	—
$PbCrO_4(s)$	−2519.2	—	—

Cobalt

$Co(s)$	−424.7	−380.3	−149.475
$Co(g)$	0	0	0

(continued)

TABLE B.14   Standard-State Enthalpies, Free Energies, and Entropies of Atom Combination, 298.15 K (continued)

Substance	$\Delta H°_{ac}$ (kJ/mol$_{rxn}$)	$\Delta G°_{ac}$ (kJ/mol$_{rxn}$)	$\Delta S°_{ac}$ (J/mol$_{rxn}$·K)
$Co^{2+}(aq)$	−482.9	−434.7	−293
$Co^{3+}(aq)$	−333	−246.3	−485
$CoO(s)$	−911.8	−826.2	−287.60
$Co_3O_4(s)$	−3162	−2842	−1080.3
$Co(NH_3)_6^{3+}(aq)$	−7763.5	−6929.5	−3018
Copper			
$Cu(s)$	−338.32	−298.58	−133.23
$Cu(g)$	0	0	0
$Cu^+(aq)$	−266.65	−248.60	−125.8
$Cu^{2+}(aq)$	−273.55	−233.09	−266.0
$CuO(s)$	−744.8	−660.0	−284.81
$Cu_2O(s)$	−1094.4	−974.9	−400.68
$CuCl_2(s)$	−807.8	−685.6	−388.71
$CuS(s)$	−670.2	−590.4	−267.7
$Cu_2S(s)$	−1304.9	−921.6	−379.7
$CuSO_4(s)$	−2385.17	−1530.44	−869
$Cu(NH_3)_4^{2+}(aq)$	−5189.4	−4671.13	−1882.5
Fluorine			
$F_2(g)$	−157.98	−123.82	−114.73
$F(g)$	0	0	0
$F^-(aq)$	−411.62	−340.70	−172.6
$HF(g)$	−567.7	−538.4	−99.688
$HF(aq)$	−616.72	−561.98	−184.8
Hydrogen			
$H_2(g)$	−435.30	−406.494	−98.742
$H(g)$	0	0	0
$H^+(aq)$	−217.65	−203.247	−114.713
$OH^-(aq)$	−696.81	−592.222	−286.52
$H_2O(l)$	−970.30	−875.354	−320.57
$H_2O(g)$	−926.29	−866.797	−202.23
$H_2O_2(l)$	−1121.42	−990.31	−441.9
$H_2O_2(aq)$	−1124.81	−1003.99	−407.6
Iodine			
$I_2(s)$	−213.676	−141.00	−245.447
$I_2(g)$	−151.238	−121.67	−100.89
$I(g)$	0	0	0
$HI(g)$	−298.01	−272.05	−88.910
$IF(g)$	−281.48	−250.92	−103.38
$IF_5(g)$	−1324.28	−1131.78	−646.9
$IF_7(g)$	−1603.7	−1322.17	−945.6
$ICl(g)$	−210.74	−181.64	−98.438
$IBr(g)$	−177.88	−149.21	−97.040

Substance	ΔH°_{ac} (kJ/mol_{rxn})	ΔG°_{ac} (kJ/mol_{rxn})	ΔS°_{ac} $(J/mol_{rxn}\cdot K)$
Iron			
$Fe(s)$	-416.3	-370.7	-153.21
$Fe(g)$	0	0	0
$Fe^{2+}(aq)$	-505.4	-449.6	-318.2
$Fe^{3+}(aq)$	-464.8	-375.4	-496.4
$Fe_2O_3(s)$	-2404.3	-2178.8	-756.75
$Fe_3O_4(s)$	-3364.0	-3054.4	-1039.3
$Fe(OH)_2(s)$	-1918.9	-1727.2	-644
$Fe(OH)_3(s)$	-2639.8	-2372.1	-901.1
$FeCl_3(s)$	-1180.8	-1021.7	-533.8
$FeS_2(s)$	-1152.1	-1014.1	-463.2
$Fe(CO)_5(l)$	-6019.6	-5590.9	-1438.1
$Fe(CO)_5(g)$	-5979.5	-5582.9	-1330.9
Lead			
$Pb(s)$	-195.0	-161.9	-110.56
$Pb(g)$	0	0	0
$Pb^{2+}(aq)$	-196.7	-186.3	-164.9
$PbO(s)$	-661.5	-581.5	-267.7
$PbO_2(s)$	-970.7	-842.7	-428.9
$PbCl_2(s)$	-797.8	-687.4	-369.8
$PbCl_4(l)$	-1011.0	—	—
$PbS(s)$	-574.2	-498.9	-252.0
$PbSO_4(s)$	-2390.4	-2140.2	-838.84
$Pb(NO_3)_2(s)$	-3087.3	—	—
$PbCO_3(s)$	-2358.3	-2153.8	-685.6
Lithium			
$Li(s)$	-159.37	-126.66	-109.65
$Li(g)$	0	0	0
$Li^+(aq)$	-437.86	-419.97	-125.4
$LiH(s)$	-467.56	-398.26	-233.40
$LiOH(s)$	-1111.12	-1000.59	-371.74
$LiF(s)$	-854.33	-784.28	-261.87
$LiCl(s)$	-689.66	-616.71	-244.64
$LiBr(s)$	-622.48	-551.06	-239.52
$LiI(s)$	-536.62	-467.45	-232.78
$LiAlH_4(s)$	-1472.7	-1270.0	-683.42
$LiBH_4(s)$	-1401.9	-1333.4	-675.21
Magnesium			
$Mg(s)$	-147.70	-113.10	-115.97
$Mg(g)$	0	0	0
$Mg^{2+}(aq)$	-614.55	-567.9	-286.8
$MgO(s)$	-998.57	-914.26	-282.76

(continued)

TABLE B.14   Standard-State Enthalpies, Free Energies, and Entropies of Atom Combination, 298.15 K (continued)

Substance	ΔH°_{ac} (kJ/mol$_{rxn}$)	ΔG°_{ac} (kJ/mol$_{rxn}$)	ΔS°_{ac} (J/mol$_{rxn}\cdot$K)
$MgH_2(s)$	−658.3	−555.5	−346.99
$Mg(OH)_2(s)$	−2005.88	−1816.64	−637.01
$MgCl_2(s)$	−1032.38	−916.25	−389.43
$MgCO_3(s)$	−2707.7	−2491.7	−724.2
$MgSO_4(s)$	−2708.1	−2448.9	−869.1
Manganese			
$Mn(s)$	−280.7	−238.5	−141.69
$Mn(g)$	0	0	0
$Mn^{2+}(aq)$	−501.5	−466.6	−247.3
$MnO(s)$	−915.1	−833.1	−275.05
$MnO_2(s)$	−1299.1	−1167.1	−442.76
$Mn_2O_3(s)$	−2267.9	−2053.3	−720.1
$Mn_3O_4(s)$	−3226.6	−2925.6	−1009.7
$KMnO_4(s)$	−2203.8	−1963.6	−806.50
$MnS(s)$	−773.7	−695.2	−263.3
Mercury			
$Hg(l)$	−61.317	−31.820	−98.94
$Hg(g)$	0	0	0
$Hg^{2+}(aq)$	+109.8	+132.58	−207.2
$HgO(s)$	−401.32	−322.090	−265.73
$HgCl_2(s)$	−529.0	−421.8	−359.4
$Hg_2Cl_2(s)$	−631.21	−485.745	−487.8
$HgS(s)$	−398.3	−320.67	−260.4
Nitrogen			
$N_2(g)$	−945.408	−911.26	−114.99
$N(g)$	0	0	0
$NO(g)$	−631.62	−600.81	−103.592
$NO_2(g)$	−937.86	−867.78	−235.35
$N_2O(g)$	−1112.53	−1038.79	−247.80
$N_2O_3(g)$	−1609.20	−1466.99	−477.48
$N_2O_4(g)$	−1932.93	−1740.29	−646.53
$N_2O_5(g)$	−2179.91	−1954.8	−756.2
$NO_3^-(aq)$	−1425.2	−1259.56	−490.1
$NOCl(g)$	−791.84	−726.96	−217.86
$NO_2Cl(g)$	−1080.12	−970.4	−368.46
$HNO_2(aq)$	−1307.9	−1172.9	−454.5
$HNO_3(g)$	−1572.92	−1428.79	−484.80
$HNO_3(aq)$	−1645.22	−1465.32	−604.8
$NH_3(g)$	−1171.76	−1081.82	−304.99
$NH_3(aq)$	−1205.94	−1091.87	−386.1
$NH_4^+(aq)$	−1475.81	−1347.93	−498.8
$NH_4NO_3(s)$	−2929.08	−2603.31	−1097.53

Substance	ΔH_{ac}° (kJ/mol$_{rxn}$)	ΔG_{ac}° (kJ/mol$_{rxn}$)	ΔS_{ac}° (J/mol$_{rxn}$·K)
$NH_4NO_3(aq)$	−2903.39	−2610.00	−988.8
$NH_4Cl(s)$	−1779.41	−1578.17	−682.7
$N_2H_4(l)$	−1765.38	−1574.91	−644.24
$N_2H_4(g)$	−1720.61	−1564.90	−526.98
$HN_3(g)$	−1341.7	−1242.0	−335.37
Oxygen			
$O_2(g)$	−498.340	−463.462	−116.972
$O(g)$	0	0	0
$O_3(g)$	−604.8	−532.0	−244.24
Phosphorus			
P(white)	−314.64	−278.25	−122.10
$P_4(g)$	−1199.65	−1088.6	−372.79
$P_2(g)$	−485.0	−452.8	−108.257
$P(g)$	0	0	0
$PH_3(g)$	−962.2	−874.6	−297.10
$P_4O_6(s)$	−4393.7	—	—
$P_4O_{10}(s)$	−6734.3	−6128.0	−2034.46
$PO_4^{3-}(aq)$	−2588.7	−2223.9	−1029
$PF_3(g)$	−1470.4	−1361.5	−366.22
$PF_5(g)$	−2305.4	—	—
$PCl_3(l)$	−999.4	−867.6	−441.7
$PCl_3(g)$	−966.7	−863.1	−347.01
$PCl_5(g)$	−1297.9	−1111.6	−624.60
$H_3PO_4(s)$	−3243.3	−2934.0	−1041.05
$H_3PO_4(aq)$	−3241.7	−2833.6	−1374
Potassium			
$K(s)$	−89.24	−60.59	−96.16
$K(g)$	0	0	0
$K^+(aq)$	−341.62	−343.86	−57.8
$KOH(s)$	−980.82	−874.65	−357.2
$KCl(s)$	−647.67	−575.41	−242.94
$KNO_3(s)$	−1804.08	−1606.27	−663.75
$K_2Cr_2O_7(s)$	−4777.4	−4328.7	−1505.9
$KMnO_4(s)$	−2203.8	−1963.6	−806.50
Silicon			
$Si(s)$	−455.6	−411.3	−149.14
$Si(g)$	0	0	0
$SiO_2(s)$	−1864.9	−1731.4	−448.24
$SiH_4(g)$	−1291.9	−1167.4	−422.20
$SiF_4(g)$	−2386.5	−2231.6	−520.50
$SiCl_4(l)$	−1629.3	−1453.9	−589
$SiCl_4(g)$	−1599.3	−1451.0	−498.03

(continued)

TABLE B.14 Standard-State Enthalpies, Free Energies, and Entropies of Atom Combination, 298.15 K (continued)

Substance	ΔH°_{ac} (kJ/mol_{rxn})	ΔG°_{ac} (kJ/mol_{rxn})	ΔS°_{ac} $(J/mol_{rxn} \cdot K)$
Silver			
$Ag(s)$	−284.55	−245.65	−130.42
$Ag(g)$	0	0	0
$Ag^+(aq)$	−178.97	−168.54	−100.29
$Ag(NH_3)_2{}^+(aq)$	−2647.15	−2393.51	−922.6
$Ag_2O(s)$	−849.32	−734.23	−385.7
$AgCl(s)$	−533.30	−461.12	−242.0
$AgBr(s)$	−496.80	−424.95	−240.9
$AgI(s)$	−453.23	−382.34	−238.3
Sodium			
$Na(s)$	−107.32	−76.761	−102.50
$Na(g)$	0	0	0
$Na^+(aq)$	−347.45	−338.666	−94.7
$NaH(s)$	−3811.25	−313.47	−228.409
$NaOH(s)$	−999.75	−891.233	−365.025
$NaOH(aq)$	−1044.25	−930.889	−381.4
$NaCl(s)$	−640.15	−566.579	−246.78
$NaCl(g)$	−405.65	−379.10	−89.10
$NaCl(aq)$	−636.27	−575.574	−203.4
$NaNO_3(s)$	−1795.38	−1594.58	−673.66
$Na_3PO_4(s)$	−3550.68	−3224.26	−1094.75
$Na_2SO_3(s)$	−2363.96	−2115.0	−803
$Na_2SO_4(s)$	−2877.20	−2588.86	−969.89
$Na_2CO_3(s)$	−2809.51	−2564.41	−813.70
$NaHCO_3(s)$	−2739.97	−2497.5	−808.0
$NaCH_3CO_2(s)$	−3400.78	−3099.66	−1013.2
$Na_2CrO_4(s)$	−2950.1	−2667.18	−949.53
$Na_2Cr_2O_7(s)$	−4730.6	—	—
Sulfur			
$S_8(s)$	−2230.440	−1906.00	−1310.77
$S_8(g)$	−2128.14	−1856.37	−911.59
$S(g)$	0	0	0
$S^{2-}(aq)$	−245.7	−152.4	−182.4
$SO_2(g)$	−1073.95	−1001.906	−241.71
$SO_3(s)$	−1480.82	−1307.65	−580.3
$SO_3(l)$	−1467.36	−1307.19	−537.2
$SO_3(g)$	−1422.04	−1304.50	−394.23
$SO_4{}^{2-}(aq)$	−2184.76	−1909.70	−791.9
$SOCl_2(g)$	−983.8	−879.6	−349.50
$SO_2Cl_2(g)$	−1384.5	−1233.1	−508.39
$H_2S(g)$	−734.74	−678.30	−191.46
$H_2SO_3(aq)$	−2070.43	−1877.75	−648.2
$H_2SO_4(aq)$	−2620.06	−2316.20	−1021.4

Substance	ΔH°_{ac} (kJ/mol$_{rxn}$)	ΔG°_{ac} (kJ/mol$_{rxn}$)	ΔS°_{ac} (J/mol$_{rxn}$·K)
$SF_4(g)$	−1369.66	−1217.2	−510.81
$SF_6(g)$	−1962	−1715.0	−828.53
$SCN^-(aq)$	−1391.75	−1272.43	−334.9
Tin			
$Sn(s)$	−302.1	−267.3	−124.35
$Sn(g)$	0	0	0
$SnO(s)$	−837.1	−755.9	−273.0
$SnO_2(s)$	−1381.1	−1250.5	−438.3
$SnCl_2(s)$	−870.6	—	—
$SnCl_4(l)$	−1300.1	−249.8	−570.7
$SnCl_4(g)$	−1260.3	−1122.2	−463.5
Titanium			
$Ti(s)$	−469.9	−425.1	−149.6
$Ti(g)$	0	0	0
$TiO(s)$	−1238.8	−1151.8	−306.5
$TiO_2(s)$	−1913.0	−1778.1	−452.0
$TiCl_4(l)$	−1760.8	−1585.0	−588.7
$TiCl_4(g)$	−1719.8	−1574.6	−486.2
Tungsten			
$W(s)$	−849.4	−807.1	−141.31
$W(g)$	0	0	0
$WO_3(s)$	−2439.8	−2266.4	−581.22
Zinc			
$Zn(s)$	−130.729	−95.145	−119.35
$Zn(g)$	0	0	0
$Zn^{2+}(aq)$	−284.62	−242.21	−273.1
$ZnO(s)$	−728.18	−645.18	−278.40
$ZnCl_2(s)$	−789.14	−675.90	−379.92
$ZnS(s)$	−615.51	−534.69	−271.1
$ZnSO_4(s)$	−2389.0	−2131.8	−862.5

**TABLE B.15   Electron Configurations of the
First 86 Elements**

Atomic Number	Symbol	Electron Configuration
1	H	$1s^1$
2	He	$1s^2 = $ [He]
3	Li	[He] $2s^1$
4	Be	[He] $2s^2$
5	B	[He] $2s^2 2p^1$
6	C	[He] $2s^2 2p^2$
7	N	[He] $2s^2 2p^3$
8	O	[He] $2s^2 2p^4$
9	F	[He] $2s^2 2p^5$
10	Ne	[He] $2s^2 2p^6 = $ [Ne]
11	Na	[Ne] $3s^1$
12	Mg	[Ne] $3s^2$
13	Al	[Ne] $3s^2 3p^1$
14	Si	[Ne] $3s^2 3p^2$
15	P	[Ne] $3s^2 3p^3$
16	S	[Ne] $3s^2 3p^4$
17	Cl	[Ne] $3s^2 3p^5$
18	Ar	[Ne] $3s^2 3p^6 = $ [Ar]
19	K	[Ar] $4s^1$
20	Ca	[Ar] $4s^2$
21	Sc	[Ar] $4s^2 3d^1$
22	Ti	[Ar] $4s^2 3d^2$
23	V	[Ar] $4s^2 3d^3$
24	Cr	[Ar] $4s^1 3d^5$
25	Mn	[Ar] $4s^2 3d^5$
26	Fe	[Ar] $4s^2 3d^6$
27	Co	[Ar] $4s^2 3d^7$
28	Ni	[Ar] $4s^2 3d^8$
29	Cu	[Ar] $4s^1 3d^{10}$
30	Zn	[Ar] $4s^2 3d^{10}$
31	Ga	[Ar] $4s^2 3d^{10} 4p^1$
32	Ge	[Ar] $4s^2 3d^{10} 4p^2$
33	As	[Ar] $4s^2 3d^{10} 4p^3$
34	Se	[Ar] $4s^2 3d^{10} 4p^4$
35	Br	[Ar] $4s^2 3d^{10} 4p^5$
36	Kr	[Ar] $4s^2 3d^{10} 4p^6 = $ [Kr]
37	Rb	[Kr] $5s^1$
38	Sr	[Kr] $5s^2$
39	Y	[Kr] $5s^2 4d^1$
40	Zr	[Kr] $5s^2 4d^2$
41	Nb	[Kr] $5s^1 4d^4$
42	Mo	[Kr] $5s^1 4d^5$

Atomic Number	Symbol	Electron Configuration
43	Tc	[Kr] $5s^2\, 4d^5$
44	Ru	[Kr] $5s^1\, 4d^7$
45	Rh	[Kr] $5s^1\, 4d^8$
46	Pd	[Kr] $4d^{10}$
47	Ag	[Kr] $5s^1\, 4d^{10}$
48	Cd	[Kr] $5s^2\, 4d^{10}$
49	In	[Kr] $5s^2\, 4d^{10}\, 5p^1$
50	Sn	[Kr] $5s^2\, 4d^{10}\, 5p^2$
51	Sb	[Kr] $5s^2\, 4d^{10}\, 5p^3$
52	Te	[Kr] $5s^2\, 4d^{10}\, 5p^4$
53	I	[Kr] $5s^2\, 4d^{10}\, 5p^5$
54	Xe	[Kr] $5s^2\, 4d^{10}\, 5p^6$ = [Xe]
55	Cs	[Xe] $6s^1$
56	Ba	[Xe] $6s^2$
57	La	[Xe] $6s^2\, 5d^1$
58	Ce	[Xe] $6s^2\, 4f^1\, 5d^1$
59	Pr	[Xe] $6s^2\, 4f^3$
60	Nd	[Xe] $6s^2\, 4f^4$
61	Pm	[Xe] $6s^2\, 4f^5$
62	Sm	[Xe] $6s^2\, 4f^6$
63	Eu	[Xe] $6s^2\, 4f^7$
64	Gd	[Xe] $6s^2\, 4f^7\, 5d^1$
65	Tb	[Xe] $6s^2\, 4f^9$
66	Dy	[Xe] $6s^2\, 4f^{10}$
67	Ho	[Xe] $6s^2\, 4f^{11}$
68	Er	[Xe] $6s^2\, 4f^{12}$
69	Tm	[Xe] $6s^2\, 4f^{13}$
70	Yb	[Xe] $6s^2\, 4f^{14}$
71	Lu	[Xe] $6s^2\, 4f^{14}\, 5d^1$
72	Hf	[Xe] $6s^2\, 4f^{14}\, 5d^2$
73	Ta	[Xe] $6s^2\, 4f^{14}\, 5d^3$
74	W	[Xe] $6s^2\, 4f^{14}\, 5d^4$
75	Re	[Xe] $6s^2\, 4f^{14}\, 5d^5$
76	Os	[Xe] $6s^2\, 4f^{14}\, 5d^6$
77	Ir	[Xe] $6s^2\, 4f^{14}\, 5d^7$
78	Pt	[Xe] $6s^1\, 4f^{14}\, 5d^9$
79	Au	[Xe] $6s^1\, 4f^{14}\, 5d^{10}$
80	Hg	[Xe] $6s^2\, 4f^{14}\, 5d^{10}$
81	Tl	[Xe] $6s^2\, 4f^{14}\, 5d^{10}\, 6p^1$
82	Pb	[Xe] $6s^2\, 4f^{14}\, 5d^{10}\, 6p^2$
83	Bi	[Xe] $6s^2\, 4f^{14}\, 5d^{10}\, 6p^3$
84	Po	[Xe] $6s^2\, 4f^{14}\, 5d^{10}\, 6p^4$
85	At	[Xe] $6s^2\, 4f^{14}\, 5d^{10}\, 6p^5$
86	Rn	[Xe] $6s^2\, 4f^{14}\, 5d^{10}\, 6p^6$ = [Rn]

APPENDIX
C
ANSWERS TO SELECTED CORE PROBLEMS

Chapter 1

3. (a) diamond   element, nonmetal
   (b) brass   mixture, metals
   (c) soil   mixture, contains metals and nonmetals
   (d) glass   mixture, contains metals and nonmetals
   (e) cotton   mixture, nonmetal
   (f) milk of magnesia mixture, contains metals and nonmetals
   (g) salt   compound, composed of metal and nonmetal
   (h) iron   element, metal
   (i) steel   mixture, metals

5. (a) Sb   (b) Au   (c) Fe   (d) Hg   (e) K   (f) Ag   (g) Sn   (h) W

6. (a) sodium   (b) magnesium   (c) aluminum   (d) silicon   (e) phosphorus
   (f) chlorine   (g) argon

7. (a) titanium   (b) vanadium   (c) chromium   (d) manganese   (e) iron
   (f) cobalt   (g) nickel   (h) copper   (i) zinc

8. (a) molybdenum   (b) tungsten   (c) rhodium   (d) iridium   (e) palladium
   (f) platinum   (g) silver   (h) gold   (i) mercury

16. $^{52}_{24}Cr^{3+}$

17. $p = 19$, $n = 20$, $e = 18$, $Z = 19$, $A = 39$

20. $^{79}_{34}Se^{2-}$

22. ^{31}P 15   31   15
    ^{18}O 8   18   8
    $^{39}K^+$ 19   39   18
    $^{58}Ni^{2+}$ 28   58   26

24. Same number of protons. Different number of neutrons.

27. (b)

28. -1

29. $+2$

30. (a) $Mg(NO_3)_2$ (b) $Fe_2(SO_4)_3$ (c) Na_2CO_3

31. (a) (a) Na_2O_2 (b) $Zn_3(PO_4)_2$ (c) K_2PtCl_6

32. Na_3N; AlN

33. K_2O_2

34. $x = 6$

38. (a) Na, metal   (b) Mg, metal   (c) Al, metal   (d) Si, semimetal
    (e) P, nonmetal   (f) S, nonmetal   (g) Cl, nonmetal   (h) Ar, nonmetal

39. (a) N, nonmetal   (b) Sb, semimetal   (c) Sc, metal   (d) Se, nonmetal
    (e) Ge, semimetal   (f) Sm, metal   (g) Sn, metal   (h) Sr, metal

41. Elements of Group VA: nitrogen, nonmetal; phosphorus, nonmetal; arsenic, semimetal; antimony, semimetal; bismuth, metal. In general, metallic properties become more pronounced going down the group.

42. Elements of the third period: sodium, metal; magnesium, metal; aluminum, metal; silicon, semimetal; phosphorus, nonmetal; sulfur, nonmetal; chlorine, nonmetal; argon, nonmetal. In general, metallic properties decrease from left to right across the period.

Chapter 1 Appendix

1. 88.7 mL, 92.0 mL, 90.0 mL
5. 1.7 cm
6. 0.9 cm, 2, 1, 1
7. 1.65 cm, 3, 1, 1, 0.8 cm
8. 0.85 cm, 2, 0.825 cm, 3
11. (a) 48.0 cm   (c) 90.32 cm

Chapter 2

3. 79.90 amu
4. 65.4 amu
6. 8.67×10^{-17} g
8. 7.30×10^{-23} g
9. 1.2011×10^6 amu, 12.011 amu, no
10. 12.011 amu, 1.9945×10^{-23} g
15. 51.996 g
16. 2:1
17. (a) 12.011 g C   (b) 58.693 g Ni   (c) 200.59 g Hg
18. $2.25, 30 g
19. 2.7×10^{26} atoms/new mole
20. 6.0×10^{23} atoms O, 6.0×10^{23} atoms P, 6.0×10^{23} atoms S
21. 5.62×10^{21} atoms of Cl
22. (a) 2.41×10^{23} atoms O   (b) 7.5×10^{23} atoms O   (c) 1.4×10^{24} atoms O
23. 106.1 g/mol
24. 12
25. 48
26. 6.022×10^{23}
27. 2.409×10^{24}
28. 16.043 amu
29. 16.043 g
32. (d) 4 mol NH_3 and 3 mol N_2H_4
33. (c) 0.050 moles glucose
34. formic acid, 46.025 g/mol; formaldehyde, 30.026 g/mol
35. (a) 16.043 g/mol   (b) 180.155 g/mol   (c) 74.122 g/mol   (d) 75.135 g/mol
36. (a) 444.56 g/mol   (b) 46.005 g/mol   (c) 97.46 g/mol   (d) 158.032 g/mol
37. (a) 220.056 g/mol   (b) 241.859 g/mol   (c) 294.181 g/mol   (d) 310.174 g/mol

38. 162.187 g/mol
39. 169.111 g/mol
40. (a) 375.937 g/mol   (b) 284.745 g/mol   (c) 444.438 g/mol
41. 7.61×10^{-3} mol C
42. 0.2247 mol P
43. 195.1 g/mol
44. 8.95 g
45. 1.587 g/cm^3
46. 9.99 cm^3
47. (a) 76.470%   (b) 68.421%   (c) 52.000%
48. (a) 21.200% N   (b) 13.854% N   (c) 16.480% N   (d) 46.646% N
49. 47.435% C, 2.5589% H, 50.006% Cl
50. 31.352% Si
51. Since $CaCO_3$ = 40.044% Ca, $CaSO_4$ = 29.439% Ca, and $Ca_3(PO_4)_2$ = 38.763% Ca, tablets that contain $CaCO_3$ are most efficient in getting Ca^{2+} ions into the body.
53. SnF_2
54. The formula of magnetite is Fe_3O_4.
55. Pyrolusite has the formula MnO_2 (b).
56. Nitrous oxide is N_2O.
57. The empirical formula of chalcopyrite is $CuFeS_2$.
58. The empirical formula is XeF_6.
59. The empirical formula of MDMA is $C_{11}H_{15}NO_2$.
60. The empirical formula is NO_2.
61. The empirical formula is CuCl.
63. $C_{40}H_{56}$
64. $C_{20}H_{14}O_4$
65. $C_8H_{10}N_4O_2$
66. $C_{14}H_{18}N_2O_5$
68. $n = 4$
69. $C_{10}H_{14}N_2$
72. 5.99 M
73. 1.3×10^{-5} M
74. 14.8 M
75. 4.13×10^{-3} M
76. 0.0814 M
77. 10.7 g Na_2SO_4
78. The total volume was more than 1 liter. The solution was less than 1.00 M.
79. 2.5 ppm C_6H_6
80. 0.25 M
81. 6.3×10^{-8} M
82. 172 mL of 17.4 M acetic acid
83. 2.25 M HCl
85. (a) $4\,Cr(s) + 3\,O_2(g) \rightarrow 2\,Cr_2O_3(s)$
    (b) $SiH_4(g) \rightarrow Si(s) + 2\,H_2(g)$
    (c) $2\,SO_3(g) \rightarrow 2\,SO_2(g) + O_2(g)$

86. (a) $2 \text{ Pb(NO}_3)_2(s) \rightarrow 2 \text{ PbO}(s) + 4 \text{ NO}_2(g) + \text{O}_2(g)$
    (b) $\text{NH}_4\text{NO}_2(s) \rightarrow \text{N}_2(g) + 2 \text{ H}_2\text{O}(g)$
    (c) $(\text{NH}_4)_2\text{Cr}_2\text{O}_7(s) \rightarrow \text{N}_2(g) + \text{Cr}_2\text{O}_3(s) + 4 \text{ H}_2\text{O}(g)$

87. (a) $\text{CH}_4(g) + 2 \text{ O}_2(g) \rightarrow \text{CO}_2(g) + 2 \text{ H}_2\text{O}(g)$
    (b) $2 \text{ H}_2\text{S}(g) + 3 \text{ O}_2(g) \rightarrow 2 \text{ H}_2\text{O}(g) + 2 \text{ SO}_2(g)$
    (c) $2 \text{ B}_5\text{H}_9(g) + 12 \text{ O}_2(g) \rightarrow 5 \text{ B}_2\text{O}_3 + 9 \text{ H}_2\text{O}(g)$

88. (a) $\text{PF}_3(g) + 3 \text{ H}_2\text{O}(l) \rightarrow \text{H}_3\text{PO}_3(aq) + 3 \text{ HF}(aq)$
    (b) $\text{P}_4\text{O}_{10}(s) + 6 \text{ H}_2\text{O}(l) \rightarrow 4 \text{ H}_3\text{PO}_4(aq)$

89. (a) $2 \text{ C}_3\text{H}_8(g) + 10 \text{ O}_2(g) \rightarrow 6 \text{ CO}_2(g) + 8 \text{ H}_2\text{O}(g)$
    (b) $\text{C}_2\text{H}_5\text{OH}(l) + 3 \text{ O}_2(g) \rightarrow 2 \text{ CO}_2(g) + 3 \text{ H}_2\text{O}(g)$
    (c) $\text{C}_6\text{H}_{12}\text{O}_6(s) + 6 \text{ O}_2(g) \rightarrow 6 \text{ CO}_2(g) + 6 \text{ H}_2\text{O}(l)$

93. 1500 molecules O_2, 15.0 mol O_2

94. 2.25 mol O_2

95. 9.00 mol CO

97. 2.0 g O_2

98. 39.9 g O_2, 27.4 g CO_2

99. 4.90 lb S

100. 9.79 g O_2

101. CrO_3

102. 0.500 kg $\text{C}_2\text{H}_5\text{OH}$

103. 782.1 lb Al

104. 3.73 g PH_3

105. 11.9 g HCl

106. 50.0 g N_2

107. 500 molecules H_2O. If the amount of O_2 doubles, the yield remains 500 molecules H_2O. If the amount of H_2 doubles, then the yield is 1000 molecules H_2O.

108. 0.400 mol P_4S_{10}; no change; 0.500 mol

109. 0.35 mol NO_2; limiting reagent: NO; excess reagent: O_2; increase; no change

110. 10.3 g HCl; to increase the amount of HCl, add more chlorine

111. 67.7 g Ca_3N_2

112. 0.482 mol PF_5

113. 5.21 g $\text{Al(CH}_3)_3$

114. 175 g Fe

115. 0.120 M; ionic equation: $\text{K}^+(aq) + \text{Cl}^-(aq) + \text{Ag}^+(aq) + \text{NO}_3^-(aq) \rightarrow \text{AgCl}(s) + \text{K}^+(aq) + \text{NO}_3^-(aq)$; net ionic equation: $\text{Ag}^+(aq) + \text{Cl}^-(aq) \rightarrow \text{AgCl}(s)$

116. 36 ml NaI; ionic equation: $2 \text{ Na}^+(aq) + 2 \text{ I}^-(aq) + \text{Hg}^{2+}(aq) + 2 \text{ NO}_3^-(aq) \rightarrow \text{HgI}_2(s) + 2 \text{ Na}^+(aq) + 2 \text{ NO}_3^-(aq)$; net ionic equation: $\text{Hg}^{2+}(aq) + 2 \text{ I}^-(aq) \rightarrow \text{HgI}_2(s)$

117. 7.469×10^{-2} M acetic acid; ionic equation: $\text{CH}_3\text{CO}_2\text{H}(aq) + \text{Na}^+(aq) + \text{OH}^-(aq) \rightarrow \text{Na}^+(aq) + \text{CH}_3\text{CO}_2^-(aq) + \text{H}_2\text{O}(l)$; net ionic equation: $\text{CH}_3\text{CO}_2\text{H}(aq) + \text{OH}^-(aq) \rightarrow \text{CH}_3\text{CO}_2^-(aq) + \text{H}_2\text{O}(l)$

118. 0.9803 M NaOH: ionic equation: $\text{H}_2\text{C}_2\text{O}_4(aq) + 2 \text{ Na}^+(aq) + 2 \text{ OH}^-(aq) \rightarrow 2 \text{ Na}^+(aq) + \text{C}_2\text{O}_4^{2-}(aq) + 2 \text{ H}_2\text{O}(l)$ net ionic equation: $\text{H}_2\text{C}_2\text{O}_4(aq) + 2 \text{ OH}^-(aq) \rightarrow \text{C}_2\text{O}_4^{2-}(aq) + 2 \text{ H}_2\text{O}(l)$

119. 0.577 ml H_2SO_4; ionic equation: $2 \text{ H}^+(aq) + \text{SO}_4^{2-}(aq) + 2 \text{ NH}_3(aq) \rightarrow 2 \text{ NH}_4^+(aq) + \text{SO}_4^{2-}(aq)$ net ionic equation: $2 \text{ H}^+(aq) + 2 \text{ NH}_3(aq) \rightarrow 2 \text{ NH}_4^+(aq)$

120. 0.376 M $C_6H_{12}O_6$
121. 0.150 M $H_2C_2O_4$
122. 24.3 g/mol (Mg)
123. $C_2HBrClF_3$
124. $2 \, KClO_3(s) \rightarrow 2 \, KCl(s) + 3 \, O_2(g)$
126. CrO_3
127. MnO_2
129. 6.65×10^4 g/mol
130. (d) $ZnCO_3$ (MW = 125.40)   is   35.10% CO_2.
131. 39.452 g
132. 32.1 g Mg
135. 80.0%

Chapter 3

7. $17 \, \dfrac{m}{cycle}$

8. 6.0×10^{-7} m

9. $4.28 \times 10^{14} \, \dfrac{cycle}{s}$

12. 11.922 m This radiation falls in the radio wave region of the electromagnetic spectrum.

13. $4.949 \times 10^{14} \, \dfrac{cycle}{s}$ This radiation falls in the visible region of the electromagnetic spectrum.

14. $5 \times 10^{16} \, \dfrac{cycle}{s}$

16. $5.089895 \times 10^{14} \, \dfrac{cycle}{s}$ $5.084741 \times 10^{14} \, \dfrac{cycle}{s}$

19. 656.3 nm < 486.1 nm < 434.0 nm < 410.2 nm in order of increasing energy.

20. 3.027×10^{-19} J

21. 2.0×10^{-26} J

22. 4.914×10^{-7} m This radiation falls in the visible portion of the electromagnetic spectrum.

25. $IE_{Li} > IE_{He}$

26. $IE_{F^-} < IE_{Ne}$

27. Kr > Br > Rb

34. Na < Li < Be < F

35. (c) N

36. (b) Ca

39. Mg

41. Kr

44. for $n = 3$, 3 subshells; for $n = 4$, 4 subshells

45. $n = 1$, 2 electrons; $n = 2$, 8 electrons; $n = 3$, 18 electrons; $n = 4$, 32 electrons

46. 10 electrons

47. 5 unpaired electrons

48. The difference between atomic number of any adjacent pair of elements in a group of the periodic table will be 8, 18, or 32 because the change between members of a group corresponds to the number of electrons to fill a shell.

49. (b)

50. (e)

51. (b)

53. Na   [Ne] $3s^1$
    Mg   [Ne] $3s^2$
    Al   [Ne] $3s^2\,3p^1$
    Si   [Ne] $3s^2\,3p^2$
    P   [Ne] $3s^2\,3p^3$
    S   [Ne] $3s^2\,3p^4$
    Cl   [Ne] $3s^2\,3p^5$
    Ar   [Ne] $3s^2\,3p^6$ = [Ar]

54. N   ⇅$(1s)$ ⇅$(2s)$ ↑ ↑ ↑$(2p)$

55. Ni   ⇅$(1s)$ ⇅$(2s)$ ⇅ ⇅ ⇅$(2p)$ ⇅$(3s)$ ⇅ ⇅ ⇅$(3p)$ ⇅$(4s)$
    ⇅ ⇅ ⇅ ↑ ↑$(3d)$

56. $x = 2$

57. (b)

58. (d)

59. (c)

60. 11 electrons

61. 6 electrons

62. Group IVA below Pb (lead)

63. (d) Y

64. (d) P

65. (d) Fe^{3+}

68. scandium, row 4 and column 3

69. niobium, row 5 and column 5

70. Group VA (antimony)

71. Group IA

73. (c) Cr^{2+}

76. 0.1442 nm, 144.2 pm

77. (c) Al

79. (d) P (phosphorus has the largest covalent radius, $P > S > N > O > F$)

82. $H^- < F^- < Cl^- < Br^- < I^-$

85. $Fe^{3+} < Fe^{2+}$

86. N^{3-}, O^{2-}, F^-, Ne, Na^+, Mg^{2+}, Al^{3+}, Si^{4+} are isoelectronic; P^{3-}, S^{2-}, Cl^-, Ar, K^+, and Ca^{2+} are isoelectronic.

87. (a), (b), (c)

88. Al^{3+} is smaller than Mg^{2+}

89. (e) Se^{2-}

90. (d) Mg^{2+}

91. (c) Be^{2+}
92. (b) P^{3-}
93. (e) P^{4+}
95. (d) [Ne] $3s^2\ 3p^1$ (Al)
96. (f) 3rd *IE*
97. (a) Na
98. (d) Mg
99. Mg < Be < Ne < Na < Li
100. The most common ions of lead and tin are Pb^{2+}, Pb^{4+} and Sn^{2+}, Sn^{4+}.

Chapter 3 Appendix

3A.3 (a) *n* (b) *l* (c) *m* (d) *s*
3A.6. $s = +\frac{1}{2}, -\frac{1}{2}$
3A.7. 5
3A.12. 16
3A.16. *s, p, d, f*
3A.19. (c)
3A.22. 5
3A.27. (a)
3A.28. (d)

Chapter 4

2. (a) Li has 1 valence electron.
   (b) C has 4 valence electrons.
   (c) Mg has 2 valence electrons.
   (d) Ar has 8 valence electrons.
3. (a) Fe has 8 valence electrons.
   (b) Cu has 11 valence electrons.
   (c) Bi has 5 valence electrons.
   (d) I has 7 valence electrons.
4. C^{4-}, N^{3-}, S^{2-}, and I^- all have 8 valence electrons. All of these elements gain electrons to achieve an octet of electrons.
5. All have 8 valence electrons. All of these elements lose electrons to achieve an octet of electrons.
6. Filled *d* and *f* subshells are ignored because they seldom take part in the chemical reactivity of the element.
7. (a) 24 (b) 8 (c) 8 (d) 32
8. (a) 22 (b) 34 (c) 48 (d) 56
9. (a), (b), (c), and (d) have 16 valence electrons; (e), (f), and (g) have 18 valence electrons.
25. CO, CN^-, and NO^+
27. (a), (c), (d)
28. (a)
29. (d)

32. All of the ions and molecules listed have nonbonding valence electrons and can act as bases.
33. (b), (c)
37. (a)
38. (b) and (c) have no resonance structures
41. (e)
42. (b)
47. (a) +1 (b) +2 (c) +1
48. (a) 0 (b) 0 (c) 0 (d) 0
49. (a) 0 (b) +1 (c) 0 (d) −1
50. (a) end N = 0, middle N = +1 (b) 0, +1 (c) +1
51. BF_3: formal charge B = 0; F_3B—NH_3: formal charge N = +1, formal charge B = −1; NH_3: formal charge N = 0
52. end S = −1, middle S = +2
57. (a) trigonal pyramidal (b) trigonal planar (c) T shaped (d) T shaped
58. (a) tetrahedral (b) tetrahedral (c) tetrahedral (d) tetrahedral
59. (a) bent (b) trigonal pyramidal (c) tetrahedral (d) octahedral
60. (a) trigonal pyramidal (b) seesaw (c) square pyramidal (d) octahedral
61. (a) linear (b) T shaped (c) seesaw (d) square planar
62. (a) linear (b) bent (c) trigonal planar
63. (a) linear (b) tetrahedral
64. (a) XeF_3^+
65. (a) NH_2^- and H_2O are bent
66. (c) CO_2 and BeH_2
67. (a), (c), (d), and (e)
68. (a), (c), and (e)
69. (a), (b), (c), and (d)
70. (a), (b), and (d)
71. (c) C
75. (a), (b), and (d)
76. H_2CO has a dipole moment along the C=O bond.
82. (d) S
83. (d) N
84. (c) Si
85. VIIA
86. VIIA
87. IA or VIIA

Chapter 4 Appendix

4A.1. (a) sp^3 (b) sp^3 (c) sp^2 (d) sp^2 (e) sp
4A.2. (a) sp^3 (b) sp^2 (c) sp^2
4A.3. (a) sp^3d (b) sp^3 (c) sp^3d (d) sp^2
4A.6. A σ molecular orbital is one that results from the head-on overlap of atomic orbitals, whereas a π molecular orbital is one that results from the sideways or parallel overlap of atomic orbitals.

4A.11. H_2 is more stable than H_2^+ and H_2^-, and H_2^{2-}.

4A.14. Peroxide ion is not paramagnetic.

4A.16. (a) diamagnetic   (b) diamagnetic   (c) diamagnetic   (d) paramagnetic
   (e) diamagnetic

Chapter 5

1. decreasing metallic character: Na > Mg > Al > Si > P > S > Cl > Ar, increasing non-metallic character: Na < Mg < Al < Si < P < S < Cl < Ar.

2. N < P < As < Sb < Bi

3. (d) Al < B < N < F

6. (a) Ca   (b) Na   (c) K   (d) Mg

7. (a) K   (b) Na   (c) Ca   (d) Ca

9. RbBr, RbH, Rb_2S, Rb_3N, Rb_3P, Rb_2O, Rb_2O_2

13. BaF_2, BaH_2, BaS, Ba_3N_2, Ba_3P_2, BaO, BaO_2

17. (a) Mg is in Group IIA, Mg^{2+}
   (b) Al is in Group IIIA, Al^{3+}
   (c) Si is in Group IVA, Si^{4+}
   (d) Cs is in Group IA, Cs^+
   (e) Ba is in Group IIA, Ba^{2+}

18. (a) C is in Group IVA, $4 - 8 = -4$, C^{4-}
   (b) P is in Group VA, $5 - 8 = -3$, P^{3-}
   (c) S is in Group VIA, $6 - 8 = -2$, S^{2-}
   (d) I is in Group VIIA, $7 - 8 = -1$, I^-

19. magnesium

20. IVA

21. IIIA

26. AlN

27. Sr_3P_2

28. GaAs

29. (a) $2\,Na(s) + F_2(g) \rightarrow 2\,NaF(s)$
   (b) $4\,Na(s) + O_2(g) \rightarrow 2\,Na_2O(s)$
   (c) $2\,Na(s) + H_2(g) \rightarrow 2\,NaH(s)$
   (d) $16\,Na(s) + S_8(s) \rightarrow 8\,Na_2S(s)$
   (e) $12\,Na(s) + P_4(s) \rightarrow 4\,Na_3P(s)$

30. (a) $Ca(s) + H_2(g) \rightarrow CaH_2(s)$
   (b) $2\,Ca(s) + O_2(g) \rightarrow 2\,CaO(s)$
   (c) $8\,Ca(s) + S_8(s) \rightarrow 8\,CaS(s)$
   (d) $Ca(s) + F_2(g) \rightarrow CaF_2(s)$
   (e) $3\,Ca(s) + N_2(g) \rightarrow Ca_3N_2(s)$
   (f) $6\,Ca(s) + P_4(s) \rightarrow 2\,Ca_3P_2(s)$

31. (a) KF   (b) AlF_3 (c) SnF_2 or SnF_4 (d) MgF_2 (e) BiF_3 or BiF_5

36. Using the guide that ionic substances generally result when metals combine with non-metals: (a) covalent, (b) covalent, (c) ionic, (d) ionic. Using bond type triangle as a guide: (a) covalent, (b) covalent, (c) ionic, (d) covalent/ionic.

37. Using the guide that ionic substances generally result when metals combine with non-metals: (a) covalent, (b) covalent, (c) covalent, (d) ionic. Using bond type triangle as a guide: (a) covalent, (b) covalent, (c) covalent, (d) ionic

38. (d)

39. (e)
41. (b), (c), and (d)
46. None of these
47. (a), (b), and (e)
48. (a) and (d)
49. (e)
52. $\Delta EN = 2.5$
53. No, could be metallic.
54. (a) $CdHg$, $AlSb$, CH_4
    (b) Cs_2O, $BaCl_2$, MgH_2
    (c) ICl, NP, SiO_2, HgO
    (d) CsI, BaS, Mg_3N_2
    (e) $BaSi_2$, MgH_2, CO_2
55. (a) covalent
    (b) semimetal
    (c) ionic
    (d) ionic
    (e) covalent
    (f) borderline ionic/metallic
    (g) covalent with some semimetal character
    (h) ionic/covalent character
    (i) covalent
61. (a) -3 (b) $+5$ (c) $+3$ (d) $+2$
62. (a) $+3$ (b) $+3$ (c) $+3$
63. Al^{3+}
64. (e) and (g)
65. (a) 0   (b) -1 (c) $+1$ (d) $+3$ (e) $+5$ (f) $+7$
66. (a) -1 (b) -1 (c) 0   (d) $+1$ (e) $+5$ (f) $+5$ (g) $+7$ (h) $+7$
67. (a) $+2$ (b) $+4$ (c) $+4$ (d) $+6$ (e) $+6$ (f) $+6$ (g) $+8$ (h) $+8$
69. (a) $+4$ (b) $+4$ (c) $+2$ (d) $+4$ (e) $+4$ (f) -4 (g) -4 (h) 0   (i) $+4$ (j) $+2$
70. (a) 0   (b) -2 (c) -2 (d) $+4$ (e) $+6$ (f) $+4$ (g) $+6$ (h) $+4$ (i) $+6$ (j) $+4$ (k) $+6$
71. (a) $+2$ (b) $+3$ (c) $+4$ (d) $+6$ (e) $+7$ (f) $+2$ (g) $+3$ (h) $+6$ (i) $+7$ (j) $+4$ (k) $+2$
72. (a) $+2$ (b) $+4$ (c) $+3$ (d) $+4.66$ (e) $+3$ (f) $+4$ (g) $+4$ (h) $+4$ (i) $+4$
73. Prussian blue, $+3$; Turnbull's blue, $+2$.
74. (a) C in CH_3 group $= -3$
         $H = +1$
         $Br = -1$
         C in center $= +1$
    (b) $C = -2$
         $H = +1$
         $O = -2$
    (c) C in $CH_3 = -3$
         $H = +1$
         $O = -2$
         C in center $= +2$

(d) C in CH_3 = -3
     H = $+1$
     O = -2
     C in center = 0

76. (a) C in CH_3 group = -3
        H = $+1$
        O = -2
        Cl = -1
        C in center = $+3$
    (b) C = -2
        H = $+1$
        N = -3
    (c) C in CH_3 group = -3
        H = $+1$
        O = -2
        C in C=O group = $+3$
        C in CH_2 group = -1
    (d) C = -3
        H = $+1$
    (e) H = $+1$
        C in CH_2 group = -2
        C in CH_3 group = -3
        C in CH group = -1
    (f) H = $+1$
        C = -1

77. (b) Iron is reduced from $+3$ in Fe_2O_3 to 0 in $Fe(s)$. Carbon is oxidized from $+2$ in CO to $+4$ in CO_2.
    (c) Carbon is reduced from $+4$ in CO_2 to $+2$ in CO. Hydrogen is oxidized from 0 to H_2 to $+1$ in H_2O.
    (d) Carbon is reduced from $+2$ in CO to -2 in CH_3OH. Hydrogen is oxidized from 0 in H_2 to $+1$ in CH_3OH.

79. (a) The phosphorus is reduced from $+5$ to 0; Carbon is oxidized from 0 to $+2$.
    (b) The phosphorus is oxidized from 0 in P_4 to $+5$ in P_4O_{10}. The oxygen is reduced from 0 in O_2 to -2 in P_4O_{10}.

81. The reaction with cyanide involves an exchange of next nearest neighbors (coordinated ligands) about the silver ion. The second reaction involves the exchange of electrons between the silver ions in the "tarnish" and the aluminum metal. Therefore, the second reaction with aluminum involves oxidation.

84. (a) P_4S_3 (b) SiO_2 (c) CS_2 (d) CCl_4 (e) PF_5

85. (a) SiF_4 (b) SF_6 (c) OF_2 (d) Cl_2O_7 (e) ClF_3

86. (a) $SnCl_2$ (b) $Hg(NO_3)_2$ (c) SnS_2 (d) Cr_2O_3 (e) Fe_3P_2

87. (a) BeF_2 (b) Mg_3N_2 (c) Ca_2C (d) BaO_2 (e) K_2CO_3

88. (a) $Co(NO_3)_3$ (b) $Fe_2(SO_4)_3$ (c) $AuCl_3$ (d) MnO_2 (e) WCl_6

89. (a) potassium nitrate
    (b) lithium carbonate
    (c) barium sulfate
    (d) lead(II) iodide

90. (a) aluminum chloride
    (b) sodium nitride

    (c) calcium phosphide
    (d) lithium sulfide
    (e) magnesium oxide

91. (a) ammonium hydroxide
    (b) hydrogen peroxide
    (c) magnesium hydroxide
    (d) calcium hypochlorite
    (e) sodium cyanide

92. (a) antimony(III) sulfide
    (b) tin(II) chloride
    (c) sulfur tetrafluoride
    (d) strontium bromide
    (e) silicon tetrachloride

93. (a) CH_3CO_2H (b) $HCl(aq)$ (c) H_2SO_4 (d) H_3PO_4 (e) HNO_3

94. (a) H_2CO_3 (b) HCN (c) H_3BO_3 (d) H_3PO_3 (e) HNO_2

95. sodium sulfite, Na_2SO_3; sodium hydrogensulfite, $NaHSO_3$

96. thiosulfate ion, SSO_3^{2-} or $S_2O_3^{2-}$

97. (a) fluorite, calcium fluoride
    (b) galena, lead(II) sulfide
    (c) quartz, silicon dioxide
    (d) rutile, titanium(IV) oxide
    (e) hematite, iron(III) oxide

98. (a) calcite, calcium carbonate
    (b) barite, barium sulfate

Chapter 6

1. Ar, Co, CH_4, Cl_2

4. (a) pressure, (b) average kinetic energy, and (d) number of molecules per container.

7. $\dfrac{14.7 \text{ lb}}{\text{in}^2}$, $1.01 \times 10^5 \dfrac{\text{newtons}}{\text{m}^2}$

8. 0.9813 atm, 9.941×10^4 Pa

10. 29.9 in Hg, 1.01×10^5 Pa

11. 29.0 psi

12. 3×10^{-6} atm

13. 1.2×10^{19} lb

14. 324 mmHg

15. The volume of the balloon will decrease to 0.349 L.

16. 13 L

17. 500 L

18. 5.29 atm

19. 2.8×10^2 °C

21. 17.9%

22. 0.332 L

24. 2:1

25. 0.690 L N_2, 2.07 L H_2

26. 15.0 L NO

27. 18.0 L
30. (b)
31. N_2O
34. (e)
35. (b)
36. $1.21 \times 10^3 \dfrac{mL \cdot psi}{mol \cdot K}$
37. $\dfrac{\$0.00014}{g\ N_2}$
38. 1.5 atm
39. 253 K
40. 8.0 kg O_2
41. 3.13 atm
42. Yes, the cork will blow out because the internal pressure is 7.1 atm
43. 1.3 L
44. 3.4×10^2 atm
45. 0.994, 1.99 atm
46. 24.6 L
47. 6.85 L
48. 2.3×10^2 K
49. 12 atm
50. $3.30\ \dfrac{g}{L}$ The density of the liquid is ~ 400 times greater than the density of the gas.
51. $0.825\ \dfrac{kg}{m^3}$
52. $0.179\ \dfrac{g}{L}$ = density helium
53. 0.0631
54. Kr
55. $28.95\ \dfrac{g}{mol}$
56. 424 mmHg
57. 2.5 atm
58. 19.0 atm
59. 0.452 atm
70. C_3H_6
71. CH_2N_2
73. 0.542 g Mg
74. CrO_3
75. 3.66×10^4 L CO_2
76. 2.20×10^3 L O_2
77. 2.45 L CO_2
78. zinc

Chapter 6 Appendix

6A.1. $SF_6 < CF_2Cl_2 < Cl_2 < SO_2 < Ar$

6A.2. 2.236

6A.6. 13 s

6A.7. 16.0 g/mol

6A.8. The gases will meet at row 22.

6A.9. $234.8 \dfrac{g}{mol}$

6A.16. $\dfrac{He}{Ne} = 1.1$, $\dfrac{Ar}{Ne} = 1.2$

6A.17.

$PV = nRT$		van der Waals	
$V = 1.0$ L	$P = 2.2 \times 10^1$ atm	$V = 1.0$ L	$P = 2.2 \times 10^1$ atm
$= 0.10$ L	$= 2.2 \times 10^2$	$= 0.10$ L	$= 2.3 \times 10^2$
$= 0.010$ L	$= 2.2 \times 10^3$	$= 0.010$ L	$= 20$

The calculation fails for the van der Waals equation ($V = 0.010$ L) because the b term is larger than the volume. This would indicate that at this small volume the molecular size becomes very significant.

Chapter 7

1. Temperature is a quantitative measure of whether an object can be labeled "hot" or "cold." Heat is that which causes a change in temperature.

2. As a balloon filled with helium becomes warmer, the particles within become more agitated. This makes a greater pressure on the walls, causing the balloon to expand.

5. The Kelvin temperature of an ideal gas is directly proportional to the energy of the gas. For more complex systems, detection of changes in energy require that both temperature and work done by or on the system be monitored.

12. The energy of the system will decrease.

15. The standard state is defined by a unique set of conditions. There is only one standard state. Therefore, the standard-state enthalpy change must be unique, whereas the enthalpy change for the many other possible states of the system can vary over an extremely wide range.

17. State functions of a trip are (g) location of the car, (h) elevation, (i) latitude, and (j) longitude.

18. The state functions are (a) temperature, (b) energy, (c) enthalpy, (d) pressure, and (e) volume.

19. Because bonds between atoms must be broken.

20. For $+\Delta H$, energy must be provided to the chemical reaction; for $-\Delta H$, energy is given off by the chemical reaction.

22. (a) endothermic
    (b) exothermic
    (c) exothermic

23. $\Delta H = 0$

24. Yes, the ΔH to form 2 mol of CH_4 would be twice the ΔH to form 1 mol of CH_4.

25. Positive

27. (a), (b), (c)

31. -1.12×10^3 kJ/mol$_{rxn}$
32. -186 kJ/mol$_{rxn}$
33. Heat is absorbed ($\Delta H° = +178.3$ kJ/mol$_{rxn}$).
34. $+49.9$ kJ/mol$_{rxn}$
35. -1284.4 kJ/mol$_{rxn}$
36. -1076.87 kJ/mol$_{rxn}$
37. -5313.97 kJ/mol$_{rxn}$
38. -905.5 kJ/mol$_{rxn}$
39. -410.7 kJ/mol$_{rxn}$
40. -189.32 kJ/mol$_{rxn}$
41. -851.5 kJ/mol$_{rxn}$
42. -536.0 kJ/mol$_{rxn}$; the iron reaction evolves more heat.
47. -1.01×10^5 kJ/mol$_{rxn}$

Chapter 7 Appendix

7A.2. $+1.90$ kJ/mol$_{rxn}$
7A.3. -189.32 kJ/mol$_{rxn}$
7A.4. $+180.6$ kJ/mol$_{rxn}$
7A.5. $+22.8$ kJ/mol$_{rxn}$
7A.6. -2043.96 kJ/mol$_{rxn}$
7A.7. -53.9 kJ/mol$_{rxn}$
7A.8.   (a) P_4 and (f) F_2
7A.9. $+178.3$ kJ/mol$_{rxn}$
7A.10. $+49.9$ kJ/mol$_{rxn}$
7A.11. -1284.4 kJ/mol$_{rxn}$
7A.12. -1076.87 kJ/mol$_{rxn}$
7A.13. -5313.9 kJ/mol$_{rxn}$
7A.14. -905.5 kJ/mol$_{rxn}$
7A.15. -410.6 kJ/mol$_{rxn}$
7A.16. -91.1 kJ/mol$_{rxn}$
7A.17. The iron reaction ($\Delta H° = -851.5$ kJ/mol$_{rxn}$ gives off more heat than the chromium reaction ($\Delta H° = -536.0$ kJ/mol$_{rxn}$).

Chapter 8

7. solid $= 2.68 \times 10^{22}$ molecules, liquid $= 2.16 \times 10^{22}$ molecules, and gas $= 2.69 \times 10^{19}$ molecules
11. (e)
21. The larger surface area will allow a greater total amount of bromine to enter the gas phase, but the actual vapor pressure will remain the same as long as the temperature is not changed. The vapor pressure in a closed container would not change if more liquid were added or if some of the liquid were removed.
29. 38°C
30. 18 mmHg
31. (c)

32. Nitrogen

33. (c)

36. The surface tension of water acts to hold the surface together even with the mass of the steel needle attempting to force its way through.

43. The strength of a hydrogen bond is related to the electronegativity difference between the hydrogen atom and the two electronegative atoms involved in the hydrogen bond. This observation follows from the order of hydrogen bond strengths given.

46. The anesthetics and olive oil have low molecular polarity.

48. (e) 1-heptanol

49. If carbon tetrachloride, a nonpolar solvent, were added to the reaction mixture, I_2 would tend to dissolve in the nonpolar solvent. Response (b).

56. (c)

57. (e)

58. (e)

Chapter 8 Appendix

8A.3. Colligative properties refer to solutions not pure substances.

8A.8. The freezing point of the ice is lowered, thus causing melting.

8A.9. The k_f values are larger than the k_b values. Thus, for a given number of moles of solute the freezing point depression is greater than the boiling point elevation.

8A.10. $-0.192°C$

8A.14. $-1.80°C$

8A.16. 86% dissociates

Chapter 9

1. (a), (d), (e)

5. (a), (b), (d), (e)

7. (e) diamond

8. (d) dispersion forces

9. (d) dispersion forces

10. (c) ionic solids

12. (d)

13. (a) LiF

24. (a) CN = 12   (b) CN = 12   (c) CN = 8   (d) CN = 6

25. The larger the number of next nearest neighbors, the larger the number of induced dipole–induced dipole interactions. Both (c) cubic closest packed and (d) hexagonal closest packed structure types allow for 12 next nearest neighbors.

26. (b) body-centered cubic structure

27. $V_{unit\ cell} = 2.40 \times 10^{-23}$ cm^3, $r_{Cr} = 1.25 \times 10^{-8}$ cm

29. $r_{Ar} = 1.933 \times 10^{-8}$ cm

30. 10.50 g/cm^3

31. body-centered cubic

32. face-centered cubic

33. body-centered cubic

34. (d) Cu
35. 4.352×10^{-8} cm
36. 0.3451 nm
37. 0.3635 nm; 0.148 nm
38. 0.176 nm
39. $N = 4$
40. $N = 4$
41. r_{Fe} (room temperature) $= 1.24 \times 10^{-8}$ cm; d (910°C) $= 15.7$ g/cm^3. The face-centered cubic structure of iron is more dense, suggesting that iron contracts on changing from the body-centered cubic structure to the face-centered structure.

Chapter 10

9. (d) $K_c = \dfrac{[ClF_3]^2}{[Cl_2][F_2]^3}$

10. (d) $K_c = \dfrac{[NO]^2[O_2]}{[NO_2]^2}$

11. (a) $K_c = \dfrac{[OF_2]^2}{[O_2][F_2]^2}$ (b) $K_c = \dfrac{[SO_3]^2}{[SO_2]^2[O_2]}$ (c) $K_c = \dfrac{[SO_2Cl_2]^2[O_2]}{[SO_3]^2[Cl_2]^2}$

12. (a) $K_c = \dfrac{[N_2][H_2O]^2}{[NO]^2[H_2]^2}$ (b) $K_c = \dfrac{[NO]^2[Cl_2]}{[NOCl]^2}$ (c) $K_c = \dfrac{[NO_2]^2}{[NO]^2[O_2]}$

13. (a) $K_c = \dfrac{[CO_2]^2}{[CO]^2[O_2]}$ (b) $K_c = \dfrac{[CO][H_2O]}{[CO_2][H_2]}$ (c) $K_c = \dfrac{[CH_3OH]}{[CO][H_2]^2}$

14. $5.72 \times 10^{-6} = K_c$

15. trial I   59.686

    trial II   59.45

    trial III   59.21

    K_c is a constant within the limits of experimental error.

17. $K_c(c) = 1.5 \times 10^{12}$
18. $K_c(c) = 0.78$
19. (b)
20. (c)
21. (c)
22. $Q_c > K_c$; the system is not at equilibrium and must shift to the left to form reactant.
26. $\Delta(CO) = 0.250\ M$, $\Delta(Cl_2) = 0.250\ M$
27. $\Delta(N_2) = 0.117\ M$, $\Delta(H_2) = 0.351\ M$
28. (b) $\Delta(NO_2) = 2\Delta(O_2)$
29. (e) $\Delta(F_2) = 3\Delta(Cl_2)$
30. (b) $1.5\Delta C$
31. $[N_2] = 0.922\ M$, $[H_2] = 0.766\ M$, $[NH_3] = 0.16\ M$
32. $K_c = 0.062$
36. $[PCl_5] = 0.96\ M$, $[PCl_3] = 0.036\ M$, $[Cl_2] = 0.036\ M$; % decomposition of PCl$_5$ = 3.6%.
37. $[PCl_5] = 0.99\ M$, $[PCl_3] = 6.5 \times 10^{-3}\ M$, $[Cl_2] = 0.21\ M$; % decomposition of PCl$_5$ is 0.65%

38. $[NO_2] = 0.098\ M$, $[NO] = 0.0019\ M$, $[O_2] = 9.5 \times 10^{-4}\ M$
39. $[NO_2] = 0.10\ M$, $[NO] = 2.6 \times 10^{-4}\ M$, $[O_2] = 0.050\ M$
41. $[SO_3] = 0.40\ M$
42. $[N_2O_4] = 0.099\ M$, $[NO_2] = 2.4 \times 10^{-3}\ M$
43. $[N_2O_4] = 0.50\ M$
44. $[NH_3] = 0.01\ M$, $[N_2] = 0.15\ M$, $[H_2] = 0.24\ M$
45. $[CO] = [H_2O] = 0.54\ M$, $[CO_2] = [H_2] = 0.46\ M$
46. The initial concentrations of CO and H_2O would have to be 1.18 M.
47. $[N_2] = 0.100\ M$, $[O_2] = 0.090\ M$, $[NO] = 1.7 \times 10^{-6}\ M$
48. $[SO_2Cl_2] = 0.049\ M$, $[SO_2] = [Cl_2] = 8.4 \times 10^{-4}\ M$
49. $[NOCl] = 0.20\ M$, $[Cl_2] = 2 \times 10^{-4}\ M$, $[NO] = 0.30\ M$
51. (a) The reaction will shift to the right.
    (b) The reaction will shift to the right.
    (c) The reaction will shift to the right.
52. (a) The reaction will shift to the right.
    (b) The reaction will not shift.
    (c) The reaction will shift to the left.
53. (a) Increasing the concentration of $NO_2(g)$ will shift the reaction to the right until equilibrium is reached.
    (b) Increasing the concentration of $O_2(g)$ will shift the reaction to the left until equilibrium is reached.
    (c) Increasing the concentration of $PF_5(g)$ will shift the reaction to the right until equilibrium is reached.
54. (a) Decreasing the concentration of $NO(g)$ will cause the reaction to shift to the right, forming more NO.
    (b) Decreasing the concentration of $O_2(g)$ will cause the reaction to shift to the left forming more $O_2(g)$.
    (c) Decreasing the concentration of $F_2(g)$ will cause the reaction to shift to the left forming more $Cl_2(g)$.
59. $K_{sp} = [Ag^+]^2[CrO_4{}^{2-}]$
60. $K_{sp} = [Bi^{3+}]^2[S^{2-}]^3$
61. $K_{sp} = 2.47 \times 10^{-9}$
62. $K_{sp} = 5.145 \times 10^{-3}$
63. $K_{sp} = 1.2 \times 10^{-10}$
64. $C_s = 6.2 \times 10^{-16}$ g Ag_2S in 100 mL of water
65. (a) $\dfrac{1.4 \times 10^{-22}\ \text{g }Hg_2S}{100\ \text{mL soln}}$

    (b) $\dfrac{2 \times 10^{-25}\ \text{g HgS}}{100\ \text{mL soln}}$

Chapter 10 Appendix

10A.2. As the reaction system approaches more closely the equilibrium state, the concentrations of all components will tend to change less and less as Q_c gets closer to K_c.

10A.3. The assumption of a small change in concentration is doomed to failure when the initial concentrations are very different from the equilibrium concentrations. By being "very different" the changes must become "very much larger" to accomplish the adjustments required.

10A.5. (c)

Chapter 11

2. $HCl(aq) + NaOH(aq) \rightarrow NaCl(aq) + H_2O(l)$
   $H_2SO_4(aq) + 2\,NaOH(aq) \rightarrow Na_2SO_4(aq) + 2\,H_2O(l)$
   $HCl(aq) + NH_3(aq) \rightarrow NH_4Cl(aq)$
   $H_2SO_4(aq) + 2\,NH_3(aq) \rightarrow (NH_4)_2SO_4(aq)$

6. (a), (b), (c), (e)

7. (a), (c), (d)

8. $HBr(g) + H_2O(l) \rightarrow H_3O^+(aq) + Br^-(aq)$
   $H_2O(l) + NH_3(g) \rightarrow OH^-(aq) + NH_4^+(aq)$

13. (a), (e)

14. (a), (c)

15. CH_4 and CH_3^+

16. (a) $\underset{\text{Brønsted acid}}{HSO_4^-(aq)} + \underset{\text{Brønsted base}}{H_2O(l)} \rightarrow \underset{\text{Brønsted acid}}{H_3O^+(aq)} + \underset{\text{Brønsted base}}{SO_4^{2-}(aq)}$

   (b) $\underset{\text{Brønsted acid}}{CH_3CO_2H(aq)} + \underset{\text{Brønsted base}}{OH^-(aq)} \rightarrow \underset{\text{Brønsted base}}{CH_3CO_2^-(aq)} + \underset{\text{Brønsted acid}}{H_2O(l)}$

   (c) $\underset{\text{Brønsted acid}}{CaF_2(s) + H_2SO_4(aq)} \rightarrow \underset{\text{Brønsted acid}}{CaSO_4(aq) + 2\,HF(aq)}$
   The fluoride ion of CaF_2 is a Brønsted base.
   The SO_4^{2-} ion of $CaSO_4$ is a Brønsted base.

   (d) $\underset{\text{Brønsted acid}}{HNO_3(aq)} + \underset{\text{Brønsted base}}{NH_3(aq)} \rightarrow NH_4NO_3(aq)$
   The NH_4^+ ion of $NH_4NO_3(aq)$ is a Brønsted acid.
   The NO_3^- ion of $NH_4NO_3(aq)$ is a Brønsted base.

   (e) $LiCH_3(l) + NH_3(l) \rightarrow CH_4(g) + LiNH_2(s)$
   CH_3^- from $LiCH_3(l)$ is a Brønsted base. CH_4 is a Brønsted acid.
   $NH_3(l)$ is a Brønsted acid in this reaction. NH_2^- is a Brønsted base.

17. (a) $\underset{\text{acid}}{HCl(aq)} + \underset{\text{base}}{H_2O(l)} \rightarrow \underset{\text{acid}}{H_3O^+(aq)} + \underset{\text{base}}{Cl^-(aq)}$

   (b) $\underset{\text{base}}{HCO_3^-(aq)} + \underset{\text{acid}}{H_2O(l)} \rightarrow \underset{\text{base}}{OH^-(aq)} + \underset{\text{acid}}{H_2CO_3(aq)}$

   (c) $\underset{\text{base}}{NH_3(aq)} + \underset{\text{acid}}{H_2O(l)} \rightarrow \underset{\text{base}}{OH^-(aq)} + \underset{\text{acid}}{NH_4^+(aq)}$

   (d) $\underset{\text{base}}{CaCO_3(s)} + \underset{\text{acid}}{2\,HCl(aq)} \rightarrow Ca^{+2}(aq) + 2Cl^-(aq) + \underset{\text{base}}{H_2CO_3}\underset{\text{acid}}{(aq)}$

18. (c)

21. (a) H_2O (b) OH^- (c) O^{2-} (d) NH_3

22. (a) PO_4^{3-} (b) HPO_4^{2-} (c) CO_3^{2-} (d) S^{2-}

23. (a) OH^- (b) H_2O (c) H_3O^+ (d) NH_3

24. (a) NaH_2PO_4 (b) H_2CO_3 (c) $NaHSO_4$ (d) HNO_2

25. (b)

27. (a) $Al_2O_3(s) + 6\ HCl(aq) \rightarrow 2\ AlCl_3(aq) + 3\ H_2O(l)$
    (b) $CaO(s) + H_2SO_4(aq) \rightarrow CaSO_4(aq) + H_2O(l)$
    (c) $Na_2O(s) + H_2O(l) \rightarrow 2\ Na^+(aq) + 2\ OH^-(aq)$
    (d) $MgCO_3(s) + 2\ HCl(aq) \rightarrow H_2CO_3(aq) + MgCl_2(aq)$
    (e) $Na_2O_2(s) + H_3PO_4(aq) \rightarrow H_2O_2(aq) + Na_2HPO_4(aq)$

34. Strong acids are (c) and (e); all others are weak.

37. $CH_4 < NH_3 < H_2O < HCl$

38. $PH_3 < H_2S < H_2O < NH_3$

39. (a)

40. (e)

41. (e)

42. (c)

43. (e)

44. (e)

47. (a)

48. (a)

49. (e)

51. The fraction that dissociates will increase.

54. 6.02×10^{13}

55. pH = 1.5, pOH = 12.5

56. pH = 1.2, pOH = 12.8

57. $[H_3O^+] = 1.9 \times 10^{-4}\ M$, $[OH^-] = 5.3 \times 10^{-11}\ M$

59. (a)

60. (d)

61. (a)

62. $HOAc < HNO_2 < HF < HOClO$

65. 1.3

66. 1.000; 1.002; to correct significant figures they are equal

70. (a)

71. 1.3%, 4.2%, 13%; the % ionization increases as the solution becomes more dilute

72. $[HCO_2H] = 9.6 \times 10^{-2}\ M$, $[H_3O^+] = [HCO_2^-] = 4.2 \times 10^{-3}\ M$

73. $[HCN] = 0.174\ M$, $[H_3O^+] = [CN^-] = 1 \times 10^{-5}\ M$

74. $[H_3O^+] = 1.3 \times 10^{-6}\ M$

75. 0.22 M

76. 8.1×10^{-5}

77. 5.4×10^{-4}

78. (c)

80. $[HCO_2H] = 0.080\ M$, $[OH^-] = [HCO_2H] = 2.1 \times 10^{-6}\ M$

81. $[OBr^-] = 0.10\ M$, $[OH^-] = [HOBr] = 6.5 \times 10^{-4}\ M$

82. pH = 9.3

84. $K_b = 5.0 \times 10^{-4}$; methylamine is stronger

85. 0.057 M

86. pH = 11.8

87. 1.2×10^{-6}

88. 8.6

91. (e)

92. (c)

93. (d)

94. (c)

101. $[H_3O^+]$ in rain $= 4 \times 10^{-3}$ M; $[H_3O^+]$ in 0.10 M acetic acid $= 1 \times 10^{-3}$ M

Chapter 11 Appendix

11A.1. $[H_2CO_3] = 0.100$ M, $[H_3O^+] = [HCO_3^-] = 2.1 \times 10^{-4}$ M, $[CO_3^{-2}] = 4.7 \times 10^{-11}$ M

11A.2. $[H_2M] = 0.250$ M, $[H_3O^+] = [HM^-] = 1.9 \times 10^{-3}$ M, $[M^{-2}] = 2.1 \times 10^{-8}$ M

11A.4. $[H_2G^+] = 1.9$ M, $[HG] = [H_3O^+] = 9.3 \times 10^{-2}$ M, $[G^-] = 2.5 \times 10^{-10}$ M

11A.5. (b) and (d)

11A.6. (b)

11A.7. $[H_2C_2O_4] = 1.01$ M, $[H_3O^+] = [HC_2O_4^-] = 0.24$ M, $[C_2O_4^{2-}] = 5.4 \times 10^{-5}$ M

11A.8. (d)

11A.9. pH = 8.4

11A.10. $[CO_3^{-2}] = 0.150$ M, $[OH^-] = [HCO_3^-] = 5.6 \times 10^{-3}$ M, $[H_3O^+] = 1.8 \times 10^{-12}$ M, $[H_2CO_3] = 2.2 \times 10^{-8}$ M

11A.12. (d)

11A.15. acidic

Chapter 12

3. One of the carbons change from -2 in succinate to -1 in fumarate

5. It is a redox reaction: Fe is oxidized from the 0 to the $+2$ oxidation state; oxygen is reduced from 0 to -2.

6. (a) $+6$ (b) $+6$ (c) $+4$

7. $+3$ in all the compounds

8. (c)

9. $CH_4 < CH_3OH < H_2CO = C < CO < CO_2$

11. (a) -2 (d) $+6$ (g) $+6$
    (b) -2 (e) $+6$ (h) $+2$
    (c) -2 (f) $+4$

13. As the oxidation number of Cl increases the strength of the acid increases.

14. (b)

15. (a)

16. (a) The reaction involves oxidation–reduction; CH_3OH is oxidized.
    (b) The reaction involves oxidation–reduction; $CH_3CH_2CH_2OH$ is oxidized.

24. (a) and (d); (e) is an oxidizing agent only under very extreme conditions

25. (c); (b) only under extreme conditions

26. All of them

27. CrO_4^{-2} and P_4

28. (b)

29. (a)

30. (c) and (d)
32. (a) NO, Mn^{+2}, H_2O
    (b) $MgCl_2$ and $CrCl_2$
33. (a) and (d)
34. (c)
49. (a) and (b)
50. (a), (b), and (c)
51. $Cl_2 > Br_2 > H_2O_2 > Fe^{3+} > I_2$
53. (a) and (b)
54. (a) and (c)
56. 0.56 V, yes
57. −1.67, no
58. 1.93 V, yes
59. 0.364 V, yes
60. −0.770 V, no
64. (a)
65. (b)
66. (c)
67. (a) H^- (b) Fe^{2+} (c) Au^+
68. 0.150 M
69. 80.1%
70. 0.467 M
71. 6.1 g
72. basic
73. 0.2385 M

Chapter 12 Appendix I

12AI.3.   −1.36 V
12AI.4.   (c)
12AI.8.   (c)
12AI.11. (b)
12AI.13. 1.1180×10^{-3} g
12AI.14. 0.100 kW·h
12AI.15. 5.4 h, 11 h, 16 h
12AI.16. H_2 and Br_2, 7.4 g Br_2, 0.094 g H_2
12AI.17. 2.722×10^9 C
12AI.18. 13 g
12AI.19. 8:1
12AI.20. 1.18 g
12AI.22. (e)
12AI.23. (a)
12AI.24. 22.4 g, 6.64 g, 3.89 g
12AI.25. $x = 1$, $y = 4$ ($CeCl_4$)

12AI.26. +4

12AI.27. +3

12AI.31. The cell potential for the reaction is positive, +0.81 V.

12AI.34. The zinc is more active than the iron. Thus the zinc is oxidized before the iron.

Chapter 12 Appendix II

12AII.4. (a)

12AII.5. $E = -0.414$ V

12AII.6. E becomes smaller; weaker

12AII.7. $E = -1.68$ V

12AII.8. 0.35 V

12AII.11. decrease

12AII.14. 2×10^{37}

12AII.15. (a) $2\,NH_3(g) + 3\,CuO(s) \rightarrow N_2(g) + 3\,H_2O(l) + 3\,Cu(s)$
(b) $3\,NH_3(aq) + 3\,Cl_2(g) \rightarrow N_2(g) + NH_4Cl(aq) + 5\,HCl(aq)$

12AII.16. (a) $2\,H_2O_2(aq) \rightarrow O_2(g) + 2\,H_2O(l)$
(b) $3\,NO_2(g) + H_2O(l) \rightarrow 2\,HNO_3(aq) + NO(g)$

12AII.17. (a) $HIO_3(aq) + 5\,HI(aq) \rightarrow 3\,I_2(aq) + 3\,H_2O(l)$
(b) $8\,SO_2(g) + 16\,H_2S(g) \rightarrow 3\,S_8(s) + 16\,H_2O(l)$

12AII.18. (a) $6\,HCl(aq) + 2\,HNO_3(aq) \rightarrow 2\,NO(g) + 3\,Cl_2(g) + 4\,H_2O(l)$
(b) $2\,HBr(aq) + H_2SO_4(aq) \rightarrow SO_2(g) + Br_2(aq) + 2\,H_2O(l)$
(c) $10\,HCl(aq) + 6\,H^+(aq) + 2\,MnO_4{}^-(aq) \rightarrow 5\,Cl_2(g) + 2\,Mn^{2+}(aq) +$
    $8\,H_2O(l)$

12AII.20. $8\,H_2S(g) + 8\,H_2O_2(aq) \rightarrow S_8(s) + 16\,H_2O(l)$

12AII.21. $3\,P_4(s) + 10\,KClO_3(s) \rightarrow 3\,P_4O_{10}(s) + 10\,KCl(s)$

12AII.22. (a) $6\,I^-(aq) + 2\,CrO_4{}^{2-}(aq) + 16\,H^+(aq) \rightarrow 3\,I_2(aq) + 2\,Cr^{3+}(aq) +$
    $8\,H_2O(l)$
(b) $6\,Fe^{2+}(aq) + 14\,H^+(aq) + Cr_2O_7{}^{2-}(aq) \rightarrow 6\,Fe^{3+}(aq) + 2\,Cr^{3+}(aq) +$
    $7\,H_2O(l)$
(c) $24\,H_2S(g) + 16\,CrO_4{}^{2-}(aq) + 80\,H^+(aq) \rightarrow 3\,S_8(s) + 16\,Cr^{3+}(aq) +$
    $64\,H_2O(l)$

12AII.24. $Br_2(aq) + SO_2(g) + 2\,H_2O(l) \rightarrow H_2SO_4(aq) + 2\,HBr(aq)$

12AII.26. (a) $NO(g) + MnO_4{}^-(aq) \rightarrow NO_3{}^-(aq) + MnO_2(s)$
(b) $2\,NH_3(g) + 2\,MnO_4{}^-(aq) \rightarrow N_2(g) + 2\,H_2O(l) + 2\,MnO_2(s) + 2\,OH^-(aq)$

12AII.28. (a) $2\,Cr(s) + 2\,OH^-(aq) + 6\,H_2O(l) \rightarrow 2\,Cr(OH)_4{}^-(aq) + 3\,H_2(g)$
(b) $4\,OH^-(aq) + 2\,Cr(OH)_3(s) + 3\,H_2O_2(aq) \rightarrow 2\,CrO_4{}^{2-}(aq) + 8\,H_2O(l)$

12AII.30. (a) $2\,OH^-(aq) + Cl_2(g) \rightarrow OCl^-(aq) + H_2O(l) + Cl^-(aq)$
(b) $6\,OH^-(aq) + 3\,Cl_2(g) \rightarrow ClO_3{}^-(aq) + 3\,H_2O(l) + 5\,Cl^-(aq)$

12AII.32. $2\,NH_3(aq) + ClO^-(aq) \rightarrow N_2H_4(aq) + H_2O(l) + Cl^-(aq)$

Chapter 13

2. $\Delta S_{universe}$ must be positive.

7. (e)

8. (d)

9. (b)

10. (c)

11. (b)

12. $\Delta S° = -332.2$ J/mol$_{rxn}$·K; not favorable.

13. $\Delta S° = -1538$ J/mol$_{rxn}$·K

19. (b)

20. (a) $\Delta S° < 0$, $\Delta H° < 0$ (c) $\Delta S° < 0$, $\Delta H° < 0$
    (b) $\Delta S° < 0$, $\Delta H° < 0$ (d) $\Delta S° > 0$, $\Delta H° < 0$

21. $\Delta S° < 0$, $\Delta H° < 0$; for $\Delta H° < 0$ an increase in temperature shifts the reaction to left

22. $\Delta H° = 178.3$ kJ/mol$_{rxn}$, $\Delta S° = 160.5$ J/mol$_{rxn}$·K; entropy is the driving force

23. $\Delta H° = -851.5$ kJ/mol$_{rxn}$, $\Delta S° = -38.58$ J/mol$_{rxn}$·K; enthalpy is the driving force

24. $\Delta H° = -3310.3$ kJ/mol$_{rxn}$, $\Delta S° = -307.7$ J/mol$_{rxn}$·K, enthalpy is the driving force

25. $\Delta H° = -602.8$ kJ/mol$_{rxn}$, $\Delta S° = +579.7$ J/mol$_{rxn}$·K; enthalpy and entropy drive the reaction

26. $\Delta H° = -220.8$ kJ/mol$_{rxn}$, $\Delta S° = 68$ J/mol$_{rxn}$·K; enthalpy, entropy, and Le Chatelier's principle all favor product formation

27. None are spontaneous; the $\Delta G°$ values are as follows:
    (a) $\Delta G° = 4.31$ kJ/mol$_{rxn}$, (b) $\Delta G° = 63.75$ kJ/mol$_{rxn}$, (c) $\Delta G° = 231.04$ kJ/mol$_{rxn}$, (d) $\Delta G° = 29.11$ kJ/mol$_{rxn}$.

28. $\Delta H° = 24.9$ kJ/mol$_{rxn}$, $\Delta S° = 307.6$ J/mol$_{rxn}$·K; $\Delta G° = 24.9 - T(0.3076)$ kJ/mol$_{rxn}$, at high temperature the reaction is favored

32. $\Delta H° = -34.3$ kJ/mol$_{rxn}$, $\Delta S° = 75.58$ J/mol$_{rxn}$·K; both $\Delta H°$ and $\Delta S°$ are favorable and therefore the reaction occurs at all temperatures.

33. (c)

37. $\Delta H° = -1076.87$ kJ/mol$_{rxn}$, $\Delta S° = -56.66$ J/mol$_{rxn}$·K, $\Delta G° = -1060.01$ kJ/mol$_{rxn}$

42. (c)

43. (c)

44. $K_p = 0.0192$

45. $P_{COCl_2} = 0.11$ atm, $P_{CO} = P_{Cl_2} = 1.8 \times 10^{-2}$ atm

46. $P_{SO_3} = 0.490$ atm

47. $P_{SO_3} = 0.471$ atm, $P_{SO_2} = 1.9 \times 10^{-2}$ atm, $P_{O_2} = 9.7 \times 10^{-3}$ atm

48. $P_{N_2} = 0.60$ atm, $P_{H_2} = 0.90$ atm

49. to the left

50. -1.2×10^5 J/mol$_{rxn}$

51. $K_p = K_c$ and $[N_2] = 0.40$ M, $[O_2] = 0.60$ M, and $[NO] = 3.2 \times 10^{-5}$ M

52. -3×10^4 J/mol$_{rxn}$

53. (a)

54. $K_p = 2.2 \times 10^4$

55. $K_p = 9.3 \times 10^{-5}$

56. $K_p = 9.7 \times 10^{33}$

57. For acetic acid, $\Delta H° = -0.25$ kJ/mol$_{rxn}$, $\Delta S° = -92.0$ J/mol$_{rxn}$·K, $\Delta G° = 27.14$ kJ/mol$_{rxn}$, and $K_a = 1.8 \times 10^{-5}$; K_a for HCl $= 2 \times 10^6$

58. $\Delta H° = 3.62$ kJ/mol$_{rxn}$, $\Delta S° = -78.7$ J/mol$_{rxn}$·K, $\Delta G° = 27.0$ kJ/mol$_{rxn}$, $K_b = 1.8 \times 10^{-5}$

59. 977.08 K

62. 516 K

63. (b) and (c)
65. $\Delta H° = 64.77$ kJ/mol$_{rxn}$, $\Delta S° = -2.1$ J/mol$_{rxn}$·K; because $\Delta H° > 0$, K will increase with increasing temperature
66. 63°C

Chapter 14

8. time^{-1}
9. M^{-1} time^{-1}
10. M^{-2} time^{-1}
11. 0.0014 M s^{-1}
12. (e)
13. (c)
14. 0.0296 M s^{-1}
17. 0.027 M s^{-1}
18. 0.0114 M s^{-1}
25. Rate $= k(NO)^2(H_2)$
26. (c)
27. (a) and (c)
29. Rate $= kK_{eq}(N_2O_5)$
32. $K_{eq} = 3.5 \times 10^4$
33. Rate $= k(NO)^2(Cl_2)$
35. Rate $= k(NO)^2(O_2)$
37. Rate $= k(CH_3I)(OH^-)$
38. Rate $= 3.3 \times 10^{-7}$ M s^{-1} $k = 6.5 \times 10^{-5}$ M^{-1} s^{-1}
40. 55°C
41. Yes, 26 s
43. 12 s
44. 24.6 s
46. 1.1×10^8 s
47. 1.19×10^{10} y
48. 0.845
49. 1.90×10^3 y
50. 0.15
51. 2.4×10^4 y
52. 3.91×10^3 y
54. 4.2×10^4 y
55. Rate $= k$ (N$_2$O)
56. 1.44×10^{-3} s^{-1}, 0.026 M
58. 1.3×10^2 h
59. first order
61. first order
64. first order
70. 218 kJ/mol$_{rxn}$
71. 51.6 kJ/mol$_{rxn}$

72. 33.3 kJ/mol$_{rxn}$
74. 0.014 M^{-1} s^{-1}
77. first order, 1.4×10^{-2} mL s^{-1}

Chapter 15

4. Both techniques separate mixtures into their components utilizing a stationary and a mobile phase. The stationary phase (typically a liquid) is contained in a column through which the mobile phase passes. GC is used to separate mixtures of volatile components and employs a gaseous mobile phase, whereas HPLC separates mixtures that are soluble in a liquid mobile phase.

6. Hexane is nonpolar and therefore will interact more strongly with the nonpolar C_{18} stationary phase than with the highly polar methanol and water mobile phase.

7. (b) and (e)

8. Heptanol, butyl propyl ether, heptane; heptane, butyl propyl ether, heptanol

10. The purpose of the GC is to separate the mixture into its components. The mass spectrometer then identifies each of the components.

11. (b)

14. quantitative, (c) and (d); qualitative, (a) and (b)

15. Each technique uses a different type of electromagnetic radiation, therefore causing different energy changes within the molecule and yielding different information about the molecule.

17. (a) absorbance increases; (b) absorbance decreases; (c) absorbance decreases

18. spectrum a, isobutyric acid; spectrum b, butyraldehyde; spectrum c, 3-pentanol; spectrum d, octane

19. butyraldehyde

21. (c); (a)

22. 576 nm is the wavelength of maximum absorbance in this region of the spectrum. Using the wavelength of maximum absorbance increases the sensitivity of the measurements and increases the concentration range over which a Beer's law plot of absorbance versus wavelength is linear.

23. 0.37

28. elemental: (d); molecular: (a), (b), (c), (e)

29. Ginger ale is a mixture; therefore, the first step is to separate it into its components using a chromatographic method. HPLC is the technique of choice. A chromatogram is also run of a normal ginger ale sample, and the two chromatograms are compared. If a component exists in the suspect ginger ale that is not in the normal sample, it should be isolated from the sample and identified using IR, NMR, or mass spectroscopy.

30. (a) GC   (b) ion-selective electrode   (c) atomic absorption   (d) IR   (e) IR and NMR   (f) GC

APPENDIX
D
ANSWERS TO CORE CHECKPOINTS

Chapter 1

- •(p. 6) The symbol 8 S represents 8 individual atoms of sulfur. The symbol S_8 represents 8 atoms of sulfur bonded together to form a single molecule.
- •(p. 13) In 10,000 atoms of lithium you would expect to find approximately 742 ^6Li atoms and 9258 atoms of ^7Li.
- •(p. 16) NaOH: Na^+ and OH^-; K_2SO_4: 2 K^+ and SO_4^{2-}; $BaSO_4$: Ba^{2+} and SO_4^{2-}; $Be_3(PO_4)_2$: 3 Be^{2+} and 2 PO_4^{3-}.
- •(p. 18) F: 9; Pb: 82; 24: Cr; 74: W.

Chapter 2

- •(p. 35) K atomic weight 39.098 amu; molar mass 39.098 g/mol; U atomic weight 238.03 amu; molar mass 238.03 g/mol.
- •(p. 37) 2.2×10^{22} Al atoms
- •(p. 42) There are two atoms of carbon in one molecule of C_2H_2; two moles of carbon atoms in one mole of C_2H_2; and 1.20×10^{24} atoms of carbon in one mole of C_2H_2.
- •(p. 45) CaC_2. The molecular weight must be known to find the molecular formula.
- •(p. 52) $H^+(aq) + Cl^-(aq) + Na^+(aq) + OH^-(aq) \longrightarrow Na^+(aq) + Cl^-(aq) + H_2O(l)$
- •(p. 56) $x = 2$
- •(p. 60) 4 moles of H_2, 2.41×10^{24} H_2 molecules, 4.82×10^{24} H atoms
- •(p. 66) In the balanced chemical equation I_2 and Mg react in a one to one mole ratio. Therefore you have just enough I_2 (4 molecules) to react with the Mg (4 atoms). Both I_2 and Mg are the limiting reagent for these reaction conditions.
- •(p. 68) ionic equation: $CaCO_3(s) + 2 H^+(aq) + 2 Cl^-(aq) \longrightarrow$
$$Ca^{2+}(aq) + 2 Cl^-(aq) + CO_2(g) + H_2O(l)$$
  net ionic equation: $CaCO_3(s) + 2 H^+(aq) \longrightarrow Ca^{2+}(aq) + CO_2(g) + H_2O(l)$

Chapter 3

- •(p. 83) The radius of a typical atom would have to be 10,000 cm. This is equivalent to 100 meters.
- •(p. 85) The frequencies of microwaves cover a range of approximately 3×10^{10} to 3×10^{12} Hz. This range of frequencies is well outside the visible spectrum that we can see.
- •(p. 89) Yes, the color of a heated metal bar can be used to determine its temperature. The higher the temperature of an object the higher the frequency of electromagnetic radiation emitted by the object.
- •(p. 92)Using Figure 3.5 we see that a change in energy of 1166 kJ/mol corresponds to an energy change between the first ($n = 1$) and third energy levels ($n = 3$). Electrons in the third energy level have an energy of -145.8 kJ/mol. The electrons would be in the $n = 3$ energy level.

- •(p. 95) Most of the first 20 elements can be arranged in their appropriate positions in the periodic table using first ionization energies. However, there are a few elements that deviate from the expected trends. B, O, Al, and S have ionization energies which are slightly lower than the elements that immediately precede them in the periodic table.
- •(p. 96) Carbon has a core charge of +4 and fluorine has a core charge of +7.
- •(p. 97) The first ionization energy of sodium is much less than the first ionization energy of neon. If there were nine electrons in the second level we would expect the electrons to be held more tightly by the 11 protons in the nucleus of sodium than the eight electrons are held by the 10 protons in the nucleus of neon. This would give sodium a larger first ionization than neon rather than the observed smaller ionization energy.
- •(p. 98) Sodium has three shells whereas lithium only has two. The first ionization energy of sodium is much less than the first ionization energy of lithium. Therefore, the radius of the valence shell of sodium is larger than that of lithium.
- •(p. 100) The number of peaks in a photoelectron spectrum is determined by the number of different energies that the electrons in an atom have.
- •(p. 101) There is only one peak in the PES of He because both of its electrons have the same energy.
- •(p. 104) Neon has 2 electrons in its $n = 1$ shell and 8 electrons in its $n = 2$ shell.
- •(p. 107) The energies and the relative intensities of the PES of Sc are given below:

Energy (MJ/mol)	Relative Intensity
433	2
48.5	2
39.2	6
5.44	2
3.24	6
0.77	1
0.63	2

Electrons are removed from the subshells of Sc in the following order: 4s, 3d, 3p, 3s, 2p, 2s, 1s. The subshells of Sc are filled with electrons in the following order: 1s, 2s, 2p, 3s, 3p, 4s, 3d. There is a difference in the order in which electrons are added to and taken away from Sc. Electrons are added to the lowest energy subshell (most stable subshell closest to the nucleus) first and then to subsequent subshells in order of increasing energy until the 3p subshell is filled. After the 3p subshell is filled the 4s is filled followed by the 3d subshell. Electrons are removed from the highest energy subshell (subshell farthest from the nucleus) first and then from lower energy subshells.

- •(p. 108) Li, B, and N all have at least one unpaired electron and would therefore be expected to have magnetic properties. Be has no unpaired electrons and would therefore have no magnetic properties.
- •(p. 113) C, 2; N, 3; Ne, 0; F, 1
- •(p. 116) A covalent radius is one half the distance between two adjacent nuclei of two like atoms in a covalent compound. A metallic radius is one half the distance between two adjacent nuclei of two like metal atoms measured in a solid crystal lattice.

•(p. 119) The ionic radius of Si^{4+} would be less than the ionic radius of Al^{3+}, so it would be at the bottom of Table 3.8. Si^{4+} and Al^{3+} have the same number of electrons, but Si^{4+} has one more proton that pulls the electrons more closely to the nucleus.

•(p. 121) The maximum positive charge on atoms in Group IVA is +4. Sn has the electron configuration $[Kr]\ 5s^2\ 4d^{10}\ 5p^2$. It can lose four electrons to form a Sn^{4+} ion with the electron configuration $[Kr]\ 4d^{10}$.

•(p. 123) The most easily removed electron (lowest ionization energy) of an atom is located in the highest energy level (energy level farthest from the nucleus). Therefore, there is an inverse relationship between the ionization energy (energy to remove an electron) and the energy associated with an electron in a particular energy level (potential energy of the electron relative to the nucleus). If Figure 3.28 were a plot of relative electron energies rather than ionization energies the plot would be inverted. The lines would decrease in energy from left to right across the plot rather than increase.

•(p. 124) AVEE values increase from left to right across a period because each subsequent atom in a period has one more positive charge in the nucleus which attracts the electrons more tightly. As you move from the bottom to the top of a group there are fewer shells containing electrons. The electrons are closer to the nucleus and therefore held tightly by the nucleus. This is the same trend followed by first ionization energy. The trend for AVEE values is the reverse of the trend followed by atomic radii.

•(p. 135) Three quantum numbers are required to describe three dimensions in space. The numbers describe the following: n, size; l, shape; and m, orientation.

Chapter 4

•(p. 143) An atom in Group IVA has 4 valence electrons.

•(p. 145) H_2^+ has only one electron to form the bond between the two hydrogen atoms. A covalent bond is formed by the attraction between the nucleus of one atom and the electron(s) of another atom. Because there is only one electron in H_2^+ both nuclei share the single electron. In addition there is less electron density to shield the repulsion between the two nuclei in H_2^+. Therefore the nuclei in H_2^+ are not pulled as closely together as the nuclei in H_2.

•(p. 147) In CO_2 the nonbonding electrons are located in the nonbonding electron domains of the oxygen atoms. In CO_2 each oxygen has two nonbonding domains. There are two electrons in each nonbonding domain but four electrons are located in each of the bonding domains between carbon and oxygen. Four electrons are required to form a double bond.

•(p. 148) A skeleton structure shows the relative positions of the atoms. A Lewis structure shows the positions of the electrons.

•(p. 150)

$$: \overset{..}{\underset{..}{Cl}} :$$
$$: \overset{..}{\underset{..}{Cl}} - \overset{|}{\underset{|}{C}} - \overset{..}{\underset{..}{Cl}} : \qquad [: \overset{..}{\underset{..}{O}} - H]^-$$
$$: \overset{..}{\underset{..}{Cl}} :$$

•(p. 151) The structure for BF_3 given in the checkpoint satisfies the octet rule for all of the atoms and therefore appears to be a good Lewis structure. However, B and F are not in the list of atoms that typically form multiple bonds. In addition, experimental evidence of the reactivity of BF_3 suggests that the Lewis structure shown in the text with only six electrons around the B is probably the Lewis structure that best matches the actual structure.

•(p. 152) In order for an atom to expand its octet it must have free unoccupied orbitals with an energy close to the energy of its valence electrons (within the same shell). Nitrogen and oxygen have valence electrons in the second shell. The second shell contains one s orbital and three p orbitals. These four orbitals can contain a maximum of eight total electrons. Because there are no more subshells in the second shell, there are no more orbitals available to hold any additional electrons.

•(p. 155) The nitrogen–nitrogen bond length would be greatest in H_2N—NH_2 because single bonds are longer than multiple bonds.

•(p. 156) The two Lewis structures for acetate are resonance structures because they differ only in the location of a multiple bond. Isomers have the same molecular formula but differ in the arrangement of the atoms. All of the atoms in the two structures shown for acetate are in the same positions.

•(p. 157)

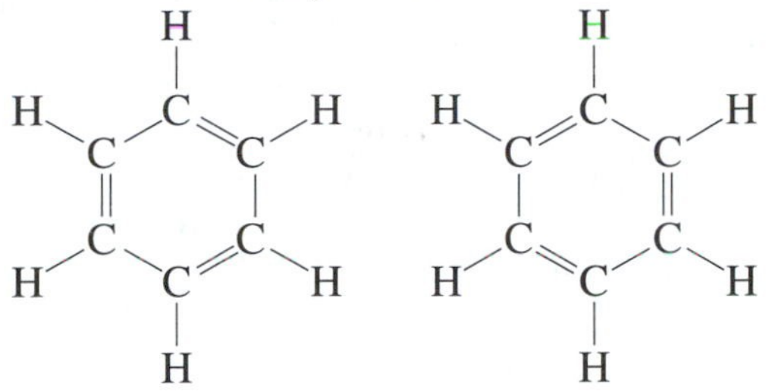

The two resonance Lewis structures of benzene have alternating single and double bonds in two different arrangements. The actual structure of benzene is an average of these two structures. Therefore, we would expect each bond in benzene to behave like a bond and a half (an average of a single bond and a double bond). The bond length of each carbon–carbon bond will be greater than the C=C double bond of ethylene but less than the C—C bond length of ethane.

•(p. 159) The bond in NO would be polar covalent because N and O have different electronegativities. The bond in O_2 would be pure covalent because both oxygen atoms have the same electronegativity.

•(p. 161) The most electronegative atom will have the negative partial charge. If the electronegativity of B is greater than that of A, B would have the negative partial charge.

•(p. 163) The formal charge on S is +1, on the single bonded O is −1 and on the double bonded O is zero.

•(p. 164) The formal charge on Cl in each of the acids is as follows:

$$HClO_4 \;\; +3$$
$$HClO_3 \;\; +2$$
$$HClO_2 \;\; +1$$
$$HClO \;\;\; +0$$

The Cl in $HClO_4$ has the largest formal charge, +3. The hydrogen with its positive formal charge will be most easily removed from the $HClO_4$ which has the largest formal charge on Cl. The Cl atom in HClO has a formal charge of zero and therefore does not lose its hydrogen so easily.

•(p. 169) Each carbon in the Lewis structure of benzene has one C—C single bond, one C—C double bond, and one C—H single bond. Therefore, each carbon atom has three electron domains.

•(p. 170) A molecule with five electron domains around the central atom will have one of four possible molecular geometries: trigonal bipyramidal, seesaw, T-shaped, or linear. Which of the geometries that the molecule will have is dependent on how many of the domains are bonding and how many are nonbonding.

•(p. 174) The shape of a molecule can be found using Table 4.3. The shape of a molecule with three bonding domains and one nonbonding domain around the central atom is trigonal pyramidal.

•(p. 175) The C_1—C_2—C_3 is 109°. The C_1—C_2—H_g is 109°. The H_e—C_3—O is slightly less than 109°.

•(p. 175) There are three bonding electron domains around each carbon. Therefore, the bond angles would be approximately 120°. The hybridization of the carbon atom would be sp^2.

•(p. 177) Yes, chloroform has a dipole moment because the polarity of the C—H bond is different than that of the three C—Cl bonds. The polarities of the bonds do not cancel and therefore result in a dipole moment for the molecule.

•(p. 190) N_2 has the Lewis structure :N≡N:.

Valence bond theory describes the bonding in N_2 as one sigma and two pi bonds formed by head-to-head and side-to-side overlap, respectively, of half-filled $2p$ orbitals. The number of bonds is three in both models, but the Lewis structure depicts all the bonds as being the same.

Chapter 5

•(p. 202) Ca < K < Rb

•(p. 205) Atoms in Group IIA have two electrons in their valence shell that can be more easily lost than their inner shell electrons. Group IA elements have only one valence shell electron. Therefore, an electron must be removed from an inner shell in order to lose a second electron. These electrons are closer to the nucleus and are more difficult to remove.

•(p. 211) The most common ion of Sr is Sr^{2+}. The peroxide anion is O_2^{2-}. The formula is SrO.

•(p. 214)

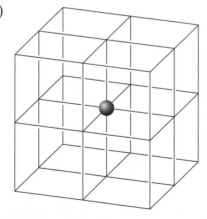

•(p. 216) The bond in $CuAl_2$ is primarily a metallic bond whereas NaCl has a primarily ionic bond. The electrons in the metallic bond are delocalized. An electron is transferred from sodium to chloride in the ionic bond. This electron now belongs to the chloride and is therefore localized. The ionic bond is held together by the attraction of the positive sodium ion for the negative chloride ion.

•(p. 219) The difference in electronegativity between Al and Cl is 1.26. Solely on the basis of the difference in electronegativity we would predict that $AlCl_3$ is polar covalent. It has a difference in electronegativity similar to that between oxygen and hydrogen atoms in water.

•(p. 223) Using the bond type triangle we find $AlCl_3$ located directly on the dividing line between covalent and ionic bonding. We would predict that $AlCl_3$ would have approximately equal covalent and ionic character.

•(p. 231) Formal charge and oxidation numbers are tools that have been developed by chemists. Formal charge treats all bonds as if they were pure covalent bonds. Oxidation numbers treat all bonds as if they were pure ionic bonds. Both of these are useful tools for making predictions about chemical structures and changes which occur during reactions.

•(p. 233) The first reaction is not an oxidation–reduction reaction. None of the atoms undergo a change in oxidation state. The second reaction is an oxidation–reduction reaction. The oxidation number of N changes from -3 in NH_3 to $+2$ in NO. The oxidation number of O changes from zero in O_2 to -2 in both NO and water. The third reaction is an oxidation-reduction reaction. The oxidation number for carbon in CH_3OH changes from -2 to -3 while the oxidation number of carbon in CO changes from $+2$ to $+3$.

Chapter 6

•(p. 248) If the liquid water and gaseous water are both at the same temperature, then the average kinetic energy of the water molecules should be the same.

•(p. 253) The pressure exerted on the floor by the spike heel will be greater. If the same person wears both shoes then the same total force is exerted on the floor. In the case of the spike heels, however, that force is exerted over a much smaller area of the floor.

•(p. 257) No, the diameter of the tube does not change the relationship between the pressure and volume of the gas.

•(p. 260) $PV \propto T$

•(p. 262) Boyle's law describes the relationship between pressure and volume, and Amontons' law describes the relationship between pressure and temperature. By combining the laws we can obtain a relationship between temperature and volume. This is Charles' law. In other words, yes, Charles' law could be predicted from Boyle's and Amontons' laws.

•(p. 269) If we assume that the 1 L of wet air and 1 L of dry air are at the same temperature and pressure, then there must be the same number of moles in both containers. The gas in the container of dry air has an average molecular weight of 29.0 g/mol. The container of wet air contains both air and water vapor. Because that container contains both air and water vapor but has the same total number of moles of gas as the dry air, there are fewer moles of dry air (29 g/mol) and more moles of water vapor (18 g/mol) than in the container of dry air. Therefore, the container of wet air is lighter than the container of dry air because the container of wet air has fewer heavy molecules and more light molecules of gas than the container of dry air.

•(p. 275) As the temperature of a gas increases the molecules of gas begin to move more rapidly, resulting in their having more kinetic energy.

Chapter 7

•(p. 299) No bonds are broken in the reactants. One bond is formed in the products. Therefore, more bonds are made than broken. This reaction will release energy because it is a bond making process.

•(p. 300) When heat enters a balloon the temperature of the gas particles increases. This results in an increase in the motion of the gas particles, causing them to strike the sides of the balloon more frequently and with more force. If the balloon is elastic it will expand, increasing the volume of the gas while the pressure remains constant.

•(p. 302) According to the first law of thermodynamics the total energy change must equal zero. This requires that the sum of the change in energy of the system and the change in energy of the surroundings equals zero. If we assume the brick to be our system and the water to be the surroundings, the brick can gain energy only if the surroundings supply that energy. Therefore, the first law does not prevent the brick from becoming hotter and the water from becoming colder. The first law requires only that what is gained by one part of the universe must be lost by another part of the universe.

•(p. 304) The temperature of the gas will increase the most for the gas in a fixed volume piston and cylinder. All of the heat transferred to this gas is used to increase the temperature of the gas. Some of the heat transferred to the movable piston and cylinder is used to do work.

•(p. 306) While at rest on the tenth floor the crate will have only potential energy. During the fall the crate will have both potential and kinetic energy. As the crate falls its velocity increases, thus increasing its kinetic energy. As it moves closer to the ground its potential energy in relation to the ground decreases. When the crate strikes the ground it can release its energy in the form of sound and heat or by doing work on the ground (making a hole or indentation).

• (p. 308) Table sugar can provide or give off energy. The bond making process must provide more energy than is required by the bond breaking process. The products must therefore be more stable than the reactants. The sum of the strengths of the bonds formed are stronger than the corresponding sum for the bonds broken.

•(p. 310) All of the reactions listed in Table 7.2 release or give off energy. Therefore, they are exothermic reactions, and the sign of the change in enthalpy is negative.

•(p. 314) ΔH°_{373} represents the change in enthalpy for a process that occurs under standard conditions (i.e., all gases at 1 atm pressure) and at 373 K.

•(p. 316) ΔH°_{ac} for $N(g)$ and $H(g)$ are zero because both substances are isolated atoms in the gaseous state. The ΔH°_{ac} for NH_3 is -1171.76 kJ/mol$_{rxn}$. ΔH°_{ac} for NH_3 represents the formation of NH_3 from its gaseous atoms. No bonds are broken in the reactants but bonds are formed by the products, and therefore ΔH°_{ac} is negative.

•(p. 320)

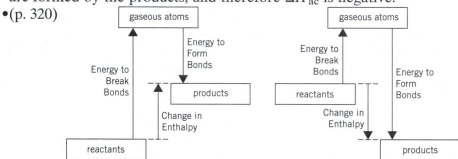

•(p. 322) The reaction is exothermic. We can visualize the reaction proceeding by breaking two triple bonds and one double bond in the reactants and forming four double bonds in the product. More energy is released in the bond making process than is used in the bond breaking process.

•(p. 324) The enthalpy of atom combination of liquid ethanol is greater than the enthalpy of atom combination of gaseous ethanol. More energy is released in the formation of liquid ethanol than in the formation of gaseous ethanol from the gaseous atoms. The change in state from liquid ethanol to gaseous ethanol requires an input of energy. There are forces of attraction between the molecules of liquid ethanol that are not present in gaseous ethanol.

•(p. 325) When similar atoms are bonded together the enthalpy of atom combination decreases with increasing bond length. BrF would be expected to have weaker bonds (a smaller enthalpy of atom combination) than ClF, because BrF is a larger molecule with a larger bond length.

Chapter 8

•(p. 349) As an atom becomes larger the electrons farther from the nucleus are more easily induced into a temporary dipole. As the size of an atom increases there is an increase in the dispersion force between atoms.

$$He < Ne < Ar < Kr < Xe < Rn$$

•(p. 350) Hydrogen bonds between NH_3 molecules are weaker than those between H_2O molecules because the oxygen is more electronegative than nitrogen. This gives a larger partial charge to the H in the OH bond.

•(p. 353) (a) The three substances have very different molecular weights. We would predict that the boiling point should increase with an increase in molecular weight due to increasing dispersion forces. Therefore, the boiling point of tetrabromobutane (373.6 g/mol) is greater than the boiling point of carbon tetrachloride (154 g/mol), and acetone (58 g/mol) has the lowest boiling point.

(b) The three substances have very similar molecular weights (142 to 144 g/mol). When comparing the boiling points of substances with similar molecular weights it is necessary to examine the intermolecular forces. Octanoic acid is capable of hydrogen bonding and will therefore have the highest boiling point. Nonanal is polar and therefore has dipole–dipole intermolecular forces. Decane is nonpolar and will have only dispersion forces. Therefore, decane is expected to have the lowest boiling point.

•(p. 354) The phase change from liquid to gas requires more energy (enthalpy of vaporization 44 kJ/mol_{rxn}) than the phase change from solid to liquid (enthalpy of fusion 6 kJ/mol_{rxn}). The solid state has the strongest intermolecular forces. Some of the intermolecular forces must be overcome to melt the solid, and the remaining forces must be overcome to cause the liquid to boil.

•(p. 357) If the volume of the container is decreased the ideal gas law tells us that the pressure of the vapor in the container should increase. However, if the vapor in the container is at its vapor pressure (the maximum pressure of vapor for a given temperature) the pressure cannot increase. Therefore, as the volume is decreased condensation of the vapor will occur so that the pressure of the vapor remains constant at the vapor pressure. On the molecular level some of the gaseous molecules enter the liquid phase.

•(p. 363) Specific heats are based on mass of a substance, and molar heat capacities are based on moles. One mole of iron has a much larger mass than one mole of aluminum. Given two substances with similar structures and properties, the one with the higher molecular weight will have a higher molar heat capacity and a lower specific heat compared to the substance with the lower molecular weight.

•(p. 371) The bond breaking process for one mole of $BaCl_2(s)$ requires +1282 kJ/mol_{rxn}. The attractive forces which form between Ba^{2+} ions and water release 718 kJ/mol_{rxn}. The attractive forces which form between 2 Cl^- ions and water release (2) × (289) kJ/mol_{rxn}. The change in enthalpy associated with the dissolution of $BaCl_2(s)$ is −14 kJ/mol_{rxn}.

•(p. 372) The bond breaking process for $AgCl(s)$ requires 533.30 kJ/mol_{rxn}. The attractive forces which form between Ag^+ ions and water release 178.97 kJ/mol_{rxn}. The attractive forces which form between Cl^- ions and water release 288.838 kJ/mol_{rxn}. The change in enthalpy associated with the dissolution of $AgCl(s)$ is +65.49 kJ/mol_{rxn}.

•(p. 377) For a side chain to be hydrophilic it must be able to interact with water. This means that the side chain should have intermolecular forces similar to the intermolecular forces of water (hydrogen bonding). The side chain on alanine contains carbon and hydrogen atoms that have similar electronegativities. This side chain is nonpolar. The side chain on cysteine is somewhat polar due to the presence of sulfur with its two nonbonding pairs of electrons, but it is not capable of hydrogen bonding. These two side chains are therefore hydrophobic (they do not readily interact with water). Lysine contains an $-NH_3^+$ group and serine contains an $-OH$ group. Both lysine and serine are capable of forming hydrogen bond intermolecular forces. Therefore, the side chains of both will interact with water.

Chapter 9

• (p. 403) On the atomic scale molecular solids are composed of individual molecules held together by intermolecular forces. Network covalent solids are composed of a three-dimensional array of atoms held together by covalent bonds. On the macroscopic scale network covalent solids have high melting points and molecular solids have relatively low melting points.

• (p. 404) On the atomic scale molecular solids are composed of individual molecules held together by intermolecular forces. Ionic solids are composed of cations and anions in a three-dimensional array held together by ionic bonds. Ionic solids have higher melting points than molecular solids. Ionic solids conduct an electric current when melted; molecular solids do not.

• (p. 405) The lattice energy of MgF_2 is greater than that of $MgCl_2$ because F^- is a smaller ion than Cl^-. The smaller fluoride ion is held more closely and tightly to the Mg^{2+}.

• (p. 407) Both the strength of the attraction of an atom for its valence electrons and the energy difference between valence subshells within an atom contribute to the AVEE and electronegativity. Atoms that have a relatively small attraction for their valence electrons and have small differences in energy between valence subshells have low AVEE and electronegativity. Atoms with those characteristics have metallic behavior.

• (p. 408) Metallic solids are composed of positively charged metal cations in a three-dimensional array. The metal cations are surrounded by a sea of delocalized electrons that form metallic bonds. Ionic solids are composed of cations and anions in a three-dimensional array held together by localized electrons in ionic bonds. Metallic solids have a wide range of melting points compared to ionic solids, which have relatively high melting points. Metallic solids can conduct electricity in the solid state but ionic solids cannot.

• (p. 412) According to the electronegativity data in Appendix B.7, B_4C has an average EN of 2.3 and a ΔEN of 0.4. This places it in the covalent region of a bond type triangle (see Figure 5.10). MoC has an average EN of 1.97 and a ΔEN 1.1 and is in the ionic region. B_4C could be used as an electrical insulator, and MoC could be used as heat insulator.

• (p. 419) In order for a substance to be classified as a body-centered cubic cell, all of the positions in the unit cell must be occupied by the same type of atom or ion.

• (p. 420) When X rays strike the surface of a crystal they are diffracted into repeating patterns. The only way that this can occur is for the X rays to remain in phase. For the diffracted X rays to remain in phase they must be diffracted by an ordered arrangement of planes of atoms. Thus, solids must consist of an ordered arrangement of particles.

• (p. 422) The three factors that contribute to the density of a metal are the atomic weight of the metal atoms, the size of the metal atoms, and the type of unit cell formed by the metal atoms.

• (p. 424) The simple cubic cell has the simplest relationship between the radii of the ions and the length of the edge of the unit cell. In a simple cubic cell the length of the edge of the cell is equal to $r + r$. In a face-centered cube $2(r + r) = a2^{1/2}$ where a is the length of the edge of the unit cell. For a body-centered cubic cell the relationship is $2(r + r) = a3^{1/2}$.

Chapter 10

• (p. 431) After the reaction has reached equilibrium both N_2O_4 and NO_2 will be present in the container.

• (p. 433) Using Table 10.1 we see that at equilibrium the ratio of *trans*-2-butene to *cis*-2-butene is 0.559 to 0.441 moles.

• (p. 436) An increase in the rate constant of a reaction will result in an increase in the rate of the reaction. Therefore, the rate of the reaction will increase with temperature.

- (p. 439) If the rate of the reverse reaction is greater than the rate of the forward reaction at a given moment in time, this does not mean that the reverse rate constant is greater than the forward rate constant. The rate of a reaction is dependent on both the rate constant and the concentrations of the reactants. If the rate of the reverse reaction is greater than the forward rate, this may be due to a higher concentration of products (reactants for the reverse reaction) than reactants for the forward reaction.
- (p. 441) The equilibrium constant can be determined from the division of the forward rate constant by the reverse rate constant. For the reaction described the equilibrium constant would be 2.
- (p. 448) The stoichiometry of the reaction is 1:1:1. For every mole of PCl_3 that reacts one mole of PCl_5 is produced and one mole of Cl_2 is used up. Therefore, if the concentration of PCl_3 decreased by 0.96 mol/L, the concentration of Cl_2 also decreased by 0.96 mol/L leaving only 0.04 mol/L. The concentration of PCl_5 must increase from 0 to 0.96 mol/L.
- (p. 454) ΔC will be small when the equilibrium constant is significantly smaller than one. ΔC would be small for K values of 1.0×10^{-5} and 1.0×10^{-10}.
- (p. 459) If more P_2 is added to the system at equilibrium the system will shift to use up the added P_2. The system will shift to the right to produce more P_4. If the pressure increases while the temperature remains constant, a reaction will shift in a direction to relieve the increase in pressure by decreasing the number of moles of gas. For the reaction $2 P_2(g) \rightleftharpoons P_4(g)$, the equilibrium will shift to the right to produce P_4 and use up P_2.

- (p. 462) $K_{sp} = [Li^+][F^-] = 6 \times 10^{-3}$

    $6 \times 10^{-3} = (\Delta C)(\Delta C)$

    $\Delta C = 8 \times 10^{-2}\ M = [Li^+] = [F^-]$
- (p. 462) The K_{sp} expression is $[Mg^{2+}][OH^-]^2$. Calculate a value Q for the given concentrations. $Q = (1.7 \times 10^{-4}\ M)(2.4 \times 10^{-4}\ M)^2 = 9.8 \times 10^{-12}$; Q (9.8×10^{-12}) $< K_{sp}$ (1.8×10^{-11}). No precipitant will form.

Chapter 11

- (p. 485) HNO_3: acid $Mg(OH)_2$: base $C_2H_3O_2$: acid
- (p. 488) $H_3PO_4(aq) + H_2O(l) \rightleftharpoons H_2PO_4^-(aq) + H_3O^+(aq)$. The conjugate base is $H_2PO_4^-$. Phosphoric acid is a polyprotic acid, meaning that it can further dissociate to lose an additional proton and produce additional H_3O^+. $C_6H_5NH_2(aq) + H_2O(l) \rightleftharpoons C_6H_5NH_3^+(aq) + OH^-$. The conjugate acid of aniline is $C_6H_5NH_3^+$.
- (p. 494) The K_w will still equal 10^{-14}. The addition of acid will result in a higher overall concentration of H_3O^+ even though some of the hydronium ion from the dissociation of water may shift back to the left. What is most significant about the shift in the equilibrium of water back to the left is the decrease in the concentration of hydroxide ion. The concentration of the hydroxide ion multiplied by the concentration of the hydronium ion will therefore equal K_w, 10^{-14}.
- (p. 498) Because $pH = -\log[H_3O^+]$, as the concentration of hydronium ion increases there will be a decrease in the pH.
- (p. 502) $K_w = [H_3O^+][OH^-]$

    $$K_a = \frac{[H_3O^+][OH^-]}{[H_2O]}$$

    The K_a expression includes a concentration of H_2O for H_2O acting as a reactant.
- (p. 503) Both solutions have a low concentration of acid. The acid with the larger K_a will be a stronger acid and will dissociate more than the acid with the small K_a. The solution with the larger K_a will produce more hydronium ion and will therefore have a smaller pH.

- (p. 505) A strong acid produces a weak conjugate base. HOCl is the stronger acid; therefore, its conjugate base, OCl^-, will be the weaker base. OBr^- is the stronger base.
- (p. 508) H_2S is a stronger acid than H_2O. Sulfur is a larger atom than oxygen. The sulfur–hydrogen bond is longer and therefore weaker than the oxygen–hydrogen bond. The sulfur–hydrogen bond is more easily broken than the oxygen–hydrogen bond.
- (p. 510) The only difference between the acids HOCl and HOI is that the oxygen is bonded to Cl in one acid and to I in the other. Because Cl is more electronegative than I, it will pull electron density away from the oxygen–hydrogen bond more than I. Thus the oxygen–hydrogen bond will be weaker for HOCl than for HOI, and the hydrogen will more easily be lost from HOCl. Therefore, HOCl is a stronger acid than HOI.
- (p. 512) When a strong acid is added to water it is assumed to completely dissociate to produce hydronium ion. For a monoprotic acid the H_3O^+ concentration will be equal to the initial concentration of strong acid.
- (p. 515) It is not possible to determine which of the two solutions has the higher pH. Solution 1 contains a stronger acid and therefore dissociation is greater than for the acid in solution 2, but the concentration of the acid in solution 2 is greater than the concentration of the acid in solution 1. It is possible that solution 2 could produce more hydronium ion because it is more concentrated even though it is a weaker acid.
- (p. 517) The 1.0 M solution with a pH of 2.4 is the most acidic.
- (p. 518) A large K_b indicates a stronger base that would react with water to accept more hydrogen ions and produce more hydroxide ions. Since both bases have the same concentration, the solution of the base with the larger K_b is more basic.
- (p. 521) NO_2^- is the conjugate base of HNO_2. Therefore, $K_b = K_w/K_a$. $K_b = (1.0 \times 10^{-14})/(5.1 \times 10^{-4}) = 2.0 \times 10^{-11}$.
- (p. 526) If acid is added to a buffer the pH of the buffer will decrease slightly, but the decrease will not be as great as when the same amount of acid is added to water. If base is added to a buffer the pH of the buffer will increase slightly but not as much as when the same amount of base is added to water. The resulting pH is dependent on how much acid or base is added.
- (p. 527) Lactic acid produced in the body can react with HCO_3^- by the following reaction.

$$HC_3H_5O_3(aq) + HCO_3^-(aq) \rightleftharpoons C_3H_5O_3^-(aq) + H_2CO_3(aq)$$

- (p. 530) As NaOH is slowly added to a solution of acetic acid the pH will gradually increase until just enough NaOH has been added to completely react with the acetic acid. At this point the pH will increase rapidly over several pH units. As additional NaOH is added, the pH will resume a more gradual increase. A plot of pH versus volume of NaOH will yield a typical titration curve.

Chapter 12

- (p. 537) Oxidation is the loss of electrons, reduction is the gain of electrons. One process cannot occur without the other.
- (p. 558) Fe = +3, Cl = −1; C = −4, H = +1; O = 0; Mn = +7, O = −2; C = +2, H = +1, O = −2, N = −3.
- (p. 563) The anode is negative and the cathode is positive.
- (p. 565) The number of atoms is balanced but the charge is not. There is a charge of +2 on the reactant side and a charge of +3 on the product side of the chemical equation. The equation is not balanced.
- (p. 566) Sn(s) is oxidized and nitrogen is reduced. The oxidizing agent is HNO_3, and the reducing agent is Sn.

•(p. 571) If the reaction is reversed the sign (+ or −) of the cell potential is changed.

•(p. 572) The standard potential corresponds to 1 M concentrations of both Cu^{2+} and Ni^{2+}. A concentration of 1.5 M Cu^{2+} on the reactant side and a concentration of 0.010 M Ni^{2+} on the product side will result in a greater tendency for the reaction to proceed toward the products. This will result in a greater (more positive) value for the potential (E_{cell}) under these conditions than at standard conditions.

•(p. 605) The student received no credit because the reaction was balanced in base rather than in acid.

Chapter 13

•(p. 619) Dissolving salt in water is a spontaneous process. Even though heat is absorbed during the process, the salt dissolves because the system (water and salt) becomes more disordered as the salt dissolves.

•(p. 621) The second law states that for a spontaneous process the entropy of the universe must increase. The brick cannot get hotter nor the water colder.

•(p. 622) Water freezing does not violate the second law of thermodynamics. In the process of water freezing water is considered the system. Water becomes more ordered, and therefore the surroundings must become more disordered so that the entropy of the universe can become more positive (disordered).

•(p. 624) It is possible for a process to proceed if the entropy of the system decreases as long as the entropy of the surroundings increases even more, causing the entropy of the universe to increase.

•(p. 625) At absolute zero, motion at the molecular and atomic level approaches a minimum. A substance would then have minimum entropy.

•(p. 629) Glucose would have greater freedom of motion (more disorder) in its open chain form. The sign of ΔS would be positive.

•(p. 630) Both molecules consist of the same kind and number of atoms. Their reactions of atom combination would involve the same gaseous atom reactants. Because the entropy of atom combination of isobutane is more negative than the entropy of atom combination of butane, the formation of isobutane results in the loss of more freedom of motion.

•(p. 634) If ΔH for a spontaneous process, such as dissolving table salt in water, is positive the entropy of the process must be positive enough to make the change in Gibbs free energy negative. The increase in disorder must be enough to make up for the endothermic process.

•(p. 637) No, the hardening of cement has a negative ΔS because the cement becomes more ordered. In order for this to be a spontaneous process it must therefore be exothermic (releasing heat).

•(p. 644) If $Q_p > K_p$ the reaction must proceed toward the reactants. In order for Q_p to decrease to K_p the numerator (product concentrations) of the expression must decrease and the denominator (reactant concentrations) must increase.

•(p. 651) If ΔH is > 0 (endothermic) the reactants are favored. If ΔS is > 0 the products are favored. A change in temperature will change the equilibrium constant. In the relationship between K, ΔH, and ΔS, it is the enthalpy term $\left(\dfrac{\Delta H}{RT}\right)$ which is dependent on temperature. As temperature decreases the term $\left(\dfrac{\Delta H}{RT}\right)$ becomes more negative and therefore K becomes smaller. Thus the equilibrium is shifted toward the reactants.

If ΔH is > 0 (endothermic) the reactants are favored. If ΔS is < 0 the reactants are favored. These circumstances will have the same result for a decrease in temperature as in the previous question because ΔH is still > 0. Therefore, a decrease in temperature will shift the equilibrium toward the reactants.

Chapter 14

• (p. 663) The kinetics of the reaction is so slow that during a human lifetime there will be no observable change of diamond to graphite.

• (p. 666) 18.2 seconds is required for the concentration of phenolphthalein to decrease from 0.00300 to 0.00250 M. 69.4 seconds is required for the concentration of phenolphthalein to decrease from 0.00100 to 0.00050 M. As the reaction proceeds it takes more time for the concentration to decrease by a given amount. As the reaction proceeds the rate of reaction decreases.

• (p. 667) If a large value for dt is used there will be a significant change in rate during the time interval. The resulting rate would then be more of an average rate during the time period rather than an instantaneous rate at some moment in time.

• (p. 668) As the reaction progresses the concentration of phenolphthalein decreases as it reacts. The concentration term in the rate law decreases, and therefore the rate of the reaction decreases. As the reaction proceeds phenolphthalein is used up and there is a decrease in the probability of a collision between phenolphthalein molecules and base molecules. Thus the reaction rate decreases.

• (p. 669) The rate describes how quickly the reactants are used up and the products are produced with respect to time. The rate constant is a proportionality constant used in the rate law. The rate of most reactions decreases with respect to time and concentration of reactants. The rate constant, however, remains constant with respect to time and concentration.

• (p. 670) The compressed air provides a constant pressure pushing on the liquid in the buret and therefore a constant rate of drainage. Without the compressed air the pressure exerted by the column of liquid at the buret tip decreases as the column of liquid decreases.

• (p. 672) The rate of disappearance of HI is twice as fast as the rate of formation of H_2: $\text{Rate}_{HI} = 2(\text{Rate}_{H_2})$.

• (p. 679) The rate of a zero-order reaction is independent of the concentration of the reactants. The rate of reaction of a first-order reaction is dependent on concentration. As a reaction proceeds the rate of reaction of a zero-order reaction will remain constant while a first-order reaction will slow down.

• (p. 681) In trials 2 and 3 both the concentrations of NH_4^+ and NO_2^- change. Therefore there will be two unknowns (m and n) in the equation. One of the concentrations must remain constant between two trials so that its effect will cancel out of the equation. Trials 1 and 2 cannot be used to find the order with respect to NH_4^+ because the concentration of NH_4^+ does not change in those trials. However, the concentration of NO_2^- does change. Any change in the rate will therefore be due to NO_2^- not NH_4^+.

• (p. 681) For trial 1 the following are known: $(NH_4^+) = 5.00 \times 10^{-2}$, $(NO_2^-) = 2.00 \times 10^{-2}$, $m = 1$, $n = 1$, and rate $= 2.70 \times 10^{-7}$. Solving the rate law equation for the rate constant, k, we get $k = 2.70 \times 10^{-4} \ M^{-1} \text{s}^{-1}$.

• (p. 683) The rate constant for the decay of ^{15}O is less than the rate constant for ^{19}O. Therefore, the rate of decay of ^{15}O will be slower, and it will have a longer half-life.

• (p. 690) No, it is not true for an endothermic reaction.

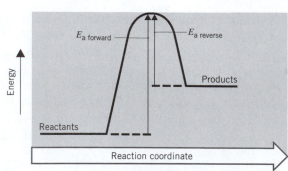

•(p. 696) Because the external pressure is less at the top of a mountain the water will boil at a lower temperature than at sea level. Therefore, the egg will not be as hot in the boiling water. Cooking the egg to make it hard-boiled involves a chemical reaction. According to the Arrhenius equation the rate constant for the reaction will be smaller at the lower temperature of the boiling water at the top of the mountain.

Chapter 15

•(p. 716) There is a much higher concentration of LLL in the suspect sample than in the olive oil sample. There is a much lower concentration of OOO in the suspect sample than in the good olive oil sample.

•(p. 720)

•(p. 725) The base T binds to A, and G binds to C. The probe must contain the sequence AAGC.

•(p. 727) Both the second and the third electron configurations correspond to a zinc atom which has absorbed energy because one or more electrons have moved to a higher energy level than in the ground state zinc atom. The first electron configuration represents the ground state of the zinc atom.

•(p. 731) Of the three wavelengths given (500, 540, and 590 nm), the hemoglobin–oxygen complex has the greatest absorbance at 540 nm. Therefore, the molar absorptivity must be greatest at that wavelength.

•(p. 734) NO and NO_2 are radicals containing an odd number of electrons. The best Lewis structure for NO has a nitrogen–oxygen double bond. The best Lewis structure for NO_2 has one nitrogen–oxygen single bond and one nitrogen–oxygen double bond. However, the actual NO_2 structure will be an average of two resonance structures, which results in each nitrogen–oxygen bond behaving as a bond and a half. Therefore, the nitrogen–oxygen bond strength in NO will be greater than in NO_2. Table 15.3 shows that when comparing a carbon–carbon double bond to a carbon–carbon triple bond, the triple bond has a smaller value for wavelength, a larger value for wave number, and would therefore have a larger frequency of absorption. The same pattern is observed when comparing a C=O bond to a C—O bond. We would therefore expect the stronger nitrogen–oxygen double bond of NO to have a smaller wavelength of absorption, a larger wave number, and a larger frequency than NO_2.

•(p. 742) Acetone would show one peak and acetic acid would show two. In acetone there are six equivalent hydrogens in the same environment, and therefore they absorb at the same frequency. In acetic acid the three hydrogens bonded to carbon are equivalent to one another, whereas the hydrogen bonded to oxygen is in a different environment and therefore absorbs at a different frequency.

PHOTO CREDITS

Chapter 1: *Page 7:* Andy Washnik. *Page 9:* Courtesy of International Business Machines Corporation.

Chapter 2: *Page 44:* Ken Karp/Omni-Photo Communications, Inc. *Page 46:* Courtesy Bayer Corporation. *Page 64:* Bob Burch/Bruce Coleman, Inc.

Chapter 3: *Page 82:* Corbis-Bettmann.

Chapter 4: *Page 166:* Courtesy American Association for the Advancement of Science. *Page 199:* Ken Karp/Omni-Photo Communications, Inc.

Chapter 5: *Page 203:* Ken Karp/Omni-Photo Communications, Inc. *Page 214:* George Bodner.

Chapter 6: *Page 251:* Ed Braveman/FPG International. *Page 256:* Ken Karp/Omni-Photo Communications, Inc.

Chapter 7: *Page 301 and 309:* Andy Washnik.

Chapter 8: *Page 366:* Andy Washnik.

Chapter 9: *Page 414–416 and 418:* George Bodner.

Chapter 10: *Page 463:* Courtesy of BASF Corporation.

Chapter 11: *Page 499:* Michael Watson.

Chapter 12: *Page 564:* Courtesy Bakken Library. *Page 589:* Ken Karp/Omni-Photo Communications, Inc.

Chapter 13: *Page 619:* E.R. Degginger. *Pages 626 and 628:* Andy Washnik.

Chapter 14: *Page 664:* Michael Watson. *Page 683:* Douglas Burrows/Gamma Liaison.

Chapter 15: *Page 724:* Courtesy Lawrence Livermore Laboratory. *Page 726:* From R. Saferstein, DNA Fingerprinting, *Chem. Matters:* 9, 12 (1991). Reproduced with permission.

INDEX

Table of Atomic Number and Average Atomic Weight

Element	Symbol	Atomic Number	Mass (amu)
Actinium	Ac	89	227.03*
Aluminum	Al	13	26.982
Americium	Am	95	241.06*
Antimony	Sb	51	121.76
Argon	Ar	18	39.948
Arsenic	As	33	74.922
Astatine	At	85	209.99*
Barium	Ba	56	137.33
Berkelium	Bk	97	249.08*
Beryllium	Be	4	9.0122
Bismuth	Bi	83	208.98
Bohrium	Bh	107	262.12*
Boron	B	5	10.811
Bromine	Br	35	79.904
Cadmium	Cd	48	112.41
Calcium	Ca	20	40.078
Californium	Cf	98	252.08*
Carbon	C	6	12.011
Cerium	Ce	58	140.12
Cesium	Cs	55	132.91
Chlorine	Cl	17	35.453
Chromium	Cr	24	51.996
Cobalt	Co	27	58.933
Copper	Cu	29	63.546
Curium	Cm	96	244.06*
Dubnium	Db	105	262.11*
Dysprosium	Dy	66	162.50
Einsteinium	Es	99	252.08*
Erbium	Er	68	167.26
Europium	Eu	63	151.97
Fermium	Fm	100	257.10*
Fluorine	F	9	18.998
Francium	Fr	87	223.02*
Gadolinium	Gd	64	157.25
Gallium	Ga	31	69.723
Germanium	Ge	32	72.61
Gold	Au	79	196.97
Hafnium	Hf	72	178.49
Hassium	Hs	108	(265)**
Helium	He	2	4.0026
Holmium	Ho	67	164.93
Hydrogen	H	1	1.0079
Indium	In	49	114.82
Iodine	I	53	126.90
Iridium	Ir	77	192.22
Iron	Fe	26	55.847
Krypton	Kr	36	83.80
Lanthanum	La	57	138.91
Lawrencium	Lr	103	262.11*
Lead	Pb	82	207.2
Lithium	Li	3	6.941
Lutetium	Lu	71	174.97
Magnesium	Mg	12	24.305
Manganese	Mn	25	54.938
Meitnerium	Mt	109	(266)**
Mendelevium	Md	101	258.10*
Mercury	Hg	80	200.59
Molybdenum	Mo	42	95.94
Neodymium	Nd	60	144.24
Neon	Ne	10	20.180
Neptunium	Np	93	237.05*
Nickel	Ni	28	58.693
Niobium	Nb	41	92.906
Nitrogen	N	7	14.007
Nobelium	No	102	259.10*
Osmium	Os	76	190.2
Oxygen	O	8	15.999
Palladium	Pd	46	106.42
Phosphorus	P	15	30.974
Platinum	Pt	78	195.08
Plutonium	Pu	94	239.05*
Polonium	Po	84	209.98*
Potassium	K	19	39.098
Praseodymium	Pr	59	140.91
Promethium	Pm	61	146.92*
Protactinium	Pa	91	231.04
Radium	Ra	88	226.03*
Radon	Rn	86	222.02*
Rhenium	Re	75	186.21
Rhodium	Rh	45	102.91
Rubidium	Rb	37	85.468
Ruthenium	Ru	44	101.07
Rutherfordium	Rf	104	261.11*
Samarium	Sm	62	150.36
Scandium	Sc	21	44.956
Seaborgium	Sg	106	263.12*
Selenium	Se	34	78.96
Silicon	Si	14	28.086
Silver	Ag	47	107.87
Sodium	Na	11	22.990
Strontium	Sr	38	87.62
Sulfur	S	16	32.066
Tantalum	Ta	73	180.95
Technetium	Tc	43	98.906*
Tellurium	Te	52	127.60
Terbium	Tb	65	158.93
Thallium	Tl	81	204.38
Thorium	Th	90	232.04
Thulium	Tm	69	168.93
Tin	Sn	50	118.71
Titanium	Ti	22	47.88
Tungsten	W	74	183.85
Uranium	U	92	238.03
Vanadium	V	23	50.942
Xenon	Xe	54	131.29
Ytterbium	Yb	70	173.04
Yttrium	Y	39	88.906
Zinc	Zn	30	65.39
Zirconium	Zr	40	91.224

Source: These data are based on the definition that the mass of the ^{12}C isotope of carbon is exactly 12 amu. The atomic weight of an element is the weighted average of the masses of the isotopes of that element. Data from "Atomic Weights of the Elements," *Pure and Applied Chemistry,* **63,** 975 (1991).

*Radioactive element has no stable isotope. The mass is for one of the element's common radioisotopes.

**Mass number of the most stable isotope of this radioactive element.

Electronegativities of the Elements

H 2.30																	He 4.16
Li 0.91	Be 1.58											B 2.05	C 2.54	N 3.07	O 3.61	F 4.19	Ne 4.79
Na 0.87	Mg 1.29											Al 1.61	Si 1.92	P 2.25	S 2.59	Cl 2.87	Ar 3.24
K 0.73	Ca 1.03	Sc 1.2	Ti 1.3	V 1.4	Cr 1.5	Mn 1.6	Fe 1.7	Co 1.8	Ni 1.9	Cu 1.8	Zn 1.6	Ga 1.76	Ge 1.99	As 2.21	Se 2.42	Br 2.69	Kr 2.97
Rb 0.71	Sr 0.96	Y 1.0	Zr 1.1	Nb 1.3	Mo 1.4	Tc 1.5	Ru 1.7	Rh 1.8	Pd 1.9	Ag 2.0	Cd 1.5	In 1.66	Sn 1.82	Sb 1.98	Te 2.16	I 2.36	Xe 2.58
Cs 0.66	Ba 0.88										Hg 1.76						